Proceeding of the 24th International Conference on Industrial Engineering and Engineering Management 2018

George Q. Huang · Chen-Fu Chien ·
Runliang Dou
Editors

Proceeding of the 24th International Conference on Industrial Engineering and Engineering Management 2018

Springer

Editors
George Q. Huang
Department of Industrial and Manufacturing Systems Engineering
The University of Hong Kong
Hong Kong, Hong Kong

Chen-Fu Chien
Department of Industrial Engineering and Engineering Management
National Tsing Hua University
Hsinchu, Taiwan

Runliang Dou
Department of Information Management and Management Science
Tianjin University
Tianjin, China

ISBN 978-981-13-3401-6 ISBN 978-981-13-3402-3 (eBook)
https://doi.org/10.1007/978-981-13-3402-3

Library of Congress Control Number: 2018963180

This Springer imprint is published by the registered company Springer Nature Singapore Pte Ltd.
The registered company address is: 152 Beach Road, #21-01/04 Gateway East, Singapore 189721, Singapore

Contents

Smart Manufacturing

A Hybrid Algorithm for Facility Layout Problem of Mixed Model Assembly Line

Shi-jun Yang and Ling Zhao(✉)

School of Mechanical and Electric Engineering,
Soochow University, Suzhou 215000, China
zhaoling@suda.edu.cn

Abstract. For solving the facility layout problem of mixed model assembly line (MMAL-FLP), the multiobjective model of MMAL-FLP was built for optimizing logistics and production efficiency according to characteristics of MMAL. For minimizing logistics cost and maximizing line balance as the index of objectives, the new hybrid algorithm named nondominated sorting genetic algorithm 2 with tabu search (NSGA2-TS) was proposed to solve this model. NSGA2-TS apply the powerful ability for local search of TS to settle the premature convergence matter of NSGA2. The practical case study proved the effectiveness and feasibility of MMAL-FLP model and the validity and stability of the NSGA-TS.

Keywords: MMAL-FLP · Multiobjective · NSGA2-TS

1 Introduction

A mixed model assembly line (MMAL) is a type of production line where a variety of product models similar in product characteristics are assembly. The products usually involve different assembly tasks, methods and standard time and require different facilities, components and raw materials [1]. With the growing trend of more variety and minor batch, there are more and more companies apply MMAL in manufacture field, but the product system facility layout problem of MMAL was researched rarely. There were many objectives for optimizing the facility layout and there were the impacts of many factors on these objectives, the multiobjective model of MMAL-FLP was built for optimizing the logistics and the production efficiency as the key points of the facility layout problem, now the logistics cost and the line balance was adopted as the index of the logistics and the production efficiency respectively according to characteristics of MMAL.

The nondominated sorting genetic algorithm 2 was developed by Ded [2], and this algorithm was one of the best popular algorithms for solving the multiobjective optimization problem [3]. For improving NSGA2 in terms of convergence, spread and computation effort, some researchers optimized the genetics operator [4], others proposed the hybrid algorithm which was the popular direction and effective way [5–9]. In fact, Glover who proposed TS have presented the theoretically possible to combine GA

G. Q. Huang et al. (eds.), *Proceeding of the 24th International Conference on Industrial Engineering and Engineering Management 2018*,
https://doi.org/10.1007/978-981-13-3402-3_1

and TS [10]. Now the new way was proposed to combine NSGA2 and TS for taking advantage of two algorithms in the paper. Now the new hybrid algorithm named nondominated sorting genetic algorithm 2 with tabu search (NSGA2-TS) was proposed to solve this model. NSGA2-TS apply the powerful ability for local search of TS to settle the premature convergence matter of NSGA2.

The remainder of the paper was organized as follow. Section 2 presented the model definition of MMAL-FLP and MMAL-FLP model. Section 3 described the new hybrid algorithm NSGA2-TS for solving MMAL-FLP. The case study and the detail analysis were presented in Sect. 4. The paper was concluded in Sect. 5.

2 MMAL-FLP Model

MMAL-FLP model mainly considered the difference of production type, process, operate time and machines of MMAL. The research scope was adopted by the minimum part set (MPS) Jonathan F. Bard proposed [11]. The assumptions of MMAL were set as follow.

1. The position and distance of MMAL was determined;
2. The number of machines was same as the number of position machines matched;
3. The size and profile of all machines were same.

For the MMAL-FLP, the logistics factor was the preferred target for optimal layout [12]. Assembly line balancing problem was an important decision problem in any case [13]. So the mathematical model was established for minimizing the logistics cost and maximizing the line balance, it was convenient for the hybrid algorithm that the problem of maximizing the line balance was transformed into the problem of minimizing the line balance loss. The MMAL-FLP model was presented as follow.

2.1 Relevant Parameters in the Model

1. Global variable
   i, j and α, β represented the machine code and position code in the production system, the number of the machines and positions was N, p represented the product type, and the number of total type was M.
2. Decision variable
   $X_{pi\alpha}$ and $X_{pj\beta}$ were the expression of decision variables and have the same meaning. The following description was based on $X_{pi\alpha}$. $X_{pi\alpha}$ was determined by the input parameter X_{pi} and the decision variable $X_{i\alpha}$. $X_{i\alpha}$ was the only variable that need to be solved in the mathematical model, indicating whether the machine i was matched with the position α.
3. Related parameter
   d_p represented the demand of production in the MPS of the product p, c_{pij} represented the coefficient of logistics cost per unit distance between i and j of product p, and $l_{\alpha\beta}$ represented the distance between the positions α and β, CT_P represented the cycle time of the product p.

4. Procedure parameter
   t_{pw} represented the time of workstation w belong product p considered the transfer time of work in process. For each product p, the principle how to partition the workstation was that maximizing the time of workstation according the order of process as long as each t_{pw} did not beyond the CT_P.

2.2 Objectives and Constraints

Refer to (1), the first objective was that minimizing the sum of the logistics cost for all products. Refer to (2), another objective was that minimizing the sum of the line balance loss for all products. Refer to (3), $X_{pi\alpha}$ was belong to variable (0 or 1), $X_{pi\alpha} = 1$ only if $X_{pi} = 1 \wedge X_{i\alpha} = 1$, otherwise $X_{pi\alpha} = 0$. Refer to (4), X_{pi} was belong to variable (0 or 1),$X_{pi} = 1$ only if product p required the machine i, otherwise $X_{pi} = 0$. Refer to (5), $X_{i\alpha}$ was belong to variable (0 or 1), $X_{i\alpha} = 1$ only if machine i was matched with the position α, otherwise $X_{i\alpha} = 0$. Equation (6) showed that each machine can only be matched with one position; Eq. (7) showed that each position can only be matched with one machine.

$$f_1 :$$

$$Min \sum_{p=1}^{M} \sum_{i=1}^{N} \sum_{j=1}^{N} \sum_{\alpha=1}^{N} \sum_{\beta=1}^{N} d_p c_{pij} l_{\alpha\beta} X_{pi\alpha} X_{pj\beta} \tag{1}$$

$$f_2 :$$

$$Min\left(1 - \sum_{p=1}^{M} \frac{\sum_{w=1}^{\Gamma_p} t_{pw}}{CT_P \Gamma_p} * \frac{d_p}{\sum_1^M d_p}\right) \tag{2}$$

$$S.t :$$

$$X_{pi\alpha} = \begin{cases} 1, \; X_{pi} = 1 \wedge X_{i\alpha} = 1 \\ 0, \; X_{pi} = 0 \vee X_{i\alpha} = 0 \end{cases} \tag{3}$$

$$X_{pi} \in \{0, 1\}, i = 1, \cdots, N, p = 1, \cdots, M \tag{4}$$

$$X_{i\alpha} \in \{0, 1\}, i = 1, \cdots, N, \alpha = 1, \cdots, N \tag{5}$$

$$\sum_{i=1}^{N} X_{pi\alpha} = 1, \; \alpha = 1, \cdots, N, p = 1, \cdots, M \tag{6}$$

$$\sum_{\alpha=1}^{N} X_{pi\alpha} = 1, i = 1, \cdots, N, p = 1, \cdots, M \tag{7}$$

3 3 NSGA2-TS

NSGA2 was of elitist strategy which suit for solving large-scale project optimization. However, due to its limit local search function, it is difficult to find globally optimal solution [14]. However, TS has the powerful ability as the extension of local search based on the idea of greed to search the neighborhood of the current solution. Therefore, the new hybrid algorithm NSGA2-TS was designed to take advantage of NSGA2 and TS. Comparing with the way that applying TS to mutation operator, the core of NSGA2-TS: the NSGA2 was used as the main algorithm to search globally, and monitoring the process of NSGA2 running, TS would run for stepping out local optimal solution when there was the premature phenomenon in NSGA2.

3.1 Definition of Fitness Value

MMAL-FLP was the multiobjective problem so that there was the definition of fitness value to direct how TS run. For Pareto solution set, the scalarization of the vector optimization problem is performed by assigning to each visited solution $\bar{x}_i$ the fitness values as follows [15]:

$$fit(\bar{x}_i) = \frac{1 + N(x_d \prec \bar{x}_i)}{1 + N(x_{nd} \succ \bar{x}_i)} \tag{8}$$

3.2 Step of Hybrid Algorithm

The hybrid algorithm NSGA2-TS flow chart was presented in Fig. 1 and all steps was described as follow

Step.1 inputting the variable and parameter of MMAL-FLP model and setting the population size S and maximum iteration time T.

Step.2 initial population P_0 including S individual was created by random arrangement. For optimizing the current layout of MMAL-FLP, the initial population contained the arrangement on behalf of current layout scheme. In the stage of population iteration, P_t and Q_t represented the parent population and progeny population in the stage of population iteration t.

Step.3 determining the rank by nondominance sorting and calculating the crowding distance for individual in P_t. Then Q_t was created by crossover operator and mutation operator.

Step.4 R_t which population size was 2S was created by combining P_t and Q_t and the new parent population P_{t+1} was selected by elitist strategy for R_t.

Step.5 comparing P_{t+1} with P_t, if the population was evolving, return step 3, if not, goes on to step 6.

Step.6 TS population was created by running the TS algorithm based on Pareto front of P_{t+1}.

Step.7 temp population was created by combining P_{t+1} and TS population, the new parent population P'_{t+1} was selected by elitist strategy for temp population.

Step.8 comparing P'_{t+1} with P_{t+1}, if the population was evolving, return step.3, if still not, goes on to step.9.
Looping the step.3 through step.8, when the population was still not evolving thought step.8 or the maximum iteration time T was reached, population iteration was end.

Step.9 output the result of NSGA2-TS.

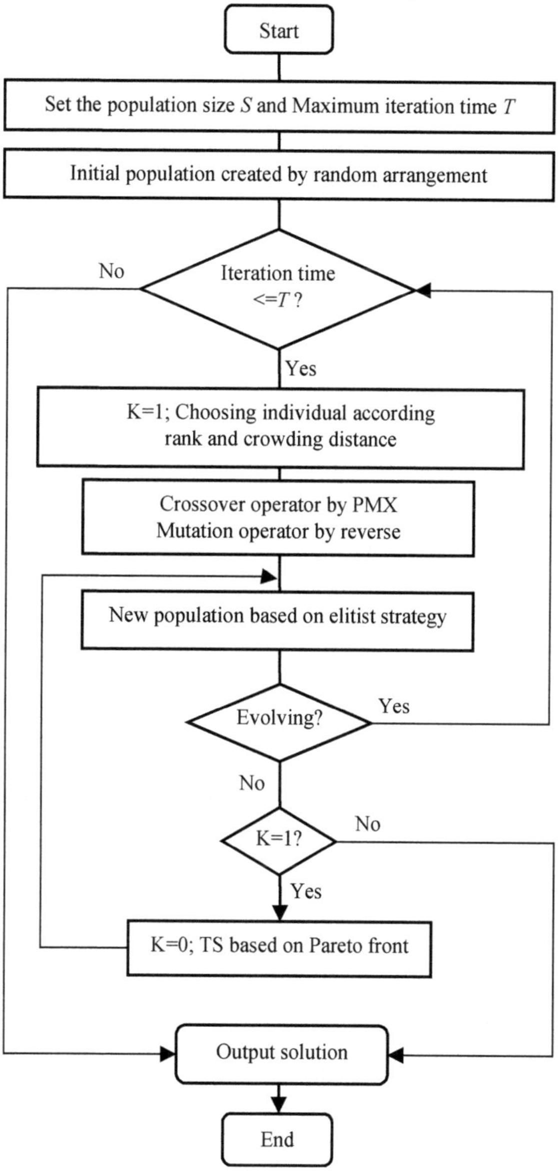

Fig. 1 NSGA2-TS flow chart

4 Case Study and Analysis

4.1 Case Introduction

In order to verify the feasibility and effectiveness of MMAL-FLP model and the validity and stability of the NSGA-TS, the case that was the booster assembly line in the automotive company showed the process how to build the MMAL-FLP model and how NSGA2-TS work. This booster assembly line for five product families was a classical mixed manufacture assembly line which was made up of sixteen machines. According to the survey of actual production situation, the minimum sequence of demand for the five product families A, B, C, D and E in the MPS was (3: 1: 2: 3: 1), and the logistics cost and line balance loss of current layout were that $f_1 = 547CNY, f_2 = 21.26\%$. The size and profile of all machines were regard as the same though there was minor difference, and the possible position of the machine was identified. Figure 2 show the code of position from 1 to 16, and the distance of position could be calculated according the information in Fig. 2. Each full permutation of machine cord represented the scheme when the every machine has a unique code. For example, the full permutation for current layout was (16, 15, 14, 13, 12, 11, 10, 9, 1, 2, 3, 4, 5, 6, 7, 8) in the MMAL-FLP model, which was represented by the series variable including $X_{16,1} = 1$, $X_{15,2} = 1$ and so on. The process time was summarized in Tables 1, 2 show the coefficient of logistics cost per unit distance that was calculated for MPS, and the cycle time was presented in Table 3.

4.2 Hybrid Algorithm and Analysis of Result

The MMAL-FLP model which was established according the actual data of the booster assembly line was solved by NSGA2-TS. The parameter was set in the process of NSGA as follow: population size $S = 40$, Maximum iterations time $T = 500$, crossover probability $p_c = 0.9$, mutation probability $p_m = 0.1$. For TS, the full field was composed of candidates by 2-optimization, and the variation of component was taboo for sixteen steps, the candidate would replace the current solution when this candidate dominated the current solution absolutely.

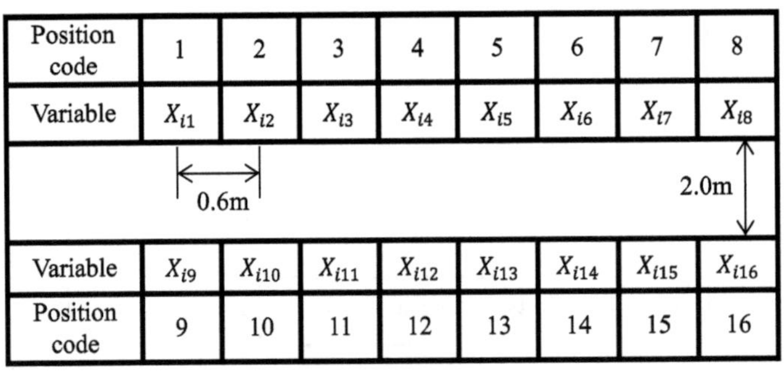

Fig. 2 Position code and Variable

Table 1 Process time of five family products (in seconds)

Machine code Product	1	2	3	4	5	6	7	8	9	10	11	12	13	14	15	16
A	0.0	5.4	6.7	0.0	5.9	0.0	0.0	9.5	3.6	4.5	0.0	9.4	0.0	3.8	0.0	8.1
B	9.8	5.2	6.7	0.0	0.0	3.3	4.9	8.2	0.0	0.0	2.6	9.3	0.0	0.0	0.0	8.5
C	9.8	5.0	7.3	0.0	6.2	0.0	0.0	9.8	0.0	4.5	0.0	0.0	9.8	0.0	1.8	8.2
D	0.0	5.8	7.6	8.2	0.0	3.4	6.0	0.0	3.6	0.0	0.0	9.5	0.0	3.2	0.0	8.2
E	0.0	5.8	0.0	8.4	0.0	3.3	4.8	0.0	0.0	0.0	2.1	0.0	9.0	2.7	2.3	8.4

Table 2 Coefficent of logistics cost per unit distance (in CNY/m)

Machine	1	2	3	4	5	6	7	8	9	10	11	12	13	14	15	16
Coefficient of logistics cost per unit distance	2				3				5				10			

Table 3 Cycle time of five family products (in seconds)

Product	A	B	C	D	E
CT	16.5	16.4	16.8	16.1	16.0

The process of NSGA2-TS was monitored to judge if the population was evolving every 50 generations in software MATLAB. The NSGA2-TS run for ten times, now there was a presentation by taking third process as example, Fig. 3 show the trend of Pareto front direction was obvious, the asterisk represented that TS run effectively when the premature phenomenon happened to NSGA2. The solution set of optimization was these points in Fig. 4, the point that horizontal and vertical line through point 'O' enclosed dominated the point 'O' which represented objectives value of current layout. Figures 5 and 6 show that the minimum and the average of the logistics cost and the line balance loss were descending obviously and stabilized gradually.

Considering the logistics route and other factors, the permutation of best solution was (15, 16, 14, 12, 13, 10, 9, 11, 1, 2, 3, 5, 6, 4, 8, 7), and the logistics cost and line balance loss of current layout were that $f_1 = 494CNY, f_2 = 17.83\%$, there was the significant improvement compared to current layout.

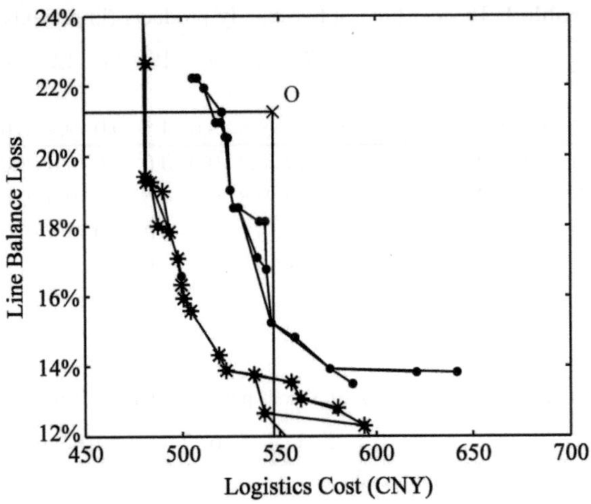

Fig. 3 Trend of Pareto front direction (*Note* * represented TS run)

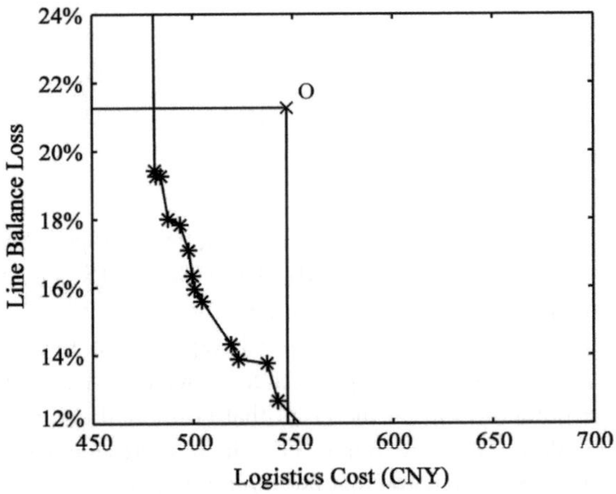

Fig. 4 Solution set of optimization

There were three points about validity and stability of NSGA2-TS when running this hybrid algorithm for ten times as follow.

1. The average of iteration time was 465 and all iteration time was behind 1000 and the rate of convergence was faster than NSGA2.
2. The average of time TS run was 2.8, it was proved that TS run effectively when the premature phenomenon happened to NSGA2.
3. The average of solution sets was 10.5, in general, the solution set of NSGA2-TS contain the solution set of NSGA2, for the single running, the solution set of NGA2-TS dominated the solution set of NSGA2.

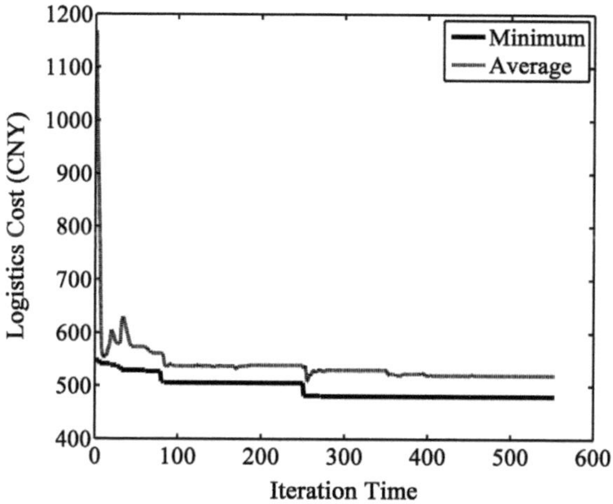

Fig. 5 Trend of logistics cost

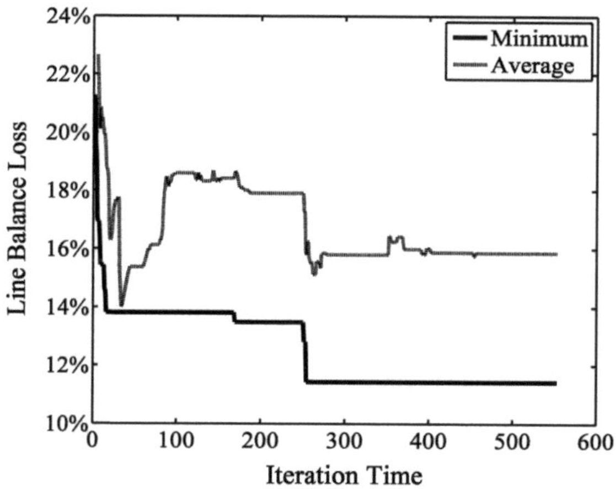

Fig. 6 Trend of line balance loss

5 Conclusion

The result showed that effectiveness and feasibility of MMAL-FLP model. The NSGA2-TS has improved the convergence and ability of local search compared to NSGA2, and the validity and stability of the NSGA-TS were proved by running the hybrid algorithm for ten times. Meanwhile the new concept of NSGA-TS can solve other multiobjective problem. The next phase of work will focus on how to solve all solution set when the initial population was completely random.

References

1. C.J. Hyun, Y. Kim, Y.K. Kim, A genetic algorithm for multiple objective sequencing problems in mixed model assembly lines. Comput. Oper. Res. **25**(7–8), 675–690 (1998)
2. K. Deb, A. Pratap, S. Agarwal et al., A fast and elitist multiobjective genetic algorithm: NSGA-II. IEEE Trans. Evol. Comput. **6**(2), 182–197 (2002)
3. M.G. Gong, L.C. Jiao, D.D. Yang et al., Research on evolutionary multi-objective optimization algorithms. J. Softw. **20**(20), 271–289 (2009)
4. S. Halelfadl, A.M. Adham, N. Mohd-Ghazali et al., Optimization of thermal performances and pressure drop of rectangular microchannel heat sink using aqueous carbon nanotubes based nanofluid. Appl. Therm. Eng. **62**(2), 492–499 (2014)
5. M. Delgado, M.P. Cuellar, M.C. Pegalajar, Multiobjective hybrid optimization and training of recurrent neural networks. IEEE Trans. Syst. Man & Cybern. Part B Cybern. Publ. IEEE Syst. Man & Cybern. Soc. **38**(2), 381 (2008)
6. C.M. Kwan, C.S. Chang., Timetable synchronization of mass rapid transit system using multiobjective evolutionary approach. *IEEE Press* (2008)
7. H.C.W. Lau, T.M. Chan, W.T. Tsui et al., A fuzzy guided multi-objective evolutionary algorithm model for solving transportation problem. Expert. Syst. Appl. Int. J. **36**(4), 8255–8268 (2009)
8. C. Zhang, W. Li, P. Jiang et al., Experimental investigation and multi-objective optimization approach for low-carbon milling operation of aluminum. *ARCHIVE Proc. Inst. Mech. Eng. Part C J. Mech. Eng. Sci. 1989–1996*. **203–210** (2016)
9. Q. Liu, W. Cai, J. Shen et al., A speculative approach to spatialtemporal efficiency with multiobjective optimization in a heterogeneous cloud environment. Secur. & Commun. Netw. **9**(17), 4002–4012 (2016)
10. F. Glover, J.P. Kelly, M. Laguna, Genetic algorithms and tabu search: hybrids for optimization. Comput. Oper. Res. **22**(1), 111–134 (1995)
11. J.F. Bard, E. Dar-Elj, A. Shtub, An analytic framework for sequencing mixed model assembly lines. Int. J. Prod. Res. **30**(1), 35–48 (1992)
12. B.H. Ulutas, A.A. Islier, A clonal selection algorithm for dynamic facility layout problems. J. Manuf. Syst. **28**(4), 123–131 (2009)
13. C. Becker, A survey on problems and methods in generalized assembly line balancing. Eur. J. Oper. Res. **168**(3), 694–715 (2006)
14. J. Li, B. Yang, D. Zhang et al., Development of a multi-objective scheduling system for offshore projects based on hybrid non-dominated sorting genetic algorithm. Adv. Mech. Eng. **7**(3) (2015). https://doi.org/10.1177/1687814015573785
15. S. Carcangiu, A. Fanni, A. Montisci, Multiobjective tabu search algorithms for optimal design of electromagnetic devices. IEEE Trans. Magn. **44**(6), 970–973 (2008)

Assessment of the Vulnerability of Emergency Rescue System in Coal Enterprises Based on GGAHP

Tian-bo Xu(✉)

School of Mechanical Engineering, Xi'an University of Science and Technology, Xi'an, China
xutianbo@sohu.com

Abstract. To identify the weak links of the emergency rescue system in coal enterprises, this study assessed the vulnerability of this system with the Group Gray Analytical Hierarchy Process (GGAHP), which combined the Group Decision Making and Grey Theory. By setting up the index system about vulnerability and judgment matrix, then whitening the matrix, the vulnerability data at the bottom of the index system could be calculated to spot the weak links of the system.

Keywords: Coal mines · Emergency rescue · GGAHP · Vulnerability

1 Introduction

The concept of vulnerability stems from the study of natural disasters to describe the vulnerabilities and disruptions of related systems, the lack of resistance to interference, and the ability to restore their structures and functions, which are now widely used in the fields of disaster science, ecology, sociology, economics and Engineering [1–2]. Production conditions in coal mines are complex and changeable, prone to all kinds of unexpected disasters. Good emergency rescue system can control the disaster situation in time, ensure the safety of human life and property, and start reconstruction as soon as possible [3]. Therefore, it is necessary to evaluate the vulnerability of coal mine emergency rescue system, analyze its weak links, improve and perfect the system in order to improve its anti-interference and recovery ability.

The "1-9 scaling method" proposed by T. L. Saaty is always chosen when Analytic Hierarchy Process (AHP) is applied to compare different systems, by which values are first compared in pairs and then a definite value is given [4]. By doing so, experts tend to give an numerical interval rather than a specific value in the evaluation, that is, the comparison value is ambiguous or grey. In addition, the decision-making of a complex system often requires the participation of several experts from different professions and the judgment of each expert is often of preference. In order to evaluate the vulnerability of coal mine emergency rescue system which is a such complex system and features multiple hierarchies, grey index and multiple expert decision-making process, this

G. Q. Huang et al. (eds.), *Proceeding of the 24th International Conference on Industrial Engineering and Engineering Management 2018*,
https://doi.org/10.1007/978-981-13-3402-3_2

paper constructed a Group Gray Analytic Hierarchy Process (GGAHP) based on AHP and combined with Group Decision-making Theory and Grey Theory. At present, the research and application of GGAHP are few. Tang and Li [5] used this method to evaluate the vulnerability of railway traffic accident emergency rescue system; Cheng and Song [6] applied it to study the vulnerability of Beijing Subway under the impact of large passenger flow and Xu and Tian [7] used it to evaluate the vulnerability of coal mine enterprise emergency management system.

2 Methodology

2.1 Establishing the Index System of Vulnerability Evaluation

According to the new Safety Production Law and the implementation process of coal mine emergency rescue, combined with the analysis of man, material, environment and management which lead to the frangibility of emergency rescue system, the paper constructs the evaluation Index system of the vulnerability of coal mine emergency rescue [8–9]. As shown in Fig. 1.

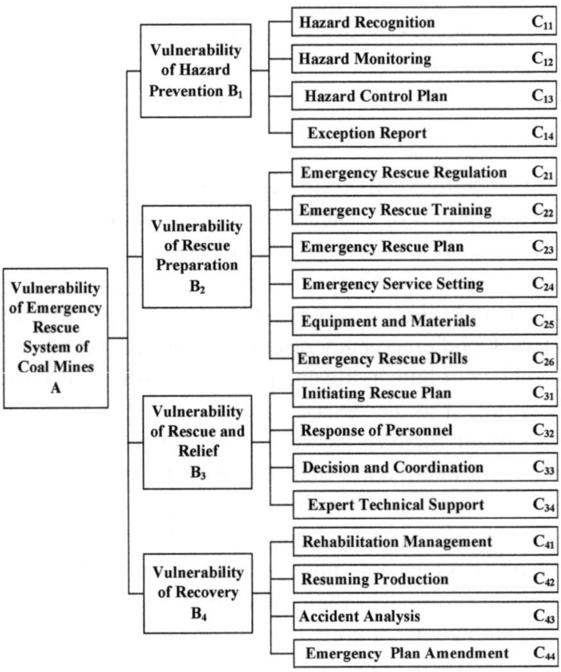

Fig. 1 Evaluation index system of vulnerability of emergency rescue system in coal mines

2.2 Constructing the Judgment Matrix of Group Grey

(1) *Constructing a single expert grey number judgment matrix*. Suppose a group of experts are invited to make decision by using the 1-9 scaling method to compare each level index in pairs, then give a numerical interval which is the scale gray number, thus the grey number judgment matrix of each level index $A^{(k)}(\otimes)$ [7] is obtained [10]. As in (1).

$$A^{(k)}(\otimes) = \begin{bmatrix} a_{11}^{(k)}(\otimes) & a_{12}^{(k)}(\otimes) & \cdots & a_{1j}^{(k)}(\otimes) & \cdots & a_{1n}^{(k)}(\otimes) \\ a_{21}^{(k)}(\otimes) & a_{22}^{(k)}(\otimes) & \cdots & a_{2j}^{(k)}(\otimes) & \cdots & a_{2n}^{(k)}(\otimes) \\ \vdots & \vdots & \vdots & \vdots & \vdots & \vdots \\ a_{i1}^{(k)}(\otimes) & a_{i2}^{(k)}(\otimes) & \cdots & a_{ij}^{(k)}(\otimes) & \cdots & a_{in}^{(k)}(\otimes) \\ \vdots & \vdots & \vdots & \vdots & \vdots & \vdots \\ a_{n1}^{(k)}(\otimes) & a_{n2}^{(k)}(\otimes) & \cdots & a_{nj}^{(k)}(\otimes) & \cdots & a_{nn}^{(k)}(\otimes) \end{bmatrix} \tag{1}$$

where *a(k) ij*(⊗) is interval given by the expert by comparing index i and j, among which k∈[1,m], m is the number of experts; i∈[1,n], j∈[1,n], n is the number of matrix participating in the comparison; *a(k) ii* (⊗)=1;

$a_{ij}^{(k)}(\otimes) \cdot a_{ij}^{(k)}(\otimes) = 1$; $a_{ij}^{(k)}(\otimes) \in \left[\underline{a}_{ij}^{k}, \overline{a}_{ij}^{k}\right]$, $\underline{a}_{ij}^{k}$, $\overline{a}_{ij}^{k}$ is the upper and lower limit of grey numbers.

(2) *Constructing the group grey number judgment matrix*. Grey number judgment matrices obtained by experts should be considered comprehensively to construct group grey number judgment matrix A(⊗), as in (2).

$$A(\otimes) = \begin{bmatrix} a_{11}(\otimes) & a_{12}(\otimes) & \cdots & a_{1j}(\otimes) & \cdots & a_{1n}(\otimes) \\ a_{21}(\otimes) & a_{22}(\otimes) & \cdots & a_{2j}(\otimes) & \cdots & a_{2n}(\otimes) \\ \vdots & \vdots & \vdots & \vdots & \vdots & \vdots \\ a_{i1}(\otimes) & a_{i2}(\otimes) & \cdots & a_{ij}(\otimes) & \cdots & a_{in}(\otimes) \\ \vdots & \vdots & \vdots & \vdots & \vdots & \vdots \\ a_{n1}(\otimes) & a_{n2}(\otimes) & \cdots & a_{nj}(\otimes) & \cdots & a_{nn}(\otimes) \end{bmatrix} \tag{2}$$

Where $a_{ij}(\otimes)$ is the grey number of comparison between i and j after synthesizing all experts' result, $a_{ij}(\otimes) \in \left[\underline{a}_{ij}, \overline{a}_{ij}\right]$. $\underline{a}_{ij}$, $\overline{a}_{ij}$ are obtained by Weighted Geometric Mean Method (WGMM), as in (3).

$$\underline{a}_{ij} = \prod_{k=1}^{m} \left(\underline{a}_{ij}^{k}\right)^{w_k}, \ \overline{a}_{ij} = \prod_{k=1}^{m} \left(\overline{a}_{ij}^{k}\right)^{w_k} \tag{3}$$

Where w_k is the degree of influence of the expert k on overall judgment.

2.3 Whitening of Group Grey Number Judgment Matrix

Suppose the positioning coefficient of the group grey number judgment matrix $A(\otimes)$ is ρ_{ij}, then $\rho_{ij}\in[0,1]$, $\rho_{ij}= 1 - \rho_{ji}$. $A(\otimes)$ could be whitening by (4) in order to get its whitening matrix $\tilde{A}(\otimes)$, as in (5).

$$\tilde{a}_{ij}(\otimes) = \underline{a}_{ij}^{\rho_{ij}} \times \bar{a}_{ij}^{1-\rho_{ij}} \tag{4}$$

$$\tilde{A}(\otimes) = \begin{bmatrix} \tilde{a}_{11}(\otimes) & \tilde{a}_{12}(\otimes) & \cdots & \tilde{a}_{1j}(\otimes) & \cdots & \tilde{a}_{1n}(\otimes) \\ \tilde{a}_{21}(\otimes) & \tilde{a}_{22}(\otimes) & \cdots & \tilde{a}_{2j}(\otimes) & \cdots & \tilde{a}_{2n}(\otimes) \\ \vdots & \vdots & \vdots & \vdots & \vdots & \vdots \\ \tilde{a}_{i1}(\otimes) & \tilde{a}_{i2}(\otimes) & \cdots & \tilde{a}_{ij}(\otimes) & \cdots & \tilde{a}_{in}(\otimes) \\ \vdots & \vdots & \vdots & \vdots & \vdots & \vdots \\ \tilde{a}_{n1}(\otimes) & \tilde{a}_{n2}(\otimes) & \cdots & \tilde{a}_{nj}(\otimes) & \cdots & \tilde{a}_{nn}(\otimes) \end{bmatrix} \tag{5}$$

2.4 Consistency Test of Whitening Matrix

By means of (6) and (7), the consistency of the whitening matrix $\tilde{A}(\otimes)$ is verified. If the consistency ratio of C.R. is less than 0.1, the inconsistency of the judgment matrix is acceptable; otherwise the judgment matrix should be corrected.

$$C.I. = \frac{\lambda_{\max} - n}{n - 1} \tag{6}$$

$$C.R. = C.I./R.I. \tag{7}$$

where C.I. is the consistency index of the matrix; λ_{max} is the maximal eigenvalue of the whitening matrix; n is the matrix order number; C.R. is the consistency ratio; R.I. is the mean random consistency index, obtained from the table.

2.5 Calculating the Vulnerability of Index

Calculate the eigenvector w corresponding to the maximal eigenvalue λ_{max} and deal with normalization to get the vulnerability values. From the bottom index to the top level index, the vulnerability value of the lower level index relative to the upper index is multiplied layer by layer to get the comprehensive vulnerability values. According to the evaluation result, the index with higher comprehensive vulnerability value is the weak link of the emergency rescue system.

3 Implementation

Using the above-mentioned method to evaluate the vulnerability of a coal mine emergency rescue system, 4 experts, with full knowledge of its current emergency rescue system, were invited to compare the vulnerability of the system, and the expert number was M_1–M_4.

Table 1 Grey number judgment matrix of B_1–B_4 obtained by expert M_1

A	B_1	B_2	B_3	B_4
B_1	1	(1/7,1/5)	(1/5,1/3)	(1/5,1/3)
B_2	(5,7)	1	(2,4)	(2,4)
B_3	(3,5)	(1/4,1/2)	1	(1,2)
B_4	(3,5)	(1/4,1/2)	(1/2,1)	1

Table 2 Grey number judgment matrix of B_1–B_4 obtained by expert M_2

A	B_1	B_2	B_3	B_4
B_1	1	(1/6,1/4)	(1/5,1/3)	(1/4,1/2)
B_2	(4,6)	1	(2,3)	(2,4)
B_3	(3,5)	(1/3,1/2)	1	(1,3)
B_4	(2,4)	(1/4,1/2)	(1/3,1)	1

Table 3 Grey number judgment matrix of B_1–B_4 obtained by expert M_3

A	B_1	B_2	B_3	B_4
B_1	1	(1/7,1/5)	(1/4,1/3)	(1/4,1/2)
B_2	(5,7)	1	(1,3)	(2,3)
B_3	(3,4)	(1/3,1)	1	(1,2)
B_4	(2,4)	(1/3, 1/2)	(1/2,1)	1

Table 4 Grey number judgment matrix of B_1–B_4 obtained by expert M_4

A	B_1	B_2	B_3	B_4
B_1	1	(1/6,1/5)	(1/4,1/3)	(1/3,1/2)
B_2	(5,6)	1	(2,3)	(2,4)
B_3	(3,4)	(1/3,1/2)	1	(1,2)
B_4	(2,3)	(1/4,1/2)	(1/2,1)	1

(1) *Constructing a Single Expert Grey Number Judgment Matrix.* 4 experts respectively compared the index B_1–B_4 with A, C_{11}–C_{14} with B_1, C_{21}–C_{26} with B_2, C_{31}–C_{34} with B_3, C_{41}–C_{44} with B_4, in pairs, and gave the evaluation interval. The grey number judgment matrix of B_1–B_4 to A is shown in Tables 1, 2, 3 and 4.

(2) *Constructing Group Grey Number Judgment Matrix.* Suppose the judgment weight of 4 experts are equal, $w_k = 0.25$, based on Tables 1, 2, 3, 4, 5 and 6, establish the group grey number judgment matrix of B_1–B_4 by using (3), as in Table 5.

Table 5 Group grey number judgment matrix of B_1–B_4

A	B_1	B_2	B_3	B_4
B_1	1	0.1543, 0.2115	0.2236, 0.3333	0.2541, 0.4518
B_2	4.7287, 6.4807	1	1.6818, 3.2237	2, 3.7224
B_3	3, 4.4721	0.3102, 0.5946	1	1, 2.2134
B_4	2.2134, 3.9360	0.2686, 0.5	0.4518, 1	1

(3) *Whitening the Group Grey Number Judgment Matrix.* Suppose the positioning coefficient of the matrix $\rho_{ij}= 0.5$, $\rho_{ji}= 1-\rho_{ij}= 0.5$. Whitening the matrix shown in Table 5 by using (4), as in Table 6.

Table 6 Group whitening judgment matrix of B_1–B_4

A	B_1	B_2	B_3	B_4
B_1	1	0.1806	0.2730	0.3388
B_2	5.5358	1	2.3284	2.7285
B_3	3.6628	0.4295	1	1.4877
B_4	2.9516	0.3665	0.6722	1

(4) *Consistency Test of the Group Whitening Judgment Matrix.* From Table 6 we know $\lambda_{max} = 4.0261$, by using (6) we get C.I. = 0.0087, and by looking up the table we get R.I. = 0.89, so by using (7) we can get C.R. = 0.0098 < 0.1, which is proved to be consistent with the requirement. Similarly, we can find out the whitening matrix by comparing C_{11}–C_{14} with B_1, C_{21}–C_{26} with B_2, C_{31}–C_{34} with B_3, C_{41}–C_{44} with B_4, then do the consistency test.

(5) *Calculating the Vulnerability Values of Each Index.* As shown in Table 6, the eigenvectors corresponding to λ_{max} are (0.1239, 0.8390, 0.4247, 0.3168), after normalization, we could get the vulnerability value of B_1–B_4 compared with A, that are (0.0727, 0.4922, 0.2492, 0.1859). Similarly, we can obtain the vulnerability value of C-layer index compared with B-layer, and then multiply the vulnerability value derived by comparing C-layer index with B-layer and that derived by comparing B-layer index with A-layer, so we can get the vulnerability value between C-layer index and A-layer index. If the vulnerability value of C_{11}–C_{14} to B_1 are (0.1276, 0.2702, 0.5380, 0.0642), then that between C_1 and A is $0.1276 \times 0.0727 \approx 0.0093$. The calculation results of each index's vulnerability value are shown in Table 7.

Table 7 Calculation results of comprehensive vulnerability value of evaluation index

	Evaluation index **B**	Evaluation index **C**	Overall vulnerability
Vulnerability of emergency rescue system of coal mines **A**	Vulnerability of hazard prevention $\mathbf{B_1}$(0.0727)	C_{11} 0.1276	0.0093
		C_{12} 0.2702	0.0197
		C_{13} 0.5380	0.0391
		C_{14} 0.0642	0.0047
	Vulnerability of rescue preparation $\mathbf{B_2}$(0.4922)	C_{21} 0.0342	0.0168
		C_{22} 0.0824	0.0406
		C_{23} 0.3871	0.1905
		C_{24} 0.1669	0.0821
		C_{25} 0.2133	0.1050
		C_{26} 0.1161	0.0572
	Vulnerability of rescue and relief $\mathbf{B_3}$(0.2492)	C_{31} 0.1514	0.0377
		C_{32} 0.0752	0.0187
		C_{33} 0.1236	0.0308
		C_{34} 0.6498	0.1619
	Vulnerability of recovery $\mathbf{B_4}$(0.1859)	C_{41} 0.4592	0.0854
		C_{42} 0.1018	0.0189
		C_{43} 0.1895	0.0352
		C_{44} 0.2495	0.0464

(6) *Analysis of Evaluation Results.* By calculating the vulnerability value of evaluation index, we can get that the emergency rescue plan ($C_{23} = 0.1905$), expert technical support ($C_{34} = 0.1619$) and emergency equipment and materials ($C_{25} = 0.1050$) are the weakest links in the emergency rescue system of that coal mine enterprise, which should be given priority in reinforcement.

4 Conclusion

Based on AHP, Grey Theory and Group Expert Decision-making theory, this paper constructed the Group Grey Analytic Hierarchy Process (GGAHP) to evaluate the vulnerability of the emergency rescue system of coal mine enterprises. On the basis of establishing the evaluation index system, this paper constructed the group grey number judgment matrix and whitened it, then calculated the comprehensive vulnerability value of the underlying index in order to identify the weak links in the emergency rescue system so that they could be improved. This method can synthesize group experts' opinions and fit the fuzzy thinking mode of human decision making process, thus the evaluation results would work for the actual situation of the emergency rescue system of the coal mine enterprise.

Acknowledgements. This work was supported by Fund for Scientific Research Training of Xi'an University of Science and Technology (No. 6310214027).

References

1. Y. Shang, Review on concept model of disaster vulnerability. J. Catastrophology **28**(1), 112–116 (2013). (Chinese)
2. M.A. Janssen, M.L. Schoon, W. Ke, Scholarly networks on resilience, vulnerability and adaptation within the human dimensions of global environmental change. Glob. Environ. Change **16**(3), 240–252 (2006)
3. Z. Feng, J. Hao, Assessment on emergency rescue capacity of coal mine accident. J. UESTC (Soc. Sci. Ed.) **12**(3), 53–65 (2010). (Chinese)
4. S. Xu, *Principles of Analytic Hierarchy Process* (Tianjin University Press, Tianjin, 1988), pp. 6–10. (Chinese)
5. S. Tang, X. Li, Study on method for assessment of vulnerability of railway emergency rescue system. J. China Railw. Soc. **35**(7), 14–20 (2013). (Chinese)
6. Y. Cheng, S. Song, Research on vulnerability of Beijing subway based on GGAHP. Urban Rapid Rail Transit **28**(3), 29–33 (2015). (Chinese)
7. T. Xu, H. Tian, Vulnerability assessment of emergency management system in coal enterprises. Saf. Coal Mines **47**(12), 219–221 (2016). (Chinese)
8. L. Yang, L. Wang, Assessment index system of coal mine emergency rescue ability based on interpretative structural modeling method. *J. China Univ. Min. & Technol. (Soc. Sci.)* **17**(1), 55–61 (2015). (Chinese)
9. R. Shan, C. Wang, X. Chen, Research on emergency rescue capability of coal mine based on vague set theory. Coal Technol. **35**(2), 320–321 (2016). (Chinese)
10. N. Xie, S. Liu, Grey analytic hierarchy process and its position-solving. *J. South. Yangtze Univ. (Nat. Sci. Ed.)* **3**(1), 87–89 (2004). (Chinese)

Research on Carbon Footprint Analysis Model of Mechanical and Electrical Products from the Perspective of LCA

Hui Zheng[1,2(✉)] and Yi-Fan Yan[1,2]

[1] College of Mechanical Engineering, Tianjin University of Science and Technology, Tianjin 300222, People's Republic of China
tjzhenghui@tust.edu.cn
[2] Tianjin Key Laboratory of Integrated Design and On-line Monitoring for Light Industry & Food Machinery and Equipment, Tianjin 300222, People's Republic of China

Abstract. In light of the carbon footprint of mechanical and electronic products distribution, based on the whole life cycle, the product carbon footprint analysis model is established, the product system boundary is determined, inventory data at each stage of the life cycle is collected and analyzed, Carbon footprint calculation model is set up, Gabi software is applied to calculate product carbon footprint. Through the establishment of model sensitivity analysis, the influence factors of the sensitivity of the product carbon footprint is calculated and the key factors are extracted. Therefore, the method can reflect the whole life cycle of real product carbon footprint and follow-up low carbon design with feedback guidance. Finally, the feasibility and effectiveness of the proposed method were verified by the example of a transformer.

Keywords: Life cycle · Carbon footprint · Sensitivity analysis
Uncertainty analysis · Transformer

1 Introduction

Under the background of global warming, low carbon economy based on low energy consumption and low pollution has become a hot topic in the world [1]. In today's economic life, industrial products occupy an important position, and the whole life cycle produces a lot of carbon emissions, which is one of the main sources of carbon emission [2]. China is implementing the scientific outlook on development, striving to build a resource conserving and environment-friendly society, and taking energy conservation and emission reduction as an important task for the development of the country. At present, the domestic and foreign scholars actively explore approaches to reduce the carbon footprint of products via manufacturing system, process, material

Fund project: Ministry of Education Philosophy and Social Sciences major research project (Number:16JZD014); Tianjin Philosophy and Social Science Planning Project (Number: TJGLWT15-006)

G. Q. Huang et al. (eds.), *Proceeding of the 24th International Conference on Industrial Engineering and Engineering Management 2018*,
https://doi.org/10.1007/978-981-13-3402-3_3

preparation and energy production. For example, the corresponding demand results are calculated from the life evaluation cycle [3, 4], the carbon footprint heuristic method [5], the analytic hierarchy process [6] and the carbon footprint tolerance analysis method [7]. Or establish the energy consumption forecast model of the product [8], establish the life cycle inventory analysis parameter [9], and realize the product optimization design based on the carbon footprint [10]. It's all about to actively explore ways to reduce the product's carbon footprint. The analysis model based on life cycle complete carbon footprint is not established by the carbon footprint scholars mentioned above, thus systematically calculate the carbon footprint spanning the whole life from production to waste, and perform the sensitivity analysis on the major factors that greatly affect carbon emissions.

This paper is based on the life cycle and establish the product carbon footprint analysis model and carbon footprint calculation process with the determination of the system boundaries. The inventory data is collected and analyzed for raw materials acquisition stage, manufacturing assembly stage, stage of transport, and stage of recovery and processing. The carbon footprint calculation model established by the whole life cycle analysis method, analyses the carbon footprint throughout every stage, and then the Gabi software is used to calculate the carbon footprint of products. After the establishment of the sensitivity analysis model, the sensitivity of product carbon footprint impact factors are calculated and analyzed, followed by extracting the key influencing factors and optimizing the data collection scheme, which can truly reflect the product carbon footprint and guide the subsequent design through the feedback.

2 Carbon Footprint Calculation Process

Product carbon footprint refers to "from cradle to grave" life process of a single product, in other words, it's the emissions of greenhouse gas from a single product accounting for the use and the treatment of the fuel, which goes through the entire life process from its manufacture, its use to its waste [11]. Based on the whole life cycle theory, the emission of greenhouse gases produced from the raw materials, energy acquisition, manufacturing, assembly, transportation, use and recycling of the products are quantified and calculated. Product carbon footprint calculation includes four key parts: goal and scope determination stage, data preparation and calculation stage, the follow-up processing stage and the verification and reputation stage.

2.1 Target and Scope Determination

A product is selected as the research object. Its distribution of the carbon footprint in each stage of the whole life cycle was obtained by analyzing the inventory and calculating carbon footprint, and through the result analysis to identify on the life cycle stage which has the relatively great influence on carbon emissions and the reasons for carbon emissions.

2.2 Selection of Functional Units

Functional unit is the basis for the calculation of carbon footprint [12]. the determination of the unit function is to provide references for the data collection in the data analysis process.

2.3 Determination of System Boundaries

System boundary is the data collected by the range of product data [13]. According to the enterprise actual investigation and study data acquisition, in this paper the PAS2050 life cycle model is used to set the evaluation system boundary of product carbon emission consisted of raw materials and energy acquisition stage, parts manufacturing and product assembly stage, transport stage, use stage, recycling processing stage.

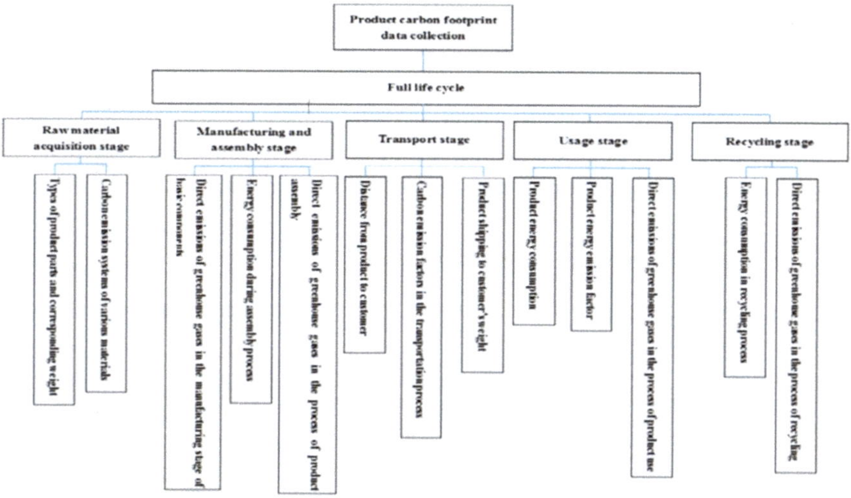

Fig. 1 Product carbon footprint data collection

2.4 Collection of Input Data

Collecting the carbon footprint data and establishing its own database are the premises on the assessment and calculation of carbon footprint of life cycle (Fig. 1). The determination of environment waste emissions requires scientific measures to collect output data [14]. At present, some developed countries have collected a large amount of carbon footprint emission data through the LCA list analysis software, but due to the regional differences, some foreign data can not be used directly for us. In order to analyze China's carbon footprint emissions data and gradually establish a database suitable for the national conditions of our country, A "carbon emissions reporting card" is established and shown in Fig. 2, so as to facilitate enterprises to fill in, and provides data support for the follow-up design of low-carbon.

Category	Type	Quantity	Remarks
Resource consumption	Water		
	Petroleum		
	Mineral		
Energy consumption	Power		
	Vapor		
Air pollutants	Carbon dioxide		
	Carbon monoxide		
	Sulfur oxide		
	Carbon hydride		
	Suspended particles		
Solid waste	Nec		
	Dangerous waste		

Fig. 2 Carbon emission report card

3 Product Carbon Footprint Calculation Model

In this paper, combined with the life cycle theory, through system analysis and definition of life cycle, using the PAS2050 as a carbon footprint calculation criterion, material consumption, energy consumption and greenhouse gas direct emissions and other factors are considered. The five stages of one product, including raw material acquisition stage, fabrication, installation and allocation stage, stage of transport, the use phase, the recovery process, are disassembled in order to quantify, and thereby the various stages of carbon footprint calculation formula can be given.

3.1 The Stage of Access to Raw Materials (G_M)

The calculation includes material consumption and energy consumption part

$$G_M = \frac{\sum_{i=1}^{n} M_i \times MEF_i + \sum_{j=1}^{m} E_j \times EF_j}{\eta_{ij}} \tag{1}$$

M_i, the quantity of class i material; E_j, the quantity of class j energy; MEF_i, the production emission factor of class I; EF_i, the energy production emission factor of class j

3.2 The Manufacturing and Assembly Stage (G_P)

The main source are energy consumption and greenhouse gas

$$G_P = \frac{\sum_{i=1}^{n} E_i \times EF_i + \sum_{j=1}^{m} O_j \times GWP_j}{\eta_{ij}} \tag{2}$$

E_i, the class energy consumption in the manufacturing assembly process; O_j, the class J greenhouse gas volume; EF_i, the energy emission factor; GWP, the global warming potential.

3.3 Transport Stage (G_T)

Mainly includes the direct discharge of transportation tool consumption and greenhouse gas

$$G_T = \sum_{t=1}^{p} M_t \cdot D_t \cdot EF_t + \sum_{t=1}^{p} O_t \cdot GWP_t \tag{3}$$

M_t, the quality of transport products; D_t, the transport distance; EF_t, the transport of carbon emission factors; O_t, greenhouse gas emissions directly.

3.4 Usage Stage (G_U)

The main consideration of product run time power consumption of carbon emissions and the use of the process of direct emissions of greenhouse gases

$$G_U = E \cdot T_w \cdot 365 \cdot EF + \sum_{j=1}^{m} O_j \cdot GWP_j \tag{4}$$

3.5 Recycling Stage (G_R)

Mainly by the material consumption and energy consumption of two parts

$$G_R = \sum_{i=1}^{n} E_i \times EF_i - \sum_{j=1}^{m} M_j \times MEF_j \tag{5}$$

The carbon footprint of products can be expressed as a quantitative model

$$G = G_M + G_P + G_T + G_U + G_R \tag{6}$$

3.6 Sensitivity Analysis

By product carbon footprint calculation, the most significant life cycle stages of carbon footprint can be acquired and the carbon footprint impact factors of the stages in the life cycle can be found out. In order to compare the sensitivity of carbon emissions by applying different factors, the greatest influence factor on carbon footprint is extracted. Through sensitivity analysis, calculate the sensitive data of various factors on the carbon footprint in different life cycle, we can get the following matrix:

$$\begin{bmatrix} S_{m1} & S_{m2} & \cdots & S_{mi} \\ S_{p1} & S_{p2} & \cdots & S_{pi} \\ S_{t1} & S_{t2} & \cdots & S_{ti} \\ S_{u1} & S_{u2} & \cdots & S_{ui} \\ S_{r1} & S_{r2} & \cdots & S_{ri} \end{bmatrix} = \frac{\Delta I_i}{G \cdot I_i} \begin{bmatrix} \Delta G_{m1} & \Delta G_{m2} & \cdots & \Delta G_{mi} \\ \Delta G_{p1} & \Delta G_{p2} & \cdots & \Delta G_{pi} \\ \Delta G_{t1} & \Delta G_{t2} & \cdots & \Delta G_{ti} \\ \Delta G_{u1} & \Delta G_{u2} & \cdots & \Delta G_{ui} \\ \Delta G_{r1} & \Delta G_{r2} & \cdots & \Delta G_{ri} \end{bmatrix} \tag{7}$$

Among them, Sm1, Sp1, St1, Su1, Sr1 respectively represent the raw material acquisition stage, manufacturing and assembling stage, transport stage, usage stage, recovery stage of the class i factors on the product carbon footprint sensitivity; ΔG_{m1}, ΔG_{p1}, ΔG_{t1}, ΔG_{u1}, ΔG_{r1} respectively these stage of the class i factor changes in the unit value of the carbon footprint of products under minimal change.

The sensitivity factor of carbon emission influencing factors: G-Product carbon footprint by factor value; I_i-The existing value of the class i carbon emission influencing factor; ΔI_i-The change value of the class i carbon emission influencing factor; S_{mi}-The degree of sensitivity of the class i factor to the carbon footprint of the product during the raw material acquisition stage; ΔG_{mi}-The change value of the class i factor product carbon footprint under the minimum change of the unit during the raw material acquisition stage.

4 Case Analysis

The following model, SFSZ-240000/330 transformer, is employed as the product instance to analyze the full life cycle carbon footprint. In this paper, the function unit of the carbon footprint calculation of the transformer is set up for 25 year operation. Through the function structure mapping, the transformer can be divided into 5 modules: the device body, the cooling device, the voltage regulator, the protection device and the outlet device. This transformer is three-phase, air cooling, forced oil circulation, load voltage regulating transformer, voltage level of 330 kV, capacity of 240 thousand kVA. After collecting the list of components and materials of transformers, the carbon footprint inventory is analyzed. After the collection of the transformer parts and materials list, the carbon footprint of these inventory analysis. Through the calculation of the transformer life cycle carbon footprint, and based on Gabi software to verify the transformer carbon footprint, the data shown in Table 1.

Table 1 Transformer in each life cycle carbon footprint

Functional unit	Carbon footprint calculation results/(kgCO2e)	Confirm calculation results on Gabi (kgCO2e)
Raw material acquisition stage	$5.573 * 10^5$	$5.767 * 10^5$
Manufacturing and assembly stage	$4.831 * 10^5$	$4.659 * 10^5$
Transport stage	$1.504 * 10^7$	$1.598 * 10^7$
Use stage	$5.385 * 10^9$	$5.394 * 10^9$
Recycling and processing stage	$-3.797 * 10^5$	$-3.581 * 10^5$
Total carbon footprint per unit transformer	$5.401 * 10^9$	$5.412 * 10^9$

Through the calculation of the carbon footprint of transformer, the most significant contribution to the life cycle stage are the use of stage and stage of transport. Power consumption/energy transport mode and system quality are the main factors that respectively affect the carbon emission during the use and transportation stage. The data from Gabi database isn't in accord with the current circumstance in China by reason of regional difference. The most significant stage of the carbon footprint of the transformer is the stage of use, which is the carbon footprint of the electricity production. In our country, the power generation is slightly different from the foreign power generation model, so the calculation results are relatively different. Regarding the existing carbon footprint of the transformer as an indicator, the carbon emission factor of electricity production, average operating power, transport energy options, recycling rate and quality of the system was taken as carbon emission factors for the sensitivity analysis. Set factors influencing changes in the range of 10%, the results as shown in Table 2.

Table 2 Sensitivity analysis of influencing factors of carbon footprint

Influence factors	Carbon footprint changes	Sensitivity
Electricity production emissions factor	$5.2374 * 10^8$	$9.6972 * 10^{-1}$
Average operating power	$5.2369 * 10^8$	$9.6963 * 10^{-1}$
Transportation energy options	$1.8033 * 10^6$	$3.339 * 10^{-3}$
Recycling rate	$3.1708 * 10^5$	$5.871 * 10^{-4}$
quality of the system	$1.8211 * 10^6$	$3.371 * 10^{-3}$

From the above results, the carbon footprint is influenced by the electricity production emission factors and the average operating power with the most sensitivity, followed by systems quality, transport, energy, and recycling rate sensitivity with the minimal sensitivity. The carbon emission factor is an important part of the carbon emission of the transformer and also affects the whole life cycle of the transformer. Therefore, reducing the emissions of power production factors will not only reduce the carbon footprint of transformer using stage, but also reduce raw materials acquisition, manufacturing, and assembly stage and power consumption in the process of recycling of carbon footprint. Due to the huge power of the transformer, the change of the parameters can also cause a significant change in carbon emissions. The power is also the key factor affecting the carbon footprint of the transformer.

5 Conclusion

In this paper, the model of SFPZ-240000/330 transformer, based on the whole life cycle of thinking, its systematic carbon footprint analysis, The calculation results show that carbon emission has the highest contribution ratio of the total carbon footprint in the whole life cycle of the transformer, reaching 99%, followed by transport phase. Carbon emissions from the use of the transformer were mainly produced in the process of electricity production. Based on the sensitivity calculation method results, effects of electric power production and emission factor and the average operating power of carbon footprint is the most sensitive, and electricity production emission factors, compared with the uncertainty of the factors that influence the transport of energy, is more significant. At the same time, in order to analyze China's carbon footprint emissions data and gradually establish a database suitable for the national conditions of our country, providing data support for the follow-up design of low-carbon.

References

1. L. Sun, Product structure regenerative design and feedback for low-carbon. Zhejiang University, 2014
2. X. Bian, S. Zhang. A review study of the carbon footprint. *Environ. Prot. Recycl. Econ.* **10**, 16–18 (2010)
3. S. Ross, D. Evans, Use of cycle assessment in environmental management. *Environ. Manag.* **29**(1), 132–142 (2002)
4. T. Yoshika, K. Aruga, T. Nitami, H. Kobayashi, H. Sakai, Energy and carbon dioxide balance of logging residues as alternative energy resources. System analysis based on the method of a life cycle inventory analysis. *J. Forset Res.* **10**(2), 125–134 (April 2005)
5. J. Toby, T.A. Okrasinski, W. Schaeffer, Estimating the carbon footprint of telecommunications products a heuristic approach. J. Mech. Des., Trans. ASMS **132**(9), 0945021–0945024 (2010)
6. X.F. Zhang, S.Y. Zhang, Z.Y. Hu et al., Identification of connection units with high GHG emissions for low-carbon product structure design. *J. Clean. Prod.*
7. G. Ameta, M. Mani, S. Rachuri et al., Carbon weight analysis for machining operation and allocation for redesign. *Int. J. Sustain. Eng.* **2**(4), 241–251 (2009)
8. K.R. Haapala, K.N. Khadke, J.W. Sutherland, Predicting manufacturing waste and energy for sustainable product development via WE-Fab software. *Proceedings Global Conference on Sustainable Product Development and life Cycle Engineering* (September 29–October 1, 2004), pp. 243–250
9. C. Zhang, G. Pu, C. Wang, Comparison of Life Cycle Assessment Between E-bike and Motorbike. *Machine Design and Research* **04**, 69–71 (2003). (In Chinese)
10. H. Bao, G. Liu, J. Wang, Optimal design of products with low-carbon footprint analysis. *J. Comput.-Aided. Des. & Comput. Graph.* **25**(2), 264–272 (2013). (In Chinese)
11. Z. Deng, Study on carbon footprint analysis and assessment model of the model of the molded bra product in the production process. Xi'an University of Engineering, 2013. (In Chinese)
12. G. Jiang, The method of extension knowledge classification and its application in products low carbon design. Zhejiang University of Technology, 2014. (In Chinese)
13. J. Wang, Research on the method of calculation for carbon emissions based on life cycle assessment. *Hefei Univ. Technol.* (2012). (In Chinese)
14. H. Dai, P. Dai, Study on data collection and inventory analysis of life cycle assessment. *Chongqing Technol. Bus. Univ.* **03**, 1–3 (2003). (In Chinese)

Ontology-Based Knowledge Representation Approach for Packaging Cost Estimation of Mechanical and Electrical Products

Yun-qi Gong[1(✉)], Xiao-jing Yu[2], and Li-xin Ma[1]

[1] College of Light Industry, Harbin University of Commerce, Harbin, China
gongyunqil@sohu.com
[2] Harbin Electric Power Equipment Plant, Harbin, China

Abstract. In the process of packaging design, the estimation of product packaging cost can provide support for the technical and economic evaluation of the mechanical and electrical products packaging design scheme and the reduction of the packaging cost. Information support for packaging cost estimation from multiple sources was analyzed, the application of ontology technology makes it more clear to estimate the cost of packaging, and to share the concept of packaging cost in different packaging design stage, first-order logic was used in common sense reasoning of the packaging cost. To solve the problem that multiple information integration, knowledge sharing and utilizing solution was proposed to support cost estimation in different packaging design stage, in which estimation methods including case-based reasoning, activity-based costing and simulation technology were used.

Keywords: Ontology · Packaging cost · Estimation · Knowledge representation

1 Introduction

A reasonable product packaging quotation should be given in time for the customer's inquiry or order when the packaging enterprises participate in the market competition. Packaging cost estimates could be used for technical and economic evaluation in the packaging design process of product, for support of the reduction of packaging cost also [1].

The current study on cost estimation mainly focus in the design stage, such as the development of the construction project bidding software based on BIM [2], Staub-French studies on the effects of design conditions on the cost in [3]. The cost estimation result was used for construction scheme selection in [4].

Existing cost estimation needs the support of many information sources, for different packaging design phase, such as scheme design stage, detail design stage and

---

Sponsor: Project supported by the Heilongjiang Provincial Department of Education, China (No.12541203)

G. Q. Huang et al. (eds.), *Proceeding of the 24th International Conference on Industrial Engineering and Engineering Management 2018*,
https://doi.org/10.1007/978-981-13-3402-3_4

construction stage, correspondingly, case-based reasoning [5], activity-based costing analysis [6] and simulation technology [7, 8] could be used in packaging cost estimation, different methods need the source of information is not the same, it is necessary to solve the integration problem of knowledge representation in the process of cost estimation and multivariate information effectively.

There has been a lot of research on ontology constructed [9–11], but there are few applications for packaging cost estimation. Ontology itself is not universal, that is to say, well-built domain ontology does not apply to other fields [12]. Therefore, the ontology constructed in this paper is the packaging cost field of electromechanical products, which is not applicable to other fields. The purpose of ontology model construction is not to establish a detailed concept list, but to build an effective domain application ontology based on packaging cost estimation. Ontology construction requires a combination of theory and practice in order to establish ontological relationships and axioms better.

In this paper, the analysis of cost estimation for the scheme packaging design stage, the detailed packaging design stage and packaging construction stage are discussed. On this basis, the concepts and relationships related to packaging cost are clarified, and the research of packaging cost estimation knowledge representation based on ontology is carried out.

2 Cost Estimation Analysis for Different Packaging Design Stages

Compared with the general lifestyle consumer goods, mechanical and electrical products packaging was almost not affected by the concept of consumption. Storage and transportation packaging delivered it as the core goal, in order to protect the product accuracy, easy for picking up and field assembly is for the principle, according to the product traits (such as shape, size, material, quality, etc.) to implement one or more times, partial or whole packaging to moisture proof, rust protection, impact resistance and other purposes [13]. Mechanical and electrical products of spare parts and whole machine packaging is different, the scheme determination of complete sets of mechanical and electronic products packaging is also closely related with the dissolution of the product process, it has important influence on the writing of the packing list and packaging cost control [14].

2.1 Cost Estimation Analysis of Package Scheme Design Stage

Mechanical and electronic products packaging under double constraints of product property and customer needs, the information analysis of product, such as the whereabouts of the goods, the mode of transportation and the specifications of the work pieces, plays a decisive role in the selection and implementation of the packaging forms, among them, packing method, packing grade, technical implementation, technical requirements determine the packaging cost, it can be used in cost estimation in the packaging scheme design phase, for example as the parameter applied to the case retrieval and reuse in case-based reasoning for product cost estimation. In the package scheme design phase, the cost estimation influence factors and their levels are shown in Fig. 1.

Product packaging shall be designed according to the nature of the product components, ensure that the products are not damaged, lost, rusted in the process of loading and unloading, transportation and warehouse storage. Proper packing methods should be chosen and reasonable packing size and tonnage should be taken into consideration.

From the perspective of packaging scheme design, the components of mechanical and electrical products can be considered in terms of shape, weight, property and production management, the classification of hydropower equipment as example is shown in Table 1.

In classification according to shape and weight: large-sized parts are often the main body of the unit, so it is necessary to attach importance to it when packaging. Heavy large part is a component with large shape, width or weight, which can not be transported in whole. The design should be divided into two parts, which should be carefully protected. Thin-walled parts are easy to deform, and vibration, impact and collision are unavoidable during transportation. Therefore, they need to be fixed with tension bars or bracing bars.

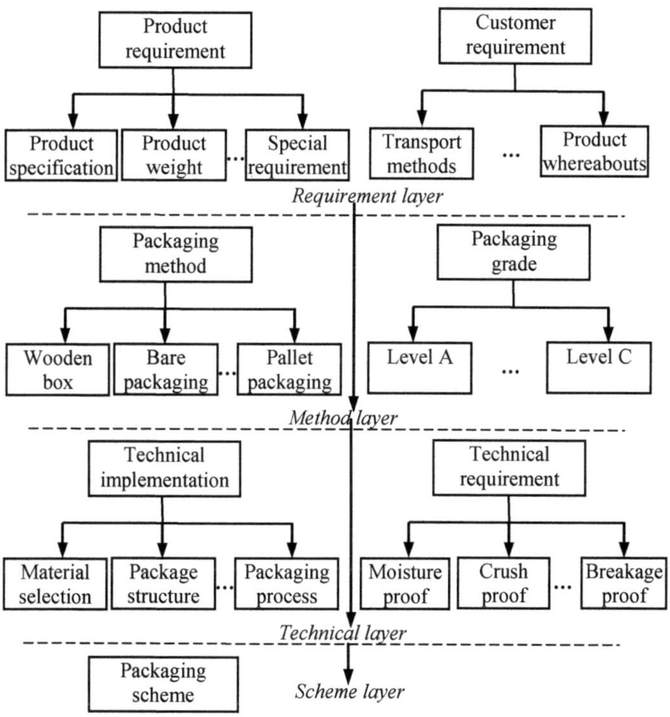

Fig. 1 Packaging cost influence factors and their architecture in scheme design stage

Table 1 Classification of some typical parts

Classification method	Classification	Typical parts
Shape weight	Heavy large parts	Framework, volute
	Large-sized parts	Main shaft, field spider
	Thin-walled parts	Shell, net cover
	Long thin parts	Oil pipeline, copper pipe
	Extra small piece	Plug screw, bolt
Property	Precision parts	Axle, bush
	Shock proof	Pressure gauges, other instruments
	To prevent broken	Glass tube for indicator
	To prevent pressure	Brush, Bakelite pieces
	Moisture proof	Coil, heating element
	Oil proof	Transmission belt
	Heat preserving	Mica band, silicone rubber belt
Production management	Missing parts	Valve, elbow
	Auxiliary material	Wool felt, filler
	Tool reserve parts	hand reamer, square broach
	Technical document	Drawing specification

The long thin parts are suitable in method of bundles packaging. Extra small pieces are often easy to lose, although the economic losses are not large, but it could affect the installation schedule, it should be classified into a special box, packaged in the instrument package.

In accordance with the classification of parts properties: precision parts should be strict rust prevention and storage according to the relevant packaging specifications. Instruments and gauges that need shock proof shall be carried out by instrument packing standard, to prevent broken parts as fragile packaging. The parts that need moisture proof shall be shipped in damp proof packing cases. The analysis of the other is similar to this and no longer to discuss.

Product packaging should be based on the classification of components, transportation methods, lifting methods, transportation routes, storage time and other factors to choose different packaging methods. Medium and small hydropower equipment should be installed by open, bundled, local packaging, boxes and a sealing tank etc.

Product parts transport from the factory to the installation of power station between motor transport, rail transport and maritime transport (or container) and other methods, the size and weight of parts of packing box should be selected, and in accordance with the relevant provisions. Sealed packing box is usually divided into three kinds: ordinary box, sliding wooden box and frame.

2.2 Cost Estimation Analysis of Detailed Packaging Design Stage and Packaging Construction Stage

Mechanical and electrical products packaging and traditional packaging or general products have similar life cycle, unit processes including demand analysis, packaging design, packaging production, packaging logistics and recycling reuse consist of packing product life cycle. From the cost estimation point of view, the cost categories and cost parameters corresponding to the unit process are shown in Table 2.

Table 2 Cost parameter per unit process

Cost category	Factor	Cost parameter
Package cost	Material	Quantity, unit cost
	Structure	Quantity, size
Packaging production cost	Process	Category, unit cost
	Material	Second processing cost
	Packing	Equipment cost, working hour cost
Packaging logistics cost	Load & unload	Equipment usage and work hour cost
	Logistics	Mode of transportation/distance, unit cost
	Storage	Storage mode/time, unit cost
Recycling, reuse cost	Recycling	Recycling & processing mode, processing costs, offset costs
	Reuse	Reuse or reprocessing type/mode, processing cost, offset cost
	Waste	Waste type/process/disposal cost/offset cost

2.3 The Concept System of Packaging Cost of Mechanical and Electrical Products

Product packaging costs are related with packaging mode, packaging production, packaging logistics and recycling reuse, packaging scheme is decided by the demand of packaging products, products related packaging variables including shape, properties, weight and so on, these characteristic variables determine the packaging mode and the packaging materials, in the process of packaging production and packaging logistics, activity-based costing may be used in indirect cost calculation, concept system related to the packaging cost was shown in Fig. 2.

Ontology is a concept derived from philosophy, introduced in the field of artificial intelligence, especially the explicit description of conceptualization. Ontology could be used to express various concepts and the relationship between them explicit and formal in the domain, there are a lot of researches on ontology, but study on the domain of packaging is still very weak, it is necessary to study knowledge representation of the packaging cost estimation of mechanical and electrical products based on ontology.

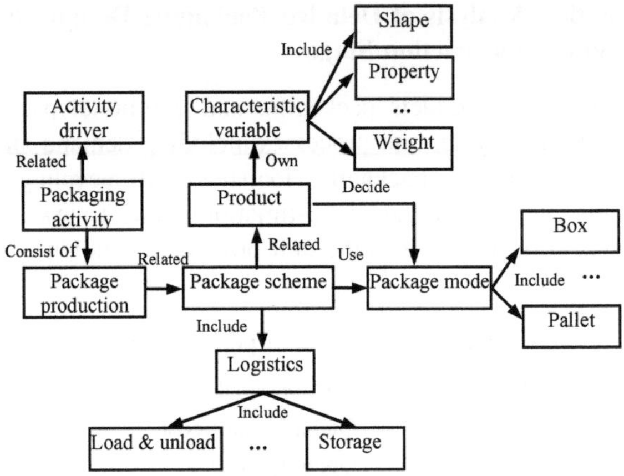

Fig. 2 Concept system related to product packaging cost

3 Ontology-Based Knowledge Representation of Packaging Cost Estimation

Ontology conceptual description describes the concept of ontology through multiple knowledge representation elements, Perez et al. summed up five basic modeling primitives with ontology taxonomy organization. Packaging cost was taken as the research object in this paper, activity cost analysis of packing detail design phase and construction phase as an example, indicating the ontology modeling and representation for packaging cost estimation knowledge, among them, simulation technology can be used in the measurement of cost drivers.

3.1 Classes or Concepts

Semantically, classes represent collections of objects, such as resource used in packaging activities. The attribute of concept reflects the static semantic features, they have no relationship with context, such as the static properties of resource, mainly the production characteristics of the resource management information and resources of their own description, management characteristics is a collection of related resource management information, such as equipment serial number, type of equipment and specifications of the equipment, machine-hour fee. The production feature refers to the collection of information related to the processing of equipment, such as processing type, machining accuracy, machining range, and replaceable equipment sequence. Concept and relation to support activity-based costing was shown in Fig. 3.

The related parts of activity-based costing are implementation class, resource class, product class and cost information submodel, in which cost information model is the core [15]. Implementation class is executable activity, resource driver class represents cost driver for a variety of resources, and cost center class is used for the merger of the same kind activity.

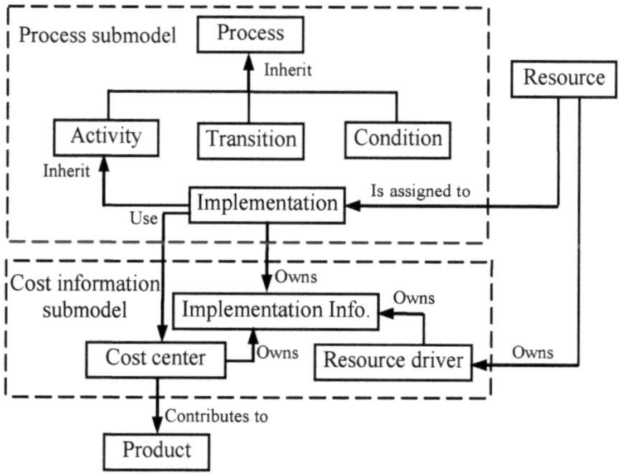

Fig. 3 Concept system to support activity-based costing

3.2 Relations

Relations refer to the interaction between concepts in a field, and the relation between concepts reflects a dynamic semantic feature. The activity-based costing analysis has the following key links between the four elements of the process, resources, activity-based costing information and products. Take relation "is assigned to" for example, it describes the distribution and support of the relationship between the elements, such as "Resource, Implementation", represents the distribution of resources to support the execution of an activity. In the actual modeling process, we can define the corresponding relationship according to the specific situation of the field.

3.3 The Other Modeling Primitives

Function is a special kind of relationship. The first *n-1* elements of the relationship can only determine the *n*th element

Axioms represent true assertions, for example, if there is a synonym for relationship concept of C1 and C2, such as resources decision making and resources choice is actually the same concept, the axiom expressed by first-order logic is as follows:

$$\forall(C_1, C_2) syn - concepts(C_1, C_2) \Rightarrow \\ syn - concepts(C_2, C_1)$$

Instances represent elements. Semantically speaking, instances represent objects

3.4 Packaging Cost Knowledge Query Using Ontology

Facing the packaging cost estimation process, the different requirements of packaging cost knowledge in different packaging design phases are defined. For the packaging cost knowledge query, in addition to the query and retrieval method of precise matching, namely the traditional retrieval methods, but also through the following ways:, retrieval concept with synonyms, inherit, the whole and the part of semantic relationship by using ontology; semantic retrieval by using the concept of correlation and similarity analysis of knowledge ontology; according to different stages of the design process of the packaging, the packaging cost and retrieval of knowledge push different versions.

4 Conclusion

Through the cost estimation analysis of different packaging design stage, and the study on representation of packaging cost knowledge, semantic information and constraint relationship between the concept of mechanical and electrical products packaging costs can be clearly expressed for the use of the packaging cost ontology query. But for different types and distribution in different locations of packaging cost knowledge, collaborative management at the logical level needs further study.

References

1. G. Gong, M. Lan, K. Yin, Research on comprehensive evaluation AHP model of total packaging cost (in Chinese). Packag. Eng. **34**(3), 100–103 (2013)
2. Z. Ma, Z. Wei, X. Zhang, Semi-automatic and specification-compliant cost estimation for tendering of building projects based on IFC data of design model. Autom. Constr. **30**(3), 126–135 (2013)
3. S. Staub-French, M. Fischer, J. Kunz, An ontology for relating features with activities to calculate costs. J. Comput. Civ. Eng. **17**(4), 243–354 (2003)
4. S. Lee, K. Kim, J. Yu, BIM and ontology-based approach for building cost estimation. Autom. Constr. **41**(5), 96–105 (2014)
5. Y. Gong, Research on packaging cost estimation of product based on CBR (in Chinese). *Packag. Eng.* **33**(15), 64–67, 87 (2012)
6. Y. Chen, J. Liu, Cost estimation of product packaging by activity-based-cost method (in Chinese). Packag. Food Mach. **28**(5), 30–34 (2010)
7. C. Li, Simulation and optimization of one kind of manufacturing & packaging processes (in Chinese). Packag. J. **5**(1), 10–14 (2013)
8. Wang C, D. Mu, Manufacturing enterprise's logistics operational cost simulation and optimization based on system dynamics (in Chinese). Syst. Eng.-Theory & Pract. **32**(6), 1241–1250 (2012)
9. X. Liu, S. Jiang, Z. Li, Ontology-based knowledge representation approach for building cost estimation. J. Eng. Manag. **29**(3), 19–24 (2015)
10. G. Wu, Y. Shi, H. Jiang, S. Liang, Support method used to assist decision making in packaging design via ontology (in Chinese). Packag. Eng. **36**(3), 59–64 (2015)
11. S. Gao, P. Wu, S. You, Introduction to researches on ontology-based product design knowledge representation. J. Intell. **30**(11), 156–161 (2011)
12. T. Walter, F.S. Parreiras, S. Staab, An ontology-based framework for domain-specific modeling. Softw. Syst. Model. **13**, 83–108 (2014)

13. F. Xu, Y. Ji, G. Qi, X. Gu, Packaging conceptual design of mechanical and electrical products based on low-carbon and cost constraint (in Chinese). J. Mech. Eng. **50**(10), 199–205 (2014)
14. Z. Ma, Export packaging design of large-scale mechanical and electrical products (in Chinese). China Packag. **30**(6), 51–54 (2010)
15. Y. Zheng, Y. Fan, Activity-based costing based on workflow meta model (in Chinese). Comput. Integr. Manuf. Syst. **13**(1), 178–184 (2007)

A Smart Framework for Intelligent Manufacturing in the Aluminum Profile Industry Using IoT Technologies

Xin Wang[1,4](✉), Yong Zhong[2], Anquan Zou[1], and Hepu Deng[3]

[1] Department of Management, Foshan University, Foshan, China
wangxinfosu@163.com
[2] Department of Internet of Things Engineering, Foshan University, Foshan, China
[3] Business and Law School, Foshan University, Foshan, China
[4] School of Business Information Technology and Logistics, RMIT University, Melbourne, Australia

Abstract. Intelligent manufacturing is becoming a critical strategy in "Made in China 2025. As a result, Chinese manufacturers are encouraged to effectively adopt IoT technologies for improving their competitiveness in the global market. This paper presents a smart framework for facilitating the adoption of IoT technologies in the aluminum profile industry. Such a framework can be used for help individual manufacturers in the aluminum profile industry adequately address various critical factors for product design and production planning with respect to the unpredictable requirements of customers and the availability of much production data. In particular, mould management, warehouse management and their effective integration in aluminum manufacturing can be considered with the use of IoT Technologies. This leads to much better performance in individual manufacturing organizations in the aluminum industry.

Keywords: IoT technologies · Mould management · Warehouse management Information systems · Intelligent manufacturing

1 Introduction

Contemporary manufacturing environments are changing rapidly with the development of information and communication technologies (ICT) across the world. Examples of such innovative environments are reflected with the adoption of latest ICT technologies worldwide including "Industries 4.0", "Strategic Innovation Promotion Programs, and "Advanced Manufacturing Partnership 2.0" [1]. The development of such intelligent manufacturing environments provides individual manufacturing organizations with numerous opportunities through better product design and production planning to improve their competitiveness in the world market, therefore leading to the rapid improvement in the national productivity of individual countries.

G. Q. Huang et al. (eds.), *Proceeding of the 24th International Conference on Industrial Engineering and Engineering Management 2018*,
https://doi.org/10.1007/978-981-13-3402-3_5

Following the trend of the whole world, China has been being committed at the development of intelligent manufacturing across the whole country. A specific national strategy referred to as "Made in China 2025" has been formulated. Such a strategy highlights the importance of intelligent manufacturing in the development of the manufacturing industry in China [2]. It effectively encourages Chinese manufacturers to adequately adopt IoT, cloud computing, big data, and mobile Internet in their product design and production planning so that the overall competitiveness of these manufacturing organizations can be improved.

There are various technologies that can be adopted for making the manufacturing process intelligent. IoT technologies, cloud computing and big data technologies are the priority selection for the Chinese manufacturing organizations. To decide which intelligent technologies are to be adopted in a specific situation, three critical issues have to be considered including: (a) the degree of digitalization of the manufacturing environment, (b) the development of a cyber-infrastructure with analytic capability, and (c) the adequate integration of the manufacturing environment with the cyber-infrastructure that leads to smart manufacturing [3].

This paper presents a smart framework for facilitating the adoption of IoT technologies in the aluminum profile industry. Such a framework can be used for help individual manufacturers in the aluminum profile industry adequately address various critical factors for product design and production planning with respect to the unpredictable requirements of customers and the availability of much production data. In particular, mould management, warehouse management and their effective integration in aluminum manufacturing can be considered with the use of IoT Technologies. This leads to much better performance in individual manufacturing organizations in the aluminum industry.

The paper is organized as follows. Section 2 identifies existing problems in the production process in the aluminum profile industry. Section 3 provides a smart framework for intelligent manufacturing in order to effectively solve these problems. Section 4 explains in details how IoT technologies can be adequately incorporated into the intelligent manufacturing framework. This is followed by the conclusion and the discussion about the possible future research.

2 Problems and Issues

The aluminum profile industry is related to the manufacturing of various aluminum profiles with respect to various requirements of the customers. The aluminum profile manufacturing is an extrusion process with thousands of moulds frequently shifted between work stations in an organization. As a result, the product quality, the mould life and the organizational productivity are very much affected by the effective management of moulds in the organization.

The current situation in the aluminum profile industry is that most of the manufacturing organizations are still at the lower stage of informatization with respect to their core manufacturing processes. These organizations are still remained in the traditional management process. As the production capacity grows, many manufacturing organizations have implemented various management information systems such as enterprise resources planning and customer relationship management. These systems are developed one after another in order to improve different parts of the manufacturing process in the organization [4]. This leads to a huge improvement of the production capacity and the productivity of the manufacturing organizations. Such improvements, however, are still not enough. The digitization evolution of the production process is still sluggish due to the frustrating bunch of obstructions, such as numerous influencing factors on the production plan, unpredictable customer requirements, the availability of the production data, and the difficulty in data collection in the manufacturing organization.

To improve the production efficiency and the product quality, aluminum profile manufacturing organizations have to adopt latest communication and information technologies and advanced manufacturing technologies in their process processes. This moves the whole organization to a higher level of digitalization. This calls for more developments in data collection and information processing [5]. In this situation, the critical parts of the production process that need to be digitized most are mould management, warehouse management and the integration of both.

Mould management is about the acquisition of various moulds and the adequate use of moulds in the production process in an aluminum profile manufacturing organization. With the increasing competition, the complexity of mould management is growing. This is due to increase of the number of moulds and their categories. This leads to various problems in managing modules including waste of time in fixing the position of a suitable mould, the use of wrong raw material in production, the failure of returning the mould to its position, the failure of counting the number of uses, and the use of a manual process for maintaining moulds [6].

Warehouse management is related to the storage of moulds in an aluminum profile manufacturing organization. To improve the warehouse management, latest warehouse management systems need to be adopted. With a large variety of modules, large aluminium profiles manufacturers are always immersed in inefficient warehouse management.

The effective integration of mould management and warehouse management is critical for improving the competitiveness of aluminum profile manufacturing organizations. Without an adequate integration of mould information and warehouse information, it is difficult, if not impossible, to effective schedule the production in the aluminum profile manufacturing organization.

An analysis of the production process in the current aluminum profile manufacturing organization shows that there are several factors that impede the fully automatic production with the incorporation of intelligent technologies. These critical factors include the presence of an information black box and the existence of an information island in the production planning process. This leads to inefficient production in aluminum profile manufacturing organizations. To effectively address these problems, the use of an intelligent system in the production planning in aluminum profile manufacturing organizations is becoming necessary.

3 An Intelligent Manufacturing Framework

An intelligent manufacturing environment consists of several modules. Figure 1 presents an overview of such an intelligent manufacturing framework.

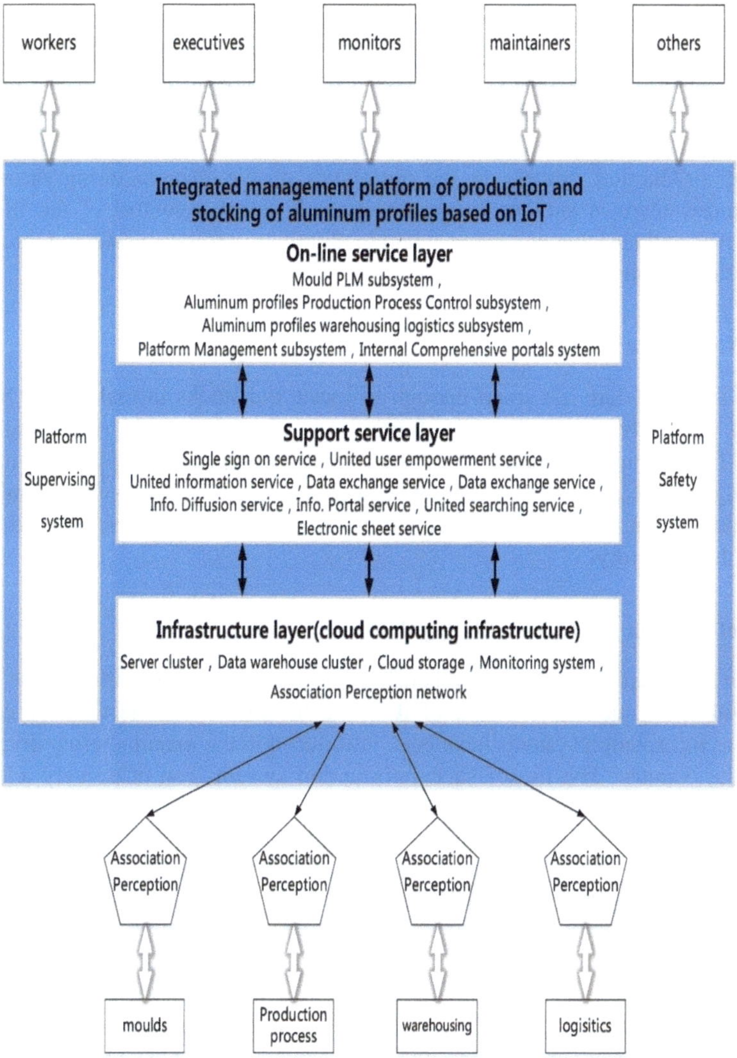

Fig. 1 An intelligent manufacturing framework using IoT technologies

3.1 Mould Management

There are two critical issues that needs to be adequately addressed in mould management including locating the proper moulds in real-time to promote the production efficiency and supervising the lifetime of the moulds to control the production quality [7]. To adequately address these issues, more dynamic data needs to be collected automatically and transferred to the database for achieving informatization which is the basis for intelligent mould scheduling.

Usually, the bar code technology is adopted in mould scheduling for streamlining the tools handling procedures. There are, however, several problems including misplacement, misuses and confused number of uses that cannot be adequately addressed with the use of this technology. This is due to the nature of the random order and the changing production process in the aluminum profile manufacturing organization. Furthermore, there is another issue related to the precise control of the lifetime of moulds before their interconnection with the aluminum profiles production process and the warehouse management process.

3.2 Warehouse Management

Large warehouses are not smart enough to locate the stocks in real time. They often lack the capability to calculate the shortest way of entry-and-exit in managing their stocks. To adequately address these issues, an appropriate adoption of intelligent technologies is required. The adoption of such technologies allows the manufacturing organizations to use advanced production theories such as just in time for reducing the cost of the inventory.

3.3 Integration of Information Systems

Large manufacturer need a comprehensive information system with the use of latest ICT. With the rapid development of ICT and the growth of the Chinese economy, numerous information systems have been introduced in the manufacturing organization across the country. The use of such information systems can effectively address the problem of the presence of information islands.

4 Application of IoT Technologies

The development of an intelligent manufacturing environment needs to solve all the problems above while meeting the demand of management at the same time. IoT is a leading and practical technology that can be adopted for adequately tackling challenges discussed above. There are five essential IoT technologies that have been broadly tested and innovated for the development of an intelligent manufacturing environment in manufacturing organizations. These IoT technologies include the radio frequency identification (RFID), the wireless sensor networks (WSN), the middleware, the cloud computing and the IoT application software [8].

A smart manufacturing environment consists of IoT-enabled monitoring and control systems for gathering data on equipment performance, positions of tools and goods, condition of critical parts and environmental conditions. Such systems can intelligently adjust the production process with respect to the data collected, therefore leading to an optimized production process in the aluminum profile manufacturing organization.

With such a great potential in the implementation of IoT technologies, it is sensible for manufacturing organizations to consider the challenges before investing. As a disruptive technology, the IoT brings multiple challenges to the adopting manufacturing organizations. RFID, for example, is the core strength of IoT as well as the weakness for its higher cost compared to the bar code technology. This calls for individual manufacturing organizations to adequately evaluate their own situations before committing to the adoption of latest IoT technologies in the development of intelligent manufacturing environments.

There are different characteristics in various manufacturing organizations. Such characteristics often lead to the formation of different manufacturing environments with unique managerial control styles in the aluminum profile manufacturing organization. As a result, the range and the depth of the application of IoT technologies in these organizations are very much different. In contrast with logistics and supply chains, manufacturing processes require more strictly on RFID in all directions including the product, the technique, and the information system. This means that the RFID system has no universal version. It have to be carefully designed with respect to the specific manufacturing environment in order to reduce the negative effects from the equipment, the information system, the workshop condition, and the assembly characteristics in the production process in the aluminum profile manufacturing organization.

4.1 Information Technology Adoption Study

The implementation of IoT technologies is a complicated system engineering activity. This requires a close cooperation between multi-departments. Often such cooperation is even extended between multi-firms. As a result, the adoption of IoT technologies is a risk investment. To minimize the risk in the development of intelligent manufacturing environments in the aluminum profile manufacturing organization, the capability maturity model can be used for evaluating the feasibility of such an immediate investment with respect to the strategic direction of the organization.

The information technology adoption theory can be used to analyze the prerequisite and capability of the adoption of RFID. Such an analysis can help the manufacturing organization to comprehend the internal expansion process with the adoption of the RFID technology. Four innovations can be used to improve the module management, warehouse management and the integration of these two. These innovations include (a) a fusion of RFID, bar code and DPM to control the cost, and with RFID middleware to process increasing timely data, (b) information encoding management throughout the total lifecycle of the production and the module, (c) middleware to fully integrate producing components based on module production lifecycle management, and (d) cloud storage and backup for sheer volume of IoT data.

4.2 A Module Lifetime Management System Based on IoT

Moulds for aluminium profiles are extrusion die. They work at a temperature between 500 and 600 °C. To achieve a lifetime management of moulds, it is crucial to study the mould identification technology by exploring the adaptability, reliability, and practicability of all the identification technology available, like bar code, RFID, and DRM during the life cycle of the mould.

There are various stages of the lifetime of a mould including design, production, inventory, employment and rejection. These stages require careful studies in order to develop an intelligent streamline production process in aluminum profile manufacturing organizations. Firstly, the routines of mould management, such as daily incoming check of new mould, the write-off of the worn mould, daily examination and condition of services need to be examined. Secondly, the key links including the prediction of the life span of moulds, the early warning from mould examination, and the statistical analysis of all the items produced by a certain mould need to be optimized.

The RFID technology is introduced in common stereoscopic warehouse to achieve an automatic collection of the mould stock data in a real time. The use of such a technology also automates the acquisition and transportation of moulds in the production process in aluminum profile manufacturing organizations. In this situation, the IoT technologies are used to improve the process control and the management of aluminum profiles manufacturing for enhancing the production quality and the monitoring of the production process.

Aluminum profiles manufacturing has special requirements for production identification and data collection. RFID is applied with the bar code technology to better manage the data, quality tracking, and monitoring producing at a lower cost. The hardware construction is built to collect and upload data quickly, precisely and in a real time. The key quality parameters are predetermined. Various quality theories are used in the software to analysis relevant data to figure out the head stream of the quality problem. Meanwhile, the real-time online data also reveal the actual producing situation and facilitate the production engineering.

4.3 Aluminum Profiles Manufacturing Warehouse Management Using IoT

The special make-to-order manufacturing mode turns the boom of the aluminium profiles categories, which is hard to be handled with one kind of identification technologies no matter it is bar code or RFID. A syncretic identification approach is presented to collect the basic data of the made-up articles. By using IoT technologies, warehouse management is redesigned to fulfill inbound, outbound and goods handling. To achieve the target more efficiently in a super large stereoscopic warehouse, an optimal path is automatically computed when enter-and-exit orders are auto-generated on the liftfork. This allows the liftfork to be navigated to the exact position of the racks in the production.

4.4 A Fusion Technique of Cloud Computing and IoT

Cloud computing is used to process and store data as a back-end infrastructure to help IoT support the production process and intelligent warehouse management. The combination of both new technologies can integrate the scattering information portals and data bases of the manufacturing organizations, therefore improving the performance of the aluminum profile manufacturing organization.

5 Conclusion

Large manufacturers with isolated information systems have great potential to become smarter in their factories and warehouses. This paper shows that IoT technologies can be used to reshape the mould management, inventory management and global information system in an aluminum profiles enterprise to improve its overall performance by providing greater agility and flexibility in production.

Acknowledgements. This article is supported by the National Social Science Foundation of China:'Carbon emission measurement and control study of ceramic supply chain in low carbon manufacturing'(17BJY008)

References

1. H. Lee, Framework and development of fault detection classification using IoT device and cloud environment. J. Manuf. Syst. **43**, 257–270 (2017)
2. J. Zhou, Intelligent manufacturing—main direction of "made in China 2025". Chin. Mech. Eng. **26**(17), 2278–2284 (2015)
3. M.M. Herterich, F. Uebernickel, W. Brenner, The impact of cyber-physical systems on industrial services in manufacturing, *Proc CIRP 2015*, Cape Town, SA, pp. 323–328
4. Y. Zhong, Z. Zhen, D. Lin, X. Qin, A method of fair use in digital rights management, *Proc. of the 10th International Conference on Asian Digital Libraries* (LNCS 4822, Hanoi, Vietnam, 2007), pp. 160–164
5. K.L. Choy, W.B. Lee, V. Lo, Design of case based on intelligent supplier relationship management system–the integration of supplier rating system and product code system. Expert Syst. Appl. **25**(1), 87–100 (2003)
6. D. Liu, J. Lv, B. Wang, M. Ji, Research on process of mold management system based on RFID. Inf. Technol. **32**(1), 120–123 (2017)
7. D. Liu, M. Ji, B. Wang, Design of mould dynamic detection model based on the internet of manufacturing things. Manuf. Technol. & Mach. Tool **4**(2), 29–34 (2017)
8. I. Lee, K. Lee, The internet of things (IoT): applications, investments, and challenges for enterprises. Bus. Horiz.S **58**, 431–440 (2015)

Two-Wheel Drive Practice Led by Lean Thinking

Yu-chun Wang[1(✉)], Li-fang Wang[2], and An-qing Zuo[2]

[1] Faw Car Co., Ltd, Changchun, Jilin, People's Republic of China
wangyuchun@fawcar.com.cn

[2] Huzhou Vocational & Technical College, Huzhou, Zhejiang, People's Republic of China

Abstract. Lean thinking is the core and soul of lean production; with technological and management innovations, the two-wheel drive is an important carrier for lean thinking. Lean thinking plays an important role in promoting lean production of enterprises by two-wheel drive. The work, based on the progress of lean production in the United States, Japan and Germany, proposed the two-wheel drive model led by lean thinking. Meanwhile, through the cases, it analyzed the practice and innovation of two-wheel drive led by lean thinking in detail. The results showed the two-wheel drive led by lean thinking is the theoretical basis, ideological basis and effective method for enterprises to achieve the supply lateral structure reform as well as take the road of "quality to efficiency".

Keywords: Lean thinking · Lean culture · Management innovation
Technical innovation

1 Background, Purpose and Meaning

The purpose of lean production is to minimize the resources occupied by the production of enterprises as well as reduce management and operation costs, with excellence, self-transcendence and continuous improvement. Two-wheel drive requires the lean thinking should be used in the enterprises [1]—enterprises should minimize the waste while providing customers with satisfactory productions and services. The application of intelligent manufacturing and Internet+ will bring lean management changes, taking the lean production as a prerequisite [2]. Over the past 30 years, China has been learning the industrial engineering application and taken lean production as the core for management innovation, thus achieving many results. E.g., FAW Car has established "Hongqi production system" after ten years of efforts. Taking FAW Car as an example, the work established the two-wheel drive model led by lean thinking in automobile manufacturing enterprises. It has the important practical significance for the maintenance and appreciation of values to state-owned enterprises, waste reduction in the whole value chain, adaption to the new normal economy, optimizing the state-owned enterprises in the supply lateral structure reform, independent innovation and breakthrough.

G. Q. Huang et al. (eds.), *Proceeding of the 24th International Conference on Industrial Engineering and Engineering Management 2018*,
https://doi.org/10.1007/978-981-13-3402-3_6

2 Establishment of Two-Wheel Drive Model Led by Lean Thinking

Value is the key of lean thinking, only determined by the end users—the value of the product (with a specific function at a particular price) is precisely defined by the user's conversation [3]. The five principles proposed in Lean thinking are the important theoretical basis for the construction of the two-wheel drive model. The contents are as follows: Redefining the value according to the customer requirements; identifying the value stream and re-establishing the enterprises' activities; activating the value stream; pulling the value stream based on customer requirements; continuous improvement and pursuit of perfection.

Innovation is the behavior to obtain a certain beneficial effect by improving or creating new things in a particular environment (including but not limited to various methods, elements, paths, and environment). Technological innovation refers to a series of activities (e.g., efficiency and quality improvement, and cost reduction) under the guidance of lean thinking through intelligent manufacturing to optimize design and manufacturing process. It runs through the complete process from the design of new products and technology to the market applications. Management innovation is to introduce new management elements (such as methods, means and modes) or element combination into the enterprise management system, thus effectively achieving the objectives of organization. Technological innovation leads the future, which is the fundamental of lean thinking; management innovation system comes first, which is the protection [4]. Lean thinking accelerates the two-wheel drive, whose innovation practice further enriches lean thinking.

FAW Car is market demand-oriented, taking "high quality, low cost and high efficiency" as the goal. Besides, it uses lean thinking to guide corporate culture change, and six "characteristics" to constitute two wheels (technological and management innovations). The two-wheel drive led by lean thinking embodies the pull creates value and provides solutions to problems, thus enabling enterprises to upgrade from "quantity" to "quality" (See Fig. 1).

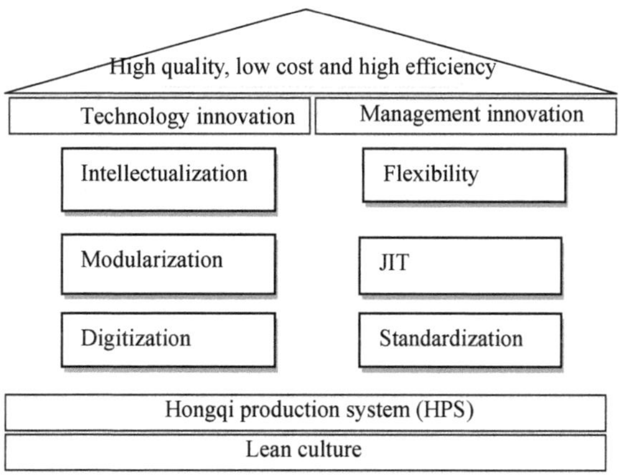

Fig. 1. Two-wheel drive model

3 Case of Two-Wheel Drive Achieved by Lean Thinking

The two-wheel model of lean production has been applied to the daily production and new car production of FAW Car, with the use of various means, tools and methods from technology and management. It has achieved remarkable results in many innovation cases, overall upgrading the management level of lean production.

Technological innovation is the fundamental of lean, reflecting the technology, information, transformation and service process of value stream. That is, more values are created for enterprises with less manpower and equipments, the shortest time and smaller sites. Its core is to approach users with satisfactory services.

3.1 Intellectualization of Lean Design

Intellectualization of lean design is create and increase the added product value by applying sensing, Internet, artificial intelligence, etc. in the automotive design based on user requirements. Its essence is to create values for users.

Besturn X40 is the first intelligence-interconnected SUV at level A0 in SUV market segment, as well as the first SUV with social feature. It is equipped with an advanced car network D-Life system, which makes customers travel more secure, convenient and stylish (with the greatest added value). FAW-Besturn, drived by Besturn X40, has a good sale due to its excellent performance, advanced configuration, fashionable appearance and strong power.

3.2 Modularization of Lean Development

Modularization of lean development is the simplified, scientific re-combination of complex and large automotive products. Standardized modules are integrated to generate modular effect. Therefore, the car assembly can become convenient and reliable as building blocks—the vehicle can be assembled with different modules. Modularity means each module can be designed independently, and the smaller subsystems playing an integral role are used to construct complex products or service.

The visible parts to users (display) can be re-modeled according to different modeling of cars. However, the invisible parts can use modular design, with mobile Internet and voice recognition. The development cycle is shortened through simultaneous development. Modularization is the important way to realize lean thinking, reflecting less investment, moderate station density, collinear multi platform, reduced car production time, and enhanced rolled yield. It simplifies the production process and shortened the production line and the cycle of inter-module synchronization development.

3.3 Digitization of Lean Production Arrangement

Digitization of lean production arrangement is to improve technology reachability and leaneness, shorten the production preparation cycle and reduce the stoppage and rework of production line through the extensive application of 3D analysis, virtual evaluation and offline programming in the engineering design and production preparation [5].

In the process of production preparation of welding, physical tooling rectification is an important factor troubling craftsmen and affecting production preparation cycle. Thus, in the tooling design of manual production line, FAW Car takes the following measures: Key stations such as 100% main work station with the whole process of man-machine simulation (See Fig. 2); key areas such as 100% simulation of welding spots in the launder area of side-wall rear light (See Fig. 3); the recognition rate of problem welding spots controlled in more than 98%; the main line of manual line in side wall, through man-machine simulation, optimizes the station layout and process

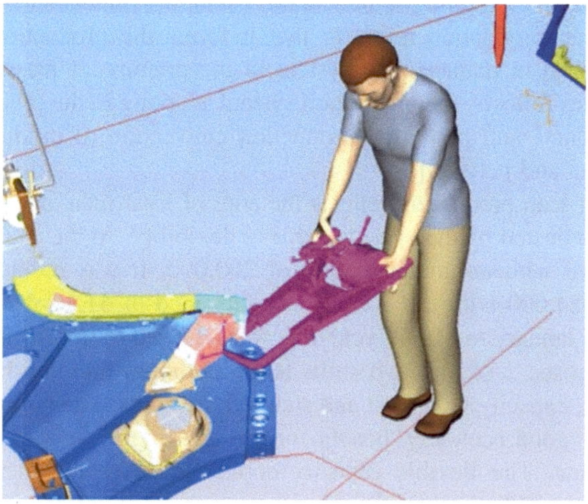

Fig. 2. The whole process of man-machine simulation

Fig. 3. 100% simulation of welding spots

arrangement; enhancing 10% of working performance. Digitization of lean production preparation reflects the lean principle—the maximizing efficiency, which is the effective tool to shorten production preparation cycle and speed up the replacement of produce life cycle.

Management and innovation drive the lean operation, which is an important guarantee for lean implementation as well as an important manifestation of lean system.

3.4 Flexibility of Lean Production

Flexibility of lean production shows the lean thinking, and embodies the principle of value flow. It is the premise of punctuality, and the fundamental of automation. Flexibility of lean production refers to that it forms the production of variety and flexibility with lean in management system in the premise of mass production. The flexibility includes follows: the production system adapting to the process requirements of different products and parts; the production conversion of production systems in different products and parts [6].

Flexibility of lean production reflects the core of least man, and increase/decrease of the output is adjusted by the cycle to achieve flexibility. At the end of 2015, Factory One was planned with an annual output of 200,000. It was predicted the full-year production was 84,000 with the average capacity load of 31.4% based on the actual production from January to May. Cycle time was adjusted to improve the capacity load and save labor costs. 2.16 min two shifts is the optimum program by analyzing production capacity, capacity load, cost and staff requirements. The program has achieved positive effects—good economy, benefit for the stability of the staff, and the increase per capita income. The flexible capacity coping leads to the increase of average capacity load from 77% (in the original play) to the actual 115%.

3.5 JIT of Lean Logistics

JIT of lean logistics refers to the transportation of goods (including raw materials, WIP and finished products) from the supplying to receiving place with minimum total cost according to users' requirements. Therefore, the source of lean appreciation is to provide necessary quantity of necessary goods at necessary time. JIT of lean logistics reflects the characteristics such as diversified, personalized customer requirements, short delivery cycle and production according to customers' expected delivery date [7].

FAW Car uses the logistics mode of sequential distribution to transport the engine, tire parts or assemblies. These parts or assemblies have characteristics including large volume, weight, high quality and multiple types or colors. In addition, the supplier or warehouse with sequential distribution ability is close to the company. E.g., five kinds of main, deputy and rear seats are provided by three suppliers. FAW Car achieves obvious economic benefits by reducing the number of suppliers to one for sequential distribution. The man hour for each car decreases by 30 s; the logistics area of workshop by 600 m^2; the number of staffs by 4.

3.6 Standardization of Lean Production

Standardization of lean production refers to the reduction of waste production and continuous improvement of production process by focusing on value-added items. In each link of the process, the current best implementation method is set as the standard, and constantly improved by relevant personnel. As the basis of lean appreciation, it reflects the perfect principle of lean thinking.

Standardization of lean production is the establishment, implementation and continuous improvement of the standards, reflecting the basic essentials of staff operation. FAW Car has developed a series of manufacturing standards, such as team management standards, basic operation essentials of operating units and staffs (bar chart, combinative table, SOP, operation standard and standard operation card). Through the formulation, implementation and development of standards, we improve production skills and comprehensive qualities of team members, thus reducing quality defects and rework.

Evaluation standard of star team management is formulated based on safety, personnel, quality, cost and production. After that, the evaluation standard is regularly revised to train section and team leaders (a total of 1,345 people trained within 6 years). Performance of star rating, team, section, workshop, factory level evaluation, excellent teams is recommended by step-by-step evaluation of team, section, workshop and factory. Meanwhile, workshop self-assessment results are randomly detected by IE department in accordance with the 20% sampling rate at regular intervals to verify the accuracy of workshop self-assessment, forming star team evaluation management closed-loop.

In FAW Car, team management levels are comprehensively improved through star team evaluation practice for 6 years. Using star team evaluation standard, team leader management abilities are continuously promoted to provide adequate personnel protection for the enterprises.

Under the guidance of lean thinking, FAW Car insists on two-wheel drive (of management and technology innovation) for continuous promotion of six leans. The system management abilities are constantly improved to provide powerful guarantee for the promotion of core competitiveness of enterprises.

4 Law of Two-Wheel Drive Practice Led by Lean Thinking

4.1 Two-Wheel Drive Led by Lean Thinking

The development concept determines the direction of development [8]. Two-wheel drive determines the business result. For FAW Car, five principles of value-oriented lean thinking constitute the core of enterprise survival and development. The whole process of business is driven by technology and management innovation of "six leans". The magic weapon of enterprise management is the organic combination of lean business concept, technology, management tools and enterprise culture.

4.2 Two-Wheel Drive Supported by Lean System

Lean system is the guarantee of achieving lean thinking. Ten elements of Hongqi production management system is the foundation of technology, management innovation and enterprise development. “Target management” is an effective tool to realize enterprise operation strategy. “Level production and JIT logistics” are important pillars of lean thinking. “Standard operation, process quality control and efficient maintenance” are effective guarantee for lean production. “Team management, complete staff improvement, 5S management and talent cultivation” are the basic methods of lean production. Hongqi Production Management System (HPS) is the core of two-wheel drive practice led by lean thinking of FAW Car.

4.3 Two-Wheel Drive Promoted by Lean Culture

Lean culture aims at pursuing “zero defect, stop and inventory”, which is the soul of lean culture. Lean culture consists of 5 pillars: “Prevention” in safety, “correctness” in quality, “personnel-oriented cultivation” in talent, “user orientation” in product and “excellence” in manufacturing. Five elements (challenge, improvement, respect, responsibility and innovation) form the mental foundation of lean culture.

4.4 Enterprise Innovation Mechanism Changed by Two-Wheel Drive

FAW car updates enterprise operation mechanism to enhance operation performance by two-wheel drive of technology and management innovation. Through the learning, digestion, absorption, application and innovation of two-wheel drive lean tools and methods, we form the continuously improved atmosphere with uniform direction, clear objective and definite responsibility to promote the innovation of whole system, personnel and process. The above tools and methods include intellectualization, modularization, platformization, digitization (synchronous engineering, virtual simulation, etc.), flexibility, standardization, punctuality, matrix project team management, personnel quality model, big calendar management, five-step work method, star team evaluation and quality management promotion [9].

5 Conclusions

The development of automobile industry has different driving forces. Two-wheel drive practice and innovation led by lean thinking of FAW Car are of great significance to the development of “window period” autonomous passenger vehicle enterprise [10]. Research result contributes to guiding lean production of other enterprises. China’s automobile manufacturing enterprises should follow the five principles of lean thinking to win in the fierce international competition. Based on “two-wheel drive” of technology and management innovation, lean production is developed by continuous innovation. In the nature, the season changes to bring about the new weather. However, the change of thinking generates the confusion. Only the concept and cultural gene heritage are the basis for long-term development of enterprises.

References

1. J.P. Womack, *Lean Thinking* (Commercial Press, 2001), p. 8
2. T. Ohno, *Toyota Production System* (China Railway Publishing House, 2006), p. 25
3. Z. Xiao, *Lean Production JIT* (Shenzhen Haitian Publishing House, 2003), pp. 12–21
4. Q. Shi, *Practice of Toyota Manufacturing Management* (China Machine Press, 2012), p. 38
5. C. Su, *Modeling and Simulation of Manufacturing System* (China Machine Press, 2008), pp. 43–51
6. Q. Wang, *Flexible Organization* (Shanghai University of Finance and Economics Press, 2007), pp. 22–47
7. A. Jose, M. Tollenaere, Modular and platform methods for product family design: literature. J. Intell. Manuf. (3) (2005)
8. J. Liu, *Promotion Strategy of Innovation Capability of Independent Brand Automobile Enterprises in China* (Fujian Normal University, 2015), pp. 10–30
9. X. Tian, Development status and countermeasure of independent brand passenger cars in China. China Price **57** (2015)
10. L. Deng, *Innovation Capacity System of China's Independent Brands* (Harbin Engineering University, 2007), pp. 75–98

Controlling and Optimization of Maintenance Production System for EMU Bogies

Yi-fan Ge, Bao-zhang Qu, Liang Zhang, and Bi-hong Lu(✉)

School of Mechanical Engineering, Dalian Jiaotong University, Dalian, China
bhlu@djtu.edu.cn

Abstract. Aiming to increase production capacity and efficiency, also avoid production imbalance, bogie of Electric Multiple Unit (EMU) maintenance production line was optimized. The 5th level wheel set maintenance line was analyzed and its process flow chart was drawn. Then use Plant-simulation software to model and simulate. The results showed that there are five bottlenecks and three blocks in the production line currently. The reasons were analyzed combination with the industrial engineering theories. The process layout adjustment was carried out on eight procedures respectively. The working hour of single train reduces by 13.3% after optimization. In order to further improve the production efficiency, the longest CRH380A wheel set maintenance line was simulated with mixed-flow production scheduling. Two scheduling methods were conducted, one was the motor trucks (D) and tow trucks (T), and the other was the 3rd, 4th and 5th level of maintenance. When the quantity ratio of the three levels of maintenance is 1:1:1, the optimal scheduling order is five-four-three. Considering single 5th level maintenance, the best order of trains is TDDDTDDD, the working hour of the maintenance decreases by 3.8%. The research accomplishments above provide technical support for process layout optimization, mixed-flow scheduling of EMU bogie maintenance and production controlling.

Keywords: Bogie maintenance · EMU · Mixed flow production
Production controlling · Simulation

1 Introduction

With the development of EMU, the quantity of trains is increasing rapidly. Also, the increasing speed of them results that the reliability and the performance of trains are requested more and more. Maintenance for trains has to be conducted timely. As an important part of the trains, the demand for bogie maintenance production increased greatly. The research on the optimization of bogie workshop can be of great significance to further development of Chinese railway freight transportation.

At present, production efficiency of the bogie maintenance workshop in most locomotive and rolling stock works is low, and their capacity is small, which is caused by multiple factors. Many scholars have carried out a large amount of researches. Beijing Jiaotong University, in view of the Shanghai South Maoyuan, conducted researches on the 3rd level maintenance for the bogie of CRH2 type motor train. They

G. Q. Huang et al. (eds.), *Proceeding of the 24th International Conference on Industrial Engineering and Engineering Management 2018*,
https://doi.org/10.1007/978-981-13-3402-3_7

improved the workshop's logistics facilities planning and optimized bottles based on Flexism software [1]; Researchers from Southwest Jiaotong University improved production efficiency by analysis the maintenance process with the bogie of passenger trains of Kunming Railway Bureau train segment as the object. They used intelligent tooling to take place of man-made operation [2]; Dalian Jiaotong University adopted SLP layout method to optimize the workshop layout of bogie assembly workshop and carry out data analysis. Meanwhile, optimized the production process from three aspects: low cost, highest productivity and maximum profit. The optimization results were evaluated by using two evaluation methods: balance rate and smoothness index [3]. A mathematical model with minimum total penalty cost was established for load balancing and solved by heuristic algorithm in 1967 [4]. In 2003, a new heuristic algorithm based on the uniform consumption rate of the assembly line was proposed by Jin MZ, which improved the target tracking method [5]. Bukchin J studied the mixed-flow line balance problem based on order production conditions [6]. A multi-level mixed-line sorting model that minimizes labor and conversion costs, equalization of parts consumption rate, and sub assembly workload were researched [7]. Reference [8] studied the sorting frame sorting model in the single-item delivery mode, which can improve the quality of delivery, reduce the errors and omissions of parts selection, and balance the consumption rate of components during assembly. Reference [9] studied the product conversion rate, total product conversion time, and improved minimization of workstation overload and total idle costs. LU Jianxia aimed to minimize overloading, idle time and total tuning time [10]. N Manavizadeh took reducing the production cycle time, work load and waste for each workstation as the goal under the constraint of buffer capacity [11]. Reference [12] studied the assembly line scheduling problem with the goal of minimizing the average unit cost and completing time, and designed a new multi-objective optimization genetic algorithm to solve the problem and optimize the sorting. Reference [13] studied two sorts of production scheduling problems with limited buffering, studied the sorting of prefabricated lines and the sorting of foaming line molds, which made the operation of prefabricated line and foaming line match as much as possible to reduce foaming. And the number of empty molds and the number of products in stock. Most studies focus on the optimization of logistics and process adjustment. But mixed-flow scheduling with variety types of vehicles, multilevel maintenance is less involved.

Mixed-flow production refers to the production mode which arranges a number of products with same production methods and processes in the same pipeline scientifically. Implementation of mix-flow production conforms the idea of industrial engineering theory and lean production concept, which effectively improves the bogie maintenance capacity and production efficiency. According to the field survey, the factory schedules production based on experience. However the bogie maintenance processes contain variety of procedures and complex production workshop layout, the scheduling based on experience cannot meet the large production requirements with short delivery cycle, resulting in production backlog in maintenance workshop and lack of railway running vehicles. The search is meaningful in the study of mixed flow scheduling in bogie workshop.

The research is based on Plant-Simulation, a simulation platform written in C++ aiming to analyze production layout, resource utilization, production capacity and production efficiency of different production lines and production systems. Plant-Simulation usually applies to automobile assembly line, mostly in assembly production. Modeling and simulation were applied to the maintenance of EMU bogies. Bogies maintenance production is different from traditional manufacturing production. Traditional manufacturing production is the process of converting raw materials into products, whereas bogie maintenance is the process of decomposing to maintenance, then getting the original product after assembly.

Objects of the EMU bogie maintenance can be divided into motor truck and tow truck. There are a variety of EMU types in maintenance production. According to the different operating time/mileage, maintenance level can be divided into the 3rd, the 4th and the 5th level of maintenance. The bogie maintenance production process consists of three different lines, which are wheel set maintenance line, structure maintenance line and small and medium maintenance line. Plant-Simulation was used to model and simulate bogie maintenance production line to analyze production line balance and bottlenecks. It is based on the basis of current production equipment and logistics layout state. Take the wheel set maintenance line as the research object since it has the longest working hours. Two scheduling method are conducted to improve the production efficiency and reduce the obstruction and waiting time. One is the motor trucks and tow trucks, the other is the 3rd, 4th and 5th level of maintenance.

2 The Process Flow Analysis of Bogie Maintenance

CRH380A model is selected as the research object, whose speed is up to 350 km/h. It is used in China Passenger Dedicated Lines. It is composed of 6 motor trucks and 2 tow trucks. In the following section, to conduct the analysis more clearly, use the letter T on behalf of the tow truck and the letter D on behalf of the motor truck.

2.1 Maintenance Procedures Analysis

There are 5 levels of bogie maintenance. Among them, the 1st and the 2nd level of maintenance is basic maintenance, while the 3rd, 4th and 5th level of maintenance are more complex. The 1st level of maintenance have to be carried out after each operation or every two days; the 2nd level of maintenance is carried out every month or after 30,000 km running distance; the 3rd level of maintenance has two phases. The initial 3rd level of maintenance is carried out every 1.5 years or 600,000 km running distance, while the forth level is carried out after every 4 years or 600,000 km running distance. The 3rd level of maintenance for the second time is 1.5 years after the 4th level of maintenance or another 600,000 km running distance. Then 600,000 km running distance or 1.5 years after, the 5th level need to be conducted. The 3rd, 4th and 5th level of maintenance happens in cycle.

With the field survey of the company's EMU bogie maintenance workshop, based on the actual process plan and production layout, the entire maintenance production line is divided into three main maintenance lines, successively named as wheel set maintenance line, structure maintenance line and small and medium maintenance line. These three lines share the same initial procedures, conducting maintenance after decommission, and with assembly process at the end. Among them, as the most important one, the wheel set maintenance line has most procedures and longest working hour. For the following detailed analysis of the entire manufacturing process, the 5th level wheel set maintenance process flow chart was drawn, as shown in Fig. 1.

The 5th Level Wheel Set Maintenance Process Flow Chart		Items	Frequency
Names	Wheel Set Maintenance	Processing○	24
Start Procedure	Bogie Protection	Inspection □	4
End Procedure	Bogie Test	Transportation→	6
Researcher	Yifan Ge	Waiting D	3

Procedure	Time/h	Processing	Inspection	Transportation	Waiting
1.Bogie Protection	1	●	□	→	D
2.Bogie Dismounting	1	●	□	→	D
3.Wheel Axle Box Dismounting	1	●	□	→	D
4.Wheel Transportation	0.1	○	□	→	D
5.Wheel Cleaning	1	●	□	→	D
6.Wheel Size Measurement	0.33	○	■	→	D
7.Wheel Surface Repair	1	●	□	→	D
8.Wheel Surface Flaw Detection	1	●	□	→	D
9.Gear Case Cleaning	2	●	□	→	D
10.Transportation	0.05	○	□	→	D
11.Bearing Dismounting	2	●	□	→	D
12.Axle Depainting	1	●	□	→	D
13.Wheel Dismounting	2	●	□	→	D
14.Wheel and Axle Flow Detection	2	○	■	→	D
15.Axle Magnetic Powder Inspection	0.33	●	□	→	D
16.Transportation	0.05	○	□	→	D
17.Gear Case Cleaning	1	●	□	→	D
18.Gear Case Disassemble	1	●	□	→	D
19.Detection/Clearance Measurement	1.5	○	■	→	D
20.Cleaning	0.5	●	□	→	D
21.Gearbox Running at High Speed	4	●	□	→	D
22.Waiting in Constant Temperature	8	○	□	→	◗
23.Transportation	0.05	○	□	→	D
24.Wheel Set pressing	4	●	□	→	D
25.Waiting for Inspection	48	○	□	→	◗
26.Wheel Manometric Assessment	0.5	●	□	→	D
27.Wheel Dynamic Balance	0.03	●	□	→	D
28.Ultrasonic Flaw Detection	0.5	●	□	→	D
29.Wheel Size Measurement	0.5	○	■	→	D
30.Transportation	0.05	○	□	→	D
31.Wheel Coating	1	●	□	→	D
32.Axle Box Body Assembly	3	●	□	→	D
33.Waiting in Constant Temperature	8	○	□	→	◗
34.Transportation	0.05	○	□	→	D
35.Bogie Assembly	4	●	□	→	D
36.Bogie Wiring	2	●	□	→	D
37.Bogie Test	4	●	□	→	D

Fig. 1. Wheel set 5th level maintenance process flow chart

2.2 Modeling the Bogie Maintenance Production Line

(1) *The establishment of object type library*: The production line contained of the 3rd, 4th and 5th level of maintenance. CRH380A has two tow trucks and six motor trucks. Use DC3, DC4 and DC5 represent the 3rd, 4th and 5th level maintenance of tow trucks separately, while TC3, TC4 and TC5 represent the three levels of maintenance of motor trucks separately. The Source is representing incoming and the Drain means leaving the factory.
(2) *Build a basic model*: The SingleProc objects in the software are used to represent manufacturing procedures. The Buffer objects are used to represent waiting in constant temperature procedures, and the line objects are used for handling procedures. Each procedure is connected by a connector. Every procedures are named in figures according to the inspection process flow chart (Fig. 1) and Table 1. Build a basic maintenance model in the software (Fig. 2).
(3) *Setting the properties and logical strategy*: Because of the differences between motor truck's maintenance and tow truck's maintenance, it is necessary to use the software method (Method Debugger) to write programs in Simtalk to control the

Table 1. Other procedures number

Number	Procedures	Number	Procedures	Number	Procedures
41	Frame disassemble	45	Framework maintenance	49	Outsource2
42	Frame cleaning	46	Framework assembly	50	Small and medium part maintenance
43	Depainting detection	47	Inspection		
44	Framework coating	48	Outsource1		

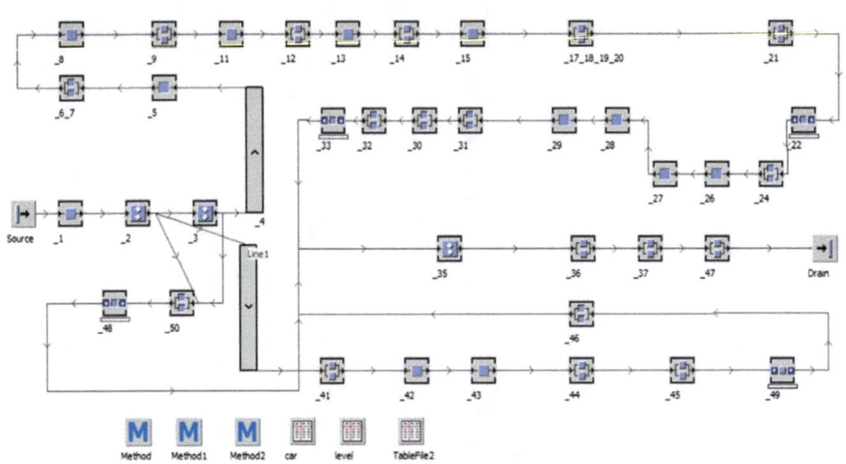

Fig. 2. Basic model of bogie maintenance production line

flow of objects. Set the properties of each station, including processing time, recovery time and so on. Considering the actual situation, the processing time is normally distributed with a ratio of 5%.

2.3 Simulation of Current Maintenance Production

Take the 5th level maintenance as an example. Every truck has a pair of bogies, which is four wheel sets. The simulation results show the total working time is of one train is 150 h. The utilization of equipment is shown in Fig. 3.

The bottleneck procedure refers to the procedure limiting the output of the entire process. In a broad sense, the bottleneck refers to various factors that restricts the output. For the production process, bottleneck procedure is the procedure with the longest working time. It is reflected in the procedure with the largest proportion of working in Fig. 3. Select the top five for optimization respectively; that is, wheel disassembly, wheel set press-fitting, gearbox maintenance, axle box assembly and gear box running at high speed.

(1) *Wheel disassembly*: There are four wheel-sets per truck. When the procedure is carried out, the production quantity is doubled while there existing one only equipment, so two additional equipment can be added to increase the production efficiency.

(2) *Wheel set press-fitting and gear box running at high speed*: During a production process, after the completion of the wheel press, the following procedure has to wait for the 48 h in constant temperature. Similarly after running gearbox high speed, eight hours waiting is needed. To optimize the process, the wheel coating, bearing press-fitting and axle box assembly that needn't to be carried out immediately after the check-up procedure can be brought forward, which shorten the waiting time.

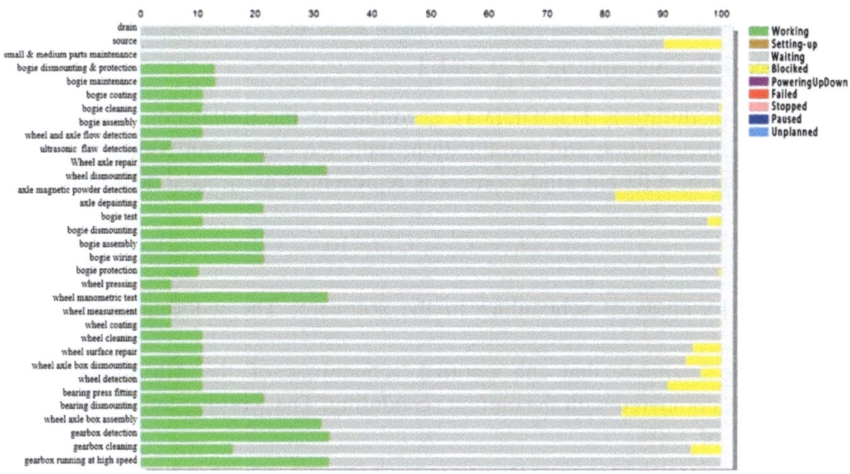

Fig. 3. Production equipment utilization currently

(3) *Gearbox maintenance and axle box assembly*: These two procedures rely mainly on manual operation, it may be appropriate to increase the number of labors.

Blocking procedure refers to the procedure leading to the subsequent procedure of material accumulation, in Fig. 3, reflected as a large proportion of blocking. Select the three largest procedures to optimize, that is, the framework assembly, the axle paint stripping and bearing off.

(1) *Axle paint stripping and bearing off*: The following procedure of these two procedures is the wheel disassembly, the long working time of which causes the two procedures being blocked. Optimize wheel disengagement will ease blockage.
(2) *Framework assembly*: The two previous procedures of the framework assembly are the framework painting and framework maintenance. The working time of the framework painting is short, so the number of stations can be reduced by one, which benefits balancing the production and saving space; frame maintenance occupied large area, mostly used as the storage. Completed framework can be subcontracting to save space for the framework assembly.

2.4 Simulation Results After Optimization

After adjusting the above procedures in the model, the simulation result shows that the working hour of the 5th grade maintenance were reduced from 150 h to 129.88 h, decreased by 20.12 h with ratio of 13.3%. The balance rate were reduced from 59.6% to 68.9%, increased by 9.3%.

3 Mixed-Flow Scheduling of the Bogie Maintenance

The analysis above shows that the wheel maintenance line takes the longest working time, with the most procedures and problems. Therefore it is separately extracted as an object of mixed-flow scheduling research.

3.1 Mixed-Flow Scheduling of D and T

The current scheduling is TDDDDDDT according to the installation structure of the train. Use the permutation and combination, with semi-row train conducting maintenance, numbers of different kinds' orders is represented as I_1, as shown in Eq. (1)

$$I_1 = C_4^1 \times C_4^1 = 16 \tag{1}$$

Simulate 16 kinds of scheduling in the software. The working time shows in Table 2.

The Table 2 shows that the optimal scheduling of the 5th maintenance is TDDDTDDD, compared with the current scheduling. The working time reduces by 7 h and decreases of 3.8%.

Table 2. Working time of 16 mixed-flow scheduling

Scheduling	Working time (D:H:M)	Scheduling	Working time (D:H:M)
TDDDDDDT	4:14:42	DDTDDDDT	4:15:32
TDDDDDTD	4:12:42	DDTDDDTD	4:13:32
TDDDDTDD	4:10:42	DDTDDTDD	4:11:32
TDDDTDDD	4:07:42	DDTDTDDD	4:09:32
DTDDDDDT	4:14:42	DDDTDDDT	4:15:32
DTDDDDTD	4:12:42	DDDTDDTD	4:13:32
DTDDDTDD	4:10:42	DDDTDTDD	4:11:32
DTDDTDDD	4:08:42	DDDTTDDD	4:11:32

3.2 Mixed-Flow Scheduling of Three Levels Maintenance

To research on three levels of maintenance production, firstly assuming that there is one maintenance vehicle of each levels of the maintenance, using the full permutation method to list all kinds. The number is represented as I_2, as shown in Eq. (2)

$$I_2 = A_3^2 = 6 \tag{2}$$

Simulate 6 kinds in Plant-simulation by setting delivery table in circular sequence, the results show in Table 3.

The Table 3 shows the optical scheduling with short working time is five-four-three (number 5), which means the 5th level of maintenance has the highest priority, then the 4th level and the 3rd level.

Table 3. Working time of 6 scheduling with 3 levels

Number	Scheduling	Working time (D:H:M)
1	345	5:13:12
2	354	5:11:12
3	435	5:06:42
4	453	5:05:42
5	543	5:03:02
6	534	5:03:42

4 Conclusions

A method was presented for the analysis and optimization of the EMU bogie maintenance production line with variety of production procedures and different kinds of trains. The main outcomes are as follows

(1) The program flow chart of the wheel set maintenance line was drawn based upon the analysis method of industrial engineering.
(2) The production system was simulated in Plant-Simulation. Results shows that there are five bottleneck procedures and three blocking procedures in the current production line. After adjusting on eight procedures respectively, the working hour of the 5th grade maintenance are reduced from 150 h to 129.88 h, decreases by 20.12 h with ratio of 13.3%. The balance rate are reduced from 59.6 to 68.9%, increases by 9.3%.
(3) The wheel-set maintenance production line were simulated by scheduling maintenance levels and vehicle types, the priority order given to the maintenance level is a five-four-three, the optimal sequence of vehicle types is TDDDTDDD.
(4) The research results above provide technical support for process layout optimization, mixed-flow scheduling of EMU bogie maintenance and production controlling as well.

References

1. J.I. Siwen, A maintenance strategy and simulation of maintenance line for high speed train bogie. M.E. thesis, Mechanical Engineering, University Beijing Jiaotong, Beijing, China (2012)
2. L.V. Liang, Research on bogie check and maintain line for railway passenger car. M.E. thesis, Mechanical Engineering, University Southwest Jiaotong, Chengdu, China (2007)
3. L. Tang, Bogie assembly workshop layout analysis and simulation optimization. M.E. thesis, Mechanical Engineering, University Dalian Jiaotong, Dalian, China (2016)
4. N.T. Thomopoulos, Line balancing-sequencing for mixed-model assembly. Manag. Sci. **14** (2), 59–75 (1967)
5. M.Z. Jin, S.D. Wu, A new heuristic method for mixed model assembly line balancing problem. Comput. Ind. Eng. **44**(1), 159–169 (2003)
6. J. Bukchin, E.M. Dar-El, R. Jacob, Mixed model assembly line design in a make-to-order environment. Comput. Ind. Eng. **41**(4), 405–421 (2004)
7. M. Zheng, M. Dong, Sequencing problem for a mixed model assembly line considering switching constraints. Ind. Eng. Manag. **17**(2), 85–89 (2012)
8. Y. Lin, Z. Zhao, Research on sequencing mixed model assembly lines based on set part supply. Ind. Eng. J. **16**(2), 112–116 (2013)
9. Q. LIU, Z. Fan, C. Zhang, Mixed model assembly line sequencing problem based on multi-objective cat swarm optimization. Comput. Integr. Manuf. Syst. **20**(2), 333–342 (2014)
10. L.U. Jianxia, W.E.N. Yaowei, L.I. Xiulin, Application of hybrid artificial bee colony algorithm in mixed assembly lines sequencing. Comput. Integr. Syst. **20**(1), 121–127 (2014)

11. N. Manavizadeh, M. Rabbani, F. Radmehr, A new multi-objective approach in order to balancing and sequencing U-shaped mixed model assembly line problem: a proposed heuristic algorithm. Int. J. Adv. Manuf. Technol. **79**(1), 415–425 (2015)
12. W.A.N.G. Binggang, Sequencing engine mixed-model assembly lines under buffer size constraints. Ind. Eng. J. **19**(1), 81–85 (2016)
13. X. Jiang, P. Su, L. Qi, A study of mixed-model assembly line sequencing problem of freezer door. Ind. Eng. J. **20**(6), 54–59 (2017)

Design and Development of the Personalized Customized Platform of Electric Motor Based on J2EE

Jian-guo Yu(✉), Wei-wei Zhang, Jian-ming Xiang, and Feng Zeng

School of Mechanical and Electrical Engineering, Jiangxi University of Science and Technology, Ganzhou, China
weier_ww@163.com

Abstract. With the rapid change of the times, the advent of the "Internet+" era, consumer demand continues to be personalized, and gradually promote the transformation of the production mode of motor manufacturing enterprises. Under the premise of satisfying customer needs and increasing participation, based on the characteristics of the production of motor companies, and combined with the status quo of motor manufacturing, through the company's on-site inspection and enterprise needs analysis, in order to build a personalized manufacturing management system for motor manufacturing manufacturers. At the same time, the physical data model of the system and the structure and module of the personalized customization platform system were respectively developed and designed using Power Designer modeling software and J2EE technology. Through the practical application of the company, the customer's experience with personalized design and the real-time interaction between the company and the customer are realized. The distance between the enterprise and the customer is shortened and take the customer as the center of the production mode is implemented.

Keywords: Motor production · Production mode · Personalized customization Personalized design

1 Introduction

With the continuous development of internationalization, as a large manufacturing country, the motor plays an important role in China's national economy. In the development of computer technology, to improve manufacturing level of today's market economy; as has been widely used electrical equipment manufacturing, should raise the level of design and innovation of marketing mode, to enhance the customer experience and the enterprise competitiveness, promote the development of China to make power [1].

Author: Jian-guo Yu, Born in 1979, male, Anlu Hubei, Associate Professor, master's degree; Direction Research: Enterprise Informatization.

G. Q. Huang et al. (eds.), *Proceeding of the 24th International Conference on Industrial Engineering and Engineering Management 2018*,
https://doi.org/10.1007/978-981-13-3402-3_8

Under the condition of economic globalization of the market economy, living in the era of "Internet+" the new generation of consumer groups of universality of life taste also from the traditional to individuation, differentiation, the standardization of traditional corporate interests centered, scale of production methods have been unable to meet the increasing personalized needs of consumers, to cater to the market changes, motor companies should keep pace with The Times, through personalization way to provide customers with personalized products and services, in order to attract customers, improve the enterprise core competitive ability, make the enterprise in the competition. According to the research on personalized customization, literature [2] summarizes the mass customization and personalization, and compares and analyzes the differences between mass customization and personalization, relevant technologies and practical applications. Literature [3] in order to identify and resolve problems when enterprises implement mass customization product, according to the characteristics of customization system and diagnosis of intentions, based on the fuzzy mathematics theory and set covering technology of multi-level enterprise customization diagnosis model was set up, customize the consistency of the diagnosis model and coverage is analyzed, and gives the solving strategy; In the literature [4], aimed at product customization, for the entire product family design and customize to create a consistent formal method, the integrated design model, and provide customized industrial application in a manufacturing company for case study; Through various cooperative mode, literature [5] was to implement mass customization production of cluster supply chain (CSC), based on the characteristics of MS building model, put forward the method of using various market under the condition of MSC, finally, according to the CSC actual production conditions, the simulation experiment to solve the problem; Literature [6], through the increase of the complexity of mechatronics products, was to adapt to the personalized and diversified demands of customers, the traditional design method of "product-centric" to "customer as the center", the design method of the proposed modeling method for multiple customer requirements, obtain and describe the needs of customers, in high-speed train design, for example, to verify the effectiveness of the proposed complex mechanical and electrical products development and the rationality and feasibility of multidisciplinary requirements modeling method; Literature [7] used CAD/CAM/CAE software parametric functions and parametric design of series hardware modeling, through the parameters related to realize data update, so as to ensure its accuracy, for the mass customization furniture series hardware modeling new solutions are put forward; Literature [8] proposed a real-time reliable physiological data based on human-computer interaction interface and the personalization of three-dimensional fitness system design method, in order to design a set of use exercise bikes, the device body cameras, feeling the heart rate sensor devices such as fitness system, and through the field investigation and test to get a better evaluation; Literature [9] breakthrough the limitation of traditional model car design, draw lessons from the Internet, try to the existing auto appearance design, configuration, personalized collocation to carry on the design of choose and buy, let consumer personalized fun in design experience, at the same time, the change of mode led industry development.

Modeling method based on the above research, it was the personalized customization of the study was more, some furniture in terms of system design, 3D fitness areas such as obtains a good effect of experience, but the personalization system research in the field of motor is relatively small, with the wide application of motor is manufacturing, to meet the personalized needs of customers, customization was carried out on the field of motor system study is very necessary. Customization production is the "single or mixed flow" type of production, is in recent years, with the improvement of economic level and the information technology and the development of new production mode, its characteristic is the number of small, high quality, quick market response ability, customer participation higher [10], the early stage of the custom at the customers demand, however, enterprise issued by the designer design stage, although it is the custom, but it can't reflect the customer's individuation.

This article embarks from the higher customer participation, and promote the real-time interaction with the user of the enterprise, to maximize meet customer custom experience for the purpose. Based on the J2EE technology, proposed personalization system was studied for the electrical products, to meet the personalized needs of customers.

2 Manufacturing Status of Motor Enterprises

Through on-the-spot investigation and analysis of some motor manufacturing enterprises, and found that the motor industry in our country have formed a complete system of business, product varieties, specifications, performance and production also has basically meet the requirements of the development of the national economy, the industry growth rate is on the rise as a whole. However, from the current policy and market, the motor enterprises have the following problems:

(1) Though motor product variety in China, but its efficiency was generally not high, there was the "big horse-drawn cart" phenomenon, thus design and application of high efficiency motor become the inevitable choice of enterprise;
(2) As customer's requirements and the differentiation of increased demand for motor, the motor needs to be customized: A new generation of consumers has been gradually tends to the trend of personalized customization, customer to improve energy efficiency and specialization of products of the demand was higher and higher, prompting companies according to the actual needs of customers design and manufacture of related parts and components of the motor parameters, with customers to design research and development into enterprise enhance motor is a new way to research and development ability;
(3) Customer requirements for the reliability of the electrical product quality and after-sales service more and more high: Motor product quality and after-sales service level and fault repair ability gradually becomes a key indicator of evaluation of enterprises for the customer.

3 System Function Design

According to the investigation and analysis of some motor manufacturing enterprises, the modular function of the motor personalized customization system is as follows

(1) Access system: Users can register on the access interface through the system entry, enter the motor personalized customization system, and then the system was accessed.
(2) Personnel management system: The function modules according to different users set different permissions, according to the information submitted to the backend database, through the system management review, to assign different access permissions.
(3) Spare parts library management system: This module can see the appearance of all accessories and corresponding functions, and even browse to the origin of accessories.
(4) Order information management system: The platform provided customers with the delivery of orders, queries, modifications and deletions of orders, and provides technical support for the management of the order.
(5) Product display system: The system can display the 3D model of existing products, can also according to the customer choice of parts in the system generated in the product design drawings, in 3D model, system interface for custom products have a more intuitive display.
(6) Design evaluation system: The system through the platform will complete the products sent to customers design background evaluation system database, in accordance with the relevant indicators of product performance and security of the objective evaluation, it was concluded that weight range, the final assembly of customer expectations through the system customization product model [11].

4 System Data Model Design

Personalization system in this paper, the motor design mainly includes the personnel management, parts inventory management, order information management, product information, evaluation system and so on several parts, based on the physical database model is established. Power Designer modeling software was adopted in this system was designed and developed the system database relational model, a complete function and convenient application of the computer aided modeling tool, it can provided different models with different development environment, and between the model and can be independent, and then connect [12]. The physical data model of the motor personalized custom system (partial module representation model) designed by Power Designer was shown in Fig. 1.

5 System Architecture Design

J2EE was the Sun's launch of a new concept for enterprise application development model, it used a multi-layer distributed model, according to the function could be divided into components, application layer distribution of the various application components according to their application in different machines. A multiple stratification application was abled to provide different services independent layer, typical

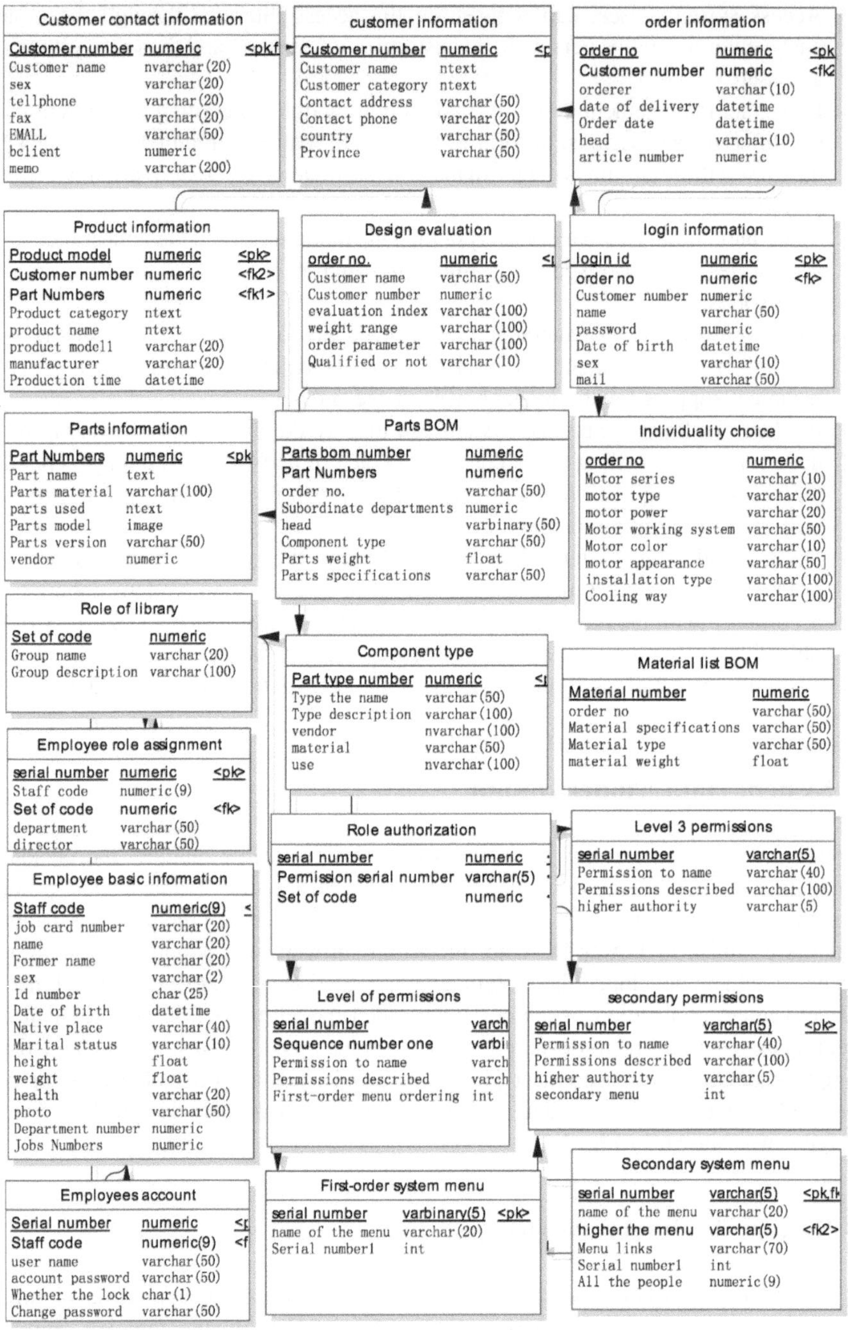

Fig. 1. Physical database model of personalized custom system

J2EE architecture was divided into the following four layer structure: the Web client and a Web component layer, business logic layer components, enterprise information system software, based on the components design, development, assembly and deployment of enterprise applications, high portability, scalable, easy to maintenance, apply to the design and development of the enterprise information system, a good development environment for enterprise application developers [13].

System integrated use of computer network technology, J2EE platform and SQL Server database technology, motor product customization design based on J2EE platform, designed for real-time interaction between customers and enterprises and the view was to provided fast and convenient communication platform. The system was divided into four layers: client layer, business layer, access layer and data resource layer, as shown in Fig. 2. Based on the characteristics of J2EE "write once, run around", easy access to database technology and the ability to secure data in Internet applications, solve complex problems in system development effectively [14, 15].

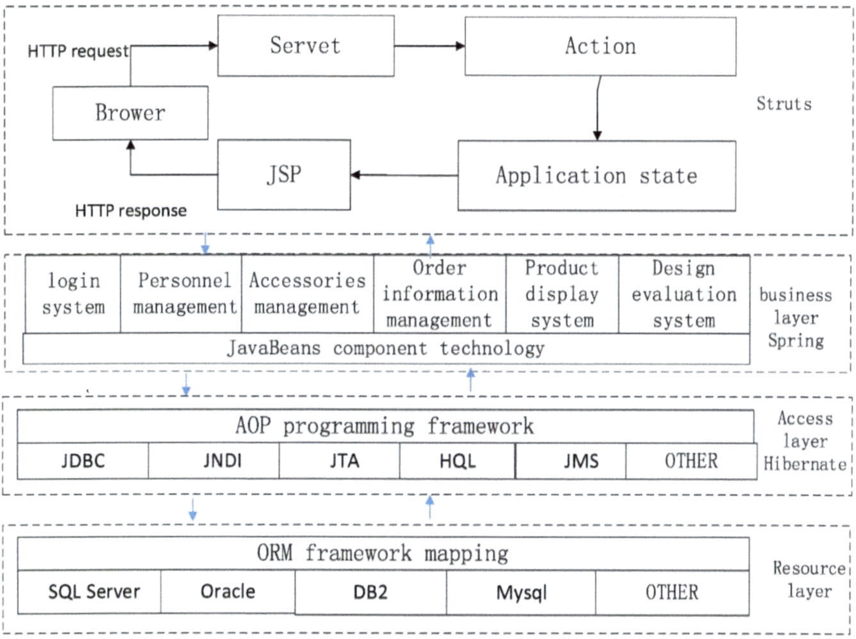

Fig. 2. System diagram

6 System Development

The system used tomcat as the Server, SQL Server as the background database management system, and the personalized customization system was developed. The key functions in the system development were brieflied introduced as follows:

(1) Login system
The login system was the authorization of users who accessed the system. Through the authentication, the user can customize the product and interact with the product designer
(2) Main interface of system
The main interface of the motor personalized customization management system was shown in Fig. 3, which mainly shown the types of motor, main parts of the motor and national standards
(3) Customize the system
Motor personalization system interface as shown in Fig. 4, the module to visual display system of personalized customization, set the important parameters of the motor for customers to choose according to individual be fond of, and select the results on the same page.

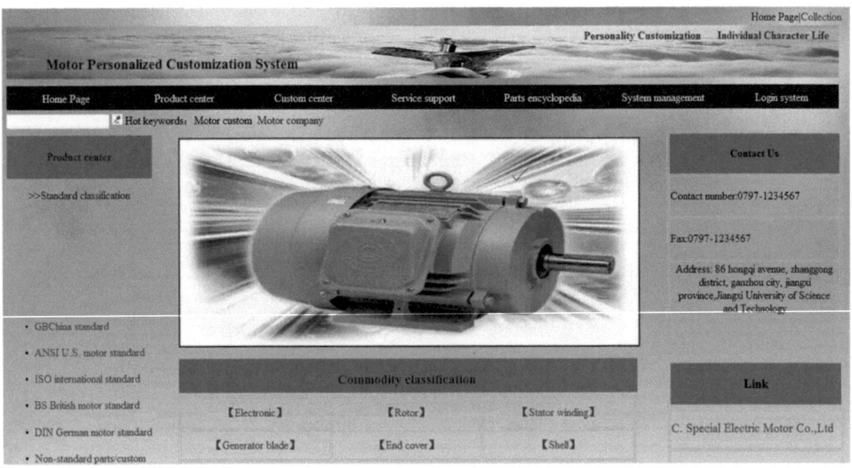

Fig. 3. Main interface of the motor personalized customization system

Fig. 4. Motor personality customization system

7 Conclusion

In order to improve the market competitiveness, motor manufacturing enterprises with the development of the modern market economy, put forward motor product customization platform based on J2EE technology, analyzes the architecture of the platform and related functions, to customer satisfaction and participate as the ultimate purpose, effectively meet the personalized needs of customers to provide support for the enterprise, reducing production cost, improve enterprise in the position in global market competition; After completion of system development, after some motor enterprise applications, make customers to be able to realize personalized experience, let customers fully enjoy the pleasure of design, the realization of the maximum customer satisfaction, as well as motor enterprise development provides a new train of thought.

Acknowledgements. Project Funding: Jiangxi Province Colleges and Universities Humanities and Social Science Research Project, GL1544.

References

1. H. Jia, Y. Wang, C. Guo, Development and development of personalized customized platform for hardware products based on web. Light Ind. Mach. **30**(2), 100–103 (2012)
2. J. Tiihonen, A. Felfernig, An introduction to personalization and mass customization. J. Intell. Inf. Syst. **2**, 1–7 (2017)
3. W. An, W. Yang, W. Guo et al., Research on enterprise customization diagnosis for mass customization. Int. J. Adv. Manuf. Technol. **76**(1–4), 669–674 (2015)
4. M. Bonev, L. Hvam, J. Clarkson et al., Formal computer-aided product family architecture design for mass customization. Comput. Ind. **74**(C), 58–70 (2015)

5. X. Xue, Z.Z. Liu, S.F. Wang, Manufacturing service composition for the mass customised production. Int. J. Comput. Integr. Manuf. **29**(2), 119–135 (2016)
6. X.J. Ma, G.F. Ding, S.F. Qin et al., Transforming multidisciplinary customer requirements to product design specifications. Chin. J. Mech. Eng. 1–12 (2017)
7. Z. Ye, S. Wang, P. Liu, A series of hardware modeling methods for mass customization. Furniture **2**, 20–25 (2013)
8. H. Zhao, G. Meng, L. Han et al., Design and implementation of personalized and customized virtual fitness system. J. Chin. Graph. Graph. **20**(7), 953–962 (2015)
9. Z. Pan, M. Cui, Explore the personalized customization mode of automobile in the Internet era. Packag. Eng. **2**, 21–25 (2015)
10. Z. Zhou, K. Zhao, J. Yu et al., The realization of personalized customized system based on networked manufacturing. Mech. Des. Manuf. (3), 87–88 (2008)
11. Y. Duan, *Research on Personalized Customization System Based on Web* (Mechanical College of Tianjin University, Tianjin, 2007), pp. 43–44
12. J. Liang, W. Zhao, B. Chu, A database design forward project and its application based on PowerDesigner. Comput. Knowl. Technol. **12**(22) (2016)
13. Y. Tao, Z. Wu, D. Tang et al., Research on application framework based on J2EE. Comput. Eng. Des. **28**(4), 826–828 (2007)
14. J. Tian, Development platform of Java development language and research on J2EE programming technology. Inf. Technol. Inf. Technol. **4**, 112–113 (2016)
15. J. Yu, M. Feng, Study on the integrated production planning and scheduling optimization system of tungsten powder processing. Manuf. Autom. **35**(3), 58–62 (2013)

Joint Optimization of Flexible Periodic Preventive Maintenance and Nonidentical Parallel Machines Scheduling with Consideration of Delivery on Time

Ji-shan Zhang(✉), Hang-yu Lou, and Ya-run Liu

Department of Industrial Engineering, Northeastern University, Shenyang, China
389520623@qq.com

Abstract. This paper targeted at multiobjective integration optimization of equipment maintenance and production scheduling. We take all speed parallel machines as an example, providing that flexible periodic preventive maintenance and repair maintenance is adopted as the maintenance strategy. A mixed integrated mathematical model is constructed simultaneously including makespan, maintenance cost, total tardiness time and considering JIT strategy. A Pareto-based multiobjective hybrid variable neighborhood genetic algorithm is designed to solve the model. The computational results show that it is efficient, and is advantageous over standard genetic algorithms. Finally, the effectiveness of the model and the algorithm are validated via a given case.

Keywords: Nonidentical parallel machines · Preventive maintenance
Variable neighborhood genetic algorithm · Delivery on time

1 Introduction

Non identical parallel machine scheduling problem belongs to a parallel machine scheduling problem [1]. In the actual processing workshop, due to different machine purchase time and age lead to different wear and tear of the equipment, resulting in different processing rates. Thus, these functions are similar, but the processing rate different nonidentical parallel machines scheduling problem has a broader application.

Modern manufacturing enterprises to use more automation and precision equipment, once the equipment failure will result in processing disruption, equipment breakdown and other issues in the production process, resulting in low processing efficiency, delivery delay, so companies pay more attention to the reliability of the equipment and running stability [2].

Scholars carry out a lot of studies on the joint optimization of production scheduling and machine maintenance recently. According to different production systems, different assumptions are put forward to construct the different model and design corresponding algorithms to solve the model. Mohammadreza and Lee & Kim studied the scheduling maintenance problem of periodic preventive maintenance considering JIT [3, 4]. Based on the limitation of the number of assumed tasks, a

G. Q. Huang et al. (eds.), *Proceeding of the 24th International Conference on Industrial Engineering and Engineering Management 2018*,
https://doi.org/10.1007/978-981-13-3402-3_9

multiobjective particle swarm optimization algorithm was developed to solve the model, to achieve the completion time and earliness/tardiness to minimize.

Sun & Li and Jie He et al. take two parallel machines as the research object and the minimization of the maximum makespan as the optimization goal [5, 6]. The author transforms the fixed periodic preventive maintenance problem into the bin packing problem and then uses the heuristic algorithm to solve the problem. C. Low et al. study single machine scheduling problem with flexible periodic maintenance strategy which the problem is also first converted to bin packing problem, and then the heuristic algorithm based on the first fit rule (DFF) is used to solve the model [7]. Some of the other studies have considered machine scheduling problem considering maintenance periodic. For example, Fitouhi & Nourelfath proposed a joint optimization decision model for nonidentical periodic preventative maintenance and production scheduling of single machine production system [8]. Cui et al. and Low suggested a single machine scheduling problem considering flexible periodic preventative maintenance and considered job release time at the same time, where they research two cases of recoverable and nonrecoverable [9, 10]. Yin et al. and Eduardo & Stefan have studied the parallel machine scheduling where consider deteriorating job scheduling problem with potential machine failure and assume that machine after repair as good as new [1, 11].

The joint optimization of machine maintenance and production scheduling usually focus on single machine or parallel machine, however, less researches concentrate on nonidentical parallel machine [12]. Equipment maintenance focuses on the consideration of fixed periodic preventive maintenance strategies, and less research on flexible periodic preventive maintenance or mixed maintenance strategies. In this paper, we proposed a joint optimization model of hybrid maintenance and nonidentical parallel machine considering JIT strategy. Numerical results show that the effectiveness of the model and the performance of the algorithm.

The rest of the paper is organized as follows: Sect. 2 details describe the problem. Mixed nonlinear programming models are developed, and the related denote parameters of the problem in Sect. 3. Section 4 describes a solution algorithm for solving the problem, and then Sect. 5 analysis the performance of the model and the effect of the algorithm via computational results, and compare with accurate algorithm NSGA-II. In Sect. 6, we conclude our paper and some directions for future research.

2 Problem Description

In this section, we put forward a multiobjective model that accords with the actual practice, to provide a basis for enterprise production. Due to the interaction of production scheduling and machine maintenance in the real production system, it is vital to coordinate relation- ships between them. JIT is a production management strategy was proposed by the Toyota and is very popular around the world. The key to implementing JIT is to achieve punctual delivery, meet customer needs and improve customer satisfaction. Therefore, in this paper, we consider JIT in model to ensure that the proposed model more universally.

Nonidentify parallel machine system consists of m machines that processing rate different, equipment performance similar $M = \{M_j | j = 1, 2, \ldots, m\}$, manufacture receives J_i jobs (i = $\{1, 2, ..., n\}$). The processing rate of M is S_j (i = $\{1, 2, ..., n\}$). The basic processing time of J_i is P_i^o (i $= \{1, 2...n\}$), and the actual working time of J_i on machine M_j is $P_i^j = P_i^o / s_j$. Equipment life obeys Weibull Distribution in production area, so in this paper we assume that equipment life fits two parameter Weibull distribution (shape parameter $\beta > 1$, size parameter η) to determine the reliability of processing machine and the distribution of failure rate. Then the reliability function of machine $R(t)$ and failure distribution $\lambda(t)$ as follows:

$$R(t) = \exp\left[-\left(\frac{t}{\eta}\right)^{\beta}\right]$$

$$\lambda(t) = \frac{\beta}{\eta}\left(\frac{t}{\eta}\right)^{\beta-1} \quad \text{or} \quad \lambda(a_{j(k-1)}) = \frac{\beta}{\eta}\left(\frac{a_{j(k-1)}}{\eta}\right)^{\beta-1}$$

Equipment maintenance uses two maintenance methods which are flexible periodic preventive maintenance and minor repair. We assume that after preventive maintenance, the performance and reliability of the machine will be improved, and prolong the life of machine, however, can't restore to the new initial state [13]. Therefore, we introduce preventive maintenance improvement factor γ to amend the machine remainder service life. In the function μ represents the cost adjustment coefficient, v represents the time adjustment coefficient. w is the learning effect adjustment factor of the equipment maintenance, l represents the preventive maintenance index.

If machine breakdown between two preventive maintenance, we repair it with minor repair. Minor repair does not affect equipment life, reliability and process performance. The preventive maintenance time and cost to Machine M_j were t_j^{pm} and C_j^{pm}, respectively, and the time and cost of a minor repair were t_j^r and C_j^r.

3 Problem Formulation

In this section, we aim at formulation the problem model with consideration the JIT strategy as a mixed integer programming. We use production time index, cost index, and delivery on time index as optimization objective to formulate mathematical model of the problem. Our problem assumes the following:

(a) All Jobs irrelevant and arrive on time at zero
(b) During the processing of job production and the maintenance of machine are not allowed to be interrupted
(c) The resources of preventive maintenance are adequate and the machine can maintain at any time immediately
(d) Machine setup time is negligible after maintenance activity
(e) Parallel machines are independent of each other. That is not failure correlation
(f) Machine failure may only take place during the operation of a machine, which belongs to ODF (operation dependent failure) failure model. The related parameters and variables were denoted as follows:

Indices and parameters:

Rt_j^k	The idle time of machine M_j precedent process position $K-1$ to next process position K
T_j^k	The completion time of machine M_j at position K
C_i	Expected completion time of job J_i
$E(C_{\max})$	Expected completion time of machining system
a_{j0}	Initial service life of machine M_j
$\bar{a}_{jk}$	Remainder service life of machine M_j before position K
a_{jk}	Remainder service life of machine M_j after position K
C_m	Maintenance cost in a processing cycle.
d_i	Delivery date of job J_i
w_i	Weighted of job J_i
T_i	Tardiness of job J_i
TDT	Total tardiness of machining system

Decision variables:

$$x_{ijk} = \begin{cases} 1, & \text{if job } J_i \text{ is assigned to machine at position } k; \\ 0, & \text{otherwise} \end{cases}$$

$$y_{jk} = \begin{cases} 1, & \text{if position } K \text{ of machine } M_j \text{ has been preventive maintenance;} \\ 0, & \text{otherwise} \end{cases}$$

$$z_{jk} = \begin{cases} 1, & \text{if the machine } M \text{ processes the last job of the machine at position } K; \\ 0, & \text{otherwise} \end{cases}$$

Objective function (Minimize):

$$E(C_{\max}) = \max\left\{ \sum_{k=1}^{n} \left[\sum_{i=1}^{n} \begin{matrix} x_{ijk} \cdot \dfrac{P_i^o}{s_j} + y_{jk} \cdot t_j^{pm} \\ + E\left\{N_{jk}^r\right\} \cdot t_j^r + Rt_j^k \end{matrix} \right] \right\} \tag{1}$$

$$C_m = \sum_{j=1}^{m} \left(\sum_{k=1}^{n} y_{jk} \cdot C_j^{pm} + \sum_{k=1}^{n} E\left\{N_{jk}^r\right\} \cdot C_j^r \right) \tag{2}$$

$$TDT = \sum_{i=1}^{n} w_i \cdot [\max(0, C_i - d_i)] \tag{3}$$

Subject to:

$$\sum_{j=1}^{m}\sum_{k=1}^{n} x_{ijk} = 1, \quad \forall i = 1,2,\ldots,n \tag{4}$$

$$\sum_{i=1}^{n} x_{ijk} \le 1, \quad \forall j = 1,2,\ldots,m; k = 1,2,\ldots,n \tag{5}$$

$$\bar{a}_{jk} = a_{j(k-1)}\left(1-\gamma\cdot y_{jk}\right), \quad \forall j = 1,2,\ldots,m; k = 1,2,\ldots,n \tag{6}$$

$$a_{jk} = \bar{a}_{jk} + P_j^k, \quad \forall j = 1,2,\ldots,m; k = 1,2,\ldots,n \tag{7}$$

$$\begin{aligned} E\left\{N_{jk}^r\right\} &= E\{N(t = a_{jk} - \bar{a}_{jk})\} = \int_{\bar{a}_{jk}}^{a_{jk}} \frac{\beta}{\eta}\left(\frac{t}{\eta}\right)^{\beta-1}dt \\ &= \left(\frac{a_{jk}}{\eta}\right)^{\beta} - \left(\frac{\bar{a}_{jk}}{\eta}\right)^{\beta}, \quad \forall j,k \end{aligned} \tag{8}$$

$$z_{jk} = \sum_{i=1}^{n} x_{ijk} - \sum_{i=1}^{n} x_{ij(k+1)}, \quad \forall j = 1,2,\ldots,m; k = 1,2,\ldots,n \tag{9}$$

$$z_{jn} = \sum_{i=1}^{n} x_{ijn}, \quad \forall j = 1,2,\ldots,m \tag{10}$$

$$\sum_{k'=1}^{k-1}\sum_{\substack{i'=1,\\ i'\neq i}}^{n} x_{i'jk'} \ge (k-1)x_{ijk}, \quad \forall i = 1,2,\ldots,n; j = 1,2,\ldots,m; k = 1,2,\ldots,n \tag{11}$$

$$C_i = \sum_{j=1}^{m}\sum_{k=1}^{n} x_{ijk}\cdot E\left(T_j^k\right), \quad \forall i=1,2,\ldots,n \tag{12}$$

$$E\left(T_j^k\right) = \sum_{k'=1}^{k}\left\{P_j^{k'} + y_{jk'}\cdot t_j^{pm} + t_j^r\cdot E\left\{N_{jk'}^r\right\} + Rt_j^{k'}\right\} \quad \forall j = 1,2,\ldots,m \tag{13}$$

$$T_i = \max[0, C_i - d_i], \quad \forall i = 1,2,\ldots,n \tag{14}$$

$$E\left(T_j^{k-1}\right) \le E\left(T_j^k\right) - y_{jk}\cdot t_j^{pm} - \frac{P_i^o}{s_j}\cdot x_{ijk}, \quad \forall j = 1,2,\ldots,m; k = 1,2,\ldots,n \tag{15}$$

Objection function (1) is the minimization of expected completion time of the machining system. Objection function (2) shows that the total maintenance cost of the machining system is minimization. Objection function (3) minimizes the total tardiness time. Constraints (4) to (5) indicates that each job is processed on one machine and only processed exactly once, moreover one machine at the same time at most process one job. Constraints (6) and (7) show that preventive maintenance can improve machine performance and reliability to a certain extent, introducing improvement factor γ to determine the machine remainder service life of machine before and after machining at position K. Constraints (8) is the expected failure number of the machine during the production process at position K. Constraints (9) and (10) indicate that if the machine position K processes the last job of the machine, no job will be assigned to later position. Constraints (11) show that if jobs are processed at position K of machine, it means all of the fronts of $(K-1)$ position already have jobs to process. Constraints (12) to (14) implies tardiness time produced by the jobs delayed delivery. Constraints (15) indicates that machine M_j is allowed to have idle time after position $(K-1)$ and before position K, and machine must finish the job at position $(K-1)$ before starting to process K.

4 The Hybrid Variable Neighborhood Genetic Algorithm Procedure (VNSGA)

In this section, we are going to present an improved algorithm based on genetic algorithm (GA) to solve the problem. GA derived the simulation of the natural evolution process and was proposed by J. Holland in 1975. Variable neighborhood search (VNS) can gradually exchange searching range to avoid local optimal. The VNSGA combines the advantage of GA and VNS, and it is a Pareto algorithm with well global and local research capability at the same time.

Initial population contains N chromosomes that generate randomly, each of length n + m − 1. The sequence of jobs coding range 1–9 with real number coding, and number code −1 ~ −m + 1 denote different machine, that is −1 ~ −m + 1 is the machine identification symbol. The preventive maintenance (PM) of machine is encoded by binary, which 1 represent machine should repair before machining job. However, if binary code 1 corresponds to the machine symbol −1 ~ −m + 1, then convert 1 to 0 compulsory. For example, there are nine jobs arrive at zero time and need to process on three parallel machines. The chromosome coding showed as the Fig. 1.

1	3	5	2	-1	7	4	-2	9	6	8

0	1	0	0	0	0	1	0	0	0	1

Fig. 1. Chromosomes coding

In multiobjective optimization problem, it is difficult to evaluate the solution by simple sort. However, the strategy based on Pareto dominance can evaluate the solution advantage and disadvantage effectively, that is assessing the solution with Pareto method. In this paper, we have used three important parameters of Pareto, which are fast nondominated sorting, crowding distance, and crowding comparative factor. Crowding distance presents the intensity of individual and others around it. The individual with bigger can ensure the population diversity, and individual with smaller crowding comparative factor guarantee the elite of population.

There are three operation designed for VNSGA. In selection operation, we select individuals with two methods to avoid algorithm premature. Selecting an individual from population randomly and selecting an individual by tournament method. In crossing operation, the sequence code of jobs processed cross by using linear order crossing, and the code sequence of preventive maintenance by using single point crossing. In mutation operation, because the jobs sequence code is real value coding, select randomly two mutation points on two individuals, and exchange gene code each other between two mutation points. Preventive maintenance use binary encoding, so the first choose one mutation point and then convert 1 to 0 or 0 to 1. Based on the above mentioned discussions, the detail steps about the algorithm are given as follows:

- Step1 Determine three neighborhood structure N_k, $K = 1, 2, 3$. Input initialization parameters: Cycle times p, initial solution. Let $i = 0$, optimum individual $BX = X^*$.
- Step 2 If $i > p$ then output optimum individual BX, else let $K = 1$.
- Step 3 Basis for X a new solution X^* is generated randomly, and then evaluation value of new solution and the initial solution.
- Step 4 If X^* better than X, output X^*, and then let $BX = X^*$, continue searching within neighborhood structure, else $K = K + 1$.
- Step 5 If $K > 3$, let $i = i + 1$ then go to step 2, else go to step 3, enter the next neighborhood structure search.

5 Computational Results and Discussion

In this section, in order to evaluate the model performance and algorithm effectiveness some numerical experiments are conducted. Production system including two parallel machines that performance similar but efficiency differently. The basic parameters of the machine showed as the Table 1. Eight machined jobs arriving at the same time need to be assigned to a certain machine to process. The basic data of the processing operations is given in Table 2. We set algorithm parameters as follows: Initial population p = 100, the maximum number of iteration G = 300, mutation probability $P_m = 0.1$, and crossover probability $P_c = 0.15$. Assume that the Weibull distribution shape parameter $\beta = 2$ and the size parameter $\eta = 100$.

The running platform for this example is MatlabR 2014a, which generates 21 Pareto solutions by iteration, as shown in Table 3. Decision maker can set weighted to normalize the value of the objective function in the nondominated solutions [14]. The optimal solution sequence of jobs on machines is $M1$: 8-1-4-6 and $M2$: 5-7-3-2. The preventative maintenance moment is before $M1$ processing job 6 and $M2$ processing

Table 1. Basic date of machine

M_j	s_j	a_{j0}	c_j^{pm}	c_j^r	c_j^{T}	t_j^{pm}	t_j^r
*M*1	1.11	40	100	800	800,000	8	40
*M*2	0.91	46	130	1000	1,500,000	10	50

Table 2. Basic date of processing operations

J_{i}	P_{i}^{o}	P_{i}^{1}	P_{i}^{2}	d_{i}
1	25	22.5	27.5	77
2	15	13.5	16.5	59
3	34	30.6	37.4	92
4	45	40.5	49.5	88
5	30	27	33	68
6	48	43.2	52.8	97
7	55	49.5	60.5	113
8	28	25.2	30.8	64

Table 3. Pareto front target value

Job processed sequence		PM code	E(Cmax)	C_m	TDT
1	5 8 3 7–1 2 6 4 1	000100010	160.08	685.68	182.44
2	2 5 6 7–1 8 4 3 1	000100010	162.11	723.84	178.34
3	2 5 1 3–1 8 4 7 6	010100000	207.54	667.09	188.21
4	8 2 6–1 1 5 7 4 3	001000001	197.39	584.56	218.78
5	5 1 3 7–1 2 8 4 6	010101000	173.54	878.53	155.46
6	5 8 3 7–1 1 2 4 6	000100010	170.24	888.84	147.29
7	8 1 5 7 2–1 4 6 3	000100001	173.19	715.52	167.60
8	5 8 2 7–1 4 3 6 1	100100000	170.98	595.35	209.24
9	2 8 5 3 6–1 4 7 1	000100001	204.08	504.24	247.85
10	8 1 7 4–1 5 2 3 6	010100001	166.63	927.33	147.16
11	8 1 4 6–1 5 3 7 2	000100001	158.76	629.05	208.28
12	4 5 6 1–1 3 8 7 2	100001000	158.14	596.06	297.27
13	7 8 5 1–1 3 4 6 2	100100000	157.56	532.37	332.81
14	8 1 3 7–1 5 4 6 2	000100001	163.16	619.46	204.06
15	5 4 3 7–1 2 8 1 6	000100001	178.62	969.25	141.07
16	1 5 6 7–1 8 2 3 4	000100011	166.45	949.16	147.66
17	5 8 1 6–1 2 4 7 3	100000001	180.89	668.80	179.38
18	8 5 3 1–1 4 7 6 2	000100010	190.66	441.26	282.20
19	2 1 4 7–1 5 8 6 3	100100001	170.99	752.20	163.01
20	2 5 4 6–1 8 3 7 1	000100010	169.98	659.04	193.06
21	2 8 4 1–1 5 6 7 3	000100001	200.69	566.48	206.59

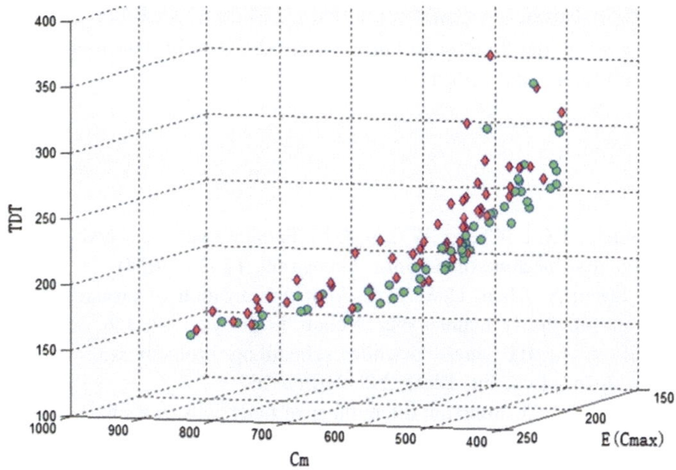

Fig. 2. Comparison about VSNGA and NSGA-II

job 2. The maximum makespan $E(C_{max}) = 158.76$, equipment total maintenance cost $C_m = 629.05$, and the total tardiness $TDT = 208.28$.

$$\min J = w_1 \cdot \frac{E(C_{\max}) - E(C_{\max})^{\min}}{E(C_{\max})^{\max} - E(C_{\max})^{\min}} + w_2 \cdot \frac{C_m - C_m^{\min}}{C_m^{\max} - C_m^{\min}} + w_3 \cdot \frac{TDT - TDT^{\min}}{TDT^{\max} - TDT^{\min}}$$

We compare our algorithm with NSGA-II, get the approximate solution in a set of iterations as the Fig. 2. The red point representation the VSNGA and the performance of the NSGA-II is shown as the green point. Easy to know that variable neighborhood search GA algorithm compares with the standard NSGA-II closer to the optimal solution.

6 Conclusion

This paper considers combinatorial optimizations of machine maintenance and scheduling problem with considering JIT strategy. Insufficient maintenance may lead to machine breakdown if repair machine with fixed periodic, conversely, frequent repairs will result in resource waste. To solve this problem, combining two maintenance policies (flexible periodic preventive maintenance and minor repair) and job delivery on time are considered in the model, and then mixed integer nonlinear programming models are suggested. To solve the problem, a variable neighborhood search genetic algorithm is derived. The numerical result proves the effectiveness of the model and the performance of the algorithm. In another word, the algorithm is quite good to solve multi objective models and find the good quality solutions.

In the future research, it would be interesting to consider other constraints, such as maintenance resource insufficient, job arrive time differently, and machine maintenance interruption is allowed, and so on.

References

1. Y. Yin, Y. Wang, T.C.E. Cheng, W. Liu, J. Li, Parallel-machine scheduling of deteriorating jobs with potential machine disruptions. Omega **69**, 17–28 (2017)
2. L. Xiao, S. Song, X. Chen, D.W. Coit, Joint optimization of production scheduling and machine group preventive maintenance. Reliab. Eng. Syst. Saf. **146**, 68–78 (2016)
3. M. Shahriari et al., JIT single machine scheduling problem with periodic preventive maintenance. J. Ind. Eng. Int. **12**(3), 299–310 (2016)
4. J.-Y. Lee, Y.-D. Kim, Minimizing the number of tardy jobs in a single-machine scheduling problem with periodic maintenance. Comput. Oper. Res. **39**(9), 2196–2205 (2012)
5. K. Sun, H. Li, Scheduling problems with multiple maintenance activities and non-preemptive jobs on two identical parallel machines. IJPEconomics **124**(1), 151–158 (2010)
6. J. He, Q. Li, X. Dehua, Scheduling two parallel machines with machine-dependent availabilities. Comput. Oper. Res. **72**, 31–42 (2016)
7. C. Low, M. Ji, C.-J. Hsu, S. Chwen-Tzeng, Minimizing the makespan in a single machine scheduling problems with flexible and periodic maintenance. Appl. Math. Model. **34**(2), 334–342 (2010)
8. M.-C. Fitouhi, M. Nourelfath, Integrating noncyclical preventive maintenance scheduling and production planning for multi-state systems. Reliab. Eng. Syst. Saf. **121**, 175–186 (2014)
9. W.-W. Cui, L. Zhiqiang, Minimizing the makespan on a single machine with flexible maintenances and jobs' release dates. Comput. Oper. Res. **80**, 11–22 (2017)
10. C. Low, C.-J. Hsu, S. Chwen-Tzeng, A modified particle swarm optimization algorithm for a single-machine scheduling problem with periodic maintenance. Expert Syst. Appl. **37**(9), 6429–6434 (2010)
11. E. Lalla-Ruiz, S. Voß, Modeling the parallel machine scheduling problem with step deteriorating jobs. EJOR **255**(1), 21–33 (2016)
12. J. Yoo, I. Lee, Parallel machine scheduling with maintenance activities. Comput. Ind. Eng. **101**, 361–371 (2016)
13. D.T. Nguyen, Y. Dijoux, M. Fouladirad, Analytical properties of an imperfect repair model and application in preventive maintenance scheduling. EJOR **256**(2), 439–453 (2017)
14. H. Mokhtari, A. Hasani, An energy-efficient multi-objective optimization for flexible job-shop scheduling problem. Comput. Chem. Eng. **104**, 339–352 (2017)

Mean Shifts Monitoring of Auto-correlated Manufacturing Processes Using Long Short-Term Memory-Based Recurrent Neural Network

Shu-mei Chen and Jian-bo Yu(✉)

School of Mechanical and Energy Engineering, University of Tongji, Shanghai, China
jbyu@tongji.edu.cn

Abstract. The traditional statistical process control (SPC) technique is used as a monitoring tool to recognize mean shifts of discrete manufacturing processes. The fundamental assumption using SPC is the independence of observed process data. Such assumption is usually violated in practical industries, which proceeds the development of alternative schemes to monitor auto-correlated processes. This paper approaches a long short-term memory (LSTM)-based model to recognize mean shifts of auto-correlated processes with different autocorrelation coefficients and shift parameters. The performance of the proposed method is evaluated by an average run length of time series residual control chart in comparison with other monitoring schemes. It was observed that the LSTM-based monitoring approach could improve significantly upon the early detection of out-of-control operating cases in auto-correlated processes.

Keywords: Auto-correlation · Long short-term memory · LSTM
Statistical process control · Shifts recognition

1 Introduction

Statistical process control (SPC) technique was developed to identify whether the manufacturing system is operating in an in-control state or deviates from the normal condition, which has witnessed a great success in discrete manufacturing operations. However, these techniques are not applicable in continuous and batch process industries [1]. In these processes, the value of process parameter at current time depends on the previous state. The observations of those continuous industries, including the manufacture of food, chemicals, paper and wood, are dependent from each other at varying time lags, which violates the assumption of SPC technique [2, 3]. Data collected at regular time intervals by process operators through automatic sensors are serially correlated [4]. Consequently, the effectiveness using traditional control charts, such as $\overline{X}$ and R control charts to identify variability of auto-correlated manufacturing processes will be greatly weakened. It promotes the development of various detecting methods to identify mean shifts in auto-correlated processes.

G. Q. Huang et al. (eds.), *Proceeding of the 24th International Conference on Industrial Engineering and Engineering Management 2018*,
https://doi.org/10.1007/978-981-13-3402-3_10

Deep neural networks (DNNs) are extremely powerful machine learning models that can achieve excellent performance on difficult problem [5], such as speech recognition [6] and visual object recognition [7]. Long short-term memory (LSTM) and conventional recurrent neural networks (RNNs) have been successfully applied to sequence prediction and sequence labeling task [8]. However, LSTM-based approach outperforms RNNs on learning context-free and context-sensitive languages [9]. LSTM network is capable of sequentially accepting each observation in time series model and mapping it into the current process state together with the historical data [10]. The architecture of LSTM is applied to deal with general sequence to sequence issues [11]. In this paper, we proposed a LSTM-based model to monitor mean shifts of auto-correlated manufacturing processes with different auto-correlation coefficients and shifts magnitude. The experimental results demonstrate that the proposed method provides a significant improvement for identifying mean shifts in auto-correlated manufacturing process over traditional statistical and regular machine learning techniques.

2 Methodology

2.1 The Basic Architecture of LSTM Neural Network

LSTM networks (LSTMs) are widely used in speech recognition, handwriting recognition, and machine translation in natural language processes. LSTMs are proposed as a special kind of RNN to learn long-term dependencies. The information is remembered by memory blocks through many time-steps learning (over 1000 time steps). Figure 1 presents the architecture of LSTMs. As shown in Fig. 1, LSTM consists of repetitive and complex network modules. An LSTM layer contains one or more recurrently connected blocks, known as memory blocks [12]. Calculations are carried out through three multiplicative units (i.e., forget gate, input gate, and output gate). The previous cell values h_{t-1} is calculated via forget gate as shown in Fig. 2; the input X_t is calculated via input gate as shown in Fig. 3; the output is calculate via output gate as shown in Fig. 4. The interaction among cells can only be carried out by gates.

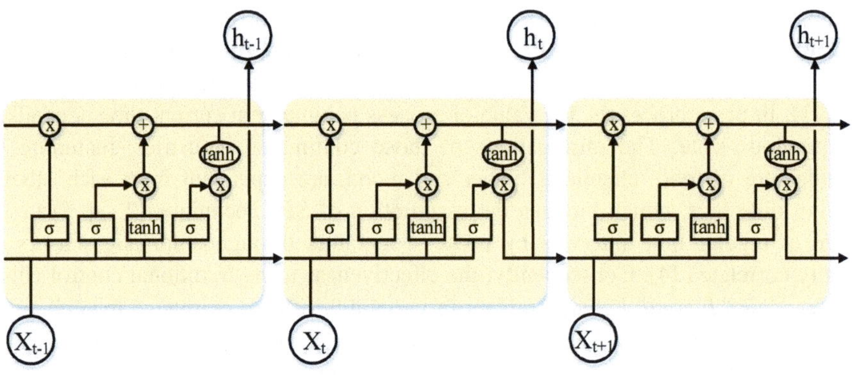

Fig. 1. The basic architecture of LSTM model

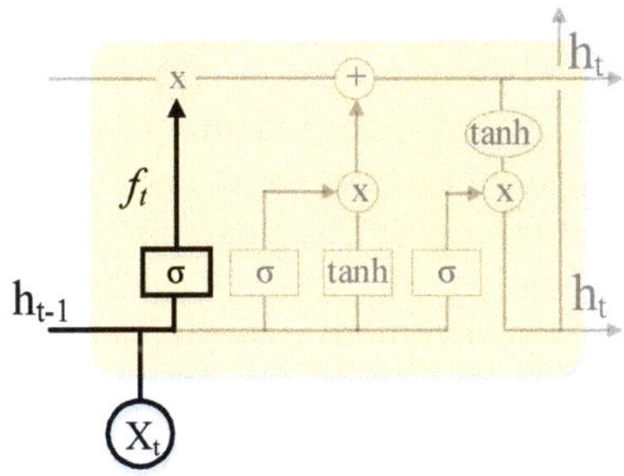

Fig. 2. The basic architecture of forget gate in LSTM

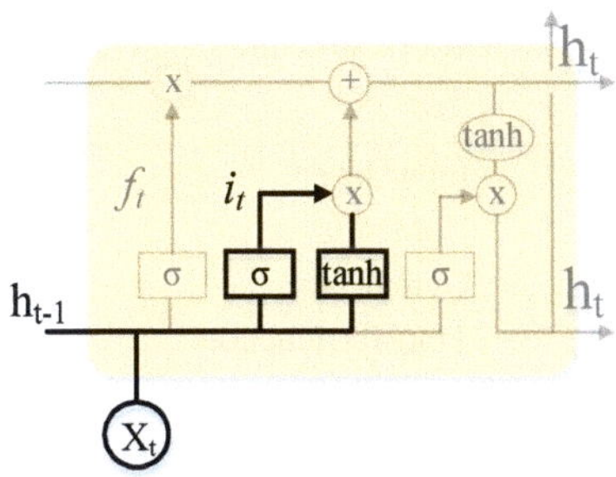

Fig. 3. The basic architecture of input gate in LSTM

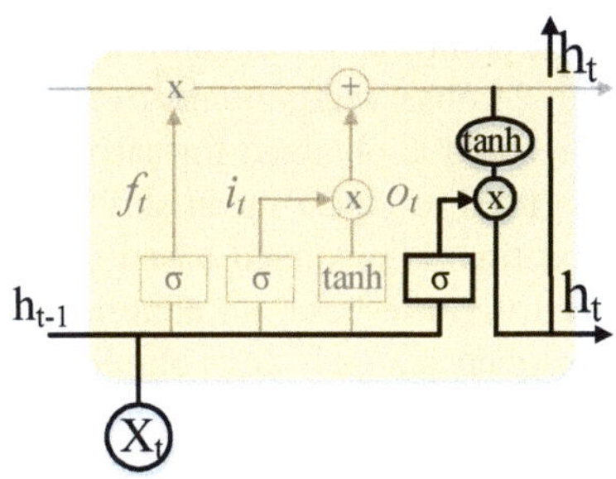

Fig. 4. The basic architecture of output gate in LSTM

2.2 LSTM-Based Model Construction

Training and testing data are generated through Monte Carlo simulation method with different auto-correlation coefficients and shift parameters. In this study, the mean shifts of auto-correlated manufacturing processes generate the out-of-control signals. The input vectors of the LSTMs are presented based on a moving window method with n observed points at varying time lags. The window size has a great influence on the performance of LSTMs. In this study, a window size of 24 is selected for data sets with different parameters. Table 1 presents the structure parameters of LSTM networks. As shown in TABLE I, the number of nodes in input layer are 24; the dimension of hidden layer is 28; the size of output layer is 1. The "sigmoid" function is selected for transfer function.

In training phase, we conducted 1000 experiments to obtain thresholds for each case with different process parameters to maintain an in-control average run length (ARL_0) of 360. In testing phase, the out-of-control average run length (ARL) is also computed from 1000 simulation. The out-of-control alarm should be given once the observed statistics exceed the cut-off.

Table 1. The network structure parameters of LSTMs

LSTMs		
Structure parameters	Input layer	24
	Hidden layer	28
	Output layer	1
	Transfer function	Sigmoid
Learning parameter	Accuracy	2.5×10^{-7}

Figure 5 presents the procedure for LSTM to identify mean shifts in auto-correlated processes. As shown in Fig. 5, data sets with different auto-correlation Φ and shift magnitudes δ are generated by Monte Carlo simulation method. The out-of-control signal is supposed to occur at point t = 71. Second, the data is presented based on a moving window method with 24 observation points over the time series stream. In off-line modelling phase, the threshold is obtained for each case. In on-line monitoring, a well-trained LSTM model is used to detect mean shifts of various measurement data sets. The process data at next time is predicted based on previous 24 observations. If the observed statistics exceed the threshold, an abnormal signal should be given for process operators to investigate causes and eliminate them from the process immediately.

3 Results

3.1 The Model of Process Data: AR (1) with Step Shifts

Autoregressive process of lag 1 (i.e., AR (1)) is very common in practical manufacturing operations [13]. The observation at varying time lags depends on the previous state in auto-correlated processes [14]. The process shift in an AR (1) model is represented by (1):

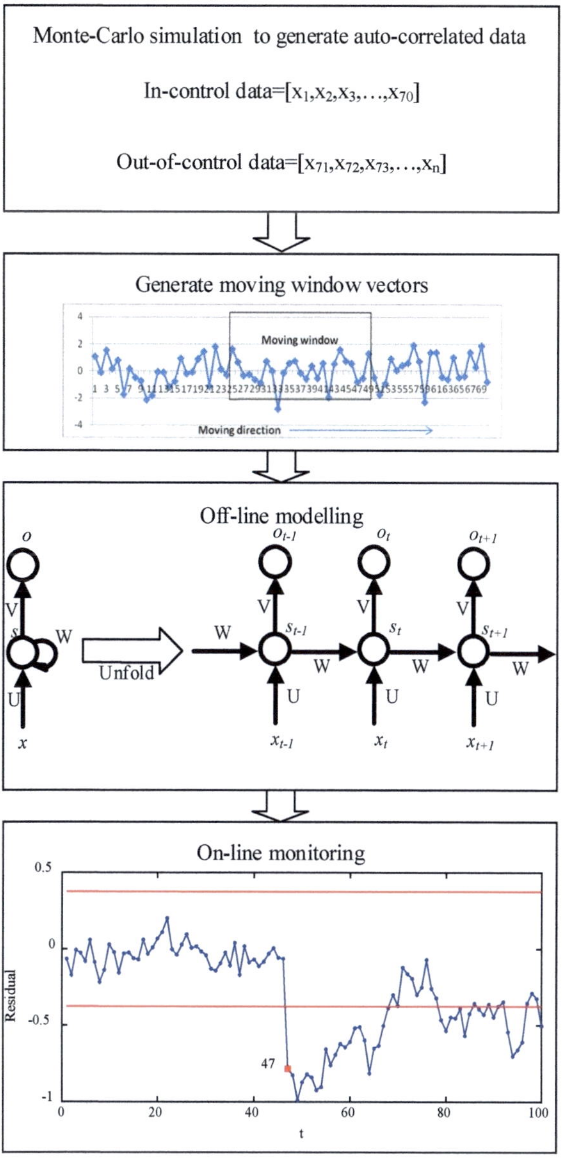

Fig. 5. Schematic diagram of the LSTM-based method for identifying mean shifts in auto-correlated processes

$$Z(t) = \mu + \Phi * (Z(t-1) - \mu) + \varepsilon_t. \tag{1}$$

where $Z(t)$ and $Z(t-1)$ represent the particular value of time series at time t and $(t-1)$, respectively; μ is the mean of process data; ε_t is an error term normal subjecting to a normal and independent distribution; and Φ is the coefficient of autoregressive process restricted to a value between −1 and 1.

When shift δ occurs at time t, the value of $Z(t)$ is represented by (2). The mean of process data has shifted at time t whereas the value of μ before t is in-control.

$$Z(t) = \delta + Z(t-1). \tag{2}$$

Therefore, the expected difference between the shifted value of $Z(t)$ and the non-shifted value of $Z(t-1)$ is δ [14]. The difference of shifted value at time $t+k$ is supposed to be $(1-\Phi)\delta$, for $k \geq 1$. It becomes more difficult to identify the shifted variable as more observations enter into monitoring processes. Hence, the best time to recognize a shift in process parameter is immediately after the shift has occurred [3].

3.2 Performance Comparison with Other Monitoring Schemes

We consider a normal process with a mean (μ) of zero and variance (σ^2) of one to generate training data sets of 500 observations for each parameter Φ (i.e., 0.0, 0.25, 0.5, 0.75, 0.9). In order to obtain an in-control threshold, we conducted 1000 simulations to maintain an ARL_0 of 360 for different Φ. Table 1 shows the threshold in the last column.

In the case, process shifts $\delta = 0.5$, $\delta = 1.00$, $\delta = 2.00$ and $\delta = 3.00$ are considered to generate testing data sets based on (1) and (2) for different Φ. In time series modeling, n moving window vectors that consists of m in-control window vectors and n–m out-of-control window vectors are fed to LSTM. The shift occurs at point 71 (the moving window number is 47) as shown in Fig. 6, when $\Phi = 0.95$, $\delta = 3.00$. After the first out-of-control window vector manifests itself in processes, the remaining out-of-control window vectors enter into LSTM.

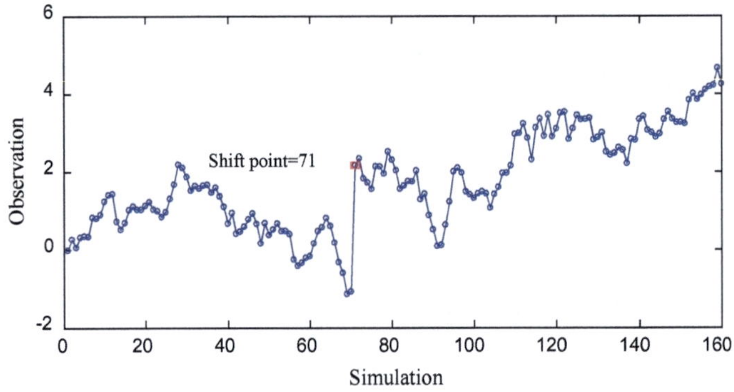

Fig. 6. Simulation data. Shift $\delta = 3.00\sigma$, $\Phi = 0.95$, $t = 71$

To further verify the performance of LSTM to identify mean shifts of in auto-correlated processes, the real-time detection using proposed method in comparison with other special-causes control (SCC) chart is also investigated. As shown in Fig. 7a,

LSTM gives an out-of-control signal when the first shifted window vectors enter into model. An alarming signal occurs at window number 47, which indicates the process has shifted to an out-of-control state at point t = 71. In Fig. 7b, the abnormal points detected by SCC chart are at point t = 29 and t = 71, which shows that SCC chart is capable of identifying large shift magnitudes $\delta = 2.00\sigma$ immediately. However, the first signal is a false alarm, which results in costly over-control for process operators to address these false out-of-control conditions.

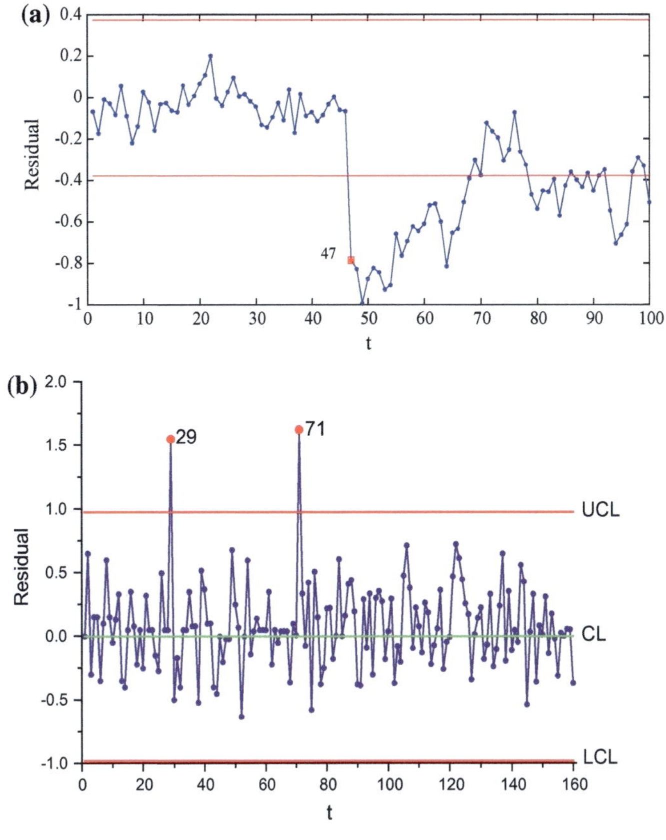

Fig. 7. In real-time detection: Shift $\delta = 2.00\sigma$, $t = 71$. **a** LSTM-based method with CL = ± 0.3765, auto-correlation coefficient $\Phi = 0.5$, **b** SCC chart with CL = ± 0.9764, auto-correlation coefficient $\Phi = 0.8981$ [14]

Table 2 presents the ARL and SRL (the standard deviation of ARL) of testing data sets with different Φ and δ. Each ARL and SRL is calculated from 1000 simulation runs for different cases. The performance of LSTM-based method is compared with other monitoring schemes (i.e., SCC chart [15], BPN-based scheme [16], and *LRProb* chart [14]). As show in Table 2, LSTM outperforms other monitoring schemes for shifts $\delta \geq 1.00$ when $\Phi = 0.00, 0.25$. As Φ value increases, our proposed method is

Table 2. ARL of LSTM-based scheme for detecting shifts of auto-correlated compared with SCC, BPN-based monitoring scheme and LRProb chart

Φ	δ	SCC[a]		BPN[b]		LRProb[c]		LSTM		Threshold
	Shift	ARL	SRL	ARL	SRL	ARL	SRL	ARL	SRL	
0.00	0.00	370.40	369.88	372.96	370.10	379.86	324.27	362.47	171.28	0.36
	0.50	152.22	154.72	25.38	17.93	20.53	14.58	91.50	69.25	
	1.00	43.89	43.39	8.29	5.76	8.92	4.49	5.03	17.47	
	2.00	6.30	5.78	2.47	1.51	3.95	1.68	1.003	5.47e−02	
	3.00	2.00	1.41	1.29	0.61	2.65	1.01	1.00	0.00	
0.25	0.00	370.40	N/A[d]	371.23	385.70	374.06	364.2	361.45	182.88	0.3855
	0.50	206.04	N/A	32.46	24.96	35.09	31.66	94.76	70.88	
	1.00	75.42	N/A	11.87	9.03	11.69	6.65	7.31	25.95	
	2.00	12.24	N/A	3.39	2.19	5.64	2.21	1.002	6.32e−02	
	3.00	2.85	N/A	1.63	0.99	3.87	1.62	1.00	0.00	
0.50	0.00	370.40	N/A	371.3	373.57	375.68	352.01	360.57	173.90	0.3765
	0.50	258.42	N/A	52.07	45.74	43.24	39.71	101.86	67.58	
	1.00	123.82	N/A	16.74	12.88	15.45	9.13	19.11	44.57	
	2.00	24.22	N/A	4.84	3.26	6.42	3.63	1.04	0.27	
	3.00	4.14	N/A	2.22	1.52	3.86	1.95	1.00	0.00	
0.75	0.00	370.40	N/A	370.6	368.36	373.23	345.21	361.95	173.28	0.336
	0.50	311.23	N/A	91.72	94.81	70.91	70.95	133.89	41.03	
	1.00	197.74	N/A	35.42	32.71	25.53	20.08	109.16	60.01	
	2.00	40.24	N/A	8.95	9.23	11.37	6.36	13.22	31.49	
	3.00	3.01	N/A	3.52	3.28	6.83	4.21	1.06	0.51	
0.95	0.00	370.40	369.88	370.37	374.11	379.20	336.06	360.07	181.75	0.213
	0.50	330.96	357.37	152.09	148.81	130.12	142.58	129.32	49.69	

(continued)

Table 2. *(continued)*

Φ	δ	SCC[a]		BPN[b]		LRProb[c]		LSTM		Threshold
	Shift	ARL	SRL	ARL	SRL	ARL	SRL	ARL	SRL	
	1.00	138.84		77.00	69.16	54.30	62.40	113.43	62.76	
	2.00	1.08	6.21	32.07	27.57	16.97	15.49	47.95	66.78	
	3.00	1.00	0	10.17	1.00	8.99	7.33	4.95	22.06	

Note [a]Results taken from [15]
[b]Results taken from [16]
[c]Results taken from [14]
[d]Not reported in the original paper

more sensitive to detect large shifts. The value of ARL is nearly equal to one for large shift $\delta \geq 2.00$ except when $\Phi = 0.75, 0.95$, which indicates that LSTM model can immediately detect the out-of-control window vector when one or more shift points enter into the processes. Compared with SCC chart, LSTM model presents a better performance than SCC chart in all cases except when $\delta \geq 2.00$ of $\Phi = 0.95$. It should be noted that the SRLs of LSTM-based method are smaller than other schemes in major case with middle and large shift magnitudes except high auto-correlation $\Phi = 0.95$, which indicates the robustness of LSTM over other schemes. It is noteworthy that it took very little time for LSTM to obtain the thresholds, ARL, as well as SRL for different cases, which is conducive to lower the time cost for quality engineer to investigate a large number of observations. Figures 8 and 9 summarize the ARL and SRL of different autocorrelation coefficients and shifts magnitudes. We can conclude that the performance of LSTMs to monitor out-of-control condition with different mean shifts improves as Φ decreases and is in particular robust when $\delta \geq 2.00$ of $\Phi = 0.75, 0.95$.

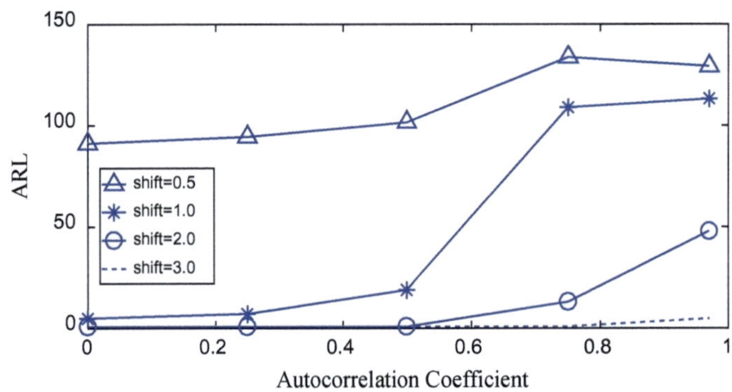

Fig. 8. ARL of different shift magnitudes and autocorrelation coefficients

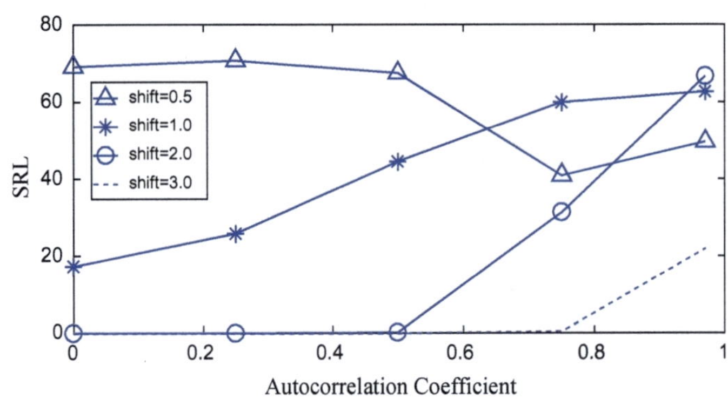

Fig. 9. SRL of different shift magnitudes and autocorrelation coefficients

4 Conclusion

In this study, we proposed a LSTM-based method to monitor mean shifts in an auto-correlated manufacturing process. The LSTM model is capable of predicting the values at next state based on previous observations to assess whether the monitoring process has shifted to an abnormal state. In time series modeling, it is difficult for regular recurrent neural network to remember information in a long-term period, while the LSTM network can deal with the long-term dependence to historic data. In addition, the experimental results indicate that the proposed LSTM model outperforms the chart-based techniques and supervised ANN monitoring scheme with regarded to the accuracy and calculation time in major cases. Therefore, we can reach a conclusion that LSTM-based process state monitoring approach is an effective, robust and economical tool to recognize mean shifts in auto-correlated manufacturing processes.

Acknowledgements. This research was supported by the National Natural Science Foundation of China (No. 51375290, 71777173), the Fundamental Research Funds for Central Universities and Shanghai Science.

References

1. C.W. Zobel, D.F. Cook, Q.J. Nottingham, An augmented neural network classification approach to detecting mean shifts in correlated manufacturing process parameters. Int. J. Prod. Res. **42**(4), 741–758 (2004)
2. D. Montgomery, D.C. Friedman, *Statistical Process Control in Computer Integrated Manufacturing* (Dekker, New York, 1989)
3. D.F. Cook, C.-C. Chiu, Using radial basis function neural networks to recognize shifts in correlated manufacturing process parameters. IIE Trans. **30**(3), 227–234 (1998)
4. D.F. Cook, C.W. Zobel, Q.J. Nottingham, Utilization of neural networks for the recognition of variance shifts in correlated manufacturing process parameters. Int. J. Prod. Res. **39**(17), 3881–3887 (2001)
5. I. Sutskever, O. Vinyals, Q.V. Le, Sequence to sequence learning with neural networks. Adv. Neural Inf. Process. Syst. 3104–3112 (2014)
6. G.E. Dahl, D. Yu, L. Deng, A. Acero, Context-dependent pre-trained deep neural networks for large-vocabulary speech recognition. IEEE Trans. Audio Speech Lang. Process. **20**(1), 30–42 (2012)
7. D. Ciresan, U. Meier, J. Schmidhuber, Multi-column deep neural networks for image classification, in *CVPR* (2012), pp. 3642–3649
8. H. Sak, A. Senior, F. Beaufays, Long short-term memory based recurrent neural network architectures for large vocabulary speech recognition, no. Cd (2014)
9. F.A. Gers, J. Schmidhuber, LSTM recurrent networks learn simple context free and context sensitive languages. IEEE Trans. Neural Netw. **12**(6), 1333–1340 (2001)
10. H. Palangi et al., Deep sentence embedding using long short-term memory networks: analysis and application to information retrieval. IEEE/ACM Trans. Audio Speech Lang. Process. **24**(4), 694–707 (2016)
11. H. Sepp, J. Schmidhuber, Long short-term memory. Neural Comput. (1997)

12. A. Graves, J. Schmidhuber, Framewise phoneme classification with bidirectional LSTM networks, in *Proceedings of International Joint Conference on Neural Networks*, vol. 4 (2005), pp. 2047–2052
13. T.P. Longnecker, M.T. Ryan, Charting correlated process data (1993)
14. J. Yu, J. Liu, LRProb control chart based on logistic regression for monitoring mean shifts of auto-correlated manufacturing processes. Int. J. Prod. Res. **49**(8), 2301–2326 (2011)
15. D.G. Wardell, H. Moskowitz, R.D. Plante, Run-length distributions of special-cause control charts for correlated processes. Technometrics **36**(1), 3–17 (1994)
16. H.B. Hwarng, Detecting process mean shift in the presence of autocorrelation: a neural-network based monitoring scheme. Int. J. Prod. Res. **42**(3), 573–595 (2004)

Exploring a New Lean Operation Mode for Chinese Manufacturing Enterprises Based on Comparing Major "Industry 4.0" Strategies

Run Zhao[1(✉)] and Haixia Sang[2]

[1] Department of Industrial Engineering, JiangSu University of Science and Technology, Zhenjiang, China
zhaorun1982@hotmail.com

[2] Department of Industrial & Management System Engineering, Waseda University, Tokyo, Japan
sanghaixia@aoni.waseda.jp

Abstract. The term "Industrie 4.0" originating from the German government evokes a new era of industrial revolution. To catch up with the world industry development trend, different countries has separately launched the corresponding strategies to define their own core meanings and terms for the 4th Industrial Revolution. In this paper, some obvious differences are summed up by comparing major "Industry 4.0" strategies. Based on the current Chinese manufacturing conditions and development history, a new lean operation mode for Chinese manufacturing enterprises is put forward to realize and deepen China's "Industry 4.0" strategy, entitled "Made in China 2025".

Keywords: Industry 4.0 · Lean operation mode · Chinese manufacturing enterprises

1 Introduction

Manufacturing is an important means for the development of human society to creates materials wealth, human cultures and provide society and its members with those things that are needed or desired [1]. In order to improve the human's living quality and standard from three aspects technologically, economically, and historically, manufacturing has been promoted by three Industrial Revolutions. In each revolution stage, an advanced manufacturing mode or management strategy is proposed to revolve the various new problems faced by manufacturing industry. In Fig. 1, with each industrial revolution, three milestones for the manufacturing mode has been put forward by three outstanding representative figures separately. Each revolution has greatly promoted the development of the industry productivity and changed the world economic landscape and country's comprehensive strength.

However, especially, since entering into the 21st century, with the development of globalization, regionalization and personalization, the manufacturing mode has been more complex and faced a variety of serious challenges, such as demand diversification, rapid response to market changes and high flexibility. Additionally, a certain

G. Q. Huang et al. (eds.), *Proceeding of the 24th International Conference on Industrial Engineering and Engineering Management 2018*,
https://doi.org/10.1007/978-981-13-3402-3_11

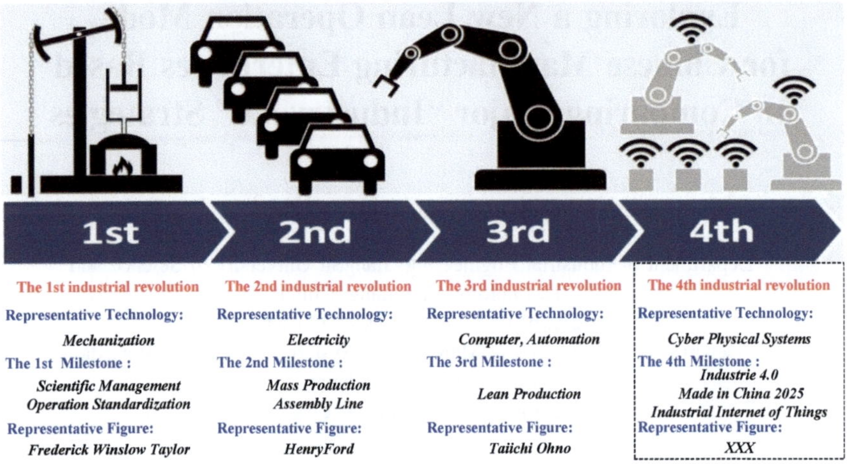

Fig. 1. Four industrial revolutions and related milestones for the manufacturing modes

manufacturing industries and processes has caused large natural resources consumption and environment pollution. Whereas, the efficiency and interests benefited from manufacturing has been narrowed as showed in Fig. 2.

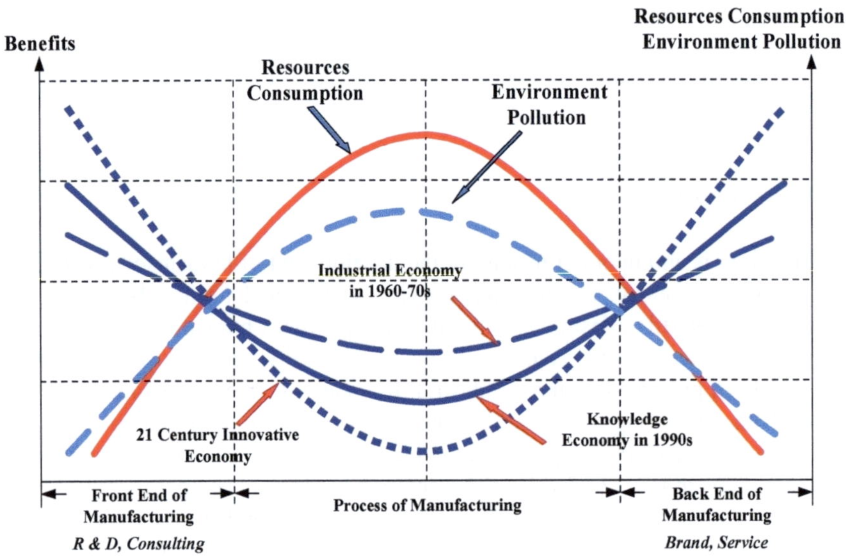

Fig. 2. The challenges from manufacturing industry and processes

Therefore, as a pillar industry, which can boom economy, increase employment and improve social material condition, the current traditional manufacturing mode should be promoted to deal with serious problems. With the rapid progress of science and technology, especially advancing in artificial intelligence, robotics, and digital manufacturing technology [2], developed or developing countries have put forward their own national industrial strategies, trying to take the opportunity to return to dominance in manufacturing industry. This development trend can be widely called the 4th Industrial Revolution as showed in Fig. 1. But, based on different backgrounds, characteristics and advantages of industrial structure, different developed or developing countries has separately defined their own core meanings and terms for the 4th Industrial Revolution. "Industrie 4.0" proposed by Germany is an intelligent manufacturing-led, designed to keep the manufacturing powerhouse of the world. "Made in China 2025" program is launched by China's government and it emphasis on innovation-driven high-quality manufacturing and domestically raising the produced content of key manufacturing components. "Industrial Internet of Things" strategy put forward by the Industrial Internet Consortium of America focuses on enabling and accelerating the adoption of internet-connected technologies across industries, both manufacturing and non-manufacturing. Nevertheless, which advanced manufacturing mode can maximize its effectiveness and lead the trend of the times, showed in Fig. 1, currently, can not be predicted distinctly.

Table 1. Comparison on major "industry 4.0" strategies

Title	Industrie 4.0	Made in China 2025	Industrial internet of things
Time	2013	2015	2014
Key authors	German government	Chinese government	Large multinationals
Background	Post crisis era	Big manufacturing industry but not strong manufacturing capacity	Financial crisis and manufacturing hollowing-out
Aim	To improve the competitiveness of the german industry and seize the initiative in the new round of industrial revolution	To promote and update from manufacturer to make power and shorten the international capacity gap in the high-end manufacturing field	To achieve a revival in the new manufacturing industry and reshape leadership
Technological focus	Supply chain coordination, cyber-physical system, information and communication technology, automation, robots, intelligent technology	Integrating information technology, automation, green manufacturing technology, intelligent technology, new materials, modern equipments	Device communication, data flows, device controls and integration, predictive analysis, industrial automation, virtualization and visualization technology
Essence	Reindustrialization strategy	Transformation and upgrading of manufacturing strategy	Reindustrialization strategy
Overall business approach	Reactive	Reactive	Proactive

2 Comparison with “Industrie 4.0”, “Made in China 2025” and “Industrial Internet of Things”

“Industry 4.0” has been firstly put forward by the German. Nowadays, a few industrial powers in the world have already begun to implement their own advanced industrial strategies respectively, and to improve the industrial level by integrating information and industrial technologies. These development strategies for industry dominated by the different governments will profoundly affect the future global industrial distribution. By comparing major “Industry 4.0” strategies from German, China, and the U.S. A., there exists some obvious differences [2–6].

For German, with the help of the information industry, the existing advanced industrial mode will be intelligentized and virtualized. The strategy, through big data and virtualization tools, emphasizes on smart factory and intelligence production to enhance product flexibility. On the other hand, the setting and extension of new standards is also put in the primary position of development. The information technologies are the medium for achieving integration and communication among advanced equipments.

For China, comparing with other developed countries, becaused of relatively weak industrial bases, in the strategy, the some basic problems need to be solved, such as the improvement of product quality, enhancement of industrial base capacity, promotion and transformation of the manufacturing sector, and so on. It calls for further integration of informatization and industrialization, as a way to boost manufacturing. The breakthrough point is the “Smart factory” or “Internet factory”, in which, on the basis of unified information management system, planning, scheduling and production are integrated into MES to achieve a high match degree between human and machines and mass customized production at lower cost.

For America, the idea of “Reindustrialization” is fully defined. The development of advanced digital manufacturing technologies, including advanced production technology platform, advanced manufacturing technology and design, and data infrastructure, is put forward. Its core is to encourage innovation, which can activate the traditional industries and reshape the industry structure through information technologies. It is called as a top-down mode for remodeling manufacturing by applying the analysis of big data tools from the CPU, systems, software, internet and other information ends. management system, planning, scheduling and production are integrated into MES to achieve a high match degree between human and machines and mass customized production at lower cost.

The differences of major “Industry 4.0” development strategies from these countries are shown in Table 1.

3 Exploring a New Lean Operation Mode for Chinese Manufacturing Enterprises

As known to all, in major developed countries, such as German and U.S.A., the manufacturing has fully experienced three times revolution before the “industrie 4.0”, which is based upon “industrie 1.0”, “industrie 2.0” and “industrie 3.0”. The related

"Industry 4.0" strategies is proposed on the basis natural development of upgrade. However, Chinese manufacturing industries still have to catch up on industry 2.0 and industry 3.0 before reaching the stage of 4.0. To avoid serious pressure from both newly emerging low-cost producers, and more effectively cooperate and compete with advanced industrialized countries, the efforts for transformation and upgrading of manufacturing mode is far broader, as the efficiency and quality of Chinese producers are highly unbalanced, and multiple challenges need to be met in a short time. So, an implicit specific aim of "Made in China 2025" is emphasis on creating more advanced manufacturing modes that is key to the most successful economic development.

On the one hand, the level of manufacturing or management mode for many Chinese enterprises is low and extensive, which can not reach a certain fine level required by "industrie 2.0" and "industrie 3.0". The implement and improvement on the standard operating procedure (SOP) and fundamental industry engineering (IE) have to be still undertaken. On the other hand, facing the increasingly fierce market competition and the urgent requirements of green economy, the advanced manufacturing mode should attach more importance to the ideas and actions in reducing pollutants discharge, lowering consumption and saving energy. Therefore, in view of the actual situation of Chinese manufacturing industry, a new lean operation mode drawing on the rich experience of Japanese companies should be considered to promote the management level and performance.

Actually, according to the current management characteristics of Chinese enterprises, lean operation mode can not completely copy the experience of Japanese enterprises, but should be improved and reconstructed. In this paper, two core parts for implement lean operation mode are considered.

3.1 A Corresponding Organizational Structure

A corresponding organizational structure for long run implementation need to be firstly set up. In this structure, three basic layers should be embodied.

Management Monitoring Layer (MML): The top leader and branch leaders should take part. In the layer, the idea of long run must be carried out, continuous implementation and improvement must be emphasized on, and long term policies also should be drawed up without just guiding short run effects. Furthermore, this layer need to reasonly balance and allocate the resource, work together to accomplish shared goals, and also evaluate the process and effect of key projects in certain period.

Project Promoting Layer (PPL): It comprises department directors. This layer should break down goals, set up detailed plans to reach the overall objective, guild plan promotion, strengthen communication and cooperation among all layers and departments, organize the related training, and deepen propagation and implement of ideas of lean operation. Moreover, recording and summarizing problems and assessing performances are the key to success.

Project Implementing Layer (PIL): It is composed by various active lean improvement teams, which are from grass-roots units, with a certain authorization. The function of these teams include: coordinate related business individuals and assign tasks; collect and analyze problems and data, and put forward some corresponding measures; organize related meetings to promote the lean improvement projects.

For Chinese enterprises, the authority of the leadership is very strong, and the business staffs of bottom layers are more dependent on managers' urges. Therefore, a top-down manner instead of the bottom-up manner is more suitable to promote the lean operation mode. In addition, these three layers are coherent and interdependent. With the aid of information technologies, mobile terminal visualization device, human-computer interaction technologies and other advanced tools serving for "Industry 4.0", smooth flow of real-time executive instruction without information Silo can be achieved easily to ensure a fast and efficient feedback to various questions about the operation. On the other hand, these three layers should also avoid over administration and instruction, but contribute to advocate and develop initiative and creativity rooted in members' potentiality. For the second layer (PPL), the comprehensive and flexible operation performances evaluation should be executed timely, and the corresponding measurements should be effective. Meanwhile, inter-departmental coordination and arrangement also should be meticulously designed and timely carried out.

3.2 A Corresponding Propulsion System Framework

After establishing the organizational structure and defining adaptable functions, a corresponding propulsion system framework for lean operation mode, including execution steps and contents, is built as shown in Fig. 3. It is obvious that three steps or stages are moved sequentially. First, for many Chines manufacturing enterprises, the fundamental Industrial Engineering (IE) improvement is the most important task. All operations and business within enterprises should be standardized. On this basis, step 2, fundamental production improvement, can be implemented to improve entire production facilities and basic management essential factors. Finally, core technologies of lean operation can be only just deployed. This management transform must be implemented step-by-step.

Ignoring basic managements in various levels and executing leap-step management changing can easily lead to failure. Furthermore, with steady progress of propulsion, some related management methodologies, including Problem Improvement, Knowledge Training, Target Management, Team Management, and Human Management, are carried out according to the order of propulsion steps.

For Chinese enterprises, with the process of implementing "Industry 4.0" Strategies, the related technologies, management measures and equipments, which do not meet the requirements of rules rooted in "Industry 1.0", "Industry 2.0", "Industry 3.0", are needed for continued transformation or improvements. It is an inevitable process while realizing great-leap-forward development, which can make up for the lack of natural consequence development experienced in the first three industrial revolutions in other developed countries. In the era of "Industry 4.0", the full application of new technologies, management measures and equipments are based on previous continuous improvements on standardization, fundamental IE, fundamental production mode, and so on. Therefore, under the background of industry "Industry 4.0", for this new lean operation mode, the new technologies, such as intelligent manufacturing, smart factory, information interconnection and so on, can not be blindly used without completing the basic management improvements, but their adaptability and application conditions need be fully considered.

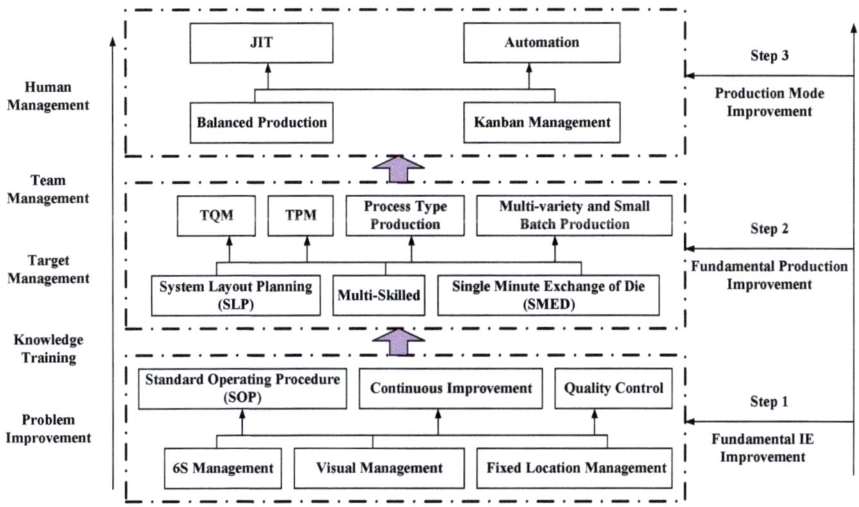

Fig. 3. A corresponding propulsion system framework for a new lean operation mode

4 Conclusion

At present, a new round of industrial revolution, "Industry 4.0", is being carried on like wildfire at a fast speed in world. Most countries are taking their own new strategies, which is compatible with national manufacturing conditions, to seize the future development highland and keep the sustainable competitive by making full use of information technologies, digital network and intelligent integrated innovation. However, in China, the manufacturing industry foundation is big but out strong, cohesion is lack of original innovation, low resource utilization, high costs, and other many basic management problems. So, in the face of the new industrial revolution, Chinese manufacturing enterprises must head-on straight on the tide of "industrialization", make greater efforts to strengthen basic management improvement and innovation by applying a new and reasonable lean operation mode while shifting to digital, networked, intelligent development. This mode can fit China's current industrial conditions and development history, which did not experience a complete process of the first three industrial revolutions, also meet the step-by-step transformation requirements from "Made in China 2025" program.

Acknowledgements. The authors wish to express their sincere gratitude to Prof. S. Takakuwa of Chuo University for his kindly suggestion and help. This research was supported by the Chuo University Joint Research "Paradigm shift of the manufacturing business in IoT/industry 4.0 environment" and the JiangSu University of Science and Technology Startup Research Fund for High-Level Talents (1042921513).

References

1. M.P. Groover, *Fundamentals of Modern Manufacturing: Materials, Processes, and Systems*, 3rd edn. (Wiley, 2007)
2. L.L. Wang, Comparative research on Germany "industrie 4.0" and "made in China", in *2nd International Conference on Humanities and Social Science Research* (2016), pp. 27–30
3. M. Hermann, T. Pentek, B. Otto, *Design Principles for Industrie 4.0 Scenarios*. Accessed 4 May 2016
4. M.A. Habib, http://www.linkedin.com/pulse/made-china-2025-inspiration-form-industry-40-md-ahasanul (2015)
5. S. Klaus, *The Fourth Industrial Revolution* (World Economic Forum, 2016), pp. 51–59
6. L.D. Bernardini, http://www.automationworld.com/industry-40-or-industrial-internet-things-whats-your-preference (2015)

Legal Guarantee of Green Manufacturing in China Under the Background of Ecological Civilization

Hui-qing Wen[1] and Jian-wei Niu[2(✉)]

[1] Department of Secretary and Law, Beijing Youth Politics College, Beijing, China

[2] School of Mechanical Engineering, University of Science and Technology Beijing, Beijing, China

niujw@ustb.edu.cn

Abstract. In traditional manufacturing industries, the protection of the natural environment was usually neglected. People did not take actions of pollution control and environmental restoration until industrial production had led to pollution and greenhouse gases. Green manufacturing emerged in this context as a revolutionary manufacturing model. The goal is to minimize the impact on the environment and to maximize the efficiency of resource use in the entire product life cycle. Green manufacturing depends on a perfect social support system, and relevant legal protection should be considered. This paper is aimed to elucidate the legal protection of a new era of China Green Manufacturing, on the basis of the ecological significance of the green manufacturing. We elaborated the connotation of green manufacturing in new era, and explained the legal systems related to green manufacturing in China in details. Finally we put forward the new requirement of green manufacture to law.

Keywords: Green manufacturing · Ecological civilization · Legal guarantee Legal systems

1 Introduction

Since the 60s of the last century, the global economy has experienced a rapid development. However, the protection of the natural environment caused by environmental pollution, global climate change, ozone depletion, acid rain and other consequences are neglected. With the economic growth and improvement of people's living standards, the life cycle of a large number of consumer products has been shortened, and the quantity of product waste has soared. People finally realized the importance of pollution control and environmental protection in the industrial sector. Nevertheless, the traditional method of industrial pollution control is end-of-pipe management. People do not take actions of pollution control and environmental restoration until industrial production has led to pollution and greenhouse gases. The remedial measures cannot achieve the objective of environment protection from the root. In order to completely

G. Q. Huang et al. (eds.), *Proceeding of the 24th International Conference on Industrial Engineering and Engineering Management 2018*,
https://doi.org/10.1007/978-981-13-3402-3_12

solve the problem of environmental pollution and high carbon emission, we must take actions from the source of the problem. Green manufacturing emerged in this context as a revolutionary manufacturing model advocated by modern civilization. The so-called green manufacturing means a modern manufacturing model considering the environmental impact and resource efficiency [1]. The goal is to minimize the impact on the environment and to maximize the efficiency of resource use in the entire product life cycle, from conception, design, manufacture, assembly, transportation, sale, use and scrap of industrial products.

As a reflection of sustainable development in manufacturing, the research and implementation of green manufacturing depend on a perfect social support system. To establish this system, first of all relevant legal protection should be considered. Therefore, this paper is aimed to elucidate the legal protection of a new era of China Green Manufacturing, on the basis of the ecological significance of the green manufacturing.

2 The Connotation of Green Manufacturing in New Era: Eco-change

As a big manufacturing country in the world, manufacturing is the largest industry in China's economy that sustains the development of China. The GDP of China in 2017 is 82.7122 trillion RMB, in which the manufacturing sector contributes 24.2707 trillion RMB, accounting for 29.34% of the Chinese economy. Therefore, in order to achieve pollution prevention and low-carbon emission in China industry, the first thing to consider is China's manufacturing industry. On March 11, 2018, the 13th China People's Congress adopted the revised "Constitutional Amendment", in which the "ecological civilization" was written into the Constitution. It is to advocate "ecological civilization" become the embodiment of the national will. For China's manufacturing industry, "ecological civilization" into the Constitution means both challenges and opportunities. This provides a fundamental legislation guarantee for the environmental protection which will benefit China's manufacturing industry. It puts forward basic requirements of a "closed loop" pattern for China's industry. Meanwhile, it is also a backstop for the Chinese green manufacturing. In the new era, it was given a kernel of "ecology" to green manufacturing.

Considering the connotation of green manufacturing, it is fundamentally necessary to study the issues from the perspective of the development of the times. If the traditional "green manufacturing" is concerned with environmental pollution prevention and energy efficiency use, then the green manufacturing in a new era, that is, in ecological construction background civilization, "green manufacturing" means to put the manufacturing into the entire ecosystem. It not only focuses on green product lifecycle process itself, but also pays more concern to whether such behavior really follows the social harmony laws of the people, the nature, and the society. It is concerned whether the manufacturing truly play a positive role for healthy ecosystem. It can be said that green manufacturing under the background of ecological civilization is a green manufacturing 2.0 upgrade that incorporates the concept of ecosystem protection on the basis of traditional "green". We can predict that, in the context of socialism with Chinese characteristics "Five in One" overall layout, China's manufacturing industry will eventually embark on the path of green low-carbon, from unsustainable modern industrial civilization to a sustainable future ecological civilization. The transition will ultimately achieve harmony and organic integration of the human society and the Earth's biosphere cycle [2]. This is the goal of China's manufacturing industry and the connotation of green manufacturing, which can be called "the ecological transformation of green manufacturing".

3 Legal Systems Related to Green Manufacturing

Currently in China, laws related to green manufacturing include: "Environmental Law", "Circular Economy Promotion Law", "Cleaner Production Law", "Renewable Energy Law", "Energy Conservation Law", "Environmental Protection Tax Law" and related natural resources legislation and pollution prevention legislation.

3.1 Renewable Energy Law and Environmental Protection Priority

The Renewable Energy Law of People's Republic of China released on January 1, 2015 is regarded as the most stringent environmental law in history. The law states: "The state shall adopt economic and technical policies and measures of conservation and recycling of resources, protection and improvement of the environment, the promotion of human and nature to make economic and social development and environmental protection in harmony." In the legacy law, it claims that environmental protection should be coordinated with the economy development. In contrast, in the latest lay, it claims that economy development should be coordinated with environmental protection. This means that economic development cannot go beyond the limits which the environment can bear. Manufacturing industry needs to give priority to the environment, and adapt to the environment. In addition to the Constitution, the Renewable Energy Law is the most direct and fundamental legal regulation for green manufacturing. The new Renewable Energy Law put forward the requirement of establishing the concept of "environmental protection as the center" for the manufacturing industry. In the product life cycle, all the designers, manufacturers, transporters, sellers and users as well as recycling and reuse of waste materials need to keep in mind "green", "cycle", "low carbon" as a primary consideration. When the conflict of economic interests and environmental interests occurs, priority should be given to environmental interests.

3.2 Circular Economy Promotion Law and Comprehensive Utilization of Resources

The Circular Economy Promotion Law of People's Republic of China was released since January 1, 2009. The purpose of the law is to promote the development of circular economy, improve the efficiency of resource use, protect and improve the environment, and achieve sustainable development. The Circular Economy Promotion Law takes low consumption, low emission, and high efficiency as the basic features of "circular economy". Although the current "Circular Economy Promotion Law" is criticized for its legislative logic disorder and dislocation [3], the status of this law as a basic norm of green manufacturing is self-evident. The basic system of the Circular Economy Promotion Law include: circular economy development planning system, total control system, evaluation system of circular economy, extended responsibility system of producers, supervision and management system of key enterprises, industrial policies of circular economy, and incentives measures [4]. Whoever in the manufacturing sector, regardless of natural persons, legal entity and other organizations should be in accordance with the relevant provisions of Circular Economy Promotion Law, to fulfill legal obligations according to the law, to comprehensive utilize the resources, to reduce the amount of pollutants and to improve the utilization of resources efficiently.

With the development of society, China has now promoted its environmental resources optimization into a national strategy. The fundamental problem to be solved by circular economy is not only economic development mode, but the way in which resources should be used [5]. Therefore, we should re-examine "circular economy" from the perspective of ecological civilization. Only in this way can we help people better fulfill their legal obligations. The "circular economy" in the sense of ecological concept means that in the big system of people, natural resources, and science and technology, and in the entire process of resource input, enterprise production, product consumption, and disposal, the traditional economy with linear dependence on resource consumption should be transformed into an economy that develops on the basis of an eco-friendly resource cycle.

3.3 Cleaner Production Promotion Law and Source Reduction/Pollution Prevention

Cleaner Production Promotion Law came into force on January 1, 2003. Cleaner Production Promotion Law was revised in 2012 and took effect on July 1, 2012. The promulgation of the new law signifies that the strategy of source prevention and full-process control has been integrated into the comprehensive strategy for economic development [6]. The purpose of this law is to promote cleaner production, improve the efficiency of resource use, reduce and avoid the generation of pollutants, protect and improve the environment, protect human health, and promote sustainable economic and social development. The law clearly stipulates the main ways of cleaner production, governance priorities, prohibited scope, and development direction [7].

Although the Cleaner Production Promotion Law has been released for many years, there are still many problems in the field of cleaner production. For example: the investment for research and development of the cleaner production technology is inadequate; standards and indicators associated with cleaner production industry lag; new technologies related to cleaner production cannot be promptly applied to production; manufacturing enterprises lack understanding of cleaner production, or unwilling to spend time and economic costs on cleaner production audits. However, in the construction of ecological civilization, governments, enterprises and individuals all have the responsibility to contribute to the reduction of sources and prevention of pollution. This is the basic requirements of the current development of China's green cleaner production.

3.4 Renewable Energy Law/Energy Conservation Law and New Energy/Energy Conservation and Utilization

The Renewable Energy Law of the People's Republic of China was released on January 1, 2006. In December 2009, the law was revised and released on April 1, 2010. The Energy Conservation Law of the People's Republic of China was released on October 28, 2007. In July 2016 the law was amended, and came into force on September 1, 2016. For manufacturing, the two laws lay down obligations of the relevant body to protect the environment and achieve sustainable economic and social development from the perspective of energy use and conservation provisions.

China's industrial energy consumption accounts for 70% of the national energy consumption. The adjustment of China's energy structure depends on the use and conservation of new energy sources in the industrial sector. Also, because green manufacturing takes into account the three major areas of manufacturing, environment and resources, promoting the use of renewable energy and energy conservation and improving energy efficiency are the inevitable requirements and historic mission of the ecological development of the manufacturing industry. The Renewable Energy Law and the Energy Conservation Law are important legal guarantees for China's manufacturing industry to change its energy structure and achieve green development.

3.5 Environmental Protection Tax Law and Pollution Regulation

On January 1, 2018, the Environmental Protection Tax Law of People's Republic of China came into effect. The legislative purpose of this law is to protect and improve the environment, reduce pollutant emissions, and promote the construction of ecological civilization. Enterprises, institutions and other production operators applicable to environmental emissions of pollutants taxable should be committed to an obligation to pay the environmental tax. In fact, the environmental protection tax is China's first specialized single-line tax law of the "green tax" to promote the construction of ecological civilization. It established an environmental restraint mechanism that uses tax leverage to force reductions in manufacturing-related units and operators [8]. The essence of the environmental protection tax is to reduce the emission of pollutants and protect the environment by internalizing the external costs of the enterprise and increasing the cost of the pollutants emitted by the company to change its emission behavior [9].

4 The New Requirement of Green Manufacture to Law in the New Era

All along, the industrial emissions of pollution are the main source of pollution in China, and industrial energy consumption accounted for more than 70% of social consumption. Environmental and resource constraints have become the bottleneck of

China's industrial development. Faced with the dilemma of environmental crisis and development demand, green manufacturing in the new era has become an urgent need for sustainable development and an inevitable choice for Chinese manufacturing to move toward the high end. It means that Chinese manufacturing industry must consider the impact of products throughout their life cycle on the environment and climate change in the production, storage, transportation, marketing and other aspects, maximize the use of raw materials and energy, reduce harmful emissions, waste and greenhouse gases, improve operational procedures, minimize pollution to the environment, mitigate and respond to climate change. In the new era of ecological civilization construction, it requires for a much higher demand on green manufacturing.

In order to achieve green development of China's manufacturing industry, "Made in China 2025" requests: in 2020, one thousand green demonstration plants and one hundred green demonstration zones will be established, part of the heavy chemical industry appears inflection point for energy resources consumptions, emissions of major pollutants in key industries drops by 20%. By 2025, the green development of the manufacturing industry and the unit product consumption of major products will reach the world's advanced level, and the green manufacturing system will be basically established. This can be understood as the basic goal of achieving ecological green manufacturing. This can be seen as new demands of the law in a new era of green manufacturing. In order to achieve these objectives, to ensure the smooth development of green manufacturing, the relevant law should respond positively.

5 Conclusion

Legal construction of green manufacturing is to promote the development of green manufacturing, to achieve the most favorable protection by the march of industrial civilization to ecological civilization. Specifically, first, the relevant laws and regulations should be revised and improved in the legislation. The Circular Economy Promotion Law and the Cleaner Production Law of Low-carbon should be revised. The related norms of recycling and green industry should be set up. The specific rules or regulations on green manufacturing practices should be developed. Based on the principles of low-carbon, recycling and green, the principle of full regulation of the manufacturing sector, including design, production, assembly, transportation, sales, use, and reuse should be laid down. Second, a clear law enforcement powers and obligations in the law enforcement body should be clarified to demand them handle issues in accordance with the green law. Third, it should be practicable and specialized in the administration of justice. The provisions of the public or the victim can pursue the legal responsibility for enterprises in non-compliance with statutory obligations and serious waste of resources through litigation, environmental pollution, and greenhouse gas emissions. The specialized environmental courts should be established, and the manufacturing and the green environment-related cases can go into the court trial. Fourth, it should be stated clearly and firmly as a requirement that various main obligations (including design, manufacturing, assembly, transportation, sale, use, reuse and utilization of renewable subject) should confirm to the principle of green, low-carbon, and recycling in the production, operation and consumer activities.

Acknowledgements. This research is supported by Social Sciences Program of Beijing Municipal Education Commission (SM201211626004) and Doctoral Scientific Fund of Beijing Youth Politics College (BS201403).

References

1. Q. Gao, J.F. Zhang, The Construction of the legal system of green manufacturing. Mark. Mod. **06**, 285–286 (2008)
2. J.X. He, The revolution of new energy system is the only way to ecological civilization: a commentary on Jeremy Rivkin's third industrial revolution. J. China Univ. Geosci. (Soc. Sci. Ed.) **14**(02), 1–10 (2014)
3. F. Peng, Resources, waste or industry promotion: reflection on the revised path of China's circular economy promotion law. Polit. Sci. Law **09**, 98–109 (2017)
4. D. Lin, The main system of circular economy promotion law. Renew. Resour. Recycl. Econ. **2** (2), 23–28 (2009)
5. Z. Yong's, *View on the Amendments of Circular economy Promotion Law of the People's Republic of China*. http://www.sohu.com/a/31275522_200899. Accessed 30 Mar 2018
6. C.B. Zhou, Z. Li, J.J. Liu, Situation, problems and countermeasures of the cleaner production development in China. Environ. Prot. **44**(10), 27–32 (2016)
7. T.B. Qin, On the role of environmental law in the process of China's new industrialization: a case study of cleaner production promotion law. Law Rev. **05**, 81–85 (2005)
8. F. Gao, Environmental taxes will bring a better environment. Tax Levy **05**, 7–8 (2017)
9. Y. Zhou, Environmental tax levy is beneficial to ecological civilization. Tax Levy **12**, 7–8 (2017)

Scheduling Optimization for Multi-AGVs in Intelligent Assembly Workshop

Ji-hong Yan(✉), Ming-yang Zhang, and Zi-min Fu

Department of Industrial Engineering, Harbin Institute of Technology, Harbin, China
jyan@hit.edu.com

Abstract. In the era of Industry 4.0, enterprises aim to make products personalized concerning consumer requirements. The logistics in intelligent assembly workshop is more sophisticated due to the variability and complexity of products. In order to cope with that, a three-layer structure was proposed to manage the dispatch of automated guided vehicles. In the first layer, Floyd algorithm was implemented to generate the route scheme between any two workstations. In the second layer, a mathematical model was established to describe the delivery time spent by AGVs. In addition, Particle Swarm Optimization algorithm and rescheduling strategy were integrated for the task assignment and dynamic scheduling. In the third layer, according to running status of AGVs, heuristic rules were proposed to prevent the collision and deadlock among AGVs. Both the feasibility and effectiveness of the proposed structure and methods were validated by the example in an assembly workshop.

Keywords: Automated guided vehicle scheduling · Intelligent workshop
Mathematical model · Optimization

1 Introduction

The increasingly fierce competition in the market and the development of advanced technology as the Information and Communication Technologies (ICT), the Internet of Things (IoT) and cloud computing, have promoted the traditional mode of manufacturing to the next industrial innovation, which is commonly known as Industry 4.0. This initiative strategy aims to make products personalized concerning each single customer requirements through the construction of intelligent factories, the development of intelligent production and the deployment of intelligent logistics [1, 2]. Caused by sophisticated demand for high-individualized products and services, both quantity and variant of parts or components factories need are so numerous that they cannot be handled with pervious planning or delivery methods [3, 4]. The inbound and outbound logistics must ensure the materials are transported to customers precisely, duly and stably. The logistics problems have been one of the research hotspots in this environment especially in modern assembly manufacturing. Parts in workshops are delivered or transferred to workstation by various vehicles, for example forklift, crane, conveyor and automated guided vehicle (AGV) [5]. AGVs have been introduced to

G. Q. Huang et al. (eds.), *Proceeding of the 24th International Conference on Industrial Engineering and Engineering Management 2018*,
https://doi.org/10.1007/978-981-13-3402-3_13

many different types of industry since its emergence in the 1950s and also developed into reliable and efficient equipment [6, 7]. Besides the convenience and efficiency AGVs bring, constraints like conflicts among themselves are also involved into production process. In an assembly workshop, once parts in the workstation are running out, there will be a line-stoppage with workers being idle [8]. Therefore, it is crucial to assure that AGVs are scheduled properly and efficiently for the high performance of production system in intelligent workshop.

Many models and methods aimed at scheduling problem in the warehouse have been considered to date. Among these papers, route optimization, task assignment and collision avoidance are more frequently discussed. And the layout of route for AGV is generally regular network. Zhu et al. [9] has established a mathematical model concerning multi-tasks and multi-AGVs. An improved genetic algorithm was implied to get optimal task assignment and route for AGV. Qiu et al. [10] analyzed the reason for deadlock of AGV and prevented them from bumping against others by optimizing the scheduling and the control of resource locks. Combinational optimization problem under the non-square network route was discussed by Li et al. [11]. Both operating and waiting time were under consideration in their research. Liao et al. [12] established an open queuing network model in flexible flow shop with non-equivalent parallel AGV. Results of an example were compared with the statistic results to validate the accuracy of the improved state space decomposition. Nageswararao et al. [13] dealt with binary particle swarm vehicle heuristic algorithm (BPSVHA) to minimize mean tardiness for simultaneous scheduling of AGVs and machines. Hu et al. [14] proposed a methodology that solving scheduling problems in storage system by updating the time window for AGVs whose priority are relative low.

Most of papers mentioned dealt with scheduling problems in warehouse or flexible manufacturing system (FMS) containing determined tasks. Few papers have referred to the impact of task properties on manufacturing or assembling. For example, different orders have different degree of urgency. However, solutions that only tackling with predetermined scenario are no longer adaptable to the intelligent manufacturing because of the variability of orders. Scheduling methods for route whose layout is large-area regular network in the warehouse could not be applied to the intelligent assembly workshop. On the other hand, there are more constraints caused by processing technology on the route of AGV in the assembly shop than those in the warehouse. In this paper, the structure for multi-AGVs scheduling system in intelligent workshop was proposed. Intelligent algorithms and heuristic strategies were implemented to optimize the route planning and task assignment. The flexibility and timeliness for processing of tasks could be guaranteed. Meanwhile, collision or deadlock among AGVs were avoided. Through the practice in an intelligent assembly workshop, the feasibility and effectiveness of the method were proved.

2 Structure of Multi-AGVs Scheduling System

Generally, the purposes of AGV operations in an assembly workshop are classified into two categories. One is material delivery, like delivering parts to workstations or finished products to the warehouse. The other is supporting system operation, like conflict-free between AGVs. Regardless of those purposes, the motion of AGV can be described as follows: for a task, AGV leave the station 1 to the station 2 on the appropriate route. After that, the AGV stops at the station 2 and unloads there until completed. Through the exploration on the inherent relationship among issues above, this paper presents a three-layer structure with strong adaptiveness for multi-AGVs scheduling system, as illustrated in Fig. 1.

In the off-line route scheme layer, stations of AGV are setup according to the workshop environment, then generate functional and distance information of these stations. The functional info refers to parking, charging and so on. The distance info, apparently, is the distance between stations. Finally, these data are put into the off-line algorithm to generate the route scheme. The scheme contains all stations that AGVs pass through when traveling in line with the shortest distance from one station to another. The functional and distance information are uploaded to the database. As for on-line task assignment layer, delivery tasks in workshop are obtained in real-time by the scheduling system. Actually, not only the task acquired from production system, the task engendered by AGV itself should also be considered into this process. When the battery capacity of an AGV is relative low, the AGV might move to the nearest charging station first rather than executing the delivery. Therefore, final scheduling

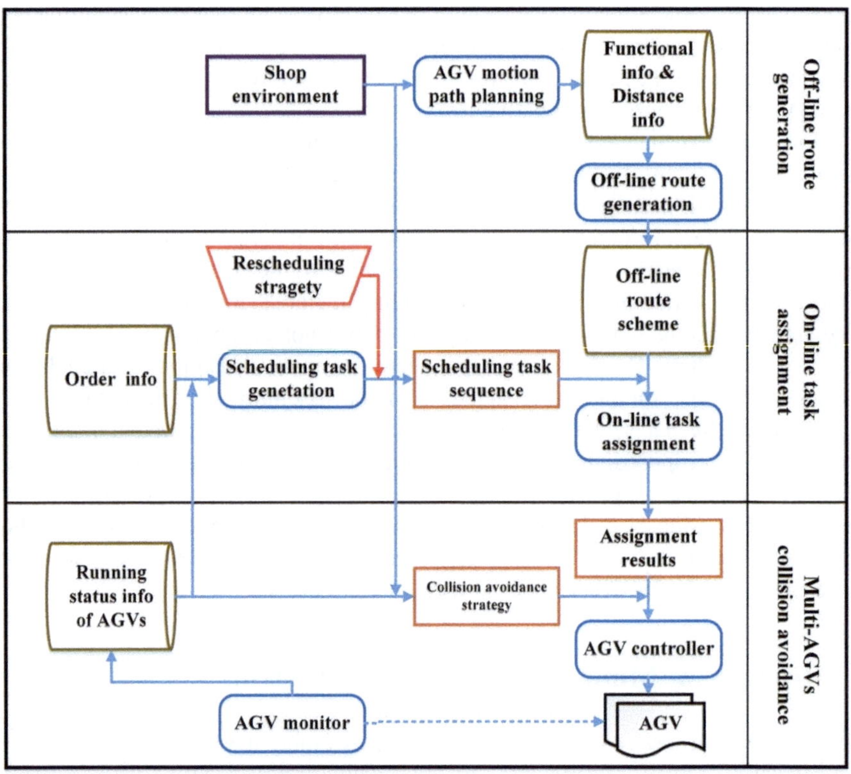

Fig. 1. Structure of multi-AGVs scheduling system

tasks are determined according to some rescheduling strategies based on running status of AGV. Intelligent algorithms are implemented to dispatch those tasks to limited quantity of AGVs. In multi-AGVs collision avoidance layer, the information of running status including the location, battery remaining capacity and so on, which is acquired by sensors installed on AGVs should be updated to the database. The scheduling system is able to generate heuristic collision-free strategies like waiting strategy according to the priority of AGVs. Then, the controller applies strategies to prevent collision.

Tasks managed by the scheduling framework achieve higher processing speed by separating scheduling stages. It is applicable to different kinds of workshops. By acquiring information including the basic environment and stations of AGV in the workshop, dynamic scheduling for multi-AGVs could be accessible. Besides, the separation of collision-free accelerates the response of the system. In an environment with relative restrict constraints process technologies cause, there are few alternatives about the routes scheme. There is great discrepancy in the different routes between two stations. The time that AGVs consume might be increased dramatically if not traveling in line with the scheme. If the route scheme is rearranged by methods like time windows, not only the time that the system costs for calculation is increased, but the result may not fulfill the time restrictions for delivery. Therefore, with the close integration of these three layers, the scheduling structure in this paper can satisfy requirements on the response speed of the system and processing capability of both complex and changeable orders in the intelligent assembly environment.

3 Solution to Multi-AGVs Scheduling

3.1 Mathematical Model

In many hybrid flow assembly workshops, materials are usually delivered from the batching area to assembly workstations. Due to the variety of products, part demand is different in each workstation, which causes the demands for parts have different urgent degrees. If the system handles all scheduling tasks by the same strategy, that leads to untimely delivery of some urgent tasks, resulting in the shortage of parts in the line side inventory and workers being idle. For multi-AGVs scheduling, minimizing the time of completing all the tasks is set as the optimization target. Within this paper, the multi-AGVs scheduling problem is limited by the following assumptions: (1) A variety of parts can be delivered by an AGV, and the capacity of AGV is non-limit. Actually, in order to assure the efficiency of production, more stations in the assembly line usually means more AGVs serving for this system. Due to the rather short cycle times, this seems like a tolerable simplification [9]. (2) No unexpected failures occur during the execution of the task, such as sudden stop. (3) The AGV runs at a constant speed v throughout the process. (4) The AGV initials in the park area and come back to this area after accomplishing the task. (5) The number of AGV accessible for scheduling is determined to be N.

Based on these assumptions, the multi-AGVs scheduling tasks in the workshop are converted into the route planning problem between AGV sites. Sites in the workshop are denoted as $i, j \in \{1, 2, \cdots, M\}$, all AGVs are denoted as $k \in \{1, 2, \cdots, N\}$. According to the description of the problem, the optimization model is established as follows (Table 1):

$$Min\ max\ \{D_k/v + T_k\}, \quad \forall k \tag{1}$$

$$\text{s.t. } D_k = \sum\nolimits_i \sum\nolimits_j x_{ijk}\, y_i\, y_j\, d_{ij}, \quad \forall i, j, k; \tag{2}$$

$$T_k = \sum\nolimits_i \sum\nolimits_j x_{ijk}\, y_i\, y_j\, d_{ij}, \quad \forall i, j, k; \tag{3}$$

$$\sum\nolimits_k \sum\nolimits_i x_{ijk}\, y_i\, y_j = 1, \quad \forall i, k; \tag{4}$$

$$\sum\nolimits_k \sum\nolimits_j x_{ijk}\, y_i\, y_j = 1, \quad \forall j, k; \tag{5}$$

$$x_{ijk} = \begin{cases} 1 & \text{AGV } k \text{ moves from site } i \text{ to } j \\ 0 & \text{Task from site } i \text{ to } j \text{ is not executed by AGV } k \end{cases} \tag{6}$$

$$y_i = \begin{cases} 1 & \text{There is a stop at site } i \\ 0 & \text{There is no stop at site } i \end{cases} \tag{7}$$

Objective function (1) seeks to minimize the running time of the last AGV who accomplishes tasks among the fleet. The total time are added up with two parts. One is motion time, and the other is waiting time. Constraint (2) calculates the running distance of each AGV in the system. Constraint (3) adds up all the waiting time during one task. (4) and (5) enforce that all the tasks are dispatched and one time only. (6) and (7) ensure the consecutiveness of those tasks.

3.2 Optimization

In the off-line route scheme layer of the structure proposed in Sect. 2, the Floyd algorithm is employed to generate the route scheme. The algorithm is a method to find the minimum path among multiple sources in a given weighted graph using dynamic programming [15]. The content of the scheme is the shortest distance between any two stations and the stations passed when AGV moves from one station to another. Those

Table 1. Notation

Variants	Specification
D_k	The total distance for AGV k to finish task
d_{ij}	The minimum distance from site i to j
T_k	The total waiting time for AGV k to finish task
tij	The waiting time in the site j from k
xijk	The task of AGV k
yi	The existence of stop at site i

are updated to the database in the form of matrixes. Scheduling tasks can be distributed on–line through Particle Swarm Optimization (PSO).

The PSO algorithm can be denoted as follows:

Suppose the dimension of the swarm is M and the quantity of particle is N. The coordinate of particle i can be denoted as $Xi = (xi1, xi2, \ldots, xid)$, and the coordinate of the best position in flying history of particle i is *pbest*, also denoted as $Pi = (pi1, pi2, \ldots, pid)$. The best one in $Pi(i = 1, 2, \ldots, N)$ is denoted as *gbest* or *pg*, whose meaning is the coordinate of global optimum in all particles' flying history. The speed of particle i is defined as $Vi = (vi1, vi2, \ldots, vid)$. Then iterations are applied to update the position of every single particle according to the formulas as follows [16]:

$$\begin{aligned} vid(t+1) = w \times vid(t) + c1 \times r1 \times (pid(t) - xid(t)) \\ + c2 \times r2 \times (pg(t) - xid(t)) \end{aligned} \tag{8}$$

$$xid(t+1) = xid(t) + vid(t+1) \tag{9}$$

where $1 \le i \le N$, $1 \le d \le M.w$ is defined as momentum, and $c1$, $c2$ are accelerating factors. $r1$, $r2$ are random numbers in the range of $(0, 1)$. The dimension of the swarm $d(1 \le d \le M)$ fluctuates in the range of $[-xd, \max, xd, \max]$ and the speed $[-vd, \max, vd, \max]$. Once the boundary of interval is extended, the boundary-value will be selected to replace the original one. Parameters $c1$ and $c2$ are chosen as integer 2, and w is also a constant which equals to 1 [17, 18]. In this paper, the priority of each task is considered. So a space matrix containing three vectors whose length are L is established as the encoding of particle. L is the quantity of tasks Each task matches a three-dimensional vector: the ID of AGV k executing this task, its order r and its urgent degree s. The space matrix can be divided into three L-dimensional vectors as Xv, Xr, Xs. When the swarm is initialized, every single variable in Xv will be assigned an integer from 1 to K, and Xr a real between 1 to L. Variables in Xs will be 1 or 0 according to their urgent degree. Integer 1 represents that the task is relative urgent while 0 is not. Variables in Vector Vx are integers in the interval of $[-(K-1), (K-1)]$, and Vr real numbers in the range of $[-(L-1), (L-1)]$.

Based on the running status of the system and the direction of the movement of AGV, two types of possible collisions are summarized shown in Fig. 2. The dash line indicates the route of AGVs. The solid point in the figure is the representative of stations. The black arrows mark the moving direction of AGV. In Fig. 2a, AGV① and AGV② are heading to stations M through different routes. They might arrive at station M simultaneously. Therefore, collision might happen between AGVs ① and ②. In Fig. 2b, AGVs ① and ② are running in the same direction. They are both toward to station N. If the AGV② stops at the station N to delivery parts to the production system and AGV① still wants to pass the station, there will be another collision. Then the AGV monitor mentioned in the third layer is implemented into the system to acquiring the status info of AGVs through many sensors installed on AGVs. Considering the constraints in the environment, waiting strategy based on the priority of each AGV is implemented to deal with the possible collision and even deadlocks.

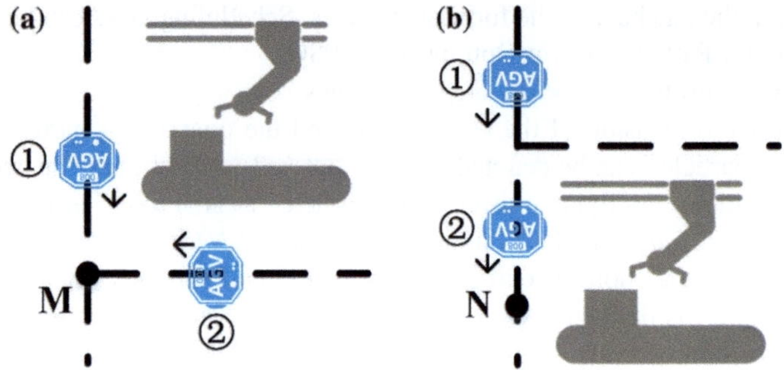

Fig. 2. Types of collision

4 Verification Results

The proposed structure and method were verified by solving the AGV scheduling problem in an actual intelligent assembly workshop. The presentation of the shop and AGV stations are shown in Fig. 3. In addition, Fig. 4 shows the AGVs used in the environment. There are 31 AGV stations in the workshop, whose routing information is stored in the system database. The workshop is consisted of 5 parts, as raw material sorting area, part storage area, hybrid flow assembly line, quality inspection area and end product inventory area. There are 3 AGVs in total accessible for the scheduling tasks.

The proposed algorithms are validated by a set of different examples. One of these examples is shown in Tables 2 and 3. There are total two phases in this example. When urgent tasks in Phase 1 were finished, the rescheduling strategy was activated and the

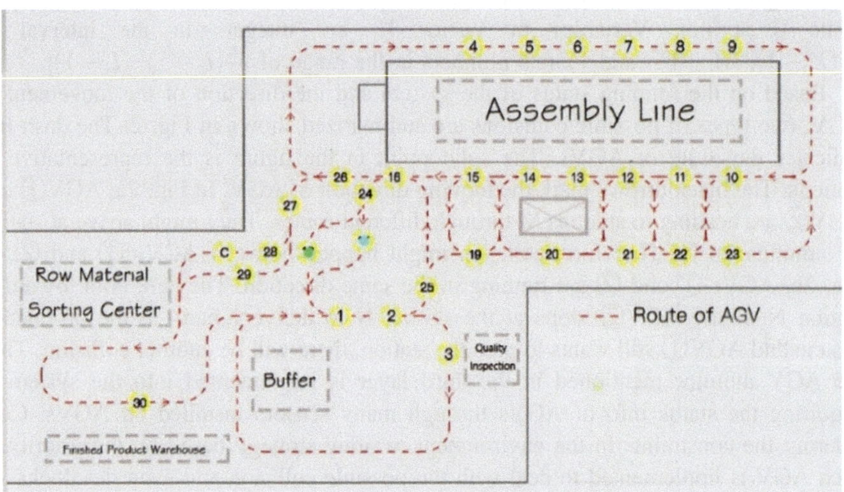

Fig. 3. The layout of the assembly shop

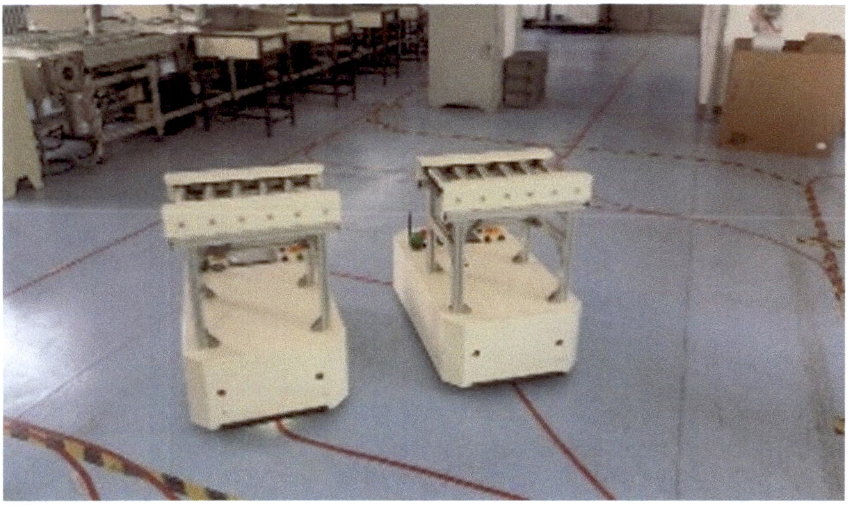

Fig. 4. AGVs employed in the verification

Table 2. Stations demanding parts

Phase	Stations
1	4, 6, 7, 8, 9, 10, 13, 16, 23
2	6, 7, 8, 10, 12, 14, 19, 22, 26,

Table 3. Results of task assignment

AGV ID	Task assigned in Phase 1	Task assigned in Phase 2
1	4-7-10	19-22-26
2	23-8-15	6-8-12
3	9-13-16	7-10-14

remain tasks with the new ones from the system were integrated as the elements in Phase 2. Tasks in Phase 1 is represented as stations need a stop in set {4, 6, 7, 8, 9, 10, 13, 16, 23}. Numbers in that set is the station ID. There are totally 9 tasks to be accomplished, while tasks from stations in the set {4, 9, 13, 23} are relative urgent. Tasks in Phase 2 is represented as stations in set {6, 7, 8, 10, 12, 14, 19, 22, 26}. The urgent tasks are marked with underline in the Table 2. Similarly, there are 9 tasks in Phase 2. The result of the assignment of tasks and the curve of fitness value during its iteration process are illustrated in Table 3 and Fig. 5 separately. All urgent tasks are executed directly. Besides, the problem of collision between AGVs has been solved properly by forecasting possible conflicts based on the monitoring of their running status.

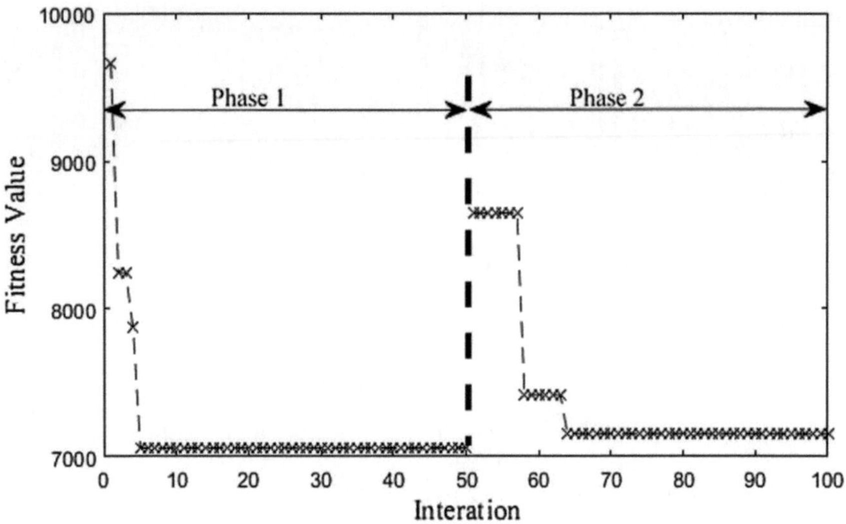

Fig. 5. Fitness value during Iterative process

5 Conclusions

The work is a step forward in scheduling problem for multi-AGVs to deal with the variability and complexity of products in the intelligent assembly workshop. On the basis of the analysis about characteristic of manufacturing in that environment, a structure for multi-AGVs scheduling system has been proposed. Utilizing the mathematical model constructed to describe the makespan, Floyd algorithm and PSO algorithm were implied to plan the route and dispatch tasks. Furthermore, this paper summarized all probable forms of collisions in our route and designed a method to resolve conflict or deadlock avoidance. As we can see in the practice, application of this method in the field of automatic assembly workshop is as efficient as we expected. This structure and method provided the theoretical supports for the construction of the unmanned factories in the intelligent manufacturing environment. For further works, we will study the performance of the method by applying it into different assembly shops.

Acknowledgements. The work is funded by NSF-NSFC (Grand No. 51561125002).

References

1. R.Y. Zhong, X. Xu, E. Klotz, S.T. Newman, Intelligent manufacturing in the context of industry 4.0: a review. Engineering **3**(5), 616–630 (2017)
2. W. Li, German industry 4.0 (industry 4.0) 'China made 2025' innovation driven research. Sci. Manage. Res. **35**(5), 71–79 (2017) (in Chinese)

3. B. Marco, F. Emilio, G. Mauro, P. Francesco, F. Maurizio, Assembly system design in the industry 4.0 era: a general framework. IFAC-PapersOnLine **50**(1), 5700–5705 (2017)
4. L. Barreto, A. Amaral, T. Pereira, Industry 4.0 implications in logistics: an overview, in *Manufacturing Engineering Society International Conference 2017*, Vigo (Pontevedra), Spain, pp. 1245–1252
5. N. Boysen, S. Emde, M. Hoeck, M. Kauderer, Part logistics in the automotive industry: decision problems, literature review and research agenda. Eur. J. Oper. Res. **242**(1), 107–120 (2015)
6. B. Li, H. Liu, D. Xiao, G. Yu, Y. Zhang, Centralized and optimal motion planning for large-scale AGV system: a generic approach. Adv. Eng. Softw. **106**, 33–46 (2017)
7. H. Neradilova, G. Fedorko, Simulation of the supply of workplaces by the AGV in the digital factory, in *International Scientific Conference on Sustainable, Modern and Safe Transport*, vol. 192, pp. 638–643
8. N. Boysen, S. Emde, Scheduling the part supply of mixed-model assembly lines in line-integrated supermarkets. Eur. J. Oper. Res. **239**(3), 820–829 (2014)
9. L. Zhu, X.M. Fan, Q.C. He, Scheduling optimization for multi-AGVs in batching area of flexible production system. Comput. Integr. Manuf. Syst. **18**(6), 1168–1175 (2012) (in Chinese)
10. G. Qiu, C.X. Ou, Y.R. Han, Map-building and scheduling algorithm for multi-AGV intelligent warehouse systems. Ind. Control Comput. **29**(12), 117–119 (2016) (in Chinese)
11. J.J. Li, B.Z. Xu, Y.S. Yang, H.F. Wh, Guided ant colony particle swarm optimization algorithm for path planning of AGVs. Comput. Integr. Manuf. Syst. **23**(12), 2758–2767 (2017) (in Chinese)
12. Y. Liao, Q.X. Chen, N. Mao, A.L. Yu, X. Li, Modelling and analysis of queening network in flexible flow shop with non-equivalent parallel AGVs. Comput. Integr. Manuf. Syst. **23**(9), 1950–1961 (2017) (in Chinese)
13. M. Nageswararao, K. Narayanarao, G. Ranagajanardhana, Simultaneous scheduling of machines and AGVs in flexible manufacturing system with minimization of tardiness criterion, in *International Conference on Advances in Manufacturing and Materials Engineering*, vol. 5 (2014), pp. 1492–1501
14. B. Hu, B. Wang, C.X. Wang, M. Yang, Dynamic routing of automated guided vehicles based on time window. J. Shanghai Jiaotong Univ. **46**(6), 967–971 (2012) (in Chinese)
15. H. Stefan, The Floyd-Warshall algorithm on graphs with negative cycles. Inf. Process. Lett. **110**(8), 279–281 (2010)
16. M. Orides, C.M.T. Carlos, F.C. Vinicius, C.P. Emerson, H.T. Roberto, PSO in 2D-space to solve reactive scheduling problems in FMS to reduce the make span, in *2013 IEEE International Symposium on Industrial Electronics*, Taipei, Taiwan (2013), pp. 1–6
17. M. Maryam, J.Y. Hwa, N.M. Siti, T. Farzad, Z.M.D. Siti, Multi-objective AGV scheduling in an FMS using a hybrid of genetic algorithm and particle swarm optimization. PLoS ONE **12**(3), 1–24 (2017)
18. C.W. Reynolds, Herds. A distributed behavioral model. Comput. Graph. **21**(4), 25–34 (1987)

Artificial Intelligence

Random Forest Classifier for Distributed Multi-plant Order Allocation

Si-han Wang[1], Wen-di Ren[2], Yi-fan Zhang[3], and Feng Liang[1](✉)

[1] Department of Management Science and Engineering, Nankai University, Tianjin, People's Republic of China
liangfeng79@nankai.edu.cn

[2] School of Data and Computer Science, Sun Yat-Sen University, Guangzhou, People's Republic of China

[3] Department of Mathematics, Nankai University, Tianjin People's Republic of China

Abstract. This paper focuses on the problem of multi-plant order allocation. It proposes a solution combining a machine learning algorithm and an optimization algorithm. It mainly concentrates on how to apply machine learning classifier to expedite the process of solving this problem in high accuracy. Random Forest classifier and an instance are used to illustrate this method, and the process of the experiment is also represented. Moreover, the result of classification by random forest is analyzed and compared with three other classifiers. The comparison approves that the proposed approach can achieve the problem more efficiently and reasonably.

Keywords: Order allocation · Random forest · Multi-plant · Accuracy Speed

1 Introduction

After INDUSTRY 4.0 was raised on the Hannover Messe in 2013, a mass of opportunities and challenges has been brought to each enterprise and each country around the world. Accordingly, when implementing Industry 4.0, enterprises may encounter various problems from many aspects. In order to facilitate the development, it is imperial to clarify and remove these obstacles.

Under such a magnificent epoch, human beings' demands can be satisfied to a larger extent. Naturally, traditional mass production is transforming into multi-variety and small batch production, and make-to-order production, to fulfill diversified needs from customers. Thus, on the aspect of order allocation, enterprises tend to deploy several distributed plants in different districts and allocate orders to specific plants with considering the goal of maximizing the profit and customer satisfaction. However, due to the large demands and the lack of high-efficient quantitative optimization decision-making, enterprises always complain about the current imbroglio of low speed and diseconomy of order allocation. To tackle the above two problems, we propose a two-stage solution approach. In the first stage, heuristic algorithms, optimization algorithms and rule bases can be used to handle the diseconomy and improve the profit. Then in

G. Q. Huang et al. (eds.), *Proceeding of the 24th International Conference on Industrial Engineering and Engineering Management 2018*,
https://doi.org/10.1007/978-981-13-3402-3_14

the second stage, machine learning algorithms would be applied to enhance the allocation speed with recorded data in the first stage. This paper mainly focuses on the second stage. To illustrate the approach, random forest algorithm is used in an instance after comparing with other machine learning algorithms.

The rest of this paper is organized as follows:

In Sect. 2, a few similar and related studies done by previous scholars is discussed, which includes the problem of multi-plant order allocation, the study and application of random forest algorithm. In Sect. 3 the instance problem is presented in detail and the data composition is explained clearly. The random forest algorithm is introduced by comparing with Decision Tree (DT), Support Vector Machine (SVM), and k-Nearest Neighbor (KNN) in Sect. 4. In Sect. 5, the experiment of implementing random forest on the instance is expounded, and the result is analyzed by a comparison with DT, SVM, and KNN. In the final section, the main points of this paper are summarized, and the superiority of random forest solving such problems is emphasized.

2 Literature Review

The problem of multi-plant order allocation has been discussed fiercely in the past two decades. After considering customer's location, transportation distance, the due date of each order, production capability, the location of each plant, and etc, enterprises would dispatch large quantities of orders to distributed plants within a short time with the objective of minimizing the production cost or time. However, such problems have never been an easy task. On the aspect of minimizing problem cost, Yang et al. [1] implements genetic algorithm effectively to solve this problem under the model of the different production cost of each plant. Besides, Zhou and Wang [2] studies and models the problem of earliness and tardiness production planning with the due date for multi-location OKP manufacturing systems, which involves earliness and tardiness penalties. To put it another way, time is regarded as a crucial measurement in the above research. Furthermore, Wang et al. [3] establishes a model innovatively which considers production cost, delay time, production load balancing at the same time. Accordingly, he puts forward a method of combining heuristic rules and multi-level decision-making process to settle it, whose reasonability is shown successfully by means of a numerical case study. Similarly, Ma [4] brings forward an approach combing a fuzzy decision-making method and a preferential assignment heuristic scheduling algorithm with taking multi-objective into account, where the coefficient weight of each objective in the multi-objective group is determined by Analysis Hierarchy Process (AHP).

Another similar problem, called supply chain order assignment and scheduling, received a lot of attention from scholars. Chen [5] summarizes current research status of supply chain scheduling problems, which proves to be insufficient and needs to be further studied. Jiang [6, 7] optimizes a three-stage process which includes multi-plant supply chain order distribution, production scheduling, and batch transport scheduling. Moreover, he applied Hybrid Taboo Search Algorithm and an integrated algorithm combining taboo search and dynamic programming method in two experiments insightfully and gains upbeat results respectively. Considering using the maximum delay of the work-piece and the number of misplaced work-pieces as the ordering target, Bai and Tang [8] establishes two mathematical model successfully and employs dynamic programming algorithm in spite of the flaw of low efficiency.

As for random forest, Breiman [9] puts forward this method formally by combining the essence of "bootstrap aggregating" and "random subspace method" on the basis of previous studies, and elaborates this method systematically, which lays a solid foundation for later related research and also promotes the development of machine learning. In the following sixteen years, other scholars improve and apply random forest in multiple domains, which cover medicine, biology, internet, finance, business and etc. Muhammad and Eiji [10] uses random forest and data recorded by various sensors to export a model to classify the transportation mode by provided features, like age, district, acceleration and etc, which proves to have high accuracy. Aung and Hla [11] employs random forest to classify web pages from Yahoo into multi categories and highly recommends the reliable accuracy and training speed of this method. Medicine is an important field of applying random forest, Warda et al. [12] utilizes random forest present a new approach to boost the accuracy of the diagnosis of mental disorders, where two datasets and classifying models are involved.

Based on previous studies, we bring forth a new two-stage solution to solve such multi-plant order allocation problems. In this paper, random forest method is innovatively applied to demonstrate the second stage, which aims to accelerate the process of order allocation.

3 Problem and Data Description

3.1 Problem Description

The problem of multi-plant order allocation in this study is mainly discussed under the frame of make-to-order production. An enterprise has n distributed plants with different production capability and production cost. Moreover, m kinds of commodities with j optional configurations are on the production list, all of which could be yielded by any one plant. In addition, it is well worth mentioning that products in one order only can be produced from one plant. The distance and its corresponding transportation expense between customers and each plant varies, which needs to be taken into account. From the perspective of customers, they select specific kinds of products in a certain number (batch size), and a promise would be made by the enterprise that their order could be finished within t days (due date). Apparently, the length of time from accepting orders to due date change as a function of batch size to some extent.

3.2 Data Description

To help illustrate the application of random forest on multi-plant order allocation, a volume of 20140 discrete production data is sampled from enterprise M which fits the condition of the foregoing problem well. In addition, the production cost of each kind of product of each plant and transportation expenses matrix is also given. A, B, C, D, and E are five distributed plants of Enterprise M.

Table 1 Potential features for prediction model

Features	Value	Number of features
Batch size	Discrete positive integer value	1
Kind of product	1(a), 2(b), 3(c), 4(d), 5(e)	1
Due date	Discrete positive integer value	1
Customer's location	1, 2, 3, 4, 5, 6, 7, 8, 9	1
Workload of each plant	Discrete non-negative value	5
Min workload plant	1(A), 2(B), 3(C), 4(D), 5(E)	1
Delay time of each plant	Discrete non-negative value	5
Cost of each plant	Discrete positive integer value	5
Min cost plant	1(A), 2(B), 3(C), 4(D), 5(E)	1
	Total number of potential features	21

As shown in Table 1, the raw data has 21 potential features after some appropriate sorting and processing. Batch Size means the total quantity of commodities in a single order, where concrete kinds do not need to be distinguished. We use 1, 2, 3, 4, and 5 to represent 5 optional configurations, a, b, c, d, and e, of one kind of product, which can be produced from Enterprise M. What the Due Date actually means here is the length of time from accepting a specific order to the delivery term. Furthermore, each Customer's Location is classified into 9 different categories, 1, 2, 3, 4, 5, 6, 7, 8, 9, based on their geographical regionalization. Workload of Each Plant refers to the time of each plant to finish remaining work at the moment of accepting a specific order respectively. Then, Min Workload Plant is the plant with least remaining work among five plants. Moreover, the meaning of Delay Time of Each Plant is the time of each plant excessing completion term if they are chosen to be the target plant respectively. Likewise, Cost of Each Plant and Min Cost Plant have similar implication with foregoing concepts, where the costs incorporate production cost and transportation expenses.

4 Methodology

4.1 Proposed Model

In this paper, we propose a Random Forest Classifier (RF) model to solve our multi-classification task. RF is one of the most effective and versatile machine learning algorithms as they are more robust to noise. After training an RF model with the historical data, the new test data can be classified very quickly. Figure 1 shows the general architecture of the proposed system.

4.2 Feature Selection by Decision Tree Tool

Feature selection is of great importance in machine learning, especially for the problem of huge databases. When the dataset is very large with multi-dimensional features, learning might not work well before removing some redundant features [13].

Fig. 1 Proposed model

Therefore, we need to use some methods to extract the necessary features. In this paper, the decision tree is used to select proper ones.

A decision tree can be built from a set of instances. The growing process can be viewed as two parts. Firstly, the top node that represents a test on one feature needs to be determined from all the candidate predictors. Secondly, the training samples are portioned according to this feature, so each branch of the tree describes the outcome of the test and the leaf node labeled with a class.

It is obvious that a measure is essential to select the candidate attributes at the first step of constructing the tree. Then, information gain is applied to estimate how well a given feature separates the training cases according to their target classification because it is the expected reduction in entropy caused by portioning the samples according to this feature [14].

The feature selection part is able to be completed by default because at each step the top nodes, on which the decision tree is split, are considered as important variables within the remaining features.

4.3 Random Forest (RF) Classifier

Random Forest is one of ensemble classification methods that uses a bagging approach to create a bunch of decision trees. Each of the classification trees is built using a bootstrap sample of the dataset, and at each split, the candidate features are a random subset of the variables [15] rather the top node in DT. The classification result of RF is acquired by combining the output of each decision tree learners through a voting process. Finally, the forest chooses the label having the most votes [11].

The RF algorithm for classification has three main steps as described [16].

1. Select n_{tree} bootstrap samples from the training data.
2. For each of the bootstrap samples, build a decision tree. The difference from DT is that at each node, randomly test m_{try} of the features and choose the best split from among those variables.
3. Output overall prediction by using majority votes of the predictions of the n_{tree} trees.

The algorithm yields an ensemble that can achieve both low bias and low variance [15]. Figure 2 shows the construction of RF classifier.

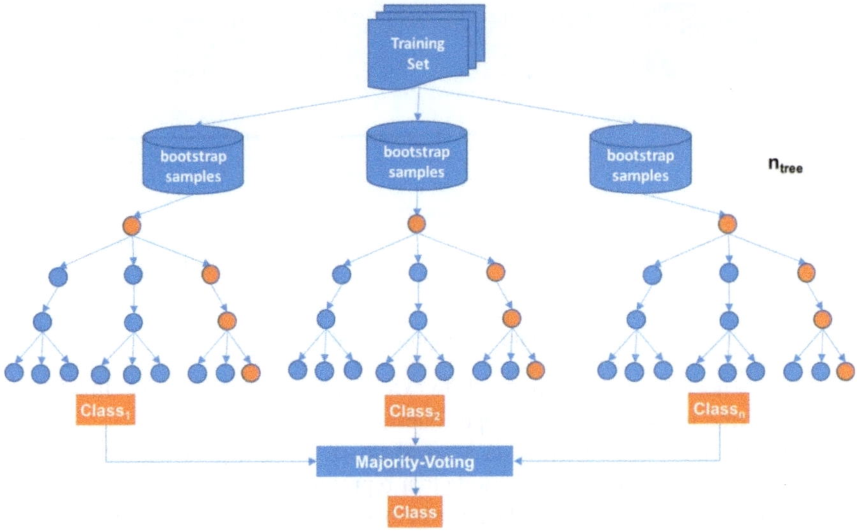

Fig. 2 Construction of RF classifier

With regard to the parameter of RF, we use a measure of the error based on the training samples to find the exact number of decision tree learners. This estimate called OOB is obtained by the following process [16]:

1. At each bootstrap iteration, predict the data not in the bootstrap sample using the tree grown with the bootstrap sample. These data are just named "out-of-bag", or OOB.
2. Aggregate the OOB predictions. Calculate the error rate which we define as the OOB estimate of error rate.

4.4 Compared Classifiers

1. k-Nearest Neighbor (k-NN) Classifier.

   The k-Nearest Neighbor (k-NN) algorithm is a non-parametric method for classification. It is one of the simplest machine learning algorithms. It has two steps; the first is to find the k nearest neighbors of a new test sample in the training set and the second is to determine the label of this instance based on a majority vote of the k neighbors in the first stage [17]. As for the parameter selection, the best choice of distance metric and k depends on the data. In this paper, the value of k is 10 and the distance metric is Euclidean.
2. Decision Tree (DT) Classifier.

   As the same construction method that is instructed in feature selection part, the DT classifier is made after computing all of the attributes. The classification rules are shown through the path from the root to the leaf node.

3. Support vector machine (SVM) Classifier.

   Support Vector Machine is a supervised machine learning algorithm. By finding a line (hyperplane), this classifier can separate the training data set into classes. When the classification problem is between two classes, in a multi-dimensional space, SVM is used to find the hyperplane owning the maximum distance from both classes. When data are not linearly divisible, no hyperplane that perfectly separates classes can be found. Thus, a hyperplane with the lowest error is the best.

As for the multiclass problem, the SVM classifier starts from many 2-class SVMs, namely the One-Vs-One (OVO) and the One-Vs-All (OVA) [18].

In this paper, we use the OVO. The m-class problem is split into $m\ (m - 1)/2$ binary classify problems. Then, a majority voting approach is employed in order to get the final classification result. Meanwhile, the cubic kernel is applied in the SVM classifier. Besides, the value of box constraint is 1.

5 Results and Discussion

5.1 Experiment

The dataset with 20,140 records is divided into one training dataset and one testing dataset. The training dataset contains 14,000 records and the testing includes the rest 6140 records.

By means of feature selection with decision tree tool, the importance of each feature in dataset is clarified. 10 of 21 potential features, Batch Size, Kind of Product, Due Date, Customer's Location, Min Workload Plant, Delay Time of Each Plant, are finalized as features in the subsequent model after dissecting and combination attempts. In order to obtain higher classification accuracy, out-of-bag classification estimate of error is conducted. As can be seen in Fig. 3, error rate plummets to 0.1 when using around 70 grown trees, and converges to approximate 0.085 at 1000 trees. Hence, 500 trees are determined to be used in the subsequent model, which can achieve sufficient accuracy without consuming much effort. In addition, the maximum number of splits are 50.

SVM, KNN, and DT are also utilized to classify the given dataset, of which results would be compared with that of random forest. Among several kinds of SVM and KNN, cubic SVM and weighted KNN performs best in this problem after repeated tentative and comparison.

5.2 Results and Discussion

The testing comprehensive accuracy of random forests reaches 91.0% in this problem. As revealed by Fig. 4, the accuracy we obtain in plant A, plant B, plant C, plant D, plant E, is 91%, 95%, 89%, 91%, 89% respectively. To illustrate, take plant A as an example. 2% of A is getting confused with B. 2% of A is misclassified into C. 2% of A are falsely regarded as D. 3% of A is mistaken for E.

It seems that 91.0% is not sufficiently high at first glance, especially when compared to some applications of random forest with excellent results, which can reach 95%, 96% or even more higher. Though 9.0% misclassified order cannot obtain

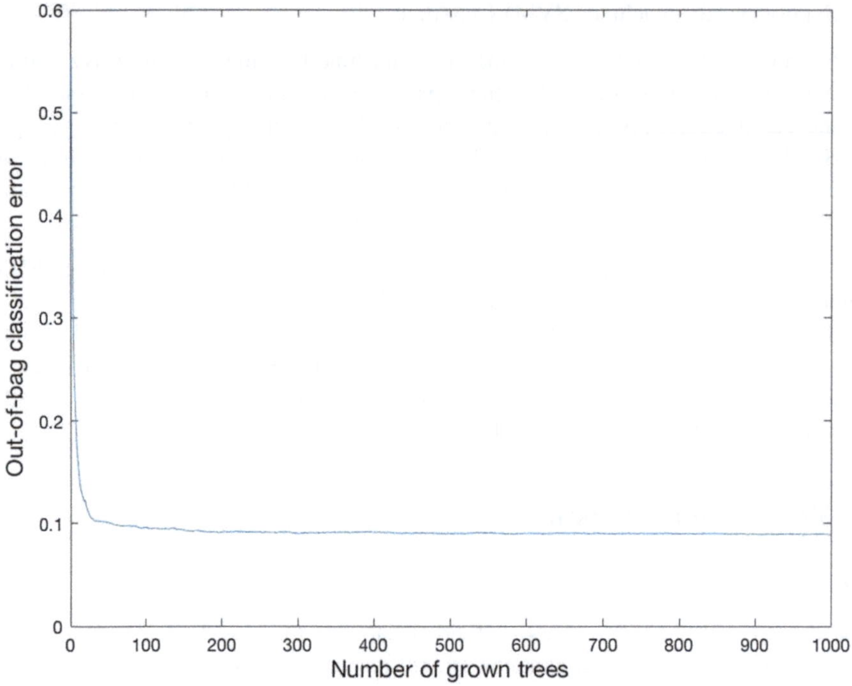

Fig. 3. Out-of-bag classification estimate of error

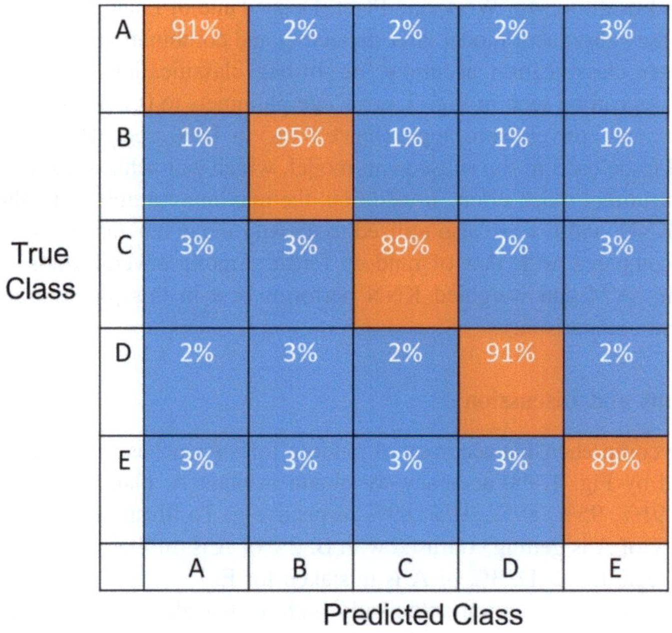

Fig. 4 Confusion matrix of random forest

Table 2 Testing accuracy of each classifier

Classifier	Accuracy (%)
Random Forest	91.00
Decision Tree	66.80
SVM	75.10
KNM	73.70

optimal solution, almost all of them take suboptimal solution after examining and contrasting the results in detail. Furthermore, as shown in Table 2, random forest is superior to DT, cubic SVM, Weighted KNN, whose testing comprehensive accuracy is 66.8%, 75.1%, 73.7% respectively.

In terms of training speed, the sequence of these four machine learning classifiers from fast to slow is DT, KNN, RF, SVM. It is suggested that there is no need to record numerical data of speed since the configuration and running speed of each computer differ. Even so, it is easy to find that SVM is significantly slower than the other three. Besides, the testing classification labels of 6140 records are outputted from model established by training dataset on a medium configuration computer within 1.3 s. Hence, it is self-evident that random forest performs much better on the efficiency of outputting the classification label of orders than optimization algorithms, like genetic algorithm, taboo search algorithm, which require a deluge of iterations and loops.

6 Conclusion

The results testify that random forest classifier could help accelerate settling the problem of multi-plant order allocation efficaciously. However, it should be noted that this study has been primarily concerned with the second stage of a proposed two-stage solution. How to render two stages compatible with each other cohesively and harmoniously requires further research. Notwithstanding its limitations, this study does prove the feasibility of applying machine learning on such problems.

Acknowledgements. This work is sponsored by a Research Grant from National Natural Science Foundation of China (71271122) and National Undergraduate Training Program for Innovation and Entrepreneurship (201610055036).

References

1. F.C. Yang, K. Chen, M.T. Wang, P.Y. Chang, K.C. Sun, Mathematical modeling of multi-plant order allocation problem and solving by genetic algorithm with matrix representation. Int. J. Adv. Manuf. Technol. **51**, 1251–1260 (2010)
2. J.H. Zhou, D.W. Wang, earliness and tardiness production planning with due-date for multi-location OKP manufacturing systems. Comput. Integr. Manuf. Syst. **5**, 52–57 (2000)

3. S.H. Wang, L.M. Yang, Y.X. Zhou, Y.F. Zhang, F. Liang, Order allocation optimization of distributed multi-plants considering multi-objectives optimization, in *Presented at the 46th International Conferences on Computers and Industrial Engineering*, Tianjin, China (2016)
4. J. Ma, Multi-objective decision analysis of multiple plants scheduling problem based on orders. Equip. Manuf. Technol. **7**, 115–119 (2010)
5. Z.L. Chen, Integrated production and outbound distribution scheduling: review and extensions. Oper. Res. **58**(1), 130–148 (2010)
6. D.K. Jianng, B. Li, Supply chain scheduling based on hybrid taboo search algorithm. J. Mech. Eng **20**, 53–59 (2011)
7. D.K. Jiang, B. Li, J.Y. Tan, Integrated optimization approach for order assignment and scheduling problem. Control Decis. **28**(2), 217–222 (2013)
8. M.Z. Bai, G.C. Tang, Integrated Production and distribution in supply chain management. Oper. Res. Trans. **1**, 113–119 (2009)
9. J. Breiman, Random forests. Mach. Learn. **45**(1), 5–32 (2001) (in Netherlands)
10. A.S. Muhammad, H. Eiji, Classification of travel data with multiple sensor information using random forest. Trans. Res. Procedia **22**, 144–153 (2017). (in Turkey)
11. W.T. Aung, K.H. Hla, Random forest classifier for multi-category classification of web pages, in *IEEE Asia-Pacific Conference on Service Computing*, Biopolis, Singapore (2009), pp. 372–376
12. H.A. Warda, N.A. Belal, Y. EI-Sonbaty, S. Darwish, A random forest model for mental disorders diagnostic systems, in *Advances in Intelligent Systems and Computing*, vol. 533 (2017), pp. 670–680
13. M. Dash, H. Liu, Feature selection for classification. Intell. Data Anal. **1**(1–4), 131–156 (1997)
14. V. Sugumaran, V. Muralidharan, K.I. Ramachandran, Feature selection using decision tree and classification through proximal support vector machine for fault diagnostics of roller bearing. Mech. Syst. Signal Process. **21**(2), 930–942 (2007)
15. R. Díaz-Uriarte, S.A. De Andres, Gene selection and classification of microarray data using random forest. BMC Bioinform. **7**(1), 3 (2006)
16. A. Liaw, M. Wiener, Classification and regression by random forest. R News **2**(3), 18–22 (2002)
17. P. Cunningham, S.J. Delany, k-Nearest neighbor classifiers, in *Multiple Classifier Systems*, no. 34 (2007), pp. 1–17
18. L.R. Quitadamo, F. Cavrini, L. Sbernini, F. Riillo, L. Bianchi, S. Seri, G. Saggio, Support vector machines to detect physiological patterns for EEG and EMG-based human–computer interaction: a review. J. Neural Eng. **14**(1), 011001 (2017)

Reliability Modeling with Application for Calibration Data

Bing-yue Liu(✉), Feng Qin, Shi-qiao Liu, and Sheng-wu Cai

Nuclear Power Maintenance Center, Research Institute of Nuclear Power Operation, Wuhan, China
liuby@cnpo.com

Abstract. In order to assess and predict the reliability of critical components, using performance degradation data for modeling has become a significant approach. In addition, degradation data often provides more information about the components' performance. The degradation can be detected directly or indirectly by records of surveillance, inspection and check. The degradation increases with time in an uncertain manner and needs to be modelled as a continuous time stochastic process. A wiener process model for modeling the calibration data has been developed. The unknown parameters of the model are obtained by the maximum likelihood estimation approach. The approach is illustrated by a real world example. This resulting model is applied to predict the reliability measure in given longer calibration intervals. The reliability index can be an input for maintenance effectiveness assessment and also be a performance indicator in a system or component monitoring program, etc.

Keywords: Calibration · Degradation · Reliability · Wiener process

1 Introduction

Reliability is a probability that an item performs its required function under given conditions for a stated time [1]. Typically, reliability assessment can be performed through reliability modeling with failure data. However, due to high reliability of critical components and comparatively conservative application of inspection and maintenance policy in the nuclear power plants, rare or even no failure data can be observed for critical components. In this case, reliability assessment with degradation data can be an effective method [2].

In past decades, degradation data plays a significant role in reliability assessment. Gertsbackh points out the value of reliability assessment with degradation data for the first time, and presents a simple linear model with both random slope and intercept [3]. Three kinds of methods for degradation data analysis are presented [4], including linear regression method, degradation path method, and stochastic process method. Degradation models in reliability assessment are reviewed and classified in certain groups [5]. Comments on their merits, applications and limitations are provided. With wiener process modeling, Hao analyses the reliability of mechanical components [6].

An analysis methodology has been described which is based on determining a statistically derived value of drift by analyzing the historical records during calibration

G. Q. Huang et al. (eds.), *Proceeding of the 24th International Conference on Industrial Engineering and Engineering Management 2018*,
https://doi.org/10.1007/978-981-13-3402-3_15

or surveillance of the transmitters in nuclear power plants [7, 8]. The mean, median, standard deviation, number of points and tolerance interval are calculated for the record data. Also, a scatter plot of the data is provided to judge whether all of the data is contained within a tolerance interval of $\pm 0.1\%$. Chen proposes an improved method which can visually describe the characteristics of instruments drift [9], and is much better to prove the stability of extended instrument calibration cycle. However, the quantitative reliability measure is not given in the literature cited above, nor does it propose any model to predict the reliability in any given longer calibration intervals.

Every component deteriorates with age and usage and eventually fails. In some cases, the degradation occurs continuously over time and is affected by operating factors (e.g., load, temperature, and material properties). This is best modeled as a continuous function of time [10]. Due to the nonmonotonic properties [6], a wiener process model is presented for representing the transmitters' degradation in this paper.

The paper is organized as follows. The Wiener process is explained in Sect. 2 and illustrated in Sect. 3. The paper is concluded in Sect. 4.

2 Preliminaries

2.1 Wiener Degradation Process

Degradation can be influenced by the random factors from the component itself and the environment. As a consequence, a good statistical model should take into account the sources of variation, and stochastic process is appropriate to describe the degradation of the system [11].

Suppose that the degradation process, $\{W(t), t \geq 0\}$, obeys a wiener process:

$$W(t) = \mu t + \sigma B(t). \tag{1}$$

where μ is the drift parameter; σ is the diffusion parameter; $B(t)$ is the standard Brownian motion.

The Wiener process has the following properties [12]:

1. $W(0) = 0$;
2. $W(t)$ has continuous sample paths;
3. $W(t)$ has independent increments;
4. $W(t) - W(s) \sim N(\mu(t-s), \sigma^2(t-s))$ for $t > s \geq 0$.

From the properties of the Wiener process, we can obtain that the degradation performance $W(t)$ is normally distributed as

$$W(t) \sim N(\mu t, \sigma^2 t). \tag{2}$$

The probability density function can be defined as

$$f(w) = \frac{1}{\sqrt{2\pi t}\sigma} \exp\left\{ -\frac{(w - \mu t)^2}{2\sigma^2 t} \right\}. \tag{3}$$

2.2 As-Found As-Left

Instruments are checked or adjusted periodically to ensure they respond to a known input signal within required accuracy limits. Generally, the transmitters' calibration check is performed across the whole instrument span. Every calibration, calibration check, or periodic surveillance has acceptance criteria for the instrument setting. If the instrument is adjusted, the as-left settings will be different from the as-found settings. The use of the terms as-found and as-left are clarified below [7]:

- As-found is the condition in which a channel, or portion of a channel, is found after a period of operation and prior to any calibration.
- As-left is the condition in which a channel, or portion of a channel, is left after a calibration or surveillance check.

It is very likely for all but the most stable instruments that the as-found setting for the present calibration will be different, even only by a small amount, from the as-left setting for the previous calibration. This variation over time represents some amount of instrument drift. We will refer to the difference between the current as-found and the previous as-left settings as the instrument drift over the time interval between calibrations.

3 Application

3.1 Background and Data

This section provides an example analysis for pressurizer level transmitters [7]. Three safety-related transmitters used to monitor pressurizer level have been combined into a single functionally-equivalent group.

All available calibration data was obtained for these transmitters at discrete points across the whole instrument span, e.g., 0, 25, 50%, etc., as shown in Table 1. For this example, these transmitters were installed in 1987 and seven calibrations have been performed since the installation.

3.2 Data Treatment

The data in Table 1 is normalized into a percent of span as follows [7]:

- Subtract the as-left value of the previous calibration from the present as-found value.
- Divide the above result by the instrument span in the same units as the calibration units.
- List the value as a percent of calibrated span.

Table 1. Calibration data

No.	Date		Data status	Interval months	Tag number	Calibration data (mA)				
	Month	Year				0%	25%	50%	75%	100%
1.	5	1993	As-found	12	LT-459	4.00	8.00	12.02	16.04	20.06
2.	5	1993	As-left		LT-459	4.01	8.00	12.02	16.00	20.02
3.	5	1992	As-Found	14	LT-459	3.98	7.99	11.98	15.93	19.96
4.	5	1992	As-left		LT-459	4.00	8.01	12.02	16.04	20.01
5.	3	1991	As-found	11	LT-459	4.01	8.00	12.00	16.01	19.99
6.	3	1991	As-left		LT-459	4.01	8.00	12.00	16.01	19.99
7.	4	1990	As-found	10	LT-459	3.96	7.95	11.90	15.92	19.91
8.	4	1990	As-left		LT-459	4.06	7.92	11.95	15.98	19.98
9.	6	1989	As-found	13	LT-459	4.00	8.00	12.02	16.07	20.02
10.	6	1989	As-left		LT-459	4.00	8.00	12.02	16.07	20.02
11.	5	1988	As-found	12	LT-459	4.05	8.06	12.05	16.12	20.15
12.	5	1988	As-left		LT-459	4.00	7.97	11.98	15.98	20.00
13.	5	1987	As-found		LT-459	New	New	New	New	New
14.	5	1987	As-left		LT-459	4.02	7.99	11.99	16.07	20.01
15.	5	1993	As-found	12	LT-460	4.02	8.03	12.02	16.01	20.00
16.	5	1993	As-left		LT-460	4.00	8.01	12.02	16.02	20.03
17.	5	1992	As-found	14	LT-460	4.00	8.00	12.01	16.02	20.03
18.	5	1992	As-left		LT-460	4.00	8.00	12.01	16.02	20.03
19.	3	1991	As-found	11	LT-460	4.01	8.03	12.02	16.04	20.06
20.	3	1991	As-left		LT-460	4.01	8.01	12.00	15.97	19.98
21.	4	1990	As-found	10	LT-460	4.01	8.01	12.00	15.98	19.97
22.	4	1990	As-left		LT-460	4.01	8.01	12.00	15.98	19.97
23.	6	1989	As-found	13	LT-460	4.04	8.05	12.06	16.03	20.04
24.	6	1989	As-left		LT-460	4.04	8.05	12.06	16.03	20.04
25.	5	1988	As-found	12	LT-460	4.06	8.03	12.02	15.99	20.00
26.	5	1988	As-left		LT-460	4.06	8.03	12.02	15.99	20.00
27.	5	1987	As-found		LT-460	New	New	New	New	New
28.	5	1987	As-left		LT-460	3.98	7.98	11.97	15.97	19.99
29.	5	1993	As-found	12	LT-461	4.02	8.03	12.02	16.03	20.04
30.	5	1993	As-left		LT-461	4.01	8.01	12.03	16.03	20.04
31.	5	1992	As-found	14	LT-461	4.02	8.03	12.03	16.02	20.03
32.	5	1992	As-left		LT-461	4.02	8.03	12.03	16.02	20.03
33.	3	1991	As-found	11	LT-461	3.99	8.01	11.99	16.05	20.06
34.	3	1991	As-left		LT-461	3.99	8.00	11.99	15.97	19.97
35.	4	1990	As-found	10	LT-461	4.04	7.97	11.95	15.93	19.90
36.	4	1990	As-left		LT-461	4.02	8.03	12.02	16.00	20.00
37.	6	1989	As-found	13	LT-461	3.96	7.94	11.95	16.00	20.07
38.	6	1989	As-Left		LT-461	3.98	7.97	11.98	15.99	19.99
39.	5	1988	As-Found	12	LT-461	3.98	7.96	11.94	15.85	19.84
40.	5	1988	As-Left		LT-461	3.99	8.00	11.98	16.03	20.00
41.	5	1987	As-Found		LT-461	New	New	New	New	New
42.	5	1987	As-Left		LT-461	4.00	7.98	12.01	16.00	20.00

The above computation represents instrument drift since the last calibration. The expression for normalized instrument drift is given by:

$$Drift_i = \frac{AF_i - AL_{i-1}}{Span}, \tag{4}$$

where,

$Drift_i$	Variation between the ith and the ith − 1 calibrations
AF_i	As-found value for the ith calibration entry
AL_{i-1}	As-left value for the previous ith − 1 calibration entry
Span	Instrument calibrated span.

3.3 Model Selection

First, investigate the drift data. It is obvious that the recorded calibration data drift in two directions.

Second, the drift data can be treated as independent and identically distributed (i.i.d.). Generally, for instrument check, calibration, adjustment, etc., the activities can be assumed to be perfect maintenance, thus the observed data are independent and identically distributed.

Then, normal probability plot is used [13]. The normal distribution probability paper is scaled so that normally distributed data will appear as a straight line. The closer the data is to a straight line, the more likely that the data is normally distributed [14]. Figure 1 shows that the plot is very straight in the middle, but deviates somewhat from the ideal line in the tails.

Thus, the wiener process model is selected.

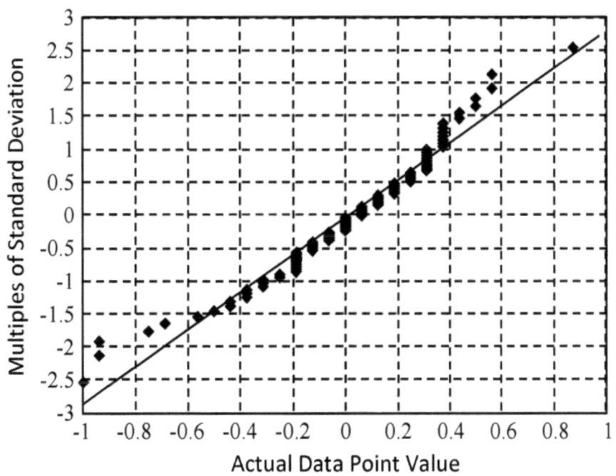

Fig. 1. Probability plot

3.4 Parameter Estimation

Totally, there are N times of calibrations. Based on the independent increment property of the Wiener process, the random variable $Drift_i$ has the following distribution:

$$Drift_i \sim N(\mu\Delta t_i, \sigma^2\Delta t_i), \ \Delta t_i = t_i - t_{i-1}. \tag{5}$$

So, the probability density function of $Drift_i$ is

$$f(Drift_i) = \frac{1}{\sqrt{2\pi\Delta t_i\sigma}}\exp\left\{-\frac{(Drift_i - \mu\Delta t_i)^2}{2\sigma^2\Delta t_i}\right\}, \tag{6}$$

where $i = 1, 2, \ldots, N$.

Thus the likelihood function of the sample is

$$\begin{aligned} L(\mu, \sigma) &= \prod_{i=1}^{N} f(Drift_i) \\ &= \prod_{i=1}^{N} \frac{1}{\sqrt{2\pi\Delta t_j\sigma}}\exp\left\{-\frac{(Drift_i - \mu\Delta t_i)^2}{2\sigma^2\Delta t_i}\right\}. \end{aligned} \tag{7}$$

Taking the logarithm of the likelihood function and the log-likelihood function as

$$\ln L(\mu, \ \sigma) = \sum_{i=1}^{N} \ln f(Drift_i). \tag{8}$$

According to the above proposed estimated method, the maximum likelihood estimations of μ and σ are obtained as:

$$\hat{\mu} = 1.27 \times 10^{-5} \text{ and } \hat{\sigma}^2 = 1.05 \times 10^{-6}.$$

3.5 Model Verification

The D' test is endorsed by ANSI N15.15-1974, to evaluate the assumption of a normal distribution for sample sizes larger than 50 [15].

(1) Order the sample data in ascending order from smallest to largest value.
(2) Compute the total sum of squares about the mean S^2 as follows:

$$S^2 = \sum x_i^2 - \frac{1}{n} \times \left(\sum x_i\right)^2 = 0.001076$$

(3) Calculate the quantity T as follows:

$$T = \sum \left[\left(i - \frac{n+1}{2} \right) \times x_i \right] = 7.6875$$

where $i = 1$ to 90.

(4) The test statistic is calculated by:

$$D' = \frac{T}{S} = 234.3549$$

(5) Compare the calculated value of D' with the D' percentage points of the distribution of this test, see Table 2. The hypothesis of normality will be accepted if the computed value of D' lies within the range 233.9–244.3. The calculated value of D' 234.3549, does lie within this interval. Therefore, it can be concluded that there is no reason to reject the hypothesis of normality (Table 3).

3.6 Reliability Measure

The calculated results are shown in Table 4.

The table shows that the transmitters' reliability decreases after transition to longer calibration intervals corresponding to the extended fuel cycles, meanwhile the failure probability increases.

Obviously, it is impossible to extend the calibration interval infinitely, a proper threshold has to be fixed. This needs a comprehensive decision making, considering the balance of reliability and availability, requirements from nuclear Technical

Table 2. Ordered sample data

−0.01	−0.00188	0	0.00125	0.003125
−0.00938	−0.00188	0	0.00125	0.003125
−0.00938	−0.00188	0	0.001875	0.003125
−0.0075	−0.00188	0	0.001875	0.003125
−0.00687	−0.00187	0	0.001875	0.00375
−0.00562	−0.00187	0	0.001875	0.00375
−0.005	−0.00187	0	0.001875	0.00375
−0.00438	−0.00187	0	0.001875	0.00375
−0.00438	−0.00125	0.000625	0.0025	0.00375
−0.00375	−0.00125	0.000625	0.0025	0.00375
−0.00375	−0.00125	0.000625	0.0025	0.00375
−0.00375	−0.00125	0.000625	0.0025	0.004375
−0.00313	−0.00125	0.000625	0.0025	0.004375
−0.00312	−0.00062	0.000625	0.003125	0.005
−0.00312	−0.00062	0.00125	0.003125	0.005
−0.0025	−0.00062	0.00125	0.003125	0.005625
−0.0025	−0.00062	0.00125	0.003125	0.005625
−0.00188	−0.00062	0.00125	0.003125	0.00875

Table 3. D′ percentage points for the 5% significance level

n	0.025	0.975
50	95.6	101.3
60	126.3	133.1
70	159.6	167.7
80	195.6	204.8
90	233.9	244.3
100	274.4	286.0
200	783.6	806.9
300	1445.0	1480.0
400	2230.0	2276.0
500	3120.0	3179.0
600	4106.0	4181.0
800	6331.0	6425.0

Table 4. Instrument reliability

No.	Interval months	Unreliability (%)	Reliability (%)
1	12	2.39	97.61
2	14	4.46	95.54
3	16	7.13	92.87
4	18	10.27	89.73
5	20	13.74	86.26
6	22	17.41	82.59
7	24	21.18	78.82
8	26	24.96	75.04
9	28	28.68	71.32
10	30	32.31	67.69

Specification, requirements of the system and component monitoring from Maintenance Rule, and potential economic losses caused by the component functional failure, etc.

4 Conclusion

In this paper, we have proposed a wiener process for representing the instruments' degradation process. Thus, the historic as-found and as-left data can be fitted to analyze and predict the reliability measure in given extended calibration intervals. In the case of no failure data, wiener process model is effective in reliability analysis and prediction considering the transmitters' degradation process. All the data required are available and easy of access in nuclear power plants.

A topic for future research is to optimize the calibration policy. A variety of performance measures can be used to evaluate and optimize the policy including availability based, reliability based and financial based. We have been working on this topic.

References

1. A. Birolini, *Reliability Engineering Theory and Practice*, 5th edn., ch. 1 (Springer, New York, Berlin, Heidelberg, 2007), p. 2
2. W.Q. Meeker, N. Doganaksoy, G.J. Hahn, Using degradation data for product reliability analysis. Qual. Prog. **34**(6), 60–65 (2001)
3. I.B. Gertsbackh, K.B. Kordonskiy, *Models of Failure*, ch. 3 (Springer, New York, 1969), pp. 86–90
4. M.J. Zuo, R.Y. Jiang, R.C.M. Yam, Approaches for reliability modeling of continuous-state devices. IEEE Trans. Reliab. **48**(1), 9–18 (1999)
5. N. Gorjian, L. Ma, M. Mittinty, P. Yarlagadda, Y. Sun, A review on degradation models in reliability analysis, in *Proceedings of the 4th Conference World Congress on Engineering Asset Management* (Marriott Athens Ledra Hotel, Athens), pp. 28–30
6. H.B. Hao, C. Su, Z.Z. Qu, Reliability analysis for mechanical components subject to degradation process and random shock with wiener process, in *Proceedings of the 19th Conference Industrial Engineering and Engineering Management* (Springer, Berlin, Heidelberg), pp. 531–543
7. E. Davis, D. Funk, Technical report. Guidelines for instrument calibration extension/reduction—revision 1: statistical analysis of instrument calibration data. EPRI, Palo Alto, CA, 1998. TR-103335-R1
8. P. Zhou, C.H. Qiu, Q.B. Chu, Application of AFAL methodology in substation of instrument calibration intervals extension in nuclear power plant. Nucl. Power Eng. **34**(5), 115–117 (2013) (in Chinese)
9. Y. Chen, L.H. Zhao, T. Yu, L.H. He, Z.J. Liu et al., Extension of instrument calibration intervals base on improved AFAL analysis. Nucl. Power Eng. **38**(2), 64–67 (2017) (in Chinese)
10. R. Jiang, D.N.P. Murthy, *Maintenance: Decision Models for Management* (Science Press, Beijing, 2008), p. 102
11. Z. Sheng, S.Q. Xie, C.Y. Pan, *Probability and Statistics* (Higher Education Press, Beijing, 2008), p. 315 (in Chinese)
12. E.A. Elsayed, *Reliability Engineering* (Publishing House of Electronics Industry, Beijing, 2013), p. 287 (in Chinese)
13. R. Jiang, M.J. Zuo, *Reliability Models and Application* (China Machine Press, Beijing, 1999), p. 34 (in Chinese)
14. B.X. Dong, S.J. Wang, M.Y. Chen, G.F. Zhai, Reliability assessment method for aerospace relay based on wiener process. Elect. Energy Manage. Technol **11**, 12–16 (2015)
15. ANSI N15.15-1974, Assessment of the assumption of normality (Employing Individual Observed Values)

Research on Long-Term Portfolio Selection Model Based on DEA Cross-Efficiency Evaluation

Chengchao Qiu(✉)

Institute for Interdisciplinary Research, Jianghan University, Wuhan, China
qcchao@126.com

Abstract. This thesis proposes a method to use Data Analysis Envelopment (DEA) for choosing a value stock which has a long-term advantage. This thesis suggests a new assurance region for a DEA model which is will prove that suitable for stock evaluation. The method is to use cross-efficiency DEA with new assurance region and exam its score and variance in several years for selecting stocks. It is a reasonable way to find a good stock for investment and focus to a durable, strong and good performance stock, not diversifying the portfolio. The Author will discuss in which input and output factors need to use in DEA model to have a result that is suitable to the purpose of long-term investment.

Keywords: DEA · RAM · Cross-efficiency · Assurance region constraints Portfolio selection · Long-term investment

1 Introduction

1.1 Background

In 1978, Chames et al. [1] created Data Envelopment Analysis (DEA) method, and then there are hundred papers researching about it. In application, Using DEA as Multi-criteria decision-making (MCDM) is one of active areas now. Basically, DMUs in DEA model correspond to multiple alternatives in MCDM, input and output factors in DEA correspond to multiple performance measures in MCDM, and the result of DEA, efficiency score, is similar to convex multi-factor performance of MCDM.

The main benefit of DEA is that it provides a way to choose reasonable weights for different input and output performance factors, and then forms a single performance measure. A DEA run determines an efficiency score for each DMU, and DEA can rank DMUs according to their efficiency scores. However, there is one famous shortcoming of DEA is its flexibility in choosing optimal weights on the input and out-put factors in favor of each DMU. A DMU can have a high efficiency score by choosing extreme high weights for some factors and zero for other factors. Doyle and Green [2] called such DMU as mavericks. This issue can give an unreasonable result or ranking. Several research has dealt with this problem such as a supper-efficiency model [3], cone-ratio model [4], and assurance region model [5]. In our proposed method, we combine Assurance region with cross-efficiency evaluation to overcome this shortcoming.

G. Q. Huang et al. (eds.), *Proceeding of the 24th International Conference on Industrial Engineering and Engineering Management 2018*,
https://doi.org/10.1007/978-981-13-3402-3_16

On the other hand, Cross-efficiency evaluation [2] uses peer-evaluation model to calculate DMU's efficiency score. It is opposite to the self-evaluation model, therefore mavericks problem has less chance to happen. The cross-efficiency evaluation also gives the better ranking result, which makes it more suitable for using DEA as MCDM. There are many applications of cross-efficiency such as project selection [6], supplier selection [7] and ABC inventory classification [8]. Although DEA cross-efficiency, enhances ranking ability of DEA much more, Cross-efficiency still has some problems that limit its use. One of those problems is the non-uniqueness of cross-efficiency, which has been improved by several approaches by introducing secondary objectives ([2, 9], etc.) or by investing the ranking range [10, 11].

A traditional use of DEA cross-efficiency evaluation in portfolio selection is to rank DMUs in a decreasing order of cross-efficiency score and choose the top n DMUs where n is the number asset we want to hold in portfolio. This method shows many advantages than the simple use of DEA. According to Lim et al. [12] it still has two problems, lack of portfolio diversification and ganging-together phenomenon. Lack of portfolio diversification is not really a problem in choosing long-term advantage stock because the main target of value investment is to focus and get the highest return in the long term. In addition to that, ganging-together phenomenon is a one-time phenomenon when we use the result of a DEA run of one-time data making a decision.

1.2 Objectives

The objective of a thesis is to find a method to select stocks for long-term investment by using Mean-Variance framework and DEA cross-efficiency. The basic idea, which has researched in the paper of Lim et al. [12], is that DMU's cross-efficiency and variety of its score represent as the DMU's return and risk characteristic. The approach of this paper is different with Sungmook Lim's paper is that it runs DEA for several finance statements in continuous years, and sum all DMU's cross-efficiency score and variance of its score. The method of this thesis is to use DEA with an assurance region and then apply multi-period cross-efficiency score and its variance to mean-variance formulation. By that way, it deals with non-uniqueness of cross-efficiency score and target on long-term investment. Author also has some discusses which input and output factors are chosen for stock evaluation in long-term aspect.

2 Methodology Development

2.1 Portfolio Selection

People have several ways to select stocks to invest, but generally there are two ways to invest in the stock market, daily trading and fundamental investing. Day trading is to buy and sell stock everyday based on supply and demand for a stock. An investor needs to analyze the trading volume and stock price trend very closely to succeed with this way. On the other hand, fundamental investors focus basically on the financial strength of stock to choose and hold it for a longer time than day trading investors. Here, the thesis uses financial ratio from financial statement to evaluate the business performance. However, how long should it take into account for analysis and how long should investors hold their investment?

Lim et al. [12] has proposed a method to use DEA with mean-variance optimization framework to evaluate and select stock in period of one year. It could be called as short-term portfolio selection. In this approach, there are some problems, which are extreme weight issue and nonuniqueness score of DEA model, time frame for evaluating business performance, and how to interpret and use variance of cross-efficiency properly. The detail will present in the next sections for these issues.

2.2 Evaluation by Financial Ratio with Assurance Region Constraints

One a big shortcoming of cross-efficiency evaluation is non-uniqueness score. This problem also is due to too flexible in choosing weights in each DMU favor. Each DMU has many optimal solutions to have the best efficiency score, therefore many cross-efficiency scores come along too. Therefore, we need to ensure that all factors need to take into account with significant weights, but still let some flexibility, because we do not know and there is no mutual best set of weights. This can be done with an assurance region [5].

Assurance region constraints are added to the model to reduce the solution region to reduce the flexibility of DEA solution, and then reduce the number of optimal weights for each evaluation. However, in this manner, we need more auxiliary information, such as prices and unit costs, to determine the lower and upper bounds properly. During developing DEA, people assume that input and output factors are technological (not monetary), and we do not know what is right price or cost for each factor, so we let each DMU choose their own weights. If we do know the range of the price or cost for these factors, we can put a constraint to them. However, in our case, we are evaluating stocks were almost factors are financial ratio, there is no way to find out this information about "price and cost" of financial ratios.

This paper proposes to set a limitation on the percent of each weight to the total weights. In that way, the DMU still has flexibility in choosing weights and each factor does not have too much or too small influence on the final efficiency score. Constraints are formed as below:

$$\frac{1}{b^*(m+s)} \leq \frac{weight\ of\ each\ factor}{(v_1 + \cdots + v_m + u_1 + \cdots + u_s)} \leq \frac{a}{(m+s)} \tag{1}$$

With: m is the number of output factors.
s is the number of input factors.
a, b are upper, lower constraint factors.

Victor [13] has proved that if we add any constraints to DEA model, it may induce free or unlimited production of output vectors in the underlying production technology, which is expressly disallowed by standard production assumptions. Therefore, we need to check the consistency of constraint whenever it is added.

For input factors:

$$\frac{1}{b^{*}(m+s)} \leq \frac{vi}{(v_1 + \cdots + v_m + u_1 + \cdots + u_s)} \leq \frac{a}{(m+s)}(i = 1, \ldots, m) \quad (2)$$

With: m is the number of output factors
s is the number of input factors
a, b are upper, lower constraint factors which are chosen approximately:
(2) is equivalence to 2 follow inequality:

$$u_1 + \cdots + u_s + v_1 + \cdots + v_m - b(m+s)^{*} vi \leq 0 \quad (3)$$

$$-a(u_1 + \cdots + u_s) - a(v_1 + \cdots + v_m) + (m+s)^{*} v_i \leq 0 \quad (4)$$

The consistent of these constraints will be checked by theorem 5 in Victor [13].
For output factors:

$$\frac{1}{b^{*}(m+s)} \leq \frac{ui}{(v_1 + \cdots + v_m + u_1 + \cdots + u_s)} \leq \frac{a}{(m+s)}(i = 1, \ldots, s) \quad (5)$$

With: m is the number of output factors
s is the number of input factors
a, b are upper, lower constraint factors which are chosen approximately
(5) is equivalence to 2 follow inequality:

$$u_1 + \cdots (1 - b(m+s))u_i + \cdots + u_s + v_1 + \cdots + v_m \leq 0 \quad (6)$$

$$(-a^{*} u_1 + \cdots + (m+s-a)^{*} u_i + \cdots - a u_s - a(v_1 + \cdots + v_m)) \leq 0 \quad (7)$$

With inequality (6), $P^T v^* = [-1]^T v^*$ always negative, because v_i is all nonnegative. According to theorem 5 in Victor [13], this inequality (constraint) is not consistent, so that I remove $\frac{1}{b^{*}(m+s)} \leq \frac{ui}{(v_1 + \cdots + v_m + u_1 + \cdots + u_s)}$ and reply by another constraint:

$$\frac{1}{b^{*}(m+s)} \leq \frac{ui}{(u_1 + \cdots + u_s)}$$

This constraint is relevant to:

$$(-a u_1 - a u_2 \cdots + (m+s-a)u_i + \cdots - a u_s) \leq 0 \quad (8)$$

In conclusion, four constraints that are (4), (5), (7) and (8) for a DEA model in this thesis. The model is run and checked the consistency of the constraint.

2.3 Long-Term DEA Cross-Efficiency Evaluation

2.3.1 RAM-DEA Cross-Efficiency Evaluation

Lack of portfolio diversification is not really a problem in choosing long term advantage stock because the main target of value investment is to focus and get the highest return in the long term. In addition to that, ganging-together phenomenon is a one-time phenomenon when we use the result of a DEA run of one-time data making a decision.

Lim et al. [12] used variance and covariance of cross-efficiencies as individual DMU risk and inter-DMUs risk of changing weights to select a high diversification portfolio. They take into account the one-period data to do so. It is the risk of using cross-efficiency DEA evaluation to select portfolio itself, it is not the risk of select a particular stock in a portfolio because the variance of cross-efficiency is created by not having a proper set of factor weights to evaluate. Addition to that, by using variance and covariance of the cross-efficiency score, we can have a portfolio of DMUs have the diversification performance strengths in the portfolio, not portfolio diversification. In the common belief of portfolio diversification, we need to minimize the system risk by choosing different assets, which are often coming from non-related industries. In this regard, using variance and covariance of DEA efficiency score will not give better diversification for selecting portfolio. Moreover, portfolio diversifying is only applicable to a large and high-turnover portfolio. It is not suitable for long-term investment.

Since both the inputs and outputs selected in our case study can have negative values, it is not appropriate to use radial DEA models and/or CRS models. When CRS models are used, data translation required for ensuring positivity changes the efficient frontier. When radial input (output)-oriented models are used, input (output) data translation will change efficiency scores, even though the efficiency classification is persevered. Moreover, since our model directly uses efficiency scores to rank DMUs, we must avoid changes in efficiency scores. As pointed out by Pastor and Ruiz [14], additive VRS DEA models are a possible alternative for dealing with negative data in both inputs and outputs. Among various additive DEA models, we use the additive VRS DEA model with a range-adjusted inefficiency measure [15] because it has several desirable properties over the others such as inclusiveness, unit invariance, and translation invariance. Following is the additive model with a range-adjusted measure (RAM) of inefficiency:

$$\begin{aligned} & Min\ \frac{1}{m+s}(R^{-}s^{-}+R^{+}s^{+}) \\ & s.t. X\lambda+s^{-}=x_k, \\ & Y\lambda+s^{+}=y_k, \\ & e^{T}\lambda=1, \\ & \lambda, s^{+}, s^{-}\geq 0 \end{aligned} \tag{9}$$

where $X=(x_{ij})\in R^{m\times n}$ and $Y=(y_{ij})\in R^{s\times n}$ denote the input and output data matrices, respectively, in which each column represents one of DMUs and each row represents the level of one of the factors of the corresponding DMU. R^{-} and R^{+} are defined as:

$$R^{-} = \left(\frac{1}{R_1^-}, \frac{1}{R_2^-}, \cdots, \frac{1}{R_m^-}\right), R^{+} = \left(\frac{1}{R_1^+}, \frac{1}{R_2^+}, \cdots, \frac{1}{R_s^+}\right),$$
$$R_i^- = \max\{x_{ij}\} - \min\{x_{ij}\}, i = 1, \cdots, m, j = 1, \cdots, n$$
$$R_r^+ = \max\{y_{ij}\} - \min\{y_{ij}\}, i = 1, \cdots, s, j = 1, \cdots, n$$

The dual program to model (5) is as follows:

$$\begin{aligned} &Max\ e_k^d = uy_k - vx_k + \xi \\ &s.t.uY - vX + \xi e \leq 0, \\ &u \geq \frac{1}{m+s} R^+, v \geq \frac{1}{m+s} R^- \end{aligned} \tag{10}$$

As discussed in Banker and Maindiratta [16] and Scheel [17] and Lim et al. [12], the output and input weight vectors u and v indicate the price of outputs and the cost of inputs, respectively. Based on that interpretation, we regard uy_k and vx_k as revenue and costs, respectively, incurred from the operations of DMU k. The ξ—adjustment is used due to the constraints of RAM DEA model and assurance region. Consequently, e_k^d is termed as the ξ— adjusted profit that DMU k attains when the price-cost vector (u, v) is used.

DMU k is efficient if and only if there is a positive price-cost vector (u, v) so that $uy_k - vx_k \geq uy_l - vx_t$ for every observed DMU l and an optimal profit of DMU k equals to 0. Let (u_k^*, v_k^*) and ξ_k^* are an optimal price-cost vector and an optimal adjustment value for DMU k. An optimal, ted profit of DMU k is $e_{kk}^* = (u_k^* y_k - v_k^* x_k + \xi_k^*)$, where * denotes an optimal solution to the model for the DMU k under evaluation. In addition, an ξ adjusted profit of DMU l, using an optimal price-cost vector that DMU k has chosen in model (6), is $(u_k^* y_l - v_k^* x_l + \xi_k^*)$, which is denoted e_{kl}^*. Notice that e_{kk}^* and e_{kl}^* correspond to the concepts of simple efficiency and cross-efficiency in conventional DEA cross-efficiency evaluation. Also note that the higher profit, the better or more efficient a DMU is. We now define the profit vector of DMU k as follows:

$$e_k = \left(e_{1k}^*, e_{2k}^*, \cdots, e_{kk}^*, \cdots, e_{nk}^*\right)^T$$

where e_{lk}^*, is an ξ adjusted profit of DMU k using an optimal price-cost vector that DMU l has chosen in model (6). Two statistical properties of e_k can be calculated: the average $\bar{e}_k = \frac{1}{n}\sum_{l=1}^{n} e_{lk}^*$ and its variance $\sigma_k^2 = \frac{1}{n}\sum_{l=1}^{n} (e_{lk} - \bar{e}_k)^2$ Average and variance are the return of DMU k and its risk related to DEA evaluation.

2.3.2 Multi-period Aggregation Mean-Variance Optimization

Lim et al. [12] said that individual DMU risk of changing weights is represented by the variance of cross-efficiencies of each individual DMU. However, this is a risk related to the DEA method in evaluating stocks, not a risk of DMU, so their method of using the variance of cross-efficiencies to find a diversifying portfolio is not quite right. Moreover, in the long-term portfolio selection we do not need to diversify our portfolio, we only focus to find down the best company that has a durable competitive advantage to survive and thrive over time, and the investor will win by holding these stocks. To do so, we should exam firm's performance in a long time where DMU will be ranked higher if it has a high efficiency score in many consecutive years. Therefore, we deploy a method to rank stocks based on multi-period data which is might longer than 2 years and we use variance of cross-efficiency scores to minimize risk of using DEA.

DEA model is run for each year and get the cross-efficiency score average and its variance, which are defined as "return" and "risk", respectively, in our evaluation. Then, we add all cross-efficiency score average of DMU in z years to get total "return" DMU l in z years:

$$\bar{E}_{l-z} = \bar{e}_{l-year1} + \bar{e}_{l-year2} + \cdots + \bar{e}_{l-yearz} \tag{11}$$

with $\bar{e}_l = \frac{1}{n}\sum_{k=1}^{n} e_{kl}$.

Total risk for evaluating each DMU is the sum of all variants of its cross-efficiency scores in z years. Because cross-efficiency scores for different years based on different data of all DMUs, it is not related to each other. Therefore, we do not include co-variance of cross-efficiency scores in our evaluation. The variance of its cross-efficiencies $\sigma_l^2 = \frac{1}{n}\sum_{k=1}^{n} (e_{kl} - \bar{e}_l)^2$ is for one year. The sum of all variances in z years of DMU l is:

$$\bar{\sigma}_{l-z}^2 = \bar{\sigma}_{l-year1}^2 + \bar{\sigma}_{l-year2}^2 + \cdots + \bar{\sigma}_{l-yearz}^2 \tag{12}$$

For all stock evaluating, we can choose S durable competitive advantage stocks which have the best results. Individual DMU is weight with $w \in R_+^n$, where

$$\sum_{i=1}^{n} w_i = s, w_i \in \{1;0\}\forall i,$$

We can find out what are durable competitive advantage stocks with an optimal weight vector w^* is determined by solving this optimization mode:

$$\begin{aligned} & Min\, w^T \sigma_z^2 \\ & s.t.\ w^T \bar{E}_z \geq (1-\gamma)^* \bar{E}_z \max \\ & \sum_{i=1}^{n} w_i = s \\ & w_i \in \{1,0\}\forall i \end{aligned} \tag{13}$$

where γ is the return-risk trade-off parameter, $\bar{E}_z \max$ is the maximum "return" we can get from S stocks in all evaluated stocks in z years. The model means that we choose S stocks who have minimum risk in evaluating with a reasonable high return (the result of evaluation) with the return-risk trade-off parameter, γ.

3 Conclusion

This paper develops a new approach of using DEA in finding a long term competitive advantage stock. There is a lot of different information that could get from financial statement, and combine that information by DEA model. The suggested assurance region with cross-efficiency RAM DEA is used to have better result. Long-term investment needs a reasonable return in many years and the firm need to perform well in many years to ensure that they have durable-competitive advantages. The paper addresses this issue by incorporating DEA cross-efficiency evaluation into multi-period data mean-variance framework, where the risk of the DEA model itself is considered by its score variance.

References

1. A. Charnes, W.W. Cooper, E. Rhodes, Measuring the efficiency of decision making units. Eur. J. Oper. Res. **2**(6), 429–444 (1978)
2. J. Doyle, R. Green, Efficiency and cross-efficiency in Dea—derivations, meanings and uses. J. Oper. Res. Soc. **45**(5), 567–578 (1994)
3. P. Andersen, N.C. Petersen, A procedure for ranking efficient units in data envelopment analysis. Manage. Sci. **39**(10), 1261–1265 (1993)
4. A. Charnes, W.W. Cooper, Z.M. Huang et al., Polyhedral cone-ratio DEA models with an illustrative application to large commercial banks. J. Econ. **46**(1–2), 73–91 (1990)
5. R.G. Thompson, L.N. Langemeier, C. Lee et al., The role of multiplier bounds in efficiency analysis with application to Kansas farming. J. Econ. **46**(1), 93–108 (1990)
6. M. Oral, O. Kettani, P. Lang, A methodology for collective evaluation and selection of industrial research-and-development projects. Manag. Sci. **37**(7), 871–885 (1991)
7. B. Marcello, P. Alberto, A quality assurance-oriented methodology for handling trade-offs in supplier selection. Int. J. Phys. Distrib. & Logist. Manag. **30**(2), 96–112 (2000)
8. H.B.S.L. Jaehun Park, Multi-criteria ABC inventory classification using the cross-efficiency method in DEA. J. Korean Inst. Ind. Eng. **37**(4), 358–366 (2011)
9. L. Liang, H. Wu, W.D. Cook et al., Alternative secondary goals in DEA cross-efficiency evaluation. Int. J. Prod. Econ. **113**(2), 1025–1030 (2008)
10. J. Alcaraz, N. Ramón, J.L. Ruiz et al., Ranking ranges in cross-efficiency evaluations. Eur. J. Oper. Res. **226**(3), 516–521 (2013)
11. F. Yang, S. Ang, Q. Xia et al., Ranking DMUs by using interval DEA cross efficiency matrix with acceptability analysis. Eur. J. Oper. Res. **223**(2), 483–488 (2012)
12. S. Lim, K.W. Oh, J. Zhu, Use of DEA cross-efficiency evaluation in portfolio selection: an application to Korean stock market. Eur. J. Oper. Res. **236**(1), 361–368 (2014)
13. M. Robaina-Alves, V. Moutinho, P. Macedo, A new frontier approach to model the eco-efficiency in European countries, J. Cleaner. prod. **103**, 562–573 (2015)
14. J.T. Pastor, J.L. Ruiz, Variables with negative values in Dea, in *Modeling Data Irregularities and Structural Complexities in Data Envelopment Analysis*, ed. by J. Zhu, W.D. Cook (Springer US, Boston, MA, 2007), pp. 63–84
15. W.W. Cooper, K.S. Park, J.T. Pastor, RAM: a range adjusted measure of inefficiency for use with additive models, and relations to other models and measures in DEA. J. Prod. Anal. **11** (1), 5–42 (1999)

16. R.D. Banker, A. Maindiratta, Nonparametric analysis of technical and allocative efficiencies in production. Econ. **56**(6), 1315–1332 (1988)
17. H. Scheel, Undesirable outputs in efficiency valuations. Eur. J. Oper. Res. **132**(2), 400–410 (2001)

Subspace Clustering Based on Self-organizing Map

Jin Tian(✉) and Mengyi Gu

College of Management and Economics, Tianjin University, Tianjin, China
jtian@tju.edu.cn

Abstract. Clustering in high-dimensional data space is a difficult task due to the interference from different dimensions. A dimension may be relevant for some clusters and irrelevant for other data. Subspace clustering aims at finding local cluster structures in certain related subspace. We propose a novel approach to finding subspace clusters based on the trained Self-Organizing Map neural network (SOM). The proposed method takes advantage of nonlinear mapping of SOM and search for subspace clusters on input neurons instead of the whole data space. Experiment results show that the proposed method performs better compared with original SOM and some traditional subspace clustering algorithms.

Keywords: Self-organizing map · Subspace clustering · High-dimensional clustering

1 Introduction

Clustering is an essential task for data mining and knowledge discovering characterized with unsupervised learning. It aims to explore intrinsic property and structure of data and provide auxiliary information for further analysis. Traditional clustering methods attempt to detect clusters composed of similar samples based on distance measurement. However, advances in information technology and Internet contribute to data with growing dimensionality, which pose great challenge for many traditional data mining methods due to the curse of dimensionality [1]. Large number of dimensions brings similarity of distances originated from sparsity and disturbs from irrelevant dimension, leading to great difficulty in finding meaningful clusters with high quality. Feature selection is a good choice to dimensionality reduction by searching optimum subspace with the most predictive information [2] and helps to improve performance in many models. But sometimes features only work for partial samples and appear as noise for others, which is a more common phenomenon in high dimensional data space. Interference to clustering will generate no matter whether to remove this kind of feature or not.

To deal with such problem, subspace clustering [3] is proposed to provide a new solution where each cluster is assigned with a subspace comprising dimensions relevant to data points in it and irrelevant to others [4]. Clustering determined in subspace reduces the computation cost and provide more targeted information of local structure in given dataset. This new topic addresses much attention from researchers and

G. Q. Huang et al. (eds.), *Proceeding of the 24th International Conference on Industrial Engineering and Engineering Management 2018*,
https://doi.org/10.1007/978-981-13-3402-3_17

different algorithms have been proposed [5–8]. Many applications of subspace clustering are found in the field of computer vision [9], bioinformatics [10], and marketing research [11]. But it still remains a big challenge to discover subspace clusters of high quality with the existence of noise and outliers.

Approaches of subspace clustering are generally divided into three major groups: cell-based, density-based and clustering-oriented [5]. Cell-based approaches, which compose an important branch of subspace clustering, divide data space into grid cells with a certain threshold and search clusters on the cells considering count of data points in these cells. Subspaces are composed of dimensions satisfying some restriction. Discretization of the data accelerates the searching process and makes contribution of efficiency, which is essential for subspace clustering. But the cell is usually in the shape regular square and may lead to information loss. The limitation may finally result in clustering with less accuracy.

Self-organizing map neural network (SOM) is an one kind of unsupervised neural networks with good approximation of the data domain [12]. High dimensional data can be mapped into a plane grid map by SOM. It is able to preserve topological structure in dataset through keeping points with high similarity in original input space close on the map. Each neuron on the map is assigned with a set of points. That is, the input space can be split into cells with arbitrarily shaped data space represented by neurons. Therefore, we proposed a novel cell-based approach called Subspace Clustering Based on Self-Organizing Map (SCBSOM). This method aims to find similar cells with related dimensions using the trained SOM and allows overlapping between clusters. Clustering will be conducted first in each dimension and then a merging procedure is followed. SCBSOM searches for clusters on cells instead of original data points, which is more efficient. And the topological preservation of SOM make contribution to higher accuracy compared with other cell-based methods.

The rest of the paper is organized as follows. Section 2 gives a brief introduction of original SOM algorithm and Sect. 3 describes the proposed SCBSOM model in details. Experimental results are presented in Sect. 4. And finally we conclude major findings in Sect. 5.

2 Self-organizing Map

SOM algorithm is first proposed by Kohonen in 1982 as an unsupervised neutral network [13]. Its good ability of dealing with high dimensional dataset and potential in visualization draw much attention from researchers. SOM has been widely used in many filed such as speech recognition, gene detection, document retrieval and satellite image classification.

An SOM is composed of input layer, output layer and connection weights. The input layer accept training data $x \in R^d$ with d neurons while the output layer is often laid out as a plane grid map with $M = m \times m$ neurons. And the weights keep connection from each input neuron to each output neuron and can be denoted as $W = \{w_i | w_i \in R^d, i = 1,..,M\}$. SOM is trained in an iterative process including competition and convergence. In the tth iterative step, SOM finds the winner in the competition, that is, the closest neuron c to the input sample $\mathrm{x}(t)$:

$$c = \arg\min_i\{\|\mathrm{x}(t) - w_i(t)\|\} \tag{1}$$

Then the convergence procedure leads the SOM model adjusting towards expected order by updating the weight vectors based on the neighborhood relationships with the winner neuron:

$$w_i(t+1) = w_i(t) + h_{ci}(t)(x(t) - w_i(t)) \tag{2}$$

where $h_{ci}(t)$ is the neighborhood function that determines the neighbor update scheme for topology-preserving nature of SOM. The function $h_{ci}(t)$ is usually in the form of Gaussian function:

$$h_{ci}(t)=\alpha(t)\exp\left(\frac{-\mathrm{sqdist}(c,i)}{2\sigma^2(t)}\right) \tag{3}$$

where $\alpha(t)$ is the learning rate that monotonically decreases with step t, $\mathrm{sqdist}(c,i)$ is the square of distance between neuron c and neuron i on the plane grid map and $\sigma(t)$ is the kernel radius that determines the range of neighborhood relationships.

During the training process, the weights adjust according to input until maximum iterations is reached. Each input sample find its winner neuron on the map. Thus every neuron on the output layer contains a set of input samples.

3 Subspace Clustering Based on Self-organizing Map (SCBSOM)

The proposed SCBSOM is a cell-based method that yields valid subspace structure based on the output map instead of on the whole dataset directly. Firstly, with the weight connections in the learned SOM, the proposed algorithm generates possible clusters of each dimension. Then a merging process is conducted to combine the neuron clusters and the corresponding subspaces with related dimensions. The final clustering result is deduced from the neuron clusters by replacing each neuron with data points in it.

3.1 Find One-Dimensional Clusters

Given a trained SOM, we can determine clusters of neurons in each dimension. We link adjacent neurons if their difference (i.e. the distance between their connection weights) is smaller than a certain threshold given in advance. Each cluster is composed of neurons that have links with other neuron in the cluster.

This clustering process is to find all connected component of undirected graph and can be solved with DBSCAN algorithm.

3.2 Merging Procedure

After finding all one-dimensional clusters, a merge process is conducted to obtain subspace clusters of neurons. Here a general method is proposed to merge two subspace clusters.

First, similar clusters with different subspace should be merged. We use Jaccard coefficient to measure the similarity of two clusters, which is computed by

$$J(E,F) = \frac{|E \cap F|}{|E \cup F|} \tag{4}$$

where E and F are two clusters of neuron. Since the two clusters to be merged can hardly be exactly the same, a rule to determine which neuron to remain in the merged cluster is necessary. Here we define the cluster after merging with neurons that appears more times than half of the size of merging subspaces in one-dimensional clusters in the two subspaces. And the cluster is assigned with a new subspace as a union of these dimensions. Then the new merged cluster is added into the result set of subspace clusters and the similar two subspace clusters are removed.

Besides, clusters with the containment relationship also need to merge. The containment relationship of cluster E and F can be measured by

$$\begin{aligned} C_1(E,F) &= \frac{|E \cap F|}{|E|} \\ C_2(E,F) &= \frac{|E \cap F|}{|F|} \end{aligned} \tag{5}$$

If $C_1(E,F)$ is close to 1 and $C_2(E,F)$ gets a smaller value, E is contained in F. The merging procedure is similar to that of similar clusters. The containing subspace cluster is maintained in the result set while the other is removed.

The merging process is executed iteratively by adding subspace clusters of one-dimension each time, which can be summarized as follows:

Step 1: set an empty set.

Step 2: choose one dimension that has not been merged and determine its one-dimensional subspace clusters.

Step 3: choose one cluster from current set of one-dimensional subspace clusters.

Step 4: compare the selected cluster with all subspace clusters in the result set to find if there is a similar cluster to the selected one. If no similar cluster exists, go to Step 7.

Step 5: merge the chosen cluster with its similar one.

Step 6: compare the selected cluster with all subspace clusters in the result set to find if there is one cluster that is contained in it. If there is, merge them. Go to Step 8.

Step 7: add the selected cluster to result set and compare the selected cluster with all subspace clusters in the result set to find if there is one that contains it. If there is, merge them.

Step 8: if all clusters of current dimension is selected, go to Step 9. Otherwise go to Step 3.
Step 9: if all dimensions is merged, return the result set. Otherwise go to Step 2.

3.3 The Clustering Result

After the merging process, we determine all subspace clusters of neurons. As we know, each neuron covers a set of data point according to the mapping of SOM.

As a result, we can obtain the final subspace clustering result through an easy transformation from neurons to data points.

4 Experiments

In our experiments, SCBSOM is compared with original SOM and some state-of-art subspace clustering methods both on the synthetic datasets from OpenSubspace framework [14] and real world datasets from UCI Machine learning Repository [15]. We adjusted parameters for each algorithm separately to obtain a good performance as much as possible. Each experiment is conducted 30 times. F1 measure is used to evaluate the performances of these clustering algorithms.

4.1 Synthetic Datasets

There are three groups in the synthetic datasets to testify the performance of SCBSOM from different aspects.

The first group includes 4 datasets with noise of different proportion. Figure 1 shows the average F1 values on 4 datasets obtained by five algorithms with different noise.

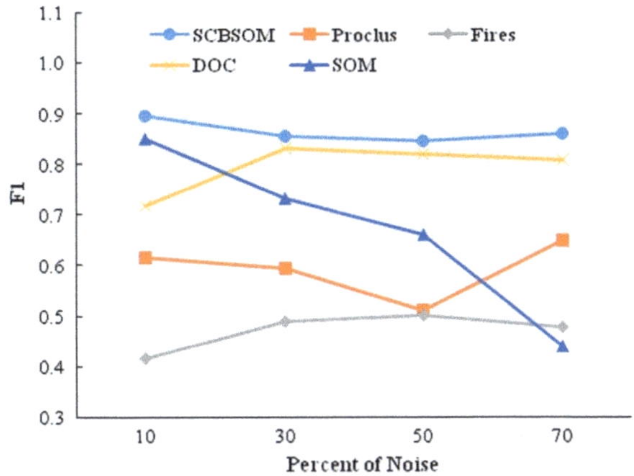

Fig. 1. Average F1 values for SCBSOM and compared algorithms on 4 datasets with different noise

As shown in Fig. 1, SCBSOM outperforms all the other four algorithms and is robust to noise while the original SOM shows poor performance when the amount of noise increases. This advantage of SCBSOM is important since in real world we can hardly obtain a dataset without noise, especially for high dimensional one.

The second group comprises 7 datasets with dimension number ranging from 5 to 75. Figure 2 illustrates the average F1 values on 7 datasets with different dimension numbers.

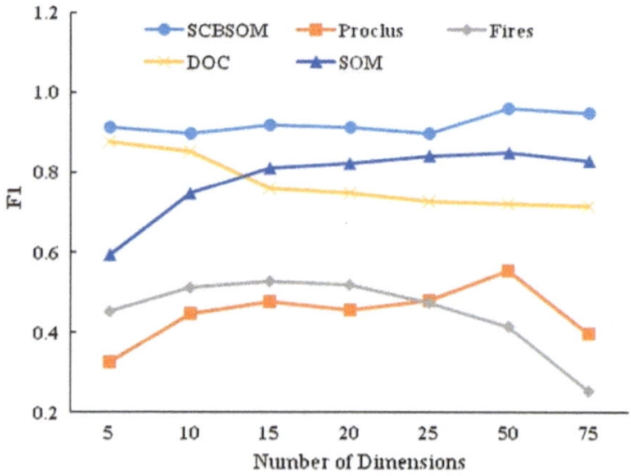

Fig. 2. Average F1 values for SCBSOM and compared algorithms on 7 datasets with different dimension number

According to Fig. 2, SCBSOM performs best among the compared algorithms. We notice that both SOM and SCBSOM get higher values of F1 with the increasing of dimensionality while the other three subspace clustering methods have decreased performance. The good ability of SOM to preserve topological structure of input data ensures the stability of SCBSOM in dimension, which enables the proposed algorithm to deal with high dimensional data with high accuracy.

The third group is 5 datasets of different size that differs from 1500 to 5500. Figure 3 presents the average F1 performances on 5 datasets with different data size.

Shown form Fig. 3, SCBSOM always achieves best clustering performances with both small size of data and large size of data, while other three subspace clustering algorithms are sensitive to the data size.

4.2 Real-World Datasets

In this subsection, the performance of SCBSOM is evaluated on 7 real-world datasets selected from the UCI Machine Learning Repository. Different from synthetic datasets with regular subspace clustering, these datasets are from real word with arbitrary structure. Table 1 shows the F1 performances of SCBSOM and other four compared algorithms. The highest F1 value in each row is outlined in bold.

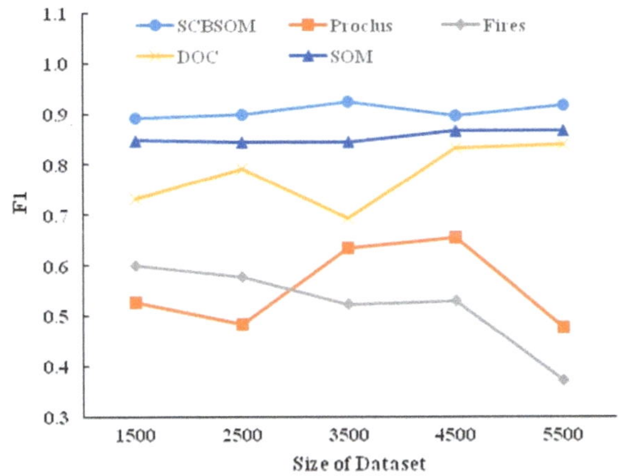

Fig. 3. Average F1 values for SCBSOM and compared algorithms on 5 datasets with different data size

Table 1. F1 values of SCBSOM and other compared algorithms on UCI datasets

Dataset	SCBSOM	SOM	Proclus	Fires	DOC
Breast	**0.7809**	0.4073	0.6639	0.4240	0.5080
Diabetes	**0.7682**	0.4739	0.4988	0.3941	0.4019
Glass	0.4729	0.4973	**0.6351**	0.0874	0.1635
Liver	**0.7125**	0.4003	0.5321	0.3489	0.3875
Shape	0.7162	**0.7211**	0.8005	0.0796	0.3156
Sonar	0.6477	0.4285	**0.8113**	0.3480	0.3480
Vowel	**0.3161**	0.3054	0.4674	0.0152	0.0816
Avg	**0.6306**	0.4620	0.6299	0.2425	0.3152

Shown form Table 1, SCBSOM outperforms other algorithms on Breast, Diabetes and Liver and produces the best clustering results on average. The superiority of SCBSOM on them makes it possible to be applied in finding complex clustering structure.

5 Conclusion

This paper presents a novel method to finding subspace clustering based on SOM called SCBSOM. It first searches for one-dimensional clusters of neurons based on the trained SOM and then a merging procedure is conduct to generated subspace clusters. Finally the neurons in clusters are replaced by corresponding data points determined by SOM map. Since clusters are discovered on neurons of SOM map with smaller size compared with whole input data, SCBSOM can be classified into cell-based approach and find subspace clusters efficiently. The SCBSOM is an extend version of SOM so that it can preserve the properties SOM, such as good ability in clustering. The experimental results show that SCBSOM can perform well with noises, high dimensionality and different size of datasets.

Many feasible extensions can be pursued relative to our work in this paper. We plan to improve iterative process since the results may be affected by the order of selected dimension. Also, a more flexible way to integrate SOM and subspace discovering is expected.

Acknowledgements. The work was supported by the General Program of the National Science Foundation of China (Grant No. 71471127, 71371135).

References

1. M. Köppen, The curse of dimensionality, in *Fifth Online World Conference on Soft Computing in Industrial Applications* (2000)
2. S. Tabakhi, P. Moradi, Relevance–redundancy feature selection based on ant colony optimization. Pattern Recogn. **48**(9), 2798–2811 (2015)
3. R. Agrawal, J.E. Gehrke, D. Gunopulos, P. Raghavan, Automatic subspace clustering of high dimensional data for data mining applications, in *Proceedings of the 1998 ACM SIGMOD* (Seattle, WA, USA), pp. 94–105
4. H.F. Bassani, A.F.R. Araujo, Dimension selective self-organizing maps with time-varying structure for subspace and projected clustering. IEEE Trans. Neural Netw. Learn. Syst. **26** (3), 458–471 (2015)
5. E. Ller, S. Nnemann, I. Assent, T. Seidl, Evaluating clustering in subspace projections of high dimensional data. Proc. VLDB Endow. **2**(1), 1270–1281 (2009)
6. C.M. Procopiuc, M. Jones, P.K. Agarwal, T.M. Murali, A Monte Carlo algorithm for fast projective clustering, in *Proceedings of the 2002 ACM SIGMOD* (Madison, WI, USA), pp. 418–427
7. H.P. Kriegel, P. Kröger, M. Renz, S. Wurst, A generic framework for efficient subspace clustering of high-dimensional data, in *Fifth IEEE International Conference on Data Mining* (Houston, TX, USA, 2005)
8. C.C. Aggarwal, J.L. Wolf, P.S. Yu, C. Procopiuc, J.S. Park, Fast algorithms for projected clustering, in *Proceedings of the 1999 ACM SIGMOD* (Philadelphia, PA, USA), pp. 61–72
9. A.Y. Yang, J. Wright, Y. Ma, S.S. Sastry, Unsupervised segmentation of natural images via lossy data compression. Comput. Vis. Image Underst. **110**(2), 212–225 (2008)
10. D. Jiang, C. Tang, A. Zhang, Cluster analysis for gene expression data: a survey. IEEE Trans. Knowl. Data Eng. **16**(11), 1370–1386 (2004)
11. P.B. Chou, E. Grossman, D. Gunopulos, P. Kamesam, Identifying prospective customers, in *Proceedings of the 2000 ACM SIGKDD* (Boston, MA, USA), pp. 447–456
12. T. Kohonen, Essentials of the self-organizing map. Neural Netw. **37**(1), 52–65 (2013)
13. T. Kohonen, Self-organized formation of topologically correct feature maps. Biol. Cybern. **43**(1), 59–69 (1982)

14. E. Müller, I. Assent, S. Günnemann, T. Seidl, OpenSubspace: an open source framework for evaluation and exploration of subspace clustering algorithms in WEKA, in *Proceedings of 1st Open Source in Data Mining Workshop, OSDM'09* (Bangkok, Thailand), pp. 2–13
15. M. Lichman, UCI Machine Learning Repository, http://archive.ics.uci.edu/ml (University of California, School of Information and Computer Science, Irvine, CA, 2013)

The R&D and Test of the Forest Harvester Control System Based CAN

Xinyue Wang[1], Qingqing Huang[1], Dian Wang[1(✉)], Jiangming Kan[1], Yang Peng[1], Pengpeng Li[1], Weimin Pan[2], Shitao Xu[2], and Tao Ma[2]

[1] Beijing Forestry University, Beijing, China
wangdian@bjfu.edu.cn
[2] Zhenjiang Zhongfuma Co., Ltd., Zhenjiang, China

Abstract. In order to intensively complete the automatic operation of logging, feeding, peeling, beating branches, making wood and so on, and satisfy the requirements of forest harvesting, the control system of forest harvesting needs to be studied urgently. The control system designed in this paper uses CAN bus, upper computer (IPC), lower computer (SPN-M*-18 controller), displayer, hand shank and diameter measuring encoder, length measuring encoder, etc. This paper introduces the system composition, hardware composition, software application, displayer design and cutting effect test. The application of CAN bus makes the communication between the host computer and the lower computer as well as other devices smooth, and the anti-interference ability is enhanced. Under the function of the controller, the operation of the equipment is realized, and the labor productivity is improved. The PID controller used in the program ensures the length of the wood in the automatic feeding process, so as to meet the project requirements. The design of the displayer screen makes the operation and maintenance of the operators easier and more convenient. After several experiments, the conclusion is reached that the control system meets the design requirements and can be used in production operations.

Keywords: Harvester · The control system · CoDeSys · Human-computer interface · PID

1 Introduction

With the continuous increase in the harvesting of artificial forests and the increase in labor costs, the traditional harvesting technologies cannot meet the requirements of forestry production. The future forestry production will gradually shift to a technologically-efficient model to increase the productivity of the unit labor force. The emergence of forest harvesting equipment will greatly increase labor productivity and provide workers with a safe and comfortable working environment, which will become an inevitable trend in forestry production.

This project has developed a control system for forest harvesting equipment. This control system is based on the CAN bus technology [1], which includes the industrial computer, display, SPN-M*-18 controller and other devices communicating with each other to complete the system work [2, 3], hydraulic actuators (hydraulic cylinders or

G. Q. Huang et al. (eds.), *Proceeding of the 24th International Conference on Industrial Engineering and Engineering Management 2018*,
https://doi.org/10.1007/978-981-13-3402-3_18

Motor) Under the control of the SPN-M*-18 controller and the hydraulic valve, the mechanism is driven to achieve operations such as felling, stripping, feeding, branching, and building materials of the harvesting equipment organization [4]. At the same time, on this basis, the operating steps could be reduced as much as possible so as to greatly increasing labor productivity. Forest harvesting equipment mechanical structure shown in Fig. 1.

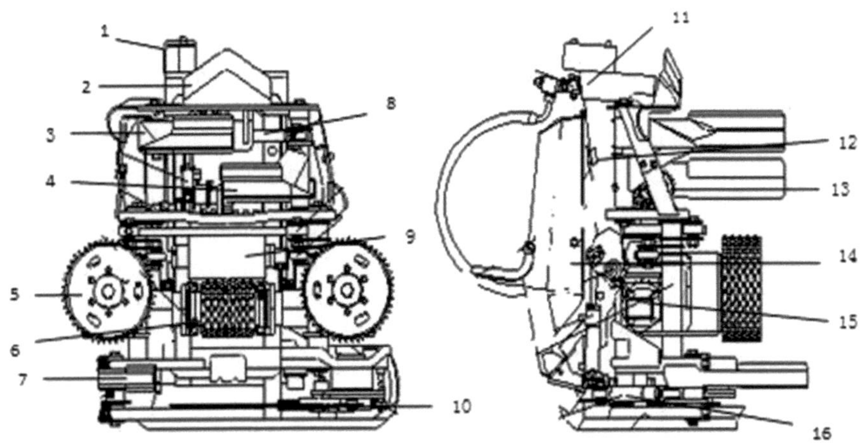

Fig. 1. The structure of Artificial forest mechanical harvester. 1 connector, 2 apron, 3, 4 top knife, 5 roller feeder, 6 belly roller feeder, 7 bottom knife, 8 top bottom knives cylinder, 9 clamp cylinder, 10 chain saw, 11 fixation frame, 12 connecting arm, 13 measuring wheel, 14 outer cover, 15 roller feeder motor, 16 chain saw frame

2 Methodology

Forest harvesting equipment control system [5] mainly includes display (included controller), handle, adapter, SPN-M*-18 controller, industrial control computer, and measuring and length measuring encoder, hydraulic valve and other components. The IPC is not only a CAN bus message sender but also a receiver. It receives messages from the SPN-M*-18 controller and continuously sends data to the CAN bus through a program written on the CoDeSys software [6].

The control system is based on the CAN open protocol, an application layer protocol established by the CAN bus protocol. The signal level is based on the ISO11898 high speed communication standard. The communication mode is set to System bus and the bit rate is 250 kbps. The signal is transmitted in a cyclic manner with a signal period of 20 ms [1]. The CAN bus adopts non-destructive priority-based bus arbitration technology, high reliability, good anti-interference performance of the communication system, and stable operation.

3 Result

The display can display the data of the tree diameter and length; the handle as an operating part, different buttons represent different functions, pressing the button will generate data information, so as to control the hydraulic valve through the SPN-M*-18 controller; SPN-M*-18 controller to receive control commands [7] from the CAN bus to control the hydraulic actuators; the length measuring encoder is mounted on the abdomen of the harvesting equipment. It is in contact with the wood during the wood feeding process. For the measurement, the caliper encoder is installed in the clamping cylinder. When the harvesting equipment is holding the wood feed, the tree diameter is measured and the corresponding data is obtained. The PI function of the SPN-M*-18 controller is used to input the system. The pulse amount is obtained through controller decoding, received by the industrial computer, and displayed on the display to form a closed loop. In this process, the operator learns the information of the tree diameter and length through the monitor and analyzes it to control the clamping conditions of the upper and lower cutting knife, the length of the automatic cutting time, the position of the harvesting equipment when harvesting the wood etc. As shown in Fig. 2, the control system diagram.

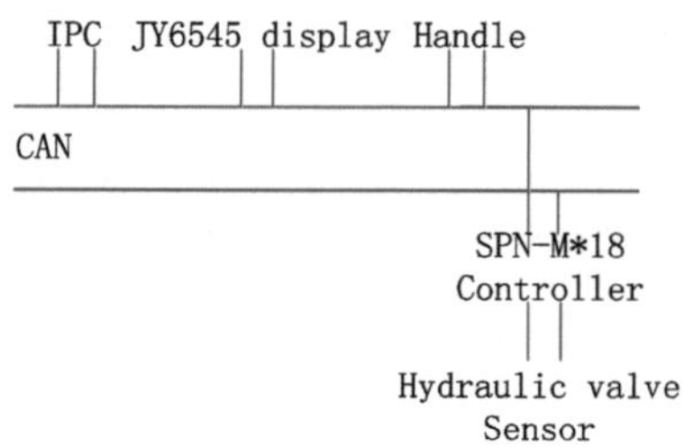

Fig. 2. Control system schematic

The display in the forest harvest control system uses the JY6545 all-in-one, as shown in Fig. 3. A programmable display based on the CoDeSys software platform supports CAN bus monitoring downloads, support for serial port monitoring downloads, and two 23-pin AMP connectors. Each connector corresponds to a handle connection, CAN bus and power cord. Only a set of plug-ins for the display are used here.

In the control system, only the right handle is used in the control system, and the eight buttons above the main button area and the two buttons on the left side of the handle are used. As shown in Fig. 4, the corresponding functions of the buttons on the handle are: Automatic feeding, claw opening, claw closing, positive rotating, reverse rotating, pitch-up and pitch-down.

The SPN-M*-18 controller is shown in Fig. 5. There are 52 output points and input points, some of which can be used as functions according to programming: DI indicates the switching output, PWM indicates the pulse width modulation output and PI indicates the pulse input. The I/O address distribution is mainly based on the function of

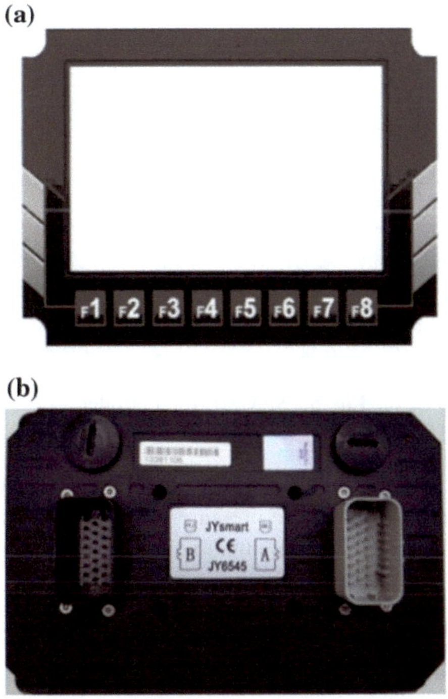

Fig. 3. **a**, **b** JY6545 integrated machine

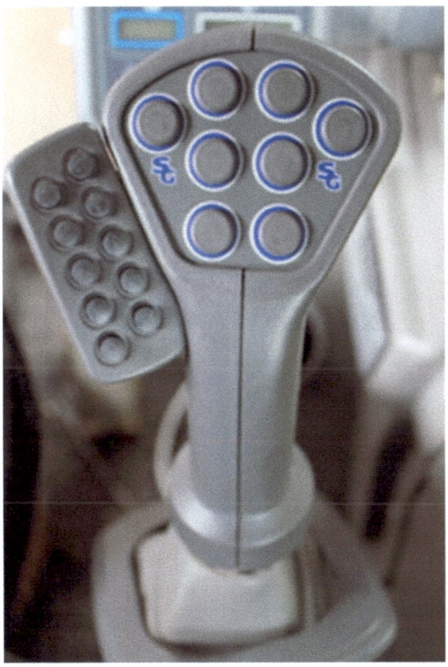

Fig. 4. Handle

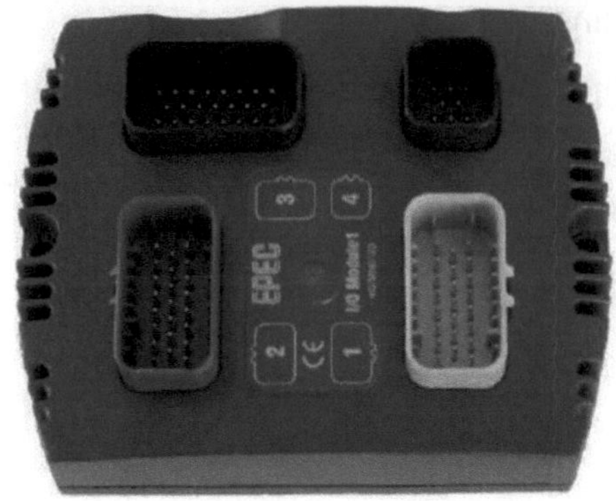

Fig. 5. *SPN-M*-18* controller

the hydraulic valve, the input and output of the encoder, the CAN bus and the switch function. The DO function is used, the pressure is PWM, and the encoder acquires the PI function. Its I/O address allocation situation is shown in Table 1.

Table 1. Location distributed of *SPN-M*-18* control

Stitch	Function	Bit address
XM1.1	DO	QX1.4
XM1.2	DO	QX1.5
XM1.3	DO	QX1.7
XM1.4	DO	QX1.6
XM1.7	DO	QX0.0
XM1.8	DO	QX0.1
XM1.14	DO	QX0.3
XM1.15	DO	QX0.2
XM1.16	DO	QX1.3
XM1.17	DO	QX1.2
XM1.23	DO	QX1.0
XM2.1	PWM	QW109
XM2.2	PWM	QW108
XM2.5	DO	QX1.11
XM2.6	DO	QX1.10
XM2.7	PWM	QW113
XM2.8	DO	QX1.8
XM2.10	PWM	QW111
XM2.16	PWM	QW121
XM2.17	PWM	QW120
XM2.22	PWM	QW122

(*continued*)

Table 1. (*continued*)

Stitch	Function	Bit address
XM3.14	PI	IW111
XM3.16	PI	IW152
XM3.17	PI	IW151
XM3.18	PI	IW155
XM3.19	PI	IW154
XM4.2		
XM4.3		
XM4.4		
XM4.6		
XM4.1		

The control system mainly uses two kinds of software:

(1) CoDeSys

CoDeSys is a complete development environment for programmable logic control PLC (CoDeSys is an acronym for Controlled Development System). In programming with PLC [8], CoDeSys provides a simple method for powerful IEC language, system editor and debugger. The function is based on a high-level programming language. The operator writes and proofs the program on the software.

In the programming process, the influence of the fixed length of the main parameters on the quality of the material is taken into account. In the sizing process, the SPN-M*-18 controller's built-in PID controller [9] (Proportional Integral Derivative Controller) was added. The PID controller outputs the steady-state error through proportional control; the integral control can eliminate the steady-state error, but may increase the overshoot; the differential control can accelerate the response speed of the large inertia system and weaken the overshoot trend. The general expression of the continuous PID controller [10] is Eq. (1),

$$u(t) = K_p e(t) + K_i \int_0^t e(\tau)d\tau + K_d \frac{de(t)}{dt} \tag{1}$$

In the formula, K_p is Proportional control gain, K_i is Control gain for integral, K_d is Differential control gain.

The PID program in the CoDeSys:

```
FUNCTION_BLOCK PID_Real
  VAR_INPUT
    MeasPoint:REAL;
    SetPoint:REAL;
    P:REAL;
    I:REAL;
```

D:REAL;
Ramp:WORD;
PAR_ZengLiangMax:WORD; PAR_ControlMAX:WORD;
Reset:BOOL;
END_VAR

According to this principle, the value of P, I, D in the controller is adjusted by the adjusted value, so that the fixed length of the harvesting equipment is greater than the standard fixed length.

(2) PCAN-View

PCAN-View is an easy-to-use CAN monitoring software for viewing and receiving messages on the CAN bus. Setting the communication mode to: System bus (CAN) [7] and the bit rate to 250 kbps. As shown in Fig. 6. The message shown in the figure is a CAN identifier in hexadecimal format.

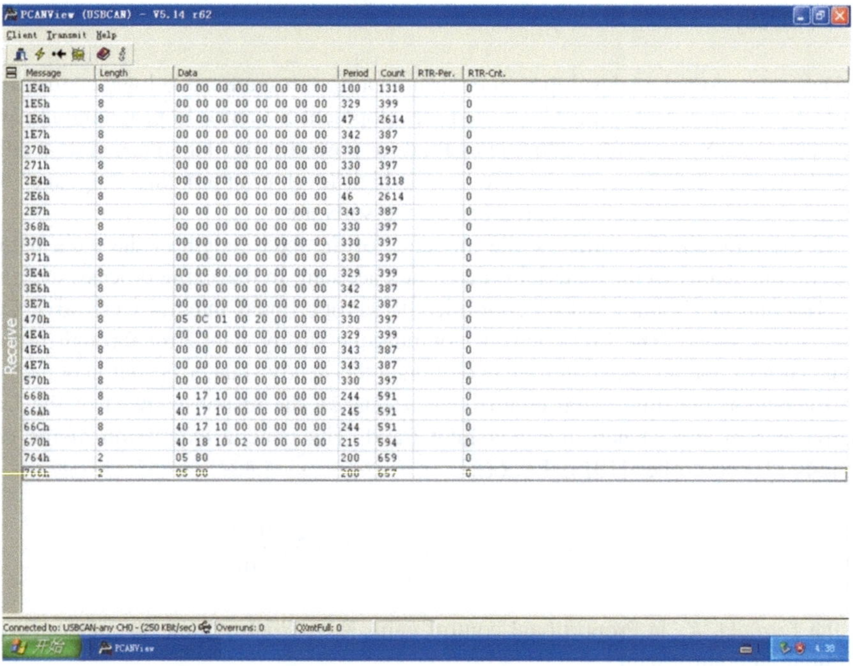

Fig. 6. PCAN-view

4 Design Display

The system has designed two interfaces, as shown in Fig. 7. The main screen shows two important parameters for the job, including the tree diameter and the fixed-length value. These two values are mainly measured by the encoder mounted on the feed roller and chain saw cylinder. When harvesting equipment is used for logging operations, the changes in these values are observed in real time, and the accuracy of the harvesting equipment encoder can be analyzed to a certain extent [11], enabling the operator to clearly understand the appropriate harvesting range. In the same way, when the material forming operation is performed, the change in length can also be observed. The operator can control the quality of the material by controlling the forward rotation and reverse rotation of the feed roller.

(a)
(b)

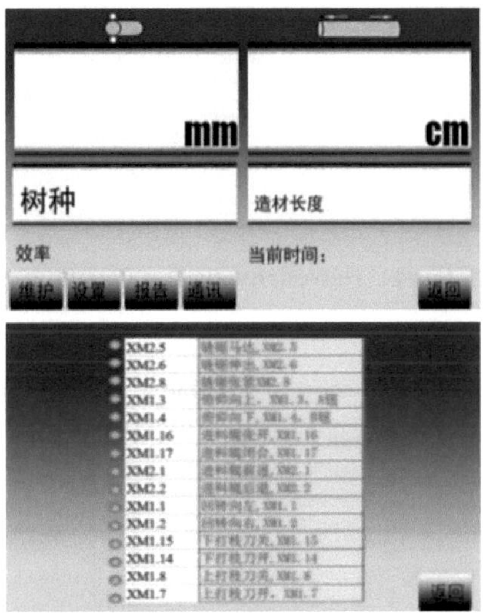

Fig. 7. Displayer interface: **a** main interface, **b** communication interface

The communication interface displays the functions corresponding to the SPN-M*-18 controller bit address, and also provides convenience for the operator in operation and maintenance.

When the forest harvesting equipment performs actual operations, the following actions are performed in turn:

(1) The harvesting equipment is in a vertical state, the angle is adjusted. The front and rear holding claws and the feed roller are opened. The trunk is held, and the sawing is performed to complete the cutting operation.
(2) The harvesting equipment is laid down and is in a horizontal state. The front and rear holding pawls and the feeding roller are still in the tight state. The feed roller is rotating forward, When it approaches the length of the material, it is properly adjusted, sawing out, sawing back and forth. And the feed roll is opened to complete the build. Recovery equipment status is shown in Fig. 8.

During the operation of harvesting equipment, the value of the tree diameter changes with the degree of clamping, and the value of the fixed length varies according to the value of the PID controller [12–14]. For the fixed length, taking different P, I, D

Fig. 8. Harvester vertical **a** upright, **b** fall

Table 2. P, I, D orthogonal test

Number	P	I	D	Test length (cm)
1		0.002		188
2	16	0.003	0	250
3		0.004		201
4		0.002		228
5	17	0.003	0	275
6		0.004		267
7		0.002		269
8	18	0.003	0	257
9		0.004		252
10		0.002		248
11	19	0.003	0	248
12		0.004		257

values, and observe the actual feed value of automatic feeding. Taking p respectively 16, 17, 18, 19, I is 0.002, 0.003, 0.004, D is 0, 12 orthogonal test, as shown in Table 2. As shown in Fig. 9, the relationship between the fixed length and the standard fixed length (250 cm) is tested.

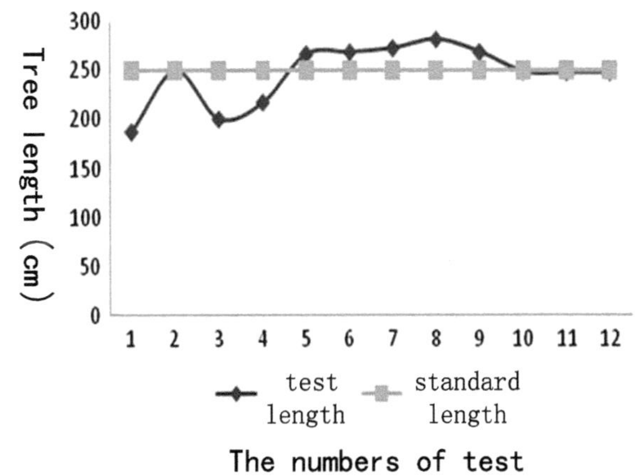

Fig. 9. Test curve of fixed length test

5 Discussion

It can be seen from Fig. 9 that the test length between the 6th and 9th test is longer than the standard fixed length 250 cm, P = 17, I = 0.003, 0.004, D = 0 and P = 18, I = 0.002, 0.003, 0.004, when D = 0, the test lengths correspond to 275, 267, 269, 257, and 252 cm, respectively, which meet the standard fixed length. The P, I, and D values in the PID controller can be referenced.

6 Conclusion

This article describes the system components of the forest harvesting equipment control system, including hardware systems and software applications. The hardware system includes a display, a controller, and a controller. Software applications include CoDesys software and PCAN-View software. In addition, the design of the display interface is also introduced. The operator can observe the data changes of the wood in real time through the display interface to facilitate the reading of orthogonal test data. Subsequent to several orthogonal tests, the corresponding values of P, I and D in the PID controller when the harvesting equipment reaches the standard fixed length are obtained, making the automatic fixed-length function of the harvesting equipment stable, which proves the design of the control system. rationality. In addition, the control system adopts CAN bus technology [12], communication is convenient, reliable, anti-jamming, and guarantees the data transmission.

Acknowledgements. This research is supported by the Fundamental Research Funds for the Central Universities (No. BLX2014-03), and Special Fund for Forest Scientific Research in the Public Welfare (No. 201504508).

References

1. X. Yang, *Fieldbus Technology and its Application* (Tsinghua University Press, Beijing, 2008)
2. L.P. Bolduc, Redundancy management system for the X-33 vehicle and mission computer, in *Digital Avionics Systems Conference, 2000. Proceedings DASC the IEEE*, vol. 1 (2000), pp. 1C3/1–1C3/7
3. X. Luo, Z. Zhang, Y. Zhao, DGPS automatic navigation and control system of Dongfanghong X-804 tractor. Chin. J. Agric. Eng. **25**(11), 139–145 (2009)
4. H. Yang, V. Cocquempot, B. Jiang, Optimal fault-tolerant path-tracking control for 4WS4WD electric vehicles. IEEE Trans. Intell. Transp. Syst. **11**(1), 237–243 (2010)
5. D. Wang, *Design and Research on Harvesting with Logging Unit* (Northeast Forestry University, Harbin, 2006)
6. Y. He, Control system of airplane rotor forming machine based on CAN bus, in *Science and Technology Forum of Ten Provinces and Districts Mechanical Engineering Society and 2007 Annual Meeting of Heilongjiang Mechanical Engineering Society*
7. Y. Yang, Data acquisition and control design of aircraft electromechanical system based on CAN, in *Academic Symposium of 2016 Aeronautical Test and Test Technology Conference*, vol. 4 (2016)
8. W. Qi, X. Zhou, Design and implementation of a distributed intelligent greenhouse computer control system. Chin. J. Agric. Eng. **20**(1), 246–249 (2004)
9. Q. Dong, Y. Wang, Development of distributed automatic control system for greenhouse computers. Chin. J. Agric. Eng. **18**(4), 94–97 (2002)
10. L. Yang, K. Lu, D. Zhang, Development of wireless intelligent control terminal based on ZigBee technology. Chin. J. Agric. Eng. **26**(3), 198–202 (2010)
11. Y. Liu, Y. Qian, Y. Xu, Design of autopilot control system for UAV based on CAN bus, in *Academic Conference on National Signal and Intelligent Information Processing and Application* (2013)
12. J. Zhao, X. Cui, Y. Zhu et al., UBot: a new reconfigurable modular robotic system with multimode locomotion ability. Ind. Robot **39**(2), 178–190 (2012)
13. Nanjing Laike Electronic Technology Co., Ltd. PCAN-View Software Help Manual [EB/OL]. http://www.njike.com
14. Y. Shen, *Research on the Implementation Mechanism and Hydraulic System of Forest Tree Combined Mining Machine* (Beijing Forestry University, Beijing, 2010)

System Modelling and Development of Accurate Feeding Control System for Forest Harvester Operation

Qingqing Huang(✉), Yang Peng, J. H. Liu, and D. Wang

School of Technology, Beijing Forestry University, Beijing, China
huangqingqing@bjfu.edu.cn

Abstract. In order to improve the efficiency of harvesters, the control system of the feed roller precision feeding has been developed and tested (we develop and test a kind of feeding roller control system which can precisely feed). This control system is based on PID control algorithm to research the automatic precision feeding control system of the harvesters. Based on the combination of the accurate automatic feeding requirement with the control process, this paper studied the application of incremental PID control algorithm aiming at automatic precise feeding control. In the experiment, an improved multi-segment adaptive PID control algorithm was proposed based on the feeding results which analyzed with different P, I and D parameters. The experimental results showed that the improved PID control algorithm can achieve automatic and accurate feeding, and its deviation is within the accepted range. This method is well-adapted which can effectively improve the operation efficiency of harvesters. But in following tasks, it is necessary to make further optimization about the characteristics of different tree species and diameter, length etc.

Keywords: Accurate feed control · Control system development
PID control

1 Introduction

With the rapid and high quality development of forest products industry, the application of forestry production and breeding equipment tends to be efficient, intelligent and informative [1]. The tending of plantations is also seasonal and must be improved by mechanization [2]. The combined mining and breeding machine for forest is a kind of high efficient tree harvesting machinery for forest operations, which is mainly used in logging, bucking and other processes [3]. It is a key step for the joint mining operation to shoot the bucking after the cutting, and the main problem to be solved in the bucking operation is how to control the log feeding length accurately.

The feed length is achieved by precise control of the flow of the hydraulic motor of the Feed roller [4], in this paper, by analyzing the structure and work flow of the working nose of the mining machine, and aiming at the load-sensitive multi-channel proportional directional valve, based on the analysis of the incremental PID control

G. Q. Huang et al. (eds.), *Proceeding of the 24th International Conference on Industrial Engineering and Engineering Management 2018*,
https://doi.org/10.1007/978-981-13-3402-3_19

algorithm in the accurate feed control. The automatic feed control system based on multi-segment adaptive PID Control strategy is developed, and accurate automatic feeding control is realized.

2 Analysis of Control Requirement of Operation Head

The working nose of the joint mining machine is mainly driven by a plurality of hydraulic cylinders and hydraulic motors which are mounted in the interior, and the movement of the mechanical claw, feeding roller and the chain saw is shown in Fig. 1. When the operation, not only required to cut inverted wood, and need to move the position of the manipulator, as well as the rotation of the head of the log to the appropriate location, and then to shoot and bucking operations.

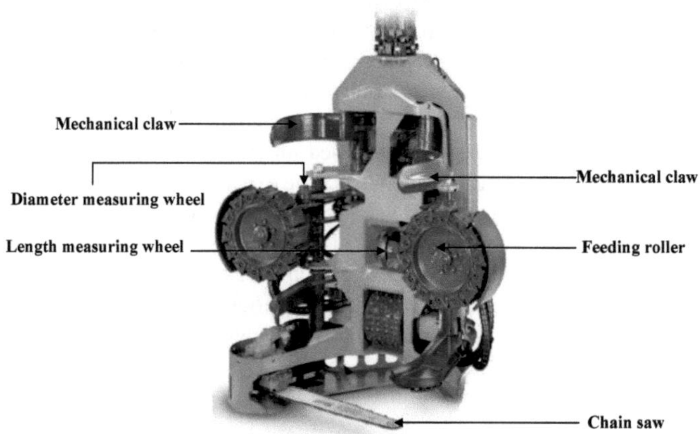

Fig. 1. The contour of the logging head of the combined mining machine

When the actual work, through the controller to control the multi-channel proportional directional valve and then drive the hydraulic motor driven feed roller rotation, so the precise automatic feed control is the valve opening of the electro-hydraulic control. Therefore, to improve the electro-hydraulic proportional system response speed and control accuracy is the key problem to be solved [5, 6]. Electro-hydraulic proportional system is a typical nonlinear time-varying system, general control algorithm is difficult to meet all the performance requirements [7]. PID controller principle simple, strong applicability, its robustness in a wide range of changes in the actual operation of the nose of the complex operation, with high stability, and the flow of multi-channel valve is difficult to accurately measure, it is difficult to establish its precise mathematical model, so we select multiple adaptive PID control in engineering [8].

3 System Composition

Figure 2 Distributed Control System Based in Can-bus of Harvester head. As shown in Fig. 2, this control system is based on can bus (Can-bus) communication protocol, which can be divided into two parts of cab and Operation Head, which can not only realize the control of working device, but also the main component of distributed control system of combined mining and breeding machine [9]. The signal receiving and transmitting device converts the electric control signal given by the control handle into a signal based on can bus, sends to the EPEC2024 controller, and sends the current feed length and the diameter level information of the controller back to the display controller for the driver to read. The EPEC2024 controller converts the control signal into an electrical signal and then controls the proportional directional valve action, at the same time, it is the core of the whole control system to collect the speed information measured by the encoder and convert it into the current feed length and diameter of the wood and return the cab. The length-measuring encoder is installed in the long wheel (see Fig. 1). When the working nose is in a lodging state, the long wheel is pressed tightly to the wood through the spring structure, the length of the measuring wheel in the wood feeding is converted to the current actual feed length, the diameter encoder is installed in the shaft of the feed roller claw, and the output data can be used to convert the current clamping log diameter when the feed roller holds the claw close. The proportional directional valve can be regarded as an amplification device from the control signal to the power execution signal, and also the interface between the control system and the hydraulic system.

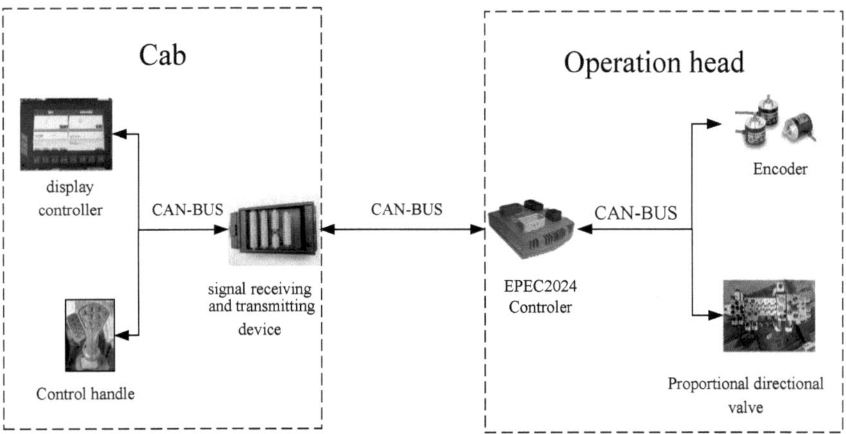

Fig. 2. The distributed Operation Head control system based on can bus

After the actual measurement, the length of the encoder measured on the display of the bucking length and the actual feed length, the error is less than 5 mm, in the actual bucking process, can be completely ignored, therefore, the display of bucking length can fully represent the current actual feed.

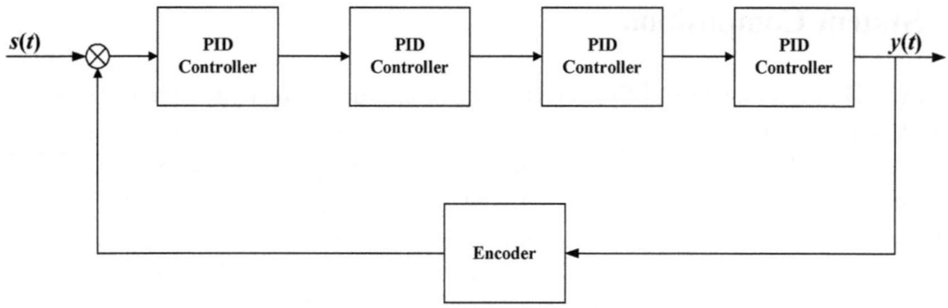

Fig. 3. Flow diagram of control system

Figure 3 flow chart the control system as shown in Fig. 3, a block diagram for an electrohydraulic proportional control system where s(t) is bucking length for a given target y(t) is the actual feed length of the log driven by the feed roller, which is fed back to the controller after the encoder is collected, compared with the given target value, and then output to the display, when the feedback value is different from the given target value. The control algorithm in the controller is used to deal with the deviation signal, and output corresponding adjustment signal, after the conversion output to the proportional directional valve, output and adjust the signal proportional flow, thus driving the hydraulic motor driving the feed roller driving the log, to reduce the deviation of the direction of movement, until the deviation is approaching 0 [10], within the allowable range, cannot drive the log into the material so far.

4 PID Controller

PID controller can be divided into analog controller and digital controller, and digital PID control algorithm can be divided into position and incremental PID. Digital PID is a kind of computer sampling control, it can only calculate the control quantity according to the deviation of sampling time, but not the continuous output control amount as analog control, and carry on continuous control [11]. Position PID because it is full output, each output is related to the past state, the workload is large, and because the output corresponds to the actual position of the executing agency, if there is a failure, it is easy to cause significant changes in the executing agencies, it may cause serious production accidents. Finally, the incremental PID control algorithm is chosen.

4.1 Incremental PID Algorithm

The so-called increment PID means that the output of the digital controller is only the increment of the control quantity, and its algorithm can be derived from the formula (1)–(5)

$$u_k = P \cdot \left[e_k + \frac{T}{I} \sum_{j=0}^{k} e_j + D \frac{e_k - e_{k-1}}{T} \right] \tag{1}$$

$$u_k = P \cdot \left[e_{k-1} + \frac{T}{I}\sum_{j=0}^{k-1} e_j + D\frac{e_{k-1} - e_{k-2}}{T} \right] \tag{2}$$

$$u_k = u_k - u_{k-1} = P \cdot \left[e_k - e_{k-1} + \frac{T}{I} e_k + D\frac{e_k - 2e_{k-1} + e_{k-2}}{T} \right] \tag{3}$$

$$u_k = Ae_k + Be_{k-1} + Ce_{k-2} \tag{4}$$

$$\begin{cases} A = P \cdot \left(1 + \frac{T}{I} + \frac{D}{T}\right) \\ B = P \cdot \left(1 + 2\frac{D}{T}\right) \\ C = \left(P \cdot \frac{D}{T}\right) \end{cases} \tag{5}$$

In the formula: P is the proportional coefficient of the controller, I is the integral time of the controller (the integral factor), the D is the differential time (differential coefficient) of the controller, (T is the sampling time period, the k is the sampling serial number, and the value is 0, 1, 2, …); The computer output value for the k-time sampling time of the u_k; e_k is the deviation value entered at the k sampling time.

4.2 Controller Software Design

The controller's programming uses the German 3S company Development CoDeSys, supports iec1131-3 standard IL (instruction table), ST (structured text), FBD (function module diagram), LD (Ladder diagram), SFC (sequential flowchart) Five kinds of PLC programming language. The calculation speed can realize the fast operation of Multitask program and realize the cycle conversion. In the case of the 128K/256K program code space, the EPEC2024 controller's program scanning speed is 10 ms, which can meet the periodical requirements of the various task programs running the logging head control process.

As shown in Fig. 4, the PID control module in the control system, wherein Pid_setpoint is set value, the bucking length is 260 cm; pid_measurepoint for the measurement return value. Pid_zengliangmax is the maximum value for a single

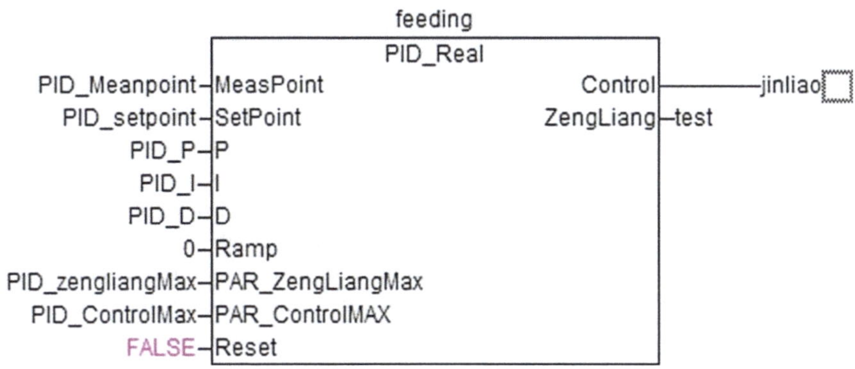

Fig. 4. PID control module

increment, Pid_controlmax is the maximum value for the control output. Control for the output of PID controller, output to variable Jinliao, the controller converted to the current signal output to the proportional directional valve, thus controlling the hydraulic motor driven feed roller.

5 Parameter Selection Description

5.1 PID Parameter Selection and Experiment

Because the controller has a bit length of 16 bits, the maximum can only reach 32767, according to the set value of the number of *Pid_setpoint*, select the initial maximum p value of 20, test between 15 and 20. If the value of P continues to decrease, the change of error cannot be tracked quickly.

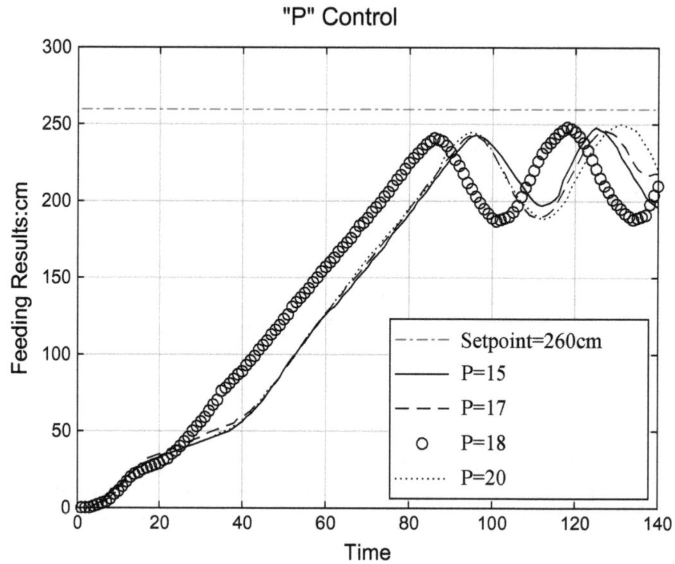

Fig. 5. The curve of actual feeding length of wood based on P control with time

As shown in Fig. 5, the horizontal axis in the figure is the feed time, unit for the number of data transmission cycle, each period of 10 ms (Figs. 6, 7 and 8, the same coordinate units in the same), when only the P link, the output can quickly respond to deviations, the controller has control, to reduce the direction of deviation change. The actual feed length increases linearly with time to a certain length and begins to produce oscillation, while the P value is 18 o'clock, the time required before oscillation is the shortest, and the amplitude changes are small, so the P value is selected as 18, and the I value is continued to adjust. Adding I is to eliminate the steady state error, but it can also reduce the response speed and increase the overshoot of the system. So from the big to the small experiment in turn, the initial setting is I = 0.005.

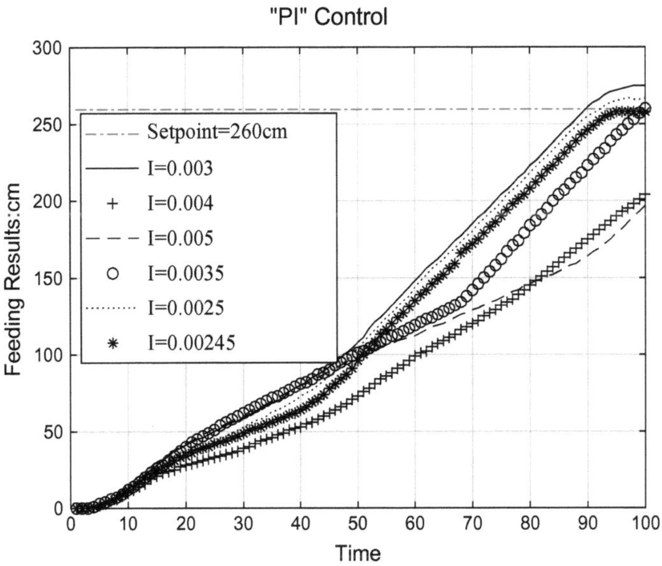

Fig. 6. The curve of actual feeding length of wood based on Pi control with time

After the field test, as shown in Fig. 6, when the I = 0.00245, the actual feed is 258 cm, the error is 2 cm, the error rate is 0.769%, within the allowable range, and has a faster response speed, so select the I value of 0.00245.

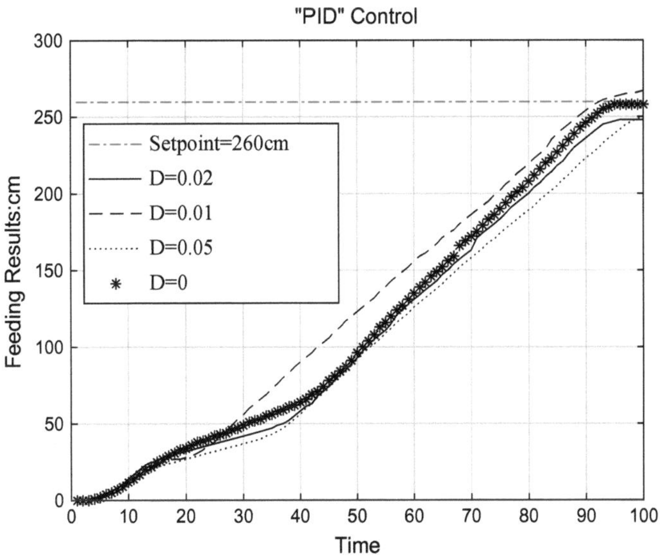

Fig. 7. The curve of actual feeding length of wood based on PID control with time

In addition to eliminating static error, the actual control system also requires speeding up the adjustment process. In order to improve the response speed, in the moment of deviation, or in the moment of deviation change, it is necessary to respond to the deviation, but also in accordance with the change in the direction of deviations in advance appropriate correction. So consider joining the D link, take D value as 0.1, 0.05, 0.02, the experiment result as shown in Fig. 7, when joins the D link, the feed all stops near 248 cm, but does not add the D link, the PI can reach 258 cm, and the response time has not obviously improved, and the differential action to the input signal noise is more sensitive, therefore, based on the actual needs of the project, choose P = 18, I = 0.0245, D = 0.

5.2 Multi-segment Adaptive PID Control Algorithm

In the actual operation process of the tree-rearing machine, the environment is complex, part of the plantation area, also contains some other woodland, its environment is highly uncertain [12, 13], there may be thickets, weeds, stones and other obstacles, when the wood cutting down, different operating environment, The resistance source of the original strip in the feeding process is different; On the other hand, the size of similar trees, lush degree is also not the same, the number of branches and other large differences, even in the same environment, feed when the size of resistance is also a certain difference [14, 15]. Therefore, in order to make the mining machine in various environments and conditions can achieve better accurate automatic feed, in order to improve the response speed, strengthen the control effect, analyze the working process of the shading Branch bucking feed and analyze the original hydraulic motor feeding process, the speed of the hydraulic motor is an acceleration. A typical feeding process that runs smoothly and slows down, so the feed speed of the original bar should be the same.

In order to solve this problem, the robustness and application range of the control are improved, and the control strategy of the hydraulic motor with multi-segment Adaptive control is designed [16, 17]. That is: In the start and occupy the length of most of the length of the smooth running section, so that the proportional solenoid valve completely open, according to the different length of the target, when close to the feeding stop position, add PID control link, can improve the speed of feed, but also to avoid environmental factors interference.

In the experiment, the target bucking length is 260 cm, so the control link is selected when the feed is in the 190, 205, 220, 235, 245 cm. As the actual results shown in Fig. 8 show, in the same time, the distance between the target value is too large, the response is slow, the distance from the target value is relatively high, the stable time is too long, when the feed reaches 235 cm, the feeding length is 264–266 cm, the error is 4–6 cm, The error rate is 2.31%, within the allowable error range. When the feed reaches 220 cm, add control link, feed length is 254–256 cm, error rate is also 2.31%, but due to insufficient target length, after bucking to Wood's further processing, often will cause the entire wood cannot use, therefore chooses when the feed reaches 235 cm, When the target length is about 25 cm, the control link is added.

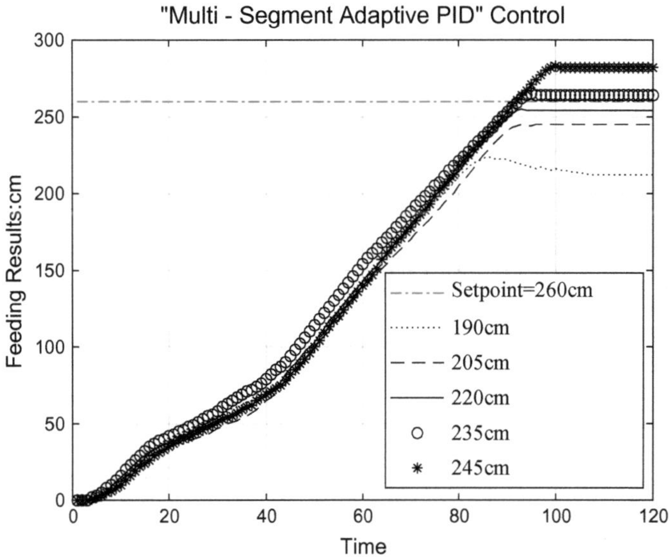

Fig. 8. Curve of actual feeding length with time based on multi-segment adaptive PID control

6 Concluding Remarks

The experimental test, the use of wood for the Metasequoia, its tree-shaped straight, the top has more branches, the middle of the trunk less, more in line with the actual situation. The timber diameter is about 153 cm and the length is about 8 m. After the commissioning of the controller, after repeated testing, when the p = 18, i = 0.0245, in distance from the target value of about 25 cm, add control links, feed error can be kept within 5 cm, the controller is stable and reliable operation, greatly reducing the feeding time and improve the accuracy of the feed. It lays a foundation for subsequent efficient harvesting operations. However, due to the limited experimental conditions, different diameter and length of wood is not able to carry out a full range of testing, in the future in the operation will need to collect more experimental data, the controller further optimization.

Acknowledgements. This research is supported by the Fundamental Research Funds for the Central Universities (No. BLX2014-03), and Special Fund for Forest Scientific Research in the Public Welfare (No. 201504508).

References

1. W.R. Zhao, J.H. Liu, Current status and development of harvesters. For. Mach. Woodwork. Equip. **36**(11), 10–12 (2008)
2. C.L. Hu, Q. Li, Research on tending management technical measures of forest in new period. J. Green Sci. Technol. (7), 146–148 (2016)

3. J.H. Liu, D. Wang, The actuality and development trend of plantation tending machines at home and abroad. For. Eng. **22**(3), 12–14 (2006)
4. B.T. Gao, Technical analysis of hydraulic control system for mobile[J]. Hydraul. Pneum. **5**, 110–112 (2012)
5. H. Duckinghaus, G. Eis, M. Pollklas, Feed control device for a harvesting machine method of controlling: US, US5901535 (1999)
6. W.L. Kaczmarski, V.C. Watts, D.S. Majkrzak et al., Feed control hydraulic circuit for wood chipper attachment: US, US6293479 (2001)
7. J. Yang, Z.X. Tong, R.P. Wang, Exploring fuzzy PID control of electro-hydraulic proportional system of hydraulic manipulator. Mech. Sci. Technol. Aerosp. Eng. **32**(6), 834–838 (2013)
8. J.H. He, X.J. Yin, C.S. Shi, The programming and implementation of PID control algorithms in siemens PLC. Ind. Instrum. Autom. **5**, 79–82 (2012)
9. L.J. Guo, H.X. Wang, J.F. Gong, A research on CAN bus-based car body network system and its control strategies. Automot. Eng. **28**(8), 774–778 (2006)
10. Q.L. Shen, *Research of the Hydraulic System Power Optimized Matching and Control Technique in Construction Machinery* (Jilin University, ChangChun, 2005)
11. H.G. Wang, H.B. Fan, G.Q. Ren, The semiconductor refrigerator temperature control system based on increasing PID controlling method. Mod. Manuf. Eng. **11**, 110–113 (2013)
12. B. Hatton, B.C. Bouzgarrou, J.C. Fauroux et al., An approach for modelling harvester head mechanism in the harvesting process of hardwood stands, in *New Advances in Mechanisms, Mechanical Transmissions and Robotics* (Springer International Publishing, 2017)
13. V. Gagnol, B. Hatton, B.C. Bouzgarrou et al., Modelling and simulation of an harvester head mechanism. Mech. Ind. **18**(3) (2017)
14. B. Hatton, V. Gagnol, B.C. Bouzgarrou et al., Modelling of the hardwood harvesting process: feeding model, in *21ème Congrès Français De Mécanique* (2013)
15. B. Hatton, G. Pot, B.C. Bouzgarrou et al., Experimental determination of delimbing forces and deformations in hardwood harvesting. Croat. J. For. Eng. **36**(1), 43–53 (2015)
16. I. Carlucho, M. De Paula, S.A. Villar et al., Incremental Q-learning strategy for adaptive PID control of mobile robots. Expert Syst. Appl. **80** (2017)
17. X. Wang, J.W. Liu, X. Zuo et al., Adaptive PID and model reference adaptive control switch controller for nonlinear hydraulic actuator. Math. Probl. Eng. **4**, 1–15 (2017)

Subspace Learning with an Archive-Based Genetic Algorithm

Kai Liu and Jin Tian(✉)

College of Management and Economics, Tianjin University, Tianjin, China
jtian@tju.edu.cn

Abstract. Feature selection is a useful technique to resolve the curse of dimensionality. Feature selection usually chooses the same feature subset for all samples. However, the different divisions of samples in local feature subsets usually have intrinsic properties in complex datasets. Subspace learning is an alternative feature selection approach by generating multiple subspaces for different classes. In this paper we proposed a subspace ensemble method based on an archive-based genetic algorithm. Experimental results show that the proposed method can outperform other conventional ensemble learning algorithms.

Keywords: Feature selection · Subspace learning · Genetic algorithm Classification

1 Introduction

Nowadays, extracting useful knowledge from dataset with high dimensions is a big challenge because of the curse of dimensionality [1].

Feature selection pertains to selecting relevant features and obtaining an optimal feature subset for classification. It is a useful technique to resolve the curse of dimensionality. There are three kinds of feature selection methods: supervised methods, semi supervised methods and unsupervised methods [2]. Supervised methods can be divided into wrappers, filters and embedded algorithms [3]. In wrapper methods feature selection is depended on the classification algorithms. The aim of wrapper methods is to find feature subsets with good classification accuracy. The filter methods attempt to delete redundant features and get feature subsets with satisfactory classification accuracy. Training examples are not need to carry out classification algorithms. The embedded feature selection methods integrate the feature selection process and the classification process into the same optimization process, that is, the feature selection is carried out automatically during the training process of the classifier. Parts of training examples with class label are needed to implement semi supervised methods. A typical method for unsupervised methods is feature clustering.

Feature selection usually chooses the same feature subset for all classes [4]. In order to avoid the losing of features with important classification competence for different classes, a feature subset is chosen for each class [5]. Moreover, the different divisions of samples in local feature subsets usually have intrinsic properties in some complex datasets. More than one feature subset is needed for some classes to get a desirable

G. Q. Huang et al. (eds.), *Proceeding of the 24th International Conference on Industrial Engineering and Engineering Management 2018*,
https://doi.org/10.1007/978-981-13-3402-3_20

classification precision [6]. Subspace learning is an alternative feature selection approach which can generate multiple subspaces for different classes. In subspace learning, each class is classified by its own feature subspaces originated from input feature space [7].

Ensemble learning can get a desirable and robust classification result by combining a number of accurate and diverse base classifiers. The first step of ensemble learning is to train some different and accurate base classifiers, which is necessary to obtain a good classification performance [8, 9]. However, there is a contradiction between diversity and accuracy in ensemble learning [10].

In this paper, we propose a subspace ensemble algorithm based on an archive-based genetic algorithm. Firstly we select different feature subsets to produce different subspaces based on niching genetic algorithm. Then we train every component classifier based on its corresponding feature subspace. We select accurate and diversity base classifiers from the iterative evolution population. The elites are kept in the archive, and the archive population influences the genetic operation of the genetic algorithm. Therefore, we can get a fine candidate set for the archive. At last, all individuals in the archive are treated as the components of the ensemble.

The contributions of this paper are concluded as follows:

(1) We present a subspace ensemble method for data with complicated feature distribution, which generates more than one feature subset for data to get good classification accuracy.
(2) In the proposed method, we improve the niching genetic algorithm to produce diverse and accurate subspaces in order to leverage the diversity and the accuracy in ensemble learning.

2 Methodology

2.1 An Overview of the Proposed Model

In this paper, we propose an archive subspace ensemble model based on genetic algorithm (ASEGA). We encode the subspace as the individuals in the evolutionary population and improve the niching genetic algorithm to produce the diverse and accurate subspaces. We keep the elites of the evolutionary population in the archive pool and aim to produce new elite subspaces in each generation that are different from elite subspaces in the archive pool to update archive pool. These final elites in the archive are the subspaces for the base classifiers in the ensemble.

Table 1 shows the framework of the proposed model ASEGA. It consists of three parts: the initialization of the evolution pool and the archive pool, the update of the evolution pool and the update of the archive pool

We output the components in the final archive pool as the ensemble

Table 1. The framework of ASEGA

Input: The training set Tr, the validation set V, the testing set Te, the evolution population size L, the archive population size M and the maximum number of iterations G
Output: The labels of the samples in Te
Procedure
1. Generate L random subspaces randomly
2. Train L classifiers based on their corresponding subspaces
3. Select M individuals from the initial evolution population to initial archive pool For $g = 1, \ldots, G$ If Archive population remains unchanged in the last k iterations, go to step 6
4. Generate the offspring of the evolution population
5. Update the archive pool
End
6. Build ensemble model with all individuals in the archive pool and predict the labels of Te with the ensemble model

2.2 The Archive Subspace Ensemble Based on Genetic Algorithm

Features are encoded as a binary chromosome in the population, and a binary gene represents the inclusion or exclusion of a feature in modeling classifiers. Thus each individual is represented as a string with d binary bits. d is the number of features. If the ith feature is chosen, the ith bit is set as 1; otherwise, the ith bit is set as 0. Thus each individual represents a subset of initial feature space, i.e. a feature subspace.

In the proposed model, we set an evolutionary pool to evolve the individuals and an archive pool to store the elites. $\mathbf{E}_l$ $(l = 1, \ldots, L)$ is the lth individual in the evolutionary pool. The size of the archive pool is M. The components in the archive pool are $\mathbf{AR} = [\mathbf{A}_1, \ldots, \mathbf{A}_M]$.

The initial individuals in the evolution population are generated randomly. Then the classification accuracies are calculated as follows:

$$accuracy = \frac{t}{m} \tag{1}$$

where t is the number of training samples and validation samples correctly classified by base classifier and m is the total number of training samples and validation samples. Based on the classification accuracy, the best M individuals are chosen to generate the initial archive pool $\mathbf{AR}^0 = [\mathbf{A}_1^0, \ldots, \mathbf{A}_M^0]$.

In the proposed method, the output diversity, the accuracy and the number of features are three objectives used to evaluate the components in the archive pool.

The output diversity measures how many samples can be classified correctly by a base classifier with a component as its input subspace but misclassified by the ensemble with the components in the archive pool as the input subspaces. We define the output diversity as follows:

$$OD_i = \mathrm{or}(f_i, fe) \quad (2)$$

where $\mathbf{f}_i = [f_{ij}]_{1\times m}$ is the classification output vector by the base classifier with the input subspace $\mathbf{A}_i$, $i = 1,\ldots,M$. $\mathbf{fe} = [fe_j]_{1\times m}$ is the classification output vector for the ensemble. If the base classifier correctly classifies the jth sample $(1 \le j \le m)$, $f_{ij} = 1$; otherwise, $f_{ij} = 0$. If $f_{ij} = 1$ and $fe_j = 0$, then $\mathrm{or}(f_{ij}, fe_j) = 1$; otherwise $\mathrm{or}(f_{ij}, fe_j) = 0$.

The second objective is the classification accuracy obtained by the base classifier with the subspace $\mathbf{A}_i$. The third objective is the number of features selected by $\mathbf{A}_i$. Then we calculate the Pareto ranking of the components in the archive pool based on the three objectives. The worst component $\mathbf{A}_k$ in the archive pool is obtained based on Pareto ranking of the multiple fitness.

As for the evolutionary pool, the output diversity, the accuracy and the feature diversity are three objectives to calculate the fitness of the individuals.

The output diversity here measures how many samples can be classified correctly by a base classifier with an individual as its input subspace but misclassified by the ensemble with the components in archive pool as the input subspaces. Similar to Eq. (2), the output diversity is computed as:

$$f_l^1 = \mathrm{or}(\mathbf{O}_l, \mathbf{fe}) \quad (3)$$

where $\mathbf{O}_l = [o_{lj}]_{1\times m}$ is the classification output vector by the base classifier with the input subspace $\mathbf{E}_l$. If the classifier correctly classifies the jth sample, $o_{lj} = 1$, otherwise, $o_{lj} = 0$.

The second objective f_l^2 is the classification accuracy. The third objective, the feature diversity is defined as:

$$f_l^3 = \min_{i=1,\ldots,M, i\neq k} \{\mathrm{ham}(\mathbf{E}_l, \mathbf{A}_i)\} \quad (4)$$

where $\mathbf{E}_l$ is the lth individual in evolutionary population and $\mathbf{A}_i$ is the ith elite in the archive population except the worst $\mathbf{A}_k$ in the archive pool. $\mathrm{ham}(\mathbf{E}_l, \mathbf{A}_i)$ is the hamming distance between $\mathbf{E}_l$ and $\mathbf{A}_i$.

We use the best individual in evolutionary population in each generation to update archive pool. Then the worst component is replaced by every individual. The classification accuracy of the ensemble with the input subspace is calculated. Finally, we select the individual with the highest ensemble accuracy to replace the worst component in the archive pool.

Then we calculate the Pareto ranking of the individuals of the evolution pool based on the three objectives. Then we select individuals based on the sort value into the mating pool. The two-point crossover is used to exchange information between two individuals that are picked randomly from the mating pool, and two offspring are produced. The mutation operation is conducted in the selected individuals by inversing the bits. That is, the bit of 0 is changed to 1, and 1 to 0. The mutation probability p is defined as:

$$p = \left(\frac{d - f_i^3}{2d}\right)^2 \tag{5}$$

where d is the feature size.

3 Results

In this paper, we use 10 datasets from UCI machine learning repository to testify the performance of the proposed method. The sample numbers of the datasets vary from 47 to 2310. The values of attribute vary from 10 to 309. The numbers of class vary from 2 to 13. The brief introduction to these datasets is shown in Table 2.

Table 2. Characteristics of datasets

Dataset	Sample	Attributes	Classes
Fertility	100	10	2
Glass	214	10	6
Hepa	159	19	2
Segment	2310	19	7
German	1000	25	2
Wdbc	569	31	2
Dermatology	366	35	6
Soybean (small)	47	35	4
Arrhythmia	452	279	13
LSVT	126	309	2

In our experiments, each dataset is separated into three parts, 50% of each dataset are set as training set, 25% of the dataset are set as validation set and the rest are set as testing set. We select Classification and Regression Tree (CART) [11] as base classifier for ensemble model. And the ensemble rule for predict result is majority voting [12, 13]. Experiment results in this paper are the average value of the 30 runs. The population size L is set as 50, and size of offspring is set as 50, too. The archive population size M is set as 25. And the Maximum number of iterations G is 200. The k is set as 10.

The compared methods include the AdaboostM1 [14], Bagging [15], Random Forest (RF) [16] and Random Subspace (RS) [17]. The comparison results among the proposed method ASEGA and other ensemble algorithms are shown in Table 3. The best result for each dataset is in bold.

Shown in Table 3, the proposed method reaches the highest for test accuracy in Glass, Hepa and Soybean. AdaboostM1 reaches the highest for test accuracy in Segment, Wdbc and LSVT. Bagging reaches the highest for test accuracy in Fertility, German and Arrhythmia. RF reaches the highest for test accuracy in Dermatology. And

Table 3. Comparison results

Dataset	AdaboostM1	Bagging	RF	RS	ASEGA
Fertility	84.66	**88.13**	87.60	88.00	88.00
Glass	75.49	74.72	76.29	73.02	**76.83**
Hepa	61.01	59.29	60.48	60.51	**61.05**
Segment	**98.16**	96.81	97.33	96.73	96.95
German	73.87	**76.75**	75.31	72.16	76.04
Wdbc	**95.99**	94.07	95.38	94.30	95.71
Dermatology	96.31	95.12	**96.87**	95.82	96.70
Soybean (small)	98.30	96.22	99.67	97.92	**99.71**
Arrhythmia	73.23	**74.94**	66.76	69.05	72.97
LSVT	**83.21**	81.55	80.87	79.61	81.67
Ave	84.02	83.76	83.65	82.71	**84.56**

the average accuracy of these 10 datasets is shown in the last row. The proposed method ranks first. We can conclude that the proposed method performs better than other ensemble methods.

We further compare the average size of selected feature subsets on UCI datasets among the proposed method, RF and RS. The comparison results are shown in Table 4. The best result for each dataset is in bold. The average rankings of the three algorithms about feature size are also shown in Table 4.

Table 4. Average size of subset versus different algorithms

Dataset	ASEGA	RF	RS
Fertility	**4.14**	6.33	5.00
Glass	**5.36**	8.30	5.00
Hepa	**7.30**	13.40	9.50
Segment	**7.78**	16.27	9.50
German	12.29	20.47	**10.50**
Wdbc	**10.41**	11.59	15.50
Dermatology	**12.51**	19.42	17.50
Soybean (small)	15.58	**4.82**	17.50
Arrhythmia	133.98	**66.95**	139.50
LSVT	142.98	**99.59**	154.50
Ranking	**1.5**	**2.2**	**2.3**

As shown in Table 4, the proposed method reaches best for average size in Fertility, Glass, Hepa, Segment, Wdbc and Dermatology. RF reaches best for average size in Soybean, Arrhythmia and LSVT. RS reaches best for average size in German. Moreover, the proposed method ranks first, which shows the proposed method can perform better than RF and RS in feature reduction.

To sum up, the experimental results in Table 3 show the proposed method performs better than other ensemble methods. The proposed method can establish a subspace ensemble model for data with complicated feature distribution by generating more than one feature subset to get high classification accuracy. The experimental results in Table 4 show that the proposed method performs well in feature reduction.

4 Conclusion

In this paper we propose an ensemble method based on the niching genetic algorithm. We improve the niching genetic algorithm to produce diverse and accurate subspaces. Moreover, we conduct feature selection in the evolutionary process. A desirable ensemble model with various precise feature subspaces is thus established which can achieve higher classification accuracies with smaller feature subsets than other conventional classification methods.

Acknowledgements. The work was supported by the General Program of the National Science Foundation of China (Grant No. 71471127, 71371135).

References

1. F. Viegas, L. Rocha, M. Gonçalves et al., A genetic programming approach for feature selection in highly dimensional skewed data. Neurocomputing 554–569 (2018)
2. Y. Li, T. Li, H. Liu, Recent advances in feature selection and its applications. Knowl. Inf. Syst. 1–27 (2017)
3. P. Yan, Y. Li, Graph-margin based multi-label feature selection, in *Machine Learning and Knowledge Discovery in Databases* (Springer International Publishing, 2016)
4. B. Seijo-Pardo, I. Porto-Díaz, V. Bolón-Canedo et al., Ensemble feature selection: homogeneous and heterogeneous approaches. Knowl.-Based Syst. (2016)
5. R. Shang, W. Wang, R. Stolkin et al., Subspace learning-based graph regularized feature selection. Knowl.-Based Syst. **112**, 152–165 (2016)
6. S. Sun, C. Zhang, Subspace ensembles for classification. Phys. A **385**(1), 199–207 (2007)
7. M. Woźniak, M. Graña, E. Corchado, A survey of multiple classifier systems as hybrid systems. Inf. Fusion **16**(1), 3–17 (2014)
8. D. Cheng, S. Zhang, X. Liu et al., Feature selection by combining subspace learning with sparse representation. Multimed. Syst. **23**(3), 285–291 (2017)
9. Y. Guo, L. Jiao, S. Wang et al., A novel dynamic rough subspace based selective ensemble. Pattern Recogn. **48**(5), 1638–1652 (2015)
10. A. Schclar, L. Rokach, Random projection ensemble classifiers, in *International Conference on Enterprise Information Systems* (Springer, Berlin, Heidelberg, 2009), pp. 309–316
11. P.L. Gutierrez, S. Siva, Classification and regression tree (CART), in *Encyclopedia of Genetics, Genomics, Proteomics and Informatics* (Springer Netherlands, 2008), pp. 370–370
12. Y.S. Chung, D.F. Hsu, C.Y. Tang, On the diversity-performance relationship for majority voting in classifier ensembles, in *International Conference on Multiple Classifier Systems* (Springer, 2007), pp. 407–420
13. G. Brown, L.I. Kuncheva, "Good" and "Bad" diversity in majority vote ensembles, in *International Conference on Multiple Classifier Systems* (Springer, 2010), pp. 124–133

14. E.A. Cortés, M.G. Martínez, N.G. Rubio, Multiclass corporate failure prediction by adaboost.M1. Int. Adv. Econ. Res. **13**(3), 301–312 (2007)
15. L. Breiman, Bagging predictors. Mach. Learn. **24**(2), 123–140 (1996)
16. L. Breiman, Random forest. Mach. Learn. **45**, 5–32 (2001)
17. M. Skurichina, R.P.W. Duin, Bagging, boosting and the random subspace method for linear classifiers. Pattern Anal. Appl. **5**(2), 121–135 (2002)

Incorporating Social Information in Recommender Systems Using Hidden Markov Model

Jia-xin Zhang and Jin Tian(✉)

College of Management and Economics, Tianjin University, Tianjin, China
jtian@tju.edu.cn

Abstract. User preference always changes over time, which makes time the strong context information in the recommender system. Many time-dependent recommender systems have been proposed to track the change of users' preferences. However, the social factor, which has been proved useful for recommender systems, is rarely considered in these models. In this paper, we consider the effects of social friends on the users' behavior and propose a dynamic recommender system based on the hidden Markov model to provide better recommendations for users. We compare the proposed model with the traditional static and dynamic recommendation methods on real datasets and the experimental results show that the proposed model outperforms the compared methods.

Keywords: Recommender system · Social information · Hidden Markov model

1 Introduction

Information overload makes it more expensive for people to discover valuable information. Recommender systems (RSs) are widely applied in various fields and help people to find products matching their preferences.

Collaborative filtering (CF) method, which makes recommendations based on users' purchase histories, has become popular in RS [1]. In addition, content-based method is another kind of widely used recommendation approaches, which recommends appropriate items for users based on analyzing the item descriptive information [2]. However, there are several limitations for the traditional recommendation approaches. For example, the similarity values, which are usually calculated based on common items, tend to be unreliable when data are sparse and the common items are few. And for content-based approaches, it is usually difficult to obtain the item information [3]. To overcome these drawbacks, more and more studies incorporate the contextual information (e.g., time, location, social, weather, etc.) into recommender systems to improve the performances [4, 5].

Generally, user preferences are changeable over time [6, 7]. Static recommender systems tend to calculate user preferences based on their purchase history which assumes that user preferences are constant. It may be able to successfully identify the

G. Q. Huang et al. (eds.), *Proceeding of the 24th International Conference on Industrial Engineering and Engineering Management 2018*,
https://doi.org/10.1007/978-981-13-3402-3_21

past preferences of users but fail to predict what people will like in future. Therefore, many researchers study on the dynamic recommendation models. Xiang et al. split rating records with a fixed time window and only considered records in current time window to model users' preferences [8]. Koren proposed time SVD++, in which the drift of both users and items bias was combined into the factor model [6]. Inuzuka et al. used matrix factorization and Kalman filtering to predict user preferences and considered the preference changes of each user by learning the order of purchase history [9]. Aghdam et al. introduced a hidden Markov model (HMM) to capture the changes of user preferences, in which the current context of the user was set as a hidden variable in the HMM model [10]. The development of social media brings new opportunities for recommender systems. However, the social factor is rarely considered in the dynamic recommendation models.

The HMM is a kind of statistical models of sequential data that have been used successfully in many applications [11]. Sahoo et al. used the HMM combined with the aspect model to identify the common transition pattern of user preferences [12], but it ignores the social effect through this process. In this paper, we propose a dynamic recommender system, SoHMM, which integrates social information into the HMM. We suppose that social friends have influence on user purchase behavior. We use the purchase records of the active user's social friends in conjunction with the user's purchase history to determine the recommendation list. Furthermore, we derive an expectation maximization (EM) algorithm [13] which is similar to the parameter estimation algorithm of traditional HMM to estimate the parameters of the framework. By comparing the performance of the proposed model with static and dynamic recommendation methods on two real data sets, we found our method shows significant improvement over the compared methods.

2 Methodology

The proposed model is based on the hidden Markov model (HMM). It has a fixed number of states corresponding to different conditional distributions of the output variables. We merge social information into it to model output sequence conditioned on it. In this section, we describe the framework and the general process of the proposed model.

2.1 Model Description

The architecture is based on standard HMMs, including a set of inputs at each time period as the condition for the emission distribution. The transition distribution is similar to the traditional HMM, Fig. 1 shows the framework of our model.

As shown in Fig. 1, the sequence $\{Z_u^1, Z_u^2, \ldots, Z_u^T\}$ represents the implicit states of user u at each time period, reflecting the trajectory of user preferences over time. These variables are said to be a Markov chain, which means that the current state only dependents on the previous state. The variable F_u^t is the input variable at time period t, representing the social information. The input variable F_u^t and the hidden state Z_u^t together determine the output I_u^t and N_u^t, that is, the observation variables. Specifically,

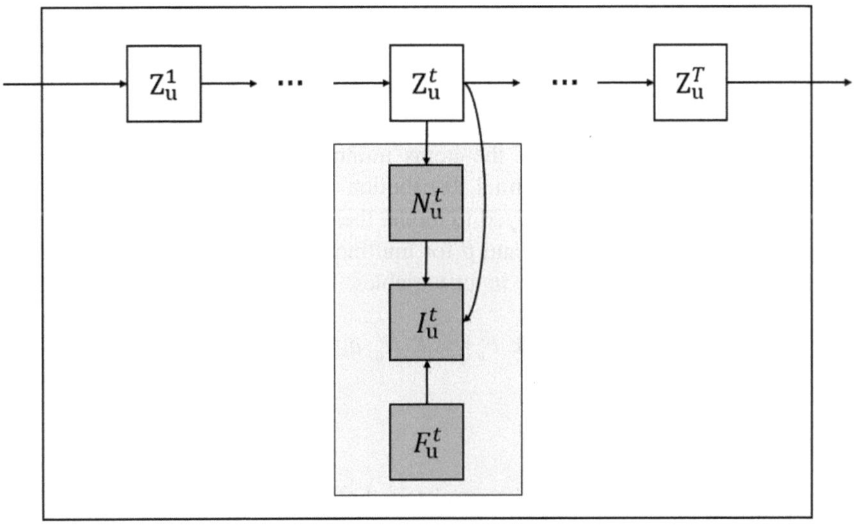

Fig. 1. The framework of SOHMM

N_u^t refers to the number of items selected by the user at time period t and I_u^t is the corresponding item set.

On this basis, the proposed SoHMM model for recommendation works as follows: At time period $t = 1$, an initial state Z_u^1 is chosen according to the initial state distribution, then the observation variables N_u^1 and I_u^1 are obtained through the emission distribution conditioned on the F_u^1. At other time periods $t > 1$, a state Z_u^t is computed at first according to the transition distribution, using the state at the previous time step Z_u^{t-1}. Then the output is generated in the same way as $t = 1$.

2.2 Model Components

The model is mainly composed of three components, which are initial state distribution, states transition distribution and the conditional output distribution models. The specific definition is as follows:

(1) *Initial State Distribution*: we specify the initial state distribution as the discrete distribution with π as the parameters, that is:

$$P\left(Z_u^1 = k\right) = \pi_k \tag{1}$$

(2) *State Transition Distribution*: in this part, hidden variable shifts from one state j to the other state i according to the transition probability matrix A, which follows the multinomial distribution.

$$P(Z_u^t = i | Z_u^{t-1} = j) = A_{ji} \tag{2}$$

(3) *Conditional Emission distribution*: The emission distribution make predict for user at each time period under the current state and input. It can be divided into two part, the distribution of the items number and the distributions of items selection. The negative binomial distribution (NBD) and the conditional multinomial distribution are employed to model these two parts respectively. We set a, b as the parameters of NBD and θ for multinomial distribution, where θ is time-variant and dependent on the input variables.

$$P(I_u^t | Z_u^t = k, F_u^t) = P(N_u^t; a_k, b_k) P(I_u^t; \theta_k, F_u^t) \tag{3}$$

2.3 Model Derivation

The most critical assumption in this model is the Markov conditional independence assumption, which means that the state variable Z_u^t summarizes sufficiently the past of the sequence, and only the current social information can affect the active user. Thus the following equality holds true:

$$P(Z_u^t | Z_u^{1:t-1}) = P(Z_u^t | Z_u^{t-1}) \tag{4}$$

$$P(I_u^t | Z_u^{1:t}, F_u^{1:t}) = P(I_u^t | Z_u^t, F_u^t) \tag{5}$$

where the $Z_u^{1:t-1}$ refers to the set $\{Z_u^1, Z_u^2, \ldots, Z_u^{t-1},\}$. These two assumptions imply that given the present state, what user would like in this period is independent of his purchase history and his friends' purchase history.

First, the whole data is set as $D_c = (I, Z, F)$. The corresponding complete data likelihood is defined as:

$$L_c(\Theta; D_c) = P(I, Z|F) = \prod_{u=1}^{U} P(I_u^{1:T_u}, Z_u^{1:T_u} | F_u^{1:T_u}; \Theta) \tag{6}$$

Using the condition independency assumption, the complete likelihood can be further factorized. Then taking the logarithm, we can obtain the final expression for the complete data log-likelihood $l_c(\Theta; D_c)$.

2.4 The EM Algorithm

Since the likelihood function depends on the unknown state variable Z, it cannot be directly maximized. We adopt EM algorithm [13] to decouple static and temporal learning. Following the EM framework, we alternate between computing the posterior distributions in the E step and updating the model parameters in the M step until the convergence.

(1) *The Estimation Step*: Estimation step computes the expected value Q of log-likelihood function with respect to the distribution of the state paths, given the data D_c and the parameters $\widehat{\Theta}$. The hat on the θ means it is calculated through last iteration.

$$Q\left(\Theta;\widehat{\Theta}\right) = E_Z\left[l_c(\Theta;D_c)|F_u^{1:T_u}, I_u^{1:T_u};\Theta\right] \tag{7}$$

To compute the Q, we use forward-backward algorithm to compute two posteriors, the state posterior probabilities and the transition posterior probabilities. The forward variables and backward variables are:

$$\alpha\left(Z_u^t\right) = P(Z_u^t|I_u^{1:t}, F_u^{1:t}) \tag{8}$$

$$\beta\left(Z_u^t\right) = \frac{P(I_u^{t+1:T_u}|Z_u^t, F_u^{t:T_u})}{P(I_u^{t+1:T_u}|I_u^{1:t})} \tag{9}$$

Using the conditional independence assumptions, these can be rewritten as recursion process:

$$\alpha\left(Z_u^t\right) = \frac{P\left(I_u^t|Z_u^t, F_u^t\right)}{P\left(I_u^t|I_u^{1:t-1}\right)}\left[\sum_{Z_u^{t-1}}\alpha\left(Z_u^{t-1}\right)P\left(Z_u^t|Z_u^{t-1}\right)\right] \tag{10}$$

$$\beta\left(Z_u^t\right) = \frac{\sum_{Z_u^{t+1}}\beta\left(Z_u^{t+1}\right)P\left(I_u^{t+1}|Z_u^{t+1}, F_u^{t+1}\right)P\left(Z_u^{t+1}|Z_u^t\right)}{P\left(I_u^{t+1}|I_u^{1:t}\right)} \tag{11}$$

Then the posterior probabilities can be expressed in terms of α and β.

(2) *The maximization step*: In this step, we update the parameters to maximize the expected log-likelihood Q.

Specifically, the initial hidden state and the transition of the hidden states both can be considered as a multinomial trails process. So for the parameters π and A, we use MAP to estimate the parameters and setting Dirichlet distribution as their conjugate prior. In addition, we model the emission distribution conditioned on the social information. For this part, we use normal MLP to get more appropriate parameter estimation.

2.5 Prediction and Recommendation

After the parameters estimation, there is two-step to generate the recommendation result. First, using the trained model, predict the possible state of the next step for each user. Then, generate the recommendation set for each user based on the predicted state and the emission distribution.

The whole process of SoHMM algorithm is shown in Table 1.

Table 1. Iterative algorithm of SOHMM

Input: $I_u^{1:T}, N_u^{1:T}, F_u^{1:T}$ for each user;
Output: I_u^{T+1} for each user;
Step 1. Randomly initialize π, A, a, b, θ;
Step 2. for each user **do**
2.1 **for** each state **do**
Compute $P(Z_u^t \mid Z_u^{t-1})$ and $P(I_u^t \mid Z_u^t, F_u^t)$ according to the transition distribution and conditional emission distribution
2.2 **for** each state **do**
Compute $\alpha(Z_u^t)$ and $\beta(Z_u^t)$ using the current value of the parameters;
Compute the transition posterior probabilities and state posterior probabilities
Step 3. for each state **do**
2.1 update the parameters π, A, a, b, θ
Step 4. Predict the next state and make recommendation

3 Experiments

3.1 Experimental Settings

In this section, we compare the performance of the proposed SoHMM model with following four methods without considering the social relationships on real datasets:

Aspect Model [14]: A probabilistic latent space model which models individual preferences as a convex combination of preference factors.

TimeSVD [6]: Incorporates time-related variables into the famous matrix factorization method. It allows for time-related user bias in matrix factorization.

HMM [12]: HMM models transitive latent states of user interest, which is the most similar algorithm to our approach, except we combine the social context into users' item selection behavior.

We evaluate the performance of our approach in the Top-N recommendation by widely used metric precision, recall, and F-measure.

We use two real-world datasets, which are Delicious dataset and Last.fm dataset. Both of them were released in the framework of the second International Workshop on Information Heterogeneity and Fusion in Recommender Systems, containing social networking, user-item records, and tagging information.

3.2 Experimental Results

In this section, we report the performances of different recommendation approaches. We set the training span to 30 months for both Delicious and Last.fm and roll the time window by month. The evaluation metrics are computed for each train-test set for the top-5 and top-10 predictions. We repeat each experiment 10 times and report the average results.

The optimal parameter settings for each method are determined by our experiments. Tables 2 and 3 show the recall, precision, and F-Measure values of all comparison approaches on two datasets. As we can see, the performance of our proposed approach is always in the group of best models, which shows that our method can predict user's interest more accurately and further verify the assumption that social information dose have great help for recommender system.

Table 2. Experiment result on delicious

	Top 10			Top 5		
	R	P	F1	R	P	F1
SoHMM	**0.0200**	**0.0039**	**0.0066**	**0.0114**	**0.0045**	**0.0064**
HMM	0.0181	0.0036	0.0059	0.0104	0.0041	0.0058
TimeSVD	0.0139	0.0027	0.0045	0.0075	0.0029	0.0042
Aspect	0.0172	0.0034	0.0056	0.0097	0.0038	0.0055

Table 3. Experiment result on Last.fm

	Top 10			Top 5		
	R	P	F1	R	P	F1
SoHMM	**0.0676**	**0.0204**	**0.0314**	**0.0403**	**0.0243**	**0.0303**
HMM	0.0655	0.0201	0.0307	0.0370	0.0227	0.0281
TimeSVD	0.0439	0.0127	0.0197	0.0284	0.0133	0.0181
Aspect	0.0511	0.0157	0.0240	0.0271	0.0166	0.0206

4 Conclusion

In this paper, we propose the SoHMM model, which is a dynamic social recommender system based on the HMM. We use it to model the social effect on the item selection of the user. The core idea of this model is that it regards social information as a useful resource of recommender system and can further improve the performance of time-aware recommend algorithm. We derive an EM algorithm similar to the parameter estimation algorithm of HMM to estimate the parameters of the proposed model. Experimental results show that the proposed model can significantly improve the performance of the recommendation.

In the future, we will further try to polish the structure of the model by introducing other dependency relationship and exploring the hybrid structure of the proposed model with traditional collaborative filtering method to improve the prediction quality.

Acknowledgements. The work was supported by the General Program of the National Science Foundation of China (Grant No. 71471127, 71502125).

References

1. A. Tuzhilin, Towards the next generation of recommender systems, in *Proceedings of the 1st International Conference on E-Business Intelligence (ICEBI2010)*, pp. 734–749
2. R.J. Mooney, L. Roy, Content-based book recommending using learning for text categorization, in *Proceedings of the Fifth ACM Conference on Digital Libraries* (2000), pp. 195–204
3. G. Guo, Integrating trust and similarity to ameliorate the data sparsity and cold start for recommender systems, in *Proceedings of the 7th ACM conference on Recommender systems (RecSys'13)* (2013), pp. 451–454
4. C. Chen, X. Zheng, Y. Wang, F. Hong, Z. Lin, Context-ware collaborative topic regression with social matrix factorization for recommender systems, in *AAAI'14 Proceedings of the Twenty-Eighth AAAI Conference on Artificial Intelligence* (2014), pp. 9–15
5. X. Ren, M. Song, E. Haihong, J. Song, Context-aware probabilistic matrix factorization modeling for point-of-interest recommendation. Neurocomputing 38–55 (2017)
6. Y. Koren, Collaborative filtering with temporal dynamics. Commun. ACM **53**(4), 89–97 (2010)
7. H. Bao, Q. Li, S.S. Liao, S. Song, H. Gao, A new temporal and social PMF-based method to predict users' interests in micro-blogging. Decis. Support Syst. **55**(3), 698–709 (2013)
8. L. Xiang et al., Temporal recommendation on graphs via long- and short-term preference fusion, in *ACM SIGKDD International Conference on Knowledge Discovery and Data Mining* (2010), pp. 723–732
9. K. Inuzuka, T. Hayashi, T. Takagi, Recommendation system based on prediction of user preference changes, in *International Conference on Web Intelligence* (2017), pp. 192–199
10. M. Hosseinzadeh Aghdam et al., Adapting recommendations to contextual changes using hierarchical hidden markov models, in *Proceedings of the 9th ACM Conference on Recommender Systems* (2015), pp. 241–244
11. L.E. Baum, Statistical inference for probabilistic functions of finite Markov chains. Ann. Math. Stat. **37**(6), 1554–1563 (1966)
12. N. Sahoo, P.V. Singh, T. Mukhopadhyay, A hidden Markov model for collaborative filtering. Manage. Inf. Syst. Q. **36**(4), 1329–1356 (2012)
13. A.P. Dempster, Maximum likelihood from incomplete data via EM algorithm. J. R. Stat. Soc. B **39**, 1–38 (1977)
14. T. Hofmann, J. Puzicha, Latent class models for collaborative filtering, in *Sixteenth International Joint Conference on Artificial Intelligence* (1990), pp. 688–693

Multivariate Process Monitoring and Fault Identification Using Convolutional Neural Networks

Xiaoyun Zheng and Jianbo Yu(✉)

School of Mechanical Engineering, Tongji University, Shanghai 200082, People's Republic of China
jbyu@tongji.edu.cn

Abstract. In multivariate process control (MPC), the conventional multivariate quality control charts (e.g., T^2) have been shown to be efficient for out-of-control signals detection based upon an overall statistic, whereas do not relieve the need for multivariate process pattern recognition (MPPR). MPPR is very beneficial to locate the assignable causes that lead to the out-of-control situation in multivariate manufacturing process. Both Support Vector Machines (SVM) and Convolutional Neural Networks (CNN) techniques have been widely applied and obtained many successes in image and visual analysis, but both methods have some weakness. Therefore, we propose a hybrid system that composes the mentioned techniques. Firstly, two different structure of CNNs were pre-trained as feature extractor due to the capability of unsupervised feature learning. The feature extracted by two CNNs were combined to train a SVM classifier. Experimental analysis show that the proposed hybrid system presents better performance than the isolated stand-alone systems.

Keywords: Multivariate process pattern recognition · Convolutional neural networks · Support vector machines · Feature learning

1 Introduction

The development of data-acquisition technology and online computers related to process monitoring have given rise to an augmented interest in the simultaneous observation of some correlated process variables or quality characteristics. An appropriate SPC charting scheme is essential to monitor and diagnose process variables, when quality feature of a product involves two correlated variables (i.e., bivariate). Particularly, process monitoring relates to process identification either in in-control or out-of-control condition, whereas process diagnosis relates to source variables identification for out-of-control situation.

It is known that the traditional SPC charting schemes for bivariate quality control (e.g., T^2 [1], multivariate cumulative sum (MCUSUM) [2], and multivariate exponentially weighted moving average (MEWMA) [3]) are effective in process monitoring. Although those control charts can successfully identify process shifts, they do not provide any information on the causes of the shifts and information on the causes of the shifts is essential for improved process control.

G. Q. Huang et al. (eds.), *Proceeding of the 24th International Conference on Industrial Engineering and Engineering Management 2018*,
https://doi.org/10.1007/978-981-13-3402-3_22

Multivariate process pattern recognition (MPPR) is important tool for monitoring multivariate processes (MP). One difficulty encountered with MPPR is the identification of the variable(s) that give rise to an out-of-control signal. Guh and Shiue [4] identified two types of fault identification techniques, namely statistics-based and learning-based methods. Some statistics-based methods for multivariate processes control already available, such as principal component analysis (PCA), partial least squares (PLS), multi-way PCA, and dynamic PCA.

With automated data acquisition and analysis techniques widely adopted in machining processes, it has become more and more popular to apply machine-learning algorithms to automated process monitoring and fault identification. Deep research has been conducted into the use of artificial neural network (ANN), such as, back-propagation network (BPN) [5], multilayer perceptron (MLP) [6], support vector machine (SVM) [7], as useful tools for abnormal signals detection or MPPR in multivariate manufacturing processes.

One disadvantage of these methods is that the process variables are directly used as their inputs and the correspondent constructed model is not reliable for MP due to the high dimension and noise in the process signals. Feature extraction from the process variables is essential in improving the performance of the recognizers through the dimension and noise reduction. However, the effective features are manually selected according to the specific MP issue, but this brings generally many difficulties for real-world applications. Thus, feature learning from process signals directly is critically needed to automatically capture the effective process characteristics.

In recent years, feature learning in the context of deep learning has attracted considerable attentions in the machine learning field [8]. Deep convolutional neural networks (CNN) [9] is feed neural networks, does not require any task-specific feature engineering and domain specific export knowledge. CNNs have recently advanced the state-of-the-art performance of image classification.

CNNs are built using a hierarchical architecture to learn deeper features presented by the image data set and scalability is one of the main advantages presented by these models. The network is normally trained using gradient backpropagation techniques. Commonly, the last layer use the Softmax function as classifier. Using the MNIST dataset and Fashion-MNIST dataset [10] as input data, the CNN with Softmax can achieve a test accuracy of $\sim$99.23% and $\sim$91.86%, respectively. However, there have been studies [11] conducted to challenge this norm. The last layers of CNNs are traditional fully-connected linear layers, which is a classifier less efficient than an SVM classifier.

In this paper, we investigate and propose a hybrid system that compose CNNs and SVM. The main idea is beginning with a relatively simple CNN (i.e., 5-layer, and 7-layer ConvNet) using back-propagation to extract feature. After this, the feature extracted are used in a second step in order to train an SVM classifier. Extensive experiments are also carried to illustrate that the proposed hybrid system presents a better accuracy than SVM or CNN used separately.

2 Methodology

Flowchart of the proposed MCC model is shown in Fig. 1. The model uses an ensemble of 5-layer and 7-layer ConvNet. The first step in order to build the hybrid model is to train the ConvNets using backpropagation algorithm. After training the ConvNets, in order to construct the proposed hybrid model, the softmax layer is dropped and the activations of the 1024 nodes fully connected previous layer are combined as the features extracted from a given image from the multivariate processes data set. Therefore, instead of feeding the SVM classifier with the row process data, the proposed hybrid system, is built by feeding an SVM classifier with the 2048 higher level features extracted from the previews trained ConvNets.

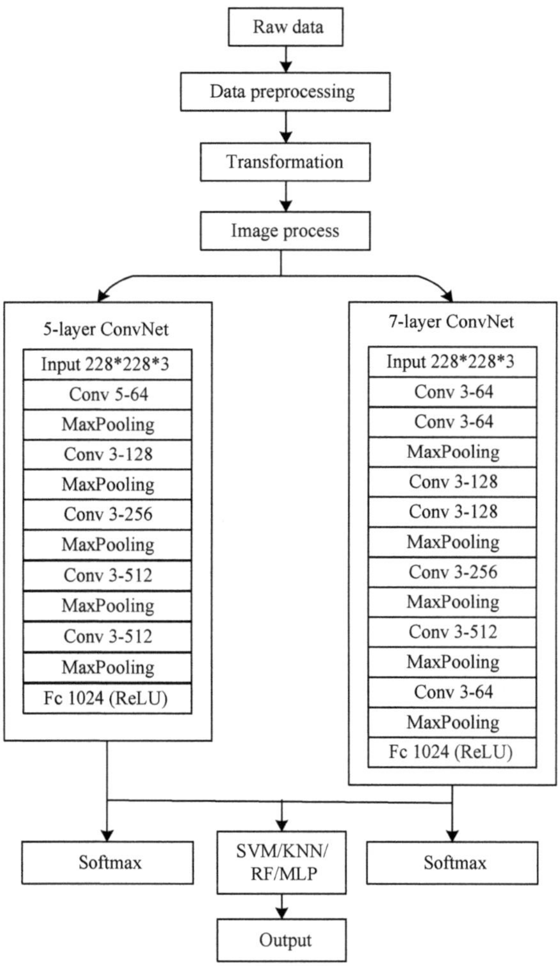

Fig. 1. Flowchart of the proposed MPPR model

After the previous procedure, the regular maximum margin criterion is used in order to optimize the resulting model. Therefore, the hybrid system has a convex error surface that uses high-level features extracted from the convent as input data.

2.1 Convolutional Neural Networks

CNNs are feed forward neural networks specialized mainly in treating image data. Taking into consideration that images are symmetrical by a shift in position, weight sharing and selective fields techniques are used in order to create filter banks that extract geometrically related features from the image data set. The process is composed hierarchically over many layers to obtain higher level features after each layer. Fully connected layers are added at the top of the architecture in order to function in a similar way of perceptron classifiers. The network is normally trained using gradient back-propagation techniques. Traditionally, softmax function is used as the classifier at the last layer of network.

CNNs are designed to treat 2-d array data and show high performances in image recognition fields. As shown in Fig. 2, a general CNN structure consists of a feature extractor and a softmax classifier, whereas the feature extractor is composed of several convolutional layers usually followed by pooling layers.

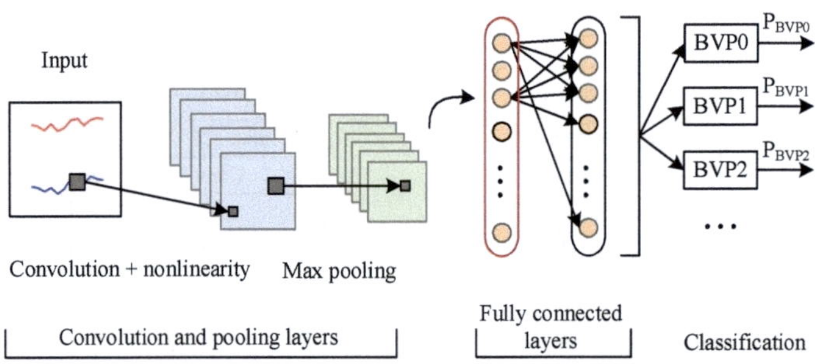

Fig. 2. General structure of CNN

Relatively simple CNN structures is used in this paper. Figure 1 shows the network structure of 5-layer CNN and 7-layer CNN. The 5-layer CNN consists of input, convolutional, maxpooling, full connection, and output layer. A set of convolutional is added to 5-layer CNN to make a 7-layer CNN. Rectangle Linear Unit (ReLU) is used as the activation function, which can prevent gradient vanishing problem and effective for training of deep networks.

2.2 SVM and Kernels

The SVM algorithm is used as the final classifier in the proposed system. This is a very successful and mathematical elegant method commonly used by machine learning community. It is shown to provide an optimization problem that presents a convex error surface. It is normally used in conjunction with kernel methods in order to enable the classification of non-linearly separate data set.

3 Results

In this paper, we used TensorFlow framework [12] to construct the CNN to extract high-level features from the images and Python Scikit-learn library [13] to process the SVM classifications and related activities.

The experiment that consists of training and testing SVM in the flat array of raw process data using Gaussian kernels and linear kernel is called SVM_L and SVM_G respectively. Our proposed hybrid system that consists of using an SVM classifier on the 2048 high-level features extracted by two simple ConvNets is labeled as ConvNetSVM. In order to demonstrate the accuracy of our hybrid model, comparisons have been extended to other class of classifiers, we trained and test Random Forest (RF), K-Nearest Neighbor (KNN), and MLP classifier on the raw process data. 2048-d labelled features are also presented to the aforementioned classifiers, labeled as ConvNetsRF, ConvNetsKNN, and ConvNetsMLP respectively.

3.1 Example with Two Variables (Bivariate)

In this study, we used Monte-Carlo simulation approach to generate the sets of example shifts for training, and testing data. Among various abnormal situations, we concerned with process mean shifts with the assumption that the not changed of the process covariance matrix. If a mean shift δ arises at time t_η, the signal $X(t)$ of a process variable at time t is expressed:

$$X(t) = \mu + Y(t) + \delta(t, t_\eta),\ t \geq t_\eta \tag{1}$$

where μ is process mean, $Y(t) \in N(0, \Sigma)$, and $\delta(t, t_\eta) = [k_1\sigma_1, k_2\sigma_2, \ldots, k_p\sigma_p]$, k_l is the shift magnitude in accordance σ_l, which is the l-th process variable.

The bivariate process is taken into consideration for performing the experimental analysis in this study. For a bivariate process, $X(t)$ follows a data distribution $N(\mu_0, \Sigma_0)$. The covariance matrix Σ_0 is:

$$\Sigma_0 = \begin{bmatrix} \sigma_{X_1}^2 & \sigma_{X_1X_2} \\ \sigma_{X_1X_2} & \sigma_{X_2}^2 \end{bmatrix} \tag{2}$$

where σ_{X_1} and σ_{X_2} denote the standard deviations of the two process variables (i.e., X_1 and X_2), respectively.

An industrial case is considered to implement the experimental analysis. The two variables of this bivariate process denote stiffness (X_1) of process target values and bending stress (X_2) for a particular product from a lumber manufacturing plant. Thus, the reference mean vector and variance/covariance matrix are $\mu_0 = \begin{bmatrix} 265 \\ 470 \end{bmatrix}$, and

$\sum_0 = \begin{bmatrix} 100 & 66 \\ 66 & 121 \end{bmatrix}$, respectively, i.e., μ_{X_1}=265, μ_{X_2}=470, $\sigma_{X_1} = 10$, $\sigma_{X_2} = 11$ and $\sigma_{X_1X_2} = 66$ (the covariance between X_1 and X_2).

As shown in Table 1, nine possible categories of bivariate patterns were considered to represent the bivariate process variation. Except for the normal pattern (i.e., BVP0), each sample for the other eight BVPs (i.e., BVP1–BVP8) starts from the normal process condition, and enters the abnormal condition at a certain time point. The window size 12 for the ConvNets model is considered in this simulation case. Six distinctive types of shifts (i.e., $\delta = 0.5\sigma$, 1.0σ, 1.5σ, 2.0σ, 2.5σ and 3.0σ) associated with the l-th process variable are considered, covering evenly the range of the possible mean shifts occurring in processes.

Table 1. Parameters for bivariate patterns

Pattern category	Mean shift	Train data	Test data
BVP0 (0, 0)	Both variables X_1 and X_2 remain in-control	3500	1500
BVP1 (1, 0)	X_1 shifted upwards, X_2 remains in-control	3500	1500
BVP2 (0, 1)	X_2 shifted upwards, X_1 remains in-control	3500	1500
BVP3 (1, 1)	Both variables X_1 and X_2 shifted upwards	175,000	7500
BVP4 (−1, 0)	X_1 shifted downwards, X_2 remains in-control	3500	1500
BVP5 (0, −1)	X_2 shifted downwards, while X_1 remains in-control	3500	1500
BVP6 (−1, 1)	X_1 shifted downwards, X_2 shifted upwards	175,000	7500
BVP7 (1, −1)	X_2 shifted downwards, X_1 shifted upwards	175,000	7500
BVP8 (−1, −1)	Both variables X_1 and X_2 shifted downwards	175,000	7500

A summary of the accuracy of each classifier is displayed in Table 2. We observe that both ConvNets are consistently more accurate than both SVM_L and SVM_G. More important, we see that ConvNetsSVM consistently improves the accuracy of the ConvNets. Therefore, our proposed hybrid system is shown to be a better classifier than both the original individual systems. In our experiments, the test accuracy of ConvNetsSVM is 96.82%, outperforms the other three hybrid ConvNets. This comparison further demonstrates that ConvNetsSVM has competitive performance.

Table 2. Performance comparison

Classifier	ACC (%)	Classifier	ACC (%)
BPN	88.62	5-layer ConvNet	94.39
SVM_L	83.85	7-layer ConvNet	95.22
SVM_G	88.25	ConvNetSVM	**96.82**
MLP	87.18	ConvNetMLP	96.01
RF	86.81	ConvNetRF	94.39
KNN	87.74	ConvNetKNN	96.04

Figure 3 further presents the combined confusion matrix for the recognition results of each classifier for the nine BVPs. The diagonal elements in this matrix are the recognition rates of each pattern. It is clear that ConvNetsSVM shows pool performance for recognizing BVP0 and BVP5. Intuitively, most recognition errors result from misclassifying a BVP5 pattern into BVP8 pattern due to the similarity of their pattern features.

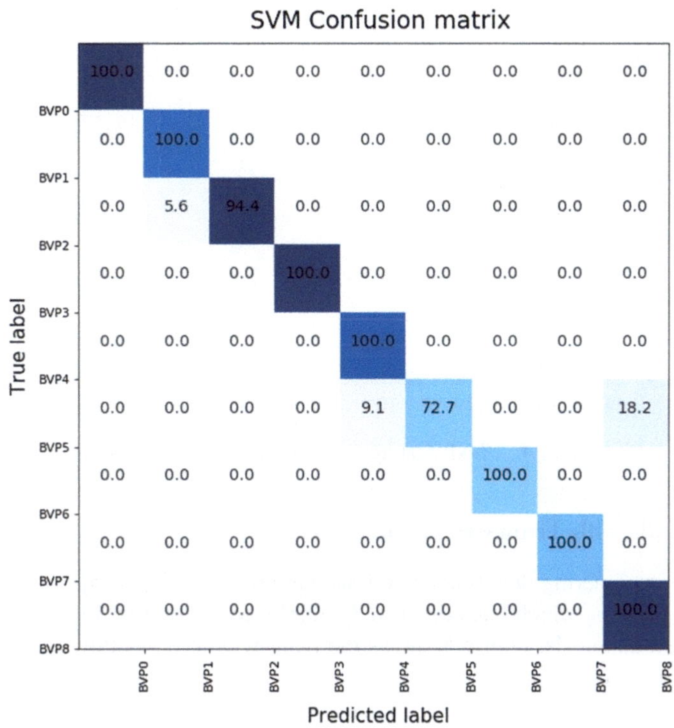

Fig. 3. Confusion matrix of the accuracy for ConvNetsSVM

Data visualization and dimension reduction is carried out by minimizing the Kullback-Leibler deviation between a data distribution and an embedded distribution [14]. The 2048-d features were transform them into a 2-d feature using t-distributed stochastic neighbor embedding (t-SNE), which is easy to visualize. Note that t-SNE is used as an informative step. Figure 4 shows the t-SNE features visualization for ConvNetsSVM. It notes that the same color (or label) points are mostly clustered together. Therefore, these visualization results indicate that the proposed model provides the powerful ability in adaptively feature learning, there is a high chance that we could use the features to train a classifier with high accuracy.

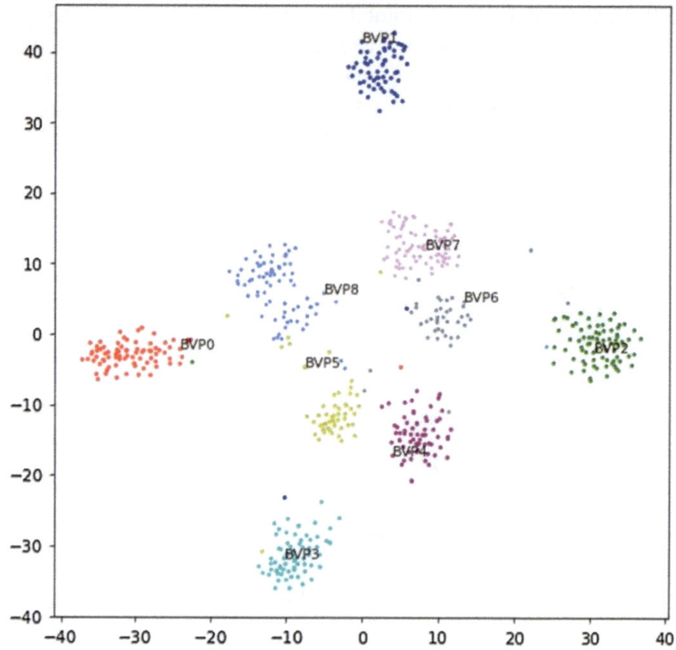

Fig. 4. t-SNE features visualization for ConvNets

3.2 Example with Tennessee Eastman Process

In chemical engineering, the Tennessee Eastman Process (TEP) [15] is a representative benchmark process, which has been widely employed to confirm the efficiency of fault detection and diagnosis. Figure 5 shows the plant-wide process flowsheet of TEP that consists of five major unit operations. The TEP contains 51 process variables, i.e., 22 continuous, 12 manipulated, and 18 compositions sampled less frequently. There are 21 types of known process faults that could happen in this process [16]. Faults 1 through faults 7 are related to a step change in a process variable. Faults 8 through faults 12 are related to an increase of several process variables variability. Faults 13 denotes a slow drift in the reaction kinetics. Faults 14, faults 15, and faults 21 are related to sticking valves.

To investigate the effectiveness of the ConvNetsSVM model for recognizing multiple fault classes that are overlapping, the eight faults 4, 5, 9, 11, 14, 15, 16 and 21 are considered in this study (labeled as TEP1, TEP2, ..., TEP8). The eight faults represent the complicated process data with overlapping distributions. For each fault, the training and testing data consist of 768 and 192 measurements, respectively. Each sample contains 52 process variables, however, only the process variable 9 (reactor temperature) and 51 (reactor cooling water valve position) are effective to distinguish these classes. Through the two process variables, Fault 4 and fault 9 data can be separated, but they overlap with fault 5 and 11. Therefore, the eight faults are misclassifying with high overlapping data.

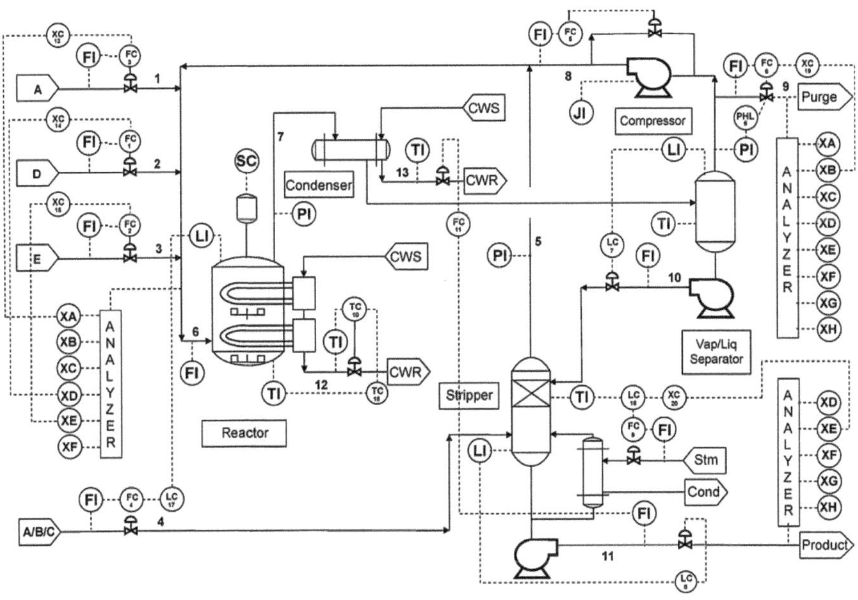

Fig. 5. Flow sheet of Tennessee Eastman process

Table 3 presents the comparisons of the accuracy for each classifier. The results indicate that ConvNetsSVM shows the best performance among all these classifiers. The other three hybrid CNNs show the better performance than that of ANNs and SVMs. This comparison further demonstrates that proposed ConvNets is very effective feature learning for MPPR in multivariate manufacturing processes.

Table 3. Performance comparison

Classifier	ACC (%)	Classifier	ACC (%)
BPN	87.79	5-layer ConvNet	97.70
SVM_L	86.55	7-layer ConvNet	96.79
SVM_G	85.65	ConvNetSVM	**98.51**
MLP	89.96	ConvNetMLP	96.89
RF	85.78	ConvNetRF	91.68
KNN	88.41	ConvNetKNN	96.21

Figure 6 shows the combined confusion matrix for the classification results of each classifier for the eight TEPs. It is clear that ConvNetsSVM shows worse performance for recognizing TEP3 and TEP7. Intuitively, most recognition errors result from misclassifying a TEP5 pattern into TEP8 pattern due to the similarity of their pattern features.

Visualization of extracted features for ConvNets is presented in Fig. 7. The feature points of each pattern are cluster together, and the features are divisible. It notes that the

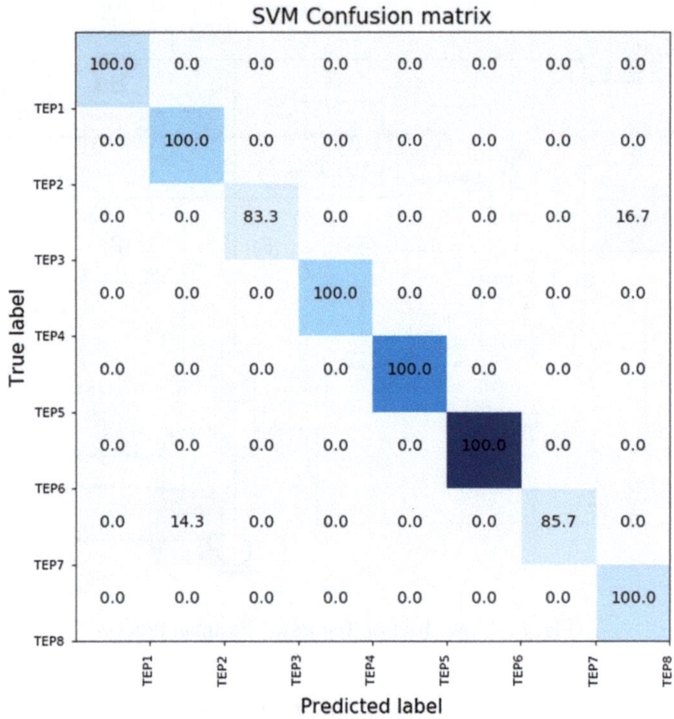

Fig. 6. Confusion matrix of the accuracy for ConvNetsSVM

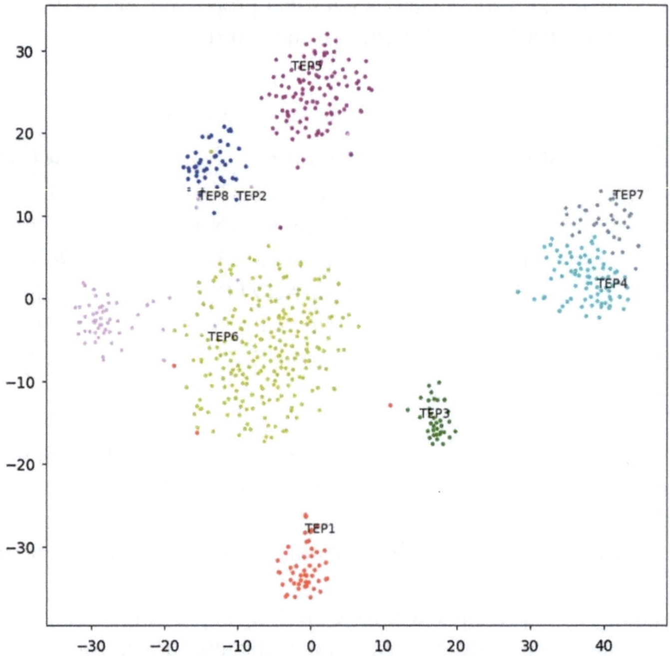

Fig. 7. t-SNE features visualization for ConvNets

proposed ConvNets provides the powerful ability in adaptively feature learning, there is a high chance that we could use the features to train a classifier with high accuracy.

4 Conclusion

MPPR is important tool for faults process detection to maintain efficient multistage process control and monitoring. In this paper, as the latest research result in the artificial intelligent field, CNN is first used as an effective feature learning. Instead of inputting process vectors into a classifier, the images are taken as input of the proposed ConvNets. The features are extracted inherently within the pre-trained ConvNets, which can learn important feature representations for a wide range of images, and provides a simple and effective procedure and achieves a good classification performance. We have shown that a hybrid system composed of an SVM classifier trained on high-level features extracted from a CNN improves consistently improves significantly the test accuracy when compared with both original systems.

Acknowledgements. This research was supported by the National Natural Science Foundation of China (No. 51375290, 71777173).

References

1. H. Hotelling, Multivariate quality control—illustrated by the air testing of sample bombsights, in *Techniques of Statistical Analysis*, ed. by C. Eisenhart, M.W. Hastay, W.A. Wallis (McGraw-Hill, NY, 1947), pp. 11–184
2. R.B. Crosier, Multivariate generalizations of cumulative sum quality control schemes. Technometrics **30**(3), 291–303 (1988)
3. C.A. Lowry, W.H. Woodall, C.W. Champ, S.E. Rigdon, A multivariate exponentially weighted moving average control chart. Technometrics **34**, 46–53 (1992)
4. R.S. Guh, Y.R. Shiue, An effective application of decision tree learning for on-line detection of mean shifts in multivariate control charts. Comput. Ind. Eng. **55**(2), 475–493 (2008)
5. R.S. Guh, On-line identification and quantification of mean shifts in bivariate processes using a neural network-based approach. Qual. Reliab. Eng. Int. **23**, 367–385 (2007)
6. S.T.A. Niaki, B. Abbasi, Fault diagnosis in multivariate control charts using artificial neural networks. Int. Qual. Reliab. Eng. **21**, 825–840 (2005)
7. C.S. Cheng, H.P. Cheng, Identifying the source of variance shifts in the multivariate process using neural networks and support vector machines. Expert Syst. Appl. **35**, 198–206 (2008)
8. Y. LeCun, Y. Bengio, G. Hinton, Deep learning. Nat. **521**(7553), 436–444 (2015)
9. A. Krizhevsky, I. Sutskever, G.E. Hinton, ImageNet classification with deep convolutional neural networks, in *Advances in Neural Information Processing Systems* (2012)
10. X. Han, K. Rasul, R. Vollgraf, Fashion-mnist: a novel image dataset for benchmarking machine learning algorithms, arXiv:1708.07747 (2017)
11. A.F. Agarap, A neural network architecture combining gated recurrent unit (GRU) and support vector machine (SVM) for intrusion detection in network traffic data, arXiv:1709.03082 (2017)
12. M. Abadi et al., Tensorflow: large-scale machine learning on heterogeneous distributed systems, arXiv:1603.04467 (2016)

13. F. Pedregosa et al., Scikit-learn: machine learning in Python. J. Mach. Learn. Res. **12**, 2825–2830 (2011)
14. L. Maaten, G. Hinton, Visualizing data using t-SNE. J. Mach. Learn. Res. **9**, 2579–2605 (2008)
15. J.J. Down, E.F. Vogel, A plant-wide industrial process control problem. Comput. Chem. Eng. **17**(3), 245–255 (1993)
16. L.H. Chiang, E.L. Russell, R.D. Braatz, *Fault Detection and Diagnosis in Industrial Systems, Advanced Textbooks in Control and Signal Processing* (Springer, London, Great Britain, 2001)

The Maintenance Evaluation Method Based on Virtual Reality

Ling-chen Zhou(✉), Qing Xue, Jia-wei Sun, Shu-lin Liu, Xiao-yue Xie, and Qian Zhang

School of Mechanical Engineering, Beijing Institute of Technology, Beijing 100081, People's Republic of China
{1360829271,895984926}@qq.com, xueqing@bit.edu.cn, {lingchen9529,zhangqian_bit}@126.com, 13121776505@163.com

Abstract. For the determination of armored vehicle repair time, the traditional method is usually to conduct maintenance experiments on the actual vehicle or full-size model to determine the maintenance time, which will lead to a long cycle and high cost for determining the maintenance time. With the development of virtual maintenance technology, it is possible to make use of virtual maintenance to indirectly determine the maintenance time. This paper cooperates with a factory in Hunan to experiment with the repair of armored track shoes. According to the experimental requirements, the comparative analysis method was used to test the actual time of the track shoe repair and the maintenance time in the virtual environment. And analyze the maintenance time under two kinds of environments, find out the relationship between them, and provide certain help for determining the maintenance time of armored vehicles.

Keywords: Comparative analysis · Repair time · Virtual reality technology

1 Introduction

At present, the maintenance time for armored vehicles is usually determined by conducting maintenance experiments on actual vehicles [1]. To a certain extent, it will consume a lot of manpower, material resources, financial resources, and time. In order to predict the repair time of armored vehicles conveniently, this paper uses virtual maintenance technology to collect the repair time in virtual environment and in reality of replacing track shoes [2], and through comparative research, the relationship between maintenance time in virtual and in reality environment can be obtained, realize the purpose of calculating the actual repair time based on the virtual repair time, make the maintenance of armored vehicle more easy.

G. Q. Huang et al. (eds.), *Proceeding of the 24th International Conference on Industrial Engineering and Engineering Management 2018*,
https://doi.org/10.1007/978-981-13-3402-3_23

2 Methodology

This paper uses the comparative study method. So for the replacement of the track shoe operation, it is necessary to simultaneously collect the actual maintenance time of the armored vehicle and the maintenance time in the virtual environment [3]. To ensure the reliability of experimental data, repeat experiments are required in this experiment.

During the experiment of collecting the maintenance time in the actual environment, the object of study is the operation time of 6 workers, and it was divided into 3 groups, and do 2 experiments, and 6 times totally [4]; what's more, during the experiment of collecting the maintenance time in the reality environment, the object of study is the operation time of 6 workers, and do 2 experiments every worker, and 18 times totally [5].

2.1 Actual Maintenance Time Collection

(1) Purpose: Collect the time of replacing track shoe operations in reality
(2) Laboratory apparatus: Armored vehicle, wrench, Afterburner, Hammer and Stopwatch, tape measure etc.
(3) Members: In order to analyze the relationship between real time and virtual time of track shoe repair fully, the workers with different ages and working hours are needed in this experiment, and grouped according to the age and length of service features to facilitate experimental analysis. The information of experiment members are shown in Table 1.

Table 1. Actual repair personnel information

Operator information			
Group	Experience category	Name	Length of service (year)
1	No	Yang hao	4
	Yes	Chen Zhiguang	9
2	No	Rao shu	3
	No	Xiao gan	4
3	Yes	Li can	9
	Yes	Long wu	11

(4) Experiment content: Subjects were in turn operated in accordance with the procedure for replacing the track shoe. The tester used a stopwatch to record the actual working time.

2.2 Collection of Maintenance Time in Virtual Environment

(1) Purpose: Collect the time of replacing track shoe operations in virtual environment.

(2) Laboratory apparatus: HTC VIVE virtual reality helmet, Base stations, 3D handle, PC, Strip, Stopwatch and so on.
(3) Members: In order to maintain the consistency of the experiment, the requirement of experiment members in the virtual environment must be consistent with the actual maintenance experiment. The information of experiment members are shown in Table 2.

Table 2. Virtual repair personnel information

Operator information			
Group	Experience category	Name	Length of service (year)
1	Yes	Xiong wei	25
2	Yes	Li Haineng	10
3	A little	Li Can	9
4	A little	Liu Guang	7
5	No	Wang Tao	5
6	No	Rao Shu	3

(4) Experiment content: At first, the testers conduct VR equipment training for the subjects. The person in charge of the test needs to wear a virtual reality helmet to complete the maintenance work in order according to the maintenance work procedure. The tester records the maintenance operation time of the test and completes the maintenance time table.

3 Results

In order to make reliable experimental results more reliable, the experiment of collecting actual time for the track shoes of the workers was a total of 2 days and was completed with a factory in Hunan [6]. According to the requirements given in the experimental plan, 6 workers were divided into 3 groups, which were performed in turn according to the maintenance procedure. The testers were responsible for recording the time of each step. The maintenance time are shown in Table 3. In order to satisfy the principle of keeping a single variable in the experiment, the maintenance time in the virtual environment is also carried out in the same factory in Hunan [7]. In the virtual environment, 6 workers conducted a total of 18 experiments. According to the requirements of the experimental program, this experiment requires the use of HTC VIVE virtual reality helmets, base stations, 3D handles and other tools combined with VMPro software to create a virtual environment [8]. Workers were required to wear a virtual reality helmet and perform virtual maintenance on the armored vehicle model presented in the VMPro. The testers recorded the virtual maintenance time of workers which are shown in Table 4.

Table 3. The actual time of track shoe repair work

No.	Operation	1	2	3	4	5	6
1	Use a sleeve, ratchet wrench to loosen the fixing bolts	15.2	29.3	25.6	28.1	22.6	23.4
	Use a punch and hand hammer to loosen the big nut	42.9	50.2	34.6	36.2	36.2	35.6
	Hold the worm head and relax the track	85.8	855	71.0	74.2	56.7	50.4
2	Remove the connected two induction gears	137	73.0	96.3	76.0	85.6	24.3
	Loosen the connecting bolts at both ends to separate the stoppers and the chain	258.	201	180	130	166	119
	Remove the track shoes	365	350	241	247	177	182
3	Remove the new track and align it with one end of the track shoe	15.2	20.2	25.3	60.1	26.7	12.3
	Install the connecting ring and limit block one by one	942	99.2	298	148	75.5	100
	Installation inducer	818	488	515	606	440	540
4	Tighten the worm head	40.8	37.3	68.5	28.8	28.3	26.8
	Tighten the big nut and lock the gear sleeve	70.3	127	55.8	80.2	47.7	51.3
	Fitting gaskets, fixing bolts	78.7	147	77.6	126	72.6	92.7

Table 4. The virtual maintenance time of workers

No.	Operation	Repair time					
0	Virtual reality environment	Experiment members numbers					
		1	2	3	4	5	6
1	Loosen the fixing bolt and remove the gasket	58.8	64.7	57.7	46.9	53.7	33.4
		48.3	109	61.3	64.1	52.7	32.0
		37.3	110	45.5	45.8	58.1	49.6
	Loosen large nut and lock gear sleeve	50.3	35.3	41.5	29.3	34.4	26.6
		35.9	26.6	31.0	29.9	23.6	21.0
		24.4	37.1	27.6	34.5	26.0	23.2
	Hold the worm head and relax the track	20.7	35.0	36.5	45.5	19.2	17.6
		40.9	31.2	30.3	37.2	33.3	30.1
		26.7	30.4	30.4	32.4	29.2	39.8
2	Demolition induction gear	40.9	48.3	49.0	28.1	35.0	30.4
		42.0	40.0	48.2	50.0	33.1	30.8
		32.6	32.5	42.1	43.5	39.1	34.7
	Loosen the connecting bolts at both ends to separate the stoppers and the chain	140	260	150	136	226	135
		196	162	164	165	181	355
		142	146	158	145	157	149
	Remove the track shoes	31.2	23.1	98.4	19.2	30.2	15.0
		15.8	18.4	18.9	14.4	22.3	18.9
		16.6	17.3	18.9	17.6	24.7	18.6

(continued)

Table 4. (*continued*)

No.	Operation	Repair time					
0	Virtual reality environment	Experiment members numbers					
		1	2	3	4	5	6
3	Remove the new track shoe and align it with one end of the track shoe	60.5	30.2	62.0	25.7	31.9	39.3
		31.7	21.0	29.2	24.4	19.7	34.7
		32.5	46.1	29.5	23.8	36.1	32.8
	Install the connecting rings, bolts and limit blocks in order	350	686	329	435	319	358
		412	364	450	592	364	419
		423	433	416	527	346	433
	Install induction gear	31.3	35.5	24.8	29.8	32.8	36.3
		23.4	30.5	33.0	38.1	26.8	37.0
		36.8	35.8	28.1	35.9	35.3	17.8
4	Tighten the worm head	27.1	32.2	25.6	23.3	24.3	28.5
		28.3	27.0	28.9	27.8	18.6	25.0
		28.8	29.9	32.0	31.8	20.5	32.1
	Tighten the big nut and lock the gear sleeve	24.1	35.2	25.6	30.8	22.5	25.0
		33.8	34.4	25.7	38.1	25.0	35.0
		27.7	26.0	25.0	32.0	20.3	28.2
	Fitting gaskets, fixing bolts	51.4	49.4	47.4	45.1	65.0	29.8
		25.2	36.8	52.8	42.6	52.0	32.3
		30.5	34.0	32.8	30.6	23.9	38.0

4 Discussion

The purpose of this paper is to find out the relationship between the actual maintenance time and the virtual maintenance time by analyzing the actual maintenance time of the track shoe repair work and the maintenance time in the virtual environment so that the actual repair time can be estimated based on the virtual repair time and reduced the cost of the track shoe repair [9]. This article has collected actual maintenance time and virtual maintenance time through experiments. Since the experimental groupings were grouped according to age and length of service during the experiment, this paper used the length of service as an influencing factor for analysis [10].

(1) The relationship between the actual maintenance time and the virtual maintenance time was studied without considering the working age variable.

Assume that the actual maintenance time obeys the lognormal distribution:

$$\ln y_r \sim N\left(\mu_r, 6r^2\right)$$

Assume that the actual virtual time obeys the lognormal distribution:

$$\ln y_v \sim N\left(\mu_v, 6v^2\right)$$

And

$$6r^2 = 6v^2$$

Through the analysis of actual maintenance and virtual maintenance time, it is can be obtained:

$$\mu_r = 1.0215\mu_v$$

Thus, the relationship between actual maintenance time and virtual maintenance time is:

$$\mu_{实} = e^{0.0215\mu_v} \cdot \mu_{虚}$$

Among them, $\mu_{实}$ is the actual average maintenance time and $\mu_{虚}$ is the average virtual maintenance time.

(2) When considering the working-age variable, analyze the relationship between the virtual repair time and the actual maintenance time.

$$\mu_r = 1.0611\mu_v - 0.0406x$$

And

$$\mu_{实} = e^{0.0611\mu_v - 0.0406x} \cdot \mu_{虚}$$

μ_v is the logarithmic mean of the virtual repair time.

5 Conclusion

In view of the fact that the cost of determining armored vehicles' maintenance time is relatively high, this article uses virtual reality technology to cooperate with a factory in Hunan to perform tracker repair work in real environments and maintenance work in a virtual environment [11], and record the time of related operations. Then use the method of comparative analysis to analyze the time in the two environments and find out the relationship between the two, so as to help determine the maintenance time of armored vehicles [12].

This paper has mainly completed several aspects of work:

(1) Carry out track shoe repair work in real environment and record the relevant operation time. Combined with the reliable track-shoe maintenance process provided by the factory and combined with the factory, the maintenance process is sorted according to the purpose of the experiment and the requirements, and an experimental scheme for repairing the track-shoe operation in a real environment is obtained [13]. For the reliable track shoe repair process, the maintenance

process is sorted according to the purpose and requirements of the experiment, and an experimental scheme for repairing the track shoe operation in a real environment is obtained. The subjects were in strict accordance with the experimental procedure and the testers recorded the relevant data.

(2) Carry out track shoe repair work in a virtual environment and record the relevant operation time. Due to the consistency of the experiment, the operation steps in the virtual environment are corrected by the operation steps in the real environment [14]. And use virtual reality helmets, VMPro and other tools to complete the experiment under the virtual environment, the tester records the relevant time.

(3) Analyze the real time and virtual time of track shoe repairs to find out the relationship between them. The relevant mathematical methods are used to analyze the experimental data [15], and the purpose of determining the real maintenance time according to the maintenance time of the virtual environment is achieved.

Acknowledgements. The author would like to thank my tutor Qing Xue for her careful guidance and thank the workers for their cooperation and support during the experiment.

References

1. Z.B. Zou, J.B. Wu, M. Ma, Virtual maintenance training system of airborne electronic equipment. Appl. Mech. Mater. **437**(437), 861–865 (2013)
2. V. Narayanan, S. Krishna, J.K.J. Jaganathan et al. Virtual reality system with haptic/auditory devices for assembly and maintenance training. Grasp (2006)
3. G.A. Figueroa, Virtual reality training system for maintenance and operation of high-voltage overhead power lines. Virtual Real. **20**(1), 27–40 (2016)
4. MIL-HDBK-470A. *Designing and Developing Maintainable Products and Systems* (Department of Defence Handbook, 1997)
5. N.I. Badler, C.A. Erignac, Y. Liu, Virtual humans for validating maintenance procedures. Commun. Acm **45**(7), 56–63 (2002)
6. A. Kaveh, A.F. Behnam, Design optimization of reinforced concrete 3D structures considering frequency constraints via a charged system search. Sci. Iran. **20**(3), 387–396 (2013)
7. R.T. Bye, A receding horizon genetic algorithm for dynamic resource allocation: a case study on optimal positioning of tugs, 114–125 (2012)
8. R. Vujosevic, J. Ianni, A taxonomy of motion models for simulation and analysis of maintenance tasks (1997)
9. M.D. Bauer, Z. Siddique, D.W. Rosen, Virtual prototyping in simultaneous product/process design for disassembly, in *Rapid Response Manufacturing* (Springer, US, 1998), pp. 141–175
10. B.S. Blanchard, W.J. Fabrycky, *Systems Engineering and Analysis* (Prentice Hall, 2002)
11. S. Mascaro, H.H. Asada, Hand-in-glove human-machine interface and interactive control: task process modeling using dual Petri nets, in *IEEE International Conference on Robotics and Automation, 1998. Proceedings*, vol. 2 (IEEE, 1998), pp. 1289–1295
12. F. Lin, C.J. Su, M.M. Tseng, An agent-based approach to developing intelligent virtual reality-based training systems, in *Proceedings of IEEE International Conference on Tools with Artificial Intelligence, 1999* (IEEE, 1999), pp. 253–260

13. K. Harada, E. Nakamae, Sampling point setting on cubic splines for computer animation. Vis. Comput. **5**(1–2), 14–21 (1989)
14. P.M. Isaacs, M.F. Cohen, Controlling dynamic simulation with kinematic constraints. Acm Siggraph Comput. Graph. **21**(4), 215–224 (1987)
15. W.W. Armstrong, M. Green, R. Lake, *Near-Real-Time Control of Human Figure Models* (IEEE Computer Society Press, 1987)

An IoT Based Solution for Cyber-Physical Fusion in Shop-Floor

Ji-hong Yan(✉), Zi-min Fu, Ming-yang Zhang, and Yan-ling Zhang

Department of Industrial Engineering, Harbin Institute of Technology, Harbin, China
jyan@hit.edu.com

Abstract. Smart manufacturing plays an important role in the transformation and upgrading of manufacturing industry and the China Manufacturing 2025 Strategy. And the cyber-physical fusion is a critical process to achieve the interconnection and interoperability between the cyber and physical world of manufacturing. Combing with IoT, a solution to realize cyber-physical fusion for heterogeneous objects in shop-floor is presented. The architecture and key technologies of the solution are investigated. Then a remote management platform is implemented. Logistics tracking, production visualization, equipment interconnection and remote operation are realized and can be remotely accessed by portable terminals via Internet. The proposed solution is verified by the actual machining and assembly workshop in a laboratory at Harbin Institute of Technology.

Keywords: Cyber model · Cyber-physical fusion · Cyber physical systems Equipment interconnection · Internet of things · Radio frequency identification

1 Introduction

With strong integrity and comprehensiveness, intelligent manufacturing is now one of the focusing points of advanced manufacturing mode [1], as well as the critical measure for the transformation and upgrading of Chinese manufacturing industry, one of the key contents of China's "13th Five-Year plan" [2]. Therefore, the research on combining advanced manufacturing with information technologies has grown significantly.

Cyber Physical Systems (CPS) is a new paradigm of manufacturing proposed to achieve the interaction between the physical and the computing processes by closed loop feedback [3, 4]. And the realization of CPS depends on the interconnection and interoperability between physical and cyber world. However, the cyber-physical fusion for shop-floor is still under development [5]. On the other hand, promoted by the growing application and reducing cost of sensors and wireless network technology, Internet of things (IoT) technologies such as RFID [6] and ZigBee [7] are increasingly applied in shop-floor to meet the needs of real time information collection, item tracking and production monitoring [8].

Aiming at the interaction of heterogeneous objects in cyber physical shop-floor system, this paper attempts to provide an IoT based solution for cyber-physical fusion in shop-floor. The architecture of the IoT system was proposed based on the definition

G. Q. Huang et al. (eds.), *Proceeding of the 24th International Conference on Industrial Engineering and Engineering Management 2018*,
https://doi.org/10.1007/978-981-13-3402-3_24

of cyber models of machining and assembly shop-floor. The key technologies and the methods of applying them are investigated. At last the system was implemented in an actual shop-floor.

2 Design of the IoT System

2.1 Cyber Models for Shop-Floor

The accuracy of cyber model reflecting physical world is decided by the comprehensiveness and typicality of physical characteristics chosen while building cyber model [9]. The feature analysis and classification for all manufacturing equipment in shop-floor is needed before the design of the IoT system because of the diversity and heterogeneity of equipment in workshop. And we also need to define a variety of Data Acquisition (DAQ) solutions for different kinds of objects so as to realize the fusion of physical and cyber efficiently and precisely. There are mainly two stages of processing in a typical shop-floor, namely, machining and assembly. Cyber models for machining and assembly are constructed through the definition of their input and output which is able to provide supports for the design of the IoT and DAQ systems.

2.1.1 Cyber Model for Machining

Machining is defined as the series of actions in the specific external working environment that are taken to change blanks into components or production according to ordered process scheme utilizing the status and ability of manufacturing equipment [10]. The input of the processing is a quadruple, (Machine info, Environment, Order info, Part status), while the output of machining could be defined as a triple, (Machine status, Quality, Efficiency). The detailed description of machining cyber model is illustrated in Fig. 1.

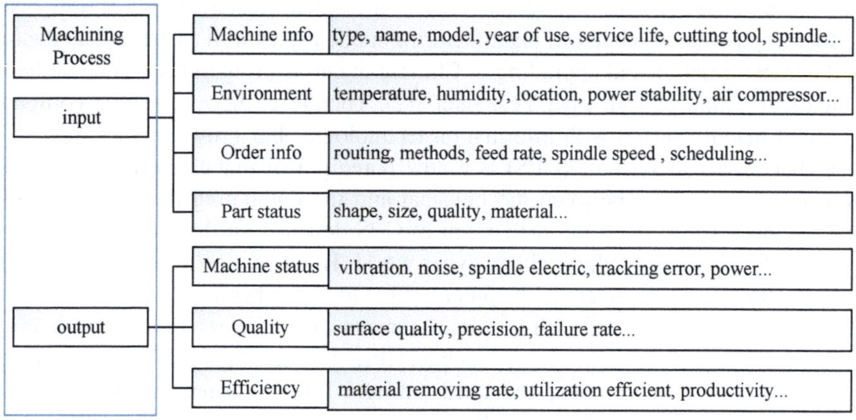

Fig. 1. Cyber model for machining

Through data acquisition and analysis of physical equipment based on machining cyber model, the relationship between status of machine tools and product quality can be discovered. This could provide guides for the design, process and maintenance of machines. And also it could provide evidences for machining technology of components, making product scheduling and analysis of power cost.

2.1.2 Cyber Model for Assembly

There are two kinds of key entities in an assembly shop-floor: warehouse and workstation. The warehouse is the starting and ending point of assembling work. And the workstation is where the components or parts assembled. They form the closed loop of assembly process. As shown in Fig. 2, the input info of warehouse could be represented as a tuple (Order, Stagey), which includes mission orders and inventory strategy. The output info is warehouse status including inventory load, cost, reliability and the like. We get these data through acquiring inventory volume timely. The input info of workstation is a triple, (Component status, Process scheme, Worker Status). And the output could be presented as (Assembly quality, Assembly efficiency). Through real time acquisition of relating characters in warehouse and workstation, the cyber model of the entire assembly line could be built, so the whole assembly process monitoring and analysis could be achieved.

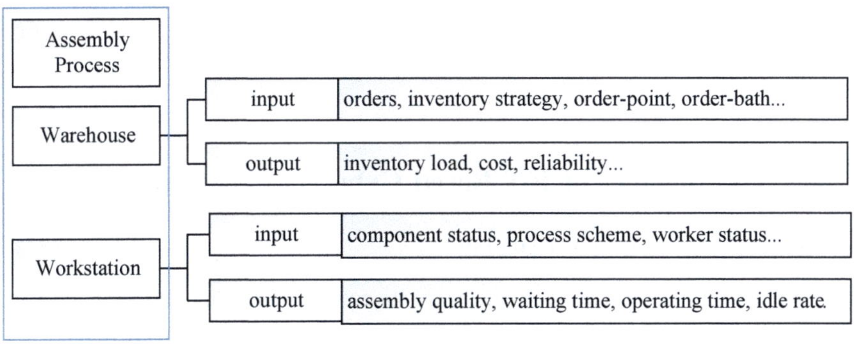

Fig. 2. Cyber model for assembly

2.2 IoT Based Architecture for Cyber-Physical Fusion

Combining the features and demand of machining and assembly shop-floor, the IoT based architecture for cyber-physical fusion is constructed as shown in Fig. 3.

The architecture consists of three layers. The bottom is the physical layer. The logistics information of assembly shops is acquired through the RFID system, while the equipment status data is acquired by the sensors and virtual instruments. The middle layer is the cyber layer, the database structure and data tables are designed and deployed in this layer, and the data are stored in the database server in real time through the LAN. The top layer is the application layer. In this layer, the database is used as medium, and the foreground applications are developed to realize remote monitoring. The three-layer system architecture enables the interconnection and communication

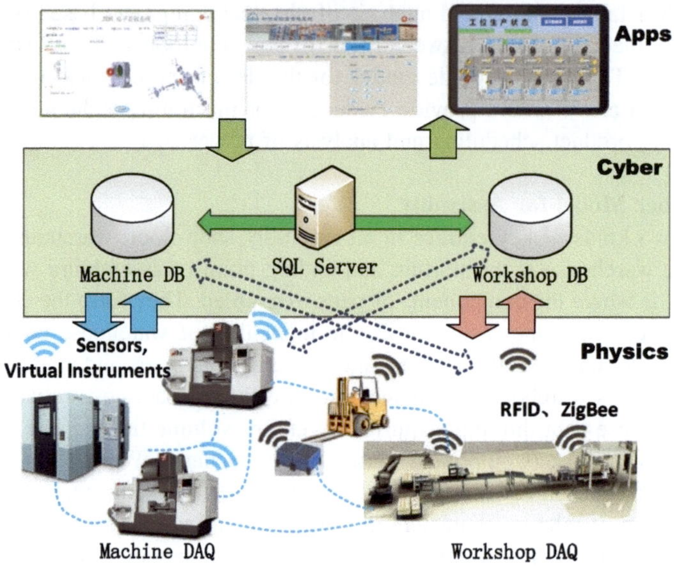

Fig. 3. IoT based architecture for cyber-physical fusion

between heterogeneous production facilities and resources, forming an IoT system that things cooperate with each other which can support efficient and personalized production.

3 Key Technologies of IoT Based Cyber-Physical Fusion

3.1 RFID Based Logistics Tracking

Radio Frequency Identification RFID is a noncontact automatic identification technology, which signals through radio frequency automatic identification and access to relevant target data, no need for manual intervention to identify job [11]. RFID systems consist of Radio Frequency (RF) tags, or transponders, and RF tag readers, or transceivers [12]. When RF tag readers are installed on each workstation, they can automatically identify and query information of products on the assembly line correspondingly. Therefore, it is convenient for workstation operation and production management.

RFID tags should be plastered on containers and the body of reducers, in order to identify and track products or components. The process of the data acquisition and transmission can be described as follows: Step 1, after the order arrives, the system sends task instructions to sorting rack. Workers get all the necessary components from the sorting rack and then put them into one container. The counter in the sorting shelf will send the number of components taken away to the server automatically through the ZigBee node. Step 2, AGV can identify the RFID tags of containers and distribute components to the corresponding stations. Step 3, when the product comes into a workstation, the RFID reader will identify the product, send the product ID to the server and display the operation animation of the process to assist operators. Step 4, the workstation sends out the instructions of the product processing, and the AGV carries the end product with the unique RFID into the warehouse after the quality inspection.

3.2 ZigBee Based Interaction of Logistics Equipment

ZigBee, based on a Wireless Sensor Network (WSN) standard, is a wireless short range transmission technology. It features a cost effective, low power and multihop wireless communication in a selforganized mesh network [13].

Logistics equipment in a typical shop-floor includes storage racks, conveyor belts, manipulators and handling equipment. In order to build the ZigBee network of logistics equipment, ZigBee nodes should be installed into PLCs of the equipment while the ZigBee server should be connected to the server computer. We use different methods to acquire data from different equipment, as described next.

(1) Manipulator. The ZigBee node is installed in the PLC of the manipulator with DAQ channels connected to the registers of PLC. The monitoring and control of the manipulator are realized by reading and writing the registers.
(2) Conveyor belt. Same as the manipulator, the ZigBee node is installed in PLC. What's more, in order to monitoring the materials conveyed by the conveyor, a photosensitive sensor is installed and connected to the ZigBee node.
(3) Storage rack. Counters are installed in the rack to collect the inventory data. And ZigBee nodes are connected to the counters in order to send the data to the server computer.

With these solutions, the ZigBee network can meet the needs of real time inventory monitoring, manipulator control, conveyor belt load monitoring and so on.

3.3 Virtual Instrument Based Machine Tool Monitoring

Virtual instrument (VM) technology is the use of high performance modular hardware, combined with efficient and flexible software to complete the application of various testing, measurement and automation applications. The basic thought of using VM is to realize measurement and control functions through software and panels in computer [14]. And LabVIEW is an integrated development environment developed by NI for industrial testing applications. It offers a graphical programming approach that empowers users to visualize every aspect of applications, including hardware configuration, data measurement, and debugging [15].

The acquisition, communication and storage of the state information of machining equipment are carried out by the LabVIEW program based on the C/S architecture. The C/S architecture realizes the separation between data acquisition terminals and data processing terminal, so that the acquisition terminals can use adequate CPU resources for acquisition, so as to reduce the cost of the acquisition terminals and achieve multisource data acquisition. The signals that we need to acquire from the processing equipment are vibration, noise and power. The DAQ system can realize the real time acquisition, processing, analysis and storage of the state data of the equipment.

Figure 4 shows the framework of the system. The machine tool vibration and noise signals are collected by sensors, then processed by PXI, finally transmitted to the computer through network. The power signals are acquired by the power meter, then transmitted to the computer through RS232. Finally the computer would process and synchronize the data from local PC to database.

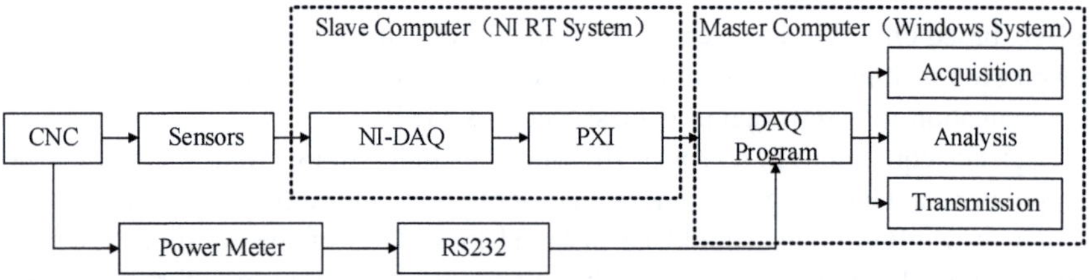

Fig. 4. Architecture of data acquisition system for machine tools

The overall design of the software framework consists of four parts: data acquisition, data receiving, data processing and database interaction, and the software is programmed using LabVIEW. The data acquisition part is the function of the server and the other three parts are the functional modules of the client. Figure 5 shows the operation workflow. The module of vibration DAQ and power DAQ collects the data. The vibration signals should be transmitted to the client through TCP by sending and receiving module. Then the computer extracts features and stores them to database. Meanwhile, the power signals are collected and stored directly by the computer. The software can be used to collect data for experiments, as well as real time online monitoring.

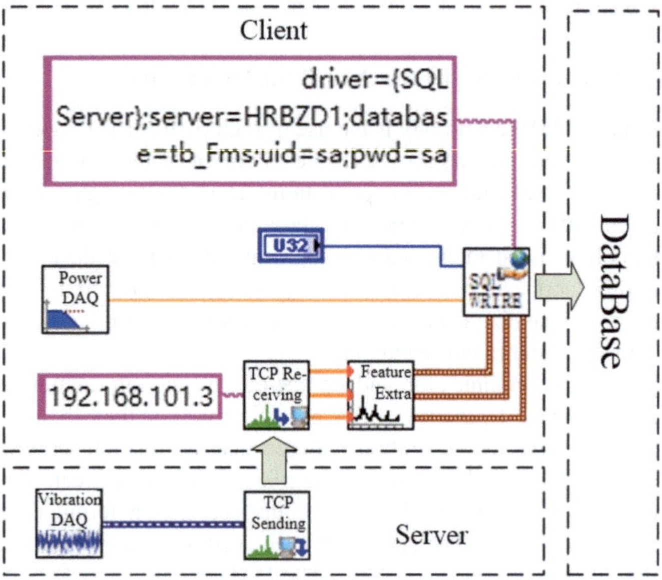

Fig. 5. Framework of the data acquisition software

4 System Implementation

4.1 Implementation of IoT

A prototype system based on the presented architecture of IoT based physical-fusion system for shop-floor was developed in order to verify the monitoring and management capability of the system for multitype and multilocation equipment involved in the typical production workshop. The structure of the IoT system implemented was presented in Fig. 6. It has two workshops connected to the IoT network over the Internet, one is an assembly workshop and the other one is a machining workshop. The assembly workshop includes an assembly line, manipulators, AGVs, storage racks and so on, while the machining workshop includes machine tools for subtractive manufacturing and 3D printers for additive manufacturing. The RFID readers, ZigBee nodes and sensors are installed in the corresponding equipment and connected to the Internet through LAN in the workshops. The collected data will be gathered to the database in the assembly workshop.

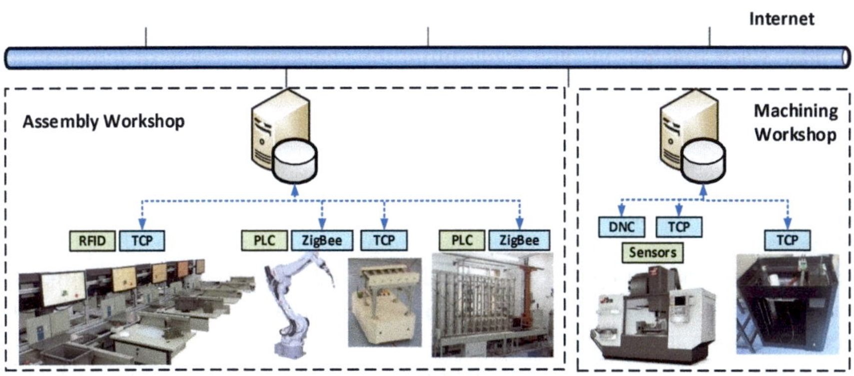

Fig. 6. Structure of the prototype system

4.2 Architecture and Mechanism of the Management Platform

Tasks in the assembly shop-floor are procuring or getting the components according to order lists, generating production plan, organizing production and delivering timely. The expected applications for assembly shop-floor are intelligent scheduling, assembly line balancing, motion analysis and so on. On the other hand, the machine tool needs to process the blanks into components according to MRP. The application of energy consumption analysis, equipment maintenance and productivity prediction need to be developed.

In order to meet the requirements of production and scientific research, a management platform was developed on the basis of the IoT system. It includes three modules: assembly line management, machine tool management and system management, which is illustrated in Fig. 7. In the module of the first one, users could get

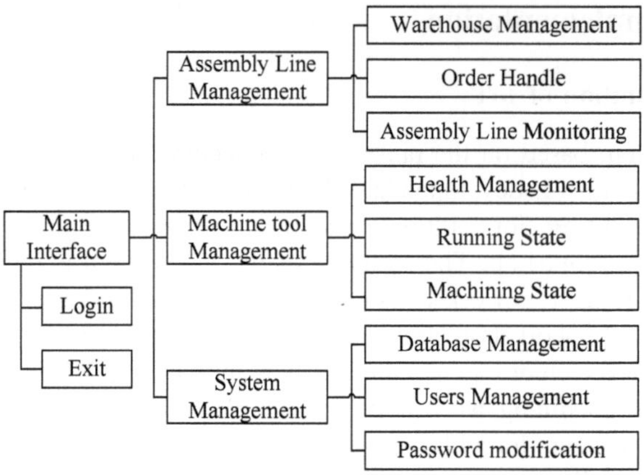

Fig. 7. Function structure diagram of the management platform

the state of assembly line through visualization interface, as well as monitor and control the logistics in real time. For the second, the health state and working state of machines are monitored and these are the fundamental for equipment maintenance and production plan. Each function module could fulfill the check and update of information in real time correspondingly.

4.3 Remote Monitoring and Control

The tablets with Windows10 operation system are chosen as the portable terminals to run the management platform. Once the equipment is networking, the platform is accessible. With accordingly accounts and passwords, users can access to the main interface. The menu shows the guidance for all sections and by clicking on the project, users can get into the selected module.

The interface of assembly line monitoring is shown in Fig. 8. The workstations displayed in the interface are workstation number 1–10 in the workshop. Each station has three parameters, workstation ID, state and number of finished products, showing the real time status of the assembly line. On the top left corner, it shows the number of Work-In-Progress (WIP) in the assembly line and the cycle time calculated according to the interval of two kinds of reducers which can be used to evaluate the efficiency. If an end product is completed in the No. 10 workstation, the system will assign a transport task to the AGV to deliver the product to warehouse.

In the equipment status monitoring module, there are three tabs which are health management, running state and machining state. Figure 9 shows the interface of the health management and it can be divided into four sections, inquiry section, basic information section, detailed data section and tool offset calculation section. In the inquiry area, users search for the data of machines based on equipment type, its location, ID and so on. When one machine tool in the equipment list is selected, sections of basic information and detailed data will show the information correspondingly. One of the main functions of running state interface is plotting the power curve of machines. When the button named Show Figure is clicked, the data of power

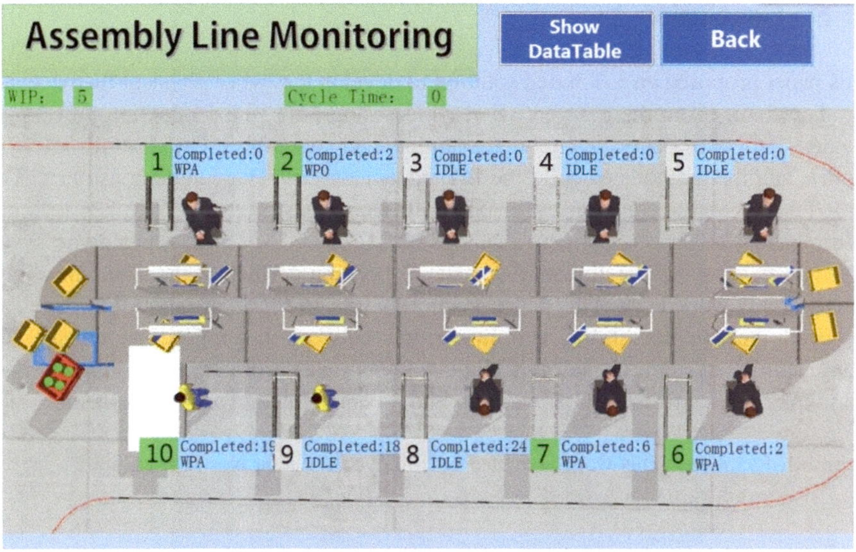

Fig. 8. Interface of assembly line monitoring

in the database will be extracted and the power curve will be plotted, so the power of machine could be monitored remotely in real time.

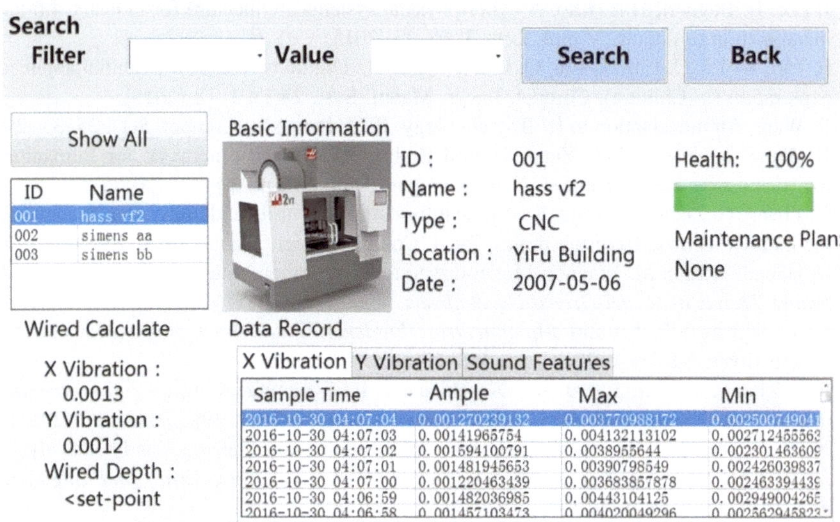

Fig. 9. Interface of equipment health status

5 5 Conclusion

This paper proposed an IoT based solution to develop the cyber-physical fusion system for shop-floor. From the perspective of typical shop-floor, the production process was divided into two stages, namely, machining and assembly. The features of cyber models and the methods to build the IoT system for each stage is demonstrated. Then an IoT based architecture for cyber-physical fusion was proposed. The key technologies related and their implement methods are discussed. At last, a prototype system based on the proposed architecture was developed and verified in the actual machining and assembly workshops. This paper practiced CPS and IoT in the shop-floor level, and achieved the interoperability and interaction between heterogeneous objects in a manufacturing system, providing technique supports to promote the production mode combining with information technology and advanced manufacturing industry.

Acknowledgements. This work is funded by NSFC-NSF (Grant No. 51561125002).

References

1. B. Chu, W.J. Tolone, J. Long, W.J. Tolone, R. Wilhelm, Y. Peng et al., Towards intelligent integrated manufacturing planning-execution. Int. J. Agil. Manuf. (in press)
2. J. Fu, Development status and trend of intelligent manufacturing equipment (in Chinese). J. Mech. Electr. Eng. **31**(08), 959–962 (2014)
3. E.A. Lee, Cyber physical systems: design challenges, in *11th IEEE Symposium on Object Oriented Real-Time Distributed Computing (ISORC)* (IEEE Computer Society, Orlando, FL, USA, 2008), pp. 363–369
4. J. Lee, B. Bagheri, H.A. Kao, A cyber-physical systems architecture for industry 4.0-based manufacturing systems. Manuf. Lett. **3**, 18–23 (2015)
5. F. Tao, W. Liu, J. Liu, X. Liu, Q. Liu, T. Qu et al., Digital twin and its potential application exploration (in Chinese). Comput. Integr. Manuf. Syst. **24**(1), 1–18 (2018)
6. R. Want, An introduction to RFID technology. IEEE Pervasive Comput. **5**(1), 25–33 (2006)
7. H. Yang, L. Yang, S.H. Yang, Hybrid Zigbee RFID sensor network for humanitarian logistics centre management. J. Netw. Comput. Appl. **34**(3), 938–948 (2011)
8. Y. Zhang, G. Zhang, J. Wang, Real-time information capturing and integration framework of the internet of manufacturing things. Int. J. Comput. Integr. Manuf. **28**(8), 811–822 (2015)
9. C. Hein, T. Ritter, M. Wagner, Model-driven tool integration with ModelBus, in *Workshop Future Trends of Model-Driven Development* (2009), pp. 50–52
10. S. Kalpakjian, S.R. Schmid, *Manufacturing Engineering and Technology* (Pearson, Upper Saddle River, NJ, USA, 2014), pp. 468–476
11. R.Y. Zhong, Q. Dai, T. Qu, RFID-enabled real-time manufacturing execution system for mass-customization production. Robot. Comput.-Integr. Manuf. **29**(2), 283–292 (2013)
12. X. Wu, Y. Wang, J. Bai, H. Wang, C. Chu, RFID application challenges and risk analysis, in *2010 IEEE 17th International Conference on Industrial Engineering and Engineering Management (IE&EM)* (IEEE, 2010), pp. 1086–1090
13. H. Christian, C. Weigand, J. Bernhard, Wireless medical sensor network with ZigBee, in *Proceedings of the 5th WSEAS International Conference on Electronics, Hardware, Wireless and Optical Communications*, Madrid, Spain, 15–17 February 2006, pp. 12–15

14. I. Orovic, M. Orlandic, S. Stankovic, Z. Uskokovic, A virtual instrument for time-frequency analysis of signals with highly nonstationary instantaneous frequency. IEEE Trans. Instrum. Meas. **60**(3), 791–803 (2011)
15. C. Elliott, V. Vijayakumar, W. Zink, R. Hansen, National Instruments LabVIEW: A programming environment for laboratory automation and measurement. J. Assoc. Lab. Autom. **12**(1), 17–24 (2007)

Research on Simulation Analysis of SRM and RGV Collaborative Operation

Wei-ping Huang[1,2(✉)] and Hua Chen[3]

[1] School of Logistics and Trade, Eurasia University, Xi'an, China
weip_h@163.com
[2] School of Management, Xi'an Jiao Tong University, Xi'an, China
[3] School of Management, Xi'an University of Science and Technology, Xi'an, China

Abstract. AS/RS becomes more and more important in logistics and manufacturing sectors in the new era. SRM and RGV are the critical subsystems in AS/RS. We show a procedure in this paper which uses simulation method to study different kinds of scheduling policy in a real world AS/RS including SRM and RGV. This procedure is demonstrated effective by a real world example. The objective warehouse is studied intensively and modeled including its entities and logic by using professional simulation software. Then two different scheduling policies are implemented in the model. The model is also simulated under many scenarios which are provided by different outbound workload to compare the effect of each policy. Future possible study is also mentioned at the end.

Keywords: AS/RS · Simulation · Scheduling policy

1 Introduction

AS/RS becomes more and more important in logistics and manufacturing sectors in the new era. Automated warehouse (automated storage/retrieval system, AS/RS) generally includes SRMs (storage and retrieve machine) and peripheral transportation systems (such as rail guided vehicle, RGV). The SRMs transport goods between the I/O points and the rack, and RGVs moves goods between the platforms (or the forklift loading platform) to the I/O points. In addition, the stacker system with single lane is rare in the general real automated warehouse, but SRM subsystem with multiple lanes and the RGV system with multiple vehicles are common in AS/RS. Since the two subsystems work together, each other is highly coupled and mutual influence, the traditional analytical method has been difficult to deal with this type of AS/RS [1]. Traditionally, the classical travel time model is the popular method for performance evaluation of SRM subsystem [2, 3], but if there is additional RGV subsystem, this analytical will been challenged. Therefore, simulation methods are used to analyze the design scheme and performance in the design phase for the compound AS/RS. When considering the stochastic of the system and the operation characteristics of the cache queues and space occupation, simulation method is almost the only choice [4]. Figure 1 illustrates a layout of AS/RS with 8 SRMs and 3 RGVs.

G. Q. Huang et al. (eds.), *Proceeding of the 24th International Conference on Industrial Engineering and Engineering Management 2018*,
https://doi.org/10.1007/978-981-13-3402-3_25

Modern professional simulation software becomes more and more powerful with IT technology, such as AutoMod software which is widely used in logistics and manufacturing field [5]. It can model and simulate the material handling system effectively with high accuracy in spatial-temporal analysis, so it has the capability to study the dynamic behavior of AS/RS.

A procedure is presented in this paper which uses simulation method to study different kinds of scheduling policy in a real world AS/RS including SRM and RGV. Firstly, the objective warehouse is studied and modeled, including its entities and logic. Then different scheduling policies are embedded in the model. Accordingly, the model is simulated under many scenarios to verify the different effect of each policies. At last, conclusion is drawn to show the effectiveness of the procedure.

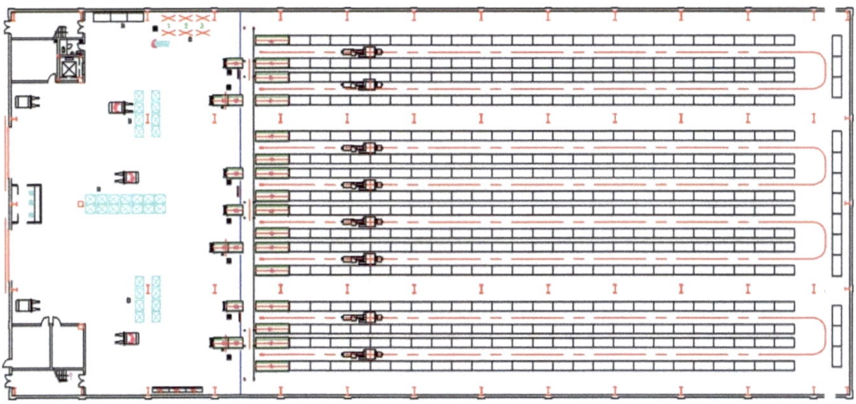

Fig. 1. A real AS/RS layout planned by the author

2 Modelling the AS/RS

In the design process of an automated warehouse, AutoMod simulation software is used to establish the model of the AS/RS, and make a comparative study of the RGV scheduling policies in the outbound operation of the warehouse. The operation of RGV is carried out entirely on the premise of the dynamic workload of the stacker system, so it can obviously reflect the characteristics of the joint operation.

2.1 Outbound Operation Logic Description

The entities in the objective warehouse are show clearly in Fig. 1, so we just focus on the operation logic of the outbound process. The scheduling policies is be used in this logic which is show in the following.

- When the WMS (warehouse management system) receives outbound orders it will produce some outbound tasks based on the inventory and some predefined rules. Then the tasks will be dispatched by WCS (warehouse control system) to different SRMs and translated into fetch tasks of SRMs.

- Each SRM will execute its fetch task and put the TUL (Transportation Unit Load) at the I/O station before its aisle.
- WCS will arrange some RGV to fetch the TUL at the mentioned I/O station then the RGV will transport the TUL to a forklift loading platform for outbound operation.
- A forklift truck will move the TUL to dock after it arrives at the platform and end the whole outbound operation cycle.

2.2 Modeling the AS/RS

AutoMod software has a well-designed UI and many powerful built-in template libraries or effective tools. It uses a true 3D graphic kernel, so it can model the AS/RS entities precisely in 3D. The static entities such as warehouse building can be built with the ACE tools bundled in the AutoMod software package while other entities can be modeled in drag-and drop way with the built-in template libraries.

As for this warehouse illustrated in Fig. 1, the rack system must be split into three parts as the SRMs because of the local Fire Code. So when modeling the rack and SRMs three AS/RS subsystem should be built and there are 8 SRMs in all 3 subsystems. In addition, although the warehouse has curved tracks for the transfer of the SRMs, it entirely aims to improve the availability of the system. In the daily operation, each aisle is still executed by the standing stacker in it, so it is still not considered when analyzing the normal throughput of the system. The AS/RS template library can be used to complete the modeling work. The conveyor template library of AutoMod software is used to complete modeling of the platforms and I/O stations. The Path mover template library is used to complete modeling of RGV system and the number of vehicles should be 3. The finished model wireframe layout is show in Fig. 2.

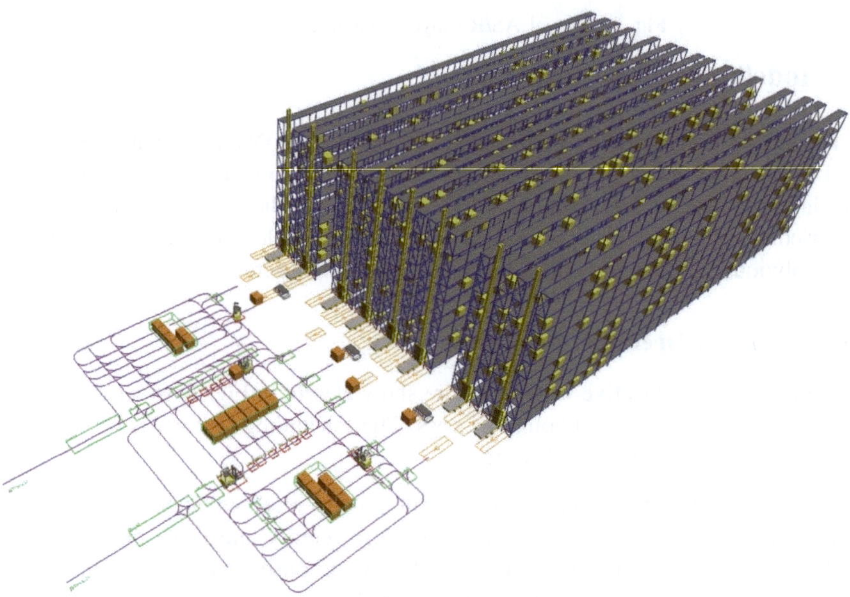

Fig. 2. Finished AS/RS model in AutoMod software

After modeling the physical entities in 3D elements all related attributes need to be set for them.

Tables 1, 2 show the running parameters for all material handling equipment (MHE) in this warehouse. From these tables it is obvious that simulation model need more parameters than analytical method for the same AS/RS. In other way, traditional analytical method cannot handle so many parameters in one solvable equation group. So it just give some approximate results for the objective AS/RS.

Table 1. SRM running parameters

SRM numbers	8	
Horizontal running parameters	Acceleration	1 m/s^2
	Deceleration	1 m/s^2
	Maxim speed	160 m/min
	Creep speed	30 m/min
	Creep distance	3 m
	Adjustment time	3 s
Vertical running parameters	Accelerate	1 m/s^2
	Decelerate	1 m/s^2
	Maxim speed	45 m/min
	Creep speed	30 m/min
	Creep distance	3 m
	Adjustment time	3 s

Table 2. Other MHEs running parameters

Frontend conveyor (I/O)	12 m/min	
RGV	Acceleration	0.3 m/s^2
	Maxim speed	120 m/min
Forklift truck	Number	4
	Acceleration	0.3 m/s^2
	Forward speed	10 m/min
	Backward speed	10 m/min
	Turn speed	4 m/min

3 Analysis with Different RGV Scheduling Policies on the Simulation Model

In order to study the effect of the RGV scheduling policies on the performance of the AS/RS, two different operational strategies were set up, and the simulation experiments were carried out under different system loads (from SRMs).

3.1 Scheduling Policies

The two policies are the following.

- FCFS (First Come First Serve) policy

The strategy is first to first come first serve strategy, which does not consider the due date of the TUL and the distance from the RGV. It is assigned RGV to carry out the outgoing task just considering the time when TULs comes into RGV's task queue. All the tasks have the same priority. Specifically, the TULs on the 8 stacker I/O is transported from the available RGV to the platform in sequence.

- NF (Nearest First) policy

The strategy is that each time the new TUL appear at the I/O port, the goods are joined in the RGV's task queue, and the task queue that has not been carried out by the RGV is sorted by the distance from the current idle RGV to the nearest cargo.

3.2 Simulation Experiment Under Different Load

Because the overall simulation strategy uses the pull strategy, the system's busy degree and workload can represented by the number of the outbound trucks at the platforms in unit time. Therefore, by changing the arrival time of trucks, the operation capability of SRMs and RGVs can be tested under different workload.

Let the simulation model run 10 h. In the experiment the arriving interval time of outbound trucks changes from 0.4 to 1.6 s. The throughput curves of the two policies under varied workload are compared in Fig. 3. The curve above (green) represents the

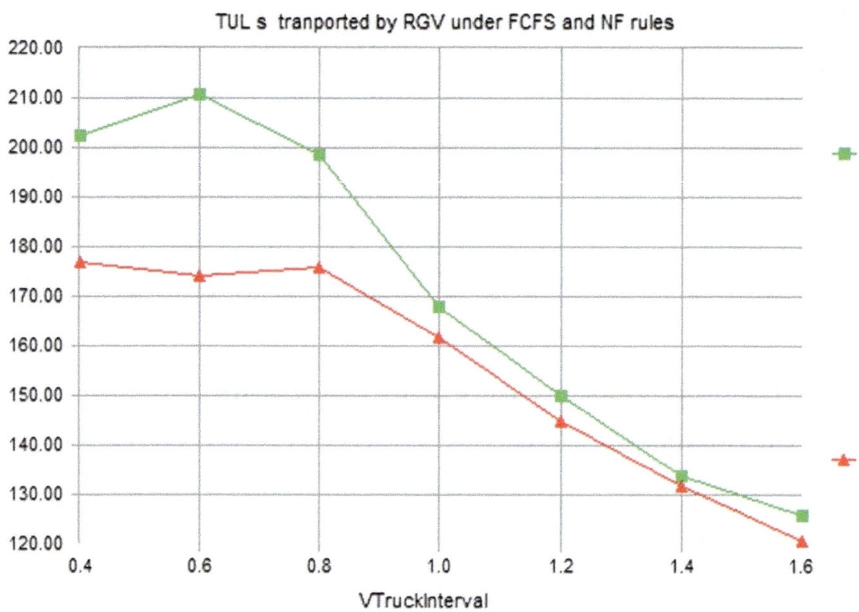

Fig. 3. Throughput comparison using scheduling policies

throughput of NF policy while the other one (red) represents the throughput of FCFS policy.

From Fig. 3, the throughput of RGV subsystem under the NF policy is better than that under the FCFS policy, but when the overall system work load is lowered, the throughput of the two policies tends to be closer.

4 Conclusion

By using simulation software we can model and simulate AS/RS dynamic behavior. So a real automated warehouse composed of the RGV and SRM can be analyzed in this way. The throughputs of different scheduling policies are compared in simulation experiment under varied workload. Deep observation of model queues and other statistics will help us understand the highly coupled and mutual influence between RGV and SRM, so we can develop better scheduling policies and improve the overall operation of the warehouse.

There are still some work can be done to the AS/RS performance evaluation research by using simulation method. For example, if characteristics of inbound or outbound orders are introduced the problem will be more complicated, then more study need to be done to solve it.

Acknowledgements. The work in this article is supported by School of Logistics and Trade, Eurasia University. I also want to show my thanks to my tutor doctor Zhili Zhou in School of Management, Xi'an Jiao Tong University.

References

1. M. Dotoli et al., Performance-based comparison of control policies for automated storage and retrieval systems modelled by coloured Petri nets (2003)
2. Y.A. Bozer, J.A. White, Travel-time models for automated storage/retrieval systems. IIE Trans. **16**(4), 329–338 (1984)
3. Y.A. Bozer, M. Cho, Throughput performance of automated storage/retrieval systems under stochastic demand. IIE Trans. **37**(4), 367–378 (2005)
4. J.-P. Gagliardi et al., A simulation modeling framework for multiple-aisle automated storage and retrieval systems. J. Intell. Manuf., 1–15 (2012)
5. W. Huang, Y. Xue, Simulation of AMHS with AutoMod, in *Proceedings of 5th International Conference on Computer-Aided Industrial Design and Conceptual Design* (CAID&CD) (2003)

Simulation and Optimization

The Framework and Methods of Quantitative Assessment for Education Reform in Industrial Engineering

Shuai Zhang, Jun-qiang Wang, Zhi-qiang Cai(✉), and Heng-yi Gao

School of Mechanical Engineering, Northwestern Polytechnical University, Xi'an, China
caizhiqiang@nwpu.edu.cn

Abstract. Industrial engineering (IE) is a system optimization technology, which is also an engineering discipline involving improvement and innovation. This paper presents the discipline development of industrial engineering in China and worldwide, as well as the key points and difficulties in the reform of industrial engineering education. A framework of quantitative assessment is then proposed for industrial engineering education reform. The multi-stage task evaluation method and importance analysis theory are introduced into educational quality management of industrial engineering for the first time. It can evaluate the whole teaching activities such as teaching, competition and practice. The author expects to give ideas of effective solutions ways to achieve successful reform for industrial engineering. It may also provide reference framework and methods to support the education reform of other related majors.

Keywords: Education reform · Framework · Industrial engineering
Innovation · Quantitative assessment

1 Introduction

Industrial engineering is a system optimization technology, which is also an engineering discipline involving planning, design, evaluation, improvement and innovation of personnel, material, equipment, information, energy and other factors [1]. Industrial engineering applies natural science, mathematics, social science, especially the theory and method of engineering technology to provide strong technical supports for the production, management, and service systems with low cost, high efficiency and high benefit [2]. As an emerging profession, industrial engineering has high recognition and high employment rate in the enterprise and society. So, the IE education plays an importance role in the society and technology developments.

Nowadays, there are mainly the following problems in the education reforms. First, the existing professional reform focuses on the cultivation of technology and management capacity, but it lacks of paying attention to the characteristics of industrial engineering. Therefore, it is urgent to study the innovation training system depending on the perspective of innovation ability [3]. Next, the comprehensive reform of industrial engineering in colleges and universities has achieved initial results, but it has

G. Q. Huang et al. (eds.), *Proceeding of the 24th International Conference on Industrial Engineering and Engineering Management 2018*,
https://doi.org/10.1007/978-981-13-3402-3_26

not yet formed a benign operation mechanism for the sustainable development. Last but not least, in the existing evaluation, the qualitative evaluation is more than quantitative evaluation, the effectiveness of industrial engineering reforms cannot be evaluated effectively for short of quantitative evaluation index system [4].

2 IE Education Worldwide

2.1 IE Education in America

After nearly a century of development, the industrial engineering education in American universities has been very mature. It has formed a large number of distinctive industrial engineering professions in America [5]. There is full-blown training plan, clear professional direction, and perfect curriculum system. Furthermore, there are high quality handouts and teaching material. The curriculum, personnel training and scientific research in American universities are coincided with regional industrial characteristics, and combine production, teaching and scientific research in close cooperation. By entrusting enterprises scientific research to colleges and universities, it provides practice base. Both sides cooperate on education closely while the goal of cultivating talents is fully embodied in the whole process of cooperative education. The two parties build engineering research center, application development research center together and send researchers work together in the meantime.

2.2 IE Education in Europe

Recently, German Production Engineering Research Association proposes a new curriculum structure, aiming to make industrial engineering education more attractive to students. The content of the courses is closer to the industrial demands [6]. Industrial engineering students need to learn basic courses in the first two semesters before they choose a professional direction in their third semester. This will help students to understand the whole subject, also to make the students develop continuously towards the direction of engineering science [7]. Students can get practice experience which related to the teaching content by completing project in groups, at the same time, students can also improve their social competence [8]. In order to meet the long-term needs of excellent personnel in engineering, there are some main approaches. Firstly, all engineering students are required to study basic courses in the first two semesters. Secondly, introduce technology courses and recombine the curriculum related to natural science [9]. In particular, the introduction of technology courses and the reorganization of the curriculum associated with natural science will be able to meet the needs of today's engineering field [10].

2.3 IE Education in Japan

In Japan, there are nearly a hundred of universities that have established industrial engineering courses, such as Waseda University and Tokyo Institute of Technology. Those first-class universities all have the department of industrial engineering where the cultivating talents are mainly engaged in the line operation and management in enterprises. The engineering education in Japan advocates the concepts that arranging a variety of technical objects in reality from the scientific level, and developing technical system's cross-correlation technique further. The industrial engineering educations in Japan aims at those talented people who have scientific attitude, engineering background, pioneering and innovative spirit, and ability of identify and solve problems by their professional knowledge. The course system which combines graduation project with practical work emphasizes experiment and practice. Therefore, it makes students master skills more practically, identify and solve problems much faster in the future.

2.4 IE Education in China

At present, there are more than 160 colleges and universities that have set up industrial engineering profession in Mainland. The colleges divide the industrial engineering profession into two categories, engineering category and management category. There are two extremes about the arrangement of industrial engineering courses. One extreme is the overweight proportion of the engineering course, on the other side is the overweight proportion of management course. In addition, comparing with foreign countries, our domestic demand for foundations of mathematics is relatively weak [11].

Hong Kong's industry is dominated by high technology in the twenty-first century, and highly trained industrial engineers become an important resource in the industrial world. Hong Kong has a high degree of international integration, where a majority of teachers has abroad experience at American universities.

As an applied discipline in Taiwan, industrial engineering applications have transcended manufacturing into services, the government and other public organizations. The University in Taiwan set up its first industrial engineering family in 1963. It has considerable influence in the international arena after the 40-year development. Tsinghua University Taiwan Hsinchu and Yuan Ze University are in the high reputation of industrial engineering education. Furthermore, Taiwan's electronic manufacturing is the mainstream industry, so the enterprises in Taiwan attach great importance to industrial engineering's application in production, education and scientific research, such as Foxconn.

2.5 IE Education in Northwestern Polytechnical University

Department of industrial engineering in Northwestern Polytechnical University (NPU) originates from mechanical engineering. At the beginning, the training project, construction of course system and other aspects all followed mechanical engineering specialty, so it gets more mechanical engineering features. In addition, the research of industrial engineering in NPU started from the information of manufacturing industry, so the initial research focused on the application of enterprise management. At present, the industrial engineering in NPU is transforming to the application base and the theoretical foundation.

NPU aims to cultivate the leading talents, who can plan, design, optimize, analyze industrial systems, and can cater to wide aperture area, too. However, many things need carrying out to cultivate industrial engineering students, who have international vision and innovative ability, such as comprehensive reform in training programs, curriculum systems, series of teaching materials, subject competitions, international cooperation and other aspects. This will play an exemplary role in the province's industrial engineering professional development and comprehensive reform, a leading role and radiation effect, too.

3 Framework of Quantitative Assessment for IE Education Reform

3.1 Main Contents

The cultivation of comprehensive innovation ability is the biggest characteristic of personnel training in industrial engineering. Based on the philosophy of system theory, this paper uses the method of quality function deployment (QFD) to support the cultivation of comprehensive innovation ability, considering stages from the top design (training program, curriculum system) to the specific implementation(teaching model, comprehensive experiment, subject competition, business practice), and makes the innovation ability training throughout the training program, curriculum system, classroom teaching, business practice, discipline competitions and other teaching links [12].

The essence of industrial engineering is that there is always a better way. The reform in this paper need to absorb essence, draw on the method of Plan-Do-Check-Act (PDAC) and the experience of the theory of importance analysis which comes from the theory of quality and reliability, identify the weak links in the comprehensive reform, and clear the key factors that affect the quality of professional education, guide the improvement and optimization of the culture system with it, optimize and improve the teaching quality with a definite object, construct continuous improvement.

The core tool of modern industrial engineering is quantitative analysis. On the basis of long-term scientific practice and research on quality management science, use the multi-stage task reliability assessment method, and divide the professional assessment into two stages: still at college and already graduated. Start with the feedback from students, teachers, administration, enterprise, graduate tutors and so on, and research on analytic hierarchy process (AHP) basing on quantitative and semi-quantitative evaluation data, then lay the foundation for the continuous optimization of training programs, curriculum system and personnel training [13].

This paper takes the undergraduate education of industrial engineering specialty in Northwestern Polytechnical University as the demonstration object of the application, and the results facilities to form colorful innovative educational activities. These activities include course content covering innovative approach, teaching mode of creating innovative thinking, subject competitions that helps to train the creative consciousness and enterprise practice that helps to exercise innovation ability. That will guide students to study basing on problems, projects and interests, and gets continuous improvement and quantitative assessment at the same time. Moreover, take Shaanxi Institute of Mechanical Engineering industrial engineering and management as communication platform, and introduce the reform process and advantages of industrial engineering specialty to colleges and universities in Shaanxi Province.

3.2 Main Objective

In order to carry out an impeccable industrial engineering education reform, the placement of industrial engineering graduates is of great significance to this paper. By analyzing the relevant literature, we can learn that industrial engineering undergraduate students have three choices. First, to engage in work related to industrial engineering; second, continue the industrial engineering learning, study in the graduate school for a master's degree; third, become a civil servant in government departments. Therefore, the employers of industrial engineering students in the future are mainly enterprises, graduate schools and government departments. The future development of industrial engineering students is described by fishbone diagram as shown in Fig. 1.

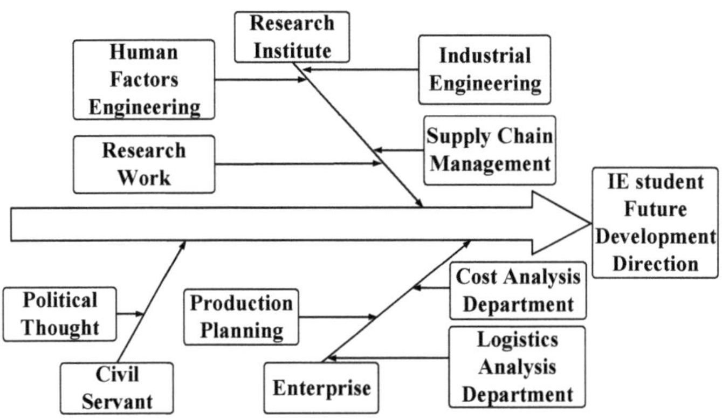

Fig. 1. The future development direction of industrial engineering graduates

According to the future development direction of industrial engineering students, we believe that industrial engineering students' qualified capability constitution is complex, as shown in Fig. 2. Industrial engineering education reform will help students to get the complicated capability constitution.

Based on the existing comprehensive reform results of industrial engineering, this paper analyzes the comprehensive reform of industrial engineering specialty in universities of developed countries, Hong Kong and Taiwan, and studies the training system of industrial engineering innovation ability, the continuous improvement of industrial engineering and the method of quantitative assessment, improves and perfects the mode of cultivating innovative talents and constructs a new system of healthy development of industrial engineering. Verifies the effectiveness of the new system through the demonstration of industrial engineering in Northwestern Polytechnical University, the promotion and application of colleges and universities in Shaanxi Province, ultimately provides effective solutions and ways to achieve the industrial engineering comprehensive reform, and also provides reference ideas and reference framework for other professional sustainable and healthy development.

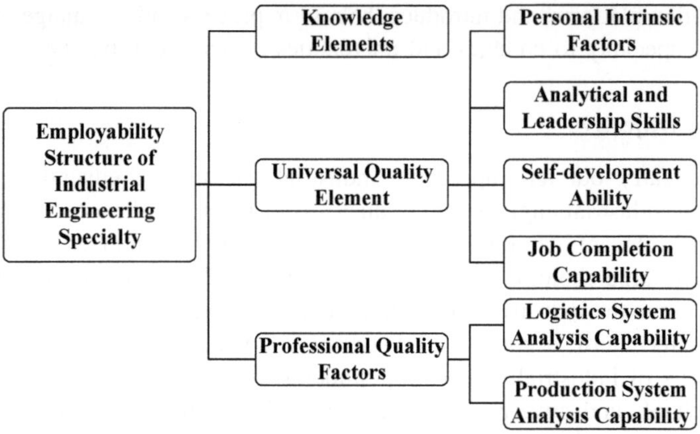

Fig. 2. Industrial engineering students' capability constitution

There is several teaching problems need solving here. First of all, when establishing the new system of cultivating innovative talents in industrial engineering, which has the main theme to train innovating talents, top level planning and system construction are lacking in the process [14]. Second, when establishing the benign operation mechanism of sustainable development of industrial engineering, which is based on the importance analysis theory and PDCA method in quality and reliability management, scientific guidance methods are lacking in the continuous problem solving process. Third, when establishing the quantitative evaluation index system, which is based on the quantitative evaluation method in system engineering, the problem is how to evaluate the efficiency of industrial engineering reform accurately. Finally, how to colligate industrial engineering reform model, how to abstract the scientific reform and development of new ideas and new models, how to form a set of overall solutions for other professional sustainable and healthy development, how to provide reference ideas and reference framework, the above problems need to be researched deeply.

4 Methods of Quantitative Assessment for IE Education Reform

4.1 Main Methods

This paper intends to in-depth analyze the industrial engineering professional comprehensive reform experience and effectiveness in domestic and foreign universities by using the method of investigation and analysis. According to the characteristics and existing problems of industrial engineering in colleges and universities in Shaanxi Province, the paper sorts out the problems in the comprehensive reform of industrial engineering. Focusing on the cultivation of innovation ability, the continuous improvement of education and the evaluation of effectiveness, locate the problems in the project research accurately.

Combine the quality of talents capacity and the quality function of the educational process, circumfusing the whole process of cultivating innovative talents in industrial engineering, basing on the QFD method. Establish a matrix to analysis the relationship between talent quality demand and cultivating project. Develop the quality of the function step by step, focusing on the education process of the quality of talent. Plan and make improvement systematically. Exchange the need of social and employing unit into the requirement of talents' quality and ability, furthermore, transform it into the corresponding countermeasures that should be taken. Then, through expanding and deriving into the various teaching processes, a scientific and systematic top-level planning and overall deployment can be formed at last.

Figure 3 shows that the QFD method can identify the needs of the stakeholders, and the AHP method introduces competitive advantages and determines the weight of competitive advantages. Then use the AHP calculation results as part of the QFD input, use the QFD method to select and determine the core attributes of the competitive advantages. Ultimately, the training program can be designed and evaluated based on the integration of QFD and AHP.

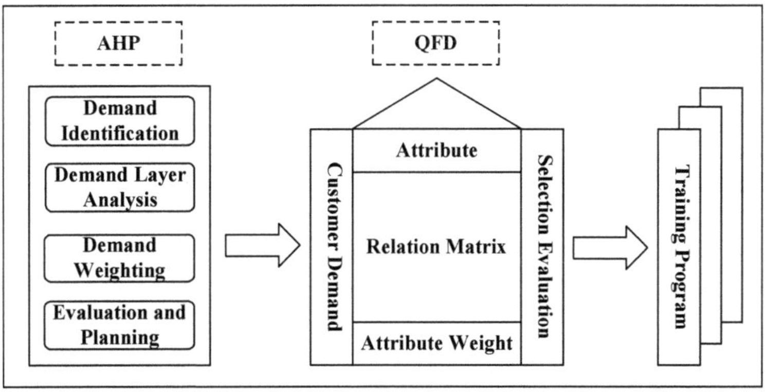

Fig. 3. Design training program based on AHP and QFD integration

Analyze the comprehensive reform and educational quality of industrial engineering, basing on the theory of importance analysis in reliability management, in accordance with the quintessence of industrial engineering and the law of learning. And then identify the weak links in the reform of industrial engineering. Next, make the core factors that influence the improvement of professional education quality clear. At last, develop improvement and optimization of core training process and links.

Carry out industrial engineering professional comprehensive persistent reform, following the PDCA scientific quality management procedures, basing on the PDCA continuous improvement model. Form the personnel training closed-loop management according to multi-source, multi-channel information feedback, including the needs of the community, the employing unit, senior instructors, students and the results of the assessment.

This paper uses the PDCA model to analyze the optimization of industrial engineering education. PDCA as a mature model has the characteristics of continuous optimization and continuous improvement. By modeling the product or event, the model is analyzed in four steps to achieve the purpose of optimization and improvement. PDCA cycle is currently more used in quality management, health care, manufacturing and other fields. The core idea of the PDCA cycle is similar to the core idea of industrial engineering, emphasizing optimization and improvement. The importance analysis is a scientific method of calculating the importance of each component in the system and the weakness of the identification system. Through the importance analysis, the importance ranking of the system components is obtained, and then the PDCA cycle is used to analyze and find the places that need to be optimized to achieve the continuous improvement of the industrial engineering education.

Summarize the evaluation index that reflects the comprehensive reform of industrial engineering, on the basis of long-term scientific practice and research on education quality management. Establish quantitative evaluation methods to evaluate the whole teaching activities such as teaching, competition and practice, based on AHP and multi-stage task evaluation methods. And finally establish a quantitative evaluation system that runs through the trinity of knowledge learning, innovation ability training and quality training process [15].

Form colorful innovative educational activities, including the course contents covering innovative approach, teaching modes of creating innovative thinking, subject competitions training the creative consciousness and enterprise practice exercising innovation ability, those above should be finished by using empirical methods, which are combined with the Northwestern Polytechnic University of industrial engineering innovation talent training process, and guide students to study, based on problems, projects and interest. Concern about the development of ability, quality, innovation and sense of responsibility, inspire thinking, broaden horizons, tap the potential, protect students' critical spirit and beneficial personality to verify, and improve the research results.

4.2 Main Advantages

The multi-stage task evaluation method and importance analysis theory in industrial engineering field are introduced into educational quality management field for the first time, carry out the improvement plan of industrial engineering professional education quality and continuous improvement, and find the way of theoretical exploration on the improvement of educational quality.

We build a trinity of personnel training model about quality goal orientation, application of process method and innovation ability training, following the process approach principle of personnel training. And we form a continuous improvement model of PDCA comprehensive education quality in industrial engineering specialty, which has a purport to improve the educational quality continuously.

This paper carries out the comprehensive reform and continuous optimization of industrial engineering, based on the continuous improvement of the closed-loop model and quantitative assessment, which is a beneficial exploration for the current professional reform, enriches the means of education and teaching reforms.

5 Conclusion

Based on the philosophy of system theory, this paper analyzes the education reform of industrial engineering in China and worldwide, considering stages from the top design to the specific implementation. The methods and innovative ability training system which supports the cultivation of comprehensive innovation ability are also studied with the method of quality function deployment. Then, this paper proposes a framework for the quantitative assessment method of industrial engineering education reform. It introduces the multi-stage task evaluation method and importance analysis theory into educational quality management. The continuous improvement model of industrial engineering education will help improve the educational quality.

Acknowledgements. The authors gratefully acknowledge the financial supports for this research from the Shaanxi Higher Education Teaching Reform Research Project (No. 15BY07) and the 111 Project (No. B13044).

References

1. G. Ksal, A. Man, Planning and design of industrial engineering education quality. Comput. Ind. Eng. **35**(3–4), 639–642 (1998)
2. T.H. Davenport, The new industrial engineering: information technology and business process re-design. Sloan. Manag. Rev. **31**(4), 11–27 (1989)
3. G. Mason, Results of an industry survey on manufacturing engineering and manufacturing engineering education. J. Eng. Educ. **87**(3), 211–214 (1998)
4. C. Makungu, K.K. Mwacalimba, A quantitative risk assessment of bovine theileriosis entering luapula province from central province in zambia via live cattle imports from traditional and commercial production sectors. Prev. Vet. Med. **116**(1–2), 63–75 (2014)
5. M. Palma, I.D.L. Rios, D. Guerrero, Higher education in industrial engineering in Peru: towards a new model based on skills. Procedia-Soc. Behav. Sci. **46**(1), 1570–1580 (2012)
6. J. Zink, Klaus, measuring universities against the european quality award criteria. Total Qual. Manag. **6**(5), 547–562 (1995)
7. J.W. Volkmann, M. Landherr, D. Lucke, M. Sacco, M. Lickefett, E. Westkamper, Engineering apps for advanced industrial engineering. Procedia CIRP **41**(3540), 632–637 (2016)
8. T.J. Gallwey, Europe needs industrial engineering degrees in order to enhance its competitiveness. Eur. J. Eng. Educ. **17**(1), 51–57 (1992)
9. E.A. Henninger, J.M. Hurlbert, Using the seven principles for good practice in undergraduate education. J. Bus. Financ. Librariansh. **12**(2), 3–15 (2007)
10. F. Alonso, D. Manrique, L. Martinez, J.M. Vines, How blended learning reduces underachievement in higher education: an experience in teaching computer sciences. IEEE Trans. Educ. **53**(3), 471–478 (2011)
11. Q.I. Er-Shi, Y. Lin, Y.J. Shi, A study on application and development of china industrial engineering in manufacturing industry. Ind. Eng. Manag. **11**(6), 1–5 (2006)
12. B. Prasad, Review of QFD and related deployment techniques. J. Manuf. Sys. **17**(3), 221–234 (1998)
13. A.S. Al-Harbi, Application of the AHP in project management. Int. J. Proj. Manag. **19**(1), 19–27 (2001)

14. G. Ostojic, S. Stankovski, L. Tarjan, I. Senk, V. Jovanovic, Development and implementation of didactic sets in mechatronic and industrial engineering courses. Int. J. Eng. Educ. **26** (1), 2–8 (2010)
15. O.S. Vaidya, S. Kumar, Analytic hierarchy process: an overview of applications. Eur. J. Oper. Res. **169**(1), 1–29 (2006)

A Feasible Method to Implement Shared Decision Making in Routine Practice

Jun-xian Wang[1(✉)], Zi-xian Liu[1], and Li-jun Liang[2]

[1] College of Management and Economics, Tianjin University, Tianjin, China
wandscape@163.com
[2] School of Management, Tianjin University of Traditional Chinese Medicine, Tianjin, China

Abstract. The paper proposes a method to incorporate the patient, physician, and general people into the shared decision making system, the method is a combination of conjoint analysis and fuzzy multi-attribute group decision making. Firstly, The preferences of various decision makers are analyzed by the conjoint analysis, secondly, the decision makers evaluate the treatment regimens with the triangular fuzzy number, then the method of ideal point is used to obtain the ideal treatment regimen, finally the relapsed or refractory chronic lymphoblastic leukemia treatment decision making is taken as an example of this method's specific application, the results show that the applicability of the method proposed by the paper is strong.

Keywords: Conjoint analysis · Fuzzy triangular number · Multi-attribute group decision making · Preference · Shared decision making

1 Introduction

With the development of economy and technology, patients' participation in medical service is increasing greater, the shared decision making, which advocates patients' greater involvement in the formulation of medical treatment, has been seen as an applicable model for the clinical practice [1]. Elywn et al. [2] has proposed an definition for Shared decision making, which attach great importance to patient's participation, preference, and the sharing of information. For the shared decision making model, there are several necessary components, including the exchange of information and preferences, the steps to achieve the optimal treatment with consensus and an agreement to implement the treatment [3].

For the problem of how to incorporate the physician and patient into the decision making system, Charles et al. [3] have described and summarized the characteristics of the model of shared decision making. Stiggelbout and Pieterse [4] have Proposed four steps to put shared decision making into practice. Elwyn et al. [5] have integrated and analyzed the existing literatures related to the shared decision making, provided a useful foundation for describing and implementing the shared decision making in future research. At the same time, scholars have proposed a series of medical decision support to motivate the application of shared decision making [6, 7]. Recently, in some

G. Q. Huang et al. (eds.), *Proceeding of the 24th International Conference on Industrial Engineering and Engineering Management 2018*,
https://doi.org/10.1007/978-981-13-3402-3_27

medical services, the shared decision making model has been put into application [8, 9].

Nowadays, for the quantitative research on shared decision making, some scholars have used the questionnaire to investigate the patient's involvement in the process of treatment formulating and analyze the factors that affect patient's involvement [10, 11], but they don't have presented a feasible model to conduct the shared decision making. Dolan has put the AHP into the practical application of shared decision making from the perspective of multicriteria decision making [12]. However, a significant drawback of this method is the consistent test cannot be passed sometimes and the method cannot bring the large group of decision makers into the shared decision making system. At the same time, the existing literatures rarely consider the patient and physician's preferences and involvement simultaneously. In view of the shortcomings of these studies, we will present a new method to conduct the model of shared decision making, we study the problem from the perspective of multi-attribute group decision making. Firstly, the differences of preferences among the patient, physician and the general people are measured by the conjoint analysis, then we put the method of fuzzy multi-attribute group decision making into the evaluation of treatment regimens and obtain the best treatment regimen finally.

The conjoint analysis can elicit the preferences of decision makers by the questionnaire [13], and the method of fuzzy multi-attribute group decision making can utilize various existing information and make a decision with the comprehensive consideration of the scheme's various attributes [14]. These two methods is simple and feasible, however, the combination of these two methods haven't been applied in the existing literatures of shared decision making. So in this paper, we combine these two method to facilitate the application of shared decision making.

2 Methodology

For a disease, there would be several different treatment regimens. these regimens form a set of options, we call the set X, $X = \{X_1, X_2, \ldots, X_I\}$, the X_i represents the i-th treatment regimen. There are multiple attributes of the treatment regimen should be considered. The set of attributes is T, $T = \{T_1, T_2, \ldots, T_J\}$, the T_j represents the j-th attribute. In the meantime, there are several decision makers incorporated into the shared decision making system, besides the physician and patient, other stakeholders would involve in the system too. Thus D is set, which represents the decision makers involved, $D = \{D_1, D_2, \ldots, D_K\}$, for the k-th decision maker, the representation is D_k. With regard to the decision makers' weights, they are different, the k-th decision maker's weight is indicated by w_k,$0 \leq w_k \leq 1$, $\sum_{k=1}^{K} w_k = 1$,$k = 1, \ldots, K$. In addition, the $\alpha_j^{(k)}$ represents the relative importance that the decision maker D_k gives to the attribute $\mathrm{T_j}$ when they evaluate the treatment regimens, $0 \leq \alpha_j^{(k)} \leq 1$, $\sum_{j=1}^{J} \alpha_j^{(k)} = 1$, $j = 1, \ldots, J$.

2.1 The Conjoint Analysis

For a disease, patient, physician, and other stakeholders may value the treatment attributes differently. Thus the paper applies the method of conjoint analysis to estimate the preference for the attributes. As a well-established, evidence-based method, the conjoint analysis is used to elicit the preferences. According to the assumption that the attributes can describe a treatment and the levels of these attributes can represent the preferences of the treatment. Participants will be asked to choose between the supposed treatments which are the combination of these attributes with different levels. With the estimation of the relative importance between the given attributes, which can be analyzed from the participants' responses, thus we can get the value of $\alpha_j^{(k)}$ [13].

2.2 The Triangular Fuzzy Number

Let $\tilde{p}$ as an example of triangular fuzzy number, $\tilde{p} = (l, q, u)$, The following formulation is the membership function:

$$f_{\tilde{p}} = \begin{cases} 0 & x \leq l, x \geq u \\ \frac{x-l}{q-l} & l < x < q \\ \frac{x-u}{q-u} & q < x < u \end{cases} \tag{1}$$

In the above formula, $x \in R$, $l < q < u$.

The decision maker evaluates the treatment regimens under the certain attribute with the linguistic term, but in the real calculation, the evaluation results should be numerical, so based on Table 1, which describes the relationship between the linguistic term and the triangular fuzzy number, we can get numerical evaluation.

Table 1. The linguistic term and the corresponding triangular fuzzy number

Fuzzy language level	Linguistic term	Triangular fuzzy number
X_0	Worst	(0, 0, 0.17)
X_1	Worse	(0, 0.17, 0.33)
X_2	Bad	(0.17, 0.33, 0.5)
X_3	General	(0.33, 0.5, 0.67)
X_4	Good	(0.5, 0.67, 0.83)
X_5	Better	(0.67, 0.83, 1)
X_6	Best	(0.83, 1, 1)

2.3 Fuzzy Multi-attribute Group Decision Making

$\tilde{e}_{ij}^{(k)}$ can represents the evaluation value of treatment regimen X_i under the attribute T_j made by decision maker D_k, the $\tilde{e}_{ij}^{(k)}$ is a triangular fuzzy number, so, through the evaluation, we can get the multi-attribute decision matrix $\tilde{E}_k$, the following is the description of the matrix.

$$\tilde{E}_k = \begin{bmatrix} \left(e_{11}^{(k)l}, e_{11}^{(k)m}, e_{11}^{(k)u}\right) & \left(e_{12}^{(k)l}, e_{12}^{(k)m}, e_{12}^{(k)u}\right) & \cdots & \left(e_{1n}^{(k)l}, e_{1n}^{(k)m}, e_{1n}^{(k)u}\right) \\ \left(e_{21}^{(k)l}, e_{21}^{(k)m}, e_{21}^{(k)u}\right) & \left(e_{22}^{(k)l}, e_{22}^{(k)m}, e_{22}^{(k)u}\right) & \cdots & \left(e_{2n}^{(k)l}, e_{2n}^{(k)m}, e_{2n}^{(k)u}\right) \\ \vdots & \vdots & \cdots & \vdots \\ \left(e_{m1}^{(k)l}, e_{m1}^{(k)m}, e_{m1}^{(k)u}\right) & \left(e_{m2}^{(k)l}, e_{m2}^{(k)m}, e_{m2}^{(k)u}\right) & \cdots & \left(e_{mn}^{(k)l}, e_{mn}^{(k)m}, e_{mn}^{(k)u}\right) \end{bmatrix} \quad (2)$$

The attributes of the treatment regimen can be divided into two types, one is the type of benefit, the other is the type of cost. For the attribute of benefit, which is the one that expects its evaluation value lager, however the cost attribute expects its evaluation value smaller. So we normalize the evaluation matrix to eliminate the impact of diverse attributes types, we can get the normalized evaluation matrix $\tilde{F}_k$.

$$\tilde{F}_k = \left[\tilde{f}_{ij}^{(k)}\right]_{m\times n}, \tilde{f}_{ij}^{(k)} = \left(f_{ij}^{(k)l}, f_{ij}^{(k)m}, f_{ij}^{(k)u}\right), 1 \le i \le I, 1 \le j \le J \quad (3)$$

For the attribute T_j, $1 \le j \le J$, we can normalize the matrix:

(1) If the attribute T_j is of benefit type, then

$$\left(f_{ij}^{(k)l}, f_{ij}^{(k)m}, f_{ij}^{(k)u}\right) = \left(\frac{e_{ij}^{(k)l}}{\sqrt{\sum_{i=1}^{m}\left(e_{ij}^{(k)l}\right)^2}}, \frac{e_{ij}^{(k)m}}{\sqrt{\sum_{i=1}^{m}\left(e_{ij}^{(k)m}\right)^2}}, \frac{e_{ij}^{(k)u}}{\sqrt{\sum_{i=1}^{m}\left(e_{ij}^{(k)u}\right)^2}}\right) \quad (4)$$

(2) If the attribute T_j is of cost type, then

$$\left(f_{ij}^{(k)l}, f_{ij}^{(k)m}, f_{ij}^{(k)u}\right) = \left(\frac{\frac{1}{e_{ij}^{(k)l}}}{\sqrt{\sum_{i=1}^{m}\left(\frac{1}{e_{ij}^{(k)l}}\right)^2}}, \frac{\frac{1}{e_{ij}^{(k)m}}}{\sqrt{\sum_{i=1}^{m}\left(\frac{1}{e_{ij}^{(k)m}}\right)^2}}, \frac{\frac{1}{e_{ij}^{(k)u}}}{\sqrt{\sum_{i=1}^{m}\left(\frac{1}{e_{ij}^{(k)u}}\right)^2}}\right) \quad (5)$$

After normalizing the matrix, for simplifying the decision making, the normalization would be taken to transform the fuzzy evaluation matrix into non-fuzzy evaluation matrix. For $\tilde{f}_{ij}^{(k)} = \left(f_{ij}^{(k)l}, f_{ij}^{(k)m}, f_{ij}^{(k)u}\right)$, the corresponding non-fuzzy number is $c_{ij}^{(k)}$.

$$c_{ij}^{(k)} = \frac{\left(f_{ij}^{(k)l} + 2f_{ij}^{(k)m} + f_{ij}^{(k)u}\right)}{4} \quad (6)$$

Thus we can get the matrix C_k composed of non-fuzzy number, $C_k = \left[c_{ij}^{(k)}\right]_{m\times n}$, $1 \le i \le I, 1 \le j \le J$.

After Combining the preference coefficient $\alpha_j^{(k)}$ and the normalized value of each attribute T_j, the matrix of each medical treatment regimen can be obtained.

$$A_k = \left[a_{ij}^{(k)}\right]_{m\times n} = \left[\alpha_j^{(k)} c_{ij}^{(k)}\right]_{m\times n}, 1 \le i \le I, 1 \le j \le J, 1 \le k \le K. \tag{7}$$

In the clinical practice, the patient, physician and other stakeholders are of different importance for the medical treatment decision making. Thus we apply the decision makers' weights in the matrix A_k, then the matrix R_k can be gotten.

$$R_k = \left[r_{ij}^{(k)}\right]_{m\times n}, r_{ij}^{(k)} = w_k \alpha_j^{(k)} c_{ij}^{(k)}, 1 \le i \le I, 1 \le j \le J, 1 \le k \le K. \tag{8}$$

After the calculation of the initial matrix, With R_k, we can determine the group's positive and negative ideal solutions, which are S^+ and S^-.

$$\begin{aligned} S^+ &= \left(y_1^+, y_2^+, \ldots, y_n^+\right), y_j^+ = max_k\, max_i\, r_{ij}^{(k)} \\ S^- &= \left(y_1^-, y_2^-, \ldots, y_n^-\right), y_j^- = min_k\, min_i\, r_{ij}^{(k)} \end{aligned} \tag{9}$$

The more the value of the treatment regimen evaluation approximates the group's ideal positive solution, the better the treatment regimen is. The more the value of the treatment regimen evaluation approximates the group's ideal negative solution, the worse the treatment regimen is. The value of p_{ik} represents the relative proximity between the treatment regimen i and the group's ideal solution, for the decision maker D_k, when the value of p_{ik} is larger, the treatment regimen i is better.

$$\begin{aligned} P_{ik} &= \frac{z_{ik}^-}{z_{ik}^- + z_{ik}^+} \\ z_{ik}^- &= \sqrt{\sum\nolimits_{j=1}^{n} \left(r_{ij}^{(k)} - y_j^-\right)^2} \\ z_{ik}^+ &= \sqrt{\sum\nolimits_{j=1}^{n} \left(r_{ij}^{(k)} - y_j^+\right)^2} \end{aligned} \tag{10}$$

The relative proximity matrix P can be constructed by the use of p_{ik}.

$$P = \begin{bmatrix} p_{11} & \cdots & p_{1l} \\ \vdots & \ddots & \vdots \\ p_{m1} & \cdots & p_{ml} \end{bmatrix} \tag{11}$$

Based on matrix P, we can determine the group's ideal solution, the positive one is V^+ and the negative one is V^-.

$$\begin{aligned} V^+ &= \left(u_1^+, u_2^+, \ldots, u_l^+\right), u_k^+ = max_i\, p_{ik} \\ V^- &= \left(u_1^-, u_2^-, \ldots, u_l^-\right), u_k^- = min_i\, p_{ik} \end{aligned} \tag{12}$$

The value of Q_i represents the relative proximity between the treatment regimen i and the group's ideal solution, when the the value of Q_i is larger, the treatment regimen i is better.

$$\begin{aligned} Q_i &= \frac{g_i^-}{\left(g_i^+ + g_i^-\right)} \\ g_i^- &= \sqrt{\sum\nolimits_{k=1}^{l} \left(p_{ik} - u_k^-\right)^2} \\ g_i^+ &= \sqrt{\sum\nolimits_{k=1}^{l} \left(p_{ik} - u_k^+\right)^2} \end{aligned} \tag{13}$$

Finally, we can determine the treatment regimen according to the value of Q_i, the optimal treatment regimen is the regimen which gains the largest value of Q_i.

3 Results

In this paper, the above mentioned shared decision making model will be applied into the medical treatment decision making of relapsed or refractory chronic lymphocytic leukemia. The decision makers are patient, physician and general people which include the relatives and friends of the patient and so on, the preferences of these decision makers for the treatment regimen attributes were obtained by the use of conjoint analysis method, we cite these preferences data from the literatures [13], and the Table 2 lists the data.

Table 2. The preference of the decision makers

Attribute	Patient	Physician	General people
Overall survival	0.369	0.436	0.329
Progression-free survival	0.09	0.111	0.111
Fatigue	0.101	0.120	0.127
Nausea	0.067	0.058	0.074
Risk of serious infections	0.192	0.170	0.227
Treatment administration	0.180	0.104	0.132

Then the different treatment regimens should be evaluated to determine the optimal regimen, which will be done with the method of fuzzy multi-attribute group decision making. Through the preliminary work, there are three treatment regimens to be chosen, these three alternatives are X_1, X_2, X_3 respectively. The decision makers D_1, D_2, D_3 are represented as patient, physician, and general people respectively.

The decision makers evaluate the treatment regimen under the attributes. In the following text, the evaluation value of the treatment regimen, which is transformed into the triangular fuzzy number, is displaced.

$$\tilde{E}_1 = \begin{bmatrix} (0.33,0.5,0.67) & (0.5,0.67,0.83) & (0.17,0.33,0.5) & (0.33,0.5,0.67) & (0.17,0.33,0.5) & (0.33,0.5,0.67) \\ (0.33,0.5,0.67) & (0.5,0.67,0.83) & (0.33,0.5,0.67) & (0.17,0.33,0.5) & (0.5,0.67,0.83) & (0.17,0.33,0.5) \\ (0.5,0.67,0.83) & (0.5,0.67,0.83) & (0.67,0.83,1) & (0.5,0.67,0.83) & (0.5,0.67,0.83) & (0.67,0.83,1) \end{bmatrix}$$
$$\tilde{E}_2 = \begin{bmatrix} (0.17,0.33,0.5) & (0.17,0.33,0.5) & (0.17,0.33,0.5) & (0.33,0.5,0.67) & (0.33,0.5,0.67) & (0.67,0.83,1) \\ (0.33,0.5,0.67) & (0.17,0.33,0.5) & (0.33,0.5,0.67) & (0.17,0.33,0.5) & (0.17,0.33,0.5) & (0.17,0.33,0.5) \\ (0.5,0.67,0.83) & (0.17,0.33,0.5) & (0.67,0.83,1) & (0.33,0.5,0.67) & (0.33,0.5,0.67) & (0.83,1,1) \end{bmatrix}$$
$$\tilde{E}_3 = \begin{bmatrix} (0.33,0.5,0.67) & (0.67,0.83,1) & (0.17,0.33,0.5) & (0.67,0.83,1) & (0.17,0.33,0.5) & (0.5,0.67,0.83) \\ (0.67,0.83,1) & (0.67,0.83,1) & (0.67,0.83,1) & (0.33,0.5,0.67) & (0.5,0.67,0.83) & (0.5,0.67,0.83) \\ (0.83,1,1) & (0.67,0.83,1) & (0.83,1,1) & (0.67,0.83,1) & (0.5,0.67,0.83) & (0.33,0.5,0.67) \end{bmatrix}$$

Then we normalize the evaluation matrix to eliminate the impact of diverse attributes types, on account of the attributes T_1 and T_2 are benefit attributes, the attributes T_3, T_4, T_5 and T_6 are cost attributes. We get the matrix $\tilde{F}_k$ and present $\tilde{F}_1$ only to keep the paper concise.

$$\tilde{F}_1 = \begin{bmatrix} (0.09,0.22,0.50) & (0.11,0.33,0.98) & (0.20,0.38,0.76) & (0.19,0.36,0.72) & (0.17,0.30,0.51) & (0.17,0.29,0.43) \\ (0.17,0.33,0.67) & (0.11,0.33,0.98) & (0.20,0.38,0.76) & (0.19,0.36,0.72) & (0.21,0.40,0.78) & (0.26,0.48,0.87) \\ (0.25,0.45,0.83) & (0.11,0.33,0.98) & (0.13,0.23,0.37) & (0.15,0.27,0.48) & (0.17,0.30,0.51) & (0.17,0.24,0.34) \end{bmatrix}$$

Then we transform the triangular fuzzy number into the non-fuzzy number, we can get the matrix constructed with the non-fuzzy number. The preferences and decision makers' status need to be considered, the preferences of decision makers for the treatment regimen's attributes have been listed in the Table 2, the weights of the patient, physician and general people in the shared decision making model are 0.3, 0.4 and 0.3 respectively, $W = (w_1, w_2, w_3) = (0.3, 0.4, 0.3)$, so the matrix R_k can be obtained.

$$R_1 = \begin{bmatrix} 0.028 & 0.012 & 0.013 & 0.008 & 0.018 & 0.016 \\ 0.042 & 0.012 & 0.013 & 0.008 & 0.026 & 0.028 \\ 0.055 & 0.012 & 0.007 & 0.006 & 0.018 & 0.013 \end{bmatrix}$$

Refer to (9), we can determine the positive and negative ideal solution, which are S_k^+ and S_k^-.

$$S^+ = (0.094, 0.019, 0.031, 0.015, 0.044, 0.028)$$
$$S^- = (0.022, 0.008, 0.007, 0.004, 0.011, 0.006)$$

We calculate the value of p_{ik} to obtain the relative proximity matrix P.

$$P = \begin{bmatrix} 0.285304 & 0.53278 & 0.13604 \\ 0.60942 & 0.69769 & 0.08118 \\ 0.65414 & 0.721998 & 0.09485 \end{bmatrix}$$

Based on the matrix P, refer to (12), we obtain the value of V^+ and V^- respectively.

$$V^+ = (0.65414, 0.72200, 0.13604)$$
$$V^- = (0.28530, 0.53278, 0.08118)$$

Refer to (13), we obtain the final relatively proximity Q_i to determine the optimal treatment regimen.

$$Q_1 = 0.12, Q_2 = 0.83, Q_3 = 0.91$$

So the order of these treatment regimens is $Q_3 > Q_2 > Q_1$, and the optimal treatment regimen is X_3.

4 Discussion and Conclusion

For the clinical treatment decision making, the doctor whose professional knowledge is richer and more proficient in the pros and cons of the various medical treatment regimens and drugs, can choose the more reasonable treatment regimen rationally for the patient. However, the patient knows their health condition and expectation better. Besides the patient and doctor, there are other stakeholders, such as the nurse, patient's family members, pharmacist and so on, who have different preferences for the patient's treatment regimen, and these preferences need to be considered too. The shared decision making model will incorporate all of stakeholders into the treatment decision making system, and trade off their different preferences and decision making status so as to reach an optimal treatment regimen finally.

To strength the applicability and practicability of shared decision making, this paper put the fuzzy multi-attribute group decision making method into application, which can settle various problems encountered in the clinical practice. On the basis of the conjoint analysis, we will obtain each decision maker's preference for each attribute of the treatment regimen respectively, in the meantime, the evaluation of each attribute of the treatment regimen will be done with the triangular fuzzy number. Finally, we rank the alternatives and obtain the optimal regimen with the ideal point method. Based on existing literatures, this paper sets the relapsed or refractory chronic lymphocytic leukemia treatment as an example to implement the method mentioned above, puts the patient, physician, and general population into the shared decision making system, and obtains the optimal treatment regimen finally.

The results of the study show that the method presented as the above can be applied in the shared decision making, which can collect the preferences of various stakeholders, make a reasonable quantification of the fuzzy language appraisal, and deal with the relationships among the preferences of decision makers, the decision makers' weights. According to the order of the alternatives, the optimal treatment regimen is obtained. From the perspective of application, this method has the characteristics of simple, efficient and convenient, it can be applied into the clinical treatment making of a variety of complex diseases and provide the stakeholders with a satisfying treatment regimen.

There are some shortcomings of this research too, first of all, in this paper, the treatment decision we made is a one-time medical decision, however, in reality, the diseases will change constantly and the treatment will be made sequentially, so there is in a need of dynamic treatment decisions. Secondly, in the clinical practice, the risk attitude and fuzzy aversion of each decision maker will have an impact on the outcome for the shared decision making. In the future research, these shortcomings should be studied to enhance the applicability of the medical treatment regimen.

Acknowledgements. This research is supported by National Natural Science Foundation of China (No. 71171142) and Tianjin Philosophy and Social Sciences Research Planning Project (No. TJGL16-013Q).

References

1. M.J. Barry, S. Edgman-Levitan, Shared decision making–pinnacle of patient-centered care. N. Engl. J. Med. **366**(9), 780–781 (2012)
2. G. Elwyn, S. Laitner, A. Coulter, E. Walker, P. Watson, R. Thomson, Implementing shared decision making in the NHS. BMJ **341**, c5146 (2010)
3. C. Charles, A. Gafni, T. Whelan, Shared decision making in the medical encounter: what does it mean? (Or it takes at least two to tango). Soc. Sci. Med. **44**(5), 681–692 (1997)
4. A.M. Stiggelbout, A.H. Pieterse, J.CJ.M. Haes, Shared decision making: concepts, evidence, and practice. Patient Educ. Couns. **98**(10), 1172 (2015)
5. G. Elwyn, D.L. Frosch, S. Kobrin, Implementing shared decision making: consider all the consequences. Implement. Sci. **11**(1), 114 (2015)
6. M.J. Barry, Health decision aids to facilitate shared decision making in office practice. Ann. Intern. Med. **136**(2), 127–135 (2012)
7. Q. Zeng-Treitler, B. Gibson, B. Hill, J. Butler, C. Christensen, D. Redd et al., The effect of simulated narratives that leverage EMR data on shared decision making: a pilot study. BMC Res. Notes **9**(1), 359 (2016)
8. D.L. Frosch, B.W. Moulton, R.M. Wexler, M. Holmes-Rovner, R.J. Volk, C.A. Levin, Shared decision making in the united states: policy and implementation activity on multiple fronts. Zeitschrift für Evidenz, Fortbildung und Qualität im Gesundheitswesen **105**(4), 305–312 (2011)
9. C. Fullwood, A. Kennedy, A. Rogers, M. Eden, C. Gardner, J. Protheroe et al., Patients' experiences of shared decision making in primary care practices in the United Kingdom. Med. Decis. Mak. **33**(1), 26–36 (2013)
10. B. Chewning, C.L. Bylund, B. Shah, N.K. Arora, J.A. Gueguen, G. Makoul, Patient preferences for shared decisions: a systematic review. Patient Educ. Couns. **86**(1), 9–18 (2012)
11. R. Say, M. Murtagh, R. Thomson, Patients' preference for involvement in medical decision making: a narrative review. Patient Educ. Couns. **60**(2), 102–114 (2006)
12. J.G. Dolan, Shared decision making—transferring research into practice: The Analytic Hierarchy Process (AHP). Patient Educ. Couns. **73**(3), 418–425 (2008)

13. E. Landfeldt, J. Eriksson, S. Ireland, P. Musingarimi, C. Jackson, E. Tweats, M. Gaudig, Patient, physician, and general population preferences for treatment characteristics in relapsed or refractory chronic lymphocytic leukemia: a conjoint analysis. Leuk. Res. **40**, 17–23 (2016)
14. Z.S. Xu, J. Chen, An interactive method for fuzzy multiple attribute group decision making. Inf. Sci. **177**(1), 248–263 (2007)

An Optimal Investment Strategy Against Information Security Risks

Bing-ning Pan and Jing Xie(✉)

College of Management and Economics, Tianjin University, Tianjin, China
xiejing@tju.edu.cn

Abstract. Information risks generally become a great challenge for individuals and organizations around the world. Managing information risks has involved various tools and approaches, among which self-protection and cyber insurance are two important methods to control the residual risk and improve security level. This paper analyzes these companies' investment strategies on self-protection and insurance respectively, and presents a company's best choice in both the weakest-link case and the partial-correlation case. The result shows that for most parameter settings Nash equilibrium could be reached and that the companies' strategies have an obvious impact on each other.

Keywords: Information security · Insurance · Interdependent risks Self-protection

1 Introduction

Internet-related products and services are widely used among IT companies and individuals around the world, and information risks have thus become much severer than ever. Consequently all of these companies seek approaches to invest on information security, and various technical solutions have been proposed to handle this challenge. However, technical approaches alone could not eliminate information risks completely. Attacking methods have been developing much faster nowadays and it is almost impossible to detect all the potential breaches ahead of time. In addition, the cost of preventing IT infrastructures from being attacked is extremely high. Companies therefore have to strike a balance between reducing potential losses by buying insurance and increasing its own self-protection level.

Cyber insurance allows companies to hedge against cyber losses. By signing an insurance contract, these companies transfer part of their information risks to an insurance company while the insurer charges a fee for the service. Policyholders are willing to reduce uncertainty effectively. The emerging of cyber insurance market gives policyholders and insurers an opportunity to achieve a win-win result. However, detailed analysis and evaluation of the market are still missing, and the current situation has been consistently lower than expected. Therefore, making balance between self-protection level and insurance level is a crucial decision for every player in the Internet.

G. Q. Huang et al. (eds.), *Proceeding of the 24th International Conference on Industrial Engineering and Engineering Management 2018*,
https://doi.org/10.1007/978-981-13-3402-3_28

Compared with traditional security risks and attack forms, information security problems have their own distinctive characteristics, among which the most significant one is the interdependence of security investment strategies among different companies.

One company's security investment decision may have a marked impact on other companies' investment choices in the market, and finally influence the overall security level. In other words, security investment has positive externalities, which may cause free riding problem, making companies invest less in self-protection. Besides, if one company suffers great losses, then other companies in the same network will also be more likely in danger. There are mainly two causes for the phenomenon. On one hand, complex physical links among IT infrastructures lead to close connections of nodes in the network. On the other hand, logical connections are equally critical for information products. The correlation among information risks results in a higher probability of being attacked, so indirect security incidents should also be taken into consideration when deciding the security investment level of a company.

In this paper, we explore a game-theoretic model particularly based on existing researches [1, 2] of the public goods field. Our purpose is to study the relationship of one company's investment incentives the other company's investment strategies, and analyze the effect of competitors on the company's behavior.

2 Literature Review

Prior research on information risks and cyber insurance has produced some interesting findings about whether market equilibrium could be achieved and how to improve social security level.

Authors in [3] [4] mathematically show that security investment produces externality with the existence of interdependent risks, which decreases the companies' investment incentives. They argue that insurance may not be completely effective, and regulation or taxation should be introduced to prevent sub-optimal social security level.

In other recent works [5–8], authors analyze participants' investment strategies with cyber insurance. Authors in [5–7] illustrate that cyber insurance can help increase the company's security investment amount and security level as well as overall social welfare. Reference [8] states that that the negative effects of interdependent security and information risks could be reduced by punishment mechanism or information sharing system.

Reference [9] shows that the effect of interdependence and imperfect detection on the companies' incentives of security investment and buying insurance are actually opposite. This work also explains the implement procedures of the insurance contract in detail. Authors in [10] trace the development of cyber insurance back to traditional policies. Both these works show that cyber insurance policies have been more sophisticated and list some relevant legislations.

Later papers have studied the problem considering the risk aversion through a numerical approach, since analytic solutions are sometimes hard to obtain. Reference

[11] uses a concave utility function to model risk aversion, while this factor is also concerned in our model.

Recent studies on information security and cyber insurance are becoming deeper. Reference [12] compares different approaches to deal with interdependent security risks. It shows the limitation of cyber insurance alone and proposes two alternative methods to help internalize the externalities of security investment. The latest work [13] discusses both regulated monopolistic and competitive markets. The results reflect that an inefficient market equilibrium exists without price discrimination, whereas price discrimination can lead to social security improvement as long as the policyholders are willing to remain part of the market.

Literatures on other fields also give us a supplementary angle to view the information security problem. Reference [14] uses case studies to explore the competitive equilibrium between rational players and address the limits of pure Nash equilibrium, arguing for a more general form of settings. Reference [15] explores the problem in the perspective of privacy and rationality. It discusses how consumers make selections between long-term privacy and short-term benefits. The lack of enough information is a severe challenge in the real market. Other works like [16] also illustrate their opinion from a economic point of view and explore the possibilities of a collaboration among different players in the whole market.

After analyzing these results and outcomes of the previous literatures, we further develop the information security investment model in this work, and discuss the company's choice in self-protection level and insurance level respectively.

3 Model Description

Companies in the network can decide their self-protection level and insurance level depending their own evaluation of the market. Each of them choose the amount of money for self-protection investment and insurance simultaneously. In addition, these companies are supposed to be identical in terms of utility functions and other aspects. To simplify the computing process, we assume there are only two different companies labeled i and j in the information network, and analyze the two identical companies' security investment decisions.

The initial wealth w of company i and j are given. The attack probability is denoted by $p \in (0,1)$ and if the company is successfully attacked, it will suffer a loss of l. Companies can take two steps to prevent or mitigate losses, deciding the insurance level $s_i \in (0,1)$ and self-protection investment level $e_i \in (0,1)$. Parameter b and c are the unit cost of these two methods respectively.

The insurance part of investment directly reduces company i's losses once being attacked with a probability p, while the effectiveness of company i's self-protection investment also depends on the investment amount of company j, which is expressed by the joint security protection function $F(e_i, e_j)$.

Each company's objective is to maximize its own expected utility of wealth. Both companies make their decisions simultaneously. The utility function takes the following form:

$$U_i(e_i, s_i) = w - pl(1 - s_i)(1 - F(e_i, e_j)) - be_i - cs_i \tag{1}$$

We will discuss two possible situations of the game in the following sections. First we learn from the existing weakest-link model in [2], and then we will explore a novel situation called partial-correlation case.

3.1 Weakest-Link Case

In this case we assume the joint protection level of company i equals to the lowest protection level among all policyholders. In other words the user with the lowest protection level will determine the whole network's self-protection effect. This assumption partly reflects the interdependence of information security investment, as the weakest node in the network will influence other companies' security degree.

As there are only two companies in our model, the protection function $F(e_i, e_j)$ becomes:

$$F(e_i, e_j) = \min(e_i, e_j) = \begin{cases} e_i, e_i \le e_j \\ e_j, e_i > e_j \end{cases} \tag{2}$$

And (1) can be written as

$$U_i(e_i, s_i) = w - pl(1 - s_i)(1 - \min(e_i, e_j)) - be_i - cs_i \tag{3}$$

Therefore the probability of being attacked depends not only on a company's self-protection alone, but also on the comparison between its own self-protection and the other company's self-protection level. If there is an extremely week node in the network, then no one can totally survive an external attack.

3.2 Partial-Correlation Case

In this scenario we change the protection function, assuming that the total protection level is achieved by multiplication. Therefore the protection function becomes $F(e_i, e_j) = e_i \cdot e_j$ and the utility function can be written as:

$$U_i(e_i, s_i) = w - pl(1 - s_i)(1 - e_i \cdot e_j) - be_i - cs_i \tag{4}$$

Company j's decision has a larger influence over company i's security level, thus affecting i's decision making process.

In repeated games company i could observe the regularity and characteristics of company j's decisions, while in a single shot game it cannot guess the other player's choice in advance. Our model discusses the first condition and company j's self-protection investment level e_j is regarded as exogenous.

4 Discussion

In this section we will proceed to determine the optimal self-protection level and insurance level of each company by analyzing Nash equilibrium.

4.1 Weakest-Link Case

According to the above analysis, the optimal self-protection investment and insurance level by company i can be achieved by calculating the partial derivative of (3).

Take the first and second derivatives of $U_i(e_i, s_i)$ with respect to s_i, and the results are:

$$U'(s_i) = pl(1 - \min(e_i, e_j)) - c \tag{5}$$

$$U''(s_i) = 0 \tag{6}$$

Thus, the company i's maximal utility can only be reached when s_i equals to 0 or 1. The first and second derivatives of $U_i(e_i, s_i)$ with respect to e_i equal to:

$$U'(e_i) = \begin{cases} pl(1 - s_i) - b, e_i < e_j \\ -b, e_i > e_j \end{cases} \tag{7}$$

Therefore the optimal utility function is only possible when e_i's value is 0 or e_j.

If company i invest nothing in self-protection while the insurance level is absolutely zero or one, then we have:

$$U_i(0, 0) = w - pl \tag{8}$$

$$U_i(0, 1) = w - c \tag{9}$$

If company i choose the same self-protection investment level as its companion, the other two conditions are:

$$U_i(e_j, 0) = w - pl(1 - e_i) - be_j \tag{10}$$

$$U_i(e_j, 1) = w - be_j - c \tag{11}$$

We then compare the utility value of these four end points and obtain the solutions under three conditions:

(1) If $pl < b$ and $pl < c$, then $(e_i^*, s_i^*) = (0, 0)$.

Proposition 1 Company i choose not to invest in self-protection or insurance when the expected loss is less than the unit cost of self-protection investment or buying insurance.

(2) If the parameters satisfy the requirements

$$c < pl < b \text{ or } c < b < pl \text{ or } \begin{cases} pl > b \\ pl > c \\ e_j < \frac{pl-c}{pl-b} \end{cases} \tag{12}$$

then $(e_i^*, s_i^*) = (0, 1)$.

Proposition 2 If the insurance expense is less than both the expected loss and the self-protection cost, or if the expected loss is greater than the unit cost of both self-protection and insurance but company j's self-protection level is relatively low, then company i will desire full insurance and invest absolutely nothing on self-protection.

(3) If the parameters satisfy the requirements

$$b < pl < c \text{ or } \begin{cases} pl > b \\ c \geq b \\ e_j > \frac{pl-c}{pl-b} \end{cases} \tag{13}$$

then $(e_i^*, s_i^*) = (e_j, 0)$.

Proposition 3 If company i's expected loss is greater than self-protection cost and less than insurance cost, or if the self-protection expense of i is rather small while j's self-protection level is relatively high, then company i will choose the same self-protection level as company j while not consider insurance at all.

4.2 Partial Correlation Case

Through a brief comparison between the results in the partial correlation case and the weakest-link case, we discover that the main difference between these two cases is that the latter one directly add company j's self-protection level to the total protection degree of company i.

According to (4) we can easily realize that the utility function in this case is twice differentiable, and that the second partial derivatives of $U_i(e_i, s_i)$ with respect to e_i and s_i both equal to zero. Therefore, the extreme value of the expected utility may be obtained if (e_i, s_i) equals to $(0,0)$, $(0,1)$, $(1,0)$ or $(1,1)$.

Let's analyze the utility of these four possibilities:

$$U_i(0,0) = w - pl \tag{14}$$

$$U_i(0,1) = w - c \tag{15}$$

$$U_i(e_j, 0) = w - pl(1 - e_i) - b \tag{16}$$

$$U_i(e_j, 1) = w - b - c \tag{17}$$

After comparing the above equations, we can draw several rules about company i's decision:

(1) If $pl < b/e_j$ and $pl < c$, then $(e_i^*, s_i^*) = (0, 0)$.

Proposition 4 Compared with the weakest-link case, companies in the partial-correlation case have fewer incentives to improve their self-protection and insurance proportion, especially when the other company in the network invests too little in its self-protection.

(2) If the parameters satisfy the requirements

$$pl > c \text{ or } 1 - e_j > \frac{c - b}{pl} \tag{18}$$

then $(e_i^*, s_i^*) = (0, 1)$.

Proposition 5 If company i's expected loss is larger than the charging rate of its insurance, while the other company in the interdependent market has a fairly low self-protection level, than company prefer to reduce the self-protection level to zero and pursue full insurance.

(3) If the parameters satisfy the requirements

$$\begin{cases} ple_j > b \\ c - b > pl(1 - e_j) \end{cases} \tag{19}$$

then $(e_i^*, s_i^*) = (e_j, 0)$.

Proposition 6 If the unit cost of insurance is much greater than the unit cost of self-protection and the gap between them exceeds the other company's unprotected losses, while the other company ensures a certain degree of security, company i will pick exactly the same level of self-protection and will not pay for insurance.

5 Conclusion

Our model presents a fundamental research of the optimal security investment strategy for information assets. We suppose that security investment has both the features of public good and private good, thus divide the investment into two parts: self-protection investment and insurance investment.

We show that market equilibrium exists in both the weakest-link case and the partial correlation case where information investments are interdependent. Our results provide some reference for Internet-related companies to decide the amount of information investment, and present the Nash equilibria in two different scenarios. From the analytical solution of our model, it becomes clear to us that a company's self-protection investment level and insurance level is largely related to the comparison of its expected loss, self-protection cost, insurance premium and the other company's self-protect decision. On the other hand, whether the other company j pay for insurance or not has nothing to do with company i's investment amount and security level.

In an information security market where there are only two identical Internet-related companies, both of them have three different strategies: investing nothing at all, arranging full insurance while no self-protection, and choosing minimal self-protection level while rejecting influence. We analyze the problem in two different cases. The interdependence factor is much stronger in the partial-correlation case than in the weakest-link case, and both two cases show that company i's decision is determined by the relationship between its own potential loss/cost parameters and the other company's state as well.

Despite the above contributions, there still remain limitations in our study. First, we give the analytical solution of the general situations, while future researchers could focus more on the numerical analysis. Second, we suppose there are only two identical policyholders in the market, while future research could further explore more complicated situations where the market includes n (n > 2) policyholders. Third, our model assumes that the player has access to perfect information when choosing its investment strategy, but in the real world it cannot observe the exact amount of the other company's investment level. Future works could consider models with imperfect information where adverse selection and moral hazard may exist.

References

1. H. Varian, System reliability and free riding. *Economics of Information Security Kluwer*, (2004), pp. 1–15
2. J. Grossklags, N. Christin, J. Chuang, Secure or insure?: a game-theoretic analysis of information security games, in *International Conference on World Wide Web*, WWW 2008, Beijing, China, April DBLP, (2008), pp. 209–218
3. H. Kunreuther, G. Heal, Interdependent security: the case of identical agents. SSRN Electron. J. (2002)
4. G. Heal, H. Kunreuther, Interdependent security: a general model, in *National Bureau of Economic Research, Inc* (2004)
5. J. Kesan, R. Majuca, W. Yurcik, The economic case for cyberinsurance, University of Illinois Legal Working Paper (2004)
6. J. Bolot, M. Lelarge, Cyber insurance as an incentive for Internet security, in *Managing Information Risk and the Economics of Security*, (Springer, US, 2009), pp. 269–290
7. M. Lelarge, J. Bolot, Economic incentives to increase security in the Internet: the case for insurance, in *Infocom IEEE*, (2009), pp. 1494–1502
8. H. Ogut, N. Menon, S. Raghunathan, Cyber insurance and IT security investment: impact of interdependent risk, in *Proceedings of Weis'* (2005)
9. H. Ogut, S. Raghunathan, N.M. Menon, Information security risk management through self-protection and insurance, in *2014 25th International Workshop on Database and Expert Systems Applications (DEXA) IEEE Computer Society*, (2005), pp. 296–300

10. R.P. Majuca, W. Yurcik, J.P. Kesan, The evolution of cyberinsurance. *ACM Computing Research Repository (CoRR)* (2006)
11. J. C. Bolot, and M. Lelarge, A new perspective on Internet security using insurance, *Proceedings—IEEE INFOCOM*, (2008), pp. 1948–1956
12. X. Zhao, L. Xue, A.B. Whinston, Managing interdependent information security risks: cyberinsurance, managed security services, and risk pooling arrangements. J. Manag. Inf. Sys. **30**(1), 123–152 (2013)
13. R. Pal et al., Will cyber-insurance improve network security? a market analysis, in *IEEE INFOCOM 2014—IEEE Conference on Computer Communications*, (2014), pp. 235–243
14. N, Christin, J. Grossklags, J. Chuang, Near rationality and competitive equilibria in networked systems, in *ACM SIGCOMM Workshop on Practice and Theory of Incentives in Networked Systems ACM*,(2014), pp. 213–219
15. A. Acquisti, J. Grossklags, Privacy and rationality in individual decision making. IEEE Secur. Priv. **3**(1), 26–33 (2005)
16. G. Danezis, R. Anderson, The economics of resisting censorship. IEEE Secur. Priv. Mag. **3** (1), 45–50 (2005)

A Measure for Modularity and Comparative Analysis of Modularity Metrics

Xian-fu Cheng(✉), Chong Wan, Hao-yang Qiu, and Jun-yi Luo

School of Mechatronics & Vehicle Engineering, East China Jiaotong University, Nanchang, Jiangxi, China
chxf_xn@sina.com

Abstract. Modularity is an effect method to response to the demand for increasing variety in the market. Modular product consists of modules, aims at decomposing complex products into simplified components by dissociated interfaces. Quantifying the degree of modularity is central to the research on modular product design. In this paper, we surveyed the literature for modularity measures, and proposed a metric for modularity to maximize cohesion degree within the modules and minimize coupling degree between the modules. At the same, comparative analysis is made on modularity metrics addressed by some literatures and this paper. It is hoped that this research will provide an approach to analyze the relationship between the two modules, which will shape future research in the field of modular product family architecture.

Keywords: Metrics · Modularity · Modular product design

1 Introduction

Modularity is an effect strategy to response to the demand for increasing variety in the market. Modular product aims at decomposing complex products into simplified components by dissociated interfaces. These simplified and relatively independent components, called modules, can be combined to create a product or product family. Using modularity can trade off increased product variety and work performance.

Compared to integral products, modular products have some advantages, such as rich product variety, rapid response to the market, relatively independent subsystem, part number minimization, and so on. Due to these benefits, modular product has been developed by many enterprises in the field of automobile, train, crane, machine tool, mechatronics, building, etc. Some approaches to modular product design have been presented by various researches. For example, Stone and Wood [1] proposed a set of heuristics to determine the modules. Bryant et al. [2] proposed a modular method from the viewpoint of recycling. Gu and Sosale [3] studied product modularization method for life cycle engineering. Yu et al. [4] presented a product modular design method incorporating product life issues and physical connection relations based on group genetic algorithm.

G. Q. Huang et al. (eds.), *Proceeding of the 24th International Conference on Industrial Engineering and Engineering Management 2018*,
https://doi.org/10.1007/978-981-13-3402-3_29

Modularity has important effect on an enterprise's productivity, development time and manufacture cost of the product. Evaluating the degree of modularity is central to the research on this subject. Some literature presented their measure for modularity, but they fit different situations. In this paper, we surveyed the literature for modularity measures, and proposed a metric for modularity to maximize cohesion degree within the modules and minimize coupling degree between the modules. It is hoped that this research will provide an approach to analyze the relationship between any two modules.

2 Measure for Modularity in Literatures

In the past literatures, some researches proposed the measure for modularity. Höltt-Otto et al. [5] compared 13 modularity metrics that have been addressed by various researches.

Algeddawy and Elmaraghy [6] proposed a simple modularity index for product architecture that is based on the 0–1 count. They expressed the degree of modularity with the number of intra-relationships among components within modules and the number of inter-relationships among modules, denoted as MI.

$$\mathrm{MI} = \mathrm{I} + \mathrm{Z} \tag{1}$$

where I is the number of '1' entries in the DSM outside modules, and Z is the zero entries inside modules. The smaller MI, the better the clustering result.

Sosa and Eppinger [7] presented a measurement for modularity on the basis of the way components share design interactions. They utilized a chi-square test to demonstrate different connectivity density between integral and modular systems. Guo and Gershenson [8] illustrated Sosa and Eppinger's method, and measure modularity with M_s, as follows.

$$M_s = \frac{\sum_{k=1}^{M}\left(\sum_{i=n_k}^{m_k}\left(\sum_{j=1}^{n_k-1} R_{ij} + \sum_{j=m_k+1}^{N} R_{ij}\right)\right)}{\sum_{k=1}^{M}\left(\sum_{i=n_k}^{m_k}\sum_{j=n_k}^{m_k} R_{ij} + \left(\sum_{i=n_k}^{m_k}\left(\sum_{j=1}^{n_k-1} R_{ij} + \sum_{j=m_k+1}^{N} R_{ij}\right)\right)\right)} \tag{2}$$

where, n_k is the index of the first element (component) in the kth cluster, m_k is the index of the last element in the kth cluster, M is the total number of clusters in the system, N is the total number of elements in the cluster, and R_{ij} is the value of the ith row and jth column element in the matrix of modularity. The metric of modularity varies from 0 to 1.

Guo and Gershenson [9] proposed a metric that measures the connectivity inside and outside clusters in a modularity matrix:

$$M_{G\&G} = \frac{\sum_{k=1}^{M}\left(\sum_{i=n_k}^{m_k}\sum_{j=n_k}^{m_k} R_{ij}/(m_k - n_k + 1)^2\right) - \sum_{k=1}^{M}\left(\sum_{i=n_k}^{m_k}\left(\sum_{j=1}^{n_k-1} R_{ij} + \sum_{j=m_k+1}^{N} R_{ij}\right)/(m_k - n_k + 1)(N - m_k + n_k - 1)\right)}{M} \tag{3}$$

As with Eq. (2), n_k is the index of the first element (component) in the kth cluster, m_k is the index of the last element in the kth cluster, M is the total number of clusters in the system, N is the total number of elements in the cluster, and R_{ij} is the value of the ith row and jth column element in the matrix of modularity. When the product is a fully integral system, its metric value is 0. When the product is composed of fully independent modules, its metric value is 1. The metric has been normalized and has nothing to do with size scale, which is a significant improvement from the early metrics.

Meng and Jiang [10] presented principles for module planning for product family designing. To guarantee the modules' function independency and minimize the physical relations and geometry relations among modules, the metrics are as follows.

The relative cohesion degree of module M_k is represented by

$$h_k = \left(\sum_{i=1}^{N_k - 1} \sum_{j=i+1}^{N_k} r_{k_i k_j} \right) \Big/ \sum_{i=1}^{N_k - 1} \sum_{j=i+1}^{N_k} 1 \tag{4}$$

In the equation, r_{kikj} is the comprehensive relation cohesion degree between the ith component and the jth component inside the module M. N_k is the numbers of components in module M_k. Assume there are U modules, the relative cohesion degree of all modules are expressed as

$$h = \sum_{k=1}^{U} h_k \tag{5}$$

The coupling degree between M_l and M_n is represented by

$$c_{\ln} = \left(\sum_{p=1}^{N_l} \sum_{q=1}^{N_n} r_{l_p n_q} \right) \Big/ \sum_{p=1}^{N_l} \sum_{q=1}^{N_n} 1 \tag{6}$$

In the equation, r_{lpnq} is the comprehensive correlation between the pth component of module M_l and the qth component inside the module M_n. The average relation coupling degree of all modules is expressed as

$$c = \sum_{l=1}^{U-1} \sum_{n=l+1}^{U} c_{\ln} \tag{7}$$

The dependency minimization criterion function is expressed as

$$F_A = \min(c - h) \tag{8}$$

3 A Module Partition Metric

In module partition, this paper aims at to maximize cohesion degree within the modules and minimize coupling degree between the modules. Cohesion degree in a single cluster shows intra-connection on physical and functional relationships between components. Coupling degree shows inter-connection on physical and functional relationships between two clusters. The larger the total cohesion degree and the smaller the total coupling degree, the better the module partition. Among the past work on measuring modularity, the metric proposed by Guo and Gershenson [9] are more comprehensive. Not only consider the asymmetric situation, but also for the non-Boolean matrix. However, the main considered coupling degree is the overall module, not suitable for calculating the coupling between the two modules, and is not conducive to the correlation analysis between the two modules. This paper proposed a metric for modularity on the basis of its study.

Suppose that the product is composed of a total of M modules, n_k is the index of the first element in the kth cluster M_k, and m_k is the index of the last element in the kth cluster, the cohesion degree of M_k may be expressed as following.

$$I_k = \sum_{i=n_k}^{m_k} \sum_{j=n_k}^{m_k} DP(i,j)/(m_k - n_k + 1)^2 \tag{9}$$

The total cohesion degree of M clusters for the product is represented by

$$I = \sum_{k=1}^{M} I_k = \sum_{k=1}^{m} \sum_{i=n_k}^{m_k} \sum_{j=n_k}^{m_k} DP(i,j)/(m_k - n_k + 1)^2 \tag{10}$$

The average cohesion degree of M clusters for the product is represented by

$$\bar{I} = \frac{I}{M} \tag{11}$$

The coupling degree between two clusters can be quantified by the total incidence degree among all elements in one cluster and that in the other cluster. Providing n_p is the index of the first element in the pth cluster M_p, and m_p is the index of the last element in the pth cluster, the coupling degree between M_k and M_p is expressed as following.

$$E_{k,p} = \frac{\sum_{i=n_k}^{m_k} \sum_{j=n_p}^{m_p} DP(i,j) + \sum_{i=n_p}^{m_p} \sum_{j=n_k}^{m_k} DP(i,j)}{2 \times (m_k - n_k + 1)(m_p - n_p + 1)} \tag{12}$$

The total coupling degree of M clusters for the product is represented by

$$E = \sum_{k=1}^{M-1} \sum_{p=k+1}^{M} E_{k,p} \tag{13}$$

According to the number of coupling comparisons, the average coupling degree of M clusters for the product is represented by

$$\overline{E} = \frac{E}{(M-1) + (M-2) + \cdots + 1} = \frac{E}{M(M-1)/2} \tag{14}$$

Thus, a metric of modularity for modular product considering the intra- and inter-cluster association is computed as

$$F_M = \overline{I} - \overline{E} \tag{15}$$

Tabel 1. Component correlation degree

Gate valve		1	2	3	4	5	6	7	8	9	10
Valve body	1	1.00	0.00	0.49	0.80	0.00	0.00	0.00	0.00	0.00	0.00
Valve stem	2	0.00	1.00	0.71	0.11	0.80	0.00	0.11	0.42	0.42	0.42
Wedge disc	3	0.49	0.71	1.00	0.00	0.00	0.00	0.00	0.00	0.00	0.00
Valve cover	4	0.80	0.11	0.00	1.00	0.00	0.51	0.00	0.42	0.42	0.31
Valve stem nut	5	0.00	0.80	0.00	0.00	1.00	0.42	0.38	0.00	0.00	0.00
Bracket	6	0.00	0.00	0.00	0.51	0.42	1.00	0.00	0.00	0.00	0.00
Hand wheel	7	0.00	0.11	0.00	0.00	0.38	0.00	1.00	0.00	0.00	0.00
Packing	8	0.00	0.42	0.00	0.42	0.00	0.00	0.00	1.00	0.80	0.71
Packing seat	9	0.00	0.42	0.00	0.42	0.00	0.00	0.00	0.80	1.00	0.27
Packing gland	10	0.00	0.42	0.00	0.31	0.00	0.00	0.00	0.71	0.27	1.00

4 Case Study

4.1 Case 1

Reference [11] on gate valve of a high pressure valve factory is considered as an instance to compare and verify the presented metric for modularity. The results of the correlation degree among all components are shown in Table 1.

The dynamic results of module division for gate valve are shown in Fig. 1. From Fig. 1, it can be seen that the clustering results vary according different value of threshold t.

To determine the best design rule of modularity, we calculated modularity metrics listed above. The metric of MI is not suitable for this case and will be discussed in the next case, since it is only fit for 0–1 value. Table 2 shows the calculated for categories 4, 5 and 7 in each metric. Results show that there was an upward trend with M_S, $M_{G\&G}$, F_A and F_M. The maximum value of every metric point to categories 7.

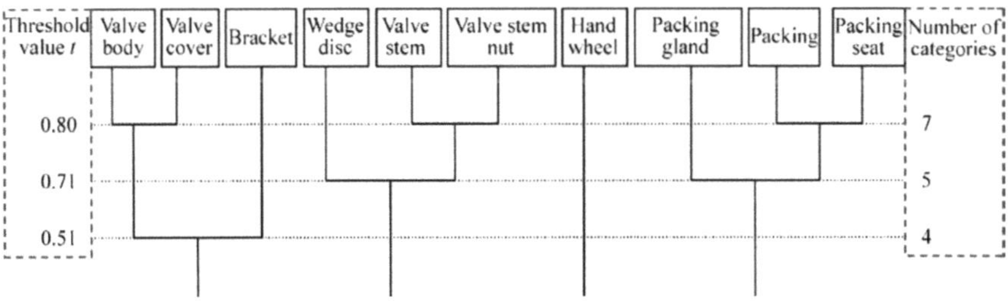

Fig. 1. Dynamic results of valve module division

Tabel 2. The calculated for gate valve in each metric

Categories	4	5	7
M_S	0.290	0.328	0.453
$M_{G\&G}$	0.654	0.750	0.818
F_A	−0.989	−0.907	0.213
F_M	0.665	0.767	0.833

4.2 Case 2

The 2nd case is from reference [6] on an automobile body-in-white (BIW). BIW is composed of large number of metal parts, as shown in Table 3. The relationships among components of BIW are depicted by DSM, as shown in Table 4. Self relationships of each element is indicated with '1' entry.

Table 3. Body-in-white components [6]

1. Adapter A-Pillar Roof Rail	15. Floor Panel	29. Reinforcement Rocker Rear
2. Adapter B-Pillar Roof Rail	16. Front Header	30. Rocker
3. A-Pillar Inner	17. Front Side Rail	31. Roof Bow
4. A-Pillar Inner	18. Front Suspension Housing	32. Roof Panel
5. A-Pillar Reinforcement	19. Heel kick	33. Roof Rail
6. Back Panel	20. Rear Floor Panel	34. Seat Cross Member Front
7. Back Panel Side	21. Rear Floor Side	35. Seat Cross Member Rear
8. Body Side	22. Rear Header	36. Shotgun
9. B-Pillar	23. Rear Panel Inner Lower	37. Spare Wheel Well
10. Channel	24. Rear Panel Inner Upper	38. Wheelhouse
11. Cowl	25. Rear Side Floor	
12. Cross track Rear Floor	26. Rear Side Rail	
13. Dash Cross Member	27. Rear Side Rail Centre	
14. Dash Panel	28. Rear Side Rail Front	

Tabel 4. The body-in-white DSM

	1	2	3	4	5	6	7	8	9	10	11	12	13
1	1	0	1	1	0	0	0	1	0	0	1	0	0
2	0	1	0	0	0	0	0	1	1	0	0	0	0
3	1	0	1	0	0	0	0	0	0	0	1	0	0
4	1	0	0	1	0	0	0	1	0	0	0	0	0
5	0	0	0	0	1	1	1	0	0	0	0	0	0
6	0	0	0	0	1	1	1	0	0	0	0	0	0
7	0	0	0	0	1	1	1	0	0	0	0	0	0
8	1	1	0	1	0	0	0	1	1	0	0	0	0
9	0	1	0	0	0	0	0	1	1	0	0	0	0
10	0	0	0	0	0	0	0	0	0	1	0	0	1
11	1	0	1	0	0	0	0	0	0	0	1	0	0
12	0	0	0	0	0	0	0	0	0	0	0	1	0
13	0	0	0	0	0	0	0	0	0	1	0	0	1
14	0	0	1	1	0	0	0	0	0	1	1	0	1
15	0	0	1	0	0	0	0	1	0	0	0	0	0
16	0	0	0	0	0	0	0	0	0	0	0	0	0
17	0	0	0	0	0	0	0	0	0	0	0	0	0
18	0	0	0	0	0	0	0	0	0	0	1	0	0
19	0	0	0	0	0	0	0	0	0	1	0	0	0
20	0	0	0	0	0	0	0	0	0	0	0	1	0
21	0	0	0	0	1	1	1	0	0	0	0	0	0
22	0	0	0	0	0	0	0	0	0	0	0	0	0
23	0	0	0	0	0	0	1	1	0	0	0	0	0
24	0	0	0	0	0	0	1	1	0	0	0	0	0
25	0	0	0	0	0	0	0	0	0	0	0	0	0
26	0	0	0	0	1	0	0	0	0	0	0	0	0
27	0	0	0	0	0	0	0	0	0	0	0	0	0
28	0	0	0	0	0	0	0	0	0	0	0	1	0
29	0	0	0	0	0	0	0	1	1	0	0	0	0
30	0	0	1	1	0	0	0	1	1	0	0	0	0
31	0	0	0	0	0	0	0	1	0	0	0	0	0
32	0	0	0	0	0	0	0	1	0	0	0	0	0
33	1	1	0	0	0	0	0	1	0	0	0	0	0
34	0	0	0	0	0	0	0	1	0	1	0	0	0
35	0	0	0	0	0	0	0	1	0	1	0	0	0
36	1	0	1	1	0	0	0	0	0	0	1	0	0
37	0	0	0	0	1	0	0	0	0	0	0	1	0
38	0	0	0	0	0	0	1	0	0	0	0	0	0

(*continued*)

Tabel 4. (*continued*)

	14	15	16	17	18	19	20	21	22	23	24	25	26
	14	15	16	17	18	19	20	21	22	23	24	25	26
1	0	0	0	0	0	0	0	0	0	0	0	0	0
2	0	0	0	0	0	0	0	0	0	0	0	0	0
3	1	1	0	0	0	0	0	0	0	0	0	0	0
4	1	0	0	0	0	0	0	0	0	0	0	0	0
5	0	0	0	0	0	0	0	1	0	0	0	0	1
6	0	0	0	0	0	0	0	1	0	0	0	0	0
7	0	0	0	0	0	0	0	1	0	1	1	0	0
8	0	1	0	0	0	0	0	0	0	1	1	0	0
9	0	0	0	0	0	0	0	0	0	0	0	0	0
10	1	0	0	0	0	1	0	0	0	0	0	0	0
11	1	0	0	0	1	0	0	0	0	0	0	0	0
12	0	0	0	0	0	0	1	0	0	0	0	0	0
13	1	0	0	0	0	0	0	0	0	0	0	0	0
14	1	1	0	1	1	0	0	0	0	0	0	0	0
15	1	1	0	1	0	1	0	0	0	0	0	0	0
16	0	0	1	0	0	0	0	0	0	0	0	0	0
17	1	1	0	1	1	0	0	0	0	0	0	0	0
18	1	0	0	1	1	0	0	0	0	0	0	0	0
19	0	1	0	0	0	1	1	0	0	0	0	1	0
20	0	0	0	0	0	1	1	0	0	0	0	1	0
21	0	0	0	0	0	0	0	1	0	0	0	1	1
22	0	0	0	0	0	0	0	0	1	0	0	0	0
23	0	0	0	0	0	0	0	0	0	1	1	0	0
24	0	0	0	0	0	0	0	0	0	1	1	0	0
25	0	0	0	0	0	1	1	1	0	0	0	1	1
26	0	0	0	0	0	0	0	1	0	0	0	1	1
27	0	1	0	1	0	0	0	0	0	0	0	0	0
28	0	1	0	0	0	1	1	1	0	1	0	1	1
29	0	0	0	0	0	1	0	0	0	1	0	1	0
30	0	1	0	0	0	1	0	0	0	1	0	1	0
31	0	0	0	0	0	0	0	0	0	0	0	0	0
32	0	0	1	0	0	0	0	0	1	0	1	0	0
33	0	0	1	0	0	0	0	0	0	0	0	0	0
34	0	0	0	0	0	0	0	0	0	0	0	0	0
35	0	0	0	0	0	0	0	0	0	0	0	0	0
36	0	0	0	0	1	0	0	0	0	0	0	0	0
37	0	0	0	0	0	0	1	1	0	0	0	0	1
38	0	0	0	0	0	0	0	1	0	1	1	1	1

(*continued*)

Tabel 4. *(continued)*

	27	28	29	30	31	32	33	34	35	36	37	38
	27	28	29	30	31	32	33	34	35	36	37	38
1	0	0	0	0	0	0	1	0	0	1	0	0
2	0	0	0	0	0	0	1	0	0	0	0	0
3	0	0	0	1	0	0	0	0	0	1	0	0
4	0	0	0	1	0	0	0	0	0	1	0	0
5	0	0	0	0	0	0	0	0	0	0	1	0
6	0	0	0	0	0	0	0	0	0	0	0	0
7	0	0	0	0	0	0	0	0	0	0	0	1
8	0	0	1	1	1	1	1	1	1	0	0	0
9	0	0	1	1	0	0	0	0	0	0	0	0
10	0	0	0	0	0	0	0	1	1	0	0	0
11	0	0	0	0	0	0	0	0	0	1	0	0
12	0	1	0	0	0	0	0	0	0	0	1	0
13	0	0	0	0	0	0	0	0	0	0	0	0
14	0	0	0	0	0	0	0	0	0	0	0	0
15	1	1	0	1	0	0	0	0	0	0	0	0
16	0	0	0	0	0	1	1	0	0	0	0	0
17	1	0	0	0	0	0	0	0	0	0	0	0
18	0	0	0	0	0	0	0	0	0	1	0	0
19	0	1	1	1	0	0	0	0	0	0	0	0
20	0	1	0	0	0	0	0	0	0	0	1	0
21	0	1	0	0	0	0	0	0	0	0	1	1
22	0	0	0	0	0	1	0	0	0	0	0	0
23	0	1	1	1	0	0	0	0	0	0	0	1
24	0	0	0	0	0	1	0	0	0	0	0	1
25	0	1	1	1	0	0	0	0	0	0	0	1
26	0	1	0	0	0	0	0	0	0	0	1	1
27	1	1	0	1	0	0	0	0	0	0	0	0
28	1	1	1	1	0	0	0	0	0	0	0	1
29	0	1	1	1	0	0	0	0	0	0	0	1
30	1	1	1	1	0	0	0	1	1	0	0	0
31	0	0	0	0	1	1	1	0	0	0	0	0
32	0	0	0	0	1	1	1	0	0	0	0	0
33	0	0	0	0	1	1	1	0	0	0	0	0
34	0	0	0	1	0	0	0	1	0	0	0	0
35	0	0	0	1	0	0	0	0	1	0	0	0
36	0	0	0	0	0	0	0	0	0	1	0	0
37	0	0	0	0	0	0	0	0	0	0	1	0
38	0	1	1	0	0	0	0	0	0	0	0	1

According to The DSM of the BIW in reference [6], a cladogram is created by NONA cladistics software. The resulting cladogram is shown in Fig. 2.

As in case 1, we calculated case 2 with every metric in Table 5. Results of MI show that level 6 scores the least MI = 160. There is an upward trend with results of M_S. Results of F_M and $M_{G\&G}$ start climbing steadily and reach a peak in level 9, then fallen. However, results of F_A are irregular. According to Table 5, we can conclude that the best level is 9. Although the least MI is 160, level 9 scores MI = 164, very closed to level 6. The metric of MI is suitable for Boolean matrix, and $M_{G\&G}$ considered symmetric DSMs only. This paper proposed a metric that suit for both non-Boolean matrix and asymmetric, also provide an approach to analyze the relationship between two modules.

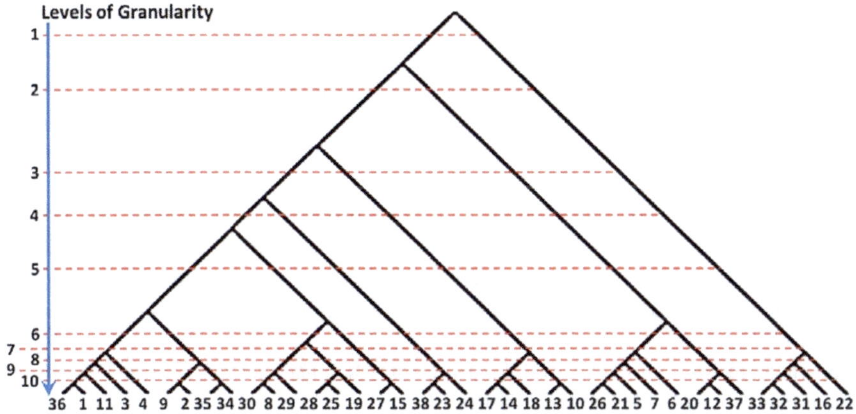

Fig. 2. Resulting cladogram of the BIW

Table 5. The calculated for an automobile body-in-white in each metric

Level	1	2	3	4	5	6	7	8	9	10
MI	884	536	380	310	218	160	166	162	164	196
M_S	0.047	0.146	0.233	0.296	0.399	0.472	0.520	0.583	0.638	0.764
$M_{G\&G}$	0.407	0.453	0.499	0.913	0.576	0.710	0.728	0.833	0.858	0.844
F_A	−0.745	−1.217	−1.722	−2.323	−2.340	−3.477	−2.177	−0.313	5.167	30.000
F_M	0.407	0.463	0.516	0.921	0.577	0.715	0.711	0.833	0.863	0.855

5 Discussion

This paper proposed a metric for modularity, aiming at to maximize cohesion degree within the modules and minimize coupling degree between the modules. And we analyzed and compared some modularity metrics that have been applied for systems or products in past work.

First of all, the cohesion degree and coupling degree proposed in this paper vary from 0 to 1. When DP is 0 in Eq. (9), the average cohesion degree of all clusters is 0 according to Eqs. (10) and (11). However, when DP is 1 in Eq. (9), the average cohesion degree of all clusters is 1. In a similar way, when DP is 0 in Eq. (12), the coupling degree of modules is 0 by Eqs. (13) and (14), and when DP is 1 in Eq. (12), the coupling degree of modules is 1. Most of other metrics vary from a larger extent. Obviously, 0 to 1 is more rational. In addition, other metrics consider a part of the DSM, they can be used to symmetric situation only. The metric proposed in this paper is more comprehensive, not only consider the asymmetric situation, but also for the non-Boolean matrix. It calculated the coupling between the two modules to analyze the correlation between the two modules, and then determine the priority for two modules.

6 Conclusion

Modularity has been proved to be an effective response to the demands for trading off increased product variety and work performance. Evaluating the degree of modularity is important to the research on this subject. Though some literature presented their measure for modularity, they are suit for different situations. This paper surveyed the literature for modularity measures and proposed a metric for modularity to maximize cohesion degree within the modules and minimize coupling degree between the modules, also suit for both non-Boolean matrix and asymmetric. It is hoped that this research will provide an approach to analyze the relationship between the two modules, which will shape future research in the field of modular product family architecture.

Acknowledgements. This work was funded by the National Natural Science Foundation of China (Grant No. 71462007 and 51765019).

References

1. R.B. Stone, K.L. Wood, Development of a functional basis for design. J. Mech. Des. **122**(4), 359–370 (2000)
2. C.R. Bryant, K.L. Sivaramakrishnan, M. Van Wie, R.B. Stone, D.A. McAdams, A modular design approach to support sustainable design, in *ASME 2004 International Design Engineering Technical Conferences and Computers and Information in Engineering Conference* (IDETC/CIE2004), Salt Lake City, Utah, USA, 28 September–2 October, 2004
3. P. Gu, S. Sosale, Product modularization for life cycle engineering. Robot. Comput. Integr. Manuf. **15**(5), 387–401 (1999)
4. S. Yu, Q. Yang, J. Tao, X. Tian, F. Yin, Product modular design incorporating life cycle issues—group genetic algorithm (GGA) based method. J. Clean. Prod. **19**(9), 1016–1032 (2011)
5. K. Höltt-Otto, N.A. Chiriac, D. Lysy et al. Comparative analysis of coupling modularity metrics. J. Eng. Des. **23**(10–11), 790–806 (2012)
6. T. Algeddawy, H. Elmaraghy, Optimum granularity level of modular product design architecture. CIRP Ann.—Manuf. Technol. **62**(1), 151–154 (2013)
7. M.E. Sosa, S.D. Eppinger, Identifying modular and integrative systems and their impact on design team interactions. ASME J. Mech. Des. **125**(2), 240–252 (2003)

8. F. Guo, J.K. Gershenson, Comparison of modular measurement methods based on consistency analysis and sensitivity analysis, in *ASME 2003 International Design Engineering Technical Conferences and Computers and Information in Engineering Conference*, American Society of Mechanical Engineers, pp. 393–401, 2003
9. F. Guo, J.K. Gershenson, A comparison of modular product design methods on improvement and iteration, in *ASME 2004 International Design Engineering Technical Conferences and Computers and Information in Engineering Conference*, pp. 261–269, 2004
10. X.H. Meng, Z.H. Jiang, Module planning for product family design. J. Shanghai Jiaotong Univ. **40**(11), 1869–1872 (2006). (chinese)
11. T.Y. Lei, W.P. Peng, J. Lei, Y.H. Zhong, Q.H. Zhang, A product module mining method for PLM database. J. Cent. South Univ. **23**(7), 1754–1766 (2016)

Analysis of Influence Factors for Reverse Recycling of Damaged Shared Bicycles Based on CRITIC-DEMATEL Method

Fu-min Deng(✉), Xian-yi Peng, and Jing Xu

Department of Industrial Engineering, Sichuan University, Chengdu, China
jingxu@scu.edu.cn

Abstract. Considering the low recovery and reuse rate in recycling damaged shared bicycle currently, this paper aims to conduct systematic studies on influence factors in terms of enterprise, product, government, and society. Combining objective weight valuation of the CRITIC (Criteria Importance though Intercrieria Correlation) method and causal classification and importance analysis of the DEMATEL (Decision Making Trial and Evaluation Laboratory) method. we overcome the shortcomings of the traditional DEMATEL method that establish initial direct influence matrix based on expert scoring or questionnaire survey, thus putting forward the CRITIC-DEMATEL method. The research example shows that the new method not only can effectively identify the key factors affecting the reverse recycling of damaged shared bicycles, but also can improve the accuracy of solving actual problems, while at the same time, it can provide scientific support for the damaged bicycle recycling.

Keywords: Damaged shared bicycles · DEMATEL · CRITIC
Influence factors · Reverse recycling

1 Introduction

According to previous reports, almost 20 brands launched about 2 million shared bicycles in 2016, and the total amount is expected to close to 20 million in 2017. These bicycles will produce almost 300 thousand tons of scrap metal when they are scrapped [1]. In the process of recycling, the present situation we are facing is the low recovery and utilization rate, the serious waste of resources, the environmental pollution, as well as poor social benefits, all of which seriously restrict the development of shared bicycle enterprises [2]. Therefore, the recycling of damaged shared bicycles has become an important issue for the enterprises, and it is also a key research area in field of reverse management of supply chain.

As a new kind of shared economy, the shared bicycle is attracting more and more attention from academic and business circles. Most scholars have centered their focus on application of recycling technology of renewable resources such as automobile, recovery mode selection, recycling system construction, etc. Generally, these research make scientific decisions on the basis of algorithm model, experimental simulation, and empirical analysis [3–6]. The shared bicycle is an emerging industry in the world, and

G. Q. Huang et al. (eds.), *Proceeding of the 24th International Conference on Industrial Engineering and Engineering Management 2018*,
https://doi.org/10.1007/978-981-13-3402-3_30

most research have been focused on the development situation, business mode, and management strategy and so on [7–9]. However, there is few work on the quantitative study on the shared bicycle while the purpose of this paper is to provide new sight on this issue taking account the shared bicycle recycling.

2 Establish Influence Factor System for Reverse Recycling of Damaged Shared Bicycles

2.1 Enterprise Dimension

In recent years, along with the development of shared economy, shared bicycles to be labeled a municipal waste on account of the output in unceasing increase. It is high time that requires a certain amount of recycling idle resources to improve efficiency and reduce the use of urban public space. In general, Recycling logistics cost of damaged shared bicycles will affect the recycling rate. If the cost of recycling logistics is too high, enterprise will give priority to manufacturing rather than recycling. Most of the shared bicycles are equipped with GPS smart locks, dismantling and reusing technology is also an important influence factor [10]. Reverse recycling problems are also affected by enterprise marketing strategy and their own environmental awareness. There is a direct relationship between the level of environmental awareness and the damage degree to the shared bicycle, it has the power to influence behavior. In addition, the recycling of damaged bicycles contribute to enhance the influence and popularity of enterprise brand.

2.2 Product Dimension

Manufacturing costs and maintenance costs are indispensable factors to the recycling of damaged bicycles. The latest data reveal that the manufacturing costs of an ordinary shared bicycle in a range of 300 yuan from 700 yuan. By contrast, the cost of shared bicycle that equipped with GPS smart lock is as high as 2,000 yuan. On the other hand, the expense of maintaining and managing based on logistics operation is high-cost, because workers need to carry out check on shared bicycles regularly, conduct a survey of market, and schedule products into the market and so on. During the operation, not the implementation of specific management measures, natural and man-made damage to more serious damage. The shortening of product life cycle become increasingly a restrictive factor for reverse recycling of damaged shared bicycles.

2.3 Government Dimension

Financial subsidies and industry policies have a critical role in the recycling of shared bicycles. Government should provide chances to develop the shared business model, so as to remedy the shortcomings of public transport. Furthermore, government departments need to establish mandatory laws and regulations to protect users' rights and enterprises' interests and endeavor to push sustainable development of the industry.

2.4 Society Dimension

Shared bicycle as a new model of shared economic in China, which fully embody the spirit of development idea and road- "innovative, coordinated, green, open and shared development". It contribute to the development of national circular economy and resource-economical and environment-friendly society, is helpful to solve the environmental pollution, traffic congestion and other urban problems, is remarkable significance to social benefits and resource and environmental benefits [11].

Through the above analysis, it can be concluded that the reverse recycling of damaged shared bicycles is affected by multiple factors (Table 1). It is extremely difficult to determine the degree of importance because of the interdependence and mutual relationship between the various factors. Therefore, the improved DEMATEL method is used to quantify the comprehensive influence relationship and accurately identify the key influence factors, which provide the scientific basis of decision-making for the recycling and disposal of the damaged shared bicycle.

Table 1. Influence factors for reverse recycling of damaged shared bicycles

Dimension	Influence factors	Number
Enterprise	Market supply and demand	S_1
	Recycling logistics cost	S_2
	Disassembling and manufacturing technology	S_3
	Environmental awareness	S_4
	Marketing strategy	S_5
	Brand value	S_6
Product	Manufacturing costs	S_7
	Maintenance costs	S_8
	Life cycle	S_9
	Damage degree	S_{10}
Government	Industry policy	S_{11}
	Laws and regulations	S_{12}
	Financial subsidies	S_{13}
Society	Environmental benefits	S_{14}
	Social benefits	S_{15}

3 Methodology

DEMATEL is a comprehensive tool for building and analyzing a structural model involving causal relationships between complex factors method. It was first conceived by the Battelle Memorial institute through its research in the 1973 [12–16]. The methodology is of great significance to solve a group of complicated and interconnected problems. But the DEMATEL method is required to access the extent of influences on other factors, it is often difficult for experts to ascertain the impact of factors and offer a precise quantitative value on the complexity of multi-factor correlation analysis. The CRITIC is the objective weight method to obtain the combination weight of the evaluation indexes, which takes into account the variability and relevance of indicators [17, 18]. To overcome the shortcomings of the methods to subjectively determine the weight of indexes, the DEMATEL method combining with the CRITIC method are adopted in this paper. It utilizes the CRITIC method as objective weighting coefficient and combines subjective weights through the DEMATEL method, and then established the normalized direct influence matrix by use of the weight ratio of each factor. The improved DEMATEL process can be summarized by the following steps:

Step 1: Identify influence factors.

The influence factors of the reverse recycling of damaged shared bikes are set as S_1, S_2, ..., S_{15} in this study, which shown in Table 1.

Step 2: Establish the initial direct influence matrix.

Experts are asked to give the degree of direct influence of each dimension on each other dimension. The numbers 0, 1, 2 and 3 represent scores that range from "no influence" to "very high influence". The initial direct influence matrix A $\left(A = \left[a_{ij}\right]_{n\times n}\right)$ is obtained with the following equation:

$$a_{ij} = \left\{ \begin{array}{ll} 3 & \text{When there are more than 10 people} \\ & \text{think that } S_i \text{ have a direct effect on } S_j \\ 2 & \text{When there are 6 to 9 people think} \\ & \text{that } S_i \text{ has a direct impact on } S_j \\ 1 & \text{When there are 1 to 5 people think} \\ & \text{that } S_i \text{ has a direct impact on } S_j \\ 0 & \text{When everyone thinks } S_j \text{ has} \\ & \text{no direct impact on } S_j \end{array} \right\} \tag{1}$$

$$A = \begin{bmatrix} 0 & a_{12} & \cdots & a_{1n} \\ a_{21} & 0 & \cdots & a_{2n} \\ \vdots & \vdots & \cdots & \vdots \\ a_{n1} & a_{n2} & \cdots & 0 \end{bmatrix} \tag{2}$$

where $1 \leq i \leq n$, $1 \leq j \leq n$; If $i = j, a_{ij} = 0$.

Step 3: Calculate the weight of influence factor.

$$C_j = \sigma_j \sum_{i=1}^{n} (1 - r_{ij}) j = 1, 2, \ldots, n \tag{3}$$

$$W_j = C_j / \sum_{i=1}^{n} C_j j = 1, 2, \ldots, n \tag{4}$$

where C_j represents the information contained in the factor j; σ_j represents the standard deviation of the factor j; r_{ij} represents the correlation coefficient between factor i and factor j; W_j represents the objective weight of the factor j.

Step 4: Normalize the matrix and form a normalized matrix.

Matrix $M\left(M = \left[m_{ij}\right]_{n\times n}\right)$ can be obtained through Eq. (5). A normalized matrix $X\left(X = \left[x_{ij}\right]_{n\times n}\right)$ can be obtained through Eq. (6), its element ranged from 0 to 1.

$$M = \frac{A}{\max\limits_{1\le i\le n} \sum_{j=1}^{n} a_{ij}}, j = 1, 2, \ldots, n \tag{5}$$

$$X = \begin{bmatrix} 0 & W_1 m_{12} & \cdots & W_1 m_{1n} \\ W_2 m_{21} & 0 & \cdots & W_2 m_{2n} \\ \vdots & \vdots & \cdots & \vdots \\ W_j m_{n2} & W_j m_{n2} & \cdots & 0 \end{bmatrix} \tag{6}$$

Step 5: Calculate the total influence matrix.

The total influence matrix $T\left(T = \left[t_{ij}\right]_{n\times n}\right)$ can be obtained by using the following formula:

$$T = X(E - X)^{-1} \tag{7}$$

where E is the identity matrix.

Step 6: Identify the total influential relations.

$$f_i = \sum_{j=1}^{n} t_{ij} \tag{8}$$

$$e_i = \sum_{j=1}^{n} t_{ji} \tag{9}$$

$$m_i = f_i + e_i \tag{10}$$

$$n_i = f_i - e_i \tag{11}$$

where f_i shows the total effects, both direct and indirect exerted by the i-th. The value e_i shows the total effects, both direct and indirect, received by factor i from the other factors. The value m_i shows the degree of importance that the i-th factor plays in the system. The value n_i shows the net effect, the i-th factor contributes to the system.

4 Empirical Analysis

4.1 Data Collection

In the process of getting data, experts and scholars are required to grade for the interactional relationship between the various dimensions and factors according to a scale of zero to three. The respondents are: shared bikes enterprises related to the responsible persons or mid-level managers who are familiar with reverse supply chain management, as well as academic experts in the field of supply chain, recycling economy, renewable resource recycling and so on. Academic experts are the teachers who strive to publish their research in internationally-recognised publications or domestic authoritative journals, so the sample data is of value in both reference and theoretical research. By using questionnaire survey, this study gathered the responses of 15 valid questionnaires, which include 7 questionnaires from business experts and 8 questionnaires from academic experts.

4.2 Data Analysis

According to the application steps of the improved DEMATEL method, the initial direct influence matrix A (Table 2) is obtained by analyzing the available questionnaire and the influence factors of the reverse recycling of damaged shared bikes. In order to obtain the improved standard direct influence matrix N (Table 3), we can utilize the EXCEL and calculate the objective weight of influence factor. Next, the total influence matrix T (Table 4) is calculated through Eq. (7). Finally, by using the formula (7–10), we can obtain the total influential relations (Table 5).

Table 2. Initial direct influence matrix *A*

	S_1	S_2	S_3	S_4	S_5	S_6	S_7	S_8	S_9	S_{10}	S_{11}	S_{12}	S_{13}	S_{14}	S_{15}
S_1	0	0	0	1	0	2	0	0	0	0	2	0	0	2	1
S_2	0	0	0	0	0	0	2	1	0	3	0	0	1	0	0
S_3	0	0	0	0	0	0	2	1	0	2	0	0	0	0	0
S_4	0	0	0	0	1	2	0	0	0	0	0	0	0	3	2
S_5	1	0	0	1	0	2	0	0	0	0	0	0	0	0	0
S_6	1	0	1	1	2	0	0	0	0	0	0	0	0	0	0
S_7	0	1	0	0	0	0	0	2	2	3	0	0	0	0	0
S_8	0	0	1	0	0	0	1	0	2	2	0	0	0	0	0
S_9	0	0	0	0	0	1	2	2	0	3	0	0	0	0	0
S_{10}	2	0	0	0	0	2	3	3	2	0	0	0	0	1	0
S_{11}	2	0	0	0	1	0	0	0	0	0	0	2	3	0	0
S_{12}	2	0	1	0	0	0	0	0	0	0	1	0	0	2	1
S_{13}	0	3	0	1	2	0	0	0	0	0	0	0	0	1	1
S_{14}	0	0	0	3	1	2	0	0	0	0	1	1	0	0	3
S_{15}	0	0	0	1	0	1	0	0	0	0	3	2	1	2	0

Table 3. Standardized direct influence matrix *N*

	S_1	S_2	S_3	S_4	S_5	S_6	S_7	S_8	S_9	S_{10}	S_{11}	S_{12}	S_{13}	S_{14}	S_{15}
S_1	0.0000	0.0000	0.0000	0.0046	0.0000	0.0092	0.0000	0.0000	0.0000	0.0000	0.0092	0.0000	0.0000	0.0092	0.0046
S_2	0.0000	0.0000	0.0000	0.0000	0.0000	0.0000	0.0095	0.0048	0.0000	0.0143	0.0000	0.0000	0.0048	0.0000	0.0000
S_3	0.0000	0.0000	0.0000	0.0000	0.0000	0.0000	0.0051	0.0026	0.0000	0.0051	0.0000	0.0000	0.0000	0.0000	0.0000
S_4	0.0000	0.0000	0.0000	0.0000	0.0047	0.0095	0.0000	0.0000	0.0000	0.0000	0.0000	0.0000	0.0000	0.0142	0.0095
S_5	0.0043	0.0000	0.0000	0.0043	0.0000	0.0087	0.0000	0.0000	0.0000	0.0000	0.0000	0.0000	0.0000	0.0000	0.0000
S_6	0.0054	0.0000	0.0054	0.0054	0.0107	0.0000	0.0000	0.0000	0.0000	0.0000	0.0000	0.0000	0.0000	0.0000	0.0000
S_7	0.0000	0.0064	0.0000	0.0000	0.0000	0.0000	0.0000	0.0128	0.0128	0.0191	0.0000	0.0000	0.0000	0.0000	0.0000
S_8	0.0000	0.0000	0.0058	0.0000	0.0000	0.0000	0.0058	0.0000	0.0117	0.0117	0.0000	0.0000	0.0000	0.0000	0.0000
S_9	0.0000	0.0000	0.0000	0.0000	0.0000	0.0046	0.0093	0.0093	0.0000	0.0139	0.0000	0.0000	0.0000	0.0000	0.0000
S_{10}	0.0171	0.0000	0.0000	0.0000	0.0000	0.0171	0.0257	0.0257	0.0171	0.0000	0.0000	0.0000	0.0000	0.0086	0.0000
S_{11}	0.0104	0.0000	0.0000	0.0000	0.0052	0.0000	0.0000	0.0000	0.0000	0.0000	0.0000	0.0104	0.0015	0.0000	0.0000
S_{12}	0.0080	0.0000	0.0040	0.0000	0.0000	0.0000	0.0000	0.0000	0.0000	0.0000	0.0040	0.0000	0.0000	0.0080	0.0040
S_{13}	0.0000	0.0144	0.0000	0.0048	0.0096	0.0000	0.0000	0.0000	0.0000	0.0000	0.0000	0.0000	0.0000	0.0048	0.0048
S_{14}	0.0000	0.0000	0.0000	0.0177	0.0059	0.0118	0.0000	0.0000	0.0000	0.0000	0.0059	0.0059	0.0000	0.0000	0.0177
S_{15}	0.0000	0.0000	0.0000	0.0052	0.0000	0.0052	0.0000	0.0000	0.0000	0.0000	0.0155	0.0103	0.0052	0.0103	0.0000

Table 4. Total influence matrix T

	S_1	S_2	S_3	S_4	S_5	S_6	S_7	S_8	S_9	S_{10}	S_{11}	S_{12}	S_{13}	S_{14}	S_{15}
S_1	−0.0001	0.0000	0.0000	0.0044	−0.0002	0.0091	0.0000	0.0000	0.0000	0.0000	0.0091	−0.0002	0.0000	0.0091	0.0044
S_2	−0.0002	−0.0001	0.0000	0.0000	0.0000	−0.0002	0.0092	0.0043	−0.0004	0.0141	0.0000	0.0000	0.0048	−0.0001	0.0000
S_3	−0.0001	0.0000	0.0000	0.0000	0.0000	−0.0001	0.0050	0.0024	−0.0002	0.0050	0.0000	0.0000	0.0000	0.0000	0.0000
S_4	−0.0001	0.0000	0.0000	−0.0004	0.0046	0.0092	0.0000	0.0000	0.0000	0.0000	−0.0002	−0.0002	0.0000	0.0141	0.0092
S_5	0.0043	0.0000	0.0000	0.0043	−0.0001	0.0086	0.0000	0.0000	0.0000	0.0000	0.0000	0.0000	0.0000	−0.0001	−0.0001
S_6	0.0053	0.0000	0.0054	0.0053	0.0107	−0.0002	0.0000	0.0000	0.0000	0.0000	0.0000	0.0000	0.0000	−0.0001	−0.0001
S_7	−0.0003	0.0064	−0.0001	0.0000	0.0000	−0.0004	−0.0007	0.0121	0.0123	0.0188	0.0000	0.0000	0.0000	−0.0002	0.0000
S_8	−0.0002	0.0000	0.0058	0.0000	0.0000	−0.0002	0.0054	−0.0005	0.0114	0.0114	0.0000	0.0000	0.0000	−0.0001	0.0000
S_9	−0.0003	−0.0001	−0.0001	0.0000	0.0000	0.0044	0.0089	0.0088	−0.0005	0.0137	0.0000	0.0000	0.0000	−0.0001	0.0000
S_{10}	0.0171	−0.0002	−0.0002	−0.0003	−0.0002	0.0168	0.0255	0.0253	0.0165	−0.0010	−0.0002	0.0000	0.0000	0.0084	−0.0002
S_{11}	0.0103	0.0000	0.0000	−0.0001	0.0052	−0.0001	0.0000	0.0000	0.0000	0.0000	−0.0001	0.0104	0.0015	−0.0002	−0.0001
S_{12}	0.0080	0.0000	0.0040	−0.0002	−0.0001	−0.0002	0.0000	0.0000	0.0000	0.0000	0.0038	−0.0001	0.0000	0.0079	0.0038
S_{13}	0.0000	0.0144	0.0000	0.0047	0.0096	−0.0002	−0.0001	−0.0001	0.0000	−0.0002	−0.0001	−0.0001	−0.0001	0.0047	0.0047
S_{14}	−0.0002	0.0000	−0.0001	0.0176	0.0057	0.0115	0.0000	0.0000	0.0000	0.0000	0.0056	0.0057	−0.0001	−0.0005	0.0176
S_{15}	−0.0003	−0.0001	−0.0001	0.0049	−0.0003	0.0050	0.0000	0.0000	0.0000	0.0000	0.0154	0.0101	0.0051	0.0101	−0.0003

Table 5. Total influential relations

Number	f_i	e_i	m_i	n_i
S_1	0.0355	0.0432	0.0787	−0.0077
S_2	0.0311	0.0203	0.0514	0.0108
S_3	0.0120	0.0145	0.0265	−0.0025
S_4	0.0362	0.0402	0.0764	−0.0040
S_5	0.0169	0.0347	0.0516	−0.0179
S_6	0.0262	0.0630	0.0893	−0.0368
S_7	0.0479	0.0530	0.1009	−0.0051
S_8	0.0330	0.0524	0.0853	−0.0194
S_9	0.0348	0.0392	0.0740	−0.0045
S_{10}	0.1072	0.0616	0.1688	0.0455
S_{11}	0.0267	0.0332	0.0599	−0.0066
S_{12}	0.0270	0.0255	0.0525	0.0014
S_{13}	0.0371	0.0111	0.0482	0.0260
S_{14}	0.0628	0.0530	0.1158	0.0098
S_{15}	0.0497	0.0390	0.0887	0.0107

Based on data collected from the aforementioned expert interviews, this study constructed a total influential relations (Table 5) using CRITIC-DEMATEL method. The influence factors for reverse recycling of damaged shared bicycles are divided into two categories: cause factor and result factor. The cause factor (n_i is positive number) is not easily influenced by other influences factors. Results showed that the damage degree (S_{10}) and the financial subsidies (S_{13}) have the greatest n_i value. It implies that factors that the damage degree have an immense impact on other factors in the system of influencing factors. Therefore, enterprises should take relevant and effective measures to decrease the man-made damage rate of shared bicycles in terms of operation and management. At the same time, the government should provide financial support for recycling and reuse of shared bicycles. In this way, shared bicycle firms have a more stable and sustainable business model, and it accord with the circular economy development. The result factor (n_i is negative number) is susceptible to other factors in a short time, which is the most obvious passive factor. Among the factors that affect the recycling of damaged shared bicycles, brand value (S_6) and maintenance costs (S_8) have the smallest n_i value, implying that it is most likely to be influenced by other factors. It is also consistent with the current social reality. The shared bicycles development in China is at preliminary stage. The enterprise should be not washed out in competition, retain competitive ability, be about to create brand core values and reinforce a brand in the user's mind. The greater the m_i value is, the stronger the effect of influence factor on the reverse recycling of damaged shared bike, the smaller the m_i value, indicating the weaker the effect of the factor. According to final results of the analysis, the study shows that in terms of the key influence factors for reverse recycling of damaged shared bicycles, the top four are including reverse cycling damage degree (S_{10}), environmental benefits (S_{14}), manufacturing costs (S_7), brand value (S_6). They are the most influential factors and the real sources which affect the other factors directly.

5 Conclusion

The CRITIC-DEMATEL method was applied in the case of analysis devoted to damaged shared bicycles. The new DEMATEL method can overcome the defects of establishing initial direct influence matrix based on expert scoring or questionnaire survey. The weight ratio of every factor can be determined by CRITIC method, and then establish the standardized direct influence matrix. Not only can reflect the causal relationship between factors, but also indicate the dynamics of the data and the degree of difference of factors. Beyond that, to a great extent, it can reduce the subjective errors and improve reliability of findings, which provide a new solution for the study of influence factors.

Application of the improved DEMATEL identify the key influence factors for reverse recycling of damaged shared bicycles. Results of the analysis indicate that reverse recycling is influenced by many kinds of factors, but the damage degree of bicycle occupies the first place among a great deal of factors. Environmental benefits, manufacturing costs and brand value are also the key factors that affect the recycling of damaged shared bicycles. Therefore, when determining how to deal with the recyclable damaged shared bicycles, enterprises should always first consider the above-mentioned key factors, and then consider other factors.

Acknowledgements. This paper is funded by the Social Science Program (SC17TJ010) of "13th Five-Year" plan for the year 2017 in Sichuan Province.

References

1. Y. Liu, Where can we place the 'sick and elderly' shared bicycle. Science and Technology Daily, June, 2017
2. J. Gan, Z. He, M. Peng et al., Based on DEMATEL method of China's damaged car recycling industry development factors analysis. Sci. Technol. Manag. Res. **36**(1), 103–107 (2016)
3. H. Lu, Automobile engine remanufacturing recovery and influencing factors, Master thesis, Dissertation, in Chinese, Tianjin University, 2011
4. M. Mahmoudzadeh, S. Mansour, B. Karimi, To develop a third-party reverse logistics network for end-of-life vehicles in Iran. Resour. Conserv. Recycl. **78**(3), 1–14 (2013)
5. Y.X. Pan, H.T. Li, Sustainability evaluation of end-of-life vehicle recycling based on energy analysis: a case study of end-of-life vehicle recycling enterprise in China. J. Clean. Prod. **131**, 219–227 (2015)
6. Y. Chen, Y. Yang, Q. Zhang, Design of reverse logistics network of waste household appliance based on third party recycling. J. Math. Pract. Theory **46**(17), 81–89 (2016)
7. K. Li, Study on the development strategy of urban shared bicycles based on shared economy. Cities (3), 66–69 (2017)
8. G. Yu, Analysis of 'shared bicycle' operation and management of the problems and measures. Knowl. Econ. (9), 87–88 (2017)
9. Y. Zhang, X. Yu, Y. Shi, Public awareness and evaluation of public economic development and its governance—taking shared bicycle as an example. Natl. Gov. (17) (2017)

10. Z. He, J. Gan, Y. Zou, Study on reverse logistics of scrapped car. Sci. Technol. Manag. Res. **14**, 162–166 (2014)
11. M. Peng, Study on influencing factors of renewable resource recycling logistics development by DEMAEL. Soft Sci. **30**(6), 140–144 (2016)
12. D. Xiao, W. Zhang, Study on key factors of green logistics development based on Dematel method. Ind. Eng. **13**(1), 52–57 (2010)
13. J. Gan, Z. He, Z. Yang, in *Influence Factors Analysis for Supplier Selection Under the Low-Carbon Economy Based on DEMATEL*. The International Conference on Management Innovation and Public Policy, 1 Jan 2012
14. C.W. Hsu, T.C. Kuo, S.H. Chen et al., Using DEMATEL to develop a carbon management model of supplier selection in green supply chain management. J. Clean. Prod. **56**(10), 64–172 (2013)
15. K. Govindan, R. Khodaverdi, A. Vafadarnikjoo, Intuitionistic fuzzy based DEMATEL method for developing green practices and performances in a green supply chain. Expert Syst. Appl. **42**(20), 7207–7220 (2015)
16. Y. Yang, S. Yang, M. Zhang et al., Evaluation system of travel reservation class app based on DEMATEL method. J. Tour. **3**(2), 64–74 (2016)
17. H. Zhang, Z. Wang, B. Zhu, Comprehensive evaluation of medical work quality based on CRITIC method. Chin. J. Health Stat. **30**(3), 454–455 (2013)
18. L. Zhang, X. Zhang, Fractive clustering method based on improved CRITIC method. Stat. Decis. **22**, 65–68 (2015)

Failure Analysis of Express Service Based on Service Blueprint and FMEA

Haolun Wang(✉) and Ranjun Deng

College of Transportation and Logistics, East China Jiaotong University, Nanchang 330013, China
haolun123@163.com

Abstract. Service failures are severely hampered the development of the current Chinese express delivery industry, and in order to avoid the aggravation of service failure, there must be our logistics service process optimization. The method using the service blueprint, which described the courier service business processes detailed and identify the destructive factors which lead to destructive failure, to build courier service quality evaluation index system. FMEA method utilizing the potential failure factors were analyzed, and the results are most likely factor to lead to service failures, at the same time, putting forward the improvement suggestion to the result of the analysis results.

Keywords: Service failure · Courier · Service blueprint · FMEA

1 Introduction

From 1990s to date, the China's private express companies were more and more, and rapid development. At the same time, it also has a lot of service failure issues, such as delivery failures, express shipments damaged or lost and poor service attitude. This has become the main factor to inhibit the sustained and healthy development of China's express delivery companies and even the courier industry. So how to avoid the failure of express service is the problem that we must consider.

More of the research at home and abroad about the quality of service delivery, Fred et al. by telephone companies, universities, and the financial industry as the research object. The effects of the relationship between perceived quality and brand image, customer satisfaction and brand loyalty and verified in the service industry service quality on brand image, customer satisfaction has a significant positive impact [1]; Gronroos regard that cannot be on the quality of service is only more objective evaluation, the gap between it by the customer to enterprise's service level expectations and actual service determined [2]. Express enterprise quality of service concept with the development of information technology and development, Chen et al. researched the service quality in logistics [3]; Yuan presented service quality evaluation index to help express company to improve its service quality [4]. For service failure, Zhang and Tao utilized the scenario simulation method to analyze the logistics service failures for customer satisfaction and loyalty [5]; Lv chose the high-grade hotels as the object of

G. Q. Huang et al. (eds.), *Proceeding of the 24th International Conference on Industrial Engineering and Engineering Management 2018*,
https://doi.org/10.1007/978-981-13-3402-3_31

study, the key factors resulting in failure of the service were in-depth studied [6]. For the service blueprint, Shostack believed that the service blueprint can be used to solve the problems in the service sector [7]; Mary et al. described that the services blueprint application in some service companies [8]. For FMEA, many researchers combine it with fuzzy comprehensive evaluation method [9, 10]. For the customer value system, Xu et al. used complex network, customer relationship management, marketing theory to establish the evaluation system of complete twitter client value and user value evaluation model [11]. And the FMEA and logistics services as the key words were searched in the academic database, the result showed that the relevant literature was less.

In the existing literatures, although the logistics service quality, service blueprint and FMEA were studied, but the combination of the three logistics of the courier service failure is not. According to the present situation of the service quality of *S* Express Company is studied in this paper. Through the qualitative and quantitative research, using service blueprint method analysis of the service process of the company, part of a service or service failure evaluation using FMEA, determine the failure process or process in the higher risk, in the light of these links put forward suggestions for improvement and prevention strategies, on failure prevention research can greatly reduce the enterprise cost and improve effectiveness, realize resource reasonable configuration, so as to improve the enterprise service quality, enhance the credibility, improve enterprise competitiveness. At the same time this paper takes *S* Company as the research object, put forward some countermeasures and suggestions on the hope of China's express industry can play certain reference significance. At the same time, to provide a theoretical basis for the future research on the failure prevention of express delivery service.

2 Express Logistics Enterprise Service Blueprint

2.1 Express Logistics Enterprise Service Blueprint

By using service blueprint to express enterprise service process optimization, improve the operation efficiency, and strive to improve customer satisfaction.

As shown in Fig. 1.

2.2 Critical Moment Analysis of Customer Behavior

In this paper, these links to carry out a critical moment analysis. Key moments analysis of customer behavior contain eight aspects: Forward consulting process, Ask for the documents, Fill in documents, packing the goods, give the package and documents to the staff, Payment of fees, Receive package and Distribution of goods. Take forward consulting process for example. The task of the main executive is to answer customer's questions. Destructive factor is to answer to the customer is not satisfied. Standard expectation value is to answer can be reassuring. Positive factors are to be enthusiastic answer customer questions.

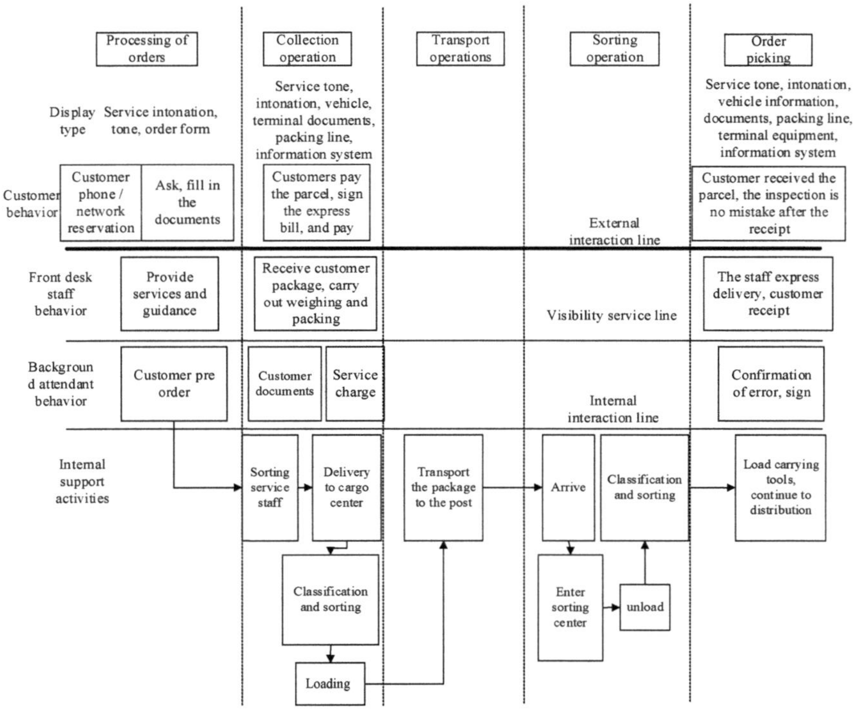

Fig. 1. Express enterprise logistics service blueprint

The failure of the *S* Company express service, that is, some of the destructive factors may cause customer resentment, leading to service failure. Mainly reflected in the following points:

(1) When the staffs answer the question, the attitude is not good or the answer is not complete.
(2) The staff is not efficient to provide the appropriate documents.
(3) The staff did not have an effective guide to the customer, to ensure that they are in the right to fill in the documents.
(4) The staff did not complete the packing of the goods to the customer.
(5) Staff at the time of the charges, the staff did not charge in accordance with the unified standard.
(6) The staff at the time of delivery attitude is poor, no advance notices.

These reasons may cause the failure of the courier service, which is from the customer point of view and more detailed, because there are a lot of details are not discussed. Therefore, the following is the use of FMEA method to the *S* Company's express service in-depth research, research and its possible failure of the specific links.

3 Failure Analysis of S Company Express Based on FMEA

3.1 Construction of Service Index System

Through the study of the existing express delivery service index, combined with the actual situation of the consumer, the express service index system (Table 1) is constructed, including 8 first level indicators and two 24 level indicators.

Table 1. Construction of express service quality index system

First order index	Two level index
Express service capability index	Completeness of goods
	Whether the goods are delivered on time
	Order tracking query ability
	Complaint resolution
Express cost index	Transport link cost control
	Other value added service cost control
Enterprise financial condition index	Express charges reasonable degree
	Abnormal rate of goods
	Insurance compensation mechanism
Error handling quality index	Error handling speed
	Compensation degree
	Compensation attitude
	Payment measures
Enterprise work efficiency index	Market survey
	Department and staff work efficiency
	Internal communication quality
Enterprise internal management ability index	Department set a reasonable level
	Reasonable degree of management method
Enterprise personnel quality index	The attitude of express delivery
	Express delivery attitude
	The ability to respond to complaints
Environmental indicator	Air pollution index
	Road traffic conditions
	Packaging recovery

3.2 Failure Factor Analysis and RPN Value

Based on the index system in Table 2, the analysis table of FMEA was established by *S* Company. The potential failure factors of S were analyzed, and its severity, detection degree, frequency and RPN were analyzed. RPN is incident severity, detecting the degree and frequency of three product (RPN = Severity $\times$ Frequency $\times$ Detection), used to measure the possible process defects, in order to take precautionary measures reduce the key changes to the process, make the process more reliable.

(1) Severity index analysis

At present, the severity of the area is less, so it is a reference to other industries to define the severity index of this area (in Table 2).

Table 2. Severity index table

Consequence	Rating according to the severity of the express activities	Severity level
Hazards without warning	When the warning is not under the circumstances, if the failure of the courier will affect the health of the customer, or is in violation of the relevant state laws or the relevant standards of the courier industry	10
Danger of warning	In the case of a warning, if the delivery failure will affect the health of the customer, or is in violation of the relevant state laws or the relevant standards of the courier industry	9
Higher	100% will result in the loss of customer property (loss of goods) or scrap	8
High	The probability of loss of customer property is very high (above 70%, less than 80%)	7
Secondary	The probability of loss of customer property is higher (less than 70%)	6
Low	The probability of occurrence of the product is less than 50%	5
Very low	The probability of occurrence of the product is less than 30%	4
Slight	The probability of occurrence of the product is less than 20% higher than 10%	3
Very slight	The probability of a product is very low (below 10%)	2
Nothing	Products do not have any problems, will not bring inconvenience to customers	1

(2) Frequency index analysis

Frequency refers to the possibility of failure occurs, in order to be consistent with the severity of the indicators, but also choose the number of 1–10 to express the frequency of failure of express (in Table 3).

Table 3. Frequency criterion table

Possibility of failure	Possible failure rate	Frequency
Very high: failure phenomenon continues to develop	$\geq$10%	10
	$\geq$5%	9
High: failure occurs more frequently	$\geq$2%	8
	$\geq$1%	7
Medium: failure occurs only occasionally	$\geq$0.5%	6
	$\geq$0.2%	5
	$\geq$0.1%	4
Low: failure occurs less	$\geq$0.05	3
	$\geq$0.01%	2
Very low: failure to occur	$\geq$0.001%	1

(3) Detection index

Detection is the possibility that a failure may be detected in advance. In this paper, based on the existing research foundation, accordance with the actual situation of *S* Company, we develop the measurement criteria based on detection index in Table 4.

Table 4. Exploration measure standard table

Detection possibilities	Criterion	Exploration measure
Scarcely possible	There is no way to know in advance the failure of the phenomenon will occur	10
Very tiny	There is almost no effective control method, which can not be effective early warning	9
Small	Effective control methods are very few, only a very small chance to make early warning	8
Less	The effective control method is very few, and the possibility of making the early warning is small	7
Small	Effective control methods are very few, there is a possibility of effective early warning	6
Secondary	Effective control methods are very few, have the opportunity to make early warning and prevent the occurrence of the phenomenon of failure	5
Middle upper	There are many ways to prevent the failure of the control method	4
High	Control method is more, have a great chance to prevent the occurrence of the phenomenon of failure	3
Higher	Control method more, can make a warning	2
Very high	Must be able to make early warning of failure, and more timely to prevent	1

The Severity, Frequency and Detection degree of each service failure mode were analyzed by the above mentioned severity, frequency and detection index.

3.3 Failure Analysis

As can be seen from the Fig. 2, the value of the risk order of the insurance payment link is up to 224, indicating that the customer is not satisfied with the link. The integrity of the goods ranked second high, the value of 144. Order tracking query ability ranked three, its value is 100. This shows that the customer service quality and service level of logistics attention more, more attention to their own rights and interests. This from the side that the roots do a good job of logistics enterprises can have a higher competitiveness, so as to obtain more market share.

Note: If service indices have several causes of failure, we will choose maximum. For example, since cargo readiness level consists of five levels: Extrusion of transport process, Handling process fall, Private holding of staff, Water immersion and Improper distribution, we will choose the maximum value 144.

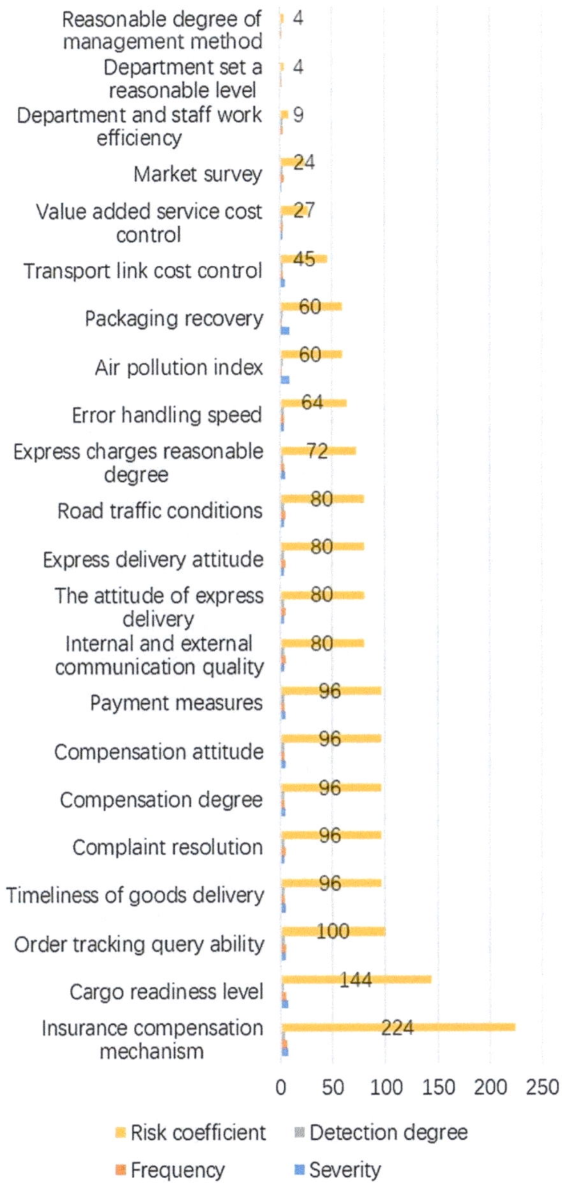

Fig. 2. FMEA analysis table

4 Improvements to the Results of Failure Analysis Are Presented

4.1 According to the Insurance Payment Link

On the one hand, *S* Company should strive to improve the quality of their transport. Do system management information system, based on improving the operation efficiency to strengthen market information gathering capabilities, to ensure that the shipment of the on time delivery and no loss; do establish the mechanism responsible for the accident, once delivery damaged or even lost, responsibility to the people; do accelerate the "last 1 km" of the development of logistics to achieve express delivery on time, in order to improve customer satisfaction.

On the other hand, it needs to establish a more perfect payment mechanism; the company should set up a margin system and actively explore the possibility of payment in advance. Without affecting the normal operation of the enterprise, the establishment of a special compensation fund, when the customer's parcel loss or damage to the event can be paid in advance.

4.2 For the Integrity of the Goods

Company to strive for the cultivation of professional talents, improve staff to enter the threshold can be choice with universities, research institutions strengthen cooperation, help to cultivate for their own business professionals, and pre post training for some time before the start of the work, in work to regularly carry out training activities; to establish enterprises in the reward and punishment mechanism, of damage to the shipment of low rate of sales incentives, of damage to the shipment of high rate of sales of punishment (in weeks); to ensure the smooth flow of complaints mechanisms, do can quickly resolve quickly resolve, can be solved quickly by a reply.

4.3 Query Ability for Order Tracking

Company to do support the optimal planning of distribution route to information technology, specification of linear guide vehicle route and ensure the timeliness of distribution; accelerate the transformation of cutting-edge technology in the application of and promote express visual simulation platform to build, let customers vision with sending process, enhancing the user experience, such as strong viscose like stick customer.

5 Conclusion

In this paper, we describe the service process of logistics enterprises by establishing a service blueprint, and the key points of the service were analyzed. At the same time, construct the evaluation index system of logistics service quality, and get the failure mode of logistics service. At the same time, the establishment of FMEA table, which appears to be the biggest failure of several links. According to the result of failure analysis, the corresponding improvement suggestions are put forward. Over all, we can see that the increase of supporting plate number can make track stiffness much smooth along train travel direction, and minimum the disruption of damping performance, this proposal is worth widely promoting.

In the future studies, the customer behavior of construction process need to be analyzed more detailed, then to dig out the failure of the key points on the cause of customer, so that the enterprises have specialized guide in the promotion process. The learned professional knowledge promotes express visual simulation system platform construction and enables them to building blocks in the current logistics industry in the forefront of the field.

References

1. S. Fred, B. Jaworski, A. Kohli, An examination of the effect of product performance on brand reputation, customer satisfaction and loyalty. Eur. J. Mark. **9**, 65–82 (1993)
2. C. Gronroos, Service management and marketing. Bus. Book Rev. Libr. **11**(4), 2–4 (2007)
3. M. Chen, M. Li, J. Mao, An empirical choice of service quality valuation indexes of tobacco logistics and its application. J. Ind. Eng. Eng. Manag. **23**(2), 99–103 (2009)
4. Q. Yuan, Quality evaluation and analysis of express service of S express company. Manag. & Technol. SME **2**, 123–124 (2015)
5. S. Zhang, N. Tao, The effect of express service failure on customer satisfaction and loyalty in the online shop—based on the theory of psychological account and attribution theory. Mod. Financ. Econ. **6**, 89–102 (2014)
6. Y. Lv, Study on the service failure factors of high grade business hotel (in Chinese), Ph.D. dissertation, Zhejiang University, Zhejiang, China, 2004
7. G. Shostack, Understanding services through blueprinting, in *Advances in Services Marketing and Management*, ed. by T.A. Schwartz, D.E. Bowen, S.W. Brown (JAI Press, Greenwich, CT, 1992), pp. 75–90
8. B. Mary, O. Amy, M. Felicia, Service blueprinting: a practical technique for service innovation, Center for Services Leadership, Arizona State University Working Paper, 2007
9. W. Cui, G. Wu, P. Sun et al., Ship safety assessment based on FMEA and fuzzy comprehensive evaluation methods (in Chinese). J. Harbin Eng. Univ. **28**(3), 263–267 (2007)
10. X. Li, Research on fuzzy comprehensive evaluation of software quality based on FMEA (in Chinese), Ph.D. dissertation, Southeast University, Hunan, China, 2005
11. X. Xu, H. Tu, J. Wang, Research on the user evaluation model of social network based on complex network (in Chinese). J. East China Jiao Tong Univ. **5**, 38–43 (2012)

Correlation Analysis Between Capital Structure and Enterprise Value of Communication Service Industry Companies—A Case Study of Shanghai and Shenzhen A-share Listed Companies

Yan-liang Zhang[1](✉), Rong-yuan Ma[1], and Xi-yu Huang[2]

[1] School of Management Engineering, Zhengzhou University, Zhengzhou, China
mars19921123@163.com
[2] School of Public Administration, Zhengzhou University, Zhengzhou, China

Abstract. Capital structure and enterprise value have a lot of contact, it is related to the overall development of the enterprise planning in practical, and related to how to find the best corporate capital structure in theoretical. At present, some scholars have conducted empirical analysis in many fields, including real estate, tourist industry, automobile industry and high-tech industry. However, there are very few researches on communication service companies. With the advent of the Internet era and the rapid development of communication services companies, this subject has the value of research. This paper mainly uses the method of empirical analysis to study the communication service companies of Shanghai and Shenzhen A shares, and finally selects the relevant data (total 115 valid samples) of 23 companies from 2013 to 2017. Excel and EViews 8.0 are mainly used on the data descriptive statistical analysis and multiple regression analysis.

Keywords: Capital structure · Corporate value · Communication service industry · DAR

1 Introduction

More and more enterprises began to choose to maximize the interests of shareholders and the value of enterprises nowadays, but no longer put the profit as the only goal of business development, so the capital structure and the value of the problem in the modern economy important Status gradually on the agenda. The choice of capital structure is directly related to the enterprise's tax situation and financial situation, will have different capital costs and governance structure, affecting the interests of different interests of the trade-offs and the total value of enterprises, and then a country or region's overall economic stability Have an impact. Therefore, it is of great theoretical significance to study.

G. Q. Huang et al. (eds.), *Proceeding of the 24th International Conference on Industrial Engineering and Engineering Management 2018*,
https://doi.org/10.1007/978-981-13-3402-3_32

2 Methodology

In this paper, after the analysis of multiple literatures, we try to minimize the bias, and have chosen four indicators that affect the value of the firm (DAR, Company Size, Flow Ratio, Ownership concentration), select a more reasonable model, take the listed companies in Shanghai and Shenzhen as the object of study, analyze the data from 2013 to 2017 with Eviews 8.0, study this problem.

3 Results

3.1 Figures and Variables

Among them, the return on net assets (ROE), also known as the rate of return on equity, is a key indicator of DuPont analysis, the value is equal to the operating net profit margin, asset turnover and equity multiplier multiplication, it is not difficult to find this data fully reflects the return on equity of investors through the formula. In addition, ROE is not limited by industry and region, so this article selects this data as explanatory variables eventually (Table 1).

Table 1. Variable statistics table

Variable type	Variable name	Variable symbol	Calculation formula
Explained variable	Net assets income rate	ROE	Net profit/average net assets
Explanatory variables	Assets-liabilities	DAR	Total liabilities/total assets
Control variable	Company size	SIZE	LN(total assets)
	Flow ratio	LIQUI	Current assets/current liabilities
	Ownership concentration	HHI5	Herfindahl Hirschman index

As to the asset structure, considering the maturity of the debt has a certain degree of ambiguity, the availability of data, the convenience of the study, this paper chooses to use the generalized capital structure, which is not only an explanatory variable but a key factor to test whether the company's financial leverage has an effective effect on its value.

Short-term solvency (LIQUI), which involves the size and quality of current assets and current liabilities. It is influenced by many factors such as market environment, policies and regulations, industry characteristics and production cycle. But it will reflect the size of short-term solvency effectively if we select the same industry in the same research period.

HHI5. Many of the literature had selected the largest shareholding ratio as a measure of the standard, in order to rule out the existence of abnormal values and improve the stability of the data, this paper chooses the sum of the square shares of the top five shareholders.

3.2 References

Early capital structure theory is the basis of modern capital structure theory, but more general. Since 1959 Modigliani and Miller's MM Theorem [1], Robichek et al. Subsequently proposed a trade-off theory in 1967 that introduced bankruptcy costs in the modified MM theory. Jensen and Meckling in 1976 proposed the agency cost theory, also known as post-trade-off theory. In 1977, Ross's signal theory argued that asset-liability ratio was positively related to firm value, and Bolton and Scharfstcin also confirmed this in 1994 and 1996 [2].

Domestic scholars on the issue of late start, is based on the views of Western scholars, taking full account of the particularity of Chinese enterprises, and empirical analysis based [3]. Because of the choice of research object, method, selection of industry, index, analysis model, there are differences, although the problem has been a hot topic of domestic and foreign scholars, but has not been conclusive [4].

3.3 Research Hypothesis

Based on the previous analysis, we present the following assumptions:

Assumption 1 Chinese communication services listed companies in the asset-liability ratio and the company's value is positively correlated [5].

Except for the net operating income theory and MM theorem analysis of the situation is extreme, most of the capital structure theory research results point of view is the existence of the optimal capital structure of enterprises, Tax Shield that the liabilities bringing will increase the cost of financial crisis, appropriate liabilities can maximize the value of the firm [6]. The optimal financing theory suggests that there is a negative correlation between the value of the firm and the asset-liability ratio only when the operating conditions are unfavorable. This assumption is made in view of the fact that the service companies do not have a tax avoidance effect on debt financing and thus can not have a positive impact on the value of the company [7].

Assumption 2 The size of the listed companies in the communications services industry is positively related to the value of the enterprise.

According to the scale effect can be drawn in the general industry [8], fixed costs and bankruptcy costs remain unchanged, with the expansion of business scale, the average cost of each product to be distributed and the proportion of bankruptcy costs to the company's assets will be reduced, so enterprises can use resources more fully and effectively, improve their operating efficiency, and have better stability [9].

Assumption 3 The current liabilities ratio is negatively correlated with the firm value. If the amount of current liabilities is large, its solvency and financial security can not be guaranteed.

Assumption 4 The distribution of equity services in listed companies is negative related to the enterprise value [10].

If the proportion of shareholders is loose, then the business decision-making more democratic.

3.4 Model Establishment

Based on the previous discussion, we choose to use the value of the company (ROE) as the explanatory variable, DAR as the explanatory variable to establish the regression model, while take company size, asset structure, short-term solvency and equity concentration into the analysis model to control the impact of other indicators on the performance of corporate value [11]:

$$\mathbf{ROE} = \boldsymbol{\beta}_0 + \boldsymbol{\beta}_1 \boldsymbol{dar} + \boldsymbol{\beta}_2 \boldsymbol{size} + \boldsymbol{\beta}_3 \boldsymbol{liqui} + \boldsymbol{\beta}_4 \boldsymbol{hhi5} + \boldsymbol{\varepsilon}$$

3.5 Sample Selection

In order to ensure the accuracy and reliability of the findings, we have processed the selected data as follows: excluding b-share enterprises; excluding small and medium-sized listed companies; excluding companies whose financial situation abnormalities had become ST class in 2013–2017. We select the enterprise 2013–2017 data, the

Table 2. Partial sample data display

Sample data	Date	ROE (%)	Capital structure	Company scale	Short-term liquidity	Ownership concentration
China Unicom	2017.03	0.35	0.61	17.9142	0.26	0.4136
	2016.03	0.18	0.61	17.9123	0.19	0.4136
	2015.03	1.34	0.57	17.809	0.22	0.4272
	2014.03	1.45	0.57	17.7748	0.2	0.4272
	2013.03	0.87	0.58	17.7415	0.17	0.4272
Citic GuoAn Group	2017.03	0.82	0.54	14.2284	0.83	0.1572
	2016.03	1.33	0.42	13.9797	1.41	0.1572
	2015.03	1.08	0.4	13.962	2.13	0.1572
	2014.03	0.75	0.5	14.0003	1.25	0.1815
	2013.03	0.69	0.5	13.9977	0.91	0.1815
Bewinner	2017.03	2.62	0.11	11.7583	4.79	0.4843
	2016.03	0.34	0.06	11.5986	6.78	0.4009
	2015.03	0.72	0.06	11.6162	7.55	0.442
	2014.03	1.08	0.03	11.5771	24.4	0.4107
	2013.03	2	0.05	10.8843	11.94	0.6143

actual number of effective enterprises is 26 (130 valid samples in total). The main source of data came from wind database and corporate financial statements (Table 2).

3.6 Descriptive Analysis

After using Excel preliminary statistical analysis of the data, we found that the capital structure of Chinese communications services listed companies in 2013–2017 mainly has the following characteristics (Fig. 1):

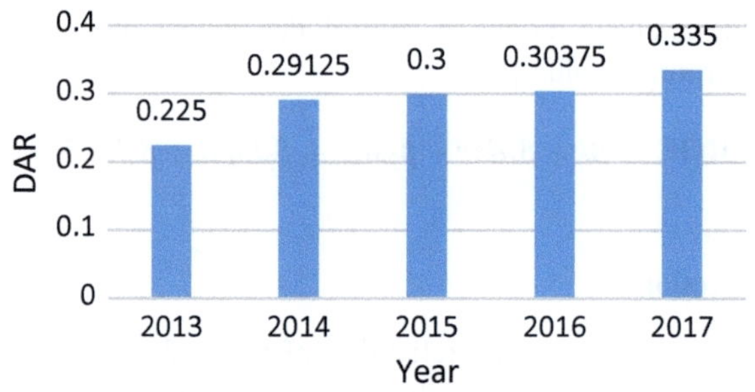

Fig. 1. Average DAR of the communications services industry listed companies in 2013–2017

The level of liabilities of listed companies in this industry is low. It is not difficult to find that the average DAR of listed companies in this industry is low (lower than the average level of Chinese enterprises), after 2013–2014 has increased, the following three years remained at 30% and had a very small change, finally got a slight increase in 2017. According to Chinese actual situation, usually the company's DAR is controlled at about 50%, if a company's data is too low, which indicating that the company has not fully utilized the advantages of debt financing; but if it's more than 60%, the phenomenon of high DAR will happen, and increase financial costs. However, the average DAR of the sample data in this industry is only 29.1%, and the annual annual averages are almost all below 35%, which is indeed lower than the theoretical value.

This may be due to: (1), the industry's listed companies own funds are still relatively abundant; (2), equity financing used properly; (3), the communications services industry is young tertiary industry, the operation process is more cautious, so do not choose the way of borrowing easily.

Chinese corporate liquidity ratio is generally located in the range of 1.5–2. However, the average sample flow ratio is 4.27, the median is 2.64, which indicates that a large part of the enterprise's liquidity ratio was significantly higher than the average level.

3.7 Regression Analysis

The sample data were analyzed by linear regression analysis with Eviews8 software. We get the result as follows:

Table 3. Data regression results

Variable	Coefficient	Std. error	t-Statistic	Prob.
C	201.1563	135.946	1.479678	0.1505
DAR?	−50.74509	13.71355	−3.70036	0.001
SIZE?	−2.902119	1.430479	−2.028774	0.0524
LIQUI?	−0.24029	0.125475	−1.915039	0.0361
HHI5?	−155.7349	140.8758	−1.105477	0.0178
Effects specification				
Cross-section fixed (dummy variables)				
R-squared		0.714574	Mean dependent var	2.701795
Adjusted R-squared		0.667178	S.D. dependent var	5.178999
S.E. of regression		4.054273	Akaike info criterion	5.88508
Sum squared resid		443.8025	Schwarz criterion	6.396945
Log likelihood		−102.7591	Hannan-Quinn criter	6.068733
F-statistic		3.182565	Durbin-Watson stat	2.058812
Prob(F-statistic)		0.006941		

4 Discussion

From Table 3 above, $R^2 = 0.71$ indicates that the overall interpretation of the model is good, the company value indeed has a linear relationship between the selected variables, the parameters we studied such as Capital Structure, Company Scale, Short-term liquidity, Ownership Concentration all have passed the significance test (p values are less than 0.1). All negative Coefficients indicate that the value of enterprises and the variables we study this time have a negative correlation.

The Explained as Follows:

1. Equity financing is generally considered the highest cost of financing in the more mature capital market of western countries, but the cost is lower in China, it is the preferred financing of most listed companies. The data tells us that this industry has a lower DAR, which indicating that the industry is very appropriate to use equity financing. We believe that the company's DAR is low, the tax shield effect and leverage effect has not yet fully play, but in such a low debt it still have a favorable impact on corporate value, indicating that the tax shield effect and leverage at this time has been moderate (le low DAR), so the related issue of decision-making in the capital structure is not highlighted in this field.
2. In normal circumstances, large-scale companies whom has better resources and a stronger ability to resist risks, the corresponding strength should also relatively strong. However, the regression results are different from our hypothesis, which may be due to the particularity of the communication service industry or the low proportion of the fixed assets. The scale effect has not yet been highlighted, and the scale effect of the scale expansion is not significant.
3. The current debt ratio is too high, it has made negative impact on the value, which is consistent with our assumptions.

4. The negative correlation between equity concentration and the company's value indicates that the current proportion of corporate shares is too concentrated, which affects the power of checks and balances and decision-making democratization.

5 Conclusion

The regression results show that from the perspective of the enterprise itself: (1) enterprises must optimize and reduce the DAR further on the base of their own actual situation. This is the core of this study; (2) the expansion of the size of the company may lead to lower value, so the company should carefully analyze the market policy and competitive environment and avoid blind expansion; (3) optimize the liabilities ratio. High value seems to be fast income in the short term, but not conducive to long-term sustainable development; (4) optimize the asset structure before full consideration of capital costs and operational risks, to achieve the expected profit while maintaining financial flexibility. The government should optimize the external environment by improving the relevant laws and regulations, strengthening the supervision.

It should be pointed out that for the enterprise, its value, capital structure, financial risk and other indicators are not static, they will change with a variety of external conditions, and thus the optimal capital structure is also in the process of changing, Therefore, enterprises should always prepare to adjust the proportion from the real environment and its own conditions, in order to achieve the maximization of corporate value.

Acknowledgements. This research is supported by Humanity and Social Science Youth foundation of Ministry of Education of China under Grant #11YJC630291 to Yanliang Zhang. Thank this financial support.

Also thank the organizers of this meeting, thank my Advisor. Zhang's and schoolmate Huang's help and guidance in the collection of literature and writing.

References

1. F. Modigliani, M.H. Miller, The cost of capital, corporation finance, and investment theory. Am. Econ. Rev. (1958)
2. P. Bolton, D.S. Scharfstcin, Optimal debt structure and the number of creditors. J. Polit. Econ. (1996)
3. A.N. Berger, E.B. Patti, Capital structure and firm performance: a new approach to testing agency and application to the banking industry. J. Bank. Financ. (2006)
4. L. Jiang, Study on dynamic optimization of capital structure of listed companies in cultural industry, Finance Research Institute, Ministry of Finance, 2013
5. X. Wu, *Chinese Listed Companies: Capital Structure and Corporate Governance* (China People's Publishing House, 2003)
6. B. Zhang, Tourism industry listed companies capital structure influencing factors empirical research. J. Tour. (2015)
7. L. Niu, X. Wang, Study on the relationship between capital structure and enterprise value of coal listed companies. J. Hebei Univ. Eng. (2016)

8. H. Zhao, Best capital structure for tourism listed companies, Dalian University of Technology, 2016
9. X. Zheng, Theoretical model of dynamic optimal capital structure of Chinese listed companies. China J. Manag. Sci. (2015)
10. Y. Cheng. Real estate enterprise capital structure and corporate value based on panel threshold model of nonlinear relationship test. Financ. Mon. (2016)
11. Tourism service industry listed companies' capital structure on the impact of corporate performance and optimization, Southwest University of Finance and Economics, 2015

An Empirical Analysis of Equity Structure and Performance of Listed Tourism Companies in China

Zan-gen Lin(✉) and Li-ming Zhao

College of Management and Economics, Tianjin University, Tianjin, China
linzangen@163.com

Abstract. The relationship between Listed tourism companies' equity structure and performance is analyzed in this paper. Select return on equity and main business profit during 2007–2015 as the index of performance. The equity structure is measured by the ratio of state-owned shares, the proportion of tradable shares, the share ratio of the largest shareholder and the proportion of the top ten shareholders. Results indicate that the ratio of state-owned shares, the proportion of tradable shares and the performance. Are not in a significant positive correlation; there was a significant negative correlation between the share ratio of the largest shareholder and performance; the proportion of the top ten shareholders is negative to performance. Finally, development proposals are proposed.

Keywords: Tourism listed companies in China · Equity structure
Performance · Empirical research

1 Introduction

Tourism, as a new service industry, is getting rapidly developed in China. The tourism enterprises are the carrier for the development of tourism, and the listed companies of tourism are the high-quality enterprises in the scope of tourism enterprises. The study of listed tourism companies can effectively promote healthy development of tourism industry. Its operating performance affects the competitive advantage and strength, and determines the listed companies to achieve healthy and stable development. Ownership structure is one of the important factors that affects the operating performance of listed companies. Different ownership structure determines the organizations of different enterprises, thus affects the performance of enterprise. The study of the relationship between ownership structure and operating performance contributes in achieving healthy development of tourism industry, especially for tourism companies.

The study of ownership structure and operating performance dates back to the 1930s. Berle and Means have shown that the weakening of shareholder's authority to restrict the company managers is due to the dispersion of the company's ownership structure [1]. Then there is more and more research on the relationship between them. Lu and Zhu select the listed companies of China in 1996 as a sample and find out that state-owned shares have negative impacts on the change of operating performance for IPO companies, are more concentrated and perform much worse in operating [2]. Chen

G. Q. Huang et al. (eds.), *Proceeding of the 24th International Conference on Industrial Engineering and Engineering Management 2018*,
https://doi.org/10.1007/978-981-13-3402-3_33

and Xu have shown that the proportion of the largest shareholder is positively correlated with the performance, and there is no significant correlation between the proportion of state-owned shares, corporate shares and performance [3]. Hu and Hu find out that when the proportion of state-owned shares is in the higher and lower range, the performance is better, the relationship between the proportion of outstanding shares and the performance is U-shaped curve, and the corporate shares aren't related to the company's performance [4]. From the current research situation, there is no conclusive conclusion on the correlation between the ownership structure and the performance of the listed companies. This paper studies the relationship between the ownership structure and performance of listed tourism companies and hopes to enrich the research results from the viewpoint of industry.

2 Research Design

2.1 Hypothesis

The contingency theory of corporate governance holds that different industries have different optimal equity structure. The ownership structures of different concentration have different mechanisms to corporate governance, and different types of equity have different roles in corporate governance. According to the agency cost theory, relative concentration of equity can avoid the problems of free ride for decentralized shareholders in the case of low concentration. Large shareholders take responsibility to effectively supervise managers, thereby enhance the effective use of corporate funds and increase the earnings. Based on the above theoretical analysis, this paper draws the first hypothesis.

H_1: The performance of tourism listed companies is positively related to the concentration of equity which increases as the concentration of ownership also increases.

Before split share reform, non-tradable shares account for the majority of the total share capital of tourism listed companies. After split share reform, the non-tradable shares are entitled to the circulation rights, the value of the circulation rights of the outstanding shares will be attributed to zero. In order to make up for the loss of shareholders of tradable shares, non-tradable shareholders consider the shareholders of tradable shares. The split share reform is to try to improve the equity composition by making the non-tradable shares get the circulation rights in order to reduce the concentration of the shares and to eliminate the problems such as the insider control caused by the dominance of the state and the seriousness of the agent. Since China's tradable shareholders are mostly small in size and complex in composition, it is difficult to act in concert to exert influence on the company. Therefore, this paper holds that:

H_2: There is no significant correlation between the performance and the proportion of outstanding shares of tourism listed companies.

When a government holds its holding position, it has the political influence that the general shareholders do not have. In order to get more cash flow, the government will oversee the management, and the government can give its holding companies preferential treatment and improve the company's performance. In China's empirical study, some scholars have also put forward similar conclusions. Zhou Ye'an has an empirical research between the equity structure and the return on net assets, and found out that it has a significant positive correlation between the proportion of state-owned shares and the return on net assets. Wu (2002) has chosen the profit of total assets to test the impact of equity structure, and found out that the proportion of state-owned shares and the performance of corporate have a curve relationship. When the proportion of state-owned shares is less, the corporate performance is negative, when the proportion of state-owned shares is more, the corporate performance is positive. Based on the above analysis, we put forward the third hypothesis.

H_3: There is a significant positive correlation between the performance and the proportion of state-owned shares of tourism listed companies.

2.2 Sample Selection

According to the classification of China Securities Regulatory Commission, there are 53 tourism listed companies to the end of 2015 including 23 in the market of Shenzhen and 30 in the market of Shanghai. This paper uses the financial data of annual report in the year 2007 to 2015 of 53 tourism listed companies as the original sample.

In order to ensure the effectiveness of the data, try to eliminate the impact of abnormal samples on the study according to the following principles to filter.

First, the company is share A and was listed before the December 31 of the year 2006.

Second, eliminate the listed companies that the financial data can not be obtained.

Third, eliminate the listed companies that were listed as ST or *ST by the SFC.

The final sample contains 23 tourism listed companies, the data is mainly from the Wind database (Table 1).

2.3 Variable Selection

The ownership structure evaluation index generally includes two aspects. First, the composition of shares, that is the number of shares held by the different backgrounds of group. Second, the shares holding by the shareholders, that is the equity concentration (CR). From the viewpoint of the two aspects, we select four index: the proportion of outstanding shares, the proportion of state-owned shares, the proportion of the largest shareholder and the proportion of the top ten shareholders.

The evaluation index of the performance can be divided into single index or comprehensive index by different scholars. Foreign scholars believe that the Tobin Q value is a good measure of the corporate performance, such as Linder and Ross, Shleifer and Vishny [5]. Chinese scholars Li and Jiang use the Tobin Q value to reflect the performance [6]. Zhu chooses the yield of net assets as an explanatory variable [7].

Table 1. The tourism listed companies

Stock code	Securities name	Time to market
600640	Hao Bai Kong Gu	1993.04.07
000524	Dong Fang Bin Guan	1993.11.18
000033	Xin Du Jiu Dian	1994.01.03
600832	Dong Fang Ming Zhu	1994.02.24
600706	Qu Jiang Wen Lv	1996.05.16
000428	Hua Tian Jiu Dian	1996.08.08
000430	Zhang Jia Jie	1996.08.29
000610	Xi An Lv You	1996.09.26
600754	Jin Jiang Gu Fen	1996.10.11
600749	Xizang Lv You	1996.10.15
000721	Xi An Yin Shi	1997.04.30
600054	Huang Shan Lv You	1997.05.06
000796	Yi Shi Gu Fen	1997.07.03
000069	Hua Qiao Cheng A	1997.09.10
000888	E Mei Shan A	1997.10.21
600138	Zhong Qing Lv	1997.12.03
000802	Beijing Lv You	1998.01.08
000809	Tieling Xin Cheng	1998.06.16
000978	Guilin Lv You	2000.05.18
600258	Shou Lv Jiu Dian	2000.06.01
600593	Dalian Sheng Ya	2002.07.11
002033	Lijiang Lv You	2004.08.25
002059	Yunnan Lv You	2006.08.10

Dai selects the net profit margin of total assets to measure the performance [8]. The comprehensive index is mainly to directly select multiple indicators or use the system method to calculate the overall score to make a comprehensive evaluation of performance. Liu selects the debt, profitability, asset management, collection capacity from four factors to evaluate performance [9]. Dong and Ban select six index system twenty specific indicators to conduct a comprehensive evaluation to measure the performance [10]. Liu et al. selects the profitability, debt, operational capacity and growth capacity to illustrate the value of Chinese tourism listed companies [11]. Hu and Chen choose the profitability, debt and operational capacity to evaluate the performance [12]. Tian and Sun use the fuzzy clustering to evaluate the performance combining with the financial indicators of the tourism listed companies [13]. Du and Gan uses the economic value-added and the yield of net assets to measure the operating performance [14]. Based on the above research literature, this paper chooses the yield of net assets and the profitability of main business to evaluate the performance.

In order to fully reflect the performance status, in this study, we choose the size and growth capacity as a control variable into the model. The size affects the performance mainly through the economic scale and management flexibility, it is represented by the natural logarithm of the total assets. The growth capacity reflects the ability and

Table 2. The index of equity structure and performance

The type of variable	The abbreviation of variable	Code	The definition of variable
The dependent variable	Rate of return on common stockholders' equity	ROE	Profit margin on net assets
	Rate of profit to main business	OPE	The main business profit is divided by the main business income
The independent variable	Percentage of outstanding shares	CS	The outstanding share is divided by the total share
	Percentage of state-owned shares	GS	The state-owned share is divided by the total share
	The proportion of the largest shareholder	CR1	The total number of shares held by the largest shareholder is divided by the total share
	The proportion of the top ten Shareholders	CR10	The total number of shares held by the top ten shareholders is divided by the total share
The control variable	The size of company	LnS	The natural logarithm of the total assets
	The growth rate of net profit	EG	The growth of net profit in this year is divided by the net profit of the last year

potential to continue to produce profit and achieve the expansion and quality improvement of the amount. It is represented by the growth of net profit. The specific indicators are in Table 2.

3 The Empirical Results and Analysis

3.1 The Descriptive Statistical Analysis

Before the specific analysis, we have a simple descriptive statistical analysis for the index data, the results are shown in Table 3. As can be seen from Table 3, the variables are basically stable and have a strong representation. It can be drawn from the following three conclusions from Table 3.

Table 3. The descriptive statistical analysis of related variables for the tourism listed companies

Variable	The maximum value	The minimum value	Mean	Median	Variance
ROE	24.00	−15.74	6.58	8.42	6.55
OPE	82.59	−2.63	41.38	47.32	14.32
CS	100.00	35.57	93.15	53.10	23.59
GS	67.83	0.00	22.54	34.85	26.90
CR1	56.15	5.74	39.28	42.28	18.73
CR10	72.07	36.48	56.49	55.45	16.45
LnS	64.26	3.71	46.23	43.29	11.45
EG	77.88	5.07	48.51	45.62	13.45

First, the proportion of outstanding shares is high in the tourism listed companies, with the average of 93.15%. It accounts for the majority of the total share. In order to change the operation mechanism, the state-owned shares are implementing the reduce reform, so the proportion has declined. Which the state-owned shares account for a large proportion of Hao Bai Kong Gu, Hua Qiao Cheng A and Lijiang Lv You, the corporate shares also decrease year by year.

Second, the average shareholding of the largest shareholder is 39.28 which indicates that the shares are relatively concentrated. With the increase of the number of shareholders, the shareholding ratio gradually increases. Until the top ten shareholders, the ratio is just over half which indicates that the degree of equity concentration is gradually reduced in the process of continuous reform of the stock market.

Third, the return of net assets is relatively large with the minimum value of −15.74 and the maximum value of 24.00. The margin profit of the main business is increasing year by year, with the average value of 41.38. But it has reached the lowest in the year 2008 because of financial crisis. It can be seen that the development of the tourism industry is easily affected by the external environment, which also illustrates the vulnerability of the tourism industry.

3.2 The Yield of Net Assets Is the Explanatory Variable

The data is smooth and there is cointegration through Eviews analysis, there is a long-term stable relationship between the variables. In contrast to the hybrid estimation model, the establishment of an individual fixed effect model can be done by F-test. The F = 7.843 is greater than F0.05(19.94) = 1.67, we reject the original hypothesis and establish an individual fixed effect model. It is shown in the Table 4.

Table 4. The estimation results of individual fixed effect model

Variable	Coefficient	Std. error	t-statistic	Prob.
C	23.0862	29.3804	0.7857	0.4340
CS	0.0332	0.0510	0.6523	0.5158
GS	0.0142	0.0563	0.2533	0.8005
CR1	0.4238	0.1686	2.5133**	0.0137
CR10	−0.0987	0.1022	−0.9648	0.3371
LnS	−1.3793	1.3572	−1.0163	0.3121
EG	0.0046	0.0009	4.9941***	0.0000
R^2	0.7612	F-statistic		11.9866
R^2adj	0.6977	Prob(F-statistic)		0.0000
D.W.	1.9191			

*indicates significant at 10% level; **indicates significant at 5% level; ***indicates significant at 1% level

The yield of net assets is explanatory variable, the R^2 is 0.7612, the adjusted R^2 is 0.6977, the regression model has a high fit degree, it indicates that the explanatory variables in the model can explain the most of the return of net assets. The P value of the probability of F is 0.0000, the linear relationship between the dependent variable and the independent variable is significant at the 5% level, that is through the general significance test of the equation. The D.W. value indicates that the model does not have autocorrelation. Through the T statistic, we can see the followings.

First, the coefficient of proportion of outstanding share and state-owned share are not through the T test, but still shows its relationship with the operating performance, the proportion of outstanding share and state-owned share are positively related with the performance. It is assumed that the Hypothesis 2 is not established, the Hypothesis 3 is established.

Second, there is a significant positive correlation between the performance and the proportion of the largest shareholder of tourism listed companies at the significance level is 0.05, the correlation coefficient is 0.4238, assuming that the Hypothesis 1 is established.

Third, the return of net assets is positively correlated with the growth rate of net profit at the significance level of 0.01, but the regression coefficient is small. It means that the company has better growth rate and the performance is higher.

Fourth, the proportion and size of top ten shareholders have failed to pass the T test, indicating that the relationship between the two and performance is not significant and yet negatively correlated.

3.3 The Profitability of Main Business Is the Explanatory Variable

The data is stable by unit root test, and the data is cointegrated by Kao test. The SSEr = 24486.12, the non-constrained model SSEu = 3435.63, the F = 31.96 is greater than the critical value, so we reject the H0, and establish the individual fixed effect model. Then use the Eviews software to regression the panel data, the results are in the Table 5.

Table 5. The estimation results of individual fixed effect model

Variable	Coefficient	Std. error	t-Statistic	Prob.
C	58.3481	47.2943	1.2337	0.2204
CS	0.0895	0.0821	1.0902	0.2784
GS	0.0835	0.0907	0.9209	0.3594
CR1	0.0730	0.2714	0.2692	0.7884
CR7	−0.0141	0.1646	−0.0857	0.9318
LnS	−1.2861	2.1847	−0.5886	0.5575
EG	0.0052	0.0014	3.5400***	0.0006
R^2	0.8931	F-statistic		31.4194
R^2adj	0.8646	Prob(F-statistic)		0.0000
D.W.	1.4932			

*indicates significant at 10% level; **indicates significant at 5% level; ***indicates significant at 1% level

From the Table 5, the R^2 is 0.8931, the adjusted R^2 is 0.8646, it indicates that the explanatory variables in the model can explain the most of the profitability of main business. The F test shows that the regression equation in overall, shows significant linearity, and the linear relationship between the dependent variable and the independent variable is significant at the 5% level. The value of D.W. indicates that the model is auto-correlated. If the correlation is considered in the model, the model is not relevant after the addition of AR(1), and the goodness is improved, while the economic significance has changed. So we still use the above model. The results show that the growth rate of net profit is positively significant with the return of net assets, but the ownership structure is not significant with the return of net assets.

4 Research Conclusion

Through the above empirical analysis, it can be concluded that the proportion of the largest shareholder is significantly positive with the performance. The proportion of the top ten shareholders is negative with the performance, so the Hypothesis 1 is not fully established. The proportion of outstanding shares has failed to pass the significant test, but still shows that the proportion of outstanding shares is positive with the performance, therefore the Hypothesis 2 is not established. There is no significant positive correlation between the state-owned shares and operating performance, so the Hypothesis 3 is established.

For the results, this paper suggests that listed companies should maintain the equity concentration in an appropriate ratio, especially the shareholding ratio of the largest shareholder. With more holdings, it has the incentive to supervise the management to protect their own interests, and also effectively motivates the management to work harder. In order to improve the performance, it should appropriately reduce the proportion of state-owned shares in the principle of "win-win for the government and investors".

It should be noted that this paper is based on the study of China's tourism listed companies, and the results are not consistent with the previous scholars. It may use different indicators to measure the performance. Since the environment of tourism industry is quite different from that of other industries, the empirical results does not represent all possible circumstances. With the development of high-tech, there may be other factors to affect the performance, such as network technology, corporate management methods, management skills and corporate culture etc. These can be the subjects of further research in the future.

References

1. A. Berle, G. Means, *The Modern Corporation and Private Property* (Macmillan, New York, 1932)
2. W. Lu, H. Zhu, A research in the relationship between the earnings pattern of IPO firms and its ownership. Study Financ. Econ. **27**(7), 45–52 (2007)

3. X. Chen, X. Xu, Equity structure, firm performance and the protection for investers' interest. Econ. Res. J. **11**(3), 11 (2001)
4. J. Hu, Y. Hu, An empirical analysis of equity structure and corporate performance in listed companies. Manag. World (3), 142–143 (2006)
5. A. Shleifer, R. Vishny, Large shareholders and corporate control. J. Polit. Econ. **94**, 461–488 (1986)
6. Y. Li, Z. Jiang, An empirical analysis of capital structure and performance of listed companies. J. Quant. & Tech. Econ. (2), 118–120 (2001)
7. Y. Zhu, A study of the performance and model of banking-securities industry cooperate. Shanghai University, Shanghai, 2003
8. Y. Dai, An empirical study on the relationship between capital structure and corporate performance of China's listed media companies. Theory Prac. Financ. Econ. **34**(181), 30–33 (2013)
9. T. Liu, An analysis and evaluation on operational performance of listed tourism companies. Tour. Trib. **20**(4), 92–96 (2005)
10. G. Dong, A study of performance evaluation system of listed tourism companies. Tour. Sci. **20**(6), 66–70 (2006)
11. L. Liu, L. Zhao, E. Duan, On the assessment of economic benefits of China's listed travelling companies. Tour. Trib. **22**(4), 79–83 (2007)
12. X. Hu, X. Chen, A study of financial situation of listed tourism companies in China based on principal component analysis. J. Hebei Tour. Vocat. coll. **15**(1), 61–65 (2010)
13. R. Tian, Y. Sun, The fuzzy clustering analysis of performance of listed tourism companies in China. Mod. Bus. Trade Ind. **6**, 5–7 (2011)
14. X. Du, S. Gan, A research study of capital structure, equity structure and performance–based on China GEM listed companies. Commer. Account. (1), 73–75 (2012)

Dynamic Analysis of Regional Scientific and Technological Innovation Capacity—A Case Study in Sichuan Province

Xue-dong Liang[1], Dong-yang Si[2], Zhi Li[3(✉)], and Tan Liu[1,2,3]

[1] Department of Industrial Engineering, Sichuan University, Chengdu, China
[2] Department of Logistics Engineering, Sichuan University, Chengdu, China
[3] Manufacturing Department, Sichuan Chengfei Integration Technology Co. Ltd, Chengdu, China
Zhili1090@163.com

Abstract. Aiming at the problem of regional scientific and technological innovation, this paper, first, adopted a dynamic model of regional scientific and technological innovation capacity, based on three aspects, environment, input, and achievement to scientific and technological innovation, and then used entropy method and improved vertical scatter degree method to respectively calculate the horizontal and vertical score of each region, finally, carried out the empirical research by taking Sichuan Province as an example, and analyzes the results.

Keywords: Entropy method · Regional scientific and technological innovation capacity · Vertical scatter degree method

1 Introduction

Regional scientific and technological innovation is the key factor to the growth of regional economy and the improvement of regional competitiveness. The regional scientific and technological innovation capacity, which reflects the basic ability of innovation elements in a certain region to bring together, integrates and promotes sustainable innovation, is an important ability to promote the sustainable development of regional economy [1], and has become an important indicator to measure the decisive position of regional economy in international competition [2].

The regional scientific and technological innovation capacity provides strategic support for national innovation, its connotation, characteristics and structure, and has become an important research object in academia. As far as the evaluation methods of regional scientific and technological innovation capacity, from the existing literature, there are static analysis methods and dynamic analysis methods. In static analysis methods, there were more used include fuzzy comprehensive evaluation [3, 4], AHP [5, 6], grey correlation [7, 8], principal component analysis [9–12] and factor analysis [13–16]. Besides, Tao Huazhi, Zhang Xin, Zhang Yan and Sui Xuejin used extension evaluation method to analyze the regional scientific and technological innovation capacity of Shandong province, which relatively well solved the problem of complex and fuzzy evaluation in the

G. Q. Huang et al. (eds.), *Proceeding of the 24th International Conference on Industrial Engineering and Engineering Management 2018*,
https://doi.org/10.1007/978-981-13-3402-3_34

comprehensive evaluation of regional scientific and technological innovation capacity and provided corresponding policy recommendations for decision-making departments at all levels [17]; Zhao Liming and Liu Meng used ideal solution to empirically analyze the development of 31 provinces and cities and 6 economic regions in China, and put forward some strategies to improve scientific and technological innovation capacity [18]. In dynamic analysis methods, Zhang Dakun, Wu Xianping, Guan Lianlong made a dynamic empirical analysis of scientific and technological innovation capacity in Jiangsu province based on factor analysis [19]; Pan Xiongfeng, Liu Qing, and Peng Xiaoxue used the entropy method to make a dynamic analysis to the innovation capacity of the Beijing-Tianjin-Hebei region, the northeast China, the Yangtze River Delta areas and the southern coastal areas, which provided a reference for the enterprises' innovation capacity in each economic zone [20]; Chen Guohong, Kang Yiping, Li Meijuan made a dynamic evaluation of regional scientific and technological innovation capacity of 31 provinces in China based on "vertical and horizontal" scatter degree method, and proposed an effective way to improve the scientific and technological innovation capacity [21].

2 Dynamic Model of Regional Scientific and Technological Innovation Index System

From the perspective of input-output, regional scientific and technological innovation capacity is scientific and technological innovative achievement acquires by scientific and technological innovative each main body based on scientific and technological innovative environment and input, which is the embodiment of regional scientific and technological innovation capacity. The main body includes government, enterprises, universities and scientific research structure, etc., and they ultimately have improved the total capacity by creating innovative environment, increasing innovative input, and enhance innovative achievement. The elements can be divided into three categories: innovative environmental factors (government environment, economic environment, educational environment, scientific and technological environment), innovative input factors (manpower input, financial resources input) and innovative achievement factors (social achievements, enterprise achievements). The model consists of two parts: the main body of scientific and technological innovation and the elements of scientific and technological innovation. In this model, the two parts constitute a dynamic system: starting from the main body, the increase of input directly leads to the improvement of achievements, the improvement of achievement improves the environment, and the improvement of environment has laid a foundation of manpower, material and financial resources for further increase of input; On the other hand, the promotion of achievement has laid a material foundation for further increase of input, the increase of input is beneficial to improve the environment, and the improvement of environment can improve the achievement [22]. Therefore, the system is not a static model, but an evaluation model of regional scientific and technological innovation capacity that is dynamically cycled over time.

Based on the above-mentioned regional dynamic model and relevant literatures of scientific and technological innovation, according to the system, scientific, quantifiable, comparable, guiding, dynamic and purpose principle, establish an index system, as shown in Table 1.

Table 1. Regional science and technology innovation capability index system

Science and technology innovation environment X_1	The number of ordinary colleges and universities X_{11}
	Number of college students per million people X_{12}
	Number of R & D activities of science and technology X_{13}
	Number of Large and Medium-sized Industrial Enterprises X_{14}
	Local fiscal general budget expenditure X_{15}
	GDP per capita X_{16}
Science and technology innovation input X_2	R&D staff equivalent at that time X_{21}
	Scientific and technical personnel X_{22}
	R&D internal expenditure X_{23}
	The proportion of sales revenue of new products in large and medium-sized enterprises in the main business income X_{24}
Science and technology innovation achievement X_3	Number of patent applications X_{31}
	Industrial output value of large and medium sized enterprises X_{32}
	The main business income of large and medium-sized enterprises X_{33}
	Value added value of large and medium sized enterprises X_{34}

3 Evaluation Steps

(1) Standardizing the data, and assume that u_{ij} is the normalized value of the i-th index of the j-th evaluation object; v_{ij} is the value of the i-th index of the j-th evaluation object; m is the number of the evaluation objects, n is the number of indicators, Pending variable P_{ij}:

$$P_{ij} = \frac{u_{ij}}{\sum_{j=1}^{m} u_{ij}} \tag{1}$$

(2) Calculating the entropy of the i-th index E_i:

$$E_i = -\frac{1}{lnm} \sum_{j=1}^{m} P_{ij} \ln(P_{ij}) \tag{2}$$

(3) W_i is the entropy of the i-th index, m is the number of objects evaluated, the entropy of the i-th index:

$$W_i = \frac{1 - E_i}{m - \sum_{i=1}^{m} E_i} \quad (3)$$

(4) Calculating the horizontal score of each region Dj:

$$D_j = \sum_{i=1}^{n} W_i \times u_{ij} \quad (4)$$

(5) In time [T1, TN], calculating the symmetric matrix:

$$\mathrm{H} = A^T A \quad (5)$$

$$A = \begin{bmatrix} v_{11} & \cdots & v_{1n} \\ \vdots & \ddots & \vdots \\ v_{N1} & \cdots & v_{Nn} \end{bmatrix}$$
$$\mathrm{t} = (1, \ldots, \mathrm{N}), v_{ij} = u_{ij} \times W_i$$

(6) k_i is the second weight of the i-th index, and the calculating eigenvalues and eigenvectors of H:

$$\begin{aligned} \max \quad & k^T H k \\ & \text{s.t. } k^T k = 1 \\ & \mathrm{k} > 0 \end{aligned} \quad (6)$$

k_i is the standard eigenvectors corresponding to the maximum eigenvalue of H.

(7) Normalizing processing k_i:

$$g_i = \frac{k_i}{\sum_{i=1}^{n} k_i} \quad (7)$$

(8) Calculating the regional longitudinal score G_t:

$$G_t = \sum_{i=1}^{n} k_i u_{ti} \quad (8)$$

4 Analysis of Regional Scientific and Technological Innovation Capacity in Sichuan Province

4.1 Data Analysis

According to the demand of selected indicators, this article analyzed and summarized the Sichuan Science and Technology Statistical Yearbook, Sichuan Statistical

Yearbook and China Urban Statistical Yearbook from 2012 to 2015. According to the geographical location of Sichuan Province, Chengdu is the central region of Sichuan Province, Guang'an City, Dazhou City, Bazhong City, Nanchong City, Suining City belongs to the eastern part of Sichuan Province, Panzhihua City, Ya'an City, Aba Prefecture, Ganzi Prefecture, Liangshan Prefecture belongs to the western region Sichuan Province, Leshan City, Yibin City, Luzhou City, Zigong City, Neijiang City, Meishan City, Ziyang City belongs to the southern region of Sichuan Province, Deyang City, Mianyang City, Guangyuan City belongs to the northern region of Sichuan Province.

By the entropy method to calculate the horizontal science and technology innovation ability score, the result was shown in Table 2. F1 is the regional science and technology innovation environment ability, F2 is the regional science and technology innovation input ability, F3 is the regional science and technology innovation achievement ability.

By the method of improved portrait scatter degree, the vertical science and technology innovation capacity score of each region was obtained, and the results were shown in Table 3.

Table 2. The horizontal science and technology innovation ability score

		Overall score	Rank	F1 score	Rank	F2 score	Rank	F3 score	Rank
2012	Central region	1.0000	1	1.0000	1	1.0000	1	1.0000	1
	Eastern region	0.0265	4	0.0375	5	0.0002	5	0.0376	3
	Western region	0.0196	5	0.0396	4	0.0099	4	0.0000	5
	Southern region	0.0603	3	0.0647	3	0.0250	3	0.0911	2
	Northern region	0.0902	2	0.0799	2	0.1113	2	0.0093	4
2013	Central region	1.0000	1	1.0000	1	1.0000	1	1.0000	1
	Eastern region	0.0186	5	0.0277	5	0.0006	5	0.0235	4
	Western region	0.0266	4	0.0515	4	0.0165	4	0.0000	5
	Southern region	0.0553	3	0.0703	3	0.0241	3	0.0650	3
	Northern region	0.0826	2	0.0757	2	0.0994	2	0.0757	2
2014	Central region	1.0000	1	1.0000	1	1.0000	1	1.0000	1
	Eastern region	0.0503	4	0.1000	4	0.0012	5	0.0284	3
	Western region	0.0620	3	0.1278	3	0.0176	3	0.0125	4
	Southern region	0.0006	5	0.0003	5	0.0015	4	0.0000	5
	Northern region	0.1204	2	0.1611	2	0.1041	2	0.0785	2
2015	Central region	1.0000	1	1.0000	1	1.0000	1	1.0000	1
	Eastern region	0.0159	5	0.0273	5	0.0000	5	0.0151	4
	Western region	0.0332	4	0.0631	4	0.0229	2	0.0000	5
	Southern region	0.0502	3	0.0772	3	0.0147	3	0.0466	3
	Northern region	0.0866	2	0.1011	2	0.0020	4	0.0581	2

Table 3. The vertical science and technology innovation ability score

		Overall score	Rank	F1 score	Rank	F2 score	Rank	F3 score	Rank
Central region	2012	6.7130	3	7.0393	2	7.5876	1	6.0115	4
	2013	6.7874	2	7.0587	1	7.3148	2	6.5607	2
	2014	5.9854	4	6.0254	4	6.3534	4	6.2028	3
	2015	6.8049	1	7.0167	3	7.3092	3	6.6985	1
Eastern region	2012	0.4319	2	0.4761	2	0.0015	3	0.2946	1
	2013	0.3161	3	0.3520	3	0.0055	2	0.2034	3
	2014	0.8273	1	1.0146	1	0.0207	1	0.2177	2
	2015	0.2886	4	0.3402	1	0.0000	1	0.1225	4
Western region	2012	0.5860	4	0.6775	4	0.1429	4	0.0000	2
	2013	0.7990	3	0.9153	3	0.2502	2	0.0000	2
	2014	1.6110	1	1.8928	1	0.1755	3	0.0877	1
	2015	1.0477	2	1.2013	2	0.3211	1	0.0000	2
Southern region	2012	0.6570	2	0.7193	3	0.2313	1	0.7057	1
	2013	0.6115	3	0.7327	2	0.2087	2	0.5491	2
	2014	0.0019	4	0.0009	4	0.0114	4	0.0000	4
	2015	0.6626	1	0.9267	1	0.1276	3	0.4006	3
Northern region	2012	0.9570	3	0.6948	3	1.6032	1	0.5620	1
	2013	0.8899	4	0.6843	4	1.4192	2	0.5460	2
	2014	1.1362	1	1.3408	1	1.2355	4	0.5260	3
	2015	0.9680	2	0.9990	2	1.2737	3	0.4332	4

4.2 Result Analysis

As shown in Table 2, whether the regional scientific and technological innovation capacity, or its' the three elements, the central region of Sichuan Province is the best, far higher than other regions, followed by the northern region of Sichuan Province, and the eastern region of Sichuan province is the weakest. Because, Chengdu is the capital of Sichuan Province, so most of the innovative resources of manpower, material and financial resources are concentrated in Chengdu; Deyang city and Mianyang city are the industrial concentrated areas of Sichuan province, bringing together most of the research institutions, which makes the scientific and technological innovation capacity of the northern region of Sichuan Province higher; the eastern region western region and southern region are far from Chengdu, the scientific and technological innovation has not been paid attention, so the innovation capacity is low.

As shown in Table 3, in the period of 2012–2015, regional scientific and technological innovation capacity and its' three elements of Sichuan Province have been developed smoothly, the scientific and technological innovation capacity in the central region has been at the leading level, the level of the eastern region. In terms of regional scientific and technological innovation capacity, the capacity of the central region declined slightly in 2014, but still far higher than other regions in Sichuan Province, the capacity of the eastern region, the western region and the northern region reached the highest level in 2014, but the southern region slightly decreased in 2014, and overall development is stable. The development trend of regional scientific and technological innovation capacity is consistent with the overall development tendency of regional scientific and technological innovation capacity. In terms of regional scientific and technological innovative input, although the northern region is lower than the central region, it is significantly higher than the eastern region, western region and southern region, the central region slightly declines in 2014, the overall development trend is stable, and the regional development trend in the eastern region, the western region and the southern region has no fluctuation. In terms of scientific and technological innovative achievement, all regions of Sichuan Province are in a stable development status.

5 Conclusions

Through the establishment of the dynamic model of regional scientific and technological innovation, this paper made the horizontal and vertical analysis to scientific and technological innovation capacity of Sichuan Province, which was based on the science and technology statistical yearbook of Sichuan Province, the statistical yearbook of Sichuan Province and the Chinese urban statistical yearbook in 2012–2015, and combined with the entropy method and the improved vertical scatter degree method. By the research and analysis, the following conclusions are drawn:

(1) On the basis of predecessors, this paper put forward the establishment of dynamic model of regional scientific and technological innovation, which enriched the content of dynamic evaluation model of innovation capacity.
(2) Using the method of horizontal and vertical combination to acquire the evaluation score of regional scientific and technological innovation capacity, which not only pays attention to the development situation at a certain time, but also pays attention to the trend of change, and provides a new idea for regional scientific and technological innovation capacity evaluation.
(3) Through the evaluation and analysis of the regions in Sichuan province, the regional scientific and technological innovation capacity of Sichuan province is steadily rising year by year, the environment, input and achievement of scientific and technological innovation interact and jointly promote the development of regional scientific and technological innovation capacity. While focusing on the development of the central Sichuan, we should pay attention to the surrounding areas and promote the steady development of the Sichuan regional regions and promote the steady development of all regions of Sichuan province.

In this paper, when establishing the index system, while considering a number of factors, the index system does not include all the factors, and there are some limitations. At the same time, the dynamic evaluation method based on timing sequence perspective can also be considered from other aspects. therefore, it is the focus of future research to further propose a more perfect dynamic model of regional scientific and technological innovation, and to study the dynamic evaluation method based on timing sequence perspective.

Acknowledgements. This research is funded by the National Nature Science Foundation of China (71302134), Foundation of Academic Leader Training in Sichuan Province ([2015]100-6).

References

1. T.L. Satty, *Muticriteria Decision Making* (RWS Publication, Pittsburgh, 1990)
2. H. Zilang, Y. Zhang, Z. Liu, The comparative study on bohai rim regional technology innovation ability. Phys. Procedia **33**, 294–300 (2012)
3. G. Gu, F. Teng, Research on operation mechanism and evaluation index system of regional scientific and technological innovation. J. Northeast. Norm. Univ. **4**, 24–30 (2003)
4. Y. Tang, Fuzzy comprehensive evaluation model and its application of regional scientific and technological innovation capacity comprehensive analysis of scientific and technological innovation capacity of Guangdong in 2001. Syst. Eng.— Theory & Pract. (2), 37–43 (2004)
5. J. Shen, Research and application of evaluation system of regional scientific and technological innovation capacity in China. On Econ. Probl. **8**, 27–29 (2005)
6. X. Yin, K. Wang, Research on evaluation of regional scientific and technological innovation capacity. Prod. Res. **6**, 99–100 (2006)
7. Z. Yang, J. Zhang, Grey correlation comprehensive evaluation of regional scientific and technological innovation capacity. J. Math. Pract. Theory **37**(9), 17–22 (2007)
8. P. Shi, Q. Li, L. Zhao, Research on evaluation of regional scientific and technological innovation capacity in China. J. Xi'an Univ. Post Telecommun. **14**(6), 122–125 (2009)
9. Y. Yang, Principal component analysis and evaluation of regional scientific and technological innovation capacity. Technol. Econ. **26**(6), 15–19 (2007)
10. Y. Dong, H. Sun, Q. Zhang, Evaluation of regional scientific and technological innovation capacity based on principal component analysis. Sci. & Technol. Prog. Policy **29**(12), 26–30 (2012)
11. X. Bai, Y. Zhang, Evaluation on innovation ability of regional scientific and technological innovation system in Liaoning province. J. Liaoning Univ. **39**(3), 268–272 (2012)
12. Y. Zhao, Y. Cai, M. Zhao, Comparative study on urban scientific and technological innovation capacity—a case study of Jiangsu province. East China Econ. Manag. **12**, 164–166 (2013)
13. T. Feng, X. Li, Evaluation of scientific and technological innovation capacity in Shaanxi province based on factor analysis. Technol. Innov. Manag. **29**(3), 221–226 (2008)
14. G. Han, Research on scientific and technological innovation capacity of Anhui province. China Dev. **9**(2), 82–85 (2009)
15. Y. Zheng, L. Zhang, J. Zheng, Empirical evaluation and analysis of regional scientific and technological innovation capacity in Fujian province. Sci. Technol. Manag. Res. **20**, 59–63 (2010)
16. D. You, Study on evaluation index system of regional scientific and technological innovation capacity. Co-Op. Econ. & Sci. **456**, 15–16 (2013)
17. H. Tao, X. Zhang, Y. Zhang, J. Sui, Research on extension comprehensive evaluation of regional scientific and technological innovation capacity. Technol. Innov. Manag. **31**(3), 256–260 (2010)
18. L. Zhao, M. Liu, Evaluation model and empirical study of regional scientific and technological innovation capacity based on entropy TOPSIS. J. Tianjin Univ. **16**(5), 385–390 (2014)
19. Y. Zhang, W. Jianping, L. Guan, Dynamic evaluation and empirical analysis of regional scientific and technological innovation capacity. East China Econ. Manag. **21**(1), 90–94 (2007)
20. X. Pan, Q. Liu, X. Peng, Dynamic evaluation and analysis of regional scientific and technological innovation capacity of China based on the global entropy method model. Oper. Res. Manag. Sci. **4**(24), 155–162 (2015)

21. G. Chen, Y. Kang, M. Li, Dynamic evaluation of regional scientific and technological innovation capacity—based on the improved "Vertical and horizontal" scatter degree method. Technol. Econ. **10**(34), 17–23 (2015)
22. R. Ma, Construction of evaluation model of dynamic bidirectional regional scientific and technological innovation capacity. Shandong Soc. Sci. **8**, 132–135 (2011)

Coordination Effects of Market Power and Fairness Preference on Supply Chain

Ju-ning Su(✉), Xiao-yu Yang, Jun-feng Jiang, and Li-yao Tan

School of Economics and Management, Xi'an University of Technology, Xi'an 710048, China
sujn@163.com

Abstract. As the core of supply chain management, supply chain coordination is influenced by several factors. The theory of behavioral supply chain management holds that in addition to economic factors, supply chain coordination is also impacted by behavioral factors. Fairness preference is a common behavioral factor. This paper examined the effect of the external factor, i.e. market power, and internal factor, i.e. fairness preference, on supply chain coordination. First, a benchmark for supply chain coordination effects was established. Next, the influences of two factors, i.e. market power and fairness preference, on supply chain coordination were separately analyzed. It was indicated that both producing negative effects on supply chain coordination. Lastly, the coordination effects of the two factors were compared, the former's influence was larger.

Keywords: Coordination effects · Fairness preference · Market power Nash game

1 Introduction

In the marketing channel, both the manufacturer and the retailer have "market power", defined as "the ability of one channel member to control the decision variables in the marketing strategy of another member in a given channel at a different level of distribution" [1]. Market power, a kind of implicit contract, determines profit distribution in the commercial process by influencing market trading behaviors. Many studies have proved that, under various market power structures, decisions made by manufacturers and retailers were different, leading to different performance, such as profit of supply chains [2–5]. This indicates that market power may exert an influence on the decisions and performance in supply chains.

In most of the existing research concerning supply chain decisions and coordination, decision-makers were assumed to be self-interested, i.e. pursuing the maximization of self-benefit while neglecting others' benefits. In recent years, lots of researchers raised objections to the absolutely self-interested assumption of the traditional economics. Experiments such as the ultimatum game, dictator game, gift exchange game, trust game and so forth laid the foundation for the proposing of the "non-self-interested" assumption, which held that people pay attention to others' interests when pursuing their self-interests, i.e. fairness preference. Based on a series of experimental

G. Q. Huang et al. (eds.), *Proceeding of the 24th International Conference on Industrial Engineering and Engineering Management 2018*,
https://doi.org/10.1007/978-981-13-3402-3_35

results, some scholars established the model of fairness preference theory. Rabin [6] proposed, the earliest, a simple linear utility function model; While Fehr and Schmidt [7] established the most representative one, i.e. inequality aversion model (FS model for short). Regarding the research of fairness preference theories, Charness and Rabin [8] assumed the value range of the disadvantageous and the advantageous unfairness preference coefficients to be [−1, 1]. According to various value ranges of fairness preference coefficients, they studies four types of fairness preference models, i.e. the narrow self-interest, competitive preference, avoiding unfair preference and social welfare preference. Considering the supply chain background, not all manufacturers and retailers, being not only self-interested but sometimes also altruistic, make decisions based on the principle of profit maximization. Kahneman and Knetsch [9] held that enterprises were also concerned about fairness, which plays a crucial role in building and maintaining the channel relationship. Through an empirical research on the marketing channel of automobiles in the USA and the Netherlands, Kumar and Knetsch [10] revealed that trust and fairness in work were essential factors in maintaining channel coordination. Loch and Wu [11] proved, through experiments, that social preference would influence trading behaviors. Besides, they also pointed out that relationship preference could promote cooperation, thus improving the performance of individuals as well as the whole supply chain; while status preference functions differently. Katok and Pavlov [12] examined, through experiments, the function of fairness in supply chain coordination, holding that people's lack of understanding, in whether they were fairly treated or not, was one of several reasons why the coordination of contract could not be realized.

Existing literature has discussed respectively, in fields of marketing channel and behavior operation, the influence of market power and fairness preference in the supply chain. This paper quantitatively studies the influence of the external economic factors and the fair behavior of the market power on the decision-making and performance of the supply chain, and compares the influence of both, so that people can use market factors and behavioral factors in the supply chain management practice, to improve the coordination level of the supply chain.

The remainder of the paper is organized as follows: Sect. 2 provides the problem description and basic assumptions. Section 3 establishes the base of comparing. Sections 4 and 5 detail the effects of market power and fairness preference on coordination performance in the supply chain respectively. Section 6 compares between the effects of market power and the effects of fairness preference. Section 7 concludes the article with a summary of the results and some future research directions.

2 Problem Description and Basic Assumptions

2.1 Problem Description

This paper examines two-stage supply chains formed by the manufacturer and retailer, producing and selling a kind of short life-cycle product. The retailer places orders with the manufacturer according to the market demand D, and the manufacturer organizes production based on MTO pattern. For the manufacture, the production cost is c and the wholesale price w; for the retailer, without consideration of the sale cost, the retail price is p, $0<c<w<p$. The manufacturer and retailer being in the imperfect competitive market, the wholesale price contract is adopted in the supply chains, with the manufacturer determining the wholesale price w and the retailer the retail price p. Supply chain members are not completely self-interested rational agents, but are finitely self-interested decision-makers with fairness preference. Supply chain members having fairness preference are concerned about the fairness of the distribution results. Both sides share information such as the cost, market demand, behavioral preference, etc.

Other parameters are described as follows: α represents the retailer's fairness preference coefficient, while β the manufacturer. π and Π indicate profits of supply chain members and the supply chain system respectively, and u means the utility of supply chain members. In the first subscript, 0 means there is no fairness preference (i.e. the fairness is neutral), S denotes that the manufacturer is with fairness preference, S that the retailer has fairness preference; in the second subscript, r represents the performance variable of the retailer, while s that of the manufacturer. A superscript, added with $*$, symbols the optimal value.

2.2 Basic Assumptions

The following assumptions are put forward, for ease of analyzing and discussing the problem.

Assumption 1 Linear market demand assumption. It is assumed that the market demand faced by the retailer is a linear form, merely influenced by the price, i.e. $D(p) = a - bp$ where a is the overall market capacity and b the sensitive coefficient of the market demand on retail price, $a > 0$, $b > 0$.

When supply chain enterprises have fairness preference, they will not aim at profits maximization anymore in making decisions; instead, they will focus on utility maximization. This paper adopts the utility function model proposed by Charness and Rabin [8]: $u = \pi + f$ where u is the decision-maker's utility, π the profits, and f the utility of fairness preference. For the fairness preferences of supply chain enterprises, this paper applies the competitive preference assumption put forward by Charness and Rabin [8], i.e. the Assumption 2 below.

Assumption 2 Competitive fairness preference assumption. For supply chain member i has fairness preference, the utility function of its fairness preference is assumed to be $f_i = \lambda_i(\pi_i - \pi_j)$, $i, j \in (s, r)$ where λ_i is the fairness preference coefficient of the supply chain member i, $0<\lambda_i<1$, indicating that for supply chain enterprises, either advantageous or disadvantageous inequality, the utility caused by inequality will be no larger than that brought by obtaining or losing equivalent economic income. In this paper, $\lambda_r = \alpha$, $\lambda_s = \beta$.

This paper examines the influence of market power and fairness preference on supply chain coordination effects. Previous research all assessed the coordination performance of the supply chain from the standpoint of the supply chain system. Based on the seller market, this concept is not accustomed to the environment of the buyer market nowadays. Under the condition of the buyer market, the supply chain providing consumers with more value will win the market competition. For this reason, the coordination performance of the supply chain should be assessed from the standpoint of consumers. In addition, as an inter-company social organization, the supply chain should also give full play to its social function, creating more social welfare benefits, to satisfy the requirements of the society. Therefore, a social perspective in assessing the coordination performance is also necessary. On this basis, this paper put forward three perspectives, i.e. the supply chain, consumers and the society, in accessing the coordination performance of the supply chain. From the angle of the supply chain, the indicator of supply chain efficiency is used to measure the coordination performance of the supply chain. As the ratio of the profits of decentralized supply chain to that of centralized supply chain, the indicator of supply chain efficiency can reflect the coordination degree of the whole supply chain. From the perspective of consumers, the consumer surplus, an indicator that can reflect benefits brought to consumers by the whole supply chain, is used to assess the coordination performance of the supply chain. In terms of the social perspective, the indicator of social welfare, sum of the supply chain profits and consumer surplus, is adopted to reflect the contribution made by the supply chain to the society. Expressions of the supply chain efficiency η, consumer surplus cs and social welfare sw, as follows: $\eta = \frac{\Pi}{\Pi_c}$, $cs = \frac{1}{2} \times \left(\frac{a}{b} - p\right) \times D$ (under the condition of the linear market demand), and $sw = \Pi + cs$.

In order to compare the influence of market power and fairness preference on the supply chain coordination effects, the following first analyzes the coordination effect of market power and fairness preference on supply chain, and then compares the relevant conclusion.

The following will first establish a benchmark for measuring the coordination effect of the supply chain: the Nash game model with balanced market power but no fairness preference, will be established in the next section. Then, the supply chain performance under different market power structures and that with balanced market power will be compared to obtain the influence of market power on supply chain coordination; lastly, through the comparison of both, we can get the inspiration of management.

3 Benchmark for Coordination Effects: The Nash Game with Balanced Market Power and No Fairness Preference

The manufacturer and the retailer making decisions to maximize their own profits in supply chain having no fairness preference. In the Nash game between each other, make decisions at the same time. Z_0 is used to represent the scenario of the Nash game without fairness preference. According to Problem Description and Assumption 1, profit functions of the retailer and the manufacturer are respectively obtained as follows:

$$\pi_r = (p - w)(a - bp)$$

$$\pi_s = (w - c)(a - bp)$$

Through simultaneous solution, we obtain: $w_0^* = \frac{a+2bc}{3b}$, $p_0^* = \frac{a-bc}{3b}$. We further obtain the supply chain coordination performance:

$$\pi_{0r}^* = \frac{(a-bc)^2}{9b}, \pi_{0s}^* = \frac{(a-bc)^2}{9b},$$
$$\Pi_0^* = \frac{2(a-bc)^2}{9b}, \eta_0 = \frac{8}{9},$$
$$cs_0 = \frac{(a-bc)^2}{18b}, sw_0 = \frac{5(a-bc)^2}{18b}.$$

4 The Coordination Effects of Market Power on Supply Chain

Ailawadi et al. [13], established American scholars in the field of food channel research, held that market power is the capacity, perceived in the market, of exerting long-term influence on the price and (or) other business conditions. In brief, market power of enterprises influences trading behaviors in the market, further determining benefits distribution in the transaction process.

4.1 Supply Chain Decisions Under Different Market Power Structures

In the two-stage supply chain formed by the manufacturer and retailer, market power structures can be divided into two scenarios, i.e. strong manufacturer with weak retailer and weak manufacturer with strong retailer. Under these two market power structures, the manufacturer-dominated and the retailer-dominated Stackelberg games are run respectively in the supply chain. Z_0^S and Z_0^R are used to represent the manufacturer-dominated and the retailer-dominated Stackelberg games respectively.

Scenario 1: The manufacturer-dominated Stackelberg game

The process of manufacturer-dominated Stackelberg game: First, the manufacturer determines the optimal wholesale price w to maximize its own profits; then, according to decisions made by the manufacturer, the retailer set the optimal retail price p so as to obtain the maximized profits.

Backward induction is applied. We obtain the optimal wholesale price and the optimal retail price:

$$w_0^{S*} = \frac{a+bc}{2b}, p_0^{S*} = \frac{3a+bc}{4b}.$$

Ultimately, we further obtain the supply chain coordination performance:

$$\pi_{0r}^{S*} = \frac{(a-bc)^2}{16b}, \pi_{0s}^{S*} = \frac{(a-bc)^2}{8b},$$
$$\Pi_0^{S*} = \frac{3(a-bc)^2}{16b}, \eta_0^S = 75\%,$$
$$cs_0^S = \frac{(a-bc)^2}{32b}, sw_0^S = \frac{7(a-bc)^2}{32b}.$$

Scenario 2: The retailer-dominated Stackelberg game

In the last past 20 years, a consensus has been formed, i.e. that market power of manufacturers has been transferred to retailers. Although the degree and speed of transferring are different in various retail formats and countries, obviously, there is a trend that the control of critical marketing decisions is transferring from manufacturers to retailers. The decision order of the retailer-dominated Stackelberg game is different from that of the manufacturer-dominated. Similarly, backward induction is applied. The equilibrium solution for the game is: $w_0^{R*} = \frac{a+3bc}{4b}$, $p_0^{R*} = \frac{3a+bc}{4b}$.Various performance indicators of the supply chain are obtained as:

$$\pi_{0r}^{R*} = \frac{(a-bc)^2}{8b}, \pi_{0s}^{R*} = \frac{(a-bc)^2}{16b},$$
$$\Pi_0^{R*} = \frac{3(a-bc)^2}{16b}, \eta_0^R = 75\%,$$
$$cs_0^R = \frac{(a-bc)^2}{32b}, sw_0^R = \frac{7(a-bc)^2}{32b}.$$

The performances of centralized supply chain are easily obtained: profit of supply chain system $\Pi_c = \frac{(a-bc)^2}{4b}$, consumer surplus $cs_c = \frac{(a-bc)^2}{8b}$, social welfare $sw_c = \frac{3(a-bc)^2}{8b}$. So the supply chain efficiency of Stackelberg supply chain with neutral fairness preference is easily calculated $\eta_0^S = 75\%$.

4.2 The Influence of Market Power on Supply Chain Coordination

Profits of the supply chain and their members under Z_0^S and Z_0^R are compared. We obtain the Proposition 1, as follows:

Proposition 1 The dominant power in the supply chain cannot change the profits of the supply chain system, but influences the distribution proportion of the channel profits, making the leader's profits 2 times the follower's.

Demonstration: $\Pi_0^{S*} = \Pi_0^{R*}$, indicating that the profits under the two scenarios, the manufacturer-dominated and the retailer-dominated, are equal. However, in Z_0^S, the manufacturer is the leader and the retailer the follower, $\pi_{0s}^{S*}/\pi_{0r}^{S*} = 2$; while in Z_0^R, the retailer is the leader and the manufacturer the follower, $\pi_{0r}^{R*}/\pi_{0s}^{R*} = 2$.

The Proposition 1 shows that market power is a beneficial tool for enterprises to obtain more profits than the trading object. The party having market power in the two-stage supply chain gains double profits than the other side.

The supply chain performance under Z_0^S and Z_0^R having market power is compared with that under Z_0 having no market power to obtain Proposition 2, as follows:

Proposition 2 Regarding the fairness of channel profit distribution and the supply chain performance, market power has a negative effect on the supply chain.

Demonstration: According to Proposition 1, in the leader-follower supply chain having market power, the profits of the leader are two times that of the follower. However, in the Nash game supply chain, the profits of the retailer are equal to that of the manufacturer $\pi_{0r}^* = \pi_{0s}^* = \frac{(a-bc)^2}{9b}$, with the two sides dividing the profits of the supply chain equally. Therefore, market power has a negative effect on the fairness of the channel profit distribution.

By comparing the supply chain efficiency, consumer surplus and social welfare under two kinds of the Stackelberg games and the Nash games, it can be seen that $\eta_0^S = \eta_0^R < \eta_0$, $cs_0^S = cs_0^R < cs_0$, $sw_0^S = sw_0^R < sw_0$. Thus, market power has a negative effect on the coordination performance of the supply chain.

5 The Coordination Effects of Fairness Preference on Supply Chain

Market power, a kind of implicit contracts, is an external market factor influencing decisions of supply chain members; while fairness preference is an internal behavioral factor of decision-makers.

5.1 Supply Chain Decisions Under Different Fairness Preference Structures

For two-stage supply chains having fairness preference, there are two scenarios concerning its fairness preference structure, i.e. the retailer and the manufacturer having fairness preference. Z_R and Z_S are used to indicate respectively the scenarios the retailer and the manufacturer having fairness preference in the Nash game.

Scenario 3: Nash game with retailer having fairness preference

When the retailer has fairness preference but the manufacturer has no fairness preference, the former makes decisions according to the principle of self-utility maximization, and the later the principle of self-profit maximization. According to Assumption 2, utility function of the retailer is:

$$u_r = \pi_r + \alpha(\pi_r - \pi_s)$$

In the Nash game, both sides make decisions at the same time, and then do simultaneous solution. We obtain the optimal decision results and various performance indicators under scenario Z_R:

$$w_R^* = \frac{a + 2bc + \alpha(a + 3bc)}{b(3 + 4\alpha)},$$
$$p_R^* = \frac{2a + bc + \alpha(3a + bc)}{b(3 + 4\alpha)},$$
$$\pi_{Rr}^* = \frac{(a - bc)^2(1 + \alpha)(1 + 2\alpha)}{b(3 + 4\alpha)^2},$$
$$\pi_{Rs}^* = \frac{(a - bc)^2(1 + \alpha)^2}{b(3 + 4\alpha)^2},$$
$$\Pi_R^* = \frac{(a - bc)^2(1 + \alpha)(2 + 3\alpha)}{b(3 + 4\alpha)^2},$$
$$cs_R = \frac{(a - bc)^2(1 + \alpha)^2}{2b(3 + 4\alpha)^2},$$
$$\eta_R = \frac{4(1 + \alpha)(2 + 3\alpha)}{(3 + 4\alpha)^2}$$
$$sw_R = \frac{(a - bc)^2(1 + \alpha)(5 + 7\alpha)}{2b(3 + 4\alpha)^2}.$$

Scenario 4: Nash game with manufacturer having fairness preference

When the manufacturer has fairness preference but the retailer has no fairness preference, the former makes decisions according to the principle of self-utility maximization, and the later the principle of self-profit maximization. According to Assumption 2, utility function of the manufacturer is:

$$u_s = \pi_s + \alpha(\pi_s - \pi_r)$$

Similarly, we obtain the decision results and various performance indicators under scenario Z_S.

$$w_S^* = \frac{a + 2bc + 2\beta(a + bc)}{b(3 + 4\beta)},$$
$$p_S^* = \frac{2a + bc + \beta(3a + bc)}{b(3 + 4\beta)},$$
$$\pi_{Sr}^* = \frac{(a - bc)^2(1 + \beta)^2}{b(3 + 4\beta)^2},$$
$$\pi_{Ss}^* = \frac{(a - bc)^2(1 + \beta)(1 + 2\beta)}{b(3 + 4\beta)^2},$$
$$\Pi_S^* = \frac{(a - bc)^2(1 + \beta)(2 + 3\beta)}{b(3 + 4\beta)^2},$$
$$\eta_S = \frac{4(1 + \beta)(2 + 3\beta)}{(3 + 4\beta)^2}, cs_S = \frac{(a - bc)^2(1 + \beta)^2}{2b(3 + 4\beta)^2},$$
$$sw_S = \frac{(a - bc)^2(1 + \beta)(5 + 7\beta)}{2b(3 + 4\beta)^2}.$$

5.2 Influence of Fairness Preference on Supply Chain Coordination

The Nash game results having fairness preference in Sect. 5.1 is compared with that having neutral fairness in Sect. 3 to obtain the following Proposition 3.

Proposition 3 Fairness preference produces negative effects on supply chain coordination performance, reducing supply chain profits, supply chain efficiency, consumer surplus and social welfare. However, the behavioral factor of fairness preference can not only promote the profits of the side having fairness preference, as a tool for gaining more profits, but also change profit distribution in the supply chain channel, making the side obtain more channel profit proportions when having fairness preference in comparison with having no fairness preference. Here, fairness preference causes an effect of "harming others and the community to benefit self".

Demonstration: By observing the profits, supply chain profits, supply chain efficiency, consumer surplus and social welfare of the side having fairness preference under scenarios Z_R and Z_S, it can be seen that when the fairness preference sensitive degree of the retailer is similar to that of the manufacturer, i.e. $\alpha = \beta$, expressions of these variables under the two scenarios are identical. Therefore, to save space, we use the results obtained under scenario Z_R in the demonstration of Proposition 3 (the results obtained under scenario Z_S are identical to those under scenario Z_R). Comparing the supply chain profits, supply chain efficiency, consumer surplus and social welfare in Z_R and Z_0, through

$$\frac{\Pi_R^*}{\Pi_0^*} = \frac{18+45\alpha+27\alpha^2}{18+48\alpha+32\alpha^2} < 1,$$

$$\frac{\eta_R}{\eta_0} = \frac{18+45\alpha+27\alpha^2}{18+48\alpha+32\alpha^2} < 1,$$

$$\frac{cs_R}{cs_0} = \frac{9+18\alpha+9\alpha^2}{9+24\alpha+16\alpha^2} < 1,$$

$$\frac{sw_R}{sw_0} = \frac{45+108\alpha+63\alpha^2}{45+120\alpha+90\alpha^2} < 1,$$

we obtain $\Pi_R^* < \Pi_0^*$, $\eta_R < \eta_0$, $cs_R < cs_0$, $sw_R < sw_0$, indicating that fairness preference has lowered the supply chain profits, supply chain efficiency, consumer surplus and social welfare, and thus causing degradation effects on supply chain coordination performance. Through $\frac{\pi_{Rr}^*}{\pi_{0r}^*} = \frac{9+27\alpha+18\alpha^2}{9+24\alpha+16\alpha^2} > 1$, we have $\Pi_{Rr}^* > \Pi_{0r}^*$, proving that fairness preference increases the profits of the side having fairness preference. Comparing the profit ratios of supply chain members in Z_R and Z_0, from $\frac{\pi_{Rr}^*}{\pi_{Rs}^*} = \frac{1+2\alpha}{1+\alpha} > 1$ but $\frac{\pi_{0r}^*}{\pi_{0s}^*} = 1$, it can be seen that fairness preference has changed the distribution proportion of supply chain channel profits, making the side obtain more channel profit proportions when having fairness preference in comparison with having no fairness preference.

Now, we have $\Pi_R^* < \Pi_0^*$, $\Pi_{Rr}^* > \Pi_{0r}^*$; from $\frac{\pi_{Rs}^*}{\pi_{0s}^*} = \frac{9+18\alpha+9\alpha^2}{9+24\alpha+16\alpha^2} < 1$, $\Pi_{Rs}^* < \Pi_{0s}^*$. Comparing the supply chains and their members' profits in Z_R and Z_0, it is concluded that: In comparison with the scenario with neutral fairness, fairness preference increases the profits of the side having fairness preference and decreases the profits and supply chain profits of the side having no fairness preference. This proves the effect of "harming others and the community to benefit self" caused by fairness preference, as in Proposition 3.

6 Comparison of Coordination Effects of Market Power and Fairness Preference on Supply Chain

Sections 4 and 5 analyzed the influences of the two factors, i.e. market power and fairness preference, on supply chain coordination performance, and revealed that they all have negative effects on supply chain coordination. Then, between the two factors, whose influence is larger? In this section, the performance of the supply chain when it has market power and that of the supply chain when it has fairness preference are contrasted and analyzed.

Proposition 4 Market power has a deeper negative influence on supply chain coordination than fairness preference.

Demonstration: In Sect. 4, through simple observation, we found that under the two scenarios, when the manufacturer and the retailer have market power, the supply chain profits, supply chain efficiency, consumer surplus and social welfare are the same. For the supply chain profits, supply chain efficiency, consumer surplus and social welfare under the two fairness preference structures in Sect. 5, we have pointed out that when $\alpha = \beta$, expressions of variables under this two scenarios were identical. Therefore, in Proposition 4, to save space, we use the supply chain performance under the manufacturer-dominated scenario Z_0^S to represent the influence of market power on supply chain coordination, and that under the scenario Z_R when the retailer has fairness preference the influence of fairness preference on supply chain coordination.

We compare the supply chain profits, supply chain efficiency, consumer surplus and social welfare under Z_0^S and Z_R. From

$$\frac{\Pi_0^{S*}}{\Pi_R^*} = \frac{27+72\alpha+48\alpha^2}{32+80\alpha+48\alpha^2} < 1,$$
$$\frac{\eta_0^S}{\eta_R} = \frac{27+72\alpha+48\alpha^2}{32+80\alpha+48\alpha^2} < 1,$$
$$\frac{cs_0^S}{cs_R} = \frac{9+24\alpha+16\alpha^2}{16+32\alpha+16\alpha^2} < 1$$
$$\frac{sw_0^S}{sw_R} = \frac{63+168\alpha+112\alpha^2}{80+192\alpha+112\alpha^2} < 1,$$

we obtain $\Pi_0^{S*} < \Pi_R^*$, $\eta_0^S < \eta_R$, $cs_0^S < cs_R$ and $sw_0^S < sw_R$, indicating that the negative effects of market power on supply chain coordination are larger than those of fairness preference.

7 Conclusion

As the core of supply chain management, supply chain coordination is influenced by several factors. This paper examined the effect of the external factor, i.e. market power, and internal fairness, i.e. preference factor, on supply chain coordination. First, the benchmark supply chain coordination effects was established. Next, the influences of two factors, i.e. market power and fairness preference, on supply chain coordination were separately analyzed. It was indicated that both producing negative effects on supply chain coordination. Lastly, the coordination effects of the both were compared, and the former's influence was larger.

This paper examines the coordination effect of market power and fair preference in the supply chain alone, and would there be the coupling effects between them? This will be a problem for us to study further.

Acknowledgements. This research is supported by the National Natural Science Foundation of China (71672143), the Project of Humanities and Social Science Research of Education Ministry of China (13YJA630077), Shaanxi Province Natural Science Foundation Research Project (2017JM7013).

References

1. K. Pan, K.K. Lai, S.C.H. Leung, D. Xiao, Revenue-sharing versus wholesale price mechanisms under different channel power structures. Eur. J. Oper. Res. **203**(2), 532–538 (2010)
2. E. Lee, R. Staelin, Vertical strategic interaction: implication for channel pricing strategy. Mark. Sci. **16**(3), 185–207 (1997)
3. A.A. Tsay, N. Agrawal, Channel conflict and coordination in the e-commerce age. Prod. Oper. Manag. **13**(1), 93–110 (2004)
4. C.H. Wu, C.W. Chen, C.C. Hsieh, Competitive pricing decisions in a two-echelon supply chain with horizontal and vertical competition. Int. J. Prod. Econ. **135**(1), 265–274 (2012)
5. J. Wei, J. Zhao, Y. Li. Pricing decisions for complementary products with firm's different market powers. Eur. J. Oper. Res. **224**(3), 507–519 (2013)
6. M. Rabin, Incorporating fairness into game theory and economics. Am. Econ. Rev. **83**(5), 1281–1302 (1993)
7. E. Fehr, K.M. Schmidt, A theory of fairness, competition, and cooperation. Q. J. Econ. **114** (3), 817–868 (1999)
8. G. Charness, M. Rabin, Understanding social preferences with simple tests. Q. J. Econ. **91**, 151–172 (2002)
9. D.J.L. Kahneman, R.T. Knetsch, Fairness and the assumptions of economics. Part 2: Behav. Found. Econ. Theory **59**(4), 285–300 (1986)
10. N. Kumar, L.K. Scheer, J.E.M. Steenkamp, The effects of supplier fairness on vulnerable resellers. J. Mark. Res. **32**(1), 54–65 (1995)

11. C.H. Loch, Y. Wu, Social preferences and supply chain performance: an experimental study. Manage. Sci. **54**(11), 1835–1849 (2008)
12. E. Katok, V. Pavlov, Fairness in supply chain contracts: a laboratory study. J. Oper. Manag. **31**(3), 129–137 (2013)
13. K.L. Ailawadi, N. Borin, P.W. Farris, Market power and performance: a cross-industry analysis of manufacturers and retailers. J. Retail. **71**(3), 211–248 (1995)

A Novel Framework for Top-N Recommendation Based on Non-negative Matrix Tri-Factorization

Jin Tian(✉) and Yingnan Qu

College of Management and Economics, Tianjin University, Tianjin, China
jtian@tju.edu.cn

Abstract. Clustering techniques have been proved effective to deal with the sparsity and scalability problems in collaborative filtering recommender systems. They aim to identify groups of users having similar preferences or items sharing similar topics. In this study, we propose an integrated recommendation framework based on matrix factorization. Firstly, users and items are clustered into multiple groups and a pair of strongly related user group and item group forms a submatrix. Then some traditional collaborative filtering technique is executed in every submatrix. The final rating predictions are generated by aggregating results from all the submatrices and the items are recommended with a Top-N strategy. Experimental results show that the proposed framework significantly improves the recommendation accuracy of several state-of-the-art collaborative filtering methods, while retains the advantage of good scalability.

Keywords: Matrix tri-factorization · Collaborative filtering · Co-clustering Personalized recommendation

1 Introduction

Clustering-based collaborative filtering (CF) approaches are a kind of model-based CF method. Although demonstrated to be efficient and scalable to large-scale data sets and able to deal with the data sparsity problem, they often suffer from relatively poor accuracy [1].

The reasons for the low accuracy of clustering-based CF are two folds. For one thing, the preferences of each user are too diverse, making it difficult to place a user into a particular group. Neglecting multiple preferences results in the loss of critical neighbors, because similar users are usually selected from the fixed size of group members in the clustering-based CF methods. For another, the relationship between user groups and items groups are less exploited and utilized in clustering tasks, and it's assumed that the preferences of a user group to different item topics are identical. However, users having similar tastes on a subset of items may have different tastes on another subset.

Cluster-level rating patterns are originally used to alleviate the data sparsity problem in CF with transfer learning methods [2]. The ratings in the auxiliary rating

G. Q. Huang et al. (eds.), *Proceeding of the 24th International Conference on Industrial Engineering and Engineering Management 2018*,
https://doi.org/10.1007/978-981-13-3402-3_36

matrix are compressed into a compact matrix referred to as a codebook [2]. In this paper, we call the codebook cluster association matrix. The cluster association matrix reflects how preferences of a user group vary on various item topics. Generally, a group of users are interested in neither a unique topic nor all the topics but just a few topics. Through the cluster association matrix, we can figure out which topics each user group prefers and build item spaces matching those topics, rather than performing CF algorithms on the whole item space, which is less accurate and time-consuming. For example, Mary is a fan of romance and adventure, while Helen not only likes romance but also watches a lot of animation because she has a little son. They may both in the group of romance fans, but we don't want to recommend an animation saw by Helen to Mary. What we expect is recommending romantic movies to them based on their historical interactive data about romantic movies.

Based on the analysis above, we propose an overlapping co-clustering (OCoC) framework to improve recommendation accuracy of CF-based recommender systems. The main contributions of our work are as follows:

(1) We discover multiple latent preferences of users and multiple latent topics of items by an interpretable non-negative matrix tri-factorization method. Moreover, the strengths of relationship between user groups and item groups are also modeled.
(2) We propose a unified recommendation framework to aggregate prediction scores from overlapping user-item submatrices, which are determined by the value in the cluster association matrices.

The proposed framework is a hybrid memory and model-based approach and has the advantages from both. The co-clustering model is used firstly to reduce dimensions, and any model-based or memory-based CF methods can be applied in each submatrix.

2 Related Work

Clustering techniques often work as an intermediate step in CF recommender systems. They are capable of alleviating data sparsity and scalability problems. However, Huang et al. [3] find the classification of users into clusters does not add any worth to the simple user-based approach. Different from one-sided clustering, co-clustering clusters both rows and columns simultaneously, which can effectively handle high dimensional data sets and discover the data structures [4]. Towards recommendation tasks, Xu et al. [5] formulate a Multiclass Co-Clustering (MCoC) model which was based on a weighted graph-cut optimization problem. Instead of using clustering as an intermediate step, Liu et al. [6] integrate rating prediction model and clustering model into a unified framework to make the clustering step and prediction step boost each other. Many previous works focus on prediction accuracy [6, 7], which cannot guarantee a satisfactory recommendation list. We focus on the top-N task, which is a more accurate reflection of how recommender systems are used in practice.

Matrix factorization-based methods have been increasingly used in data clustering. It has been demonstrated that non-negative matrix factorization is equivalent to some widely used clustering algorithms [8]. Compared to the graph-based co-clustering methods which require solving expensive eigenproblems [4], matrix factorization-based methods are computationally more efficient [9]. Recently there have been some researches using non-negative matrix tri-factorization in recommender systems. Chen et al. [10] apply orthogonal non-negative matrix tri-factorization (ONMTF) to simultaneously cluster the rows and columns of the user-item matrix, and then attain predictions within a user group. Li et al. [2] use ONMTF to generate the codebook for transfer learning task.

3 Methodology

3.1 Non-negative Matrix Tri-Factorization Based Submatrices Discovery

We define a set $\{u_1, u_2, \ldots, u_M\}$ of users, a set $\{o_1, o_2, \ldots, o_N\}$ of items. $\boldsymbol{R} \in \mathbb{R}^{M \times N}$ denotes the rating matrix and r_{ij} denotes the rating given by user u_i to item o_j. Suppose there are K user groups $\{C_u^1, C_u^2, \ldots, C_u^K\}$. The probability of u_i belonging to the pth user group C_u^p can be described as $P(C_u^p|u_i)$. Thus, we can define user group membership vector for u_i as $\boldsymbol{u}_i = [P(C_u^1|u_i), P(C_u^2|u_i), \ldots, P(C_u^K|u_i)]$ and the user group indicator matrix as $\boldsymbol{U} = [\boldsymbol{u}_1, \boldsymbol{u}_2, \ldots, \boldsymbol{u}_M]^T$.

Similarly, we define L item groups $\{C_v^1, C_v^2, \ldots, C_v^L\}$ and item group membership vector $\boldsymbol{v}_j = [P(C_v^1|o_j), P(C_v^2|o_j), \ldots, P(C_v^L|o_j)]$, which indicates how much o_j conforms to the item topics. Then we get the item group indicator matrix $\boldsymbol{V} = [\boldsymbol{v}_1, \boldsymbol{v}_2, \ldots, \boldsymbol{v}_N]^T$.

We define $\boldsymbol{S} \in \mathbb{R}^{K \times L}$ as the group association matrix which captures the strengths of relationship between user groups and item groups. Each entry s_{pq} means the overall preference of the pth user group C_u^p for the qth item group C_v^q.

The preference of u_i for o_j can be denoted as $x_{ij} = \sum_{p=1}^{K} \left(u_{ip} \times \sum_{q=1}^{L} \left(s_{pq} \times v_{jq} \right) \right)$, which is a linear combination of all user groups' preferences for o_j, weighted by the membership values of u_i with these user groups.

Based on the definition above, we can derive an approximation matrix $\boldsymbol{X} = \boldsymbol{USV}^T$ of $\boldsymbol{R}$.

The goal is to minimize the following objective function:

$$J(\boldsymbol{U}, \boldsymbol{S}, \boldsymbol{V}) = \left\| \boldsymbol{R} - \boldsymbol{USV}^T \right\|^2 \tag{1}$$

We impose non-negative constraints on the objective function to allow a more intuitive interpretation than the other matrix factorization approaches, such as Singular Value Decomposition (SVD), which allows negative values in the matrix components. Furthermore, in order to avoid the uncertainty in clustering assignments, we impose L_1 norm on rows of $\boldsymbol{U}$ and $\boldsymbol{V}$, which allows entries in $\boldsymbol{U}$ and $\boldsymbol{V}$ to be directly interpreted as cluster posterior probability.

Above all, the overall objective function is as following,

$$\begin{aligned} \min J(\boldsymbol{U}, \boldsymbol{S}, \boldsymbol{V}) = & \left\|\boldsymbol{R} - \boldsymbol{U}\boldsymbol{S}\boldsymbol{V}^T\right\|^2 + \alpha\|\boldsymbol{U}\boldsymbol{e} - \boldsymbol{e}\|^2 \\ & + \beta\|\boldsymbol{V}\boldsymbol{e} - \boldsymbol{e}\|^2 \\ s.t. \quad & \boldsymbol{U} \geq 0, \boldsymbol{S} \geq 0, \boldsymbol{V} \geq 0 \end{aligned} \tag{2}$$

where $\alpha, \beta \geq 0$ are L_1 normalization penalty parameters and e is a vector of all ones. They are set to be 0.5 through cross validation.

Using Lagrangian multipliers to enforce the constraints, we derive the following updating rules to solve this problem:

$$\boldsymbol{U} \leftarrow \boldsymbol{U}\left(\frac{\boldsymbol{R}\boldsymbol{V}\boldsymbol{S}^T + \alpha\boldsymbol{E}}{\alpha\boldsymbol{U}\boldsymbol{E} + \boldsymbol{U}\boldsymbol{S}\boldsymbol{V}^T\boldsymbol{V}\boldsymbol{S}^T}\right)^{\frac{1}{2}} \tag{3}$$

$$\boldsymbol{V} \leftarrow \boldsymbol{V}\left(\frac{\boldsymbol{R}\boldsymbol{V}\boldsymbol{S}^T + \beta\boldsymbol{E}}{\beta\boldsymbol{V}\boldsymbol{E} + \boldsymbol{V}\boldsymbol{S}^T\boldsymbol{U}^T\boldsymbol{U}\boldsymbol{S}}\right)^{\frac{1}{2}} \tag{4}$$

$$\boldsymbol{S} \leftarrow \boldsymbol{S}\left(\frac{\boldsymbol{U}^T\boldsymbol{S}\boldsymbol{V}}{\boldsymbol{U}^T\boldsymbol{U}\boldsymbol{S}\boldsymbol{V}^T\boldsymbol{V}}\right)^{\frac{1}{2}} \tag{5}$$

where E is a matrix of all ones.

Based on the factorization model, we can learn group memberships of users and items from $\boldsymbol{U}$ and $\boldsymbol{V}$ and association strengths between user groups and item groups from $\boldsymbol{S}$. Then we should decide which groups a user or an item belongs to and which users groups and item groups are tightly related.

If user the membership of u_i with a user group is larger than ε_u, we assume that u_i belongs to this group. Similarly, we assume o_j belongs to an item group if the membership is larger than ε_v. Considering each user group is interested in only a few item topics, for each user group, we select the highest-ranked k_c item groups as the interested groups based on the cluster association matrix $\boldsymbol{S}$ and the association value of the selected groups should be larger than ε_c. The threshold values are set as $\varepsilon_u = 1/K$, $\varepsilon_v = \varepsilon_c = 1/L$ and $k_c = \lceil \log_2(L) \rceil$. In this way, a group of users and a group of items can be combined together and form a compact submatrix.

3.2 Recommendation with Submatrices

In each submatrix, we can apply any memory-based or model-based CF algorithms to predict the missing ratings. This is an advantage of our framework because we can benefit from different kinds of CF algorithms adapting to different scenarios.

Since each user and item can belong to multiple groups and the relationship between user groups and item groups is many-to-many, a missing value can be predicted in several submatrices simultaneously. To decide whether an item should be recommended to a user, we need to aggregate these prediction scores.

Y_{ij} denotes the final preference score of user u_i to item o_j. $\delta_{pq} = 1$ if user cluster C_u^p is closely related to item cluster C_v^q. $pred^{(pq)}(u_i, o_j)$ is the prediction score made by CF techniques in the submatrix formed by C_u^p and C_v^q. We calculate Y_{ij} using Eq. (6).

$$Y_{ij} = \max_{p,q}\left(\delta_{pq} pred^{(pq)}(u_i, o_j)\right) \tag{6}$$

We use the maximum of preference scores among all submatrices as the final preference score, which means an item would be liked by a user as long as the item gets a high prediction score on one of the topics preferred by the user. At last, items are recommended with a top-N strategy based on the aggregated prediction scores.

4 Experiments

4.1 Experimental Setup

Table 1 summarizes the statistical properties of the data sets we use.

Table 1. Basic statistics of the data sets

	Users	Items	Ratings	Sparsity (%)
MovieLens	6040	3952	1,000,209	4.19
Last.fm	1799	1530	62,600	2.27

The Last.fm data set contains listening times of users to artists. We removed the artists that have never been listened. The elements of the user-item matrix vary in too large a range to be well used, so we recoded the data to integer numbers ranging from 1 to 5 with the method used by Langseth and Nielsen [11]. For each user, the 10% of the artists that the user has listened to the least are recoded as one. The artists with a listening time between 10th and 30th percentiles are recoded as two while the artists between 30th and 70th percentiles and the artists between 70th and 90th percentiles are encoded as three and four respectively. The artists above the 90% percentile are given five.

We choose four representative CF methods including both memory-based and model-based methods as the base methods applied to the submatrices. They are PureSVD [12], PPMF [13], UBCF [14] and MCREC [15]. PureSVD has been proved to deliver better top-N results than some RMSE-oriented latent factor models [16]. We set the dimension of latent features to be ten in PureSVD and PPMF. MCREC is a newly proposed matrix completion based method for the Top-N recommendation problem.

We adopt three accuracy-based metrics to evaluate the performance of our framework: Precision, Recall and F1 Measure, which are computed as following. We use $\boldsymbol{I}_{liked}$ to represent the items actually liked by a user and $\boldsymbol{I}_{list}$ to represent the items recommended to the user.

$$Precision = \frac{|\boldsymbol{I}_{liked} \cap \boldsymbol{I}_{list}|}{N} \quad (7)$$

$$Recall = \frac{|\boldsymbol{I}_{liked} \cap \boldsymbol{I}_{list}|}{|\boldsymbol{I}_{liked}|} \quad (8)$$

$$F_1 = \frac{2 \times Precision \times Recall}{Precision + Recall} \quad (9)$$

Considering an item's rank in a recommendation list, we also adopt a ranking-based metric Mean Average Precision (MAP), which gives more weight to the items appear earlier in the top-N list. MAP is defined as:

$$MAP = \frac{1}{|U|} \sum_{i=1}^{|U|} \frac{1}{\min\{N, |I_{liked}|\}} \sum_{j=1}^{N} prec(j) \times pref(j) \quad (10)$$

For each user u, given a ranked list with N items, we denote $prec(j)$ as the precision at the ranked position j, and $pref(j)$ as the preference indicator of the item at position j. $pref(j)$ equals to one if the corresponding item is liked by user u, otherwise equals to zero.

4.2 Comparisons on Accuracy

Tables 2 and 3 compare the performance of our framework with the base CF methods which are applied directly on the whole user-item matrix. The length of recommendation list is set to be 10. The results indicate that our framework improves the recommendation accuracy of base CF methods.

Table 2. Accuracy comparison on movielens

		PureSVD	PPMF	UBCF	MCREC
Precision@10	Base	0.262	0.091	0.224	0.229
	OCoC	0.273	0.102	0.243	0.263
Recall@10	Base	0.173	0.052	0.148	0.180
	OCoC	0.193	0.063	0.162	0.195
F1@10	Base	0.208	0.066	0.178	0.202
	OCoC	0.226	0.078	0.194	0.224

Among the base methods, all as model-based CF, PureSVD and MCREC deliver much better performance than PPMF does. This may be because PPMF is RMSE-oriented and is trained only on the known ratings, ignoring the missing entries. However, the missing values can also reflect users' preference to some degree.

Table 3. Accuracy comparison on last.fm

		PureSVD	PPMF	UBCF	MCREC
Precision@10	Base	0.147	0.042	0.123	0.142
	OCoC	0.152	0.077	0.131	0.159
Recall@10	Base	0.213	0.062	0.184	0.230
	OCoC	0.227	0.114	0.192	0.247
F1@10	Base	0.174	0.050	0.147	0.176
	OCoC	0.182	0.092	0.155	0.193

Performance of our framework relies on the recommendation quality of the base method used in the interior. The more accurate recommendation the base method yields, the higher accuracy the framework achieves.

4.3 Comparisons on Ranking Metrics

Tables 4 and 5 show the MAP comparisons of different methods with varying length of the top-N recommendation list. The results demonstrate that our framework improves the item ranking performance in most cases.

Table 4. Accuracy comparison on last.fm

	PureSVD		PPMF		UBCF		MCREC	
Top-N	Base	OCoC	Base	OCoC	Base	OCoC	Base	OCoC
5	0.253	0.257	0.069	0.077	0.214	0.227	0.203	0.239
10	0.200	0.207	0.051	0.057	0.167	0.180	0.165	0.195
20	0.170	0.181	0.040	0.047	0.139	0.152	0.148	0.172
30	0.163	0.176	0.038	0.045	0.131	0.146	0.146	0.169
40	0.162	0.176	0.037	0.045	0.129	0.145	0.148	0.171
50	0.163	0.178	0.037	0.045	0.129	0.146	0.151	0.174

Table 5. Map results with varying n on last.fm

	PureSVD		PPMF		UBCF		MCREC	
Top-N	Base	OCoC	Base	OCoC	Base	OCoC	Base	OCoC
5	0.157	0.150	0.032	0.076	0.133	0.132	0.137	0.161
10	0.130	0.129	0.026	0.061	0.111	0.110	0.125	0.142
20	0.146	0.147	0.029	0.067	0.123	0.125	0.143	0.162
30	0.153	0.156	0.031	0.069	0.128	0.131	0.152	0.171
40	0.158	0.161	0.032	0.070	0.131	0.135	0.157	0.176
50	0.161	0.164	0.033	0.071	0.133	0.138	0.160	0.180

4.4 Effect of the Number of Groups

Figure 1 shows how the number of user groups and item groups affect the recommendation performance. We only present the results on the MovieLens data set due to space limits, and the results on Last.fm data set show similar trends. We observe that generally the recommendation accuracy fluctuates in a small range which is no more than 0.03, showing our framework is not so sensitive to the number of groups. Too large number of user groups results in relatively worse performance. This is because there is not enough available information in each group for rating prediction, and the preferences of the obtained groups may be shared together.

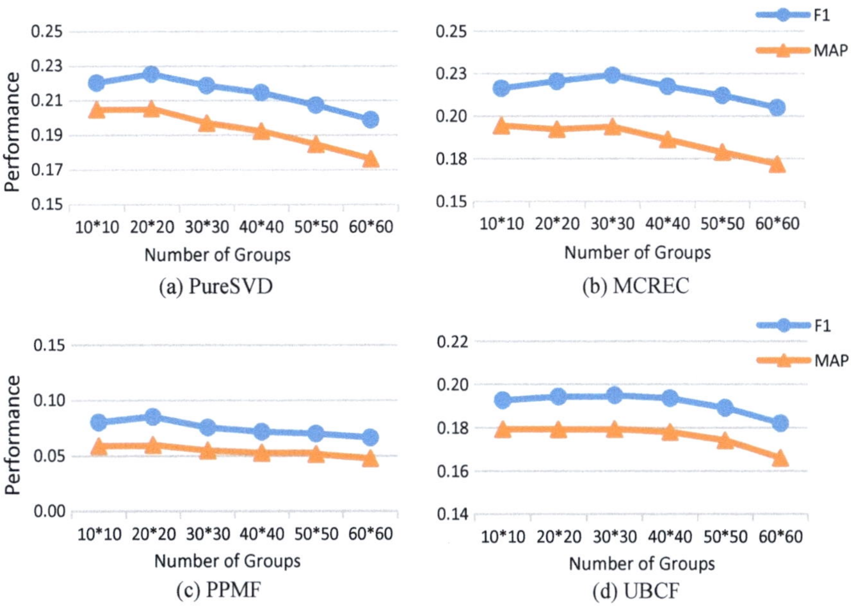

Fig. 1. F1@10 and MAP@10 with different number of groups on MovieLens

5 Conclusion

In this paper, we proposed a framework to improve upon top-N performance of CF-based recommendation methods, by exploiting the relationship between user groups and item groups, which is neglected by previous research on clustering-based recommender systems. For this purpose, we employ an interpretable matrix tri-factorization method on the rating matrix to discover the relationship between user groups and item groups and build up compact submatrices. Two real data sets based experiments demonstrate that our approach improves several state-of-the-art CF methods.

As for the future work, we would like to find better ways to combine groups and CF methods. In addition, we will testify our framework in more data sets.

Acknowledgements. The work was supported by the General Program of the National Science Foundation of China (Grant No. 71471127, 71371135).

References

1. G. Guo, J. Zhang, N. Yorke-Smith, Leveraging multiviews of trust and similarity to enhance clustering-based recommender systems. Knowl.-Based Syst. **74**(1), 14–27 (2015)
2. B. Li, Q. Yang, X. Xue, Can movies and books collaborate? Cross-domain collaborative filtering for sparsity reduction, in *Proceedings of the 21st International Jont Conference on Artifical Intelligence*, Pasadena, California, USA, pp. 2052–2057, 2009
3. Z. Huang, D. Zeng, H. Chen et al., A comparison of collaborative-filtering recommendation algorithms for e-commerce. IEEE Intell. Syst. **22**(5), 68–78 (2007)
4. Z. Li, X. Wu, Weighted nonnegative matrix tri-factorization for co-clustering, in *23rd IEEE International Conference on Tools with Artificial Intelligence*, Boca Raton, FL, USA, pp. 811–816 (2011)
5. B. Xu, J. Bu, C. Chen et al., An exploration of improving collaborative recommender systems via user-item subgroups, in *Proceedings of the 21st International Conference on World Wide Web*, Lyon, France, pp. 21–30 (2012)
6. J. Liu, Y. Jiang, Z. Li et al., Domain-sensitive recommendation with user-item subgroup analysis. IEEE Trans. Knowl. Data Eng. **28**(4), 939–950 (2016)
7. Y. Zhang, M. Zhang, Y. Liu et al., Improve collaborative filtering through bordered block diagonal form matrices, in *International ACM SIGIR Conference on Research and Development in Information Retrieval*, Dublin, Ireland, pp. 313–322 (2013)
8. C. Ding, X. He, H.D. Simon et al., On the equivalence of nonnegative matrix factorization and spectral clustering, in *Proceedings of the SIAM International Conference on Data Mining*, pp. 606–610 (2005)
9. Q. Gu, J. Zhou, C. Ding et al., Collaborative filtering: weighted nonnegative matrix factorization incorporating user and item graphs, in *Proceedings of the SIAM International Conference on Data Mining*, pp. 199–210 (2010)
10. G. Chen, F. Wang, C. Zhang et al., Collaborative filtering using orthogonal nonnegative matrix tri-factorization. Inf. Process. Manage. **45**(3), 368–379 (2009)
11. H. Langseth, T.D. Nielsen, Scalable learning of probabilistic latent models for collaborative filtering. Decis. Support Syst. **74**, 1–11 (2015)
12. B. Sarwar, G. Karypis, J. Konstan et al., Application of dimensionality reduction in recommender system—a case study, in *ACM Webkdd Workshop* (2000)
13. H. Shan, A. Banerjee, Generalized probabilistic matrix factorizations for collaborative filtering, in *IEEE 10th International Conference on Data Mining*, Sydney, NSW, Australia, pp. 1025–1030 (2010)
14. J.S. Breese, D. Heckerman, C.M. Kadie et al., Empirical analysis of predictive algorithms for collaborative filtering, in *Proceedings of the Fourteenth Conference on Uncertainty in Artificial Intelligence*, Madison, Wisconsin, USA, pp. 43–52 (1998)

15. Z. Kang, C. Peng, Q. Cheng, Top-N recommender system via matrix completion, in *Proceedings of the Thirtieth AAAI Conference on Artificial Intelligence*, Phoenix, Arizona, USA, pp. 179–184 (2016)
16. P. Cremonesi, Y. Koren, R. Turrin, Performance of recommender algorithms on top-n recommendation tasks, in *Proceedings of the Fourth ACM Conference on Recommender Systems*, Barcelona, Spain, pp. 39–46 (2010)

Comprehensive Evaluation of the Efficiency of China's Highway Freight Transport Platform Based on AHP-DEA

Wei-hua Gan, Chun-hong Cheng(✉), Ren Zhong, and Xiao-kang Li

School of Transportation and Logistics, East China Jiaotong University, Nanchang, China
729134514@qq.com

Abstract. This paper constructs the comprehensive evaluation index system of the efficiency of China's highway freight platform from three aspects: vivid level, turnover and customer satisfaction of platform. By using AHP and DEA method, the selected 18 highway freight platforms are measured in terms of overall efficiency, pure technical efficiency and scale efficiency. The results show that the efficiency of the highway freight platform is generally low, and the significant difference exist among all 18 platforms. This is mainly related to financing capability, operating mode and platform service level there is a big relationship. By analyzing the differences between the freight platforms, some suggestions are given in the conclusion of this paper, e.g. innovating the business model, improving the financing ability of the enterprises and enhancing the three capabilities (service level, timeliness and cost) of the enterprise.

Keywords: Highway freight platform · AHP-DEA evaluation method
Evaluation index system

1 Introduction

In July 2015, the State Council issued the "Guiding Opinions on Actively Promoting the 'Internet +' Action". The government actively promoted the deep integration of the Internet and economic and social fields, which referred to "Internet + Efficient Logistics". Owing to the Internet, the logistics platform has arisen account of "away from" and "away from". It provides an important way to promote the concentration of the logistics market, optimize the chain of logistics and promote the integration of industries.

With the emergence of Internet technology and the emergence of online and offline interactive economic forms, highway freight platform in China ushered in the peak of

The National Social Science Fund: Symbiosis research on influence mechanism of logistics platform from the perspective of the user enterprise development (17BJY140); Major projects Chinese Logistics Association (2015CSLKT1-002): The integration model of China's logistics industry platform; Jiangxi Provincial Academy of Social Sciences Key Project: Strategy of vertical integration of supply side reform under the background of the logistics platform (16GL02).

G. Q. Huang et al. (eds.), *Proceeding of the 24th International Conference on Industrial Engineering and Engineering Management 2018*,
https://doi.org/10.1007/978-981-13-3402-3_37

development and formed a "Internet + Logistics" upsurge. Trucks to help, Blue Rhinoceros, 58 City's "58 home", Jingdong Mall's "Jingdong home", CaiNiao Logistics' " Card swipes" and other highway freight platform flourishing.

As an emerging logistics industry, highway freight platform has an important impact on the rapid development of China's logistics industry. The way to measure the operation efficiency of highway freight transportation platform scientifically and effectively becomes an important problem. Therefore, this paper intends to take the subjective and objective combination of ideas and methods to evaluate the operational efficiency of China's highway freight platform. AHP-DEA combined evaluation model with AHP and DEA is used in the construction of evaluation system and efficiency estimation. The purpose is to make the calculation result more convincing through the organic unity of subjective evaluation and objective evaluation method, and (to more accurately reveal) the operation efficiency of highway freight platform.

2 Literature Review

As the highway freight platform is the new thing that has been produced in the past five years, the present study for Highway Freight platform mainly concentrated in the commercial mode, operation mode, system establishment, etc., the research on the efficiency of highway freight transport platform is less, but the efficiency of P2P platform is more. Since the highway freight platform and the P2P platform have some similarities and the purpose is to bring bilateral users into transactions, so this paper selects P2P platform efficiency from the platform point of view for reference. Although the academic research on the efficiency of highway freight transport platform is relatively small, but the research degree of logistics efficiency evaluation is higher, so this article will select logistics efficiency research for reference.

There is little research on the efficiency of highway freight transportation platform. Based on the quantitative data, the DEA analysis of the operational efficiency of 16 network loan platforms in China is carried out by Zheng et al. [1] with using the latest and valid data available. And he analyzes the influence factors of the efficiency of 11 net loan platform by using the analysis method of panel data. He also analyzes the relationship between popularity, technology, brand, revenue, leverage, transparency and liquidity, and he makes some suggestions. And Zhang [2] uses DEA method to measure the efficiency of P2P network lending platform in China, from the results analysis of China's network lending industry efficiency is insufficient. Through the DEA model, Zhou et al. [3] analyze the efficiency of 22 well-known P2P microfinance companies in China, and point out that the credit efficiency of P2P microfinance platform is not very high, and DEA is only 36% There is still a lot of room for improvement. Cai [4] uses the analytic hierarchy process (AHP) to construct the model of the comprehensive risk of P2P network loan platform. The weights of the five risk indicators are analyzed, and the risk situation of each platform is scored. The results show that the influence of the comprehensive risk of the platform is: transparency index, liquidity index, dispersion index, technical index and leverage index. Chen et al. [5] also use the analytic hierarchy process (AHP) to select the five indicators, including: time weighted volume, collect leverage, risk benefit ratio, concentration of borrowing and the funds, and approve to establish the standard for evaluation, analyzes the operation data of the actual volume of the 30 largest platform.

In our country, there are more achievements in the research of logistics efficiency evaluation system. Because DEA method can't consider the shortcomings of decision-maker preference and analytic hierarchy process subjective impact, Yang and Pei [6] put forward a method of combining DEA and AHP to evaluate the performance of logistics system on the basis of the logistics system comprehensive evaluation level model. Different from the problem that previous literature on the combination of the two methods of research in essence does not reflect the decision-maker preferences, they propose a method that first uses the AHP method to calculate the weight of each index, and then calculates the relative efficiency of each system by using the DEA method. Finally, combining the weight of each index and the relative efficiency value, the whole efficiency value of each logistics system is calculated and sorted. The method effectively combines the advantages of the two methods, and makes up for the shortcomings of the two methods. The final case analysis shows the applicability and operability of the method. When using DEA to evaluate the efficiency of logistics enterprises, Xu and Gao [7] propose a method to calculate the efficiency values based on the AHP/GRA method to determine the uniform common weights and then take Zhu-zhou Chang Xiang-tan Sha logistics enterprise as an example to carry on the empirical research. The result shows that the efficiency evaluation result based on DEA —AHP/GRA is more reasonable, and the decision units can be sorted effectively. The result reflects the large difference of the efficiency of each logistics enterprise in Chang Sha Zhu xiang-tan zhou.

By referring to the efficiency evaluation method of P2P platform and logistics system, this paper evaluates the highway freight platform by using AHP-DEA method. First, It construct a highway freight platform evaluation index system. And then it use the analytical hierarchy process (AHP) to calculate the weight of each index. This paper select indicators from the three indicators of the measured value by using the different weight. Finally, the index value of each freight platform is substituted into DEA model, and the efficiency difference of each platform is obtained.

3 Construction of Efficiency Evaluation Index System for Highway Freight Transport Platform

3.1 Evaluate the Data Source

At present, the domestic department on the highway freight platform data can be from the Federation of things, the major platforms of the official website and other third-party portal open access, but more indicators are difficult to obtain directly. Due to the many factors that affect the efficiency of the platform, it is difficult to evaluate the statistics on a regular basis. This paper obtains the basic data of some platforms through field investigation and interview, and forms the low-level data source of the data efficiency evaluation index system.

3.2 Evaluation System

This paper constructs the platform efficiency evaluation index system from three aspects: the vivid degree of platform, the platform transaction volume and the customer satisfaction degree of the highway freight platform. Platform activity can reflect the degree of recognition of shippers and drivers on the highway freight platform. Usually the owner of the platform and the more drivers, supply and demand matching easier to achieve on the platform at the same time, the number of daily distribution is also a platform for evaluating the freight level of a key indicator. The highway freight platform itself belongs to the information agency, its advantages and characteristics are to match the supply and demand sides of the two sides of the high efficiency and convenience. Therefore, the trading volume of the platform is a measure of its operating efficiency of the core indicators. The service object of the highway freight platform is usually the core competitiveness of the truck driver, the owner and the truck driver. The supply price of the platform is the lifeblood of the platform. Whether the service level of the platform can survive and survive on a long-term basis It is important. Therefore, the highway freight platform must coordinate the relationship between the truck driver and the owner to ensure that both the truck driver and the owner can be satisfied. Finally, under the three dimensions of platform freight, platform transaction volume and customer satisfaction, considering the availability of data, we further selected 11 third-level indicators, and constructed the efficiency of highway freight platform as shown in Table 1 Comprehensive evaluation of the index system.

Table 1. Evaluation system of highway freight platform

Objective hierarchy	Criteria hierarchy	Index hierarchy
Indicator system of highway freight transport platform	Vivid of platform	Number of platform members
		Number of truck vehicles
		Daily supply
	Turnover of platform	Total turnover
		Annual turnover
		Total volume
		Annual turnover
	Customer's satisfaction	Platform complaint rate
		Favorable rate
		Difference rate
		Cargo damage rate

In this paper, the total financing amount of highway freight platform and the number of employees as input indicators, AHP method to calculate the Taiwan in the cargo, trading volume and satisfaction of the three criteria as a result of output indicators, resulting in the platform The relative efficiency of the value, AHP calculation using the YAAHP software, DEA calculation using DEAP2.1 software.

4 The Efficiency Evaluation Result of Highway Freight Platform

4.1 The AHP Method Is Used to Solve the Index Weight of the Criterion Layer

Firstly, analytic hierarchy process (AHP) should be used to construct proportional judgment matrix. In this paper, the relative importance of indicators is determined by expert grading and questionnaire survey, then the maximum characteristic roots of weight and proportional judgment matrix are calculated by using the line and mean method. The judgment matrix of the three dimensions is passed the consistency test, therefore, the weight vector deduced by subjective judgment is reasonable. Finally, the AHP weight coefficient of comprehensive evaluation index system of highway freight platform was obtained (Tables 2 and 3). In three dimensions, the trucks number, total turnover and the rate of damage of goods are having the highest weight respectively, so this paper selects the trucks number, total turnover and the rate of damage of goods as a measure of highway freight platform activity, trading volume and users' satisfaction of three indexes.

Table 2. The index system of highway freight platform

First-class index	Second-class index	Third-class index
The index system of highway freight platform	Vivid degree	Membership
		Trucks number
		Daily supply number
	Turnover	Total turnover
		Annual turnover
		Total volume
		Annual volume
	Customer's satisfaction	Platform complaint rate
		Rave reviews
		Negative rating
		Rate of damage of goods

4.2 DEA Efficiency Evaluation of Highway Freight Transportation Platform

The index weight of each criterion layer is solved by analytic hierarchy process, using data from 18 highway freight platforms selected, the measurement values of vivid degree of platform, trading volume and users' satisfaction degree are obtained. Finally, it is used as the output result of DEA evaluation. The registered capital and number of employees of the freight transportation platform are selected as the index. According to the input and output index system, the operation efficiency of each highway freight platform is calculated, the results are shown in Tables 4 and 5.

Table 3. The weight coefficient for indexes

First-class index	Second-class index	Third-class index
1	0.24	0.17
		0.51
		0.32
	0.35	0.17
		0.28
		0.26
		0.29
	0.41	0.24
		0.23
		0.22
		0.31

Table 4. The DEA of efficiency analysis

Platform	Combined efficiency	Pure technical efficiency	Scale efficiency
Truck help	1.00	1.00	1.00
Yunniao distribution	0.78	1.00	78
Luoji Logistics	0.46	0.78	0.59
Yun manman	1.00	1.00	1.00
Supaide	0.23	0.59	0.39
No.1 truck	0.31	0.69	0.45
Yi Huo Di	1.00	1.00	1.00
Ka Xing Tianxia	1.00	1.00	1.00
Blue Rhino	0.34	0.69	0.49

Table 5. The DEA of efficiency analysis

Platform	Combined efficiency	Pure technical efficiency	Scale efficiency
Huo lala	0.46	0.67	0.69
58 Home	0.48	1.00	0.48
Bliss truck	0.79	1.00	0.79
Tianjiao	0.23	1.00	0.23
Logistics point of view	0.27	0.39	0.68
Logistics assistant	0.19	0.49	0.39
Mr. feng	0.34	0.59	0.57
Wan Ji Logistics	0.12	0.32	0.39
Express Freight	0.44	0.66	0.67

4.3 Efficiency Analysis of Highway Freight Transportation Platform

The operation efficiency of 18 highway freight platforms is analyzed by DEA method, The results indicate that the efficiency values of each platform are significantly different and generally lower. In terms of overall efficiency, the highway freight platform that reaches the efficiency of the DEA is only three, there are the truck help, the Ka Xing Tianxia and the Yunniao Distribution, the lowest platform efficiency is only 0.12. The comprehensive efficiency measure of 11 highway freight platforms is below 0.5. This shows that in the early stage of the development of "Internet + logistics" in China, many highway freight platforms are still in the initial stage of the platform, and the overall logistics efficiency of the industry has a great improvement space. From the result of the separation of technical efficiency, the efficiency of pure technology and the efficiency of scale can be improved. Tianjiao and other platforms have higher pure technical efficiency, but due to lack of scale efficiency, it finally shows inefficiency. Of the 18 platforms surveyed, three have the same pure technical efficiency and scale efficiency, with six platforms being more efficient, while the other seven platforms have higher pure technology efficiency.

4.4 Analysis of the Reasons for the Difference of Highway Freight Platform Efficiency

Firstly, the financing capacity is different. In the early stage of the highway freight transportation platform, high density investment (including manpower, technology, R & D, etc.) is needed, and with the rapid growth of the platform, the continuous capital investment will also become a key factor affecting the growth of the platform. In the eyes of capital, start-up companies' ability to finance is also an important dimension in assessing their competitiveness. Who can enter the next round, who can fully leverage the capital (including funds and funds outside the resources) to win the market, who can lay a considerable relative advantage, a greater chance of success can stand out. Therefore, the financing capacity of highway freight platform is also one of the reasons for the efficiency difference of kilometer freight platform. In the 18 highway freight platforms selected by this paper, the platform with strong financing capability has the platform of Truck Help, Yunmanman and Ka Xin Tianxia enterprises, and their comprehensive efficiency is also higher.

Secondly, the platforms operate differently. This paper selects the highway freight platform, asset light operations, a heavy asset management, a franchise model, self mode and so on. The different operation mode of the platform there will be differences in turnover, the number of vehicles and customer satisfaction, resulting in the efficiency of the different platform.

Thirdly, the types of highway freight platforms are different. Highway Freight platform at home can be roughly divided into the following five types: the road to highway port, line type, distribution, city freight and delivery type. The highway port, freight line and the distribution platform for logistics services, which is to provide services for the matching of supply and demand drivers and the owners, we call the three class platform for highway distribution platform; while the city freight class freight platform is mainly to match idle drivers and shippers who supply goods for trading, but the scope of the region is smaller, generally for urban transportation; the type of delivery platform is city freight accelerated version, regional scope is small, we call the two class platform for city distribution platform. Compared with the city distribution platform, there are two types of highway distribution platform advantage: ① Highway distribution platform driver in the use of platform before long service in the owner, have a strong sense of service, labor cost and time cost and therefore lower driver training; ② Highway distribution platform dedicated to our chance to shuttle in the country, is a platform for natural people to push a promotion and publicity to highway distribution platform, reduce highway distribution platform promotion time and cost. therefore, compared to the city distribution platform, the efficiency of highway distribution platform is relatively high.

5 Conclusion and Recommendation

This paper constructs the index system of the comprehensive evaluation of the operation efficiency of the highway freight platform, and uses the AHP-DEA model to estimate the comprehensive efficiency, pure technical efficiency and scale efficiency of the 18 highway freight platforms. The results show that the efficiency of highway freight platform is generally low, and there exists a significant difference between the efficiency levels of the platform, which is mainly related to the operation mode of the platform and the financing ability of the platform.

According to the above conclusions, in order to enhance the operational efficiency of China's highway freight platform and promote the healthy development of China's highway freight platform, the following suggestions are put forward:

First, improve the operation mode of enterprises. Platform strategy makers in the process of the operation of the platform, the platform of the operating mode needs to constantly reflect whether there is any defect, whether in the default track on the smooth forward, whether the market accepts this model of operation, what is the reason why the platform cannot expand rapidly, whether the profit point is considerable, the reasons for the rapid expansion of peers, whether there is a place worth learning from, through a series of reflection, and constantly improve the platform's operating model.

Second, strengthen financing capacity. Highway freight transport platform enterprises need to comb its own business model, build a reasonable organizational structure, search platform profit, build the platform of the core competitiveness, establish a platform for trade barriers, and outline a blueprint that will convince investors to raise their own financing capabilities.

Third, improve the platform business management, enhance the platform of the three capabilities. The highway freight platform should have three core abilities: the first is the level of service; the second is the limitation; the third is cost. For platform enterprises, whether the owner or the owner of the platform is the user of the platform, both of them play a key role in platform satisfaction, therefore, service level is one of the indispensable capabilities of the platform; the timeliness of the distribution is positively correlated with the user satisfaction, so the aging is also one of the key capabilities of the cargo platform enterprises. The cost determines the profitability of a company, and also is the most important ability of the enterprise.

References

1. S. Zheng, T. Yuan, J. Lu. Research on operation efficiency of P2P lending platforms based on DEA methods. Econ. Res. Ref. **20**, 67–77 (2016)
2. X. Zhang. Research on the efficiency and influencing factors of P2P lending platforms in China. ZheJiang University of Finance and Economics, 2017
3. P. Zhou, B. Zhang, P. Zhou et al., Research on the efficiency of P2P micro-credit platform in China. JiangSu Bus. Theory **6**, 51–53 (2014)
4. Y. Cai, Empirical analysis on the risk of P2P lending platforms based on AHP. Mod. Commer. Ind. **4**, 98–99 (2017)
5. S. Chen, C. Yu, P. Liang, et al., The choice of P2P online lending platform. Econ. J. **7**, 188–190 (2015)
6. D. Yang, J. Pei, Research on performance evaluation of logistics system based on DEA-AHP. Logist. Manag. **18**(5), 81–86 (2009)
7. X. Xiang-bing, Y. Gao, Based on the efficiency evaluation of DEA—AHP/GRA, for example. Technol. Manag. Res. **17**, 66–70 (2015)

Forecasting and Planning for Inventory Management in a Complex Rental Housing Unit Supply Chain Environment

Haixia Sang(✉) and Shingo Takahashi

Faculty of Science and Engineering, Department of Industrial and Management Systems Engineering, Waseda University, Tokyo, Japan
sanghaixia@aoni.waseda.jp, Shingo@waseda.jp

Abstract. Given the increasing demand for rental housing units, suppliers have been forced to analyze methods to optimize both their inventory levels and opportunity losses. Although inventory forecasting and planning has been studied for several decades, studies on the circulation inventory problems are extremely limited. In this context, a discrete simulation-based approach to forecast inventory levels in a complex rental housing unit supply chain was developed. An interrelationship between the forecasting method, the initial inventory level and the inventory filling indicator was identified that can help suppliers to optimize their inventory level and opportunity losses. It is suggested that this simulation-based approach is a powerful and efficient tool for mangers involved in inventory decision making.

Keywords: Rental unit · Seasonal demand · Inventory management Simulation

1 Introduction

Inventory Forecasting and planning has received considerable attention from the operational research community over the last 50 years because of its implications for decision-making at both the strategic and operational levels of organizations. Concurrently, a growing number of studies have addressed issues related to supply chain inventory management [1–4]. However, existing studies in this field tend to emphasize the cross-sector collaborations while failing to examine interactions among the various components of a supply chain. Scholars have shown that understanding the interrelationships between various factors in a complex supply chain system can help to solve the inventory problem more efficiently and effectively [5, 6].

In this study, we focus on the inventory system for a rental housing unit supply chain. Unlike most products, rental housing units can be returned to the company at the end of the rental period to meet future rental demand. However, there are several problems in the rental housing unit business. First, there is highly seasonal demand. Second, unlike rental DVDs, it is not possible to predetermine the rental period when a

G. Q. Huang et al. (eds.), *Proceeding of the 24th International Conference on Industrial Engineering and Engineering Management 2018*,
https://doi.org/10.1007/978-981-13-3402-3_38

rental house is rented to a customer as a result of the business practices of the rental company. From the standpoint of the rental housing supply chain, Since there is always a gap between lending and returning, it is recognized that an appropriate determination of the inventory level in the rental housing unit supply chain is a complex problem. Although the problems facing rental business that handle “circulating” products are an important part of the supply chain research field that urgently needs to be solved, studies on the rental business are extremely limited. Papier [7] studied the appropriate inventory size for the rail freight rental business using queueing theory. However, despite the seasonal nature of this rental business, the analysis was conducted under constant conditions that eliminated seasonal fluctuations. Endo et al. [8] proposed an approach taking into account the seasonal demand fluctuations, but instead of focusing on the inventory itself, the objective was to maximize profit.

Simulation has traditionally been used in supply chain research to provide critical support for decision-making regarding the inventory problem [9]. This paper uses a simulation-based analysis to determine the impact of demand forecasting accuracy on inventory, taking the interrelationships among components such as the forecasting method, the initial inventory level, and the inventory filling indicator for the distribution centers into consideration. This study differs from previous studies because we used real data, and thus the findings of this study are directly applicable to the rental housing unit business.

2 Rental Housing Unit Business Model

In recent years, the demand for rental housing units has gradually increased in Japan, as a result of the increase in public works such as seismic-related work and preparations for the 2020 Olympics. Figure 1 shows the business model for the rental housing unit business. The company that handles rental housing needs to understand four product flows, rental shipping, rental return, new product sales, and the sale of used rental housing.

Rental companies usually have several distribution centers and carry out their business through their local sales offices. Units are shipped from a nearby distribution center when demand arises, and are returned at the customer’s convenience. Rental housing units have durable life, For example, Japanese law stipulates that the life span of a prefabricated house is about 20 years, thus company usually turning rental houses into pre-sales markets after a certain period of production to account the maximizing rental income. The shipping, return and pre-sale of rental housing units are basically supported by the distribution center, and the number of units that enter and leave the distribution center varies each month. Figure 2 shows the difference between the number of rental units entering and leaving the distribution centers each month. It can be seen that there is a strong seasonal trend. The negative number in Fig. 2 represents insufficient demand, whereby the total number of units either shipped or pre-sold is less than the number of units returned. The demand would be filled by factory. For example, if the distribution center’s forecast dispatch quantity (rental shipping plus pre-sale) is 1000, and the filling indicator is set as 0.1, the filling quantity is $1000 * 0.1 = 100$. The inventory filling indictor will affect the distribution center’s stock level. In this paper, we set the filling inventory indicator between 0.1 and 1 to determine the change in inventory.

From Fig. 2, it can be seen, that every year from May to December, there is always high demand in excess of the factory's production capacity, and thus it is necessary for the factory to produce units in advance. Thus, the initial inventory is an important factor.

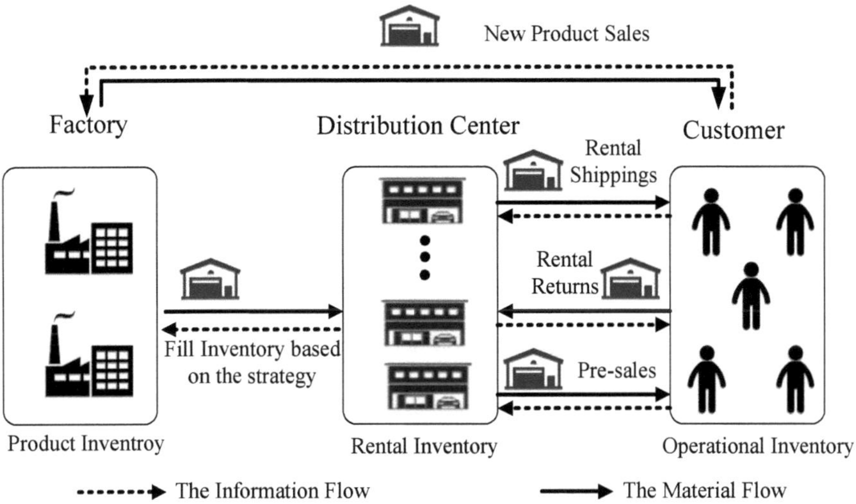

Fig. 1. Rental housing unit business model

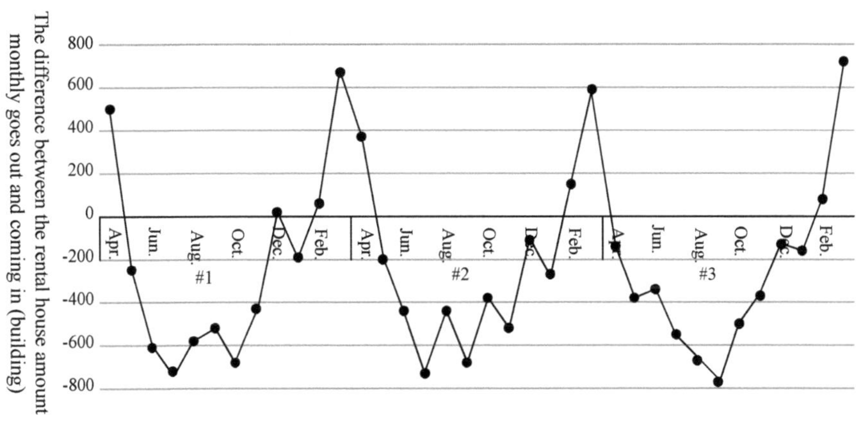

Fig. 2. Monthly differences between the numbers of rental units entering and leaving the distribution centers Difference = rental returns − rental shippings − pre-sales

3 Seasonal Forecasting Method

Problems associated with using seasonally adjusted data to estimate the parameters of econometric models have been studied for several decades. In addition to the forecasting method that is currently used by the company, we used two other popular models to examine the effect of prediction accuracy on inventory levels, one is the Holt-Winters method and the other is multiple regression analysis using a set of dummy variables.

Under the current production system, the demand can be predicted using the following function:

$$I_t' = O_t' * \left(\frac{O_{t-2}}{I_{t-2}} + \frac{O_{t-1}}{I_{t-1}}\right) * \frac{1}{2}$$

where I_t' (building) is the predicted inventory in period t, I_t (building) is the actual initial inventory in period t, O_t' (building) is the predicted initial inventory in period t, and O_t (building) is the actual initial inventory in period t.

In 1960, Winters developed an exponential smoothing method to incorporate seasonality [10], which later became known as the Holt-Winters method. The forecast for any future period ($T + \tau$) uses the following function:

$$D_{T+\tau}' = (a_t + b_t * \tau) * c_t'$$

where $D_{T+\tau}'$ is the predicted demand in period $T + \tau$, a_t is the permanent component of period t, b_t is the linear trend component of period t, and c_t is the time interval seasonal factor of period t.

For the special case of a seasonal pattern that does not change from one year to the next, we regress x_{ym}, the observation for season m in year y, on a set of dummy variables [11] as follows:

$$x_{ym} = \sum_{i=1}^{k} b_i s_{ymi} + e_{ymi}$$

where k is the number of seasons in a year.

$$s_{ymi} = \begin{cases} 1 & if \quad i = m \\ 0 & otherwise \end{cases}$$

4 Case Study

4.1 Rental Housing Unit Supply Chain

This study examines a rental housing unit supply chain located in Chiba prefecture, Japan. To satisfy the diverse demands of various customers, several types of housing unit are produced and stock is held in both the factory and in distribution centers.

To understand the interrelationships among various factors in this complex supply chain system, product A is selected as the research object to simplify the problem.

4.2 Forecasting Results

Figure 3 shows the trends in actual demand and the forecast number of rental returns based on three different forecasting methods. Table 1 shows the residual squared differences between forecast values and actual values. From Fig. 3 and Table 1, it can be seen that multiple regression analysis achieved better results than the other methods, while the current forecasting method is better than the Holt-Winters method. In the next section, we examine whether multiple regression analysis can forecast the optimum inventory level.

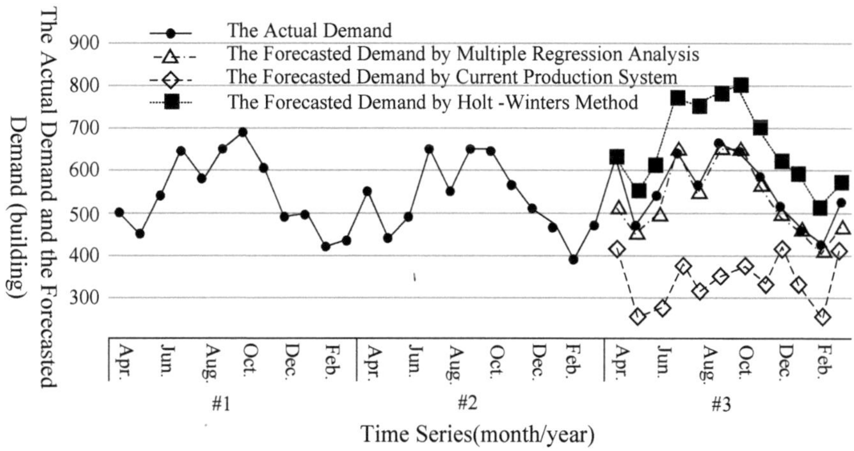

Fig. 3. Forecast demand and actual demand using various forecasting methods

Table 1. Residual squared differences between predicted and actual values

Forecasting method	Residual squared differences between predicted and actual values (buliding)
Forecasting method under the current production system	587,086
Multiple regression analysis	25,029
The Holt-Winters method	125,596

4.3 Simulation Model Construction

Based on the characteristics and structure of the actual supply chain system, a simulation model was constructed to analyze the current supply chain inventory problem. This study used the Excel-VBA simulation platform to develop a model comprising four parts, as shown in Fig. 4.

The simulation logic is shown in Fig. 5. In step 1, based on previous sales results, three forecasting methods were used to forecast demand and the monthly production master schedule was scaled to the forecast demand. However, there is always a gap between forecast demand and actual demand, thus after generating the real demand in step 2, in step 3 the production schedule was modified to fill the gap. In step 4, following production, the product is transported to the distribution centers based on the company's inventory filling strategy. The distribution center must meet customer demand in term of rental shipping and pre-sales, and thus the inventory filling indicator in the distribution center was set accordance with the total amount of rental and pre-sale units needing to be shipped. In this study, the simulation model shows the opportunity loss, average inventory level, and ratio of product utilization for various values of the inventory filling indicator from 0.1 to 1. To compare the various scenarios under the same conditions, we increase the initial inventory filling value in each scenario until there was no opportunity loss.

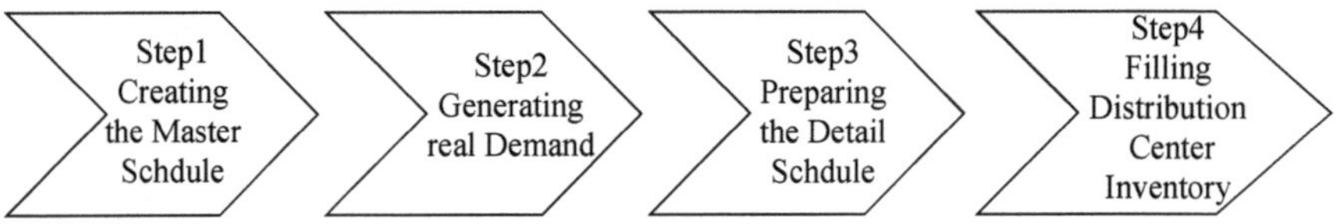

Fig. 4. The simulation model

4.4 Analysis of Simulation Results

Table 2 shows the results of the simulation model in terms of production inventory in the factory and rental inventory in the distribution center. Multiple regression analysis provides the best forecast results. However, the production in the master schedule (see Fig. 3 and Table 2. Scenario 3). However, because actual demand is higher than expected, there is an opportunity loss. In contrast, using the current forecasting method, expected returns is less than actual demand, thus the master schedule shows an increased level of production to avoid any opportunity loss. In the detailed schedule, by adjusting the actual and planned demand values, the actual production volume empty than planned. As a result, from the perspective of avoiding opportunity loss, although the forecast accuracy is less than that under multiple regression analysis, the current forecasting method achieves a suitable inventory level.

Master production schedule is based on the balance of rental returns, rental shipping, and pre-sales (balance = rental returns − rental shipping − pre-sales). If the forecast number of returns is high, which means there is a balance between the number of rental housing units arriving at and leaving the distribution center, there is no need to increase.

Figure 6 shows the changes in average rental inventory and opportunity loss under the inventory filling indicator strategy. By changing the inventory filling indicator from 0.1 to 1, the average rental inventory will be increased; however, the opportunity loss

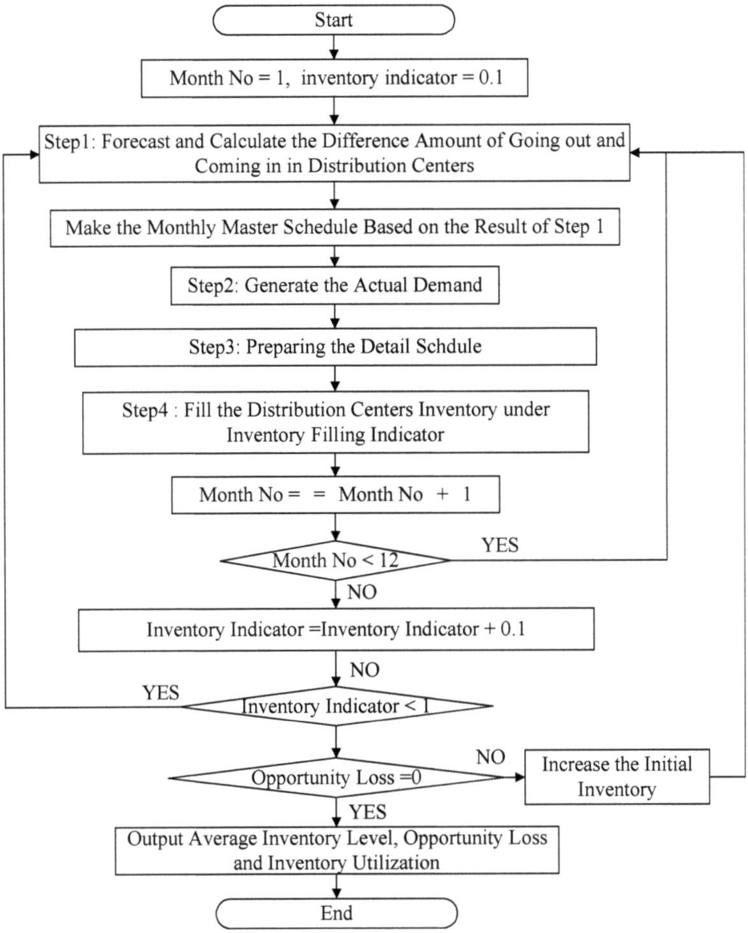

Fig. 5. The simulation logic

will not change until the inventory filling indicator is reduced to 02, 0.4 or 0.5. To avoid unnecessary transport of units among the distribution centers, it is considered that the factory must retain some inventory. Figure 7 shows the relationship among new product inventory, rental inventory, and opportunity loss under the inventory filling indicator strategy in scenario 1, while Fig. 8a, b show the relationship between the average inventory level and opportunity loss using the initial inventory level in scenario 1, which is the best scenario. From the figures, it can be seen that the initial inventory level is an important factor in reducing opportunity loss. The results of the simulation are able to help supply chain managers with both manufacturing and logistics decisions.

Table 2. Simulation results

	Factors causing fluctu ations of inventory			Master schedule forecasted value (buli ding)	The actual demand (buliding)	Production value in the detailed schedule (buliding)	The average ending inventory (buliding)	The annual amount inventory (buiiding)	The opportunity loss (buliding)	The inventory utilization rate (%)
	Forecasting method	Initial inventory level (buliding)	Inventory filling indicator							
Scenario 1	Current forecasting method	500	0.2	4242	1116	1791	422	5563	0	90
Scenario 2	Multiple regression analysis	690	0.4	1199	1130	1417	561	7425	0	91
Scenario 3	Holt-Winters method	855	0.5	386	1115	1408	642	8560	0	89

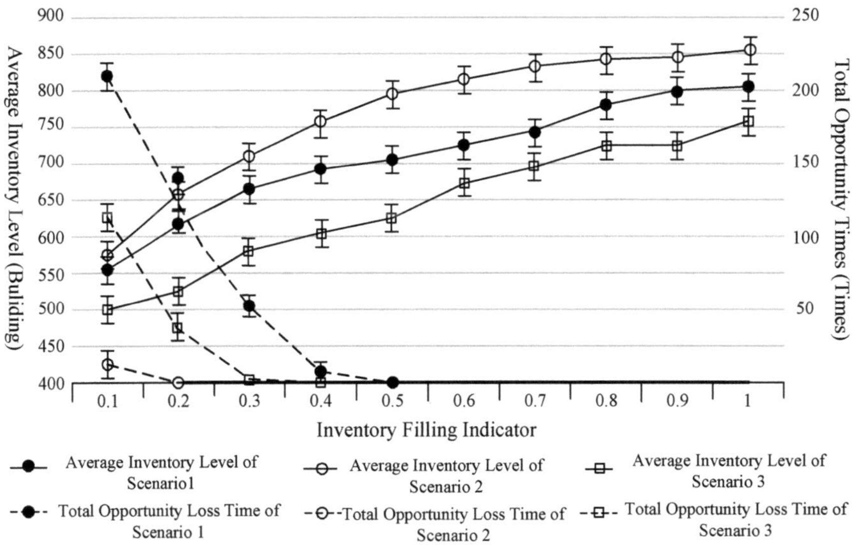

Fig. 6. Changes in the average rental inventory and opportunity loss using the inventory filling indicator strategy

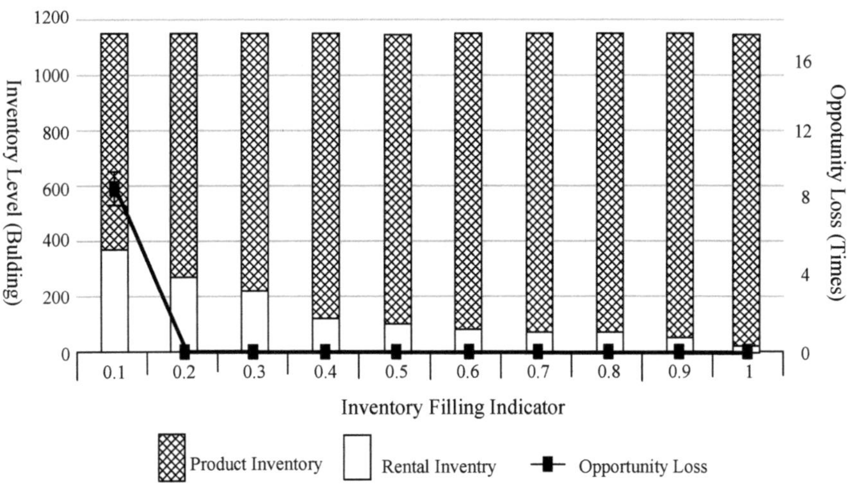

Fig. 7. Relationship among product inventory, rental inventory, and opportunity loss based on the inventory filling indicator strategy in scenario 1

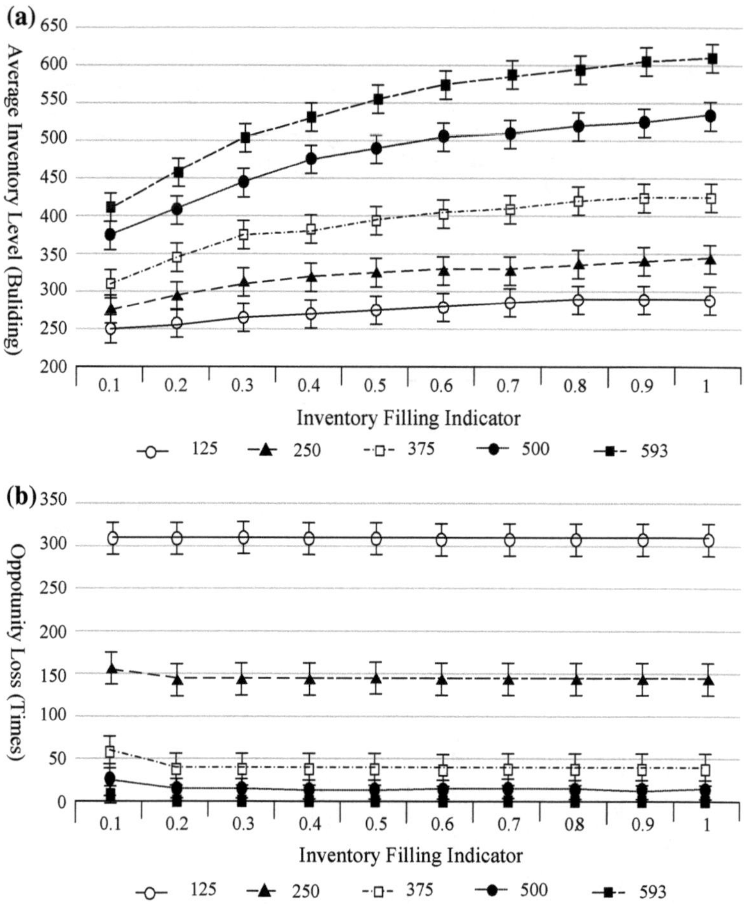

Fig. 8. **a** Average change in the rental inventory in response to changes in the inventory filling indicator in scenario 1, **b** changes in the opportunity loss in response to changes in the inventory filling indicator in scenario

5 Conclusions

(1) An efficient simulation-based modeling method is proposed for practical inventory forecasting systems.
(2) The proposed modeling method is illustrated using an actual case to demonstrate its applicability to real-world inventory forecasting problems.
(3) The flexibility of the proposed system is described. Using the proposed simulation-based modeling method, the relationships between the forecasting method and the initial inventory level, the inventory filling indicator can be identified effectively and efficiently.

Acknowledgements. This research was supported by the Grant-Aid for Scientific Research (B) (No 18K13954). The sample data relating to the rental housing unit supply chain operation are fictitious. The sample data relating to the rental housing unit supply chain operation are fictitious.

References

1. M.M. Helms, L.P. Ettkin, S. Chapman, Supply chain forecasting collaborative forecasting supports supply chain management. Bus. Process Manag. J. **6**, 392–407 (2000)
2. C. Eksoz, S.A. Mansouri, M. Bourlakis, Collaborative forecasting in the food supply chain: a conceptual framework. Int. J. Prod. Econ. **158**, 120–135 (2014)
3. U. Ramanathan, A. Gunasekaran, Supply chain collaboration: impact of success. Int. J. Prod. Econ. **147**, 252–259 (2014)
4. E. Hofmann, Big data and supply decision: the impact of volume, variety and velocity properities on the bullwhip effect. Int. J. Prod. Res. **55**(17), 5108–5126 (2017)
5. H. Akkermans, K. Van Helden, Vicious and virtuous cycles in ERP implementation: a case study of interrelations between critical success factors. Eur. J. Inf. Syst. **11**, 35–46 (2002)
6. H. Sang, S. Takahashi, R. Gaku, Big data-driven simulation analysis for inventory management in a dynamic retail environment, in *Proceedings of the 25th International Conference on IE&EM* (2018)
7. Papier F, Queuing models for sizing and structuring rental fleets. Transp. Sci. **42**(3), 302–317 (2008)
8. U. Endo, S. Morito, T. Tamayama, An approach to determine the amount of ownership in the rental business, in *Proceedings of the 2014 Japan Industrial Management Association Autumn Conference*, pp. 194–195 (2014)
9. S. Terzi, S. Cavalieri, Simulation in the supply chain context: a survey. Comput. Ind. **53**(1), 3–16 (2004)
10. P.R. Winters, Forecasting sales by exponentially weighted moving averages. Manage. Sci. **6**, 324–342 (1960)
11. M.C. Lovell, Seasonal adjustment of economic time series and multiple regression analysis. J. Am. Stat. Assoc. **58**(304), 993–1010

Simulation Research on Core Control Parameters of Electromagnetic Force in Electromagnetic Riveting System

Chang Chen[1(✉)], Yi-jian Tao[1], Zhi-kun Zhu[1], and Zeng-qiang Cao[2]

[1] Chengdu Aircraft Industry Group, Chengdu, China
caarachen@qq.com
[2] Manufacturing Engineering of Aerospace Vehicle, Northwestern Polytechnical University, Xi'an, China

Abstract. To achieve effective control of the riveting force in the electromagnetic riveting system and improve riveting efficiency, a theoretical analysis of electromagnetic riveting force is given. The Ansoft Maxwell 2D finite element software was used to establish the coil coupling model and circuit model respectively, and the numerical simulation of the influence factors of electromagnetic riveting force was studied. The results show that the electromagnetic force is perpendicular to the slave coil and is non-uniformly distributed in the radial direction, moreover, the maximum value is located at half the slave coil radius. The voltage, capacitance, inductance, and resistance in the discharge loop have a great influence on the electromagnetic force, but their effects are quite different. To verify the numerical simulation results, the electromagnetic force was actually measured and compared with the simulation results. The results show that the simulation results are credible.

Keywords: Electromagnetic riveting · Influencing factors · Electromagnetic force · Finite element · Coupling model

1 Introduction

For a long time, riveting has become the most important connection method for aircraft assembly with its advantages of simple process, good reliability and low cost. However, In order to meet the increasing requirements of new-generation aircraft, a variety of new materials, new structures are emerging, the traditional riveting process faces enormous challenges [1–3]. With the rise of large rockets and large military/civil transport aircraft, the use of large diameter rivets and interference fit fasteners has continued to increase, the traditional riveting can no longer achieve its riveting or installation, as well as meet the new stringent requirements of structural fatigue life [4, 5]. Domestic and foreign research and practice have shown that, the electromagnetic riveting is an effective means to solve the above problems and has achieved good technical results in the development of new aircraft [6, 7].

G. Q. Huang et al. (eds.), *Proceeding of the 24th International Conference on Industrial Engineering and Engineering Management 2018*,
https://doi.org/10.1007/978-981-13-3402-3_39

Electromagnetic riveting is a new riveting process. Compared with traditional riveting technology, it has the advantages of long connection life, stable process quality, good controllability, and low labor intensity. In practical applications, due to the different optimum strain rate of the material and the different waveforms of the optimal installation force of the interference fit fasteners, it is necessary to reasonably control the electromagnetic rivet force, this technique is more needed for automated electromagnetic riveting. However, current research on electromagnetic riveting mainly focuses on technical applications such as riveting of titanium alloys, riveting of composite materials, and installation of interference fit fasteners [8–10]. The study of electromagnetic riveting force is still relatively lacking. Only a few documents have studied the relationship between electromagnetic riveting discharge circuit discharge and electromagnetic riveting force [11–13]. The effects of charging voltage, capacitance, inductance and loop resistance on the electromagnetic riveting force are not yet clear. Therefore, in order to achieve effective control of electromagnetic riveting force, further improve the quality of aircraft assembly process, and expand the application of electromagnetic riveting technology, it is necessary to study the core control parameters of electromagnetic riveting force.

In this paper, the electromagnetic riveting force is theoretically analyzed, electromagnetic coupling coil electromagnetic coupling model and discharge loop model are established to analyze the influence of discharge circuit core parameters on electromagnetic riveting force, and simulation results are verified by experiments. The relevant research results can provide guidance for the control of electromagnetic riveting forces in automated electromagnetic riveting systems.

2 Theoretical Analysis

From the perspective of energy conversion, by electromagnetic induction, the electromagnetic riveting technology converts the electrical energy stored in the capacitor into the mechanical energy of the slave coil, forming rivets and releasing heat with this kinetic energy.

2.1 Electromagnetic Riveting Principle

Figure 1 is the basic principle of electromagnetic riveting. Based on the RLC discharge circuit, the master coil is connected in the pulse discharge circuit to produce a high intensity pulse magnetic field and to form the electromagnetic force with the mutual inductance coil. The electromagnetic force is reflected and transmitted in the stress wave amplifier and forms a stress wave at the incident end of the amplifier. Finally, the stress wave acts on the rivet or the high-locking bolt to install and form.

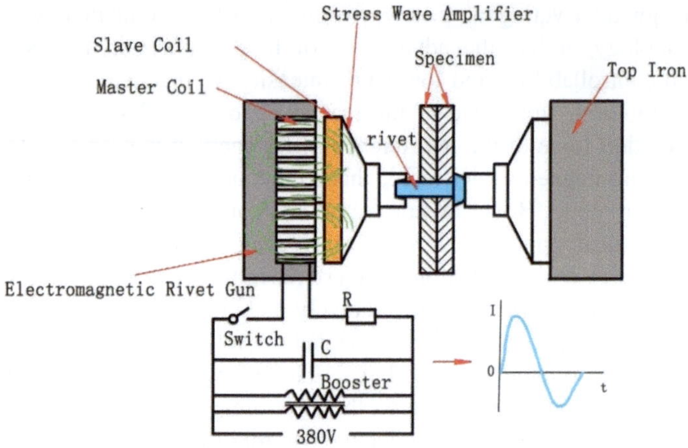

Fig. 1. Basic principle of electromagnetic riveting

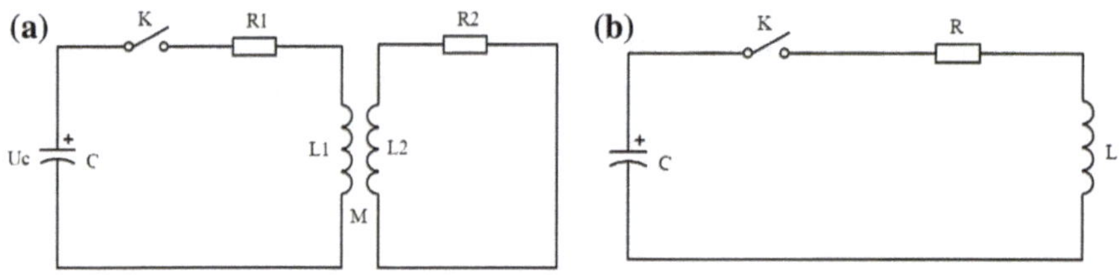

Fig. 2. Discharge circuit model

2.2 Discharge Circuit Model

Electromagnetic riveting system discharge circuit model equivalent to RLC discharge circuit. The charging process of the capacitor satisfies the loop zero state response, and the discharging process satisfies zero input response, the equivalent circuit model is shown in Fig. 2. Because the discharge process is very short, the influence of the displacement of the slave coil on the mutual inductance of the loop can be ignored, so the mutual inductance is considered as a constant, and the loop model can also be represented in the form of Fig. 2b.

C is the storage capacitor; *R* is the total resistance of the discharge loop, including the resistance of the cable and the resistance of the master coil; *L* is the total inductance of the discharge loop, including the inductance in the capacitor, cable inductance, and the mutual inductance of the master coil and the slave coil.

As the discharge circuit satisfies the transient characteristics, the second-order linear homogeneous differential equations about the current are listed:

$$LCi'' + RCi' + i = 0 \tag{1}$$

Initial Condition: $u_C(0_+) = u_C(0_-) = E$.

The magnitude of the resistance in the discharge circuit is generally small, satisfying the conditions of the underdamping oscillation circuit: $R<2\sqrt{L/C}$. Solving the above formula, the current analysis formula is:

$$i(t) = \frac{E}{L\sqrt{\omega_0^2 - \alpha^2}} e^{-\delta t} \sin\sqrt{\omega_0^2 - \delta^2 t} \tag{2}$$

From the formula, the current is a sine decay function with respect to time.

$\delta = \frac{R}{2L}$ Attenuation Coefficient;

$\omega_0 = \sqrt{\frac{1}{LC}}$ Resonant Circular Frequency;

$\omega = \sqrt{\frac{1}{LC} - \frac{R^2}{4L^2}}$ Oscillation Frequency.

As $\frac{R^2}{4L^2}$ and $\frac{1}{LC}$ differ by three orders of magnitude, so $\omega \approx \omega_0$. The current oscillation period is:

$$t = 2\pi\sqrt{LC} \tag{3}$$

Current pulse width is $\frac{t}{2}$.

According to the existing study [14], the analytical formula of the electromagnetic repulsion force between the loading coils is:

$$F = \mu_0 i_1 i_2 \cdot \frac{r_1}{\alpha} \tag{4}$$

i_1—Master coil current; i_2—Slave coil induced current. With the above expression, the analytical formula of electromagnetic force is:

$$F = \frac{r_1 \mu_0 \omega M N^2}{\alpha\sqrt{R_2^2 + (\omega L_2)^2}} \cdot I_m^2 e^{-2\delta t} \sin(\omega t)\sin(\omega t + \varphi) \tag{5}$$

In the formula:

r_1 Master coil radius;
μ_0 Vacuum permeability;
ω Oscillation round frequency;
M Coil mutual inductance;
N Master coil turns;
α Coil distance;
R_2 Slave coil resistance;
L_2 Slave coil inductance.

In the above equation, the electromagnetic force is proportional to the radius of the coil, the number of turns, the peak current, and is inversely proportional to the distance between the loading coils. The specific influence of each parameter in the discharge loop on the electromagnetic force can not be determined from the formula. The finite element method will be used in the following research.

3 Finite Element Modeling

The coil mutual inductance model and discharge loop model were established by Ansoft Maxwell 2D. The physical model of the coil is shown in Fig. 3, and the geometric parameters are shown in Table 1.

(a) Master Coil (b)Slave Coil

Fig. 3. Physical maps of master coil and slave coil

Table 1. Master coil geometry

Turns number	Inner diameter (mm)	Outside diameter (mm)	Coil thickness (mm)	Coil spacing (mm)	Thickness (mm)
16	20	100	1	1	15

The master coil is a standard spiral conductor wrapped by insulating layer and has good electromagnetic coupling performance. In the finite element analysis, the coil's helicity is neglected, and it will be considered as the concentric cylinders with equal circles.

Because of the symmetrical geometry of the loading coils, the modeling software only needs to create half of the axial section of the physical model to reduce the amount of calculation.

The finite element model is shown in Fig. 4a, and the discharge loop model is shown in Fig. 4b. The inductance in the circuit is divided into L-line and L Winding, which represent the cable inductance and load coil inductance in the discharge circuit, respectively.

The load coil inductance can be calculated by the simulation software; the inductance of the cable is measured by a digital bridge. Measurement instrument and component parameters are shown in Fig. 5

Design simulation tests:

Under the conditions of capacitance $C = 4\,\text{mF}$, inductance $L = 8\,\mu\text{H}$, resistance $R = 15\,\text{m}\Omega$, and charging voltage $U = 1000\,\text{V}$, the influence of the core parameters on the electromagnetic force is studied by the control variable method.

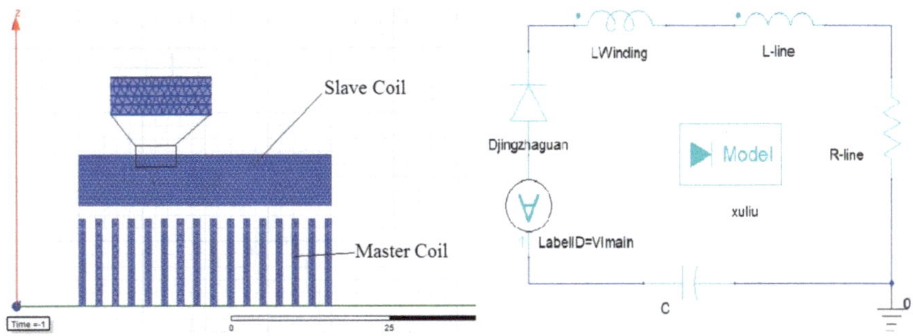

(a) Coil model (b) Circuit model

Fig. 4. Simulation model

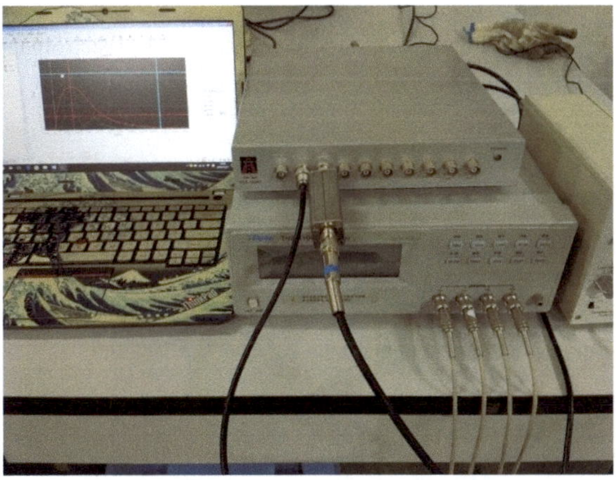

Fig. 5. Data acquisition tower and digital bridge

4 Results and Discussion

4.1 Test Verification and Electromagnetic Force

As the charging voltage is higher, the error of the test measurement is larger. Under the voltage of 500 V, the electromagnetic force simulation value and test value were compared and verified. The results of comparison of current and electromagnetic force are shown in Fig. 6.

The trend of the two types of data is roughly the same. The experimental current and the electromagnetic force reach the peak slightly later. The errors of multiple characteristic parameters shown in Table 2 are all within 10%, and the simulation results are reliable.

Therefore, the finite element simulation method can be used instead of experimental methods to analyze the specific influence of various core parameters on the electromagnetic force in the electromagnetic riveting device.

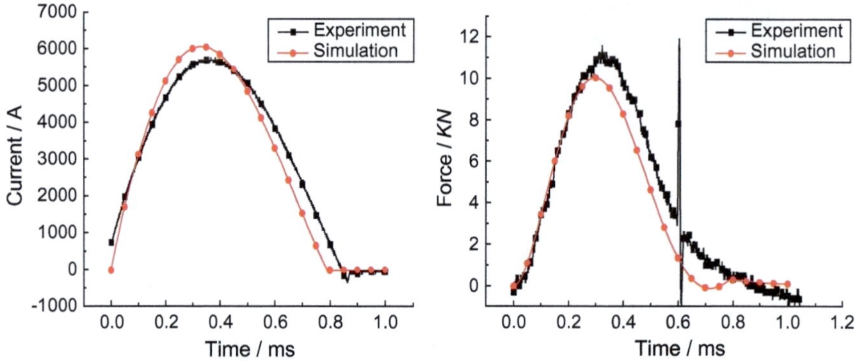

(a) Current (b) Electromagnetic Force

Fig. 6. simulation data and experiment data

Table 2. Simulation and experiment related data

	I_{max} (A)	F_{max} (KN)	Current duration (ms)	Electromagnetic force width (ms)
Simulation	6068	10	0.79	0.7
Experiment	5980	11.5	0.8	0.9

The electromagnetic rivet force is the force formed by electromagnetic coupling of the loading coils, its vector distribution is shown in Fig. 7. The induced current in the slave coil is perpendicular to the paper and faces the direction of the master coil current, which is in accordance with the basic principle of electromagnetic induction.

The coil is mainly subjected to axial force and has a slight force component in the radial direction. This is due to the uneven distribution of magnetic lines around the coil. The axial magnetic field force is distributed unevenly along the coil, and the magnetic field force is the maximum at about half of the coil radius.

4.2 Influence of Voltage

The capacitor charging voltage is an important parameter of the discharge circuit. Keep other parameters in the discharge circuit unchanged, simulate the model from a voltage of 500–1000 V to obtain the curve of voltage and electromagnetic force, as shown in Fig. 8. As the voltage increases, the electromagnetic force peaks gradually, but the electromagnetic force pulse width does not change. The higher the voltage, the greater the increase in electromagnetic force.

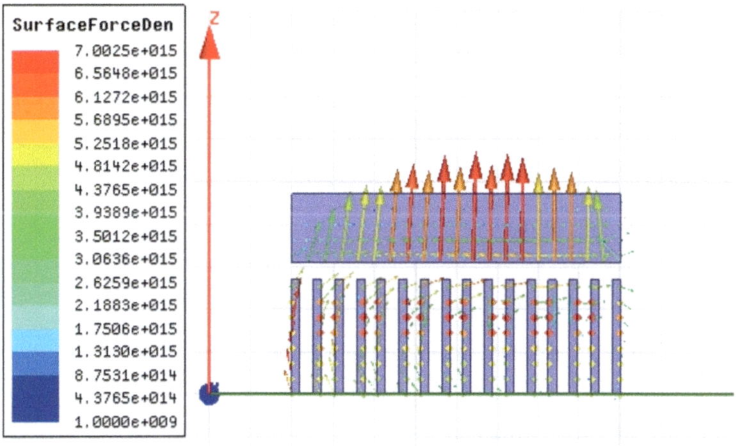

Fig. 7. Electromagnetic force vector map

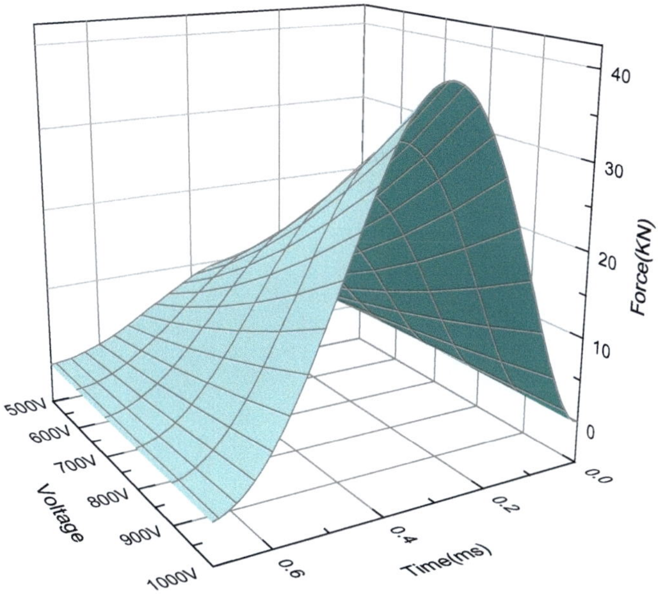

Fig. 8. Electromagnetic force-voltage

The relationship between the peak current and the electromagnetic force peak voltage is further analyzed, as shown in Fig. 9. The current peak value is linearly proportional to the charging voltage, and the peak value of the electromagnetic force is a quadratic function with the charging voltage, which is consistent with the theoretical analysis.

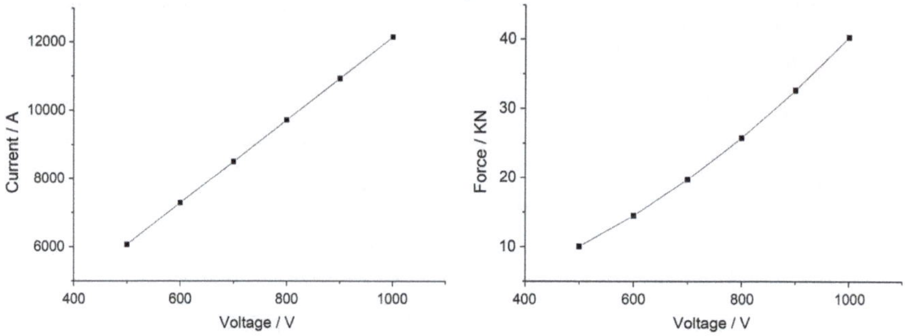

Fig. 9. Voltage—Current peak and Electromagnetic force peak

4.3 Influence of Capacitance

Seen from the formula, the current pulse width is only related to the capacitance. When the capacitance is doubled, the current pulse width is increased by $\sqrt{2}$ times.

Electromagnetic force waveforms with different capacitances are shown in Fig. 10. Compared to voltage and electromagnetic band diagrams, the band diagram of capacitance and electromagnetic force is narrower, that is, the change in capacitance has a great influence on the pulse width of the electromagnetic force. At the same voltage, the larger the capacitance, the higher the electromagnetic force amplitude and the wider the pulse width of the electromagnetic force.

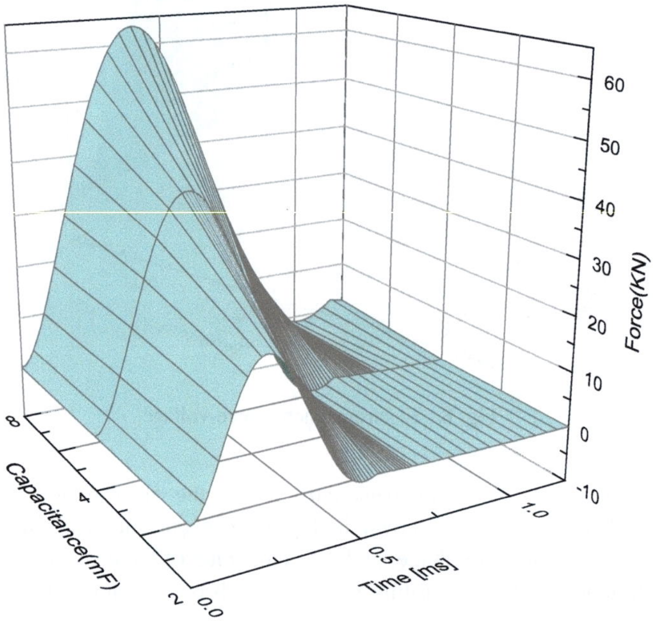

Fig. 10. Electromagnetic force—capacitance

4.4 Effect of Inductance

The inductance in the discharge circuit consists of the electrolytic capacitor inductance, the cable self-inductance, and the self-inductance of the loading coil. The total inductance of the capacitor bank is $0.5\,\mu\text{H}$ approximately

The inductance of the cable is related to its length. The inductance of two parallel laid cables per unit length is approximately $(0.5 \sim 1) \times 10^{-7}$ H/m.

Analyze the relationship between inductance and electromagnetic force, as shown in Fig. 11. As the total loop inductance increases, the magnitude of the electromagnetic force decreases, and the electromagnetic force peak point time gradually moves backward. The pulse width, amplitude, and loading rate of the electromagnetic force can be controlled by changing the inductance in the discharge loop.

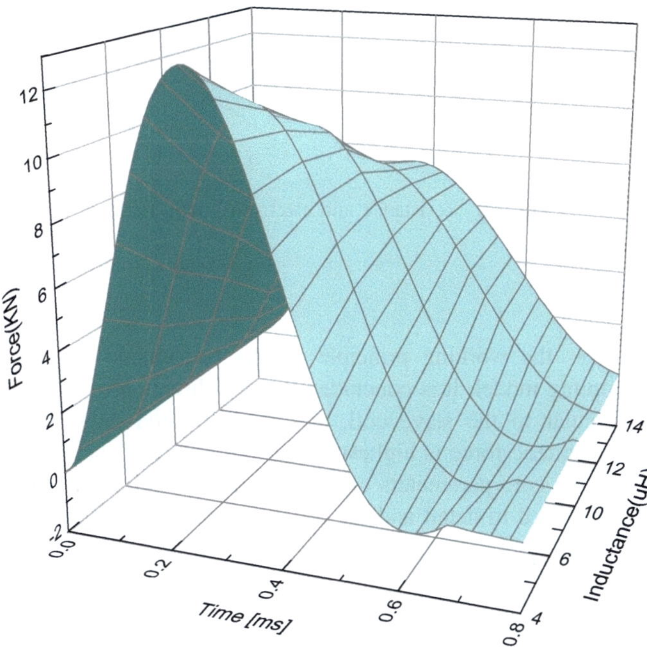

Fig. 11. Inductance and electromagnetic force

4.5 Influence of Resistance

The resistance in the discharge loop consists of the master coil resistance and the cable resistance. Using the control variable method, the relationship between the resistance and the electromagnetic force is analyzed, as shown in Fig. 12, as the resistance in the loop increases, the peak value of the electromagnetic force gradually decreases, and the time point at which the electromagnetic force reaches the peak gradually advances.

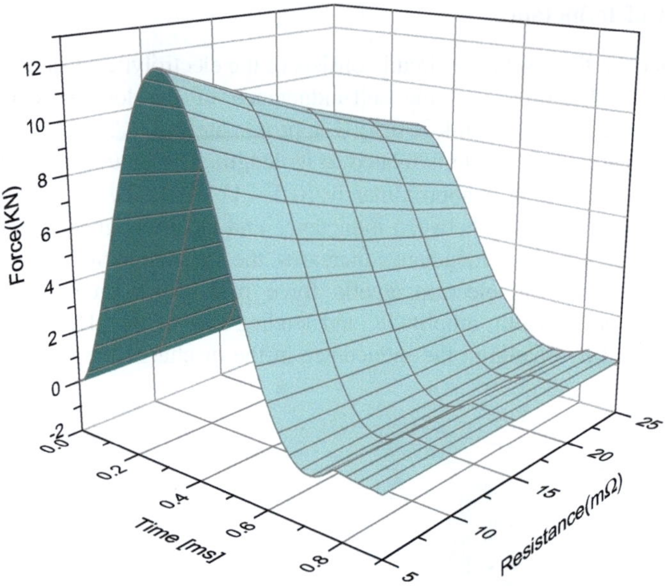

Fig. 12. Resistance and electromagnetic force

5 Conclusion

This paper analyzes the working principle of electromagnetic riveting system and simulates the coupling model. It is concluded that the direction of the electromagnetic force is perpendicular to the slave coil, and is distributed unevenly in the radial direction, electromagnetic force is highest at half the radius of the slave coil.

The impact of the core parameters of the discharge circuit on the electromagnetic rivet force was evaluated in detail and the following conclusions were obtained:

(1) As the capacitor charging voltage increases, the electromagnetic force amplitude increases and the electromagnetic force pulse width does not change
(2) The current amplitude has a linear relationship with the charging voltage; the electromagnetic force amplitude has a quadratic function relationship with the charging capacitor
(3) The electromagnetic force amplitude increases as the capacity increases. The larger the capacity, the wider the pulse width of the electromagnetic force
(4) As the total inductance of the loop increases, the amplitude of the electromagnetic force decreases, and the time point at which the electromagnetic force reaches the peak gradually shifts back
(5) As the resistance increases, the peak value of the electromagnetic force gradually decreases, and the time point at which the electromagnetic force reaches the peak gradually advances.

During the actual application process of the electromagnetic riveting system, the parameters in the discharge loop can be adaptively modified to match the optimum strain rate of the molded material.

References

1. H. Li, C.J. Yi, Q. Wang, Comparative analysis of compression riveting and hammer riveting of composite material. Aeronaut. Manuf. Technol. **537**(18), 97–99 (2017) (In Chinese)
2. Z.Q. Cao, Y. Dai, Electromagnetic riveting technology for wedge-shaped composite structures. J. Aeronaut. **30**(10), 1998–2002 (2009) (In Chinese)
3. J. Cui, L. Sun, L. Meng et al., Experimental study on electromagnetic riveting of titanium alloy rivets with carbon fiber connection structure. Forg. Stamp. Technol. **38**(2), 47–52 (2013) (In Chinese)
4. H.P. Yu, C.F. Li, Electromagnetic riveting technology for large-diameter and high-strength rivets, in *Proceedings of the 11th National Collapsing Conference on Plastic Engineering*. Plastics Engineering Society of China Mechanical Engineering Society, Beijing (2009) (In Chinese)
5. H.P. Yu, C.F. Li, Electromagnetic riveting technology of large diameter rivet with high strength, in *The 11th National Plastic Engineering Academic Annual Meeting*. Chinese Institute of Mechanical Engineering Plastic Engineering Branch, Beijing (2009) (In Chinese)
6. Z.Q. Cao, Application of electromagnetic riveting technology in the manufacture of large aircrafts. Acta Aeronautica et Astronautica Sinica **29**(3), 716–720 (2008) (In Chinese)
7. Z.Q. Cao, Exploration of electromagnetic riveting application in large aircraft manufacturing. Acta Aeronautica Et Astronautica Sinica **29**(3), 716–720 (2008) (In Chinese)
8. G. Xu, M. Gao, Q. Xiao et al., Development, equipment development and application of electromagnetic riveting technology. Aeronautical Manufacturing Technology (23), 38–41 (2010) (In Chinese)
9. J.H. Deng, C. Tang, M.W. Fu, Y.R. Zhan, Effect of discharge voltage on the deformation of Ti Grade 1 rivet in electromagnetic riveting. Mater. Sci. Eng. A **591**, 26–32 (2014)
10. C. Zengqiang, Coping with assembly connection technology for large aircraft development in China. Aeronaut. Manuf. Technol. **2**, 88–91 (2009)
11. H. Jiang, G. Li, X. Zhang et al., Fatigue and failure mechanism in carbon fiber reinforced plastics/aluminum alloy single lap joint produced by electromagnetic riveting technique. Compos. Sci. Technol. 152 (2017)
12. X. Yang, Z.Q. Cao, Y. Zuo, et al. Experimental study on mounting process of titanium interferential bolts based on stress wave loading. J. Northwest. Polytech. Univ. **35**(3), 462–468 (2017) (In Chinese)
13. Y. Qin, M. Gao, F. Zou, Simulation analysis of electromagnetic riveting discharge process. Aeronaut. Manuf. Technol. **474**(5), 68–69 (2015) (In Chinese)
14. Z.Q. Cao, Electromagnetic riveting theory and application research. Northwestern Polytechnical University, Xi'an (1999) (In Chinese)

The Application of Foreground Theory and Genetic Algorithm in Bus Dispatching

Huang Li[1(✉)], Dong-mei Luo[1], Xiao-peng Luo[2], Chen Kun[1], Zhen-li Guo[1], and Gao Tao[1]

[1] School of Mechanical Engineering, Panzhihua Universiy, Sichuan Panzhihua 617000, China
459628370@qq.com

[2] Mathematics and Computer College, Panzhihua Universiy, Sichuan Panzhihua 617000, China

Abstract. The prospect theory is used to describe and analysis the passenger behavior and the genetic algorithm is combined with it to establish the bus dispatch model in this paper. Solutions are provided to shorten the passenger waiting time and maximize the benefits of bus company. Finally, it takes the case of bus dispatching from Panzhihua University Headquarters to Xiyuan as a case study.

Keywords: Prospect theory · Genetic algorithm · Bus dispatch

1 Introduction

With the rapid development of social economy, the traffic demand of urban residents shows a sustained and rapid growth trend, and the development of intelligent transportation system has become a necessary measure to alleviate the traffic congestion. The scheduling problem of bus buss is a typical multi-objective optimization problem, which is usually optimized by the overall interests of passengers and bus companies. Genetic algorithm is an efficient algorithm for searching the global optimal solution, which has good maneuverability, parallelism and robustness, so genetic algorithm is used to optimize the bus dispatching in this paper.

2 Theoretical Basis

2.1 Prospect Theory

Kahneman and Tversky (1992) put forward the "prospect theory" to supplement the "expected utility theory" [1]. They have proved that in the face of benefits and losses,

---

Huang Li (1980.2-), female, Nan Chong, Sichuan province, master's degree in Panzhihua associate professor, school of mechanical engineering, school of mechanical engineering college of industrial engineering department, the main research direction for operation management optimization, E-mail: 459628370@qq.com.
Dong-mei Luo (1996.7-) female, person from Suining, Sichuan province, college of mechanical engineering, Panzhihua University.

G. Q. Huang et al. (eds.), *Proceeding of the 24th International Conference on Industrial Engineering and Engineering Management 2018*,
https://doi.org/10.1007/978-981-13-3402-3_40

people cannot fully grasp information, make optimal decisions. Katsikopoulos (2002) and others through a lot of route choice experiment, it is concluded that people's travel behavior is similar to "prospect theory" described [2], Zhao Lin is still 2006 prospect theory is verified by the simulation than previously expected utility theory more close to the situation of people travel route choice behavior, then the prospect theory in such aspects as traveler travel decisions has been widely research [3].

2.2 Genetic Algorithm

Genetic algorithm is a kind of random search algorithm, learning from Darwin's natural selection and genetic mechanism of the evolution of biological evolution process. It simulates the rules of natural selection and the phenomenon of genetic recombination and mutation in the process of natural genetic. In the iteration process for each part of the candidate solutions, and high fitness individuals are selected from candidate solutions, through selection, crossover and mutation, these individuals are combined and produce offspring candidate solutions. Repeating this process until some

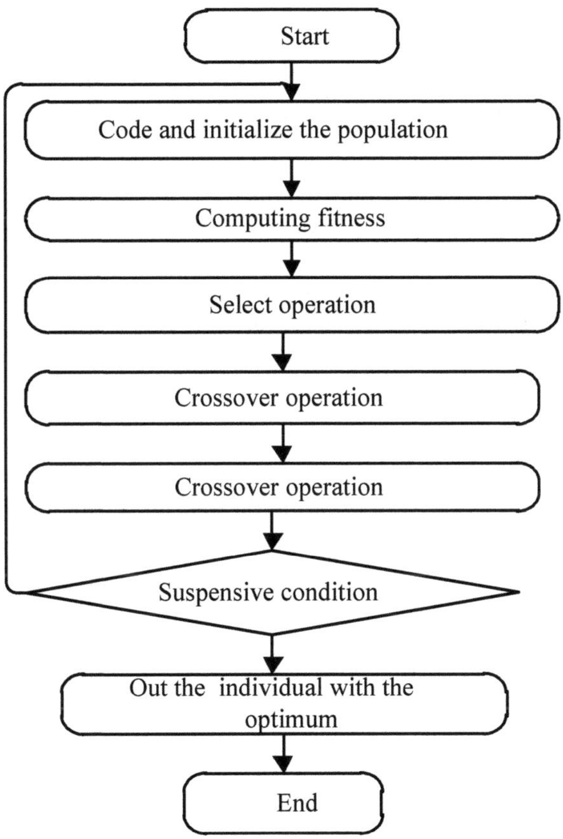

Fig. 1. Basic schematic diagram of genetic algorithm

convergence condition is satisfied [4–6]. The genetic algorithm includes five basic elements: (1) coding; (2) setting of initial group; (3) fitness function; (4) genetic manipulation; (5) control parameters, the basic principle of the algorithm is shown in Fig. 1.

3 Mathematical Model of Bus Dispatching

The research object of the bus scheduling problem includes bus companies and passengers. The bus companies always want to reduce the number of bus trips, increase the load rate and reduce operating costs. Passengers are required to wait as long as possible [7]. If only the time needed to reduce the number of passengers is met, the cost of the bus company will be increased, and vice versa. Therefore, it is a typical multi-objective optimization model to establish a mathematical model with the overall interest of the two as the objective function [8].

This paper will be divided into multiple times a day, with an average departure interval of each time period as a variable of the model, with both the average passenger waiting time and the number of bus to a minimum for scheduling goal, at least in a bus load factors and departure interval range as constraint condition to establish the mathematical model of problem solving.

3.1 Hypothesis of Traveler and Reference Point

In this paper, based on prospect theory, people are bounded rationality, evaluation of its passenger care is not the absolute time but the relative time, therefore, the selection of reference point is very important, in order to guarantee the representation of the reference point. Set the reference point for waiting for the reserved time te, because of traffic time and reserve waiting time have difference, the data processing process is simplified as shown below. On the basis of market research, the data given may represent the common people's waiting time and each time interval is used as waiting time.

3.2 Model Hypothesis

Because of the influence of the bus dispatch is affected by many environmental factors, in order to make more universal mathematical model, the following assumptions are made: (1) specific object at a particular time period, all bus route in accordance with the relevant provisions; (2) all buses are the same model; (3) there will be no passengers when each bus passing through the station; (4) the reserve waiting time is equal; (5) the departure interval is the same within the same time period; (6) accidents such as traffic jams and traffic accidents are not considered.

3.3 Objective Function

There are two objective functions: (1) the minimum number of buss, namely the least cost of the bus company; (2) passengers are the most profitable. The two objective functions are as follows:

$$\min\left(H = \sum_{i=1}^{I} \frac{T_i}{\Delta t_i}\right) \tag{1}$$

In which, I represents I periods of time,

T_i represents the length of time period of the i th period.

Δt_i represents the departure interval of the i th time period.

$\frac{T_i}{\Delta t_i}$ represents the number of times of departure for the i th period.

The total number of starts in the I time periods. The minimum is the minimum number of times a day.

According to the value function given by Kahneman and Tversky.

$$v(x) = \begin{cases} (t_e - \Delta t_i)^{\alpha}, t_e \geq \Delta t_i \\ -\lambda(\Delta t_i - t_e)^{\beta}, t_e \leq \Delta t_i \end{cases}$$

α, β the risk attitude coefficient is less than 1. λ the coefficient of loss aversion is greater than 1.

$\alpha = \beta = 0.88, \lambda = 2.25$ When $t_e \geq \Delta t_i$ by the actual waiting time traveler less than waiting for reserve of his time, is characterized by revenue, if not, for the loss.

$$\max\left(\varphi = \sum v(x) = \sum_{i=1}^{I} \sum_{j=i}^{J} m_i \times \rho_{ij} \times \left[\sigma(t_e - \Delta t_i)^{\alpha} - (1 - \sigma)\right]\right) \tag{2}$$

Among them, J represents the total number of stations in Xiyuan to the bus line.

m_i represents the number of cars that are issued during the ith period.

ρ_{ij} represents the number of passengers arriving at the jth site at the ith time period.

σ represents a variables is between 0 and 1, when $t_e \geq \Delta t_i$ is 1, otherwise 0.

So Φ is the maximum value that all passengers get when they travel to the bus in Xiyuan.

In order to consider both the interests of passengers and bus companies, Through the weighted coefficient α and β, the function H and Φ are integrated into a function F:

$$\min\left(F = \alpha \times \max \varphi + \beta \times \sum_{i=1}^{I} \frac{T_i}{\Delta t_i}\right) \tag{3}$$

where α represents the passenger average waiting time weight coefficient, and β represents the total number of times weight coefficient of the bus company.

3.4 Model Constraints

In order to improve the full capacity of buses and make full use of the bus resources, the full capacity of the model must be more than 70%. In the model, it is expressed as follows:

$$\frac{\sum_{i=1}^{I}\sum_{j=1}^{J}\mu_{ij}}{Q_{const}\times\sum_{i=1}^{I}\frac{T_i}{\Delta t_i}} > 70\% \tag{4}$$

μ_{ij} represents the number of passengers on the *j*th site in the *i*th time period, and $\sum_{i=1}^{I}\sum_{j=1}^{J}\mu_{ij}$ indicates that the total number of passengers in one day.

4 Genetic Algorithm Optimizes Bus Scheduling

4.1 Chromosome Coding

According to the characteristics of mathematical model, a binary coded form is used to describe the departure interval of each period. If the minimum departure interval is $D_{\min}$ and the maximum departure interval is $D_{\max}$, then the binary code length of each interval is x, and x must satisfy

$$2^x \geq D_{\max} - D_{\min} + 1 \tag{5}$$

Suppose the minimum departure interval of $D_{\min} = 3$ the maximum departure interval $D_{\max} = 13$, so the binary code length corresponding to the departure interval Δt_i at each time is at least 4. For example, the decimal and binary encoding processes of the departure interval of 3, 5, 7 and 13 min are: 3 min—0-0000; 5 min—2-0010; 7 min—4-0100; 13 min—10-1010. If the day is divided into *I*, then the length of the chromosome is at least 4*I*.

4.2 Chromosome Decoding

Get a binary string $b_0b_1\ldots b_{xI-1}$, which is a solution to the scheduling problem, and the integer that is turned into $[D_{\min}, D_{\max}]$ is the corresponding departure interval. Suppose that the range of the departure interval is [3, 13] and the binary string is divided into four consecutive groups, with a total of *I* groups, and the binary string into the decimal number to get the corresponding time interval Δt_i:

$$\Delta t_i = 3 + \sum_{j=0}^{3} b_j + 2^j \tag{6}$$

4.3 Selection of Initial Group

Initial population with completely random method, firstly the randomly generated a certain amount of a topic, and then choose the optimal individuals to join the initial population, iterated, this process until the initial number of individuals in a population achieves predetermined scale. At last, we randomly selected the number of K binary string of $4I$ from these initial populations as the initial population.

4.4 Fitness Function

In the genetic algorithm to calculate the fitness of the individual comparison sort, and to determine the probability of the individual selected, so the value of fitness function generally is positive, therefore the objective function is often mapped as a maximum value in the form of the objective function, the introduction of a large enough constant $C_{\max}$ to limit the current maximum of the objective function, G get new objective function as follows:

$$G = C_{\max} - \left(\alpha \times \max \varphi + \beta \times \sum_{i=1}^{I} \frac{T_i}{\Delta t_i}\right) \tag{7}$$

4.5 Genetic Manipulation

1. *Track of choice*
   By fitness proportion method, this method is also called roulette algorithm or Monte Carlo choice, set group species for n, the fitness of the ith individual is f_i, then the probability of the ith individual selected is $p_i = f_i / \sum_{j=1}^{n} f_j$. The larger the fitness, the greater the probability of being selected.
2. *Cross*
   This paper adopts a single point of intersection, the crossover rate is set to 0.78, the biggest cross one chromosome number is 12.
3. *Variation*
   Mutation operation can improve the local random searching ability of genetic algorithm. The selection of mutation probability is influenced by the population size and chromosome length, and it is usually selected with a smaller value. In this paper, the mutation rate is 0.01, and the maximum variation of a chromosome is 12.

5 Bus Dispatching of Panzhihua College Bus

5.1 Current Analysis

In order to get reliable analysis data, in this paper, In order to get reliable analysis data, in this paper, through the way of questionnaire to understand the passenger's feeling when taking bus from Panzhihua University headquarters to XiYuan campus.

240 students who often take campus bus are chosen as the subjects of investigation and 217 copies were valid. From the investigation study, the following conclusions are drawn as follows:

(1) due to the long waiting time and crowded bus, 82.52% of the total number of buses were abandoned, so it is necessary to adjust the time interval of departure.
(2) the actual waiting time is more than expected waiting time, based on prospect theory "bounded rationality" of passenger behavior description, by causing them often complain about, even give up to take the bus as shown in Fig. 2.

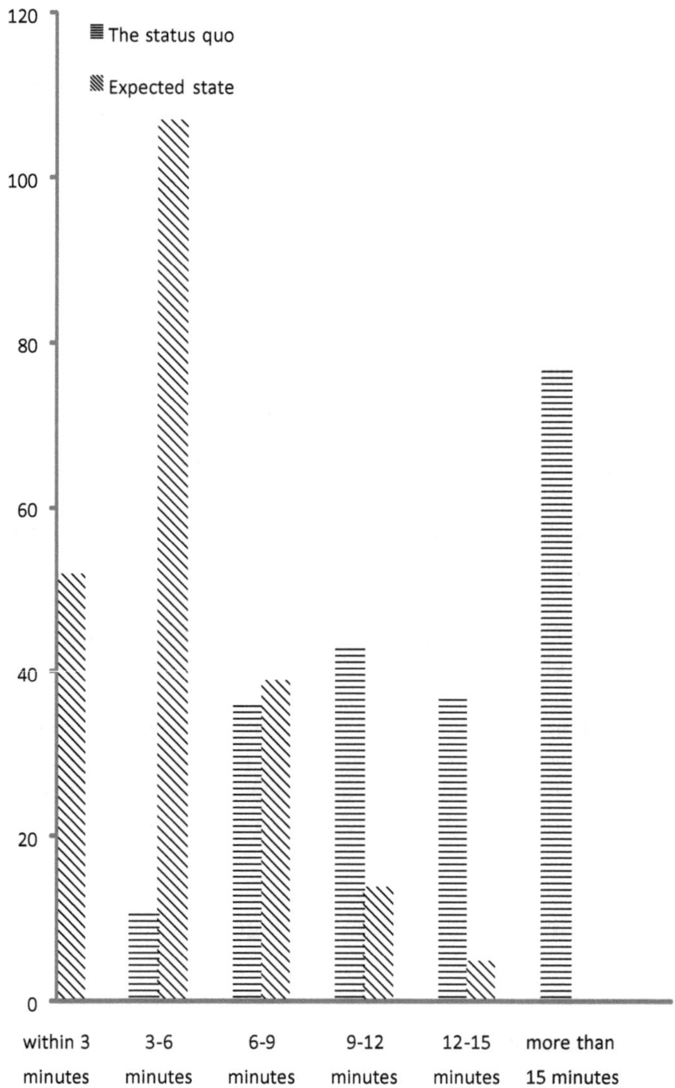

Fig. 2. Comparison of the status of waiting times and expectation

5.2 Scheduling Optimization

This paper is to optimize the grid interval, shorten the traffic waiting time and reduce the number of traffic department start times as the goal.

Set algorithm of the basic parameters: dividing the operating time of a day for nine times, minimum departure intervals for 3 min, the biggest departure intervals for 13 min. When bring traffic data into model analysis, the maximum number of iterations is 45, the minimum loading rate is 70%, population size is set to 20. The relationship between individual fitness and mutation rate was analyzed by a set of experiments.

The fixed cross rate is 0.85, set the mutation rate to 0.01. The experiment results were shown in Fig. 2. Set the mutation rate to 0.1, as shown in Fig. 3. According to the result of Fig. 4 can be found that with the increase of iteration times, the individual fitness are slowly rising, starting with 6 individual fitness increasing significantly, from the 8th generation beginning to smooth the individual fitness tends to a stable state. It explains that at this time a state diagram has been in a state of optimal solution; From Fig. 3, it can be found that when the mutation rate increases, the individual fitness convergence rate slows down, and then becomes stable in the 10th generation.

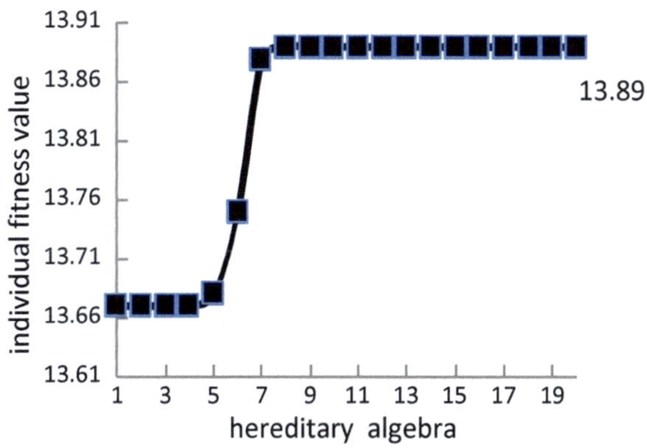

Fig. 3. Variation rate 0.01 genetic algebra and individual fitness chart

After analysis, it can be found that when the fitness reaches 14.38, it is most reasonable to maximize the benefits of passengers and bus companies.

5.3 Practical Application

In This paper prospect theory and genetic algorithm are applied in the headquarters to solve intershool public transportation scheduling problem, based on the number of the segment route hop on and off the site statistics and analysis, to improve original irregular grid random phenomenon, make the departure schedule. After investigation, the main parameters required by the algorithm are: the total number of sites is J = 4,

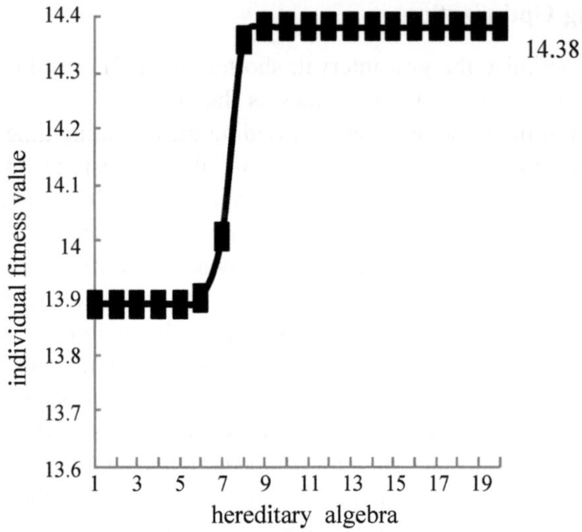

Fig. 4. Mutation rate 0.1 genetic algebra and individual fitness chart

Q = 30, the total number of times is P = 9, and the interval of departure interval [3, 13], and the value is an integer. The maximum evolutionary algebra G = 45, population Size = 20, crossover probability Pc = 0.85, variation concept Pm = 0.01, weighted coefficient of α, β respectively 1, 1, expectation waiting time $t_e = \frac{\sum_i f_i x_i}{\sum_i f_i} =$ 4.91 min. Through the above algorithm design and parameter value selection of the genetic algorithm, the simulation experiment was carried out by using the computer, and the optimization result was obtained: the time table of the departure time was shown in Table 1.

Table 1. Schedules of depaeture

Time interval	Interval	Departure interval/min	Departure flight
7:20–8:30	0	7:30	1
8:30–9:50	40	8:30 9:10 9:50	3
9:50–12:00	40	10:30 11:10 11:50	3
12:00–14:40	55	12:50 14:10 14:40	3
14:40–16:50	60	15:40	1
16:50–18:00	70	16:50	1
18:00–19:30	80	18:00 19:20	2
19:30–21:30	60	20:20	1
21:30–21:40	70	20:30	1

6 Conclusion

This paper is based on prospect theory that assumes the passenger as the "bounded rationality" combined with genetic algorithm, in view of the long waiting time of passengers, widely difference between peak and off-peak passenger flows, a multi-objective model is established, by adjusting the bus departure intervals for the traffic department. Solutions are provided to meet the maximization of benefits for both passengers and bus companies. At last take Panzhihua University as an example to analyze the status quo and provide the optimal solutions.

Acknowledgements. Fund project: 2017 cxcy049—prospect theory and genetic algorithm optimization application in bus scheduling (2017 university-level university student innovation and entrepreneurship training program project).

References

1. R. Li, A review of prospect theory (in Chinese). Soc. Sci. BBS (2), 214–222 (2014)
2. K.V. Katsikopoulos, Y. Duse-Anthony, D.L. Fisher, S.A. Duffy, Risk attitude reversals in drivers' route choice when range of travel time information is provided. Hum. Fact. **44**(3), 466 (2002)
3. L. Zhao, Research on travel decision model and ATIS simulation based on "strong view theory" (in Chinese), Beijing Jiaotong University (2007
4. S. Cui, Application of genetic algorithm in bus dispatching (in Chinese), Jilin University (2004)
5. P. Han, J. Wang, H. Feng, Z. Yao, Selection of college students' travel mode based on the prospect theory (in Chinese). J. Transp. Eng. Inf. **15**(3), 146–153 (2015)
6. Z. Yang, Study on departure time and travel path selection model based on prospect theory (in Chinese), Harbin Institute of Technology (2007)
7. Z. Yao, G. Zhao, H. Bing, Multi-population transformation genetic algorithm and its application in optimization scheduling (in Chinese). Control Theory Appl. **18**(6), 882–886 (2001)
8. Y. Xi, F. Jian, Rolling scheduling strategy based on genetic algorithm (in Chinese). Control Theory Appl. **4**, 589–594 (1997)

China GDP Prediction on Traffic and Transportation by PCA

Hui-Jun Yang[1(✉)], Yan-Xu Wang[2], Jun-Jian Xia[3], and Yan Zhang[1]

[1] Military Transportation Command Department, Army Military Transportation University, Tianjin, People's Republic of China
yanghuijun110@163.com
[2] The Fifth Team of Cadets, Army Military Transportation University, Tianjin, People's Republic of China
[3] General Teaching Department, Military Transportation University, Tianjin, People's Republic of China

Abstract. First, based on the Principal Component Analysis on a decade China' GDP, the main components in traffic and transportation contributing to the GDP are found. Secondly, the GDP is predicted by a more precision regression equation. Thirdly, some results about the GDP prediction are proposed.

Keywords: Principal component analysis · Traffic and transportation GDP prediction

GDP is influenced by many components and it is key to national development. China' GDP growth rate was about 23% in 2006 and down to 6% in 2015 [1]. So, it is meaningful to study GDP growth from the viewpoint of traffic and transportation and give some suggestions in GDP growth and prediction method.

1 To Reduce Dimensions by Principal Component Analysis

Principal Component Analysis method [2] is a statistic method to reduce indices' dimension, namely to change p, the related indices observed to m, the non-related indices, and absolutely more information of raw p indices can also be reflected.

1.1 To Make Raw Data of Principal Component Analysis Non-dimensional by Mean of Equalization

The key of Principal Component Analysis is to seek Principal Component, and its tool is covariance matrix. Because covariance matrix is easily affected by quantities and dimensions of indices, it is usual to make raw data standardization and the covariance matrix is changed to coefficient matrix. But the different information among indices is also be eliminated by the effect of quantities and dimensions [3]. To make raw data non-dimensional by equalization means, namely to obtain new sample data by dividing raw data with indices average value [4]. Then the new data series not only eliminate the

G. Q. Huang et al. (eds.), *Proceeding of the 24th International Conference on Industrial Engineering and Engineering Management 2018*,
https://doi.org/10.1007/978-981-13-3402-3_41

effect of quantities and dimensions, but also include all information of raw data, namely the covariance matrix main diagonal components of new data series is just the square of variance coefficient among indices which reasonably affect variance grade of indices, and covariance matrix non-diagonal components of new data series obviously includes the information of interrelated parts of indices.

Suppose n objects to be evaluated, p evaluation indices, the raw data $x_{ij}(i = 1, 2, \ldots, n; j = 1, 2, \ldots, p)$ reflect ith object to be evaluated to jth indices, namely raw data matrix is

$$X = \left(x_{ij}\right)_{n \times p} \tag{1}$$

to make

$$\bar{x}_j = \frac{1}{n}\sum_{i=1}^{n} x_{ij} \tag{2}$$

$$y_{ij} = \frac{x_{ij}}{\bar{x}_j} \quad i = 1, 2, \ldots, n; j = 1, 2, \ldots, p \tag{3}$$

Then y_{ij} is the new data series after non-dimensional process to raw data series by equalization method.

1.2 To Compute Components Coefficient Matrix Through Non-dimensional Process Data

Suppose the covariance matrix of new data series

$$y_{ij}(i = 1, 2, \ldots, n; j = 1, 2, \ldots, p) \tag{4}$$

Is

$$U = \left(u_{ij}\right)_{p \times p} \tag{5}$$

If

$$b_{ij} = Cov\left(\xi_i, \xi_j\right), i, j = 1, 2, \ldots, n. \tag{6}$$

is the covariance of n dimensions random variance

$$\xi = (\xi_1, \xi_2, \ldots, \xi_n) \tag{7}$$

Then

$$B = \begin{bmatrix} b_{11} & b_{12} & \cdots & b_{1n} \\ b_{21} & b_{22} & \cdots & b_{2n} \\ \cdots & \cdots & \cdots & \cdots \\ b_{n1} & b_{n2} & \cdots & b_{nn} \end{bmatrix} \tag{8}$$

is called covariance matrix of

$$\xi = (\xi_1, \xi_2, \ldots, \xi_n). \tag{9}$$

1.3 To Compute the Eigenvalue and Eigenvector of Coefficient Matrix *R*, then to Seek Orthogonal Matrix *U* and Set Principal

The related calculating steps are as follows:

We mark the coefficient matrix of non-dimensional process sample as follows:

$$R = \frac{1}{n}(X'')'X'' = \begin{bmatrix} r_{11} & r_{12} & \cdots & r_{1p} \\ r_{21} & r_{22} & \cdots & r_{2p} \\ \cdots & \cdots & \cdots & \cdots \\ r_{p1} & r_{p2} & \cdots & r_{pp} \end{bmatrix} \tag{10}$$

According to function $\left|R - \lambda_j E\right| = 0$, we may seek p eigenroots $\lambda_j (j = 1, 2, \ldots, p)$ of R and related eigenvector $U_j = \left(u_{1j}, u_{2j}, \ldots, u_{pj}\right)$, then the principal component is marked as

$$Y_j = X'' U_j \tag{11}$$

$$\text{e.g. } y_{kj} = x''_{k1} u_{1j} + x''_{k2} u_{2j} + \cdots + x''_{kp} u_{pj} (k = 1, 2, \ldots, n; j = 1, 2, \ldots, p), \tag{12}$$

then to compute the contribution rate $v_k = \frac{\lambda_k}{p}$ of variance.

1.4 To Select m Principal Enable Accumulated Contribution Rate $\sum_{j=1}^{m} \eta_j$ Exceeding 97%

To define the contribution rate of *j*th principal component as follows

$$\eta_j = \frac{\lambda_j}{\lambda_1 + \cdots + \lambda_p} (j = 1, \ldots, p) \tag{13}$$

it expresses the percentage of variance λ_j of y_j in total variance. In actual application, as accumulated contribution rata of current m principal

$$\sum_{j=1}^{m} \eta_j > 97\% \tag{14}$$

the other $(p - m)$ principal may be missed.

2 The Case Study

2.1 Let's Take China' GDP and Related Database Shown as Table 1 as Study Sample to Seek Some Related Results About GDP

Table 1. China' GDP and related components data

Year	Total population [1]	Total fixed asset investment [1]	Passenger traffic [5]	Turnover of passenger traffic [6]	Freight traffic [7]	Turnover of freight traffic [8]	GDP [1]
2006	13.14	109998.2	202.42	19197.21	203.71	88839.85	219438.5
2007	13.21	137323.9	222.78	21592.58	227.58	101418.81	270232.3
2008	13.28	172828.4	286.79	23196.70	258.59	110300.49	319515.5
2009	13.35	224598.8	297.69	24834.94	282.52	122133.31	349081.4
2010	13.41	251683.8	326.95	27894.26	324.18	141837.42	413030.3
2011	13.47	311485.1	352.63	30984.03	369.70	159323.62	489300.6
2012	13.54	374694.7	380.40	33383.09	410.04	173804.46	540367.4
2013	13.61	446294.1	212.30	27571.65	409.89	168013.80	595244.4
2014	13.68	512020.7	203.22	28647.13	416.73	181667.69	643974
2015	13.75	561999.8	194.33	30058.90	417.59	178355.90	685505.8

Annotate:

1. Table 1 was introduced from year book of China transportation & communications (2016).
2. Item unit in Table 1: population (a hundred million person); Total fixed asset investment (a hundred million yuan); Passenger traffic (a hundred million person); turnover of passenger traffic (a hundred million person-mile); freight traffic (a hundred million ton); turnover of freight traffic (a hundred million ton-mile); GDP (a hundred million yuan).

First, we discover the regression equation predicting the GDP as follows [9]:

$$\widehat{y} = -2.4682 \times 10^6 + 1.8771 \times 10^5 x_1 + 0.2510 x_2 - 3.0040 \times 10^2 x_3 + 4.7518 x_4 + 4.0288 \times 10^2 x_5 + 0.9745 x_6$$

About PCA algorithm mentioned above, we find the later four components may be missed because the accumulated contribution rate of the former two components is up to 99.3434% shown as Table 2.

Then the predicted GDP by the regression equation shown as Table 3, the actual value and predicted one of other related components shown as Table 4.

Table 2. The result of principal analysis

Principal component	Eigenroots λ_i	Contribution rate η_i (%)	Accumulated contribution rate (%)
1	4.6851	78.0852	78.0852
2	1.2755	21.2583	99.3434
3	0.0205	0.3413	99.6847
4	0.0141	0.2343	99.9190
5	0.0031	0.0511	99.9701
6	0.0018	0.0299	100.0000

Table 3. GDP data of actual value and predicted one

Year	2006	2007	2008	2009	2010	2011	1012	2013	2014	2015
Actual	219,438	270,232	319,515	349,081	413,030	489,300	540,367	595,244	643,974	685,505
Predicted	225,036	272,174	303,766	355,580	415,372	483,989	546,413	594,702	648,236	680,417
Standard deviation	5598	1942	−15,749	6499	2342	−5311	6046	−542	4262	−5088
Absolute relative deviation (%)	2.55	0.72	−4.93	1.86	0.57	−1.09	1.12	−0.09	0.66	−0.74

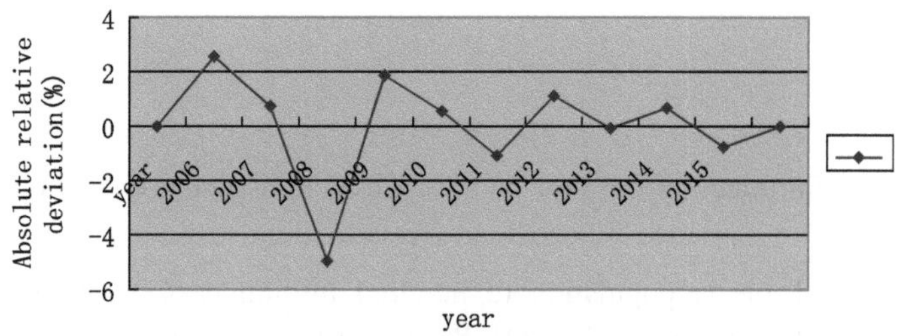

Fig. 1. Absolute relative deviation (%) of year from 2006 to 2015

2.2 Some Proposals About the Result

First, shown as Table 2, the main components affecting the GDP are as follows serially: total population, total fixed asset investment, passenger traffic, turnover of passenger traffic, freight traffic, turnover of freight traffic.

Secondly, shown as Table 2 too, the total population and total fixed asset investment are the most important components among these six components for the GDP. The other four components stem from above two most important components and have less importance. So we should pay more attention on total population and total fixed asset investment, especially the consumption of people and quality of fixed asset investment.

Thirdly, shown as Table 3 or Fig. 1, the absolute relative deviation (%) shown in Table 3 are at the range from 2.55 to −4.93, and the biggest ones is in 2006’ and 2008’.

Table 4. The actual value and predicted one of other related components

Year	Total population		Total fixed asset investment		Passenger traffic		Turnover of passenger traffic		Freight traffic		Turnover of freight traffic	
	Actual	Predicted	Actual	Predicted	Actual	Predicted	Actual	Predicted	Actual	Predicted	Actual	Predicted
2006	13.1	13.1	109,998	112,548	202.4	186.2	19,197	18,870	203	205.1	88,839	89,398
2007	13.2	13.2	137,323	136,592	222.7	243.3	21,592	21,398	227	224.7	101,418	98,781
2008	13.2	13.2	172,828	170,552	286.7	286.5	23,196	23,832	258	254.5	110,300	111,515
2009	13.3	13.3	224,598	213,027	297.6	315.4	24,834	26,060	282	290.7	122,133	126,206
2010	13.4	13.4	251,683	262,614	326.9	329.6	27,894	27,967	324	329.1	141,837	141,460
2011	13.4	13.4	311,485	317,910	352.6	328.6	30,984	29,439	369	365.6	159,323	155,886
2012	13.5	13.5	374,694	377,512	380.4	312.2	33,383	30,362	410	396.3	173,804	168,088
2013	13.6	13.6	446,294	440,017	212.3	279.9	27,571	30,622	409	417.2	168,013	176,675
2014	13.6	13.6	512,020	504,023	203.2	231.3	28,647	30,106	416	424.1	181,667	180,253
2015	13.7	13.7	561,999	568,127	194.3	166.1	30,058	28,699	417	413.1	178,355	177,429
Average absolute relative deviation (%)	0.022		2.05		10.90		4.53		1.70		1.98	

It also means that the regression equation has a better consistency with the actual GDP for the other years.

Finally, shown as Table 4, the consistency between the actual value and the expected one from high to low is as follows: total population, freight traffic, turnover of freight traffic, total fixed asset investment, turnover of passenger traffic, passenger traffic. This means the former four components and GDP meet the same rule of GDP growing approximately. The later two components have not a good enough consistency with the same rule of GDP growing.

Acknowledgements. I want to take this chance to thank to my colleagues Wu-Juan, Chun-Gan Wang, Ren-Jie, Hai-Ying Liu and Dong-Fang Huo, who gave me many academic and constructive advices, and help me to correct my paper. And also share time to discuses some interesting questions. All mentioned above are very useful for the writing of this paper.

References

1. Year book of China transportation & communications, p. 977
2. J. Zhu, Applied Multivariate Statistical Analysis (Science Press, Beijing, 2006), pp. 93–106
3. R.A. Johson, D.W. Wichem, wirte, lu xuan translate, Actual statistical multianalysis (TsingHua university publishing house, Peking, 2001)
4. S. Meng, The noticed questions in multi-indices comprehensive evaluation by principal component analysis
5. Year book of China transportation & communications, p. 986
6. Year book of China transportation & communications, p. 989
7. Year book of China transportation & communications, p. 992
8. Year book of China transportation & communications, p. 995
9. H. Yan, W. Ping, SAS Statistics Analysis and Application (China machine press), pp. 231–243

Research on Facility Location Considering Location Cost and Competition Cost

Pi-ye Yuan, Nai-xu Song(✉), and Ya-wen Xie

Department of Management Engineering, Qingdao University of Technology, Qingdao, China
snx_qwert@163.com

Abstract. This paper studies the optimal location problem of new enterprises considering the competition cost and location cost under the condition that company A already exists and occupy a favorable market position. Assume the demand points are also the facility candidate points, three optional level construction scale for each facility, and facility attraction is proportional to the infrastructure construction scale. Gravity model will be used to distribute market share. The 0–1 nonlinear integer programming model is set up with the goal of maximizing profit, and the genetic algorithm is designed to solve the problem. The results show that considering the competition cost that may happen at the beginning of site selection is very necessary, which can achieve the purpose of reducing revenue and increasing income. A large number of numerical analyses show that the algorithm is fast and accurate.

Keywords: Competitive location · Genetic algorithm · Gravity model Nonlinear programming

1 Introduction

Competition is everywhere. There is also a competition problem in location problem, that is, two or more enterprises maximize the market share through location decision under the competitive environment. The competitive location began with an article published by Hotelling (1929), and became a hot topic in the field of location research after the 80s of last century [1].

There are many researches on competitive location in China and abroad. Zhang and Rushton (2008) studied the location problem of competing facilities, regarded the customers' waiting time as constraint conditions, resumed the distribution of demand is proportional to the capacity of facilities, and is inversely proportional to the distance, after the adjusting of the location and capacity, make the space of customer utility maximum, the facility capacity is discrete, related to the service desk number [2]. Dasci and Laporte (2004) optimizes the individual design attributes, but it is still not a random type of location model [3]. Aboolian et al. (2007) proposed a competitive location design problem in discrete cases [4]. At the same time, we optimized the location and design with limited facility, and put forward three different methods to solve it. The current competitive location research mostly adopts the spatial interaction model. On the basis of considering the facilities distance, it adds the attraction factors

G. Q. Huang et al. (eds.), *Proceeding of the 24th International Conference on Industrial Engineering and Engineering Management 2018*,
https://doi.org/10.1007/978-981-13-3402-3_42

of the facilities, and distributes the demand in the form of probability, which makes the model more perfect. Zhu (2017) studied the location problem of a logistics company under the competitive environment [5]. Based on the gravity utility model, a multi facility competition location model was established and solved by LINGO software. Yang and Zhou (2011) analyzed the main and subordinate Stackelberg game problems of new enterprises and existing enterprises, established a facility competitive location model, and put forward a solving strategy combining genetic algorithm with QPADM algorithm [6]. Zhang and Tang (2015) studied the location problem with the goal of maximizing market share in the competitive environment, and also considered the chance constraint [7]. Finally, the optimal solution was obtained through the emergent search algorithm. Sheng and Wang (2014) established the Stackelberg game location model of two companies under competition conditions [8]. The optimal competitive location and distribution plan considering carbon emission cost was studied, and the conclusion was drawn that the total market profit under competition condition is smaller than monopoly. Cheng et al. (2014) studied the best policy problem of two service providers, in which they open facilities one by one, and established a bilevel programming model [9]. Gao (2017) considered the consumer's choice behavior in the competitive location model, that is, consumers have more choices, which make the model more practical [10].

The above literatures consider the competitive location problem under different circumstances, and the research method is worth learning. But in reality, when the facilities are different from each other, there will be direct competition, resulting in cost expenditure (i.e. competition cost), such as supermarket chains with different brands in the same area. Based on previous studies, this paper further studies the competitive location problem considering the location cost and competition cost of facilities. Suppose there are two similar product providers A and B, and P facilities have been set up by A (Leader) to occupy a larger market share in a region, while B (Follower) have no facilities but will set up Q facilities in a same place. How should B build facilities to get the maximum profit? In view of the above background, the following assumptions are made:

(1) There is a positive correlation between the attractiveness of facilities and the investment scale of the facilities.
(2) The market share of the follower B is 0 before entering the market.
(3) There is a straight line between every two demand points.
(4) The same demand can accommodate two facilities at the same time.
(5) Only one facility can be set up by the same company at the same demand point.
(6) Consumer demand is fully met.

2 Modeling

C_{kA} is the scale of investment of facility k of company A; d_{ij} is the distance between the demand point i and j, and d_{ik} is the distance between demand point i and k; D_i is the demand for demand i; c_{ij} is the unit freight for the demand point j to the demand point i; N is arbitrary for the set of demand points, any $i \in N$; S is the set of all candidate facility points any $j \in S$; E is the selected facility point; under this hypothesis, it is also the facility point of company A (Leader), K = 1, 2, any $k \in E$ Decision variable is

$$X_j = \begin{cases} 1, \text{Setting up facilities at the demand point j} \\ 0, Other \end{cases}$$

2.1 Utility Model

According to the gravitational model proposed by Huff [11, 12], the utility function of the demand point j to point i is

$$\mu_{ij} = \frac{F(C_j)}{H(d_{ij})} \tag{1}$$

μ_{ij} indicates that the utility value of the requirement point j to the demand point j; $F(C_j)$ is a function of C_j, indicates the attraction of the facility; $H(d_{ij})$ is the function of the distance d_{ij} between the demand point i and point j. Formula (1) is the general form of utility function. According to the extension of Nakanishi [13] to the utility function, we can refer to the expression of the utility function as follow,

$$\mu_{ij} = \frac{C_j^x}{(d_{ij} + \varepsilon)^\alpha} \tag{2}$$

There are three scale levels, 1000, 2000 and 3000, x = 1, 2, 3, indicating the scale of investment. The attractiveness C_j^x of facility j is known positively related to the scale of the facility investment, and it could be used to represent the scale of the investment of the facility j. α is a parameter related to the position of the competitor. ε is an arbitrary number greater than 0, which ensures that the denominator is not 0 at $d_{ij} = 0$. Shen [14] proves that the effect of ε is minimal and can be ignored. Then the probability of demand point i receive service of point j is

$$P_{ij} = \frac{\mu_{ij} X_j}{\sum_{j \in S} \mu_{ij} X_j + \sum_{k \in E} \mu_{ik}} \tag{3}$$

2.2 Model of Competition Cost

We also need to consider that A and B are competitor. Therefore, competition cost is exit, that is, the extra expenses incurred by A and B in order to promote products. When a homogeneous product is in a homogeneous market, if one party increases sales expenditure while the other does not carry out promotion, it will change the share of market. Therefore, according to the rational person hypothesis, when one party conducts the promotion, the other party also promotes the promotion. Assume that the two companies are paying for the competition at the same time, and the market share of the two companies will not change.

There is a strong spatial correlation in the occurrence of competition in reality, the adjacent facilities that belong to different companies will compete against each other. Therefore, the competition between the facility j and the facility k is a function of the reduction of distance d_{jk}, and the competition cost of the facility j can be described as,

$$q_j = \Phi \partial_{jk} e^{-\beta d_{jk}} \tag{4}$$

Φ is competition cost factor, which is related to the type and degree of competition. ∂_{jk} shows the relative relationship between the cost of the two competition facilities, $\partial_{jk} = \frac{\lambda C_k^x}{C_j^x}$. When $\partial_{jk} \geq 1$, it shows that facility k was more competitive than the facility j; and when $0 < \partial_{jk} < 1$, the market share of the facility j could be saved by only spending less competition cost. λ is competition parameter, and its value depends on the frequency and intensity of the competition activities. The general value $0 < \lambda < 1$ indicates that the competition cost between the two competing facilities is not sustained. β is the sensitivity coefficient of distance.

2.3 Objective Function Model

The cost of facility setting can be divided into two parts. One is a, which indicates the fixed cost of facility location, the other is the variable cost related to facility attraction, which is indicated by C_j^x The total cost of new facilities at j is $f_j^x = a + C_j^x$.

According to the above analysis, the objective function model as follows,

$$\begin{aligned} \max P = & \sum_{i=1}^{n} \sum_{j=1}^{s} D_i p_{ij} m - \sum_{j=1}^{s} f_j^x X_j - \sum_{i=1}^{n} \sum_{j=1}^{s} D_i p_{ij} c_{ij} d_{ij} \\ & - \sum_{j=1}^{s} \sum_{k=1}^{e} q_j X_j \end{aligned} \tag{5}$$

S.t.

$$\sum_{i=1}^{n} \sum_{k=1}^{e} D_i p_{ik} + \sum_{i=1}^{n} \sum_{j=1}^{s} D_i p_{ij} = D_i \tag{6}$$

$$m > c_{ij} d_{ij} \tag{7}$$

$$\mu_{ij} = \frac{C_j}{(d_{ij} + \varepsilon)^{\alpha}} \tag{8}$$

$$P_{ij} = \frac{\mu_{ij} X_j}{\sum_{j=1}^{s} \mu_{ij} X_j + \sum_{k=1}^{e} \mu_{ik}} \tag{9}$$

$$f_j = a + C_j \tag{10}$$

$$\partial_{jk} = \frac{C_k}{C_j} \tag{11}$$

$$q_j = \Phi \partial_{jk} e^{-\beta d_{jk}} \tag{12}$$

$$X_j \in \{0, 1\} \tag{13}$$

The formula (5) indicates that B has the largest profit; the (6) indicates that the demand for demand point i is all satisfied; (7) indicates that the profit per unit product is greater than the unit transportation cost; the formula (8)–(12) has been described in the previous text; the (13) is the 0–1 decision variable.

3 Algorithm Design

The problem described in this paper has been proved by Mirchandani [15] to be NP-hard problem. This kind of problem cannot get the best solution in acceptable time when the solution is large, so we consider it through heuristic algorithm. The genetic algorithm has accumulated mature application experience in solving such problems, and shows good robustness and convergence, so it is our choice. The concrete steps are as follows:

Step 1 Generate each demand point and the requirements of each demand point randomly, and select the points within each set.
Step 2 Calculates the distance between every two demand point.

Table 1. Basic information of demand points

Serial number	Coordinate	Quantity demand	Fixed investment
1	64, 98	800	8000
2	81, 75	720	7200
3	52, 54	670	6700
4	39, 42	679	6790
5	43, 70	680	6800
6	24, 67	400	4000
7	80, 16	300	3000
8	45, 10	300	3000
9	43, 37	670	6700
10	78, 43	600	6000
11	20, 48	500	5000
12	9, 33	300	3000
13	17, 95	350	3500
14	39, 92	680	6800
15	83, 5	400	4000

(continued)

Table 1. *(continued)*

Serial number	Coordinate	Quantity demand	Fixed investment
16	80, 73	400	4000
17	60, 26	500	5000
18	65, 41	630	6300
19	52, 79	700	7000
20	62, 60	700	7000
21	29, 30	300	3000
22	91, 51	300	3000
23	74, 11	300	3000
24	47, 57	650	6500
25	23, 20	600	6000

Step 3 Select q points in sequence from the set S as the new construction application point of the company B, and brings into the model calculation to find the location combination of the maximum profit.

Step 4 Stop the condition to reach the maximum number of iterations (Fig. 1 and Table 1).

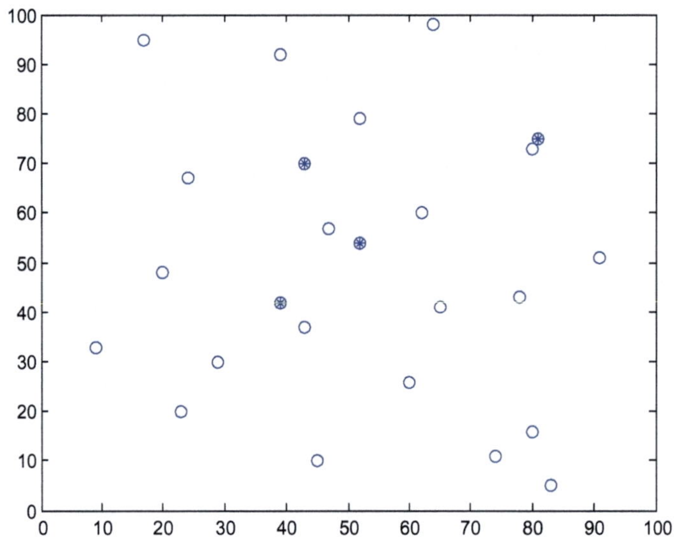

Fig. 1. Schematic map of facility location

Table 2. Sensitivity analysis of competition cost factor

Competition cost factor	Result	Construction scale	Total profit/yuan	Average distance
0	1, 2, 4, 5	3, 3, 3, 3	41905.07	61.31
5000	1, 2, 3, 9	3, 3, 3, 3	40923.66	64.22
10,000	2, 5, 18, 24	3, 3, 3, 3	37404.53	69.84
15,000	2, 3, 4, 5	3, 3, 3, 3	37264.10	74.52
20,000	2, 3, 9	3, 3, 3	35322.55	76.09
50,000	2, 3, 4	3, 3, 3	27168.78	76.91

4 Numberical Analysis

Set $\varepsilon = 1$, $\alpha = 1$, sensitivity coefficient of distance $\beta = 0.002$, competition coefficient $\lambda = 0.\ 05$, competition cost factor $\Phi = 15{,}000$, unit product profit m = 100 yuan, Unit product unit distance transportation freight c = 1 yuan. The population size is popsize = 100, the maximum iteration number is maxgen = 2000, the cross coefficient is pc = 0.9, and the coefficient of variation is pm = 0.2.

Known Company A (Leader) have facility 1 and 2 in the market, that is, p = 2, and the construction scale is 3. Company B has not been built and can be selected at all the demand points, because of the limitation of funds and other reasons, its largest location number is 4, namely q = 4. By using MATLAB, the best location of company B is 2, 3, 4, 5, and the best construction scale is 3, 3, 3, 3, and the total profit is 37264.10 yuan.

In order to know the impact of competition cost factor on the site selection results, $\Phi = 0$, 5000, 10,000, 15,000, 20,000 and 50,000 are taken, and the results are shown in Table 2.

(1) With the increase of the number of competition cost factor, the total profit of Company B decreases and the average distance increases. Because the competition cost factor is bigger, B requires the expenditure of the competition cost is higher, so the total profit is low, when the facilities of B continue to attempt to close to the facilities of A to obtain greater market share, competition costs grew faster than the growth rate of profits. Therefore company B tends to choose facilities that far from company A to set new facilities, so the average distance increases.
(2) B has chosen the largest scale of construction, which is conducive to reducing the cost of competition.
(3) When the competition cost factor is over 20,000, the number of facilities for the Company B to obtain the best profit is reduced. Due to the utility function described in this paper, every one more facility will make other facilities market share reduced. There are multiple thresholds in the competition cost factor under the discrete location conditions of this paper, which will make the marginal cost of a new facility added to B is greater than the marginal revenue, so as to increase competition cost factor the optimal location number of B is decreasing. Get the

result at $\Phi = 0$, and then reconsider competition cost, take $\Phi = 15{,}000$ in the result and the total profit is 36257.07 yuan, which is 1007.03 yuan less than the result that at the beginning of location to consider the competition cost. That means the former is better.

5 Conclusion

Competition cost and location cost are considered in the traditional competitive location problem, and the effect of the two costs on the optimal location results is analyzed. In this paper, we use gravity model to allocate market share for every facility point, and establish a mathematical model based on the profit maximization of Company B. The final example analysis proved that considering the competition cost at the beginning of facility location is very necessary, which can achieve the purpose of reducing revenue and increasing income, and has a reference for realistic decision-making.

References

1. H. Hotelling, Stability in competition. Econ. J. **39**(153), 41–57 (1929)
2. L. Zhang, G. Rushton, Optimizing the size and locations of facilities in competitive multi-site service systems. Comput. Oper. Res. **35**(2), 327–338 (2008)
3. A. Dasci, G. Laporte, Location and pricing decisions of a multi store monopoly in a spatial market. J. Reg. Sci. **44**(3), 489–515 (2004)
4. R. Abooliana, O. Bermanb, D. Krassb, Competitive facility location and design problem. Eur. J. Oper. Res. **182**(1), 40–62 (2007)
5. P.F. Zhu, Research on facility location of logistics company a under the competitive environment (in Chinese), Donghua University (2017)
6. Y.X. Yang, G.G. Zhou, Study on location model of facility competition for closed loop supply chain network (in Chinese). Chin. J. Manag. Sci. **19**(5), 50–57 (2011)
7. L.X. Zhang, H. Tang, Research on two objective competitive location based on opportunity constraint (in Chinese). Chin. Market **4**, 39–42 (2015)
8. L.L. Sheng, C.X. Wang, Competitive location of logistics distribution centers considering carbon emission (in Chinese). J. Shanghai Marit. Univ. **35**(3), 51–56 (2014)
9. C. Cheng, Y. Zhang, Z.J. Xue, M.Y. Qi, A single-level mixed-integer programming model for the competitive facility location problem (in Chinese). Ind. Eng. J. **20**(5), 21–27 (2014)
10. Y. Gao, Research on competitive location selection with customers choice behavior (in Chinese), Shenzhen University (2017)
11. D.L. Huff, Defining and estimating a trade area. J. Market. **28**(3), 34 (1964)
12. D.L. Huff, A programmed solution for approximating an optimum retail location. Land Econ. **42**(3), 293–303 (1966)
13. M. Nakanishi, L.G. Cooper, Parameter estimation for a multiplicative competitive interaction model: least squares approach. J. Mark. Res. **11**(3), 303–311 (1974)
14. R.L. Shen, J.Z. Huo, Flow-interception facility location problem under competitive conditions (in Chinese). Oper. Res. Manag. Sci. **3**, 97–101 (2013)
15. P.B. Mirchandani, R. Rebello, A. Agnetis, The inspection station location problem in hazardous material transportation: some heuristics and bounds. INFOR Inf. Syst. Oper. Res. **33**, 100–113 (1995)

A Study on Construction Project Cost Management Based on Neural Network

Jin-song Tong(✉) and Xing Bi

School of Management, Tianjin University, Tianjin 300072, China
841184923@qq.com

Abstract. Project cost management is vital to the survival and development of construction enterprises. This paper reveals the risk factors of the project cost by discussing the impact of the uncertain behavior of proprietors, construction units, designers and supervisors on the project cost management from the perspective of project management. Then, this paper establishes a project cost risk analysis model, through the use of artificial neural network algorithm to find the sensitive factors that increase the total project cost risk, so as to provide reference for project managers to prevent risks.

Keywords: Analysis · Cost management · Construction project Neural network

1 Introduction

With the rapid development of China's construction industry, its profit margins are severely squeezed due to the low-price bidding for projects, the sharply fluctuated price of raw materials, the lowered entry barrier of industries and the rising exiting cost, and the large influx of overseas builders. How to effectively control the cost of construction projects has become the focus of construction companies. A survey shows that the extensive management mode of construction enterprises in our country still has not been fundamentally changed. The operation flow and basic management of enterprises are still relatively weak. The project management level is relatively low. The project cost can be further reduced through improvement in management philosophy, means and targets [1]. Therefore, it is very important for the construction enterprises to objectively analyze the factors that cause the project cost risk and forewarn and prevent them.

Some foreign researchers have studied on the rising costs and overruns of construction projects. Most of them survey the proprietors, contractors and supervisors to analyze the reasons the construction costs exceeded the estimated costs during construction. For example, Arditi and his colleagues conducted a large-scale survey of public utilities and contractors in Turkey and found that the major reason for overruns in public construction costs between 1970 and 1980 was inflation, resulting in rising raw material prices and labor costs, delays in construction and budget errors [2]. Flyvbjerg and his colleagues found that to a large extent the cost increase depends on the project duration through statistical analysis of 258 railway, bridge, tunnel and highway projects, a total cost of 90 billion US dollars [3].

G. Q. Huang et al. (eds.), *Proceeding of the 24th International Conference on Industrial Engineering and Engineering Management 2018*,
https://doi.org/10.1007/978-981-13-3402-3_43

There are relatively fewer studies on the influencing factors of project cost in China, most of which are normative discussion. Guo Jiqiu and her colleagues systematically introduced the basic principles and methods of project cost management and their application in engineering practice [4]. Wang Li analyzed the common problems in the field of engineering cost control in China, and put forward corresponding measures for the related problems [5]. Only a tiny minority of researchers employed the method of questionnaires. Yet their questionnaires are not comprehensive enough, which are limited to the discussion of individual factors. For example, Yang Deqin conducted in-depth research on the humanistic background and cost risk of the project through questionnaires [6]. Mao Hongtao and his colleagues identified the impact of organizational environment of enterprises on project cost management [7]. In a word, the research on the factors affecting the cost management of engineering projects at home and abroad has not yet formed a relatively perfect theoretical framework.

2 Theoretical Basis

2.1 Concept of Project Cost Management

Cost management is the core part of the construction phase. Compared with the cost of traditional manufacturing products, project cost include more economic content, the composition is more complicated [8]. It is not only the cost monitoring and collecting of large amounts of data in the whole process of project construction, but more importantly, the correct analysis of all kinds of cost data and timely measures to control the final cost of the project within the target range [9].

2.2 BP Neural Network

BP neural network [10] was proposed by Rumelhant and McClelland in 1986 and is a typical multi-layer feed forward neural network. BP neural network algorithm is mainly composed of two processes: signal forward propagation and error back propagation. In the forward propagation, the input samples are passed from the input layer to the output layer after processing layer by layer through the hidden layer. If the actual output of the output layer does not match the expected output, error back propagation starts. In the back propagation, the output error is back to the input layer one by one through the hidden layer, and the units adjust the weights and thresholds according to the error signal so that the error between the network output and the expected output gradually decreases until the requirements are met.

2.3 Factors Affecting the Cost of Projects

This paper argues that multiple parties affect the project cost risk in the construction process of the project. Such being the case, the paper therefore analyses the four aspects of the proprietor, the construction units, designer and supervisor, who are the stakeholders of cost of the project, to find out the risk factors of project cost.

1. The proprietor
   The proprietor is the investor of the project, and he/she is to manage and coordinate the all the participants of the project. Therefore, the proprietor's behavior will affect the total cost of the project. Before the construction, if the proprietor fails to get the corresponding approval documents after the construction contractor has made the construction preparation; it will leave the workers and equipments unused, resulting in a cost risk. After the construction starts, in order to reduce operational risk, the proprietors usually hope that the project will be put into operation as soon as possible. To this end, the construction contractor may sacrifice the quality of the project to meet the proprietor's requirement. As a result, the quality cost will increase, and the cost risk emerged [11].
2. The contractor
   Project management competence. The complex internal and external environment of project construction often brings emergencies, which requires managers not only have solid theoretical knowledge, but also rich practical experience [12]. Some projects lose money at first, but by establishing a good relationship with the proprietors, a part of the compensation can be brought back, and it can also be turned into profit. Some construction contractors have extensive internal management and lack of collaboration between relevant departments, which can not produce effective value-added operations, and often cause delay in time and unqualified projects, and form cost risk. Moreover, improper construction method, uncertain construction period and unreasonable layout of construction organization will affect the flow process between construction processes, resulting in the rise of hidden costs and eventually the total cost of construction projects [13].
   Raw material. Material is the basis of engineering construction. Material quality is the basic guarantee for engineering quality. Only when the material quality meets the requirements, can the engineering quality meet the engineering standard [14]. In the ordering stage, the purchase mode and transportation method directly affect the material cost. In the storage phase, the factors such as the setting of the reservoir, the position of the stacking and the control of the stock will also cause the change of the material cost. For example, if the custodial rules are not strictly followed, material has been damaged or lost before being put into production, making the material cost difficult to measure, which results in cost risk and even serious waste. Finally, the total cost of the project is increased and the risk of project cost is formed.
3. The designer
   Engineering drawings. The influence of the designer on the cost of the project can never be ignored. The engineering drawings are mainly made by the designer. After the drawings are completed, he/she has controlled over 80% of the cost of the project, so the drawings are very critical [15]. At present, the domestic construction enterprises directly communicate with the designers, and the designers communicate with the proprietors. Usually, if the design is too perfect, it will cost much more to complete the project using better materials, more professional personnel and a longer period of time. If the design is too rough, the project cost may be reduced.

However, if the low quality affects the normal use of products, failure often occurs, or need to be rebuilt, and then the total cost of project actually increased.
Design alteration. A design alteration often occurs during the process of construction. If you find unreasonable construction drawings during the construction process, you need to communicate with the owner to alter the design. After a design alteration occurs, materials and mechanical equipment will need to be re-purchased, which may result in additional material costs and mechanical usage fees, which will increase the cost of the project and create cost risks.

4. The supervisor
   The national quality standard is the basic requirement of the quality of the project, and a hard one. The main responsibility of the supervisor is to supervise the whole process of the project implementation, so as to ensure the successful completion of the project and its transfer. If the supervisor is poor in responsibility, and does not strictly control the quality of the project, it will increase the cost of the project. For example, sometimes the supervisors need to supervise on spot to ensure the validity and safety of the construction process. In that time, negligence and absence without leave may lead to low quality of the project, even unnecessary casualties. Thus, the direct costs of the project, additional construction fees and management fees increased significantly, form the cost risk.

3 Research Design

3.1 Model Design

The construction project cost risk analysis model reflects the influence factors of project cost risk more intuitively. There is a three-layer logical structure in project cost risk and fault tree, yet a neural network containing one hidden layer is more than enough to represent any continuous function with arbitrary accuracy. Therefore, this paper takes the intermediate events in the fault tree as the hidden layer, and the basic events and the top events are the input nodes and the output nodes respectively.

3.2 The Learning Process of the Neural Network

The neural network model established in this paper is $6 \times 3 \times 1$ neural network. The input vector is X; the hidden layer input vector is YR; the hidden layer output vector is YC; the output layer input vector is SR, the output layer vector SC, the desired output vector is DC; the weighted value between input layer and the hidden layer is Wih; the weighted value between hidden layer and output layer is Who; the threshold of the nodal points in the hidden layer is fh; the threshold of the nodal points in the output layer is fo. The number of samples: $k = 1 - n$; $h = 1 - 4$.

(1) Network initialization. Each connection weight is assigned randomly a number within $(-1, 1)$. The error function is set as e, the calculation accuracy as ξ and the maximum number as M.
(2) To randomly select K samples and corresponding expected output DC(K).

(3) To calculate the input and output of each node in the hidden layer (Fig. 1).

$$YR_h(k) = \sum_{i=1}^{6} W_{ih}X_i(K) - f_h$$

$$YC_h(k) = f(YR_h(k))$$

$$SR(k) = \sum_{h=1}^{4} W_{ho}YC_h(k) - f_o$$

$$SC(k) = f(SR(\mathrm{k}))$$

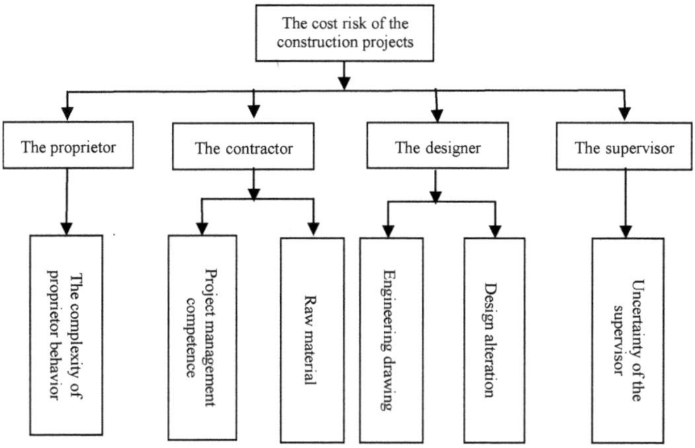

Fig. 1. Cost risk analysis model of construction project

(4) To use the actual output and expected output, to calculate the partial derivative δo (k) of the nodes of the output layer and the partial derivative δh(k) to the nodes of the hidden layer.

$$-\delta o(\mathrm{k}) \triangleq -(DC(\mathrm{k}) - SC(k))f'SC(k)$$

$$-\delta_{\mathrm{h}}(\mathrm{k}) \triangleq -(\delta o(\mathrm{k})W_{\mathrm{ho}})f'(YR_h(\mathrm{k}))$$

(5) To modify the connection weights $W_{ih}(k)$ and $W_{ho}(k)$, and to calculate the global error.

$$E = \frac{1}{2n}\sum_{k=1}^{n}(DC(k) - SC(k))^2$$

To repeat steps (3) until the calculation meets the schedule requirements. After that, the learning process completed (Table 1).

3.3 Case Study

First of all, experts evaluate the cost risk factors of the project and score them. Risk assessment indicators such as the index system established in the fault tree, the results of the expert evaluation have been collated.

Table 1. Project cost risk factor rating scale

The complexity of proprietors' behavior	Project management competence	Raw material	Engineering drawings	Design alteration	Uncertainty of supervisors
0.4689	0.6314	0.5330	0.4983	0.5412	0.5274

These data are input into the above neural network models as initial data, and computer programming aided calculation is done by MATLAB. After calculation, the total cost risk is 0.5492. Sensitivity tests were conducted for all the influencing factors: when the evaluation values of all risk factors changed +10% and −10% respectively, we apply neural network model to calculate the total cost risk comprehensive evaluation value after change.

From Table 2, it can be seen that the impact of each risk factor on the comprehensive evaluation results of the total cost risk of the project is different. The project management capability, raw materials and design alteration have the greatest impact on the comprehensive evaluation of project cost risk, and they are relatively sensitive factors in total cost risk of construction project. Therefore, we should set up the concept of precision and science, enhance professional and scientific management level, and pay attention to the direct cost management of raw materials. The designer should make a thorough investigation of the project, carry out comprehensive research on those possible problems, strengthen communication with proprietors, minimize the cost of engineering changes, and reduce the cost risk.

Table 2. Total cost risk sensitivity analysis

Participants	Factors	Rate of change (%)	Assessed value after change	Comprehensive evaluation value deviation	Sensitivity
Proprietors	The complexity of the proprietors' behavior	10	0.5158	0.5691	0.0199
		−10	0.4220	0.5083	−0.0109
Contractors	Management competence	10	0.6175	0.5804	0.0312
		−10	0.5053	0.5459	−0.0333
	Raw material	10	0.59741	0.5541	0.0049
		−10	0.48879	0.5302	−0.0190
Designers	Engineering drawing	10	0.54813	0.5541	0.0049
		−10	0.44847	0.5182	−0.0190
	Design alteration	10	0.58432	0.5814	0.0322
		−10	0.47808	0.5291	−0.0301
Supervisors	Supervision	10	0.58014	0.5741	0.0249
		−10	0.47466	0.5231	−0.0261

4 Conclusion

The paper found out the risk factors affecting the project cost through the analysis of the improper behavior of the project related parties. The paper also calculated the total cost risk evaluation value employing artificial neural network. Finally the paper got the sensitive factors of total cost risk through adjusting the various risk evaluation index, analyzing the sensitivity of the total cost of risk, providing reference for the project managers to prevent risks.

References

1. T. Xu, Application of strategy-based target cost management method in international project contracting projects, Southwestern University of Finance and Economics, pp. 6–7 (2003)
2. T. Arditi, G. Akan, Cost overruns in public projects. Proj. Manag. **3**(4), 218–224 (1985)
3. M.S. Flyvbjerg, H.S. Buhl, What causes cost overrun in transport infrastructure projects. Transp. Rev. **24**(1), 3–18 (2004)
4. J. Guo, Cost management of engineering project (China Electric Power Press, 2011)
5. L. Wang, Research on the cost control method of construction project, Shangdong University (2013)
6. Y. Deqin, An empirical study on human risk factors of engineering cost. Constr. Econ. **2**, 45–48 (2006)
7. H. Mao, B. Zhu, T. Wang, The effect of organizational environment on engineering project cost management. Nankai Bus. Rev. (1), 102–112 (2012)
8. X. Zheng, Study on cost control of construction projects. Res. Financ. Econ. Issues (11), 117–120 (2014)
9. Q. Wang, W. Xu, A survey of quality cost management in construction engineering in China. Ind. Eng. Manag. (1), 120–124 (2006)
10. B. Ou, Economic forecasting method based on back propagation neural network. J. Nanjing Inst. Technol. (2), 11–14 (2004)
11. Yu. Zhang Huanjie, H.X. Xiaozhong, Analysis on the tripartite game of owner and supervision cahoots program. Shanxi Archit. **3**, 23–24 (2008)
12. W. Xuejun, W. Xuefen, On construction project management and cost control. Value Eng. **13**, 63–65 (2014)
13. Z. Hongliang, Yu. Dong, Analysis of internal environment of construction enterprises. Constr. Econ. **11**, 24–27 (1997)
14. Z. Hongmou, Six elements of material cost management for construction projects. Constr. Enterp. Manag. **11**, 30–31 (2000)
15. T. Xuekui, Cost control at engineering design stage. Constr. Econ. **212**, 45–46 (2000)

Weighted Self-regulation Complex Network-Based Modeling and Key Nodes Identification of Multistage Assembling Process

Peng Zhu and Jian-bo Yu(✉)

School of Mechanical Engineering, Tongji University, Shanghai, China
jbyu@tongji.edu.cn

Abstract. This paper proposed a weighted self-regulation variation propagation network (WSRVPN) modeling and key nodes identification method based on the complex network for multistage assembly process. Firstly, a self-regulation weighted variation transmission network is constructed through using actual machining error, quality characteristic information and assembly process requirements. Then, the weighted LeaderRank sorting algorithm is introduced to rank the importance of nodes in the network and find the key nodes. To ensure the final assembly's quality by controlling the quality of critical nodes. The multistage assembling process of a bevel gear assembly is studied, which proves that the method can effectively model the complicated assembly deviation flow and identify the key weak points.

Keywords: Complex network · Key assembling features · Key nodes identification · Self-regulation

1 Introduction

There are various errors in the multistage assembly process, including the manufacturing error of parts and positioning errors of the assembly, which would lead to the quality deviation of the end product. Thus, it is important to model the assembly process to control the critical node so as to ensure the stability of product quality. Since the stream of variation(SOV) theory [1] has been proposed, this method is mainly used to solve the quality problem of body assembly. Then it is widely used in the error modeling, analysis of manufacturing process [2–4] and assembly process [5–7]. The state space model based on the SOV is also widely used in the modeling of multistage assembly process [8, 9]. Peng [10] reviewed the fault detection and diagnosis technology related to complex industrial process, combed the basic idea and improvement process of typical method in relevant quality inspection.

Since the complex network model [11] can abstract the feature surfaces and corresponding quality features of the actual multistage assembly process into network nodes [12]. Through the correlation relationship between nodes and the sequence of assembly process to describe the error propagation process. The topology and structure

G. Q. Huang et al. (eds.), *Proceeding of the 24th International Conference on Industrial Engineering and Engineering Management 2018*,
https://doi.org/10.1007/978-981-13-3402-3_44

characteristics of the assembly network also can be described clearly and effectively. Therefore, establishing the weighted variation propagation network model for the complex assembly process, and using the weighted LeaderRank sorting algorithm to identify the key nodes which would lead to serious quality problems. Thus, we could ensure the quality of the end product by controlling key nodes.

2 Methodology

The errors in the multistage assembly process are generally divided into two categories, namely the errors caused by the manufacturing deviation of parts and the assembly errors caused by the reference or positioning deviation in the assembling process. Each interface interaction in the process of assembly, the surface and the surface are each other's benchmark for positioning. Thus, in order to describe the relationship of the error propagation accurately, we need to construct variation propagation relation network. The nodes in the network are the feature nodes which abstracted out of the assembly process, the connection relationships are determined by the assembly sequence. Then, empowered the variation propagation relation network to construct the WSRVPN. The weighted LeaderRank sorting algorithm is used to identify the key nodes in the network. Figure 1 is the overall procedure of the proposed method to model the WSRVPN and identify the key nodes.

2.1 Constructed the WSRVPN Model

The key feature nodes (e.g., the quality features of the critical contact surfaces.) of assembly can be extracted from the component assembly information. Combined with the corresponding assembly sequence to determine the connection relationship among the feature nodes. There are three connection relations in the network, like direct contact with the surface, the relative distance between the surfaces, and the positioning based on the surface. Among them, the connection relationship between surfaces is caused by the machining error on the one surface, and this error would propagate to the next connected surface. Thus, lead to the assembly error of end product. The distance error between the surfaces would affect the performance of the component so as to affect the quality of component. The assembly error caused by the surface positioning deviation can also cause the final performance defect of the end component. The connection is all connected from former surface to the back surface.

The variation propagation relation network can be constructed though connecting the corresponding connection relationship. The WSRVPN can be expressed as $G = \{<V_i, V_j>, \omega_{ij}\}$, where V_i, V_j represent the nodes in the network, ω_{ij} represents the initial weight among the two nodes, if there is a connection between node V_i and node V_j, it's equal to 1, otherwise 0.

After relation network is constructed, we need to give weight to the influence strength among the nodes. Thus, it is necessary to model the error propagation strength between each node. Corresponding to the three connecting relationship, the three kinds of calculation methods of error weight are as follows:

(1) Direct contact with the surface

The interaction between the contact surfaces can be refined to the quality features which affect the surface like size, shape and roughness. The size error, shape error and roughness error are represented by the ε_S, ε_L, ε_R respectively. The respective expression are as follows:

$$\varepsilon_S = \frac{|e_i - e_0|}{T} W_{1j} \tag{1}$$

where e_i represents the actual measurement size, e_0 represents the rate size, T represents the tolerances for the permissible variation of this feature size. W_{1j} indicates whether the size error of node i affects the node j, if true takes 1, otherwise 0, similarly hereinafter.

$$\varepsilon_L = \frac{|t_i - T_0|}{T_0} W_{2j} \tag{2}$$

where t_i represents the actual index of shape and location, T_0 represents the manufacturing requirement of shape and location. Surface features describe the roughness and heat treatment characteristics of machining surface. This paper takes the main factor, namely roughness error as the surface characteristic error.

$$\varepsilon_R = \frac{|r_i - R_0|}{R_0} W_{3j} \tag{3}$$

where r_i indicates the machined surface roughness measured actually, R_0 represents the machining requirement of the roughness. The final weighted factor between the machining features is:

$$\omega_q = \varepsilon_S + \varepsilon_L + \varepsilon_R \tag{4}$$

(2) The distance between the nodes

$$\omega_d = \frac{|d - D|}{T} W_{ij} \tag{5}$$

where d represents the actual distance between two surfaces, D represents the requirement distance between the nodes, T is the tolerance of distance between two surfaces. W_{ij} indicates whether there is a distance requirement between node i and node j, which equal to 1 when the requirement exits, otherwise 0, similarly hereinafter.

(3) positioning and installation based on the surface.

$$\omega_p = \frac{\Delta}{T} W_{ij} \tag{6}$$

where △ denotes the deviation of the actual measured position distant away from the requirement position. T indicates the variation tolerance of the actual position allowable distance from the required position.

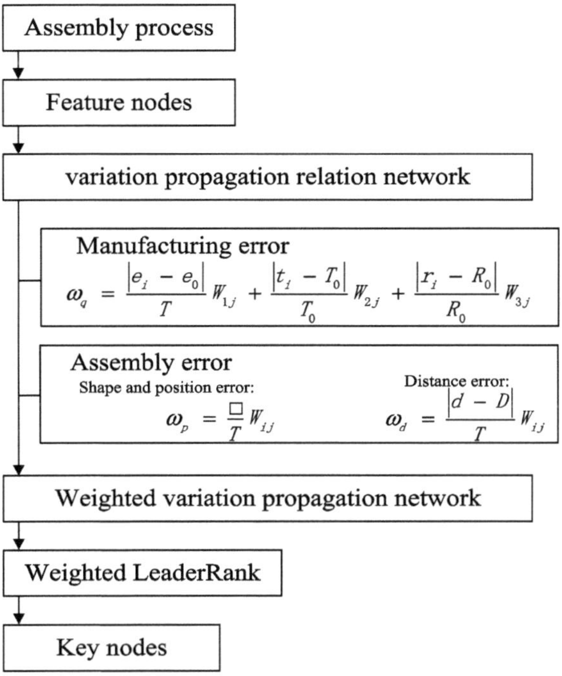

Fig. 1. The overall procedure of the proposed method to model the WSRVPN and identify the key nodes

We can construct the WSRVPN through giving weight to the variation propagation relation network, the final network can be represented as $G = \{V, E, W\}$, where $V = \{v_1, v_2, v_3, \ldots, v_n\}$ represents a node set, $E = \{e_1, e_2, e_3, \ldots, e_m\}$ represents an edge set, $W = \{\omega_1, \omega_2, \omega_3, \ldots, \omega_m\}$ denotes a weight set of the edge.

2.2 Using the Weighted LeaderRank Algorithm to Identify the Key Nodes

The key nodes in the complex network are nodes that affect the network performance and network structure more powerful than other nodes. The key nodes are monitored to realize the stability of the whole network. Therefore, the normal operation of the assembly line can be maintained by controlling the quality of the critical feature nodes in the assembly process to ensure the quality stable of the end product.

The weighted LeaderRank sorting algorithm is the optimization algorithm of the PageRank algorithm [13]. This algorithm takes into account the network characteristics of node degree, cluster coefficient, connection weight and so on. This algorithm uses a background node to realize the two-way connection with all nodes in the network, maintaining the strong connectivity of the network. And this algorithm excluded the disadvantage of all the jumps probabilities are the same in the PageRank algorithm. Compared with PageRank algorithm, this algorithm can converge faster and identify the important nodes in the network more accurately, and the key nodes which are excavated have stronger robustness [14]. The general expression of the algorithm is as follows:

$$WLR_i(t) = \sum_{j=1}^{n+1} \omega_{ji} \frac{WLR_j(t-1)}{b_j^{out}} \tag{7}$$

where $b_j^{out} = \sum_{i=1}^{n=1} \omega_{ji}$ represents the out-strength of the node j, ω_{ji} represents the weight of the node j to node i. $WLR_i(t)$ is the weighted LeaderRank (WLR) value of node i. Through this algorithm, the nodes in the network are calculated, and the nodes with larger WLR value are the key nodes. Thus, the quality of the corresponding key nodes in the assembly process should be controlled.

3 Industrial Case Study

In order to verify the effectiveness of the multistage assembly process modeling and key node identification method proposed in this paper, the actual assembly process of the bevel gear shaft assembly is studied. The assembly diagram and assembly process of the bevel gear shaft assembly are shown in Fig. 2. It can be seen from Fig. 2 that the component is consisting of 13 parts, and the contact between the surface and the surface in the connection takes the main part. It is known from the assembly sequence that the error caused by the front assembly surface can lead to the error of subsequent propagation. Thus, there are the propagation and accumulate of the error. The assembly error based on the surface will lead to the evolution of the error, i.e. the manufacturing error of the surface becomes the relative distance error. Therefore, the error propagation in the assembly process of this component is relatively complex and diverse, which can be modeled and analyzed by the complex network.

Each part has several surface contact with other parts because of the assembly process, thus, individual parts can be abstracted into different feature nodes according to the assembly information. For bearing inner ring, it's characteristic nodes have upper and lower contact surfaces and circular diameters. Thus, according to the assembly process information, a total of 48 related feature nodes are extracted. For the convenience of the text, the nodes are numbered respectively. The connection relationship of the entire network is determined by using the assembling sequence, the corresponding connection relationship is directed from the front mounting surface to the installed face. Combined with the weighted formula in the second section, three kinds of connection relationships in the variation propagation relation network are empowered, then the WSRVPN can be constructed.

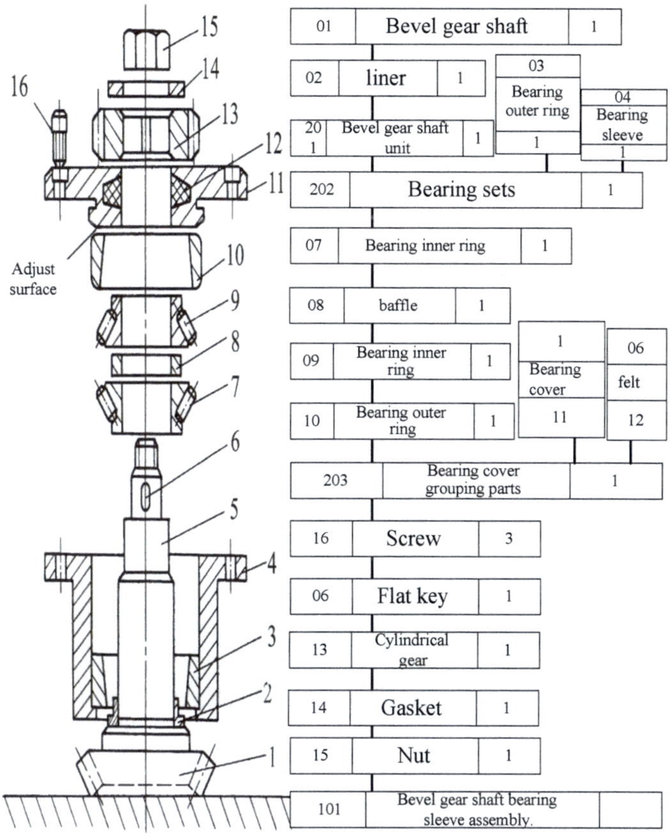

Fig. 2. Assembly process diagram of bevel gear shaft assembly

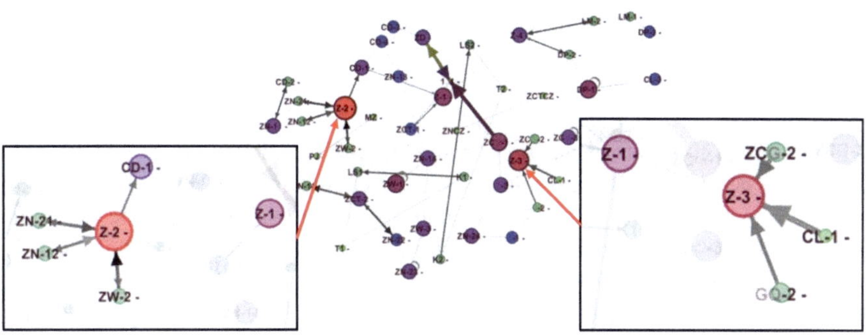

Fig. 3. The graph of WSRVPN

The WSRVPN of the bevel gear shaft assembly is shown in Fig. 3. The node size in the graph are proportional to the degree of the node, and the thickness of the edges in the network are proportional to the weight. As shown in Fig. 3, the node Z-2 (bevel gear shaft surface 2) and Z-3 (bevel gear shaft surface 3) have greater degree, thus, they may play an important role in the network.

The characteristics of weighted variation propagation network are analyzed, including node degree, betweenness, clustering coemcient and other related network characteristics [15].

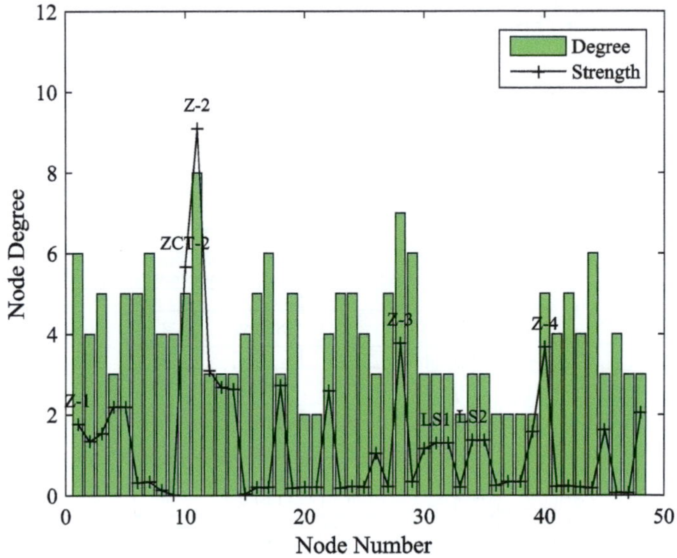

Fig. 4. The degree and degree distribution graph

As can be seen from the network node distribution diagram in Fig. 4, the degree of most nodes in the network are within a certain range, and the few nodes have a greater degree. Note that a few nodes may play an important role in the network.

Figure 4 is a graph of node degree and node degree strength. Some nodes have larger node degree and node strengths at the same time, such as node Z-2, ZCT-2 (diameter of bearing sleeve), Z-3, Z-4 (bevel gear shaft surface 4), etc., this type of nodes may play an important role in the network.

Figure 5 is a comprehensive analysis and comparison diagram of WSRVPN, which includes the node strength, betweenness, cluster coefficient and WLR of the deviation propagation network. The results of the analysis are shown in Fig. 5, the WLR index can fit well to the size of node strength and betweenness, Such as node Z-2, ZCT-2, Z-3, Z-4, etc., have a large betweenness, degree of intensity and WLR, indicate that such nodes play an important role throughout that network. Thus, they are the key nodes in the network.

Through quality monitoring of the above key nodes, the whole assembly process can be operated normally to ensure the quality of the end product.

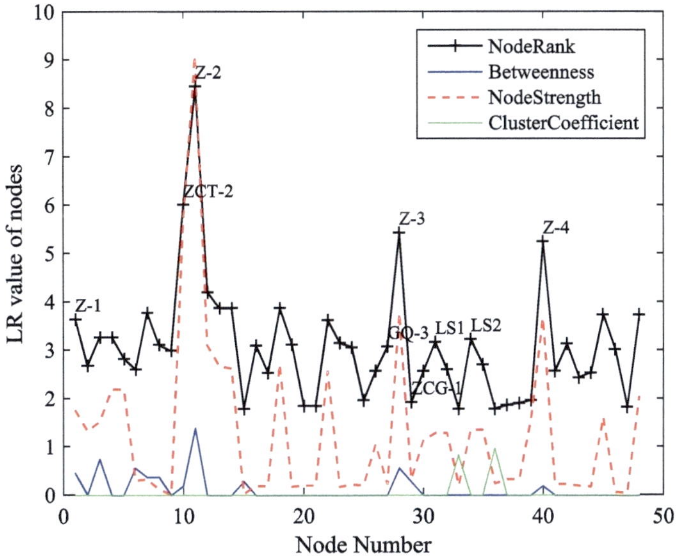

Fig. 5. Characteristics analysis of WSRVPN

4 Conclusion

In this paper, the self-regular weighted variation propagation network is constructed for the multistage assembly process, and the weighted LeaderRank sorting algorithm is used to identify the key nodes. The effectiveness of the proposed method is verified by an example of the bevel gear shaft assembly. The complex network method for multistage assembly process proposed in this paper provides a new idea for quality control in multistage assembling production. In the actual production process, a larger-scale assembly of equipment would cause the network to be too big and intricate. In the assembly process network, the network connection is tight, and the network connection between components is sparse. Therefore, it will be the focus of the follow-up study to divide the large network into different communities for analysis and search for key nodes.

Acknowledgements. This research was supported by the National Natural Science Foundation of China (No. 51375290, 71777173), the Fundamental Research Funds for Central Universities, and Shanghai Science.

References

1. S.J. Hu, Y. Koren, Stream-of-variation theory for automotive body assembly. CIRP Ann. Manuf. Technol. **46**(1), 1–6 (1997)
2. D. Ceglarek, W. Huang, S. Zhou et al., Time-based competition in multistage manufacturing: Stream-of-variation analysis (SOVA) methodology. Int. J. Flex. Manuf. Syst. **16**(1), 11–44 (2004)

3. Y. Ding, J. Shi, D. Ceglarek, Diagnosability analysis of multi-station manufacturing processes. J. Dyn. Syst. Meas. Contr. **124**(1), 1–13 (2002)
4. J. Shi, Stream of variation modeling and analysis for multistage manufacturing processes (CRC Press, 2006)
5. J. Camelio, S.J. Hu, D. Ceglarek, Modeling variation propagation of multi-station assembly systems with compliant parts. J. Mech. Des. **125**(4), 673–681 (2003)
6. H. Wang, X. Ding, Identifying sources of variation in horizontal stabilizer assembly using finite element analysis and principal component analysis. Assembl. Autom. **33**(1), 86–96 (2013)
7. J. Liu, Variation reduction for multistage manufacturing processes: a comparison survey of statistical-process-control vs stream-of-variation methodologies. Qual. Reliab. Eng. Int. **26** (7), 645–661 (2010)
8. T. Zhang, J. Shi, Stream of variation modeling and analysis for compliant composite part assembly—part II: multistation processes. J. Manuf. Sci. Eng. **138**(12), 121004 (2016)
9. J.V. Abellan-Nebot, J. Liu, F.R. Subirn et al., State space modeling of variation propagation in multistation machining processes considering machining-induced variations. J. Manuf. Sci. Eng. **134**(2), 021002 (2012)
10. K.-X. Peng, L. Ma, K. Zhang, Review of quality-related fault detection and diagnosis techniques for complex industrial processes. Acta Automatica Sinica **43**(3), 349–365 (2017)
11. Y. Shi, M. Gregory, International manufacturing networks—to develop global competitive capabilities. J. Oper. Manag. **16**(2–3), 195–214 (1998)
12. Z.H.O.U. Sheng-Xinag, Study on extraction of machining features about parts of revolution. Acta Automatica Sinica **25**(6), 848–851 (1999)
13. S. Brin, L. Page, The anatomy of a large-scale hypertextual Web search engine. Comput. Netw. ISDN Syst. **30**(1) (1998)
14. Q. Li, T. Zhou, L. Lü et al., Identifying influential spreaders by weighted LeaderRank. Physica A **404**, 47–55 (2014)
15. S. Boccaletti, V. Latora, Y. Moreno, M. Chavez, D.U. Hwang, Complex networks: structure and dynamics. Phys. Rep. **424**(4), 175–308 (2006)

Risk Identification and Assessment of the Development of Final Optics Assembly

Wei-gang Li, Yi Pan, Xin Xu, and Chao-zhong Guo(✉)

Department of Industrial Engineering, Harbin Institute of Technology, Harbin, China
cguo@hit.edu.cn

Abstract. The Final Optics Assembly (FOA is one of the most important equipment of the photoelectric control system. Its development process has great uncertainty and risk. Based on the general methods of the project risk management, this paper proposes a "DOUBLE work breakdown structure (WBS) + Risk breakdown structure (RBS)" three-dimensional risk identification model and a risk assessment model combining risk analysis with AHP. The risk identification model helps to quickly and fully identify similar risks that may exist in the current project. The proposed risk assessment model is beneficial to find out the key risks in the huge analysis data, and it has stronger practicality. Additionally, a risk source breakdown structure of the development project is also proposed. This paper provides guidance on risk identification and assessment of other large development projects.

Keywords: Final optics assembly · Risk assessment · Risk identification
Risk management

1 Introduction

Risk management is an emerging management discipline that studies risk characteristics and risk control strategies and is an important means to ensure the realization of project objectives. In risk management, risk identification and risk assessment are the prerequisites for developing a risk control strategy which are the basic work of risk management. The initial risk management mainly focus on business, finance and other related fields. However, with the development of risk management techniques, researchers find that many analytical techniques and management methods have universal applicability, such as Analytic Hierarchy Process (AHP), Project Evaluation and Review Technique (PERT) and so on. Therefore, the thought and tools for risk management have gradually been extended to various industries and various types of projects.

In 1969, the U.S. Vice Minister of Defense took the lead in pointing out that there is a lack of necessary risk assessment in the system of defense acquisition. Hence, the research of development of equipment on risk management started. Ireland introduced a management model in 1983 for identifying, assessing and managing the Management model of acquisition risk of National Defense Weapon System [1]. The model used the

G. Q. Huang et al. (eds.), *Proceeding of the 24th International Conference on Industrial Engineering and Engineering Management 2018*,
https://doi.org/10.1007/978-981-13-3402-3_45

output of WBS and network technologies such as Program Evaluation and Review Technique (PERT), Critical Path Method (CPM) and Precedence Diagramming Method (PDM) to establish the risk quantification matrix [2]. In 1989, Risk Management: Concepts and Guidance published, and it recommended a quantitative assessment method for cost risk and progress risk. This method uses the Monte Carlo simulation method based on WBS [3]. In China, the introduction of project risk management is late and the degree of emphasis is not as good as the developed countries in the world. However, in recent years, researchers have made many new progresses in the project risk management. In 2012, Zhang established a civil aircraft risk identification model by using WBE-RBS methods [4]. In 2015, Yu proposed a risk management approach based on the PDCA cycle method [5]. In 2016, Cao analyzed the typical risk sources in the development of aviation models and formulated the risk probability levels for the development of aviation models [6].

Based on the previous studies, combining with the characteristics of FOA development projects, this paper proposes a “DOUBLE WBS + RBS” three-dimensional risk identification model and a risk assessment method combining risk analysis with AHP. “DOUBLE WBS + RBS” constitutes a three-dimensional risk identification model. These three perspectives are the work nature dimension, the product structure dimension and the risk type dimension. This risk identification model helps to quickly and fully identify similar risks that may exist in the current project. AHP has the advantages of qualitative and quantitative analysis, concise structure and good consistency, which is suitable for the risk assessment of development projects lacking of historical risk data. However, the risk factors for FOA development projects are complex and diverse. Taking into account too many risk assessment objects will lead to a huge structure of the analytic hierarchy. Therefore, this paper combines AHP method with risk analysis to find out the key risks. This method will be more practical.

2 Risk Identification Method

2.1 Establishment of Risk Description Model

Accurate risk description is the prerequisite of risk identification, so it needs to define the basic risk event and the attribute of risk description. According to the definition of risk, development project risk also has three major elements, which are risk factors, risk events and risk consequences [7]. Therefore, these major elements are three basic aspects that we can describe a risk. In addition, different risks may exist in different stages of project development. For example, development projects can be divided into four basic processes: project evaluation, design and development, manufacturing process and assembly and commissioning. Clarifying the stage of occurrence of a risk event can alert stakeholders to the risk that may occur during a certain phase. The categories to which the risks belong can distinguish different sources of risk and help us recognize the nature of the risks [8]. Therefore, risk category and risk stage should also be used to describe the risk attributes.

2.2 Establishment of Risk Identification Model

Risk identification of development project recognizes and describes all the risk sources and risk events that may be encountered in the development process to determine their characteristics, analysis, classification, and ultimately forms a document. Different from general project risk, the risks of FOA development project are complex, systematic and lack of historical data. Considering this property, a three-dimensional risk identification model of "DOUBLE WBS + RBS" is proposed. WBS decomposition can press many kinds of rules. In general, there are two types WBS which are divided by stage process and by product structure plan [9]. RBS refers to the risk breakdown structure which is based on the experience and analysis of similar projects. By analyzing, classifying, refining, summarizing and summarizing various kinds of risks that may appear in the project, the main risk areas and risk sources are formed [10]. Therefore, WBS and RBS are used to establish a three-dimensional risk identification model including work quality dimension, product structure dimension, and risk type dimensions. Based on the identification of "DOUBLE WBS + RBS", a preliminary risk list is formed. Combined with the expert brainstorming method, the main risk in FOA development project can be systematically and comprehensively identified. Figure 1 shows the risk identification model.

2.3 Risk Identification for FOA Development Project

Based on the experience of similar development projects in compiling RBS, this paper combines the risk characteristics of FOA development projects and compiles the RBS of the FOA development project according to the comprehensive, streamlined, and guiding principles. This RBS is as shown in Fig. 2 which can provide a reference for the general development projects.

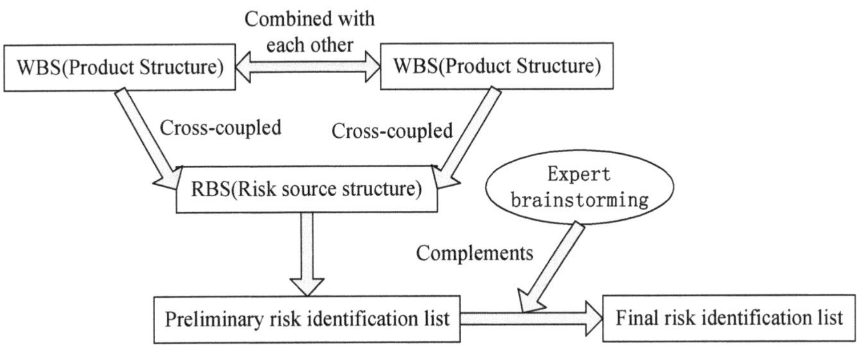

Fig. 1. Risk identification model

(1) Work breakdown structure by the nature of work

According to the working framework and processes, development characteristics and actual research situation during the FOA development process, the FOA development project can be divided into six major types of work, and each work can be divided into more detailed work packages, as shown in Fig. 3. This structure illustrates the overall workflow and scope of work on the development project. It's an important tool for project management.

(2) Work breakdown structure by product structure

Based on the relevant FOA product information, this paper compiles the work breakdown structure by product. The breakdown structure mainly includes four modules: a target window module, a frequency conversion module, a lens focusing module, and a beam measurement sampling module as shown in Fig. 4.

2.4 Establishment of Risk Identification Matrix

When using WBS-RBS method to identify risks, WBS and RBS need to be cross-linked to generate WBS-RBS matrix. The row (work package) and column (risk types) crossing checking is used to determine whether there is a corresponding risk in each unit of work. Compared with other risk identification methods, this method can systematically and comprehensively identify the risk of each working part of the project. It is not prone to omissions. It can guide the risk identification process accurately and effectively [11].

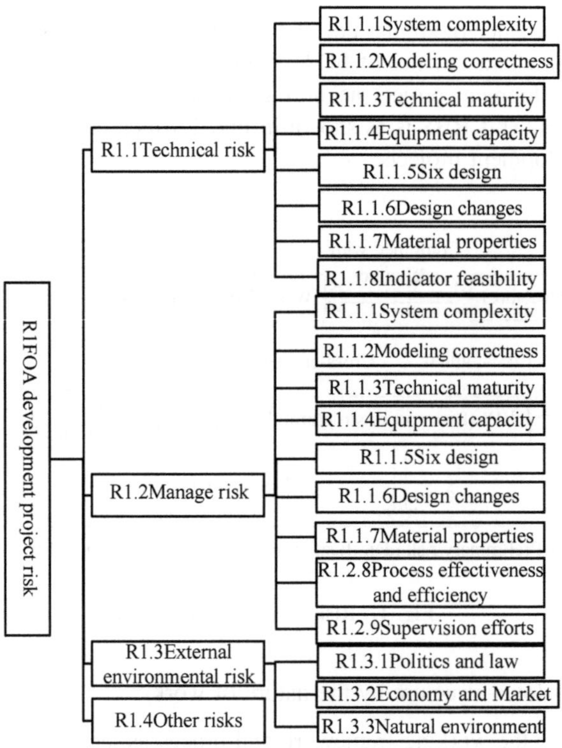

Fig. 2. FOA development project RBS

Taking WBS, RBS for FOA development projects classified according to the nature of work as an example, the coupling matrix of the WBS-RBS is shown in Table 1.

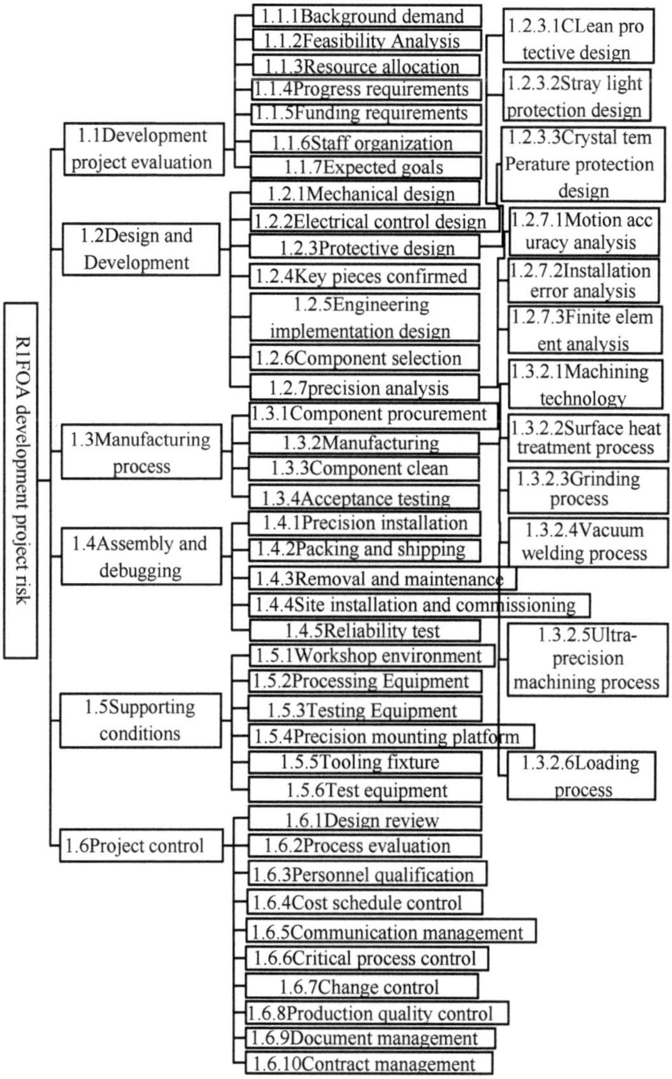

Fig. 3. FOA development project WBS (Classified by working nature)

Table 1. WBS and RBS coupling matrix

WBS			RBS			
			R1.1			
			R1.1.1	R1.1.2	R1.1.3	…
			System complexity	Modeling correctness	Technical maturity	
1.1	1.1.1	Background demand				
	1.1.2	Feasibility analysis				
	…	…				
1.2	1.2.1	Mechanical design				
	1.2.2	Electrical control design				
	…	…				

3 Risk Assessment Method

3.1 Establishment of Risk Assessment Model

Considering the complex and diverse risk of FOA development project, too many evaluation objects of risk event will generate too large structure of AHP which is unfavorable to the accuracy of calculation. Therefore, a risk assessment method based on risk analysis and AHP is explored, that is, the severity of each risk is initially determined through risk calculation, and then a few key risks affecting the project are found according to the 2–8 principles. The risk priority weight ranking is determined by using AHP quantitative analysis based on these key risks.

3.2 Risk Analysis

The size of a risk event can be characterized by a risk index. In general, the risk index can be assessed from both the likelihood of a risk event and the seriousness of the consequences which can be decomposed into three aspects: the impact on the quality, progress and cost, so that the evaluation index system of risk index can be established [12].

Suppose I represents the risk index, P indicates the probability of a risk event occurring, C indicates the severity of the risk consequence, CQ, CT, and CC represent the severity of the impact on quality, schedule, and cost, WQ, WT, and WC represent the weights of the impact of quality, cost, and schedule, the risk index formula can be expressed as:

$$I = P \times C = P \times (CQ \times WQ + CT \times WT + CC \times WC) \quad (1)$$

Due to the lack of the corresponding historical probability data, the qualitative method is used to measure the possibility of risk occurrence. The specific scale is

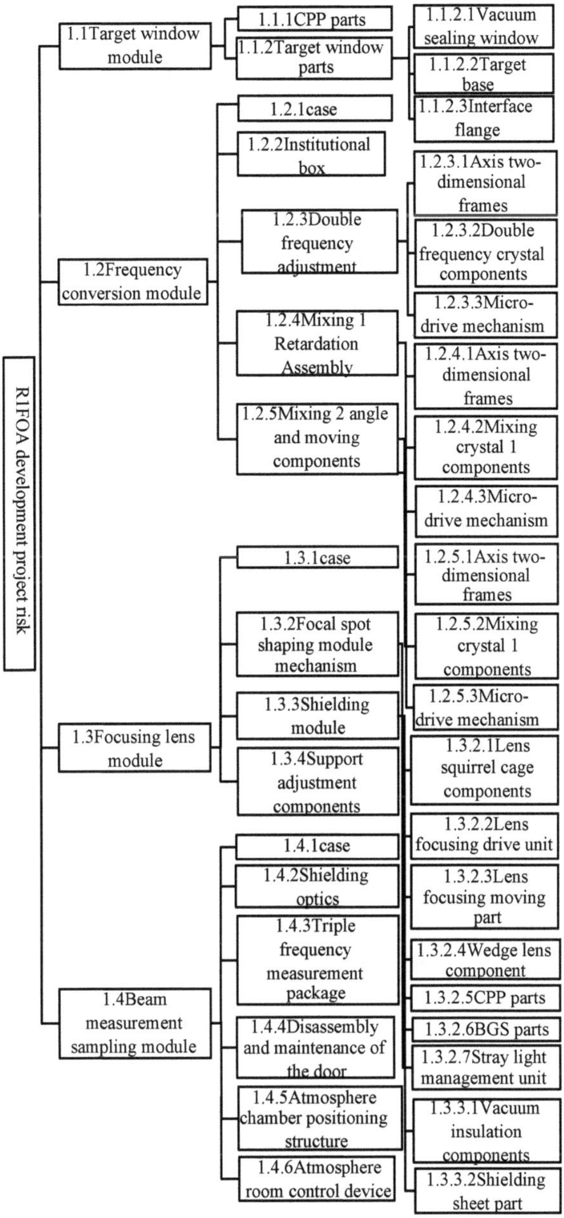

Fig. 4. FOA development project WBS (Classified by product structure)

shown in Table 2. The seriousness of the consequences of the risk can be described in terms of quality, cost and schedule, as shown in Table 3. According to the grading standard, each index of risk events is scored by Delphi method, and the size of risk is calculated according to the calculation formula of risk degree [13].

3.3 AHP Method for Risk Assessment

The AHP method combines qualitative and quantitative analysis. This method decomposes the goal from top to bottom and determines the weight of each factor from bottom to top. It is in line with human analysis-comprehensive thinking process that is simple and clear. Because of the consistency check which is required in the AHP method, it is also possible to eliminate subjective arbitrariness to some extent.

(1) Select key risks event which have larger risk index
According to the 2–8 principle, an arrangement diagram (called Plato) is drawn from the risk calculation results of risk events, and a small number of key risk events with large risks are selected so as to further quantitatively analyze these key risks.
(2) Construction of AHP index system
The purpose of using the AHP method is to conduct a more accurate assessment of a few key risks and determine the priority ranking of these risks so that risk response strategies are taken to achieve the purpose of risk management. The risk priority is arranged for three objectives: project quality, progress and cost respectively [14] (Figs. 5, 6).

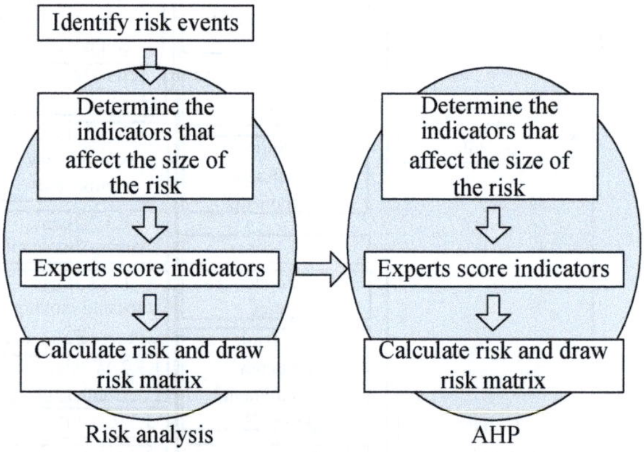

Fig. 5. Risk assessment model

Table 2. Risk probability scale

Possibility of risk	Possibility definition
0.10	Unlikely happens
0.30	Rarely happens
0.50	Possibly happens
0.70	Probably happens
0.90	Very likely to happen

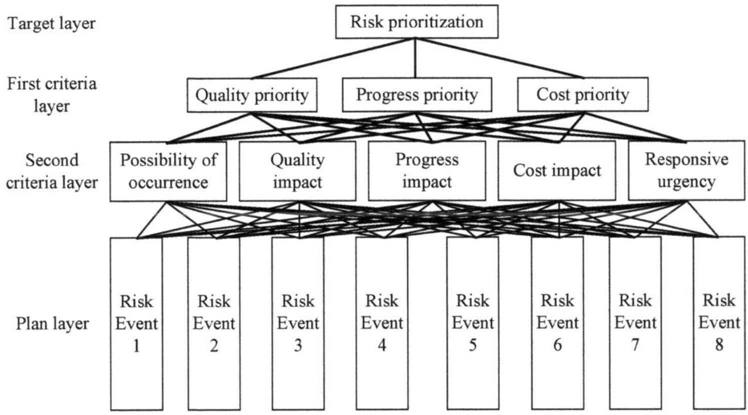

Fig. 6. AHP index system

Table 3. Risk severity rating scale

Project objectives	Very low 0.05	Low 0.10	Medium 0.20	High 0.40	Very high 0.80
Cost	Cost increase is not significant	Increased cost less than 10%	Cost increase 10%~20%	Cost increase 20% ~ 40%	Increased cost more than 40%
Progress	Delay in progress is not significant	Progress delay is less than 5%	Progress delay 5%~10%	Progress delay 10%~20%	Progress delay is more than 20%
Quality	Decline in quality is not significant	Only the most deman-ding parts are affected	Deterio-ration in quality requires approval from the initiator	The quality is reduced to the initiator can not accept	The final result of the project has no practical purpose

At the level of the guidelines, two indicators which are the probability of occurrence of the risk and the severity of the consequence are used to judge the priority of the risk. In addition, a third indicator, namely, the urgency of risk is adopted to judge the priority of the risk.

At the program level, it is the key risk event that has been identified that affects the project.

(3) Establishment of a judgment matrix

For each factor at the same level, taking the adjacent element as the criterion, according to the principle of scale, the relative importance of each element is given and a judgment and the importance judgment matrix is established:

$$A = a_{jj} \tag{2}$$

(4) Consistency check of the judgment matrix
After constructing the judgment matrix, the maximum eigenvalue $\lambda_{\max}$ and its corresponding eigenvector are solved for each judgment matrix, and the relative weights of the elements in a certain criterion layer are obtained. In order to verify the rationality of the judgment matrix, it is necessary to test whether the judgment matrix A has serious non-consistence through the value of $\lambda_{\max}$.The consistency check can be done by solving the consistency ratio *CR*.

$$CI = (\lambda_{\max} - n)/(n - 1) \tag{3}$$

$$CR = CI/RI \tag{4}$$

where *CI* in the formula is a consistent indicator; *CR* is the consistency test rate; *RI* is a random consistency indicator.

(5) Overall sorting and consistency checks
Overall sorting is to calculate the ranking weights of all the factors in the same hierarchy for the top-level goal from the top to the bottom. It can be solved according to the following formula (5).

$$C_i = \sum_{j=1}^{m} b_{ij} a_j \quad i = 1, 2, \ldots n \tag{5}$$

From the highest level to the lowest level of the overall sorting also need to make consistency test at all levels. Overall sorting consistency test can be done by:

$$CR = \sum_{j=1}^{m} b_j CI_j / \sum_{j=1}^{m} b_j RI_j \tag{6}$$

where CI_j is the individual sorting consistency index of each element of layer C to the element b_j of the previous layer B, RI_j is the random consistency index, and *CR* is the consistency ratio of the whole sorting of layer C [15].

Based on the AHP method, the priority weights of the key risk events are obtained, so as to more specifically evaluate the risks in the FOA development project and lay a foundation for the next step to formulate different risk responses measures.

4 Conclusion

FOA is a product with high-tech characteristics under the background of national defense projects. Its long development cycle, strong innovation, high unpredicted ability and maintenance cost pose higher requirements for the risk management of FOA development projects. Based on a deep understanding of the characteristics of FOA, this paper compares different risk identification methods and proposes a risk identification method suitable for the FOA development project and develops the RBS and WBS, which enables systematic and comprehensive identification of risks in development projects. After that, by studying the common methods of risk assessment, a risk assessment model and a risk assessment index system combining the characteristics of FOA development projects are established. The system further quantified the risks in the development project which provides the basis for the formulation of corresponding risk response measures and risk monitoring. However, risk management is a dynamic and continuous process. Therefore, risk identification and assessment are not an once-in-a-lifetime event. They need to be developed and perfected repeatedly as the project progresses. Only in this way one effective risk management system can be built to minimize the adverse impact of risk on the project.

Acknowledgements. This paper is supported by the Special Financial Grant from the China Postdoctoral Science Foundation (Grant No. 2016T90305).

References

1. L.R. Ireland, Risk management model for the defense system acquisition process, NTIS:AD-P002313/5/HD M (1983)
2. E.G. Ingalls, P.R. Schoeffel, Risk assessment for defense acquisition management, NTIS: AD-P002 303/6/HDM (1983)
3. W.G. Harrison, Methodology for assessing acquisition technical risk, NTIS:AD-A255130/7/HDM (1998)
4. Z. Chunxiu, Research on risk management of civil aircraft project based on WBS-RBS (in Chinese). Proj. Manag. Technol. **04**, 98–103 (2014)
5. Yu. Haichan, Application of pdca circulation method in risk management of large instrument development project Technol. Econ. **80**, 77–80 (2015) (in Chinese).
6. C. Tao, Exploration and reflection on risk management of aviation model development project (in Chinese). Aviat. Sci. Technol. **6**, 67–70 (2016)
7. T. Karaulova, S. Kramarenko, E. Shevtshenko, Risk factors in project management life cycle, *Proceedings of the International Conference of DAAAM Baltic Industrial Engineering* (2008)
8. L. Harron, R. Barlow, T. Farquhar, Enhanced leak detection risk model development. Int. Pipeline Conf. **4**, 559–564 (2010)
9. E. Siami-Irdemoosa, S.R. Dindarloo, M. Sharifzade, Work breakdown structure (WBS) development for underground construction. Autom. Constr. **58**, 85–94 (2015)
10. Z. Xingbang, Research on risk management of shenzhou X spacecraft development project (in Chinese), Harbin Institute of Technology (2014)
11. D. Hillson, S. Grimaldi, C. Rafele, Managing project risks using a cross risk breakdown matrix. Risk Manag. **8**, 61–76 (2006)
12. J. Perez, D. Weir, C. Seguin, R. Ferdous, Development and implementation of a liquid pipeline quantitative risk assessment model, in *ASME. International Pipeline Conference, Materials And Joining; Risk And Reliability*, vol. 3, V003T12A029. https://doi.org/10.1115/ipc2014-;33705
13. G. Yunfeng, Research on risk management of high pressure compressor development project (in Chinese), Jilin University (2010)
14. A. Kokangül, U. Polat, C. Dağsuyu, A new approximation for risk assessment using the ahp and fine kinney methodologies. Saf. Sci. **91**, 24–32 (2017)
15. J. Franek, A. Kresta, Judgment scales and consistency measure in AHP. Procedia Econ. Financ. **12**, 164–173 (2014)

Study on Agile Software Development Based on Scrum Method

Er-shi Qi, Hai-tian Bi(✉), and Xing Bi

College of Management and Economics, Tianjin University, Tianjin, China
hydden@sina.com

Abstract. In recent years, the rapid development of banks is more and more dependent on the function of bank software system. The demand for software development is constantly changing, which makes some banks' software development teams unable to adapt to frequent demand changes. In order to adapt to the needs of frequent changes, more and more software development teams use agile software development method. Based on the research of the basic principles of agile development, this paper studies the agile development method based on Scrum, and improves and optimizes the implementation of agile development of M software project team in C bank. Finally, the implementation of the project is evaluated and analyzed.

Keywords: Agile development · Bank software · Scrum

1 Introduction

In recent years, the rapid development of banks is more and more dependent on the function of bank software system. The demand for software development is constantly changing, which makes some banks' software development teams unable to adapt to frequent demand changes. The emergence of agile methods has pointed out the direction for the Information Department of the bank. "Rapid iteration, small step run, gradual improvement, response to needs, multiple deliveries" and other characteristics make the research and development personnel of information products better cope with the changing needs of banking users and the rapid development of science and technology society. The application of agile testing technology in the bank's IT Department has greatly released the labor productivity of the Department and improved the competitiveness of the product. In the traditional test pattern, the embarrassing situation of product release is outdated and demand development change has been effectively eliminated in the agile development.

In order to meet the needs of the market, to solve the problems existing in the software development industry,some new development methods have been created. This kind of method with quick and convenient way of thinking, quickly solve the low production efficiency of some traditional software development enterprise, and obtain a rapid promotion. We call this kind of method agile software development method. Agile software development is a lightweight development approach, against traditional, huge, and heavy processes. It advocates the importance of communicating with people,

G. Q. Huang et al. (eds.), *Proceeding of the 24th International Conference on Industrial Engineering and Engineering Management 2018*,
https://doi.org/10.1007/978-981-13-3402-3_46

replace documents with high quality software, and be able to respond to fast changing requirements quickly, such as rapid development of software capabilities [1]. As a result, agile programming is a way of developing [2], which is well suited for small and medium development teams. The core idea of agile development is to find the best balance between speed and quality, with the myriads of changes business needs, always make sure to provide customers with the greatest value, the real value of the final software can change the software technology thorough ways to provide value for the user [1].

In a word, with many advanced methods and rich experience, many advantages have made agile development seem to be the standard answer to the software crisis [3].

2 Methodologies

In 2001, Kent Beck, Martin Fowler, and Robert Martin as the leader of the Agile Manifesto for Agile Alliance released, marking the agile software development methods from the stage of history. The agile development approach is created to adapt to the rapidly changing user needs and aims to respond quickly to changes. Unlike traditional development methods, agile development does not require most of the needs in the early stage of the project and allows changes in demand. It responds to changes in demand with a positive attitude. The delivery function in the development process can be used online. The flexibility of agile development depends on iteration. It usually stipulates that a shorter period is an iteration cycle. At the beginning of the cycle, user stories are reviewed and determined. After testing, the functions of the reviews are integrated into product functions iteratively. On the one hand, iterative development and integration, can control the changes in demand caused the loss due to the iterative user stories in the short span of time (for example, only 2 weeks), to determine the function of user participation does not normally occur in such a short period of time change. On the other hand, it has no effect on the iterative demand change outside the iterative process. These changes are responded by the following iterative. As a result, the agile development approach ensures software quality, development efficiency, and user satisfaction. The agile development approach is squinted for most complex or indeterminate projects.

2.1 The Generation of Scrum Development Method

Hirotaka Takeuchi and Ikujiro Nonaka at that time by Japanese well-known enterprises involved in automobile, photo machine, computer and printer industry case conducted in-depth research, and in 1986 in the Harvard Business Review published an article for the "The New New Product Development Game" article [4]. The article emphasizes the authorization, self-organizing team, and the first use of the term Scrum, to compare a new similar to the "olive ball" style of team cooperation.

In early 1990s, Ken Schwaber and Jeff Sutherland respectively in each of the company presents a method of developing similar. In 1993, Jeff Sutherland for the first time called this method Scrum. In 1995, invited by Object Management of the Group, Sutherland Jeff and Ken Schwaber summed up years of experience of development and management and created a new methodology, and in OOPSLA'95 conference officially published jointly, first proposed the concept of Scrum [5].

The philosophical foundation of Scrum is to empower the project development team and satisfy customer needs. Its management culture is rooted in the idea of helping others accomplish their goals. The main technology tool is making decisions based on fact through learning process [6]. Scrum is a refreshing framework. It is based on the eight values of honesty, openness, courage, respect, concentration, trust, authorization and cooperation [7].

Scrum itself is not a standardized process. It does not guarantee that people can get satisfactory products according to its implementation gradually. On the contrary, Scrum is a framework for organization and management. The Scrum framework is built on values, principles and practices. It allows people to add different practices based on their industry, environment and conditions to supplement and enrich their work and form a Scrum version belonging to these organizations.

2.2 The Roles, Processes, and Artifacts of the Scrum Development Approach

(1) Scrum roles. The Scrum development project can contain one or more Scrum teams. For the Scrum main frame, each Scrum team contains three different roles, namely the product leader, Scrum Master and development team. In the process of actually pushing and applying Scrum, there are other roles. (2) Scrum Flow. In the Scrum method, first by the person responsible for the product according to the stakeholder's willingness to establish a product vision, and the vision is decomposed into a group of characteristics, to establish a list of products (also known as user stories), then, according to the characteristics of each commercial value and determine the priority of risk. At this point, not every feature in the list needs to be refined. It needs to say clearly, what effects or functions need to be achieved, and refine it until each sprint or sprint plan is completed. (3) Scrum artifacts. It includes the product backlog, the sprint planning, the burnout chart, and the definition of done.

A burnout chart is a visual view that is used to track the amount of the remaining tasks of the sprint, as shown in Fig. 1. Usually, there are two annotations for the burnout chart, one is based on the number of remaining tasks, and the other is based on the number of hours of the remaining tasks. The burnout chart is divided into the release burnout map and the sprint burnout chart, in which the sprint burnout map is also called the pending work map.

Different people have different understanding of task completion. Most programmers think that the completion of the test is finished. The product leader usually thinks that the task is completed through acceptance test. Different understanding of the word "completion" often leads to problems arising from communication between tasks. Therefore, in order to maintain consistency in terms of use and comprehension, it is necessary to unify the definition of "completion". When speaking at Chicago Scrum Gathering in 2008, Scrum co-creator suggested that the definition of done included code review, design review, reconfiguration, performance testing and unit testing [8].

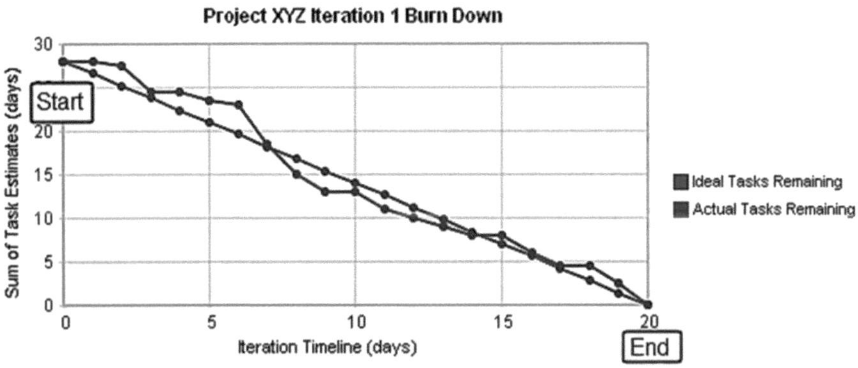

Fig. 1. Burnout chart

3 Software Development Improvement Based on Scrum Method

The theory of Scrum reveals the existing problems in the process of software project development from three aspects, namely, project team planning, project process planning and project supervision and evaluation planning, as shown in Fig. 2. Mike Cohn, one of the founder of Scrum alliance and the Agile Alliance often say a word: "Scrum is a let's focus on the delivery of high quality in the shortest time of the commercial value of the framework [9]. Combining with the Scrum agile project management framework, this paper puts forward some improved designs on the three aspects mentioned above to solve the problems in software development.

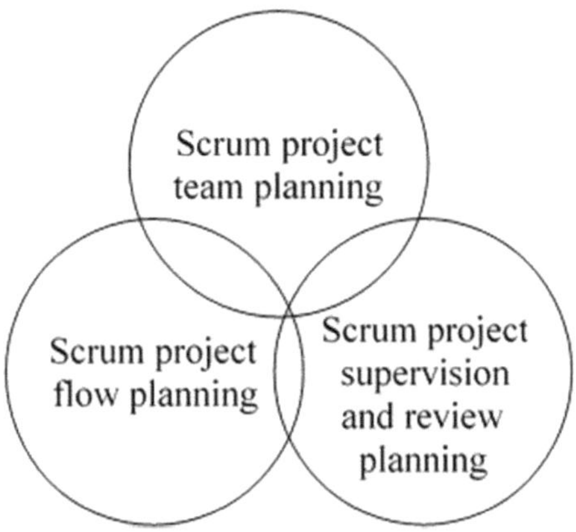

Fig. 2. Software development improvement

3.1 Scrum Project Team Planning

In order to solve the problems existing in the project team, we try to readjust the structure of the original project team according to the related theory of Scrum. We need to strengthen the customer full participation, to ensure customers and development team to maintain a high degree of consistency. On the other hand, we try to mobilize the enthusiasm of team members of the project team, to strengthen self-organization and self-management ability, to weaken the role of the project manager through the adjustment of team structure and responsibilities among members. We encourage the interaction between the product manager and the project team, and strengthen the role of the product manager in the project. Based on the above considerations, we need to create three important role in the project team, scrum Master, Product Owner and the development team, as shown in Fig. 3. There is no hierarchical relationship between these three roles. They are equal in status and need to take their respective responsibilities in the implementation of the project [10].

3.2 Scrum Project Flow Planning

The traditional waterfall development model is an ideal development state. It needs a clear, clear and fixed requirement analysis when the project started. However, in the actual implementation of the project, the customer will adjust the demand according to the changes in the market. Sometimes, the deviation of the project team to the understanding of the needs to correct in time. The non-backtracking of the waterfall development model leads to its inability to respond flexibly to this change of demand, and cannot be fed back and modified in time. Therefore, we need to make overall planning for the project process, and apply the scrum agile project management framework (Fig. 4) to solve the problems in software development.

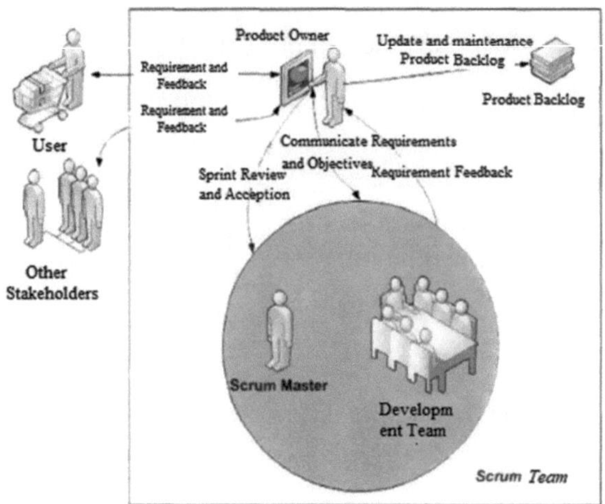

Fig. 3. Three important role in the project team

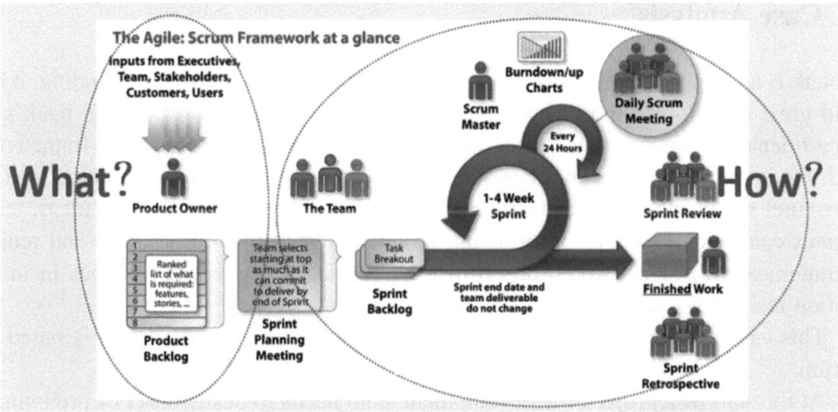

Fig. 4. Scrum agile project management framework

The common development pattern is to make a corresponding long-term work plan after determining the goal of developing a product. The agile architecture is different from the usual pattern of development. The agile architecture product owner records and manages the customer's needs by creating a product backlog. In addition, the development team decomposes it into several Sprint phases based on the priority and workload of the product backlog. Every 2–4 weeks as a Sprint phase of the cycle, every day and as a double circulation flow pattern to develop a PDCA. At the end of each Sprint phase, the Sprint review and the Sprint review will be a phased summary. A 15-min daily scrum meeting need to be held every day, and the daily research and development work is started. From a certain point of view, the dual cycle mode of Scrum is carried out every day and every Sprint phase.

3.3 Scrum Project Supervision and Review Planning

In the process of the traditional project development, the project manager needs to develop a detailed project plan at the beginning of the project. However, most of the projects are often not implemented according to the project's plan. Because we only focus on the beginning and end of the two endpoints, we cannot effectively control the long and changeable development process. Therefore, we need to decompose the function or module of the product one by one. Every Sprint cycle focuses on the realization of a function or module, decomposing the task and recording it on the Sprint backlog. In every Sprint implementation process, we use visualization tools such as task Kanban and Sprint burn out chart to monitor the daily state of the project, find out the problems in the implementation process in time and adjust accordingly.

4 Case Analysis

C bank is an important joint-stock commercial bank in China. Since its founding, it has paid great attention to the development of information technology. The C bank's IT department and the financial sector mix, build a matrix—based business framework, and maximize the integration of business and technology. Business personnel and IT personnel are in the same department. When business people have good ideas, technicians can quickly convert information products to information products and reduce communication costs between departments. C banks have great advantages in information technology and financial electronic products.

This case is an agile test of the M software project in C bank, which is based on Scrum.

At the start of a project, the development team needs to design a set of preliminary test plans based on previous experience and specific projects, and in accordance with user supplied software function demand analysis, further to the new changes and new requirements may encounter iterative testing in the process of analysis and research, for possible solutions. Scrum, the product function design, bugs are set, customer suggestions for improvement, the core features and characteristics of advanced technology such as the implementation process of the project tasks are located in the Product Backlog product records in the list, M software project in C bank corresponds to a product list. The tasks in the product record list are constructed according to the priority principle and risk assessment and system architecture factors. According to the principle of high cohesion and low coupling, the Backlog list is decomposed into different problem sets, which is completed by the second stage of Sprint development and testing.

The Sprint phase is the most critical part of the Scrum process, which is related to the successful success of the whole project. The programming and testing of all functions of the banking software system is concentrated in this stage. Every stage will form a functional and complete logically mature "finished product", and the finished product is tested by software, which is also the result of the agile test iteration.

After a number of iterations in the Sprint phase, the C bank software products meet the basic requirements of the customer. Through acceptance testing and meeting customer needs, the entire development work of the C bank software project is completed and the product is delivered to the customer.

The M software project in C bank uses the agile testing model to improve the efficiency and effectiveness of software testing, but at the same time, there are some problems.

(1) Agile development means the shortening of the time cycle. To reduce the development cycle, reduce development does not mean that the task, the task instead of heavier, larger amount of labor, so in the same development resources, inevitably bring quality risk, project management staff and is responsible for the development of people put more effort.
(2) Integrated testing requires multiple module comprehensive testing. In the early stage of Sprint test, one or more modules were not developed, and integration testing was forced to pause after the integration test was carried out according to

the test plan requirements. The reason is analyzed. On the one hand, because there is a bias in the development of test plans for the development of individual modules, on the other hand, dependencies exist in the development process of different modules, which objectively increases the development time.

(3) There are too many organizational levels in project management. Although daily meetings can ensure good communication before different teams, but because of many organizational levels, the daily communication between developers and test leaders occupy a lot of energy and time, and seriously affect the efficiency.

Through analyzing the problems of team organization, development strategy and bug management strategy of M software project in C bank, we can draw the conclusion that the key to improve the efficiency of agile development of M software project in C bank is the following three aspects.

First, team organizations need to optimize and update. The existing multi-level organization structure cannot meet the rapid iteration characteristics of agile testing. The multi-level organization framework increases the communication cost, which is not conducive to the response time of agile testing, and constantly adjust the requirements of change.

Secondly, the process and strategy of agile testing need to optimize. In the short iteration cycle of agile testing, the amount of submission of the code is much larger than that of the traditional test. Therefore, in addition to testing the function and performance of the software, agile testing is also responsible for the detection of the software development process. The agile test not only ensures the complete and correct function of the C bank software, but also needs to ensure that the software development process is not wrong.

Finally, agile development needs to improve bug management. One of the most important goals of software testing is to find bug, and to solve bug. In the current agile test fast iteration process, there will be new bug discoveries in every Sprint stage, and there may be correlation between the newly discovered vulnerabilities and the vulnerabilities that have been solved. Therefore, efficient and reasonable vulnerability management can facilitate agile testing to proceed smoothly.

5 Conclusion

With the increasingly fierce competition between domestic and foreign banks, the excellent software development model will become an important choice for the IT Department of the bank. Agile development because of its characteristics of fast development, iterative optimization, more and more banks IT department's favor, the four major state-owned bank software development center has implemented agile test model in the field of banking software development, but compared with foreign enterprises, there is still more work is needed to improve.

First, the development of the theory of software engineering cannot be separated from the change of the way of thinking. The agile development model is more reasonable than the traditional development model, and requires higher quality and ability of developers.

Second, agile development requires all people to have agile thinking. Not only developers and testers, project leaders and product leaders need to be familiar with the flow and methods of agile testing.

References

1. R.C. Martin, *Agile Software Development: Principles, Patterns and Practices* (Pearson Higher Education, London, 2013), p. 34
2. Z. Jingzhou, Q. Leqiu, Z. Sanyuan, An overview of agile methodology. Comput. Appl. Softw. **19**(6), 1–9 (2002). (In Chinese)
3. J.D. Musa, *Software Reliability Engineering* (McGraw-Hill, NY, 1998)
4. H. Takeuchi, I. Nonaka, The new new product development game. Harv. Bus. Rev. 1–14 (1986)
5. K. Schwaber, Scrum development process, in *Proceedings of OOPSLA'95 Workshop*, Austin, Texas, pp. 117–134
6. K. Schwaber, *Scrum Developer Best Practices*, 1st edn. (Microsoft Press, Redmond, WA, 2004), pp. 12–15
7. K.S. Rubin, *Essential Scrum: A Practical Guide to the Most Popular Agile Process* (Addison-Wesley Educational Publishers Inc., Boston, 2012), pp. 4–5
8. C. Sims, H.L. Johnson, X. Yi, Translation, *The Elements of Scrum* (Post and Telecom Press, Beijing, 2013), p. 34. (In Chinese)
9. L. Jin, Z. Lingling, Process design and analysis of large scale software project management. Sci. Technol. Manag. Res. **15**, 204–206 (2010). (In Chinese)
10. Z. Wenfan, Application of scrum to software project management in HM company, MBA dissertation, East China University of Science and Technology, Shanghai, China, 2012. (In Chinese)

Dispatching Rules in Assembly Job Shop with Machine Breakdown and Preventive Maintenance

Xiao-jia Yang, Jian-jun Liu(✉), Qing-xin Chen, and Ning Mao

Guangdong CIMS Provincial Key Lab, School of Electromechanical Engineering, Guangdong University of Technology, Guangzhou, China
jianjun.liu@gdut.edu.cn

Abstract. The assembly job shop is a common shop type in make-to-order companies, which satisfies with the requirement of product assembly. Compared with deterministic scheduling, the majority of researches for dynamic assembly job shop focus on dispatching rules. However, most of researches do not consider the deterioration of machine. This study extends the assumption that machine is always available. Under the dynamic assembly job shop with breakdown and preventive maintenance (PM), researches are conducted as follows: (1) Three representative PM policies are compared; (2) The performance of classic dispatching rules for assembly job shop are compared. (3) From the view of constituent attributes of dispatching rules, we analyze the influence of machine maintenance on dispatching rules, and a new dispatching rule, MALLOPN (modified allowance per operations), is proposed. The full factorial simulation experiments are carried out with different levels of machine deterioration. The results show that the policy with optimal PM period (PM policy II) is the best one and the MALLOPN rule outperforms classic rules.

Keywords: Assembly job shop · Dispatching rule · Machine reliability
Preventive maintenance

1 Introduction

With the development of industry 4.0, customization is becoming an essential production mode [1]. It is a typical make to order mode of non-repeatable production. Compared with researches of job shop scheduling [2], the researches of dynamic scheduling on assembly job shop are more satisfied with the needs of product assembly or multi-part coordination [3]. The coordination of associated parts of the same product, assuming a set of product is an order, is the key to the assembly job shop dynamic scheduling [4]. If the associated parts are not completed, the waiting time for the assembly will be prolonged, and the order tardiness will be caused in serious cases. And the early completion of parts have preempted the urgent delivery parts, which should be processed first. To the best of our knowledge, in previous dynamic assembly job shop scheduling researches, it is always assumed that the machine is available all the time. However, the machine may breakdown and carry out preventive maintenance

G. Q. Huang et al. (eds.), *Proceeding of the 24th International Conference on Industrial Engineering and Engineering Management 2018*,
https://doi.org/10.1007/978-981-13-3402-3_47

in actual production [5]. The maintenance activities are closely related to the production scheduling. If only the production activities are carried out, the reliability of the machine will be decreased continuously, which will increase the number of breakdown and affect the production progress. Although the PM action can improve the reliability of the machine and reduce the occurrence of breakdown, excessive PM action will reduce the time for production activities.

In this study, the Weibull distribution is used to describe the deterioration of machine. Three representative preventive maintenance policies, in actual production, are introduced. From the view of the attributes of the dispatching rules, we analyzes the influence of machine breakdown and PM on production scheduling, and a new dispatching rule, MALLOPN (modified allowance per operations), is proposed. Lastly, The simulation model is established, which includes five performance measures related to work-in inventory and delivery. Each measures evaluates the average level and the worst-case of the dispatching rules from mean value, maximum and standard deviation.

2 Machine Reliability and Maintenance

In actual production, with the increase of service time, the reliability of the machine will decrease continuously. Machine will inevitably fail at the end. The reliability of the machine is the probability that machine would be operational before the time of t. In many previous researches and practical applications, the Weibull distribution with two parameters is used to describe the reliability of machine [6, 7]. The distribution of Weibull is as follow:

$$R(t) = \exp\left[-\left(\frac{t}{\theta}\right)^{\beta}\right], \qquad t \geq 0 \tag{1}$$

where β is a shape parameter and θ is a scale parameter. This study assumes that the machine is repairable. Maintenance is considered as the activity to maintain or restore the machine to a better state. There are two large groups Maintenance operations: corrective maintenance (CM) and preventive maintenance (PM) [8]. CM is carry out when machine breakdown has already taken place. The time for machine repair is called repair time. This study assumes repair is minimal, i.e., the service time of a repaired machine is just like when it failed. And the PM action is maximum, i.e. it will restore the machines to the as-good-as-new state. Three representative preventive maintenance policies are introduced as follows:

policy I: The policy that does not take into account the reliability of the machine, and PM is regularly executed periodically (daily, weekly, or monthly). It is widely applied in real production [8].

policy II: Depending on the distribution of reliability, the influence of the machine breakdown and PM is balanced, and availability of machine is maximized. t_r is the number of repair time units and t_p being the number of time units the PM takes. The optimal PM interval (T_{PMop}) can be calculated by the following formula [6].

$$T_{PMop} = \theta \bullet \left[\frac{t_p}{t_r(\beta - 1)}\right]^{\frac{1}{\beta}} \tag{2}$$

policy III: For some high risk production systems or special industries with expensive raw materials, the minimum threshold of machine reliability must be high. Therefore, the minimum reliability threshold is used as a constraint to calculate the longest PM interval according to the formula (3). In subsequent studies, the reliability is set to 97%.

$$T_{PM} = \theta \bullet [-\ln R(t)]^{\frac{1}{\beta}} \tag{3}$$

3 Dispatching Rules on Assembly Job Shop

The dispatching rules are based on the status of shop or attributes of jobs, such as the current queue load, due date of the job and so on., to calculate a clear priority value for each job. According to the priority value, the processing queue is sorted. Summarizing the previous dynamic assembly job shop researches [9], constituent attributes used in dispatching rules as follows: (1) product structure; (2) production progress; (3) staging delay (assembly waiting time); (4) product due date; (5) shop congestion. For example, the TWKR rule can effectively reduce the lead time and the assembly waiting time [10], taking into account the production progress of all parts of the same product. Therefore, its constituent attributes are: (1) product structure and (2) production progress.

According to the constituent attributes mentioned above, classic rules and the rules proposed in this study are summarized in Table 1.

Among them, FCFS is the basic rule in this study; Previous researches have shown that JDD rule is the best one for minimizing mean tardiness on job [11]; TWKR-IR is a combination rule, in which IR is a tie break rule. In the case of the same TWKR, the part with high IR value is processed first; $(\text{ROPT})^2$ + SC takes into account the production progress and the shop congestion factor that is a multiple of the operation's processing time and the sum of processing times in the queue [12]; Finally, the influence of machine maintenance on the number of remaining operations will be greater than the time of remaining operations, because the more number of remaining operation is, the more times it will encounter machine maintenance. Based on the All/OPN rule, whose constituent attributes are production progress (number of remaining operations) and due date, this study proposes an improved rule MAll/OPN, which takes into account the product structure and assembly waiting time that are the two specific attributes of assembly shop.

Table 1. Dispatching rules on assembly job-shop

Rules	Attributes	Description	Calculation
FCFS	No consideration	First at workstation, first served	min r_j
TWKR-IR	(1)(2)	Total work remaining—important ratio	min $\sum_{i \in U_j} rt_{ij}$, tie break by IR
JDD (EDD)	(4)	Earliest product due date first	min d_j
MADD	(1)(3)(4)	Modified assembly due date	min if $ECT_j < d_j$, then d_j else ECT_j
$(\text{ROPT})^2$ + SC	(2)(5)	0.1* (remaining number of operation)^2 + shop congestion factor	$0.1 * (U_j - C_j)^2 + SC$
All/OPN	(2)(4)	Allowance per remaining operation	$(d_j - T)/ N_{ij}$
MALL/OPN	(1)(2)(3)(4)	Modified allowance per remaining operation	$(\text{MADD} - T)/ N_{ij}$

Where r_j = the time of product j arrives at the system; d_j = due date of product j; T = current time; ECT_j = the early completion time for product j; rt_{ij} = remaining processing time for part i of product j; U_j = set of all uncompleted operations of product j; N_{ij} = the remaining number for part i of product j

4 The Simulation for Assembly Job Shop

4.1 Simulation Settings

This simulation study follows a classic assembly job shop similar to many previous researches [13]. For sake of simplification, we consider a one-level assembly job shop with a final assembly. The shop floor consists of six machines for manufacturing parts. It is assumed that the capacity of the assembly machine is sufficient, in other words, there is no assembly bottleneck. Especially, some assembly process is carried out at location of customers. A product is assembled by 1–6 parts and a part needs to be processed from 1 to 6 operations. Orders arrive randomly and interval time of arrive according to an exponential distribution with a mean of 2.6 h. Thus the shop utilization is 90% approximately. It is assumed that all orders must accept, materials are available and the operation routing is known.

In addition, the due date of the order has a significant impact on the performance. Once the order arrives, its due date specified by the TWKCP rule, which is based on the length of the critical path in the product [14].

$$d_j = T_j + C * TWKCP_j$$

where T_j is the order j arrival time, and C is the coefficient of due date.$TWKCP_j$ is total processing time of critical path on the product j. Accordance with previous researches, the C is set 33.8 to produce 30% tardy of orders, when FCFS is used as the dispatching

Table 2. Characteristics of simulation model

Shop	Shop type	Assembly job shop with a final assembly
	Number of machine for serial operations	6
	Machine reliability	Weibull. distribution
	Shop utilization	90%
Product	Interval of product arrive	Exp. distribution
	Number of parts per product	Discrete uniform [1, 6]
	Due date of products	Product Arrival time + C × critical path processing time.
Part	Number of manufacturing operations per product	Discrete uniform [1, 6]
	Routing variability	Random routing without re-entrant
	Operation processing time	Truncated 2-Erlang (mean = 1; max = 4)

rule and 90% utilization of shop. The all characteristics of the simulation model are shown in Table 2.

4.2 Performance Measures

In this study, we choose five performance measures, which are based on system inventory and due date. The details of each performance measures are given in Table 3.

It is worth noting that an order completed before the due date cannot be delivered before due date. Thus, the flow time performance can only be used as an indicator of WIP; Effective flow time reflects the lead time for the product from arrival to delivery; Staging delay is an performance proposed by [4], to reflect the coordination of the production progress of associated parts; Percent tardy indicates how many orders are completed after the due date, which is reflection of the level of service. Tardiness can be compared when system with the same percent tardy [9];

Table 3. Performance measures

Classification	Performance measure	Calculation
Inventory based	Flow time	$f_j - rt_j$
	Effective flow time	$\max\{f_j - rt_j, d_j - rt_j\}$
	Staging delay	$(f_{ij} - f_{1j})/N_j$
Due date based	Percent tardy	NT/N
	Tardiness	$\max\{f_j - d_j, 0\}$

Where f_j = the completion time of product j; f_{ij} = the completion time of part i in product j;N_j = the number of parts in product j; N = the number of completed products; NT = the number of tardy products

5 Experimental Design and Discussion

5.1 Experimental Design

The design of the simulation experiment is based on the following questions: (1) Which is better one among three PM policies? (2) What is the performance of all dispatching rules under three PM policies and different levels of deterioration?

The factors and levels involved in experiment are summarized in Table 4. It includes seven kinds of dispatching rules, three PM policies and two Weibull shape parameters that represent two kind of deterioration of machine. When $\beta = 2$, the reliability of machine decrease more evenly. While $\beta = 3$, the reliability of machine decrease slowly at the beginning and accelerate gradually.

A full factorial experiment is carried out with 42 cells. Each experiment runs 450 days and will be executed 100 times. Avoiding the impact of the instability in the beginning, the experimental data is collected only 100 days later.

5.2 Discussion

The results of experiment are shown in Table 5. Main results and discussion are summarized as follows:

(1) In the case of the same machine deterioration (the same Weibull shape parameter), performance of each dispatching rule under Policy II outperforms other two policies. Although the utilization of the machine remains at 90% under different policies, policy II can balance the number of machine breakdown and PM to reduce the influence of maintenance on production progress.
(2) Different deterioration of the machine has an impact on the three PM policies. While it has the greatest impact on policy III. It is because that the PM period, determined by the reliability threshold, is affected by the rate of reliability change.
(3) TWKR-IR and $(\text{ROPT})^2$ + SC rule can effectively reduce percent tardy of orders. TWKR-IR rule is best one reduces flow time (an indicator of WIP) and staging delay. But the performance of effective flow time is worse than other rules. And two rules are poor in the maximum and standard deviation of all performance measures. It is that the quality of service will greatly decrease due to much delay and fluctuation.
(4) JDD and MADD produced obviously lower value of flow time than All/OPN and MAll/OPN. However, JDD and MADD have no advantage over All/OPN and

Table 4. Experimental factors and levels

Experimental factors	Levels
Dispatching rule	FCFS; EDD; MADD; TWKR-IR; Slack; $(\text{ROPT})^2$ + SC; All/OPN; MALL/OPN
PM policy	Policy I; Policy II; Policy III
Weibull shape parameter β	2; 3

Table 5. Results of experiment

	Rules	Flow time			Effective flow time			Staging delay			Tardy	Tardiness		
		Mean	Max	SD	Mean	Max	SD	Mean	Max	SD	%	Mean	Max	SD
β = 2	*Policy I*													
	FCFS	150.4	369.8	76.8	216.8	463.8	71.2	92.7	360.6	62.9	29.94	26.3	224	38.2
	TWKR-IR	65.4	2378	161.2	206.8	2378	146.2	159.7	2092	119	5.32	18.3	2092	118.6
	EDD	94.9	381.4	69.7	193.9	447.9	64.1	102.4	259.1	40.5	7.49	3.4	30.4	6.2
	MADD	92.2	619.6	72.8	192.3	601.7	68.8	101.9	468.6	46.2	6.28	1.8	186	12.4
	$(ROPT)^2$ + SC	95.2	2808	232.7	228.5	2808	211.2	173.3	2548.4	181.2	9.70	40	2548	189.6
	All/OPN	117.6	404.5	72.3	193	453	64.4	77.9	324.1	44.8	4.51	2.5	37	5.6
	MALL/OPN	115.4	544.8	73.3	191.7	541.6	66.1	77.5	458.7	48	4.14	1.2	106.9	8.4
	Policy II													
	FCFS	118.7	311.8	64.1	202.8	445.2	64.7	96.5	357.4	61.3	18.94	12.3	166.7	23.8
	TWKR-IR	54.3	1941.9	125.7	202	1941.9	116.2	159.4	1665.9	93.1	4.02	11.6	1665.9	85.4
	EDD	75.2	350.6	62	191.5	440	62.8	117.3	282.8	40.3	2.7	1.0	12.9	2.2
	MADD	74.2	431	62.7	191	517.7	64	117.4	341.2	41.7	2.19	0.5	70.1	4.1
	$(ROPT)^2$ + SC	78.9	2397.1	189	218.6	2397.1	171.3	167.9	2143.5	144.4	7.6	28.2	2143.5	146.9
	All/OPN	100.8	379.3	68.4	191.2	441.4	62.9	91.1	340.7	47.2	1.32	0.7	12	1.7
	MALL/OPN	100.2	420.6	68.5	190.7	482.3	63.2	90.8	372.1	48	1.19	0.3	38.4	2.4
	Policy III													
	FCFS	157.7	380	79.9	220.6	468.1	72.8	92.9	361.1	63.7	32.68	30.1	234.2	41.7
	TWKR-IR	66	2628.5	173.7	210.3	2628.5	158.2	164.2	2345.2	130.4	5.3	19.9	2345.2	131.5
	EDD	98.8	388.2	71.4	194.7	450.1	64.5	100.1	258.4	41.1	8.24	4.2	34.2	7.2
	MADD	94.5	604.9	74.2	192.2	604.2	69.2	99.6	450.6	47.4	7	1.8	204.9	12.8
	$(ROPT)^2$ + SC	94.9	2825.6	239.3	227.6	2827	218.2	174	2572.5	188.2	10.03	41.2	2569.8	196.7
	All/OPN	122.1	412.4	73.5	194	457.9	65.1	75.3	322.9	44.1	5.33	3.5	45.1	7
	MALL/OPN	119.1	563.5	74.5	191.8	538	67	74	472.2	47.8	4.84	1.3	101.3	9.8

(*continued*)

Table 5. (*continued*)

	Rules	Flow time			Effective flow time			Staging delay			Tardy	Tardiness		
β = 3	*Policy I*													
	FCFS	136.7	348.0	72.2	210.5	455.7	68.3	93.9	358.7	62	25.35	20	203.8	32.4
	TWKR-IR	62.9	2356	154	205.5	2356	140.1	159.6	2074.4	113.6	5.04	17.1	2074.4	112.1
	EDD	86.2	367.9	66.5	192.5	442.8	63.3	108.4	273.8	40.6	5.13	2.1	22.7	4.1
	MADD	84.7	539.1	68.8	191.8	610.9	66.8	108.4	420.9	44.5	4.32	1.3	136.2	9
	$(ROPT)^2$ + SC	86.1	2492.3	206.9	221.7	2492.3	186.9	169	2228.4	157.9	8.80	33.4	2228.4	163.8
	All/OPN	110.4	392.2	70.6	191.9	446	63.5	83	333.4	45.9	2.76	1.5	24.1	3.4
	MALL/OPN	109.2	467.5	71.4	191.2	520.4	64.6	82.8	406.8	48.1	2.49	0.8	73.9	5.3
	Policy II													
	FCFS	95.8	270.3	54.6	196.1	438.7	62.4	105.8	367.7	62.5	11.22	5.6	129.1	14.6
	TWKR-IR	46.3	1639	100.8	198.1	1639	97.2	159.4	1364.1	77.8	2.98	7.6	1359.5	62.2
	EDD	61.2	327.9	55.3	190.5	437.4	62.4	129.4	307	39.6	0.5	0.1	3.3	0.3
	MADD	61.2	342.8	55.5	190.2	450.5	62.5	129.4	314.6	39.7	0.45	0.1	21.3	0.7
	$(ROPT)^2$ + SC	65.2	1971.9	151.2	209.9	1971.9	138.1	164.2	1721.6	114.6	5.82	19.5	1719.3	110.5
	All/OPN	87.9	364.3	64.9	190.6	437.6	62.4	102.7	354.6	49.1	0.28	0.1	3	0.3
	MALL/OPN	87.9	371.6	64.9	190.1	444.8	62.5	102.7	360.3	49.2	0.24	0.1	11.2	0.5
	Policy III													
	FCFS	100.5	278.2	56.9	197.4	440.5	62.8	103.8	366	62.3	12.86	6.9	137.3	16.5
	TWKR-IR	47.6	1740.2	106	198.9	1740.2	101.1	159.7	1458.3	80.9	3.19	8.4	1456.1	66.9
	EDD	63.6	332.2	56.7	190.6	437.6	62.4	127.2	304.9	40	0.91	0.2	4.6	0.5
	MADD	63.5	359.2	56.8	190.6	463	62.6	127.3	317.8	40.2	0.75	0.1	23.5	1.1
	$(ROPT)^2$ + SC	66.7	2064.9	157.3	210.9	2064.9	143	164.8	1813.9	118.8	6.19	20.5	1813.4	116.5
	All/OPN	90.7	367.4	65.8	190.6	438.1	62.4	100.1	351.7	48.7	0.49	0.1	4.1	0.4
	MALL/OPN	90.6	391.9	65.9	190.2	462.3	62.7	100.1	373.7	49	0.43	0.1	9.7	0.9

95% confidence interval

MAll/OPN for performance of effective flow time, which is the real lead time of order in this study.

(5) Comparing with the JDD rule, MADD has significant improvements for performance of percent tardy and tardiness. The reason is that the earliest completion time of the critical path is used as the assembly time if the order will be tardy, which will reduce the priority of all associated parts and it is more conducive to urgent orders processing prior. So that the efficiency of overall shop is improved. However, we also observed that the way of modified assembly time would produce greater deviation.

(6) Although All/OPN is better than MADD for performance of percent tardy, there is no improvement for performance of tardiness. While MAll/OPN is superior to All/OPN and MADD rules for any performance except flow time. Because MALL/OPN not only considers the due date and the number of remaining processes (production progress), but also considers the assembly time.

6 Conclusion

In this study, the dynamic scheduling of the assembly job shop with machine breakdown and preventive maintenance is studied. Three representative preventive maintenance policies are introduced. The results show that PM Policy II is better than other two policies. What's more, influence of maintenance on the dispatching rules is analyzed from view of the constituent attributes and new dispatching rule, MALLOPN (modified allowance per operations), is proposed. Results show that the comprehensive performance of MALLOPN is better than classic rules in a variety of cases.

Considering the economic loss and cost of maintenance is a future study direction, which is not taken into consideration in this study.

Acknowledgements. This research was supported by the National Natural Science Foundation of China (71572049, 61573109), the Pearl River S&T Nova Program of Guangzhou (201710010004), and the Special Plan Young Top-notch Talent of Guangdong (2016TQ03X364).

References

1. G.D. Silveira, D. Borenstein, F.S. Fogliatto, Mass customization: literature review and research directions. Int. J. Prod. Econ. **72**(1), 1–13 (2001)
2. O. Holthaus, C. Rajendran, Efficient dispatching rules for scheduling in a job shop. Int. J. Prod. Econ. **48**(1), 87–105 (1997)
3. T. Fry et al., The effects of product structure and sequencing rule on assembly shop performance. Int. J. Prod. Res. **27**(4), 671–686 (1989)
4. N. Adam, J. Surkis, Priority assignment procedures in multi-level assembly job shops. AIIE Trans. **19**(3), 317–328 (1987)

5. E. Borgonovo, M. Marseguerra, E. Zio, A Monte Carlo methodological approach to plant availability modeling with maintenance, aging and obsolescence. Reliab. Eng. Syst. Saf. **67** (1), 61–73 (2000)
6. H. Seidgar et al., Simulated imperialist competitive algorithm in two-stage assembly flow shop with machine breakdowns and preventive maintenance. Proc. Inst. Mech. Eng. Part B J Eng. Manuf. **230**(5) (2016)
7. C. Richardcassady, E. Kutanoglu, Minimizing job tardiness using integrated preventive maintenance planning and production scheduling. AIIE Trans. **35**(6), 503–513 (2003)
8. R. Ruiz, J.C. García-Díaz, C. Maroto, Considering scheduling and preventive maintenance in the flowshop sequencing problem. Comput. Oper. Res. **34**(11), 3314–3330 (2007)
9. P. Philipoom, R. Russell, T. Fry, A preliminary investigation of multi-attribute based sequencing rules for assembly shops. Int. J. Prod. Res. **29**(4), 739–753 (1991)
10. K.M. Mohanasundaram et al., Scheduling rules for dynamic shops that manufacture multi-level jobs. Comput. Ind. Eng. **44**(1), 119–131 (2003)
11. K. Natarajan et al., Performance evaluation of priority dispatching rules in multi-level assembly job shops with jobs having weights for flowtime and tardiness. Int. J. Adv. Manuf. Technol. **31**(7–8), 751–761 (2007)
12. R.S. Russell, B.W.T. Iii, An evaluation of sequencing rules for an assembly shop. Decis. Sci. **16**(2), 196–212 (1985)
13. H.L. Lu, G.Q. Huang, H.D. Yang, Integrating order review/release and dispatching rules for assembly job shop scheduling using a simulation approach. Int. J. Prod. Res. **49**(3), 647–669 (2011)
14. N.R. Adam et al., Due date assignment procedures with dynamically updated coefficients for multi-level assembly job shops. Eur. J. Oper. Res. **68**(2), 212–227 (1993)

An Analysis of Strategies for Coupled Design Tasks Reviewing with Random Rework

Meng-na Wang, Xiao-ming Wang(✉), Qing-xin Chen, and Ning Mao

Guangdong Provincial Key Laboratory of Computer Integrated Manufacturing, Guangdong University of Technology, Guangzhou 510006, China
simonwang@gdut.edu.cn

Abstract. The coupled design tasks are usually need to be confirmed by customer repeatedly before they can be completed. To evaluate the merits of the coupled design task reviewing strategy with the exponentially distributed durations, the project network is firstly transformed by regarding each rework task as a potential new task. Then, the state transition process during the project execution is described by a continuous-time Markov chain (CTMC). Finally, the probability distribution of project completion time is obtained based on the phase-type (PH) distribution, and its expectation is severed as the standard for evaluating the a given reviewing strategy. To validate the constructed model and method, a calculation example is illustrated with a simple project that consists of six activities. The experimental results show that the quality of a given reviewing strategy is related to the project environment. Meanwhile, the experimental results also bring some inspiration to the design project management.

Keywords: Coupled design · Reviewing strategy · Random rework CTMC · PH distribution

1 Introduction

Faced with the fierce market competition, companies must shorten the development cycle of new product as much as possible. The design process is often subdivided into multiple phases and multiple tasks. On the one hand, there exists some coupling relationships among the adjacent tasks. On the other hand, the customer's initial product demand is usually not clear, so that some tasks can only be completed after customer confirmation. Under the influence of the above two factors, design project manager will face a variety of alternative task reviewing strategies (the combination of tasks that submitted to customer for reviewing) and need to choose one of them to minimize the project completion time.

Obviously, submitting multiple tasks to the customer for confirmation at the same time can reduce the fixed time (such as communication and preparation time). However, it will accumulate the risk of mistakes in the subsequent tasks. Conversely, submitting each task individually to the customer can decrease the cumulative risk of error, but it might increase the fixed time-consuming. In addition, the uncertainty of customer's confirmation results in the uncertainty of project network structure, which

G. Q. Huang et al. (eds.), *Proceeding of the 24th International Conference on Industrial Engineering and Engineering Management 2018*,
https://doi.org/10.1007/978-981-13-3402-3_48

making it difficult to estimate the project completion time. Therefore, accurately estimate the project completion time is essential for assisting project manager in selecting the optimal reviewing strategy.

At present, some scholars have studied the problems with random rework. In the aspect of network modeling, scholars mainly study the steady-state performance analysis with random rework. Yu and Bricker [1], Pillai and Chandrasekharan [2] considered the rework and scrap in the manufacturing process and analyzed the expected indexes based on an absorbing Markov chain. Wang et al. [3] proposed a method based on the queue length approximation and CTMC for predicting the completion time of mold orders by considering single rework. In addition, some scholars have studied the project completion time prediction without rework based on Markov theory. Kulkarni and Adlakha [4] first studied the probability distribution of project completion time based on an absorbing CTMC. Dodin [5] proposed an approximate distribution function of project completion time for a large-scale random network with arbitrary structure and task duration distribution. Azaron and Fatemi Ghomi [6] compute the lower bound of the mean completion time of a dynamic Markov PERT network by using stochastic dynamic programming. Wang et al. [7] predict the delivery time of a mold project based on the state evolution in a discrete-time Markov chain. The above research shows that the Markov theory can accurately describe the execution process of stochastic projects. As for the modeling of a project with random rework, the Markov theory has the advantages that the other methods cannot match. Up to now, however, there is still none research focus on the coupled design tasks reviewing with random rework.

We focus on the design project with random rework and coupled tasks and study the quantitative evaluation method of a given reviewing strategy. This paper is organized as follows: First of all, the basic problem is described and defined. Then, an absorbing CTMC is constructed to describe the state transitions during the project execution. Further, the probability distribution of project completion time is calculated based on the PH distribution. Finally, we give a calculation example to analyze the impact of project environment on reviewing strategies and summarize its implications for design project management.

2 Methodology

Assuming that the design project is represented by an activity-on-node $G(N, E)$, where $N = (0, 2, \ldots, n)$ denotes the set of tasks and $E = ((j, i)|j, i \in N)$ denotes the precedence constraints. Let $\varphi \subseteq N$ denotes the set of tasks that need to be submitted to the customer for reviewing. The rework probability p_j of task $j \in \varphi$ is known and is assumed to be a constant. In addition, this paper also considers the coupling relationships between adjacent tasks, which expressed by the conditional probability of rework. Let $p(i|j)$ denotes the probability that task $i \in \varphi$ will also need to rework in the case of rework of task $j \in \varphi$. All of the conditional probabilities are assumed to be known. In addition, all regular durations, rework durations, and reviewing durations are assumed to be exponentially distributed. Without loss of generality, the reviewing duration is determined by the content of the task being reviewed, which is set to

$\beta \sum_{j \in S} d_j + \varsigma$, where β is the discount factor, ς is the fixed time, $S \subseteq \varphi$ is the set of tasks being reviewed, and d_j is the mean duration of task $j \in S$.

Since all of the durations are exponentially distributed, the project execution state has the Markov property (memoryless). On this basis, we can model the problem by utilizing CTMC. It is clearly that the project will stay in the finish state forever, which means that the constructed CTMC is an absorbing CTMC. An absorbing CTMC is two-tuple $\{X, Q\}$, where X is the state space and Q is the transition rate matrix. For the studied problem, these two elements are defined as follows.

The state space $X \cdot X$ is a collection of all possible project execution states. In the research of stochastic resource-constrained project scheduling problem based on Markov decision process, the system state is composed of a task set with three possible status (finished, idle, processing) [8], or composed of an idle task set I and a processing task set P [9], or composed of a waiting task set W and a processing task set P [10]. Since, there is no waiting task set without resource constraints in this paper, we define the system state as $x = P_x$.

The transition rate matrix Q. Let $q_{xx'}$ denotes the rate at which the system transfer from state x to state x' $(x, x' \in X)$. All of the transition rates among all of the system states constitutes the transition rate matrix Q. According to the property of transition rate matrix, the diagonal element $q_{xx} = -\sum_{x' \in X, x' \neq x} q_{xx'}$. Constructing the transition rate matrix Q is equivalent to computing the transition rates from a state to other states.

Let γ_j and p_{γ_j} represent the completion result of task $j \in P_x$ (either completed successfully or rework) and the corresponding probability, respectively. The next state x' is jointly determined by the current state x, the first completed task j and its completion result γ_j, which is represented by the transition function $x' = g(x, j, \gamma_j)$. Let $q_{x,g(x,j,\gamma_j)}$ denotes the transition rate from state x to state $x' = g(x, j, \gamma_j)$ when task j is completed first and its completion result is γ_j, which is given by

$$q_{x,g(x,j,\gamma_j)} = p_{\gamma_j} / E(d_j) \tag{1}$$

Obviously, the completion of different tasks may cause system transfer to the same next state. In this case, the state transition rates need to be accumulated.

The project completion time is essentially the time from the initial state to the absorbing state of the above CTMC, which can be described by the continuous time PH distribution ($T \sim PH(\boldsymbol{\alpha}, \mathbf{K})$). Therefore, we will calculate the theoretical probability distribution of the project completion time by using the PH distribution after constructing the CTMC of the project execution. To calculate the PH distribution, we must first accurately obtain its two parameters, namely the subgenerator matrix $\mathbf{K}$ and the probability row vector $\boldsymbol{\alpha}$.

Suppose that the constructed absorbing CTMC has a total of $n+1$ states, where the $n+1$-th state is the absorbing state, and the rest states are intermediate states. Then, the probability row vector $\boldsymbol{\alpha}$ corresponding to this CTMC is a $n+1$-dimensional row vector, which represents the probability that the system is in each state at the initial time, has $\sum_{i=1}^{n+1} \alpha_i = 1$. Since the initial status of the project is usually known (for example all of the tasks are not started), we have $\boldsymbol{\alpha} = (1, 0, \ldots, 0)$. The relationship between parameter $\mathbf{K}$ and the state transition rate matrix is represented by Eq. (2).

$$Q = \begin{pmatrix} \mathbf{K} & \mathbf{K}^0 \\ \mathbf{0} & 0 \end{pmatrix} \tag{2}$$

where the $\mathbf{K}$ is a n-order square matrix representing the transition rates among the n intermediate states; $\mathbf{K}^0 = -\mathbf{Ke}$ is a n-dimensional column vector describing the transition rate from the intermediate states to the absorbing state.

The probability density function and the cumulative distribution function of the PH distribution is shown in Eqs. (3) and (4), respectively. After obtaining the two parameters of the PH distribution through the above steps, the two equations can be employed to calculate the probability distribution of the project completion time.

$$f_T(t) = P(T = t) = \boldsymbol{\alpha} e^{\mathbf{K}t}(-\mathbf{K})\mathbf{e}, \quad t \geq 0 \tag{3}$$

$$F_T(t) = P(T \leq t) = 1 - \boldsymbol{\alpha} e^{\mathbf{K}t}\mathbf{e}, \quad t \geq 0 \tag{4}$$

In addition, the moment-generating function of the PH distribution is shown in Eq. (5). Calculating the first moment can obtain the expected project completion time, which can be used to evaluate the merits of the reviewing strategy (the smaller the project completion time, the better the corresponding reviewing strategy).

$$E(T^n) = (-1)^n n! \boldsymbol{\alpha} \mathbf{K}^{-n} \mathbf{e} \tag{5}$$

3 Results

To illustrate the application of the above model and method in evaluating a given reviewing strategies more clearly, in this section we will give a calculation example with a simple design project consists of six design tasks as shown in Fig. 1. In this project, tasks 2, 3 and 4 are need to be reviewed. The mean duration of the six tasks is 5, 6, 5, 5, 6 and 5, respectively. The mean rework duration is set to half of the corresponding mean regular duration. All of the durations are exponentially distributed. In addition, there are two pairs of tasks ((2, 3) and (2, 4)) that coupling with each other.

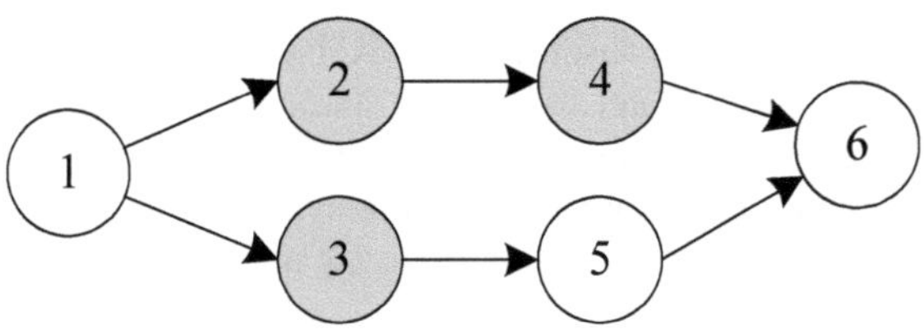

Fig. 1. A simple design project network with six nodes

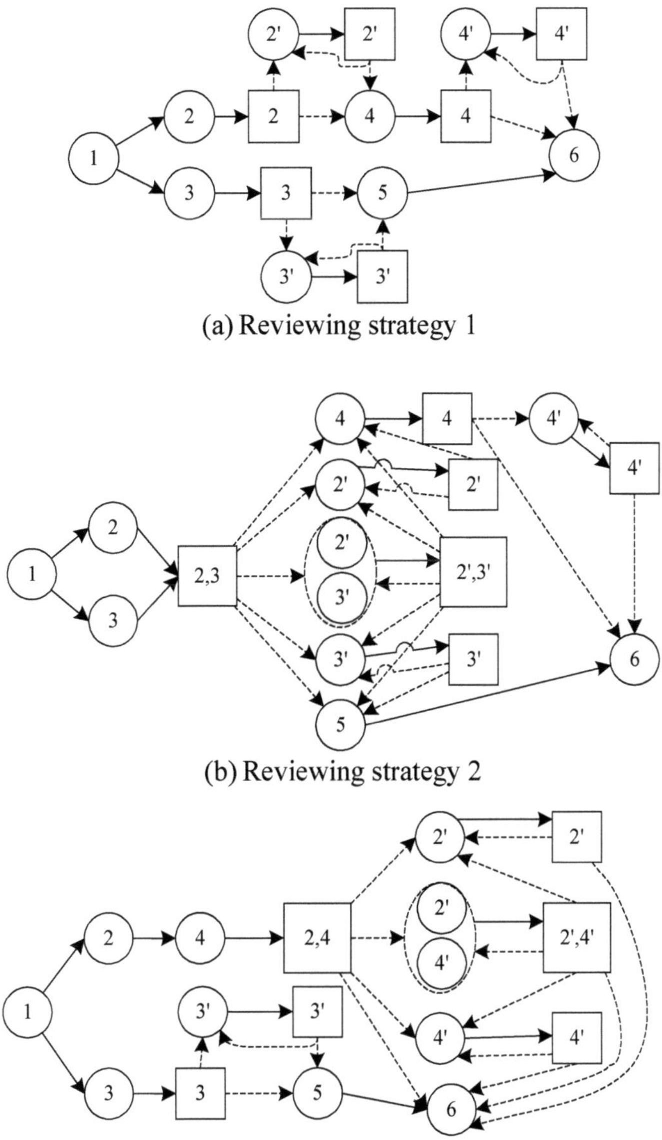

(a) Reviewing strategy 1

(b) Reviewing strategy 2

(c) Reviewing strategy 3

Fig. 2. Three reviewing strategies and corresponding project network

There are five feasible reviewing strategies in total for the above project. However, two of them are obvious not optimal. The remaining three feasible reviewing strategies are shown in Fig. 2. The first strategy (referred to as strategy 1, denoted by Π_1) indicates that task 2, 3 and 4 are submitted to the customer individually for reviewing. The second strategy (referred to as strategy 2, denoted by Π_2) indicates that task 2 and

3 are submitted to the customer together and task 4 is submitted to the customer individually for reviewing. The third strategy (referred to as strategy 3, denoted by Π_3) indicates that task 2 and 4 are submitted to the customer together and task 3 is submitted to the customer individually for reviewing. Since the probability distributions of the regular durations, reviewing durations and rework durations are different, we firstly transform the project network according to the given reviewing strategy. Afterwards, the mean project completion time can be obtained by applying the above CTMC model and PH distribution calculation method, based on which the best reviewing strategy can be determined.

To give a more comprehensive comparison of the above three reviewing strategies, we further consider a number of factors that may affect the project completion time. Intuitive analysis shows that the factors that may affect the project completion time include the reviewing duration (involves the duration discount factor and the fixed time), rework probability, and task relevance. For this reason, we construct different project environments by setting various parameter values. Among them, the duration discount factor $\beta = \{0.8, 0{,}9, 1.0, 1.1, 1.2\}$, the fixed time $\zeta = \{0, 5, 10\}$, the rework probability coefficient $c = \{0.05, 0.075, 0.1, 0.125, 0.15\}$ (the rework probability of task 2, 3 and 4 is set to $p_2 = 3c$, $p_3 = 4c$ and $p_4 = 5c$, respectively), the correlation coefficient $e = \{0.75, 1, 1.25, 1.5, 1.75\}$ (the conditional probability is set to $p(i|j) = p_{ij}/p_j$, $p_{23} = p_{32} = ec$, $p_{24} = p_{42} = ec$).

Based on the above example data, we get the absorbing CTMC by a program coded in Visual C# and compute the PH distribution by a program coded in Matlab. Due to space limitations, this paper only lists the comparison results of the probability distribution functions of the project completion time corresponding to the three strategies under a combination of parameters ($\beta = 0.8$, $c = 0.1$, $e = 1$), as shown in Fig. 3.

We will pay more attention to analyzing the impact of each parameter on the merits of the three reviewing strategies in detail by using the mean project completion time as the criterion. The comparison results are shown in Fig. 4, Fig. 5 and Fig. 6, respectively.

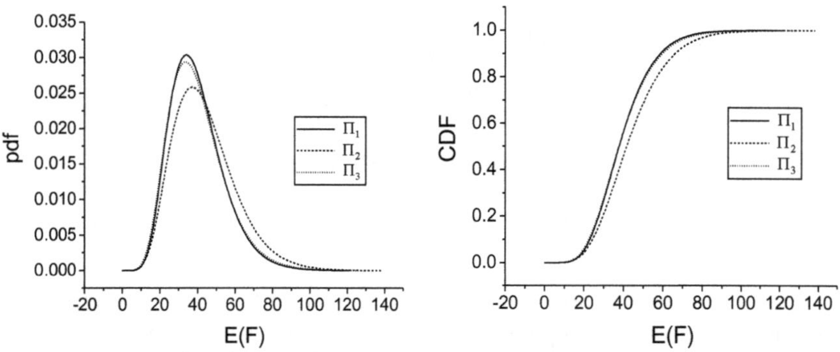

(a) Probability density function (b) Cumulative distribution function

Fig. 3. Comparison of the probability distribution of project completion time among the three strategies

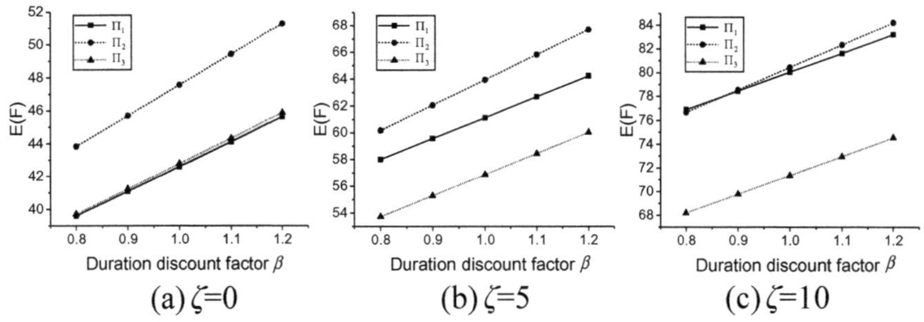

Fig. 4. Analysis of influence of β on reviewing strategies (c = 0.1, e = 1)

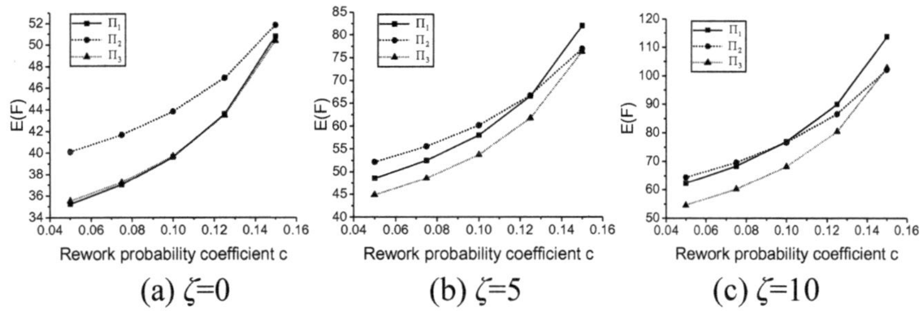

Fig. 5. Analysis of influence of c on reviewing strategies (β = 0.8, e = 1)

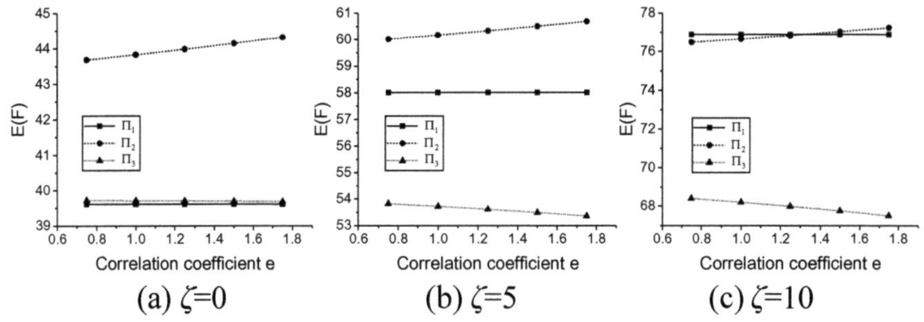

Fig. 6. Analysis of influence of e on reviewing strategies (β = 0.8, c = 0.1)

4 Discussion

From the above calculation results, it can be seen that the combination of different parameters has a significant impact on the merits of the given reviewing strategy. We divide the comparisons of the three strategies into two groups. The first group is to compare Π_1 and Π_2, which used to judge whether it is good or not to review the two parallel coupled tasks (2 and 3) together. The second group is to compare Π_1 and Π_3, which used to judge whether it is good or not to review the two serial coupled tasks (2 and 4) together. Specific analyses are as follows.

In the aspect of the duration discount factor (as shown in Fig. 4), the mean project completion time corresponding to the three strategies is approximately linear with the factor β. As for the comparison of Π_1 and Π_2, we see that in most cases the mean project completion time corresponding to Π_1 is smaller than that of Π_2, which indicated that Π_1 is better than Π_2. We also noticed that as the increase of fixed time ς, the curve of Π_1 increased significantly, while the curve of Π_2 has grown more slowly. When the fixed time is large and β is small (such as $\zeta = 10$ and $\beta = 0.8$), Π_2 is better than Π_1. The reason for this phenomenon is that when ς grows, Π_1 will increase the mean duration of three confirmation tasks, and Π_2 will only increase by two. As a result, the randomness of the project completion time is enhanced, and there is a certain probability that take large values. This advantage of Π_2 is more obvious when the influences of other factors are small. That is why the mean project completion time of the two strategies are getting closer when the duration discount factor is small. As for the comparison of Π_1 and Π_3, we see that the mean project completion time of Π_1 is smaller than that of Π_3 when $\varsigma = 0$, which means that Π_1 is better than Π_3. However, with the increase of fixed time ς, the mean project completion time of Π_3 is smaller than that of Π_1, and the advantage becomes more obvious. This indicated that Π_3 is better than Π_1 in these cases. The reason is that as ς grows, Π_1 needs to increase the mean duration in three confirmation tasks, and Π_3 only needs to increase by two, and Π_3 does not have the waiting time of coupling task for the customer to confirm.

In the aspect of rework probability coefficient (as shown in Fig. 5), the mean project completion time corresponding to the three reviewing strategies are polynomial increased with the grown of c. As for the comparison of Π_1 and Π_2, we see that Π_1 is better than Π_2 when ς and c is small. However, Π_2 is become better than Π_1 when ς and c is increased. In addition, similar to the previous results, Π_2 and Π_3 shows more stable than Π_1 when ς increased, and the internal causes is the less reviewing time. As for the comparison of Π_1 and Π_3, we see that the mean project completion time of the two strategies are very close when $\varsigma = 0$. However, Π_3 is gradually better than Π_1 with the increase of ς, and the reason is the same as it described in previous.

In the aspect of the correlation coefficient (as shown in Fig. 6), the mean project completion time corresponding to the three strategies is also approximately linear with the factor e. As for the comparison of Π_1 and Π_2, we see that the mean project completion time corresponding to Π_1 doesn't change while the value corresponding to Π_2 are increased (the growth rate decreased with the grown of ς) when e increased. Since tasks 2, 3 and 4 are reviewed separately according to strategy Π_1, the changes in the correlation coefficient has no effect on the CTMC and does not change the mean project completion time. For strategy Π_2, task 2 and 3 are reviewed at the same time. Although the reviewing time is shortened, the simultaneous assessment of concurrent tasks increases the waiting time. As a result, when the review fixed time is small, the total time for the simultaneous assessment of concurrent tasks is still greater than the single task. It will increase the transition probability in the CTMC that the system transfer to the simultaneous rework of the two tasks when e is increased, which further increase the mean project completion time. However, when the review fixed time is large (such as $\zeta = 10$, the advantage is more obvious and Π_2 is better than Π_1), that is, the slower the mean project completion time increased. As for the comparison of Π_1 and Π_3, we see that in most cases the mean project completion time of Π_3 is smaller than that of Π_1, which indicated that Π_3 is better than Π_1. In addition, we see that the mean project completion time corresponding to Π_3 is decreased when e is increased, and the decreasing speed is growing simultaneously with ς.

For Π_3, tasks 2 and 4 are reviewed at the same time, it will not only shorten the reviewing time, but also no waiting time between the serial tasks. This is the reason that Π_3 superior to Π_2 in almost all cases.

The above calculation results and analyses provide some inspirations for the design project management decisions with rework and coupled tasks. First of all, in the actual project, it is necessary to select the appropriate reviewing strategy according to the environment in which the project is located. None strategy can guarantee optimality under all circumstances, it is necessary to use the CTMC model and the PH distribution-based method to perform theoretical calculations to select the optimal reviewing strategy. Second, the rework probability has a significant impact on the project completion time and the performance of reviewing strategy. Specifically, the project completion time will grow in a polynomial manner with the increase of rework probability. Therefore, in the practical design project management, companies should reduce the rework probability of design tasks as much as possible. Finally, the experimental results also show that the fixed time for reviewing also has a significant effect on the project completion time. Since the fixed time is not directly attached to the project completion time, but constantly accumulated during the rework process. Therefore, in the practical design project management, urging customers to complete the review as soon as possible is also important to shorten the design project cycle.

5 Conclusion

This paper focus on the analyze of reviewing strategies of a new product design project that considering random rework and coupled tasks, proposes the use of absorbing CTMC to describe the transition process among the project execution states, uses the PH distribution to calculate the probability distribution of project completion time, and select its expectation as the review strategies criterion. To illustrate the effectiveness of the modeling and calculation methods described above, a simple project with six tasks is used as a calculation example. We analyzed the advantages and disadvantages of the three feasible reviewing strategies under the combination of different review duration discount factors, rework probability coefficients, the correlation of coupled tasks, and review fixed time parameters. The experimental results show that the pros and cons of a given reviewing strategy are related to the project environment, and the obtained results also bring some inspiration to the design project management.

Although the model and calculation method constructed in this paper can effectively evaluate the merits of the coupled design task reviewing strategies, there are still many related issues that deserve to be further studied. In the next phase of work, we will consider the issue of task duration that is subject to a general distribution and will also study the corresponding optimization problem.

Acknowledgements. This work was sponsored by the National Natural Science Foundation of China (No. 51505090, No. 51775120 and No. 61573109).

References

1. K.-Y.C. Yu, D.L. Bricker, Analysis of a Markov chain model of a multistage manufacturing system with inspection, rejection, and rework. IIE Trans. **25**(1), 109–112 (1993)
2. V.M. Pillai, M.P. Chandrasekharan, An absorbing Markov chain model for production systems with rework and scrapping. Comput. Ind. Eng. **55**(3), 695–706 (2008)
3. X. Wang, Q. Chen, N. Mao et al., Estimation of mold remaining duration considering reworks. J. Mech. Eng. **50**(7), 200–208 (2014)
4. V.G. Kulkarni, V.G. Adlakha, Markov and Markov-regenerative pert networks. Oper. Res. **34**(5), 769–781 (1986)
5. B. Dodin, Approximating the distribution functions in stochastic networks. Comput. Oper. Res. **12**(3), 251–264 (1985)
6. A. Azaro, S.M.T. Fatemi Ghomi, Lower bound for the mean project completion time in dynamic PERT networks. Eur. J. Oper. Res. **186**(1), 120–127 (2008)
7. X. Wang, Q. Chen, N. Mao et al., Mould projects due-date forecast methods under random environment. Comput. Integr. Manuf. Syst. **18**(2), 405–414 (2012)
8. S. Creemers, R. Leus, M. Lambrecht, Scheduling Markovian PERT networks to maximize the net present value. Oper. Res. Lett. **38**(1), 51–56 (2010)
9. S. Creemers, Minimizing the expected makespan of a project with stochastic activity durations under resource constraints. J. Sched. **18**(3), 263–273 (2015)
10. X.-M. Wang, R. Leus, S. Creemers et al., A CTMDP-based exact method for RCPSP with uncertain activity durations and rework, in *OR2017 Conference*, Berlin, Germany, 2017

Logistics Engineering

Research on Influential Factors of the Coal Mine Production Logistics System Efficiency Based on Rough Set

Yan-liang Zhang, Jun-ya Li(✉), and Yao-nan Liu

School of Management Engineering, Zhengzhou University, Zhengzhou 450001, China
642533104@qq.com

Abstract. According to the problems that so many factors that influence the efficiency of logistics system of coal mine production, and the complex effect of various factors on it, an analysis model of influencing factors of coal mine production logistics system efficiency based on rough set is proposed. Firstly, the efficiency index system of coal mine production logistics system is established, and then the rough set attribute importance theory is used to sort the influencing factors of coal mine production logistics system efficiency, and the key influencing factors are selected. Finally, the feasibility of the model is verified by empirical analysis of 15 coal mines. The results show that, the rationality of equipment layout, staff work skills, equipment mechanization and transport route planning rationality are the key factors, which can effectively improve the resource allocation efficiency of coal mine production logistics system.

Keywords: Coal mine production logistics system · Efficiency factors
Rough set theory

1 Introduction

In the development of China's national economic, the coal industry has long been in the irreplaceable basic position. The process of coal production, including coal mining, crushing and down-hole and ground transportation, is a complete logistics process, as well as water, air, material flow and equipment flow. Statistics show that the total cost of production of coal enterprises in China, the logistics cost accounted for more than 50%, thus, the efficiency of coal mine production logistics is an important factor in the safe and efficient production of coal enterprises, to some extent, affected the production efficiency of coal mines. Through what means can improve the efficiency of coal mine production logistics, and thus to promote coal production more efficient, has become the current coal enterprises are facing the most important issues.

In recent years, coal production logistics research mainly focuses on the macro strategy research, management pattern optimization, information platform construction and the construction of the index system. For example, Yang and Shao [1] stated that the reason for the low efficiency of coal production is the lack of scientific and rational

G. Q. Huang et al. (eds.), *Proceeding of the 24th International Conference on Industrial Engineering and Engineering Management 2018*,
https://doi.org/10.1007/978-981-13-3402-3_49

planning, therefore, proposed to improve the existing logistics system through the logistics information technology. Zhang [2] conducted a comprehensive analysis on traditional coal mine logistics management and control mode, realized the rational allocation of coal mine materials and the whole process of information management, and promoted the efficiency of coal mine logistics. Markovits-Somogyi and Bokor [3] proposed the data envelopment analysis combined with analytic hierarchy process to form a new tool for efficiency evaluation. Momeni et al. [4] proposed a multi-objective additive mesh DEA model for evaluating and selecting the most appropriate logistics service provider, and using case to verify the effectiveness of the model. Wang et al. [5] identified the coal mine production logistics system resource allocation efficiency evaluation index, and established the evaluation model of coal mine production logistics resource allocation efficiency by using data envelopment analysis method. Sun and Fan [6] optimized the M/D/C model, established the coal mine logistics and transportation system efficiency model, and used Origin software to simulate and predict the queuing status of the coal mine.

In summary, the research on logistics efficiency at home and abroad mainly focus on the construction of logistics efficiency evaluation index system and evaluation methods, but the research on coal mine efficiency is relatively few. There is few studies are mostly limited to the mathematical model. On the basis of previous studies, the index system of influencing factors on the efficiency of coal mine production logistics system was established. According to the rough set theory to reduce the key influencing factors, and implement more scientific configuration and input in order to improve the resource allocation efficiency of coal mine production logistics system.

2 Rough Set Theory

In 1982 Z. Pawlak first proposed the concept of rough set theory, which is used to solve the problem of uncertainty and imperfections. For the boundary problem, the classical logic can only be divided into true or false, but in daily life inevitably there will be some vague phenomenon, can not simply determine the true or false. Fuzzy sets focus on the ambiguity of knowledge and the uncertainty of the set boundary. Rough sets are different from this. Rough sets will focus on the roughness of knowledge and pay more attention to the uncertainty of the relationship between sets of objects. On the basis of the existing knowledge base, to show the uncertainty or imperfect knowledge, which can be more objective description of the knowledge. Rough set is a kind of data analysis method which mainly deals with incomplete or imprecise information. It focus on the roughness of knowledge and pay more attention to the uncertainty of the relationship between the set objects. Finding the implicit rule from incomplete or inaccurate data is the purpose of the existence of rough sets. Rough set theory have been widely used in many areas, such as financial crisis warning, construction project risk assessment, medical decision-making, logistics demand forecast, etc. The following is a few concepts of rough sets:

(1) Information system. An information system is a data set that is usually used to represent the form of a data table, and an information system can be represented as

a S = (U, A, V, f). In an information system, it usually consists of the following: the universe (U) is the whole object, U = {X_1, X_2, …, X_n}; The attribute (A) is the whole property; Range (V), V_a represents the range of attribute a; Information function (f), f: U × A→V, refers to the attribute value of object X ∈ U.

(2) Decision table. An information system that can be represented by the disjoint condition attribute C and the target attribute D, the mathematical expression can be expressed as: A = C ∪ D, and C ∩ D = ∅, such an information system can be called a decision table.

(3) Distinct Matrix. The information function assigns an information value to each attribute of each object, and the distinct matrix is a m × n order matrix, whose column i, row j, column element t_{ij} [7] is

$$t_{ij} = \begin{cases} \{c_k | c_k \in C \cap c_k(x_i) \neq c_k(x_j)\}, d(x_i) \neq d(x_j), k \leq n \\ 0, d(x_i) = d(x_j) \end{cases} \tag{1}$$

(4) Attribute importance. For an identifiable matrix T(ij) = $(t_{ij})_{m \times n}$, attribute a ∈ A, then the importance of attribute a in A [8] is

$$\sigma(a) = \sum_{i=1}^{n} \sum_{j=1}^{n} \frac{\lambda_{ij}}{|t_{ij}|} \tag{2}$$

In which $|t_{ij}|$ represents the number of attributes that t_{ij} contains, when $a \notin t_{ij}$, $\lambda_{ij} = 0$; when $a \in t_{ij}$, $\lambda_{ij} = 1$.

3 The Efficiency Factors Analysis Model of Coal Mine Production Logistics System

3.1 Construction of Indicator System of Influencing Factors of Coal Mine Production Logistics System Efficiency

Considering the dynamic complexity of coal mine production logistics system, through the analysis of related theories, including foreign academic papers, monographs and some classic books, analysis the key issues, and consult with experts and professors in the field of coal mine production logistics to select the relevant indicators, and following the principles of scientific, comprehensive, independent stability and the combination of qualitative and quantitative to build the index system. This paper puts forward the classification of the whole coal mine production logistics system from four aspects: management and technical personnel, machinery equipment and facilities planning, disaster emergency rescue, material supply and circulation, and systematically analyzes the efficiency of coal mine production logistics system, depth exploration of system efficiency and the impact of the mechanism between the various factors. The coal mine production logistics system index system of factors is shown in Table 1.

Table 1. Coal mine production logistics system index system

First grade indexes	Second grade indexes	Index meaning
Professional management and technical personnel	Staff education level C1	The education level of coal mine staff
	Staff work skills C2	Working skill level of coal mine staff
	Staff training rate C3	The proportion of trained employees in coal mines
Machinery equipment and facilities planning	Equipment mechanization C4	Mechanical level of coal mine facilities
	Equipment informatization C5	Degree of informatization of coal mine facilities and equipment
	Equipment layout rationality C6	Reasonable layout of coal mine equipment
	Equipment maintenance and repair C7	Maintenance and overhaul of coal mining and transportation equipment
Material supply and circulation	Material availability timeliness C8	Time of logistics supply in coal mine production
	Material inventory C9	Coal mine material inventory level
	Reasonable plan of transport routes C10	Reasonable degree of route planning in coal transportation
Disaster emergency rescue	Risk monitoring capability C11	Detect and monitor accidents ability before disaster
	Accident recovery capability C12	The coal mine resumed production capacity after the accident
	Emergency rescue capability C13	Emergency rescue plan and implementation of coal mine after disaster
	Emergency readiness capability C14	Emergency preparedness for accidents before a disaster accident occurs

3.2 Data Collection and Processing

Data standardization. Taking 15 coal mines as the research object, the qualitative and quantitative methods are combined to collect and sort out the index data. The quantitative indexes are divided into extremely small indexes and extremely large indexes. The former is dealt with according to formula (3), and the latter is dealt with according to formula (4).

$$y_{ij} = \frac{x_{\max} - x_{ij}}{x_{\max} - x_{\min}} \times 100 \tag{3}$$

$$y_{ij} = \frac{x_{ij} - x_{\min}}{x_{\max} - x_{\min}} \times 100 \tag{4}$$

For qualitative indexes, organize relevant experts to conduct field research and use the Delphi method quantitative indexes. The results of data standardization are shown in Table 2.

Table 2. Standardized summary of indexes

Coal mine	Evaluation index value														d
	C_1	C_2	C_3	C_4	C_5	C_6	C_7	C_8	C_9	C_{10}	C_{11}	C_{12}	C_{13}	C_{14}	
2	29	39	52	56	0	94	100	86	78	50	46	94	100	100	923
3	3	89	84	75	54	44	97	80	31	37	23	47	0	63	726
4	91	44	87	63	89	100	18	3	42	67	89	78	100	43	912
5	14	25	3	3	37	81	29	74	72	3	11	11	69	49	484
6	83	100	6	81	40	19	100	77	64	73	17	0	58	46	765
7	26	75	97	78	57	28	12	20	100	80	3	39	97	54	766
8	51	50	74	75	83	72	15	20	58	73	100	61	78	63	873
9	91	50	94	81	29	6	50	57	69	0	83	61	17	23	711
10	66	6	90	16	83	41	71	20	28	53	46	81	97	11	707
11	80	61	84	97	29	31	9	74	61	10	69	50	89	6	749
12	29	8	71	0	57	47	21	3	14	33	69	25	61	60	497
13	97	25	26	100	51	9	100	60	8	77	100	31	67	34	785
14	49	19	81	9	9	53	91	11	44	57	94	42	33	37	630
15	71	92	23	9	94	75	59	69	78	57	49	47	50	49	821

Data discretization. The method of equal distance division is used to discretize the data. Use "1" to represent the numbers in the interval [min, min + step], similarly, use "2" and "3" to represent the numbers in the interval [min + step, min + 2 * step] and [min + 2 * step, max] respectively, among them step = (max − min)/3. The data in Table 2 are discretized, and the results are shown in Table 3. The numbers "1", "2", and "3" in Table 3 indicate the indicators "general", "better" and "good".

Distinct Matrix. In order to calculate the attribute importance of coal production logistics system efficiency, the identification matrix should be established first.

According to the established decision table and formula (1), the distinct matrix is calculated, as shown in Table 4.

3.3 Attribute Importance Calculation

According to the heuristic algorithm of distinct matrix, the attribute importance of each factor is calculated according to formula (2), and the calculation results of each index importance are: $f(c_1) = 5.3562$, $f(c_2) = 5.7975$, $f(c_3) = 3.8586$, $f(c_4) = 5.5570$, $f(c_5) = 5.4169$, $f(c_6) = 5.8669$, $f(c_7) = 4.6534$, $f(c_8) = 5.1086$, $f(9) = 5.1328$, $f(c_{10}) = 5.5205$, $f(c_{11}) = 4.9701$, $f(c_{12}) = 5.2098$, $f(c_{13}) = 5.0594$, $f(c_{14}) = 3.2926$. The results show that, the impact of each factors on coal mine production logistics system efficiency degree from big to small order is equipment layout rationality staff work skills, equipment mechanization, reasonable planning of transport routes, equipment informatization, staff education level, emergency readiness capability, accident recovery capability,

Table 3. Index discretization summary table

Coal mine	Evaluation index value														d
	C_1	C_2	C_3	C_4	C_5	C_6	C_7	C_8	C_9	C_{10}	C_{11}	C_{12}	C_{13}	C_{14}	
1	1	2	3	3	1	3	3	1	2	2	3	3	1	2	2
2	1	2	2	2	1	3	3	3	3	2	2	3	3	3	3
3	1	3	3	3	2	2	3	3	1	2	1	2	1	2	2
4	3	2	3	2	3	3	1	1	2	3	3	3	3	2	3
5	1	1	1	1	2	3	1	3	3	1	1	1	3	2	1
6	3	3	1	3	2	1	3	3	2	3	1	1	2	2	2
7	1	3	3	3	2	1	1	1	3	3	1	2	3	2	2
8	2	2	3	3	3	3	1	1	2	3	3	2	3	2	3
9	3	2	3	3	1	1	2	2	3	1	3	2	1	1	2
10	2	1	3	1	3	2	3	1	1	2	2	3	3	1	2
11	3	2	3	3	1	1	1	3	2	1	3	2	3	1	2
12	1	1	3	1	2	2	1	1	1	1	3	1	2	2	1
13	3	1	1	3	2	1	3	2	1	3	3	1	3	2	3
14	2	1	3	1	1	2	3	1	2	2	3	2	1	2	1
15	3	3	1	1	3	3	2	3	3	2	2	2	2	2	3

material inventory, material availability timeliness, emergency rescue capability, risk monitoring capability, equipment maintenance and repair, staff training rate.

4 Discussion and Conclusion

The factors have different attribute importance to the efficiency of coal mine production logistics system, and have some regularity. According to the results of the attribute importance, equipment layout rationality, staff work skills, equipment mechanization, reasonable planning of transport routes are the key factors affecting the efficiency. In terms of the present situation of the coal mine production logistics system efficiency, there are some shortcomings in the above aspects. It is necessary for coal mine management to pay attention to the resource inputs in these areas and the various among factors.

A method for analyzing the factors affecting the efficiency of coal mine production logistics system based on rough set theory is proposed, and identify the key influencing factors of coal mine production logistics system. The rough set is used to deal with the characteristics of fuzzy and uncertain data, and the attribute importance of the influencing factors is identified, which makes the calculation result more objective. According to the results, the coal mine enterprises can carry out more scientific and rational allocation and input to the resources, improve the efficiency of resource allocation of coal mine production logistics system, and other similar enterprises have some reference value.

Because the coal mine production logistics system has both dynamic and complex characteristics, there are more complex factors in the actual production and factors

Table 4. Distinct matrix

·	X1	X2	X3	X4	X5	X6	X7	X8	X9	X10	X11	X12	X13	X14	X15
X1	0														
X2	C3 C4 C8 C9 C11 C13 C14	0													
X3	0	C2 C3 C4 C5 C6 C9 C11 C12 C13 C14	0												
X4	C1 C4 C5 C7 C10 C13	0	C1 C2 C4 C5 C6 C7 C8 C9 C10 C11 C12 C13	0											
X5	C2 C3 C4 C5 C7 C8 C9 C10 C11 C12 C13	C2 C3 C4 C5 C7 C10 C11 C12 C14	C2 C3 C4 C6 C7 C9 C10 C12 C13	C1 C2 C3 C4 C5 C8 C9 C10 C11 C12	0										
X6	0	C1 C2 C3 C4 C5 C6 C9 C10	0	C2 C3 C4 C5 C6 C7 C8	C1 C2 C4 C6 C7 C9 C10 C13	0									

(continued)

Table 4. (*continued*)

·	X1	X2	X3	X4	X5	X6	X7	X8	X9	X10	X11	X12	X13	X14	X15
		C11 C12 C13 C14		C11 C12 C13											
X7	0	C2 C3 C4 C5 C6 C7 C8 C10 C11 C12 C14	0	C1 C2 C4 C5 C6 C9 C11 C12	C2 C3 C4 C6 C8 C10 C12	0	0								
X8	C1 C5 C7 C10 C12 C13	0	C1 C2 C5 C6 C7 C8 C9 C10 C11 C13	0	C1 C2 C3 C4 C5 C8 C9 C10 C11 C12	C1 C2 C3 C5 C6 C7 C8 C11 C12 C13	C1 C2 C5 C6 C9 C11	0							
X9	0	C1 C3 C4 C6 C7 C8 C10 C11 C12 C13 C14	0	C4 C5 C6 C7 C8 C9 C10 C12 C13 C14	C1 C2 C3 C4 C5 C6 C7 C8 C11 C12 C13 C14	0	0	C1 C5 C6 C7 C8 C9 C10 C13 C14	0						
X10	0	C1 C2 C3 C4 C5 C6 C8 C9 C14	0	C1 C2 C4 C6 C7 C9 C10 C11 C14	C1 C3 C5 C6 C7 C8 C9 C10 C11 C12 C14	0	0	C2 C4 C6 C7 C9 C10 C11	0	0					

(*continued*)

Table 4. *(continued)*

.	X1	X2	X3	X4	X5	X6	X7	X8	X9	X10	X11	X12	X13	X14	X15
								C12 C14							
X11	0	C2 C3 C4 C5 C6 C7 C8 C9 C10 C11 C12 C13 C14	0	C4 C5 C6 C8 C10 C12 C14	C1 C2 C3 C4 C5 C6 C9 C11 C12 C14	0	0	C1 C5 C6 C8 C10 C14	0	0	0				
X12	C2 C4 C5 C6 C7 C9 C10 C12 C13	C1 C2 C3 C4 C5 C6 C8 C9 C10 C11 C12 C14	C2 C4 C7 C8 C10 C11 C12 C13	C1 C2 C4 C5 C6 C9 C10 C12 C13	0	C1 C2 C3 C4 C6 C7 C8 C9 C10 C11	C2 C4 C6 C9 C10 C11 C12 C13	C1 C2 C4 C5 C6 C9 C10 C12 C13	C1 C2 C4 C5 C6 C7 C8 C9 C12 C13 C14	C1 C5 C7 C10 C11 C12 C13 C14	C1 C2 C4 C5 C6 C8 C9 C12 C13 C14	0			
X13	C1 C2 C3 C5 C6 C8 C9 C10 C12 C13	0	C1 C2 C3 C6 C8 C10 C11 C12 C13	0	C1 C4 C6 C7 C8 C9 C10 C11	C2 C8 C9 C11 C13	C1 C2 C3 C7 C8 C9 C11 C12	0	C2 C3 C5 C7 C9 C10 C12 C13 C14	C1 C3 C4 C5 C6 C8 C10 C11 C12 C14	C2 C3 C5 C7 C8 C9 C10 C12 C14	C1 C3 C4 C6 C7 C8 C10 C13	0		
X14	C1 C2 C4 C6 C12	C1 C2 C3 C4 C6 C8 C9 C11 C12 C13 C14	C1 C2 C4 C5 C8 C9 C11	C1 C2 C4 C5 C6 C7 C10	0	C1 C2 C3 C4 C5 C6 C8 C10 C11	C1 C2 C4 C5 C6 C7 C8 C9 C10	C2 C4 C5 C6 C7 C8 C10 C13	C1 C2 C4 C6 C7 C8 C9 C10 C14	C5 C8 C9 C11 C12	C1 C2 C4 C6 C7 C8 C10	0	C1 C3 C4 C5 C6 C8 C9 C10	0	

(continued)

Table 4. (*continued*)

·	X1	X2	X3	X4	X5	X6	X7	X8	X9	X10	X11	X12	X13	X14	X15
				C12 C13		C12 C13	C11 C13			C13 C14	C13 C14		C12 C13		
X15	C1 C2 C3 C4 C5 C7 C8 C9 C11 C12 C13	0	C1 C3 C4 C5 C6 C7 C9 C11 C13	0	C1 C2 C5 C7 C10 C11 C12 C13	C4 C5 C6 C7 C9 C10 C11 C12	C1 C3 C4 C5 C6 C7 C8 C10 C11 C13	0	C2 C3 C4 C5 C6 C8 C10 C11 C13 C14	C1 C2 C3 C6 C7 C8 C9 C12 C13 C14	C2 C3 C4 C5 C6 C7 C9 C10 C11 C13 C14	C1 C2 C3 C5 C6 C7 C8 C9 C10 C11 C12	0	C1 C2 C3 C5 C6 C7 C8 C9 C11 C13	0

interact with each other. While in this paper, the analysis of the factors influencing the efficiency of coal mine production logistics system is not enough systematic and comprehensive. Therefore, more comprehensive and in-depth analysis of this problem is needed.

Acknowledgements. This research is supported by Humanity and Social Science Youth foundation of Ministry of Education of China under Grant #11YJC630291 to Yanliang Zhang.

References

1. Z.H. Yang, B. Shao, Research on coal production logistics system. Coal Technol. **01**, 271–272 (2012). (Chinese)
2. J. Zhang, Development and application of coal mine logistics control system. Mod. Min. **11**, 4–6 (2014). (Chinese)
3. R. Markovits-Somogyi, Z. Bokor, Assessing the logistics efficiency of European countries by using the DEA-PC methodology. Transport **29**(2), 137–145 (2014)
4. E. Momeni, M. Azadi, R.F. Saen, Measuring the efficiency of third party reverse logistics provider in supply chain by multi objective additive network DEA model. Int. J. Shipp. Transp. Logist. **7**(1), 21–41 (2015)
5. J.F. Wang, Y.C. Zhu, L.J. Feng, Research on resource allocation efficiency of coal mine production logistics system. Coal Eng. **08**, 114–117 (2014). (Chinese)
6. Y.L. Sun, L.B. Fan, Logistics efficiency and queuing system prediction model-case studies on coal mining enterprises. Logist. Eng. Manag. **03**:125–127+116 (2016) (Chinese)
7. B. Ding, D. Chen, The evaluation of emergency logistics plan of local government based on rough set and FAHP-FEC. Syst. Eng. **27**(4), 7–11 (2009). (Chinese)
8. L. Wu, Y.Z. Yang, Model of green supplier based on rough set and entropy theory. Ind. Eng. Manag. **16**(2), 34–41 (2011). (Chinese)

Design of Integrated Logistics Model for Ceramic Enterprises in Liling

Anquan Zou(✉) and Jun Liu

Department of Business Administration, Foshan University,
Foshan 528000, China
zaq2833@163.com

Abstract. This paper analyzes the development of ceramic industry in Liling and the current situation of its logistics, and analyzes the existing problems of ceramic Logistics in Liling as follow: 1st, enterprise's understanding and management of logistics is fall behind; 2nd, information asymmetry in procurement links; 3rd, absence of support for third party logistics; 4th, policy guidance is not clear in development of logistics. We use cost analysis and production ratio reciprocal method to optimize the production flow of ceramic products. We design logistics network information platform which constructed integrated logistics model of integrate proprietary logistics of ceramic enterprises, third party logistics and four level information platform. Our work improve the green development of ceramic enterprises in Liling to provide active exploration.

Keywords: Liling ceramic · Logistics · Pattern design

1 Introduction

Ceramic production in China has a long history, mainly in China's Guangdong Foshan, Hunan Liling, Jiangxi Jingdezhen and other regions. Liling ceramics known at home and abroad, called Liling a must, known as the "Oriental ceramic art of the peak", as well as the United States and the United States [1]. With the economic development, the ceramic industry has become a pillar industry of Liling industrial economy. Due to the large demand in developed countries, ceramic products exports increased rapidly, the development of the logistics industry has become a bottleneck restricting the development of Liling ceramics. Based on this, around the future development of Liling ceramic industry cluster, the urgent need to build to promote the green development of Liling ceramic integrated logistics model to meet the development of modern logistics.

This article is Supported by the National Social Science Foundation of China: "Carbon emission measurement and control study of ceramic supply chain in low carbon manufacturing" (17BJY008).

G. Q. Huang et al. (eds.), *Proceeding of the 24th International Conference on Industrial Engineering and Engineering Management 2018*,
https://doi.org/10.1007/978-981-13-3402-3_50

2 Analysis of Liling Ceramic Logistics

2.1 Liling Ceramic Industry Overview

Liling ceramic industry is mainly concentrated in the urban area, suburbs and 106 National Road along the southeast of the Pukou, Wang Xian, Huang Sha, Dalin, Sun Jia Wan, Ka Shui, Si Fen and other towns, the distribution range of about 40 km^2, Table 1 shows, Table 1 is from 2011 to 2016 Liling City daily ceramic production, industrial ceramics production and its growth.

Table 1. Output and growth of domestic and industrial ceramics in Liling from 2011 to 2016

	Daily-use ceramics production (1000000000 pieces)	Year-on-year growth (%)	Industrial ceramics production (10000 tons)	Year-on-year growth (%)
2011	18.32	15.9	24.02	14.47
2012	22.83	24.7	30.80	28.20
2013	26.69	16.9	37.15	20.60
2014	31.81	16.5	43.61	18.50
2015	40.20	26.40	54.51	25.00
2016	51.06	27.01	70.42	29.20

Ceramic production increased year by year [2], from the ceramic production required for the procurement of raw materials to the production process of loading and unloading handling and circulation processing, and then to the finished product storage and distribution of its logistics requirements will be further improved.

2.2 Status of the Logistics Industry in Liling

In recent years, Liling logistics demand is growing rapidly, the rapid development of logistics enterprises. At present, the Trade and Industry Bureau registered 55 transportation, warehousing and postal enterprises. Among them: express delivery business 16, storage 9, postal 1, passenger and freight and logistics information business 29; engaged in transport, distribution services, 22, 6 warehousing services, freight information services 27, its composition As shown in Fig. 1.

With the development of export-oriented economy, the ceramic industry on the efficient, professional logistics services demand gradually expanded in 2015, Liling ceramic cargo traffic more than 7 million tons. Table 2 is the 2015 Liling cargo transport volume and composition. According to the future development trend, we can sum up the Liling logistics industry there is a big room for development.

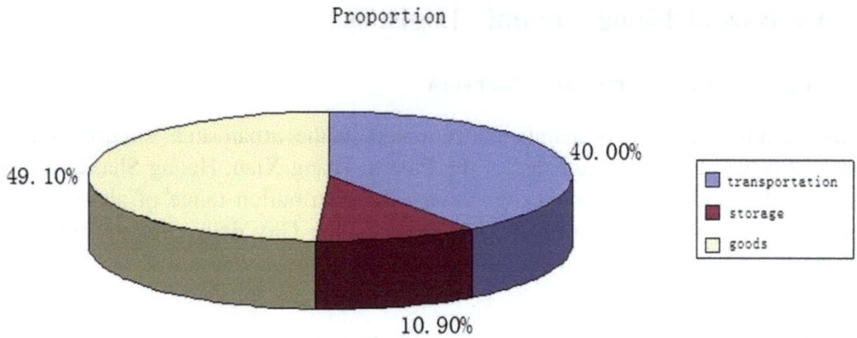

Fig. 1. Types of logistics enterprises in Liling

Table 2. Volume and composition of goods transportation in Liling (Unit: 10000 tons)

Category	Annual traffic	Remarks
Mineral products	1300	
Ceramic products	320	
Ceramic raw and auxiliary materials	290	Which Guangdong, Guangxi, Henan and other places to buy 54 million tons
Petroleum products and natural gas	40	Xinjiang
Coal	40	Which Shanxi, Shaanxi, Henan, the purchase of 310,000 tons
Color packaging and products	70	
Chemical products and products	23	
Building materials	100	
Bamboo, wood raw materials and products	50	
Agricultural means of production	4	
Agricultural products	17	Which transported 90,000 tons, shipped out of 80,000 tons
Other industry	45	
Tertiary industry	48	

2.3 Characteristics of Ceramic Production Logistics

The main features of ceramic production logistics are as follows:

(1) Liling ceramic industry, the basic form of enterprise logistics to self-run, and the existence of backward logistics equipment, business management is not clear and other issues.

(2) there is a vicious competition between the ceramic enterprises, there is no sense of win-win, lack of communication led to information asymmetry, and further lead to waste of resources, and ultimately make the ceramic product costs.
(3) ceramic logistics facilities and equipment is not perfect, and most of the third party logistics in the initial stage, can not meet the diversification of ceramic business needs. The base of the ceramic logistics is fragmented and not scale. Modern scale ceramic logistics base has not yet formed.
(4) ceramic enterprises with raw materials, complex ingredients, raw materials and production enterprises located far away from the location of raw materials changes in the larger varieties of products, the market is broad, industrial agglomeration and other characteristics, more suitable for third-party logistics [3].
(5) ceramic logistics costs relative to other industries higher. From the ceramic production process analysis that the ceramic business logistics generally include the following links. as shown in Fig. 2.

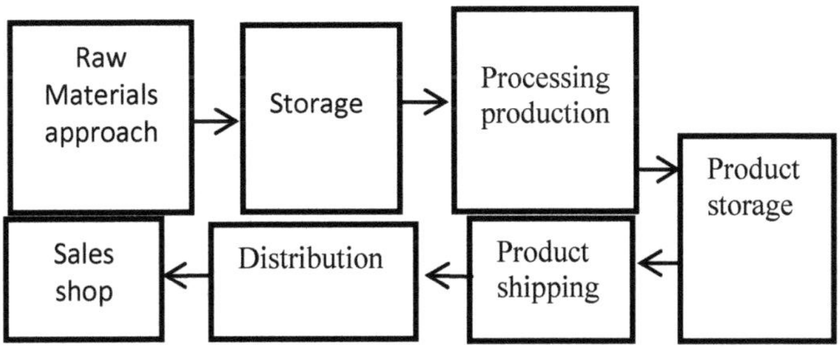

Fig. 2. Logistics link of ceramic enterprise

In the production of ceramic logistics, its ceramic production has a lower continuity, ceramic production process mechanization, low degree of automation, ceramic production cycle is longer, ceramic production needs to consume a lot of energy and so on.

2.4 Liling Ceramic Logistics of the Main Problems

(1) *the existence of ceramic enterprises themselves*: Research concluded that the majority of ceramic enterprises lack the correct understanding of modern logistics [4]. Ceramic production line layout is irrational [5], a number of enterprises of the production line confusion, resulting in the production process to produce some unnecessary logistics activities, making the cost of ceramic products increased. While unreasonable inventory management, resulting in the backlog of goods.
(2) *there is information between enterprises do not share the problem*: There is a lack of cooperation among the ceramic enterprises, especially in the area of information sharing, many companies are afraid to seize the orders by competitors, the information blockade, slow response to the market [6], resulting in great waste of logistics [7].

(3) *Third party logistics is insufficient*: Liling third party logistics enterprises are mostly based on distribution and cargo information services, service projects are more single, many transport companies do not respond to the development of the times in a timely manner, can not provide for the bulk cargo transport and effective logistics services support.
(4) *logistics industry development policy guidance is not clear*: Ceramic production enterprises "big and complete", "small and complete", that is, from the clay production, processing, glaze development to ceramic production, packaging, sales by the enterprise itself to complete the independent division of labor between the division of labor system has not yet formed [8]. Liling has not yet formulated the relevant logistics industry to guide the policy.

3 Liling Ceramic Integrated Logistics Model Design

3.1 Optimize the Self-logistics Model of Ceramic Enterprises

From the concrete situation of the ceramic enterprises, the establishment of the construction of the objectives and principles, logistics costs and organizational structure optimization, pipeline production process optimization and logistics management rationalization as the main content of the self-logistics model.

3.2 Logistics Costs and Organizational Structure Improvement

The production cost of ceramic enterprises is shown in Table 3.

Table 3 shows that all aspects of ceramic production are involved in the logistics, including the ceramic production before the procurement of raw materials, transportation and warehousing and so on logistics functions. Liling most of the ceramic enterprise management organizational structure shown in Fig. 3.

Through the above chart, we can find that there are some shortcomings, mainly the division of labor is not clear, the production workshop partition is too small and so on. This kind of organizational structure, whether from the enterprise production or management point of view is not conducive to enterprise management, so for the ceramic production of this logistics links more enterprises, this paper combines the requirements of modern logistics management, Liling ceramic enterprises. The structure has been improved, the improved organizational structure shown in Fig. 4.

The above production process, each process can not be separated from the transport of this logistics link, which means that if you can not rationally arrange the production line, will greatly increase its production costs.

Ceramic production is a process has been stereotypes, advanced technology, many types, large output, and the long process of strong demand. The process of ceramic products can be simply divided into several processes, according to the mixed pipeline to design its production line, so that in a pipeline can be produced in accordance with the order of ceramic production of a variety of ceramic products. To solve this problem generally use the production than the reciprocal method. Take the reciprocal of the ratio of the number of products as a measure of whether the product is put into production

Table 3. Compostion of cost of product and logistics of ceramic enterprise

Project cost	Cost item range	Logistics-related cost
Raw material	Raw material: including blank, glaze Outsourcing semi-finished products: including ceramic green body and fetal porcelain Gold liquid: including palladium water, electro-optical water, gold water and so on Other ceramic decorative materials: including for the product, underglaze painted blue and white pigments and seasoning agent; glaze, glaze, glaze under the glaze paper Kiln molds: including gypsum, plastic, steel mold and so on Packaging: contains packaging materials and packages that are sold separately from the product	Transportation cost Package cost Loading and unloading costs
Fuel and power	Including coal, heavy oil, wax oil, diesel	Transportation cost
Wages and benefits	Refers to the direct participation in the production of workers wages and calculated according to the provisions of the extraction of employee benefits, according to the provisions of the cost of special raw materials and fuel savings	Logistics information and management fees
Workshop funding	Including: wages, employee benefits and so on	Circulation processing cost
Enterprise management fee	Including wages and benefits, design drawing fees, rental fees, warehouse funds, transportation costs	Storage cost Transportation cost

priority criteria, the smaller the production than the reciprocal of the product output in the total output of the larger proportion of priority production.

A production line to produce X, Y, Z three kinds of ceramic products, their production were 60,000, 40,000, 20,000, according to the production than the reciprocal method to determine the order of the three ceramic products put into production.

(1) Calculate the production ratio Xi

Xx = 60000/20000 = 3
Xy = 60000/20000 = 2
Xz = 60000/20000 = 1
The sum of the production is 6, then a cycle is 6.

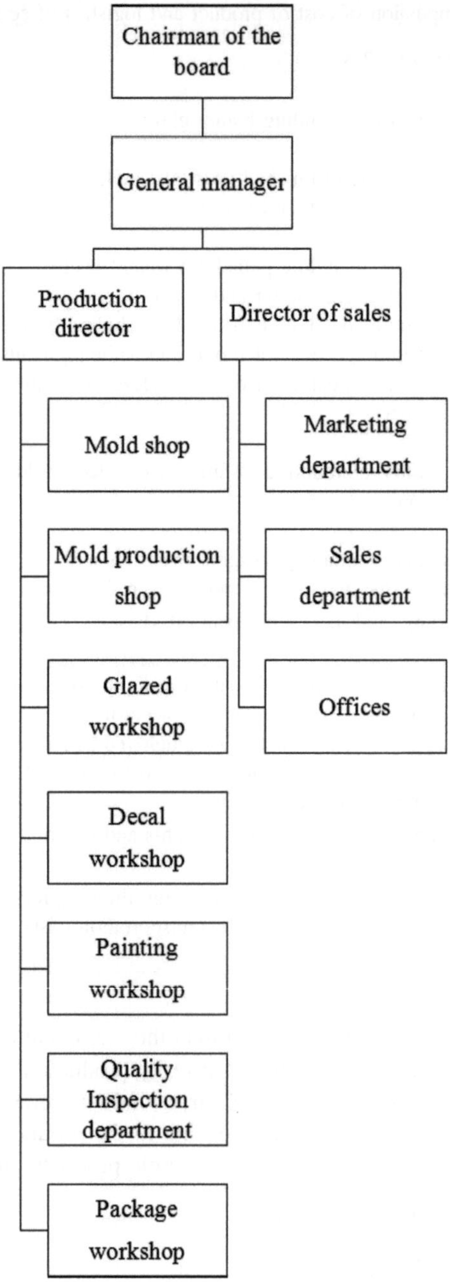

Fig. 3. Organization chart of ceramic enterprise before improvement

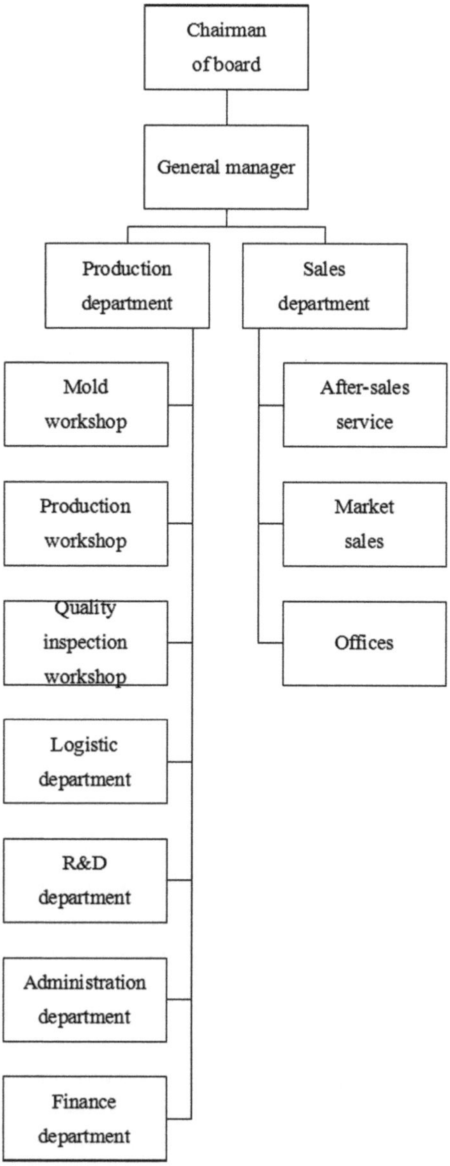

Fig. 4. Main process of ceramic production

(2) Calculate production than countdown nj

$$nx = 1/Xx = 1/3$$
$$ny = 1/Xy = 1/2$$
$$nz = 1/Xz = 1.$$

(3) According to the production of the first few lower than the first production rules to determine the order of production, and the selected ceramic products with * logo; if the minimum production than the reciprocal is the same to take the product to be identified later to vote.
(4) Update the production than the reciprocal value, select the next product line into the product. The order of the three ceramic products is shown in Table 4.

Table 4. Determine production order by production ratio reciprocal method

Counting number	X ceramic product	Y ceramic product	Z ceramic product	Production order
1	1/3*	½	1	X
2	2/3	1/2*	1	XY
3	2/3*	1	1	XYX
4	1	1*	1	XYXY
5	1*	/	1	XYXYX
6	/	/	1*	XYXYXZ

Therefore, the order of the three ceramic products is X-Y-X-Y-X-Z.

4 Conclusion

Through the analysis of the development status and logistics status of Liling ceramics industry, this paper discusses the existing ceramic logistics in Liling, which is concerned with the logistics knowledge and management. The information asymmetry in the procurement process, the lack of support for the third party logistics enterprises and the development of the logistics industry. Clear and other issues, the use of cost analysis and production than the reciprocal method to optimize the production process of ceramic enterprises, ceramic enterprises through the integrated platform, ceramic enterprises can be based on order processing system to develop production plans, while the development of material demand planning, and information feedback To the raw material suppliers, to prepare their supply, to focus on procurement, reduce the waste of raw materials, improve material utilization and reduce procurement costs, in the platform, companies can make full use of capital flow, information flow, logistics To improve the core competitiveness of ceramic enterprises.

References

1. D.J. Beebe, Signal conversion, in *Biomedical Digital Signal Processing*, ed. by W. J. Tompkins (Prentice-Hall, Englewood Cliffs, NJ, 1993), ch. 3, pp. 61–74
2. M. Akay, *Time Frequency and Wavelets in Biomedical Signal Processing* (IEEE Press, Piscataway, NJ, 1998), pp. 123–135

3. G.B. Gentili, V. Tesi, M. Linari, M. Marsili, A versatile microwave plethysmograph for the monitoring of physiological parameters. IEEE Trans. Biomed. Eng. **49**(10), 1204–1210 (2002)
4. T. Menendez, S. Achenbach, W. Moshage, M. Flug, E. Beinder, A. Kollert, A. Bittel, K. Bachmann, Prenatal recording of fetal heart action with magnetocardiography (in German). Zeitschrift für Kardiologie **87**(2), 111–118 (1998)
5. J.E. Monzon, The cultural approach to telemedicine in "Latin American homes," in *Proceedings of 3rd Conference Information Technology Applications in Biomedicine, ITAB′ 00*, Arlington, VA, pp. 50–53
6. F.A. Saunders, Electrotactile sensory aids for the handicapped, presented at the *4th Annual Meeting Biomedical Engineering Society*, Los Angeles, CA, 1973
7. J.R. Boheki, Adaptive AR model spectral parameters for monitoring neonatal EEG, Ph.D. dissertation, Biomedical engineering program, University Federal Rio de Janeiro, Rio de Janeiro, Brazil, 2000
8. J.P. Wilkinson, Nonlinear resonant circuit devices, U.S. Patent 3 624 12, 16 July 1990

Security Decisions in a Networked Supply Chain with Integration

Han-yue Zhang and Jing Xie(✉)

College of Management and Economics, Tianjin University, Tianjin, China
xiejing@tju.edu.cn

Abstract. Communication network provides an important premise for the development of supply chain integration, but also brings more and more severe information security risks. Thus, the information security of each firm depends both on the firm's own investment, as well as on the strategies of security made by supply chain firms. Using game theory model, this paper discusses the investment on security and sharing of the security information of the supply chain firms. Particularly, we analyze the impact of supply chain's integration, and inherent vulnerability of network on firms' security strategies. The results show that if a firm increases the investment on security, the other firm tends to free-riding. In addition, compared with the joint decision-making of firms, they will have less security investment when making decisions separately. Hence, firms should better form an information-sharing alliance to coordinate their security decisions.

Keywords: Security information sharing · Security investment
Integration degree

1 Introduction

Rapid innovation on information technology provides essential prerequisites for the development of supply chain integration, revealing that supply chain firms can share and integrate the business information more effectively through the Internet, therefore the operation efficiency of the supply chain is improved. However, the inherent vulnerability of the network has also increased due to the widespread popularity of the Internet and the increasing complexity of the information system. In recent times, the network attacks and security vulnerabilities are becoming a major problem, in the meanwhile, the concern of information security is increasingly serious, especially in the field of supply chain [1]. The business integration among member firms is becoming more closely with the continuous development of supply chain integration, leading to the stronger dependence of the members on the network. In this situation, each firm's information security is not only affected by his own security capital, but also associated with related firms' security strategy. Hence, the supply chain is facing with more severe security risk.

During a long time, people are focusing on the development of the advanced information security technologies e.g. firewall, access control mechanism, anti-virus software, information encryption technology and other hardware devices, so as to

G. Q. Huang et al. (eds.), *Proceeding of the 24th International Conference on Industrial Engineering and Engineering Management 2018*,
https://doi.org/10.1007/978-981-13-3402-3_51

expand the extension of information security effectively and directly, leading to an enhanced information system security and reduced loss rising from security accidents.

Nevertheless, the advanced technology can't guarantee the security of information system fundamentally [2], and high risks of network attacks still exists. People realized that besides a technical problem, the security of information is also involves many other issues in economic and management, which needs to be solved by comprehensive implements of technology, management, economic theories and methods. In early 1990s, Ross Anderson proposed that "Economics explains this better than technical factors do" [3], that is, economic analyzing can interpret many problems which seem hard for security experts to solve simply through techniques. Since then, the economics of information security began to develop and has gradually become a thriving and fast-moving discipline.

Increasing investments on security information is a straightforward way to enhance the system security for firms, including employing technical experts, updating hardware and maintenance of security management. In addition, more and more people have realized that an ideal solution to the security problem of information systems is sharing information about vulnerabilities and failed attempts of loopholes to exploit vulnerabilities, as well as developing a scheme to block with respect to detecting and repairing security vulnerabilities [4].

Different firms' information systems may suffer similar attacks in their information systems all the time. The main reason is the use of a general database or communication network [5]. Hence, the sharing of security information is not only beneficial to the firms themselves but also to the whole industry, and even the whole society. First, through the sharing of security information, firms can relatively reduce their investment in information security. Additionally, because of the reduction of duplication of efforts, firms can identify and prevent vulnerabilities in time, as well as take targeted remedies more rapidly if there has been an actual vulnerability attack, leading to minimized vulnerability losses. Second, the behavior of sharing security information between firms enhances the effectiveness of network security products, increasing the confidence of customers, leading to the spillover effect of demand, consequently, bringing the positive network externality of the entire industry. In a certain extent, information sharing is conducive to improve the whole system security level, resulting in the promotion of the overall social welfare including the industry and end users.

Therefore, growing attention has been paid to the security information sharing by governments and many private enterprises. For instance, the American government once actively promoted the development of SB/ISOs (security-based information sharing organizations), such as the CERT/CC (CERT Coordination Center), (information Sharing and Analysis Centers) [4].

Although the benefits of security information sharing are beyond any doubt, it also introduces many risks and disadvantages. For instance, sharing the unique security information will strengthen the security level of the rival, increasing the competitive edge and causing indirect harm to itself. Sharing the previous security breaches may damage its intangible assets, e.g. reputation and then lead to negative influence on the customer's trust. Therefore, a double edged sword is involved when firms determine whether to share or how much to share security information. Furthermore, if a firm finds that other firms have made security investments, it will have an incentive to take a free-riding, that is, to invest less on the security reduce their own safety investments [5]. Thus, as an effective tool to enhance the degree of information security as well as increase the social welfare, the sharing of security information still has many problems to be studied and solved.

2 Literature Review

Research on information security investments has been addressed in the literature of information economics.

From the point of view of "private supply of public goods", Varian [6] analyzed the multi-firm's security investment, and determined the free-rider effect.

Within a decision-theoretic model incorporating inherent vulnerability of one information system and potential loss of one firm, Gordon and Loeb [7] reveal that the firm may not necessarily shift its emphasis to such information set with the highest vulnerability, and demonstrate that the optimal level of expenditure to protect the given information assets is 37% of the expected loss due to the security vulnerabilities.

Besides, information sharing has also been extensively studied in the economics literature in the context of non-security related organizations, particularly in the field of TAs (trade associations), and JVs (joint ventures) [8–10]. But the research in the field of information security is still not sufficient. As an early scholar of security information sharing, Gordon et al. [4] study that the sharing of security information between firms is how to influence the level of security investment. They show that the firm will decrease its security investment if the other firm make an information sharing on security. And they also highlight the possibility of free-riding, which could result in insufficient investment on security. By a Bertrand duopoly framework, Gal-Or and Ghose [11] analyze the sharing of security knowledge is how to influences the investments on security and the competition of price between two firms. They point out that the investments as well as sharing of information on security are complementary in the Nash equilibrium, which is in contrast to Gordon's result. Liu et al. [12] analyze the property of information assets owned by the firms play a vital role in the sharing of knowledge as well as investment decisions that affect information security.

Meanwhile, the interdependence of security risks is also an important area in document research. Ogut et al. [13] point out that there exists a reduction in the motivation of security investment due to the interdependence of security risks between firms. Using a game-theoretic framework, Kunreuther and Heal [14] analyze the equilibrium level of security investment between interdependent firms. Hausken [15] examines that the firms' security investment is how to affected by the degree of substitution between different targets [16].

Nevertheless, the existing literature mainly focused on the duopoly competition environmental, and didn't systematically analyze the impact of security information sharing among supply chain firms.

3 Model

3.1 The Model Description

Considering a supply chain that makes up of two firms, that we denote as i, j. In order to prevent from being attacked, firms will simultaneously decide the optimal investments on security, denoted as t_i, t_j , as well as the optimal levels of information sharing on security with the other firm, denoted as s_i, s_j. Due to the inherent vulnerability, each firm may be directly breached by a hacker. By sharing security information, the probability of a firm's information system being breached directly will depend on its own security investment, as well as another firm's sharing of security information and the security investment. We use P to represent the probability function of the firm's security vulnerability, and taking firm i as an example, it can be expressed as: $P_i = P(t_i + \eta t_i s_j)$, where $\eta \in (0, 1)$, measures the effectiveness of information sharing. Specifically, we specify that the property of function P is a convex and two order derivable function, which decreases with the effective investment, i.e., $(t_i + \eta t_i s_j)$. In order to facilitate further analysis, we present a specific function form $P_i(t_i, y_j) = \frac{1}{1 + t_i + \eta t_i s_j}$.

Meanwhile, because the two firms' information systems are physically connected through the network, and it's vulnerable of the network that easily breached due to the spread of loophole between two firms. Thus, when a firm has been attacked by a security breach, the breach may spread through the network to the other firm, bring an indirect attack on it. We use the constant q to measure the network's inherent security vulnerability, that is, the possibility of an indirect default occurs of a firm when the other firm suffered a direct default. Thus, the probability that the firm i's information system is breached indirectly can be expressed as qP_j.

When a security breach occurs, firms incur a monetary loss, which could be tangible such as the loss of business, or intangible such as reputation and competitiveness. Due to the integration of supply chain, the two firms will share information about the business. So the two firms' losses are coupled together. Unlike the vulnerability dependence, the loss mentioned above depends on the degree of supply chain integration. And we use parameter ρ to measure the degree of SC integration of two firms' business. Thus, firm i's loss caused by a security breach can be expressed by the following two cases: (i) L if firm i is breached, including direct and indirect attacks; (ii) ρL if firm j is breached, including direct and indirect attacks.

What's more, there also exists two externalities because of the sharing of security information. For one thing, it will bring the risk of firm's information exposure, thereby generating the "leakage costs" ks_i. For another thing, it can also bring a positive influence on the efficiency or consumer demand of firms. We measure this additional benefit as $\lambda t_j s_j$ with respect to firm i, which is depends on the other firm's decision variables.

According to the above assumptions, the two firms i, j solves the following problem.

$$Min\ C_i = t_i + ks_i - \lambda t_j s_j + L(P_i + qP_j) + \rho L(P_j + qP_i) \quad (1)$$

$$Min\ C_j = t_j + ks_j - \lambda t_i s_i + L(P_j + qP_i) + \rho L(P_i + qP_j) \quad (2)$$

where C_i, C_j represent the expected total cost of investment and sharing in information security for firm i, j.

Table 1 provides a summary of all the notations.

Table 1. Summary of notations

Notation	Brief definitions
t_i, t_j	The levels of security investment
s_i, s_j	The levels of sharing on security information, $s_i, s_j \in [0, 1]$
q	The inherent security vulnerability of the network
ρ	The degree of integration between supply chain firms
L	The firms' loss by a security breach
η	The effectiveness of information sharing
k	The nonnegative coefficient of leakage costs
λ	The positive spillover of security information sharing

3.2 Comparative Statics

Both firms simultaneously determine their levels of investment and sharing. The optimal first-order conditions are as follows:

$$\frac{\partial C_{i/j}}{\partial t_{i/j}} = 0 \quad (3)$$

$$\frac{\partial C_{i/j}}{\partial s_{i/j}} = 0 \quad (4)$$

By symmetry, two firms are facing the same reaction function. Thus, we can obtain the symmetric Nash Equilibrium solutions in the situation that they separately make security decisions, as shown below.

$$s_e = \frac{1}{\eta t}\left(\sqrt{\frac{\eta L t(\rho + q)}{k}} - t - 1\right) \quad (5)$$

$$t_e = \frac{1}{1 + \eta s}\left(\sqrt{L(1 + \rho q + \eta s(\rho + q))} - 1\right) \quad (6)$$

Through comparative static analysis of each parameter in the equilibrium solutions, we can draw the following propositions.

Proposition 1.

(1) The equilibrium sharing level of information security increases with the degree of supply chain integration and the network vulnerability, but decreases with the leakage costs of sharing.
(2) The equilibrium security investment increases with the degree of integration of the supply chain system, the leakage costs of sharing, as well as the network vulnerability.

This proposition shows that the sharing levels of security information and security investment is how changing with each parameter under Nash equilibrium. For one thing, with the vulnerability of the supply chain network increasing, the probability of security breaches spreading will be increased, leading to a larger probability of indirect attack from other firms, thereby increasing the expected security loss. For another thing, the higher the degree of integration of the supply chain, the closer the business association between firms, and a firm will have more sensitive core business information which is owned by other firms. Thus, if a cooperation firm is successfully attacked, the related losses will be greatly increased. Therefore, when making security decisions, the firm should take into account not only enhancing the level of information security itself, but also try to enhance the cooperation firms' information security level, so as to reduce the influence of associated risk, and protect its own information assets more effectively. Given these considerations, the firm will increase the security investment to improve its own security level, and also share more security information with other related firms.

In addition, when the leakage cost is high, security information sharing will bring more invisible loss to the firm, such as the reduction of consumer confidence and loyalty, which can bring greater business losses. Hence, firms will tend to share less security information, but invest more in contrast, to make up for the declined information security level due to lack of security information sharing of other firms.

Proposition 2. With security information sharing, one firm increases its own investment on security will result in a reduction in the level of investment on security as well as sharing of the information on security by the competitor.

This proposition shows the interaction between the sharing of security information and investment. It reveals the phenomenon of free-riding in the field of information security. That is, if a firm can benefit from other supply chain firms' investments on security, but shouldn't be responsible for these investments, that all firms will tend to use the security information of other members freely, for the sake of taking advantage of other firms' security investment to increase its own information security level. Consequently, the behavior of free riding usually bring about an insufficient investment in the security of information, so that the information security of the entire supply chain may suffer a greater threat.

4 Model Extension

This chapter will extend the above basic model to study the situation that the supply chain firms form an information-sharing alliance (ISA), and make the security decisions jointly, namely, each firm aims at maximizing the benefit of the entire supply chain.

Thus, the two firms i, j solves the following problem jointly:

$$W = \sum_{i=1}^{2} [t_i + ks_i - \lambda t_j s_j + L(P_i + qP_j) + \rho L(P_j + qP_i)] \quad \mathrm{j} = 3 - \mathrm{i} \tag{7}$$

where W represents the expected total cost of information security of the entire supply chain.

We assume that the two companies are similar, so we now work on the symmetric solutions, that is, $t_i = t_j = t^*$, $s_i = s_j = s^*$, where t^* and s^* represent the optimal level of investment and security sharing of the entire supply chain respectively.

The information-sharing alliance decides t^* and s^* in order to minimize the total information security cost of the entire supply chains. The optimal first-order conditions about t^* and s^* are as follows:

$$\frac{\partial W}{\partial t^*} = 0 \tag{8}$$

$$\frac{\partial W}{\partial s^*} = 0 \tag{9}$$

Then, we can derive the optimal solutions when the two firms make security decisions jointly, as shown below.

$$s^* = \frac{1}{\eta t}\left(\sqrt{\frac{\eta L t(1+\rho)(1+q)}{k - \lambda t}} - t - 1\right) \tag{10}$$

$$t^* = \frac{1}{1+\eta s}\left(\sqrt{\frac{t(1+\rho)(1+q)(1+\eta s)}{1 - \lambda s}} - 1\right) \tag{11}$$

By comparing and analyzing the optimal solutions of the entire supply chain and the equilibrium solutions in the previous chapter, we can draw the proposition as below.

Proposition 3. When firms make security decisions jointly, the optimal level of investment on security as well as security information sharing is higher than that of individual decisions.

This proposition states that, when two firms make security decisions individually, the investment on security is insufficient, and the sharing of information on security has not reached the optimal level of the supply chain. This is because security investment can only create implicit value, but firms generally tend to invest more in areas where they can produce explicit value directly, thus, firms usually choose free riding, that is, have less motivation in security investment. In addition, if there is no regulation or incentive, because each firm is assumed to be rational, firms will maximize their own profits, therefore, they will not voluntarily improve the sharing level of security information. However, when firms make security decisions in order to minimize the safety cost of the whole supply chain, the externality of the network is internalized, thus eliminating the "free rider" phenomenon.

5 Conclusion and Discussion

The analysis of this paper is that the decisions of investment and information sharing on security in the supply chain firms. As we all know, the communication network can make the links between the supply chain firms more closely, which is conducive to the development of the supply chain integration, and can significantly improve the operation efficiency of the supply chain. However, communication network can also greatly increased the information security risk of the supply chain, and any node firm is likely to become a potential threat or a potential victim. Hence, the information security degree of a supply chain firm depends both on the firm's own investment, as well as on the sharing level of security information made by other supply chain firms.

Our results show that when the investment on security by a firm increases, the level of sharing and security investment on other firms will decrease, that is, the phenomenon of "free riding" caused by moral hazard. In fact, firms who choose the behavior of "free riding" will not only get benefit, but greatly increase the security vulnerability of itself and the entire networked supply chain.

In addition, this study finds that if the degree of the supply chain integration is high, firms will not only increase the investment on security but also improve the sharing's level of security information, thereby reducing the associated losses caused by the high degree of integration. But even so, firms' equilibrium level of security investment and sharing is still insufficient, namely, still below the optimal level of supply chain. Therefore, we suggest that supply chain firms should better form an information-sharing alliance to coordinate security decisions for the reason that maximize the security level of the entire networked supply chain, and each firm will also have a higher security level than the level when they make decisions individually.

Despite the above contributions, there still remain limitations in our study. First, we assume that the supply chain is composed of only two firms, thus, future research could further explore more complicated situation where the supply chain consists of n ($n > 2$) firms. Second, our model only consider that the two firms are risk neutral, and we can further consider how the risk averse attitude will affect the conclusion when the firm makes the security decisions.

References

1. T. Bandyopadhyay, V. Jacob, S. Raghunathan, Information security in networked supply chains: impact of network vulnerability and supply chain integration on incentives to invest. Inf. Technol. Manag. **11**(1), 7–23 (2010)
2. R. Anderson, Why cryptosystems fail, in *Proceedings of the 1st ACM Conference on Computer and Communications Security*, New York, USA, pp. 215–227, 1993
3. R. Anderson, T. Moore, The economics of information security. Science **314**(5799), 610–613 (2006)
4. L. Gordon, M. Loeb, W. Lucyshyn, Sharing information on computer systems security: an economic analysis. J. Account. Public Policy **22**(6), 461–485 (2003)
5. M.H.R. Khouzani, V. Pham, C. Cid, Strategic discovery and sharing of vulnerabilities in competitive environments, in *International Conference on Decision and Game Theory for Security*. Springer International Publishing, pp. 59–78 (2014)
6. H. Varian, System reliability and free riding. Econ. Inf. Secur. **2**(5799), 1–15 (2004)
7. L. Gordon, M. Loeb, The economics of information security investment. ACM Trans. Inf. Syst. Secur. **5**(4), 438–457 (2002)
8. W. Novshek, H. Sonnenschein, Fulfilled expectations in Cournot duopoly with information acquisition and release. Bell J. Econ. **13**(1), 214–218 (1982)
9. D. Fried, Incentives for information production and disclosure in a duopolistic environment. Q. J. Econ. **99**(2), 367–381 (1984)
10. E. Gal-Or, Information sharing in oligopoly. Econometrica **53**(2), 329–343 (1985)
11. E. Gal-Or, A. Ghose, The economic incentives for sharing security information. Inf. Syst. Res. **16**(2), 186–208 (2005)
12. D. Liu, Y. Ji, V. Mookerjee, Knowledge sharing and investment decisions in information security. Decis. Support Syst. **52**, 95–107 (2011)
13. H. Ogut, N. Menon, S. Raghunathan, Cyber insurance and IT security investment: impact of interdependent risk, in *Proceedings of Weis'*, 2005
14. H. Kunreuther, G. Heal, Interdependent security. J. Risk Uncertain. **26**(2–3), 231–249 (2003)
15. K. Hausken, Income, interdependence, and substitution effects affecting incentives for security investment. J. Account. Public Policy **25**(6), 629–665 (2006)
16. X. Gao, W. Zhong, S. Mei, Security investment and information sharing under an alternative security breach probability function. Inf. Syst. Front. **17**(2), 423–438 (2013)

Investigation on Musculoskeletal Disorders of the Workers in Automobile Production Logistics

Lin-hui Sun(✉), Yi-meng Zhang, Kang Shang, and An-bo Wu

The School of Management, Xi'an University of Science and Technology, Xi'an, China
45642572@qq.com

Abstract. To explore the effect of ergonomics risk factors on workers' musculoskeletal disorders in automobile production logistics. Based on the video of the four main tasks of unloading operation, storage operation, unpacking operation and SPS (set parts system) sorting operation, the authors evaluated the musculoskeletal disorders status of workers during the operation by the Rapid Upper Limb Assessment. The results show that unpacking operations need to be further investigation. Storage operations and SPS sorting operations are yet to be studied and should be improved as soon as possible. Unloading operations need to be studied and improved immediately.

Keywords: Automobile production logistics · Logistics workers
Musculoskeletal disorders (MSDs) · Rapid upper limb assessment (RULA)

1 Introduction

Automobile production logistics plays an important role in automobile logistics industry. It refers to the complex process of material flow in the process of automobile production, such as raw materials, spare parts and auxiliary materials. In the case of mechanization and intelligent has not yet fully realized, the existing traditional manual handling operations have brought heavy workload and pressure to the workers. Its health problems should arouse our high attention. Investigation of automobile production logistics workers musculoskeletal disorders (MSDs), analysis of risk factors of occupation, focus on prevention, the implementation of ergonomics intervention strategies will help to protect the staff's physical and mental health, reduce the loss of enterprises and the national economy, improve the overall competitiveness of the country.

How to correctly evaluate the ergonomic risk factors in automobile production logistics operations is a key link in the prevention of occupational musculoskeletal disorders. At present, the methods of exposure assessment of ergonomic risk factors in the workplace can be summarized into three categories: self-evaluation method, direct evaluation method and observation evaluation method. The great advantage of the observation method is that it can collect a wide range of exposure to workplace risk factors without interfering with the normal operation of the workers and without the

G. Q. Huang et al. (eds.), *Proceeding of the 24th International Conference on Industrial Engineering and Engineering Management 2018*,
https://doi.org/10.1007/978-981-13-3402-3_52

need for high cost, and the method is more suitable for static and repetitive operations evaluation. Such methods are commonly used: Ovako working posture analysis system, rapid upper limb assessment, rapid entire body assessment, checklist, quick exposure check, national institute of occupational safe and health, etc. Some of these methods only evaluate the posture of each part, some mainly evaluate the physical load, while others include the subjective evaluation of social psychological factors. Among them, RULA is the most widely used, and has been applied in many industries, such as manual machinery operation, garment production, supermarket charging, automobile manufacturing, electronic manufacturing, administrative work, ceramic production and so on [1–5]. RULA method has shown good re-test reliability and score reliability in multiple industries [6–8]. Ferdiansyah introduced the RULA into the MSDs assessment of workers to prove its reliability [9]. Automobile production logistics operation mainly takes the form of manual transportation in China, in the course of work there are likely to lead to MSDs bad posture. At present, there are few reports on MSDs caused by the bad posture of workers in the automobile production logistics industry. In this research, RULA is used to investigate the musculoskeletal injury of automobile production workers and make suggestions for improvement.

2 Objects and Methods

This article randomly selected a car logistics company in Kaifeng City, Henan Province, four long engaged in automobile production logistics operations skilled workers, which were engaged in unloading, storage, unpacking and SPS (set parts system) sorting operations. and then studied the work posture in their work process. Record the work of the four workers video, according to the video to get some static job posture screenshots, and then according to each job posture reference RULA evaluation. According to the RULA total score evaluation table, the human action posture level is divided into four levels, as shown in Table 1.

Table 1. RULA total score

Total score	Action level	Treatment scheme
1 or 2	1	Acceptable
3 or 4	2	Investigate further
5 or 6	3	Investigate further and change soon
7	4	Investigate and change immediately

3 Results

Shooting four kinds of car production logistics operations process, each job recording time were 30 min, followed by 30 s interval screenshot once, each job activity were 60 photos. This paper applies simple multi-instantaneous investigation and RULA to evaluate the action level of automobile production logistics operations. The multi-instantaneous investigation solves the RULA dynamic assessment problem and is widely used abroad. It determines the frequency of an event by random and short-term observations, which is one of the standard methods of the German labor science [10].

3.1 A Total Score (Arm, Wrist Position)

A scores were evaluated respectively for 60 images of unloading, storage, unpacking, sorting and SPS sorting of the four operations. Write the percentage of the evaluation results to Table 2, and the weighted average points are calculated and the positions are calibrated.

As can be seen from Table 2. In the upper arm posture, the labor intensity of the SPS sorting operation was the largest, and most of the workers were leaned 15°–90°, and most of them were associated with abduction of the upper arm; In the lower arm position, the SPS sorting operation labor intensity remained large, reaching a maximum of 2 points, with most of the workers crossing the median of the arm. The SPS sorting operation is shown in Fig. 1. Wrist labor intensity difference is very small; Unloading,

Table 2. Arm and wrist posture statistics

Item A	Position posture	Score	Percentage (%)			
			Unloading operation	Storage operation	Unpacking operation	SPS sorting operation
Upper arm	Forward 0–15° Backward 0–15°	1	8.3	40	35.0	0
	Forward 15–45° Backward >15°	2	43.3	53.3	65.0	45.0
	Forward 15–90°	3	48.3	6.6	0	55.0
	Forward >90°	4	0	0	0	0
	Weighted mean		2	2	2	3
	Calibration score		3	2	3	4
Lower arm	Forward 0–90°	1	86.6	95.0	53.3	3.3
	Forward >90°	2	13.3	5.0	46.6	96.6
	Weighted mean		1	1	1	2
	Calibration score		1	1	2	3
Wrist	Normal 0°	1	18.3	33.3	13.3	20
	Upward 0–15° Downward 0–15°	2	81.7	60.0	78.3	80
	Upward >15° Downward >15°	3	0	6.6	8.3	0
	Weighted mean		2	2	2	2
	Calibration score		2	2	2	2
Wrist rotation	No rotation	0	36.6	73.3	28.3	35.0
	Near the median	1	63.3	26.6	71.6	65.0
	Over the median	2	0	0	0	0
	Weighted mean		1	0	1	1

Fig. 1. SPS sorting operation

unpacking, SPS sorting have obvious wrist rotation phenomenon. According to RULA, the total score of the four jobs, including unloading, storage, unpacking and SPS sorting, that is, A score are 4, 3, 4, 4.

3.2 C (Line) Score

The use of muscle for four different jobs requires 1 points; Unloading operations and storage operations of the additional strength/load score of 1 point, the load is mostly 2–10 kg (intermittent or repeatable) of the weight, and unpacking sorting operations and SPS sorting operations without overload, So do not add points. The scores were C (line), 6, 5, 5, 5.

3.3 B Total Score (Neck, Body, Leg Position)

B scores were evaluated respectively for 60 images of unloading, storage, unpacking, sorting and SPS sorting of the four operations. Write the percentage of the evaluation results to Table 3, and the weighted average points are calculated and the positions are calibrated.

Table 3. Neck, body, leg posture statistics

Item B	Position posture	Score	Percentage (%)			
			Unloading operation	Storage operation	Unpacking operation	SPS sorting operation
Neck	Forward 0–10°	1	78.3	73.3	100	21.7
	Forward 10–20°	2	15.0	26.7	0	70.0
	Forward 20–60°	3	6.7	0	0	8.3
	Forward >60°	4	0	0	0	0
	Weighted mean		1	1	1	2
	Calibration score		2	2	2	3
Body	Backward 0–10°	1	30	23.3	63.3	41.6
	Forward 0–20°	2	13.3	20	36.7	6.7
	Forward 20–60°	3	50	56.7	0	51.6
	Forward >60°	4	6.7	0	0	0
	Weighted mean		3	3	1	2
	Calibration score		4	3	2	2
Leg	Supported or balanced	1	100	100	100	100
	Be without	2	0	0	0	0
	Weighted mean		1	1	1	1

In the SPS sorting operation, due to the special structure of the SPS car, the workers had to head down and Stretch arm to identify the parts position, resulting in increased work load of the neck posture, coupled with the SPS car unreasonable set of stops So that workers have a neck rotation phenomenon; Unloading and storage operations need to continue to bend over, the workers will be the whole box of goods moved to the tray, causing greater damage to the body, and unloading operations are also accompanied by body scoliosis or rotation phenomenon. The unloading operation is shown in Fig. 2. According to RULA, we can get the total score of four jobs, such as unloading, storage, unpacking, and SPS sorting, that is, the B scores are 5, 4, 2, 3.

3.4 C (Column) Score

Four kinds of job muscle utilization are maintained 4 times/min or more, all need to add 1 points. Additional strength/load score: four kinds of operations in the neck, body, legs are no force or load, so no extra points. The C (column) scores were 6, 5, 3, 4.

Fig. 2. Unloading operation

3.5 RULA Total Score

Calculate the final total score, unloading, storage, unpacking, SPS sorting the total score of four operations were 7, 6, 4, 5. According to Table 1, it can be seen that the posture action level of the unloading operation is AL4, which needs to be investigated and changed immediately; Storage operations and SPS sorting operations are AL3, which needs to be investigated and changed soon. The posture action level of unpacking operations is AL2, which needs further investigate.

4 Suggestion

The above four major automobile production logistics, according to the total score of the job posture action level did not reach the level of no improvement, the human body is present or light or heavy harm. The following four kinds of automobile production logistics operations in this article, according to different action posture level, gives the following 3 suggestions: Unloading and storage operations require workers to continue to bend over and accompanied by the phenomenon of body scoliosis, the greater the physical damage to the workers. Therefore, in the unloading and storage operations, we should try to avoid bending and twisting of the body, and take intermittent work to avoid long continuous operation. In addition, the auxiliary facilities can be used, such as stationary or hydraulic type DCQY, sinking platform and sheared type hydraulic elevating platform. The workers can be on the same level with the goods during unloading and storage. Workers in the SPS sorting operations, the upper arm and lower arm there are unreasonable posture led to excessive labor intensity, and the neck was forward and accompanied by a rotating phenomenon. This is mainly due to SPS car is divided into Multiple layers, the number of parts placed on each floor, large area and a certain depth. Workers need to bow or head up at the same time to straighten the arm in order to accurately place the parts in the specified position. Therefore, the SPS car should be improved, according to the ergonomics, the different parts of the layer should be set with a reasonable height and a certain slope, and neck protection measures should be taken to avoid the injury of the neck forward and the rotation of the neck. The unreasonable posture of the wrist is prevalent in the four kinds of automobile production logistics operations. In the course of operation, arm bending and twisting should be avoided, and wrist measures should be taken appropriately to reduce wrist injuries. In practice, many of the above unreasonable posture is more difficult to avoid, in addition to take the above protective measures, the need for regular staff physical examination and protective treatment. Hope that through this study, can cause the industry managers and related personnel attention, and strive for the automobile production logistics workers to create good working conditions, and Promote the healthy development of enterprises.

Acknowledgements. A. We are grateful to an anonymous referee for his/her helpful comment earlier version of this paper. This work was supported by the National Science Foundation of China [grant number 71673220].
B. The publication of this paper was supported by the relevant auto companies and the employees surveyed. The data and pictures used in the paper also received the informed consent of the relevant staff and expressed thanks to them.

References

1. L. Mcatamney, E.N. Corlett, RULA: a survey method for the investigation of work-related upper limb disorders. Appl. Ergon. **24**(2), 91–99 (1993)
2. Z. Rowshani, M. Sb, A. Khavanin, R. Mirzaei, M. Mohseni, Comparing RULA and Strain index methods for the assessment of the potential causes of musculoskeletal disorders in the upper extremity in an electronic company in Tehran. Feyz J. Kashan Univ. Med. Sci. (2013)
3. D. Sharan, P.S. Ajeesh, Correlation of ergonomic risk factors with RULA in IT professionals from India. Work **41**(Suppl1)(6), 512
4. G. Pourtaghi, A.A. Karimi Zarchi, F. Valipour, A. Assari, Ergonomic assessment using RULA technique in determining the relationship between musculoskeletal disorders and ergonomic conditions for administrative jobs in a military center. J. Milit. Center. J. Milit. Med. **17**(3), 155–162 (2015)

5. C.M.L. Rahman, Study and analysis of work posutres of workers working in a ceramic industry trough rapid upper limb assessment (RULA). Int. J. Eng. Appl. Sci. **5**(3), 14–20 (2015)
6. I.M. Zeidi, H. Morshedi, B.M. Zeidi, The effect of interventions based on transtheoretical modelling on computer operators' postural habits. Clin. Chiropr. **14**(1), 17–28 (2011)
7. L. Levanonl, Y. Lerman, A. Gefen, N.Z. Ratzon, Validity of the modified RULA for computer workers and reliability of one observation compared to six. Ergonomics **57**(12), 1856–1863 (2014)
8. S. Dockrell, E. O'Grady, K. Bennett, C. Mullarkey, R.M. Connell, R. Ruddy et al., An investigation of the reliability of Rapid Upper Limb Assessment (RULA) as a method of assessment of children's computing posture. Appl. Ergon. **43**(3), 632–636 (2012)
9. F. Syahnl, E. Sonjaya, Validity, and reliability testing by ergonomic evaluation methods for geothermal task, in *Proceedings World Geothermal Congress 2015*, Melbourne, Australia, 2015
10. Y. Lei, L. Zang, T. Decheng, A brief introduction to the investigation of labor posture, in *Industrial Health and Occupational Diseases*, 1995(gy), pp. 101–103

Comparison of Factory Layout Based on Traditional SLP Method and Improved SLP Method

Li Huang[1(✉)], Yan Yi[2], Qian Du[1], and Chuan-Chuan Mo[1]

[1] College of Mechanical Engineering, Panzhihua University, Panzhihua, China
459628370@qq.com
[2] College of Management Science, Chengdu University of Technology, Chengdu, China

Abstract. Workshop layout design is a multi-objective and multi-constrained optimization problem. It is difficult to meet the current needs only by traditional SLP method. In this paper, the improved SLP method, which is a new solving method produced by the combination of the traditional SLP method with the genetic algorithm to solve the problem of workshop layout, will be applied to the optimization layout design of the finishing workshop in the RunYing gear factory in the city of PanZhihua.

Keywords: System layout planing · The genetic algorithm · Logistics optimization · Workshop layout

1 Research Object

The RunYing gear limited liability company is a gear-based enterprises, and its products are mainly exported to Japan and other countries [1]. There are two workshops in the factory, one is for the roughing workshop and the other is for the finishing workshop. The finishing workshop has many production procedures and the internal layout is more complicated and therefore more in need of improvement.

In this case, the finishing workshop of the RunYing gear factory is regarded as the research object and the SLP method is given to priority in workshop layout design. Based on genetic algorithm and system simulation [2], this paper makes an analysis and improvement about the layout of facilities, transportation routes and the arrangements of operation of finishing workshop, and conducts a comparative analysis on the unimproved and improved layout of facilities of finishing workshop.

2 Facing Problems

Through the observation and investigation of the RunYing Gear Factory in PanZhihua, the workshop mainly has the following problems:

(1) there are a lot of cross-traffic and reverse logistics in the operating regions in the finishing workshop.

G. Q. Huang et al. (eds.), *Proceeding of the 24th International Conference on Industrial Engineering and Engineering Management 2018*,
https://doi.org/10.1007/978-981-13-3402-3_53

(2) the inconsistent problems between the workshop layout and the process flow give rise to increase of the logistics cost and lower production efficiency (Fig. 1) [3].

Fig. 1. Factory status

3 Analysis of Current Situation

The finishing workshop of the RunYing gear factory covers an area of 7665 m^2, and the length of east-west direction is 105 m, the length of north-south direction is 73 m. The total area of operating region is about 6806.25 m^2 and four doors are open in the workshop. Based on the improved SLP method [4], this paper creates new layout designs of the finishing workshop of the RunYing Gear Factory. According to the process and Place of work, the finishing workshop is divided into 31 operating units. The names and required areas of each operating unit are shown in Table 1.

The finishing workshop is divided into three lines, respectively, gear, gear ring, flywheel gear ring processing line. The main completion of the process are shown in Table 2.

3.1 Logistics Relations

The analysis of logistics intensity among operating units is a prerequisite for facility planning and an important basis for avoiding cross-cutting and detouring of logistics routes and reducing logistics costs. According to the proportion of logistics routes and the ratio of the logistics flow undertaken among the operating units, the logistics intensity can be divided into five grades of A, E, I, O, U, which respectively correspond to super-high logistics intensity, high logistics intensity, larger logistics intensity, the general logistics intensity, and negligible logistics. By the analysis of the workshop, we can conclude the distinction on levels of logistics intensity (shown in Table 3), as well as logistics interrelationship table (Fig. 2).

Table 1. Prodecting unit and area

The serial number	Name of operating unit	Length (m)	Width (m)	Area (m^2)
1	Rough turning gear	18	7	126
2	Gear ring rough turning	8	7	56
3	Gear ring cutting	14	7	98
4	Gear ring level end face	3.5	7	24.5
5	Gear ring finishing turning	10.5	7	73.5
6	Gear ring finishing hole	18	7	126
7	Finish turning I	36	21	756
8	Gear hob	14	21	294
9	Gear tooth area	17.5	19.5	341.25
10	Drilling	20.5	21	430.5
11	Burring	3	7	21
12	Gear shaving	3.5	21	73.5
13	Cleaning I	3	7	21
14	Sign	6.5	7	45.5
15	Heat treatment area	29.5	11	324.5
16	Gear grinding	6.5	7	45.5
17	Finish turning II	7	7	49
18	Make coding	7.5	15	112.5
19	Gear liner bushing	2.5	3.5	8.75
20	Gear extrusion	5	3.5	17.5
21	Inspection area	7.5	3.5	26.25
22	Working-storage section	8.5	11	93.5
23	Cleaning II	17.5	8	140
24	Packaging	13	8	104
25	The flywheel processing	36	13	468
26	Flywheel gear assembly	14	13	182
27	Flywheel gear ring inspection	17.5	13	227.5
28	Flywheel gear ring packaging	6.5	13	84.5
29	Tool stack	18	19	342
30	Electric welding workshop	18	19	342
31	Warehouse	24	73	1752

3.2 Non-logistics Relationship

Logistics analysis is an important basis for factory layout when the logistics situation has a major impact on the production of enterprises. However, the influence of non-logistics factors will also be taken seriously. Especially in the case where logistics has little impact on production and logistics is not fixed, the layout of factories can not only rely on logistics analysis, instead, the impact of other factors on the interrelation among the operating units should be taken into account. According to the survey of the actual situation of the factory, various factors affecting the production units (shown as

Table 2. Product processing procedure

Camshaft gear	Gear	Inert gear	New product	Gear fuel supply pump	Air compressor drive gear	Ring gear	Flywheel ring gear
1. Rough turning	1. Rough turning	1. Rough turning	1. Rough turning	1. Rough turning	1. Rough turning	1. Rough turning	1. Gear milling
2. Finish turning I	2. Finish turning I	2. Finish turning I	2. Finish turning I	2. Finish turning I	2. Finish turning I	2. Cut off	2. Assembly
3. Hob	3. Hob	3. Hob	3. Hob	3. Hob	3. Hob	3. Level end face	3. Inspection
4. Drill	4. Drill	4. Burring	4. Drill	4. Drill	4. Drill	4. Finish turning	4. Pack
5. Burring	5. Burring	5. Shaving	5. Burring	5. Burring	5. Burring	5. Finish hole	
6. Shaving	6. Shaving	6. Clean I	6. Shaving	6. Shaving	6. Shaving	6. Gear shaping	
7. Clean I	7. Clean I	7. Sign	7. Clean I	7. Clean I	7. Clean I	7. Drill	
8. Sign	8. Sign	8. Heat treatment	8. Sign	8. Sign	8. Sign	8. Burring	
9. Heat treatment	9. Heat treatment	9. Polish	9. Heat treatment	9. Heat treatment	9. Heat treatment	9. Clean I	
10. Polish	10. Polish	10. Finish turning II	10. Polish	10. Polish	10. Polish	10. Sign	
11. Finish turning II	11. Finish turning II	11. Code	11. Finish turning II	11. Finish turning II	11. Finish turning II	11. Heat treatment	
12. Code	12. Code	12. Extrusion	12. Code	12. Code	12. Code	12. Code	

(*continued*)

Table 2. *(continued)*

Camshaft gear	Gear	Inert gear	New product	Gear fuel supply pump	Air compressor drive gear	Ring gear	Flywheel ring gear
13. Final inspection	13. bushing	13. Final inspection	13. Bushing	13. Extrusion	13. Bushing	13. Final inspection	
14. Clean II	14. Extrusion	14. Clean II	14. Extrusion	14. Final inspection	14. Extrusion	14. Clean II	
15. Pack	15. Final inspection	15. Pack	15. Final inspection	15. Clean II	15. Final inspection	15. Pack	
	16. Clean II		16. Clean II	16. Pack	16. Clean II		
	17. Pack		17. Pack		17. Pack		

Note The 2# of the gear ring is the same as the process flow of the ring 3#, but it is only different in size

Table 3. Logistics strength level

Logistics strength level	Unit pair	Logistics volume	Percentage (%)
A	6	25490	34.6
E	7	28280	38.4
I	5	15010	20.4
O	15	4929	6.6
U	432	0	0
Total	465	73709	100

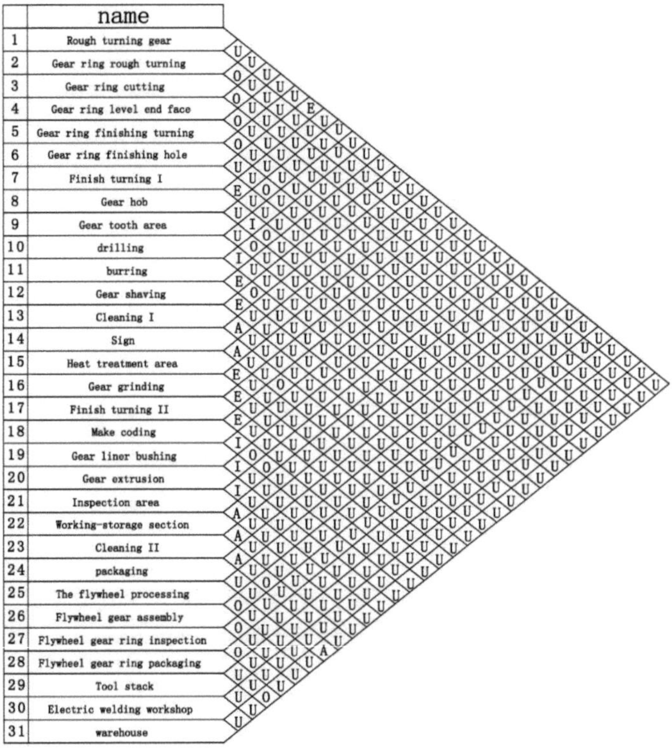

Fig. 2. Logistics interrelationship table

Table 4) were identified and a related non-logistics relationship table was drawn up (Fig. 3).

3.3 Comprehensive Relationship

The logistics relationship and non-logistics relationship in processing workshop act an effect on the degree of closeness among the operating units. In this paper, weighted proportion of logistics to non-logistics relations in the finishing workshop is assumed to be a formula: m:n = 1:1.

Table 4. Activity degree reason code

Code	Reasons
①	Craft cohesion requirements
②	Similar job properties
③	Use the same place
④	Using the similarity of processing equipment
⑤	Use the same public facilities and file files
⑥	The same group of workers
⑦	The influence of vibration and smoke
⑧	Easy to supervise and manage
⑨	Working contact frequency
⑩	The frequency and urgency of service

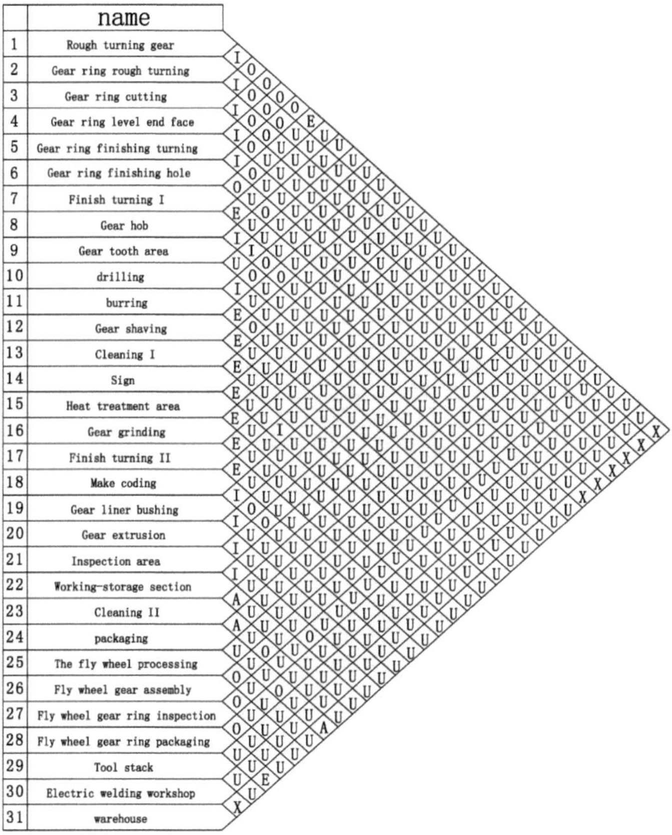

Fig. 3. Non-logistics relationship table

For each grade of degree of closeness in the table, the grade is quantified, that is, A = 4, E = 3, I = 2, O = 1, U = 0, according to the quantified grade and the determined weight, the composite correlation value of the operation unit (shown in Table 5) eventually yields a composite diagram of the operation unit (Fig. 4).

Table 5. Comprehensive hierarchy

Score	Relational grade	Unit pair	Percentage (%)
8	A	3	0.6
5–6	E	10	2.1
4	I	13	2.8
1–3	O	27	5.8
0	U	406	87.2
−1	X	7	1.5
Total		465	100

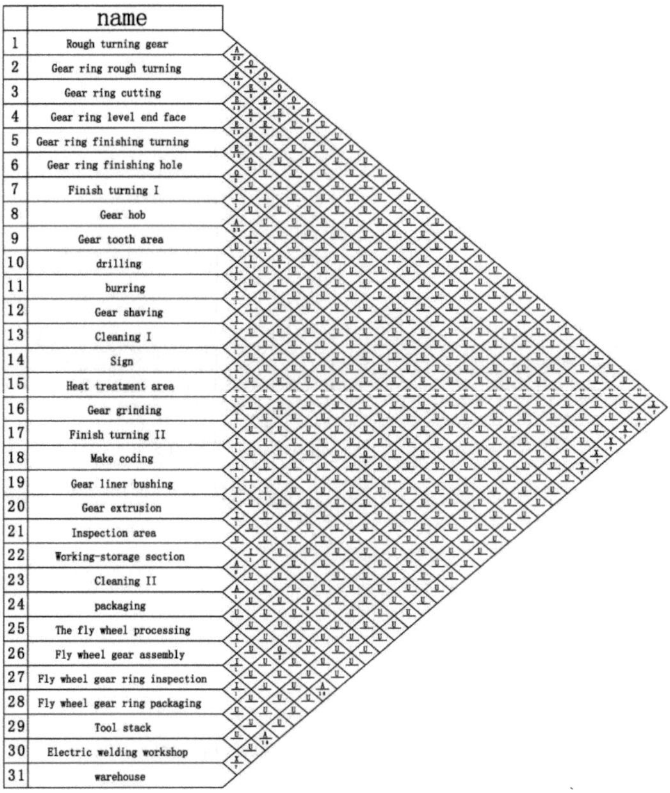

Fig. 4. Integrated operation relation table

4 Scheme Design

4.1 The Traditional SLP Method

In this paper, the location correlation map is drawn according to the comprehensive proximity. Operating units with high overall proximity scores should be placed in the center of the position-related map. On the contrary, operating units with lower score should be demonstrated at the edge of the map. When we make a arrangement, following the order of A, E, I, O, U. For the same grade of operating units, according to from high to low score of degree of comprehensive proximity, the final map about the location of the operating unit is obtained (Fig. 5). Combined with the factors of operating units, entrances and exits, staff venues, after the modification and adjustment, the layout blueprint of the factory is ultimately reached (Fig. 6).

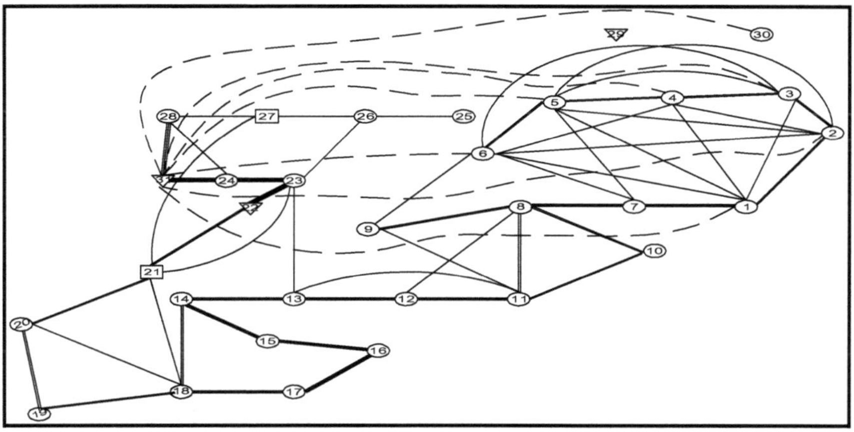

Fig. 5. Position correlation map

4.2 Improved SLP Method

4.2.1 Workshop Layout Modeling

In actual workshop layout design, the size of each work area is different, and the location of each work area is restricted by other factors.

Let's make a hypothesis: the layout scope is known and the rectangular area is S. According to the actual technological process of the workshop, the workshop is divided into n operating areas. The set of the areas is defined as N and the area of each of them is assumed as S. Make a assumption that the material transport distance is in the center of each of the operating areas. There are m positions, m > n. And the set of each position is defined as M, N = {ki. | I = 1, 2, 3 …, n}, M = {ki. | I = 1, 2, 3 …, m}, Every location and operating area is independent of each other, and if the collection of layout design is assumed as H, H = M × N, as shown in Fig. 7. The entire area of the graphical bold region is considered as the overall scope of Factory layout, a small rectangle is on behalf of one operating area. Using an example of 2 operating areas, the

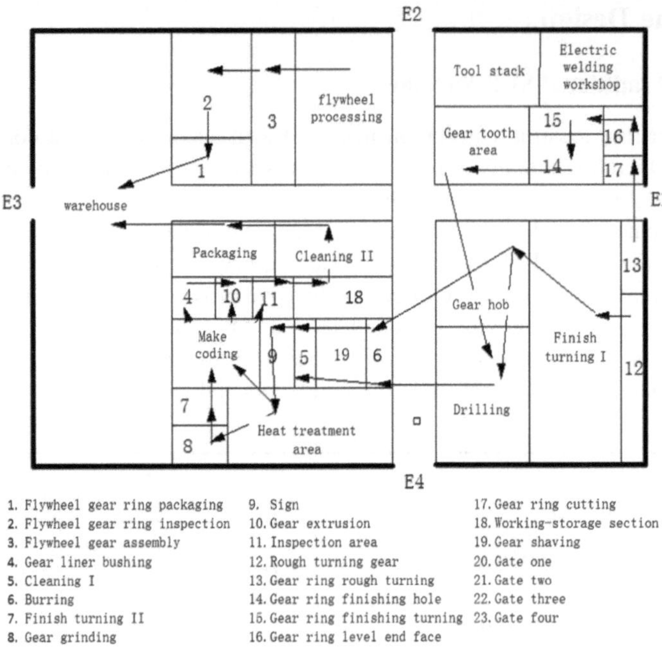

Fig. 6. Workshop layout

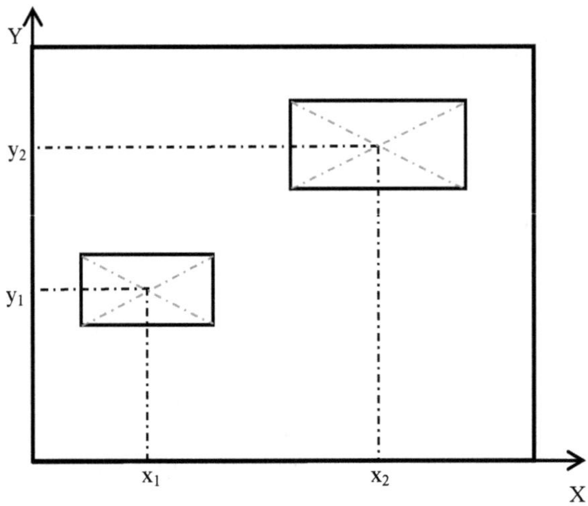

Fig. 7. Layout modeling

following figure shows how to determine the center of the work area. The intersection of two diagonal lines of 2 operating areas is the central position.

Assuming that x is one of the schemes of plant layout, x is contained in set H. on the scheme called X, the logistics volume between work zone ki and work zone kj is indicated as Qij(x), (i, j = 1, 2, 3, …, n}. The distance of center position between two work zone ki and work zone kj is defined as dij(x), (i, j = 1, 2, 3, …, n}. On the scheme called X, we can get the matrix formula of logistics quantity about Q(x) and the distance matrix expression about d(x) as shown in the following formula (1) and (2)

$$Q(X) = \begin{bmatrix} Q_{11}(x) & Q_{12}(x) & \cdots & Q_{1n}(x) \\ Q_{21}(x) & Q_{22}(x) & \cdots & Q_{2\mathrm{n}}(x) \\ \cdots & \cdots & \cdots & \cdots \\ Q_{n1}(x) & Q_{n2}(x) & \cdots & Q_{\mathrm{nn}}(x) \end{bmatrix} \tag{1}$$

$$d(X) = \begin{bmatrix} d_{11}(x) & d_{12}(x) & \cdots & d_{1n}(x) \\ d_{21}(x) & d_{22}(x) & \cdots & d_{2\mathrm{n}}(x) \\ \cdots & \cdots & \cdots & \cdots \\ d_{n1}(x) & d_{n2}(x) & \cdots & d_{\mathrm{nn}}(x) \end{bmatrix} \tag{2}$$

4.2.2 Layout Optimization Results

In this paper, the software matlab is used to solve workshop layout model of the genetic algorithm. The software runs on the ASUS notebook. Assuming that the number of sets is 50, the number of iterations is 3000, the cross probability is 0.8, and the probability of variation is 0.05, as shown in Fig. 8. Through genetic algorithm after the iteration process, it can be seen that the fitness value has reached a steady value of about 1.826 before the iteration 1000 times.

The simulation results calculated by the model can be used to obtain the coordinates of the center of each work area according to the actual situation of the factory structure. Because the number of working areas in the factory is large, it is not necessary to consider the layout of each operating area of the workshop to maintain a certain distance to simplify the layout. Each operating area is arranged adjacent to each other, and taking the actual location of four doors E1, E2, E3, E4 into account, the location correlation map of the optimal layout of the work area is drawn, as shown in Fig. 9.

5 Summary

5.1 Quantitative Analysis

The design of workshop layout directly affects the operation efficiency of the workshop, handling distance, handling costs, thus affecting the competitiveness of enterprises. In this paper, the improved SLP method is used in the finishing workshop of RunYing Gear Factory. Through the theoretical Analysis and measurement of actual data, and by means of current layout, the traditional SLP layout and the improved SLP layout, the quantitative comparison analysis on the transportation distance, the logistics cost and the non-logistics relation-ship is concluded (shown in Table 6).

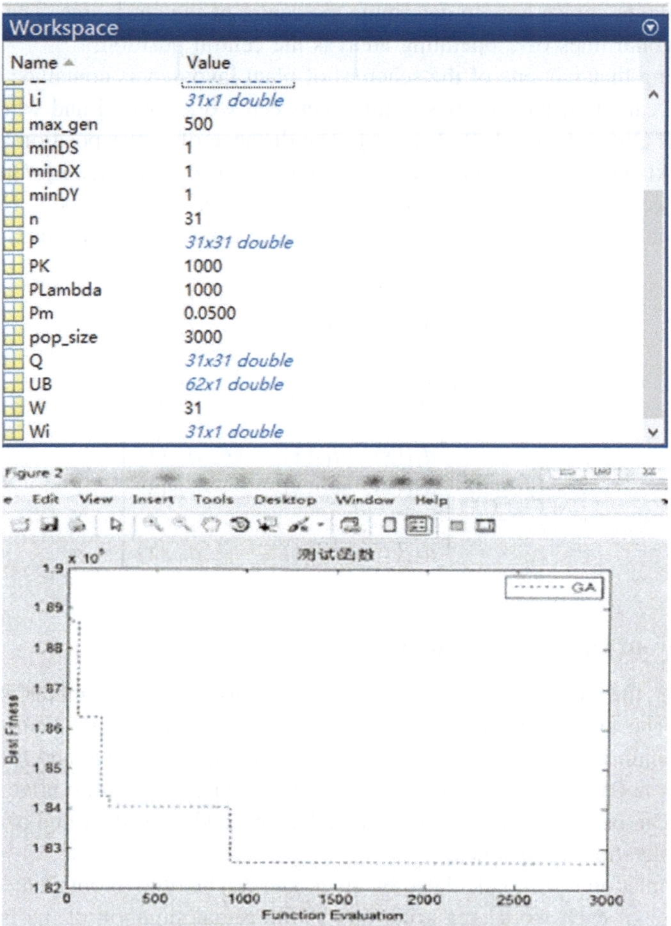

Fig. 8. Iteration process of genetic algorithm

Compared with the existing layout of the factory and the optimal layout by the traditional SLP method, it is evident that the scheme of factory layout through the improved SLP method has shorter material transportation distances, lower transportation costs, and closer non-logistics relationships, which indicate that the factory layout model based on the improved SLP method is feasible. And the improved SLP method is more effective than the traditional SLP method in the layout of the factory. Furthermore, more factors is involved in the design of factory layout. The improvement of SLP method can greatly reduce the carrying distance, handling costs to adjust and improve the layout of factory.

5.2 Qualitative Analysis

Comparing Figs. 10 and 11 with the traditional SLP method, the situation of material carrying in the workshop is obviously improved, the cross-traffic situation is obviously

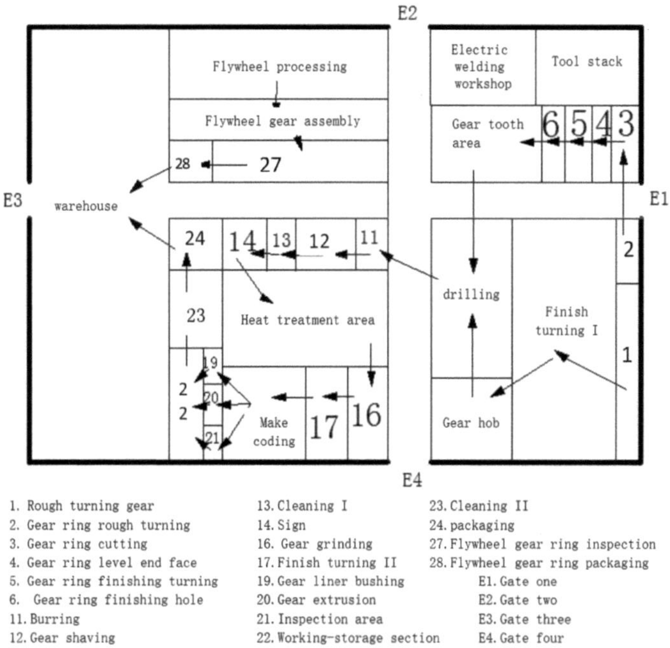

Fig. 9. Workshop layout

Table 6. Comparison and analysis of scheme

Plan	Transportation distance (m)	Handling cost (yuan)	Non-logistics relationship
Current plant layout	651.5	1475320.85	71.92
Traditional SLP layout	451.2	1021741.78	92.4
Improved SLP layout	343.196	776723.03	101.8

reduced, and the carrying distance is shortened. Comparing Figs. 11 with 12, it is clear that the layout has no cross-logistics and the process of various products changes into a pipeline in the improved SLP method. Then, the convergence of the process gets more closely by the new layout, so that it avoids unnecessary waiting and utilization of space is higher.

It is easy to find out that the improved SLP method is more practical meaningful by analyzing the optimal distance between operating units, transportation costs and non-logistics relationship. It proves that the improved SLP method is effective in solving the problems of the factory layout.

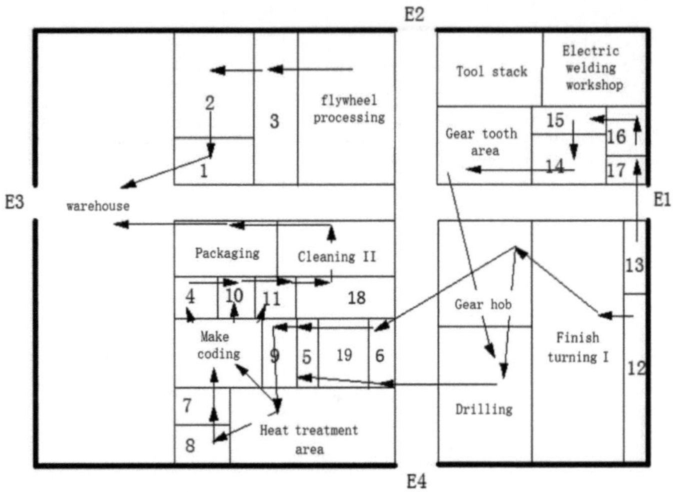

1. Flywheel gear ring packaging
2. Flywheel gear ring inspection
3. Flywheel gear assembly
4. Gear liner bushing
5. Cleaning I
6. Burring
7. Finish turning II
8. Gear grinding
9. Sign
10. Gear extrusion
11. Inspection area
12. Rough turning gear
13. Gear ring rough turning
14. Gear ring finishing hole
15. Gear ring finishing turning
16. Gear ring level end face
17. Gear ring cutting
18. Working-storage section
19. Gear shaving
E1. Gate one
E2. Gate two
E3. Gate three
E4. Gate four

Fig. 10. Current plant layout

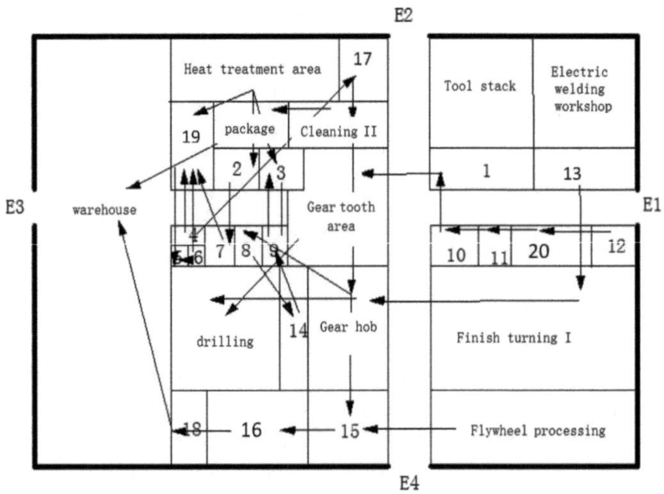

1. Gear ring finishing hole
2. Gear grinding
3. Sign
4. Inspection area
5. Gear extrusion
6. Gear liner bushing
7. Finish turning II
8. Burring
9. Cleaning I
10. Gear ring finishing turning
11. Gear ring level end face
12. Gear ring rough turning
13. Rough turning gear
14. Gear shaving
15. Flywheel gear assembly
16. Flywheel gear ring inspection
17. Working-storage section
18. Flywheel gear ring packaging
19. Make coding
20. Gear ring cutting
E1. Gate one
E2. Gate two
E3. Gate three
E4. Gate four

Fig. 11. Traditional SLP layout

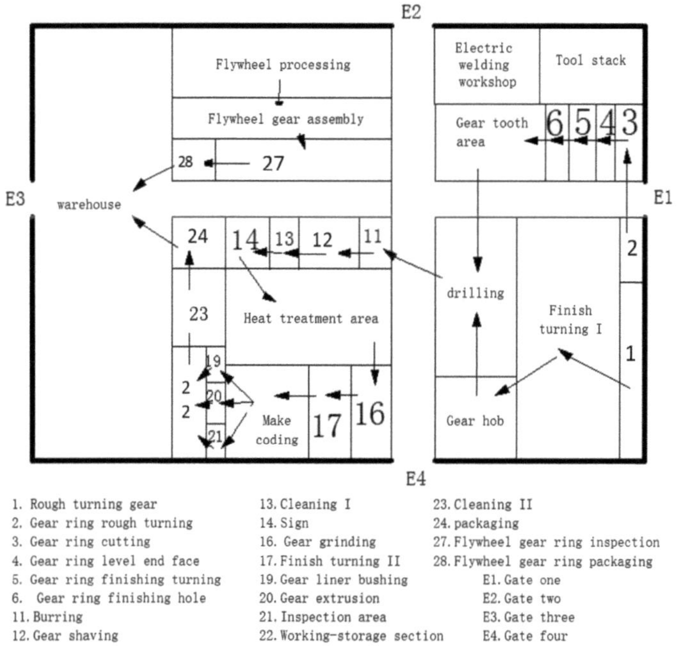

Fig. 12. Improved SLP layout

References

1. D.G. Conway, M.A. Venkataramanan, Genetic search and the dynamic facility layout problem. Comput. Oper. Res. **21**(8), 955–960 (1994)
2. K.-Y. Lee, M.-I. Roh, H.-S. Jeong, An improved genetic algorithm for multi-floor facility layout problems having inner structure walls and passages. Comput. Oper. Res. **32**(4), 879–899 (2005)
3. Y. Wang, Z. Mi, SLP and genetic algorithm are used in the layout of workshop equipment. Comput. Eng. Appl. **46**(5), 211–213 (2010)
4. Y. Mao, Based on SLP and simulation technology, enterprise production logistics system layout optimization. Chengdu University of Technology (2013)

A Pricing Model of Chinese Medicinal Materials Logistics Distribution Service Based on Consumer Environmental Preference

Chao-dong Lu[1(✉)], Xiao-Jian Hu[2], and Yan-Li Zhang[2]

[1] Department of Economy and Management, Bozhou University, Bozhou, Anhui, People's Republic of China
996697719@qq.com
[2] School of Management, Hefei University of Technology, Hefei, Anhui, People's Republic of China

Abstract. It assumes that two Chinese medicinal materials logistics distribution enterprises provide green logistics distribution service and general logistics distribution service with mutual substitution. Considering consumer purchase behavior, it divides consumers into the green consumer and general consumers. It further supposes that two types of consumer's valuations for two kinds of distribution are heterogeneous, and are uniformly distribution. We use a utility function to get the consumers' consumer behavior. Under the condition of concentration and decentralize respectively, we get optimal pricing of green logistics distribution service and general logistics distribution service by maximizing profit of the whole supply chains and the two enterprises. We identify the two Chinese medicinal materials logistics distribution enterprises' pricing is positively related to the price sensitivity of consumers in both integrated supply chain and decentralized supply chain. Under some conditions, the pricing of the two enterprises are positively related to consumer environmental awareness.

Keywords: Chinese medicinal materials · Environmental preference
Logistics distribution service · Pricing · Price sensitivity

1 Introduction

With the increasing contradiction between eco-friendly and economic development, led to a new challenge of change in economic development. Green economy is a new choice for economic development pattern in the future. "Green economy" is the essential pattern to achieve energy for sustainable and economy for recyclable. It's necessary to research green Chinese medicinal materials logistics distribution services which are less energy-consuming and less environmental pollution in the "green economy" context. Under the background of green economy, consumers with environmental awareness may predict the quality of a Chinese medicinal materials logistics distribution service based on its carbon emission level, which is considered as the assurance that the Chinese medicinal materials logistics distribution enterprises

G. Q. Huang et al. (eds.), *Proceeding of the 24th International Conference on Industrial Engineering and Engineering Management 2018*,
https://doi.org/10.1007/978-981-13-3402-3_54

provides after evaluating the advantages of Chinese medicinal materials logistics distribution services. Thereby, the carbon emission level of Chinese medicinal materials logistics distribution service becomes an important measure for encouraging low-carbon market demand by increasing the willingness to pay for consumers with environmental awareness. The Chinese medicinal materials logistics distribution enterprises have environmental responsibility to give a reasonable carbon emission level for its logistics distribution service in order to fulfill the corporate social responsibility. In addition, a reasonable carbon emission level of logistics distribution service will certainly enhance purchase willingness of consumer. However, supply chain members must make a reasonable trade-off between the profit and the investment from providing green logistics distribution services because of the cost to pay for low-carbon logistics distribution service by the Chinese medicinal materials logistics distribution enterprise. At the present stage, research and development green logistics distribution service is still in an exploring stage. Green logistics distribution service and general logistics distribution service with mutual substitution, and exists in one market. In this study, we research the pricing of green logistics distribution service and general logistics distribution service. In this paper, we use green degree describes the carbon emissions in the course of logistics distribution service. The lower the carbon emissions, the higher green degree is.

There is a lot of research and achievement of two substitutable products modeling in China and any other country. Birge et al. [1], Rajaram and Tang [2], Hopp and Xu [3], Kuyumcu and Popescu [4], Tang and Yin [5], Karakul and Chan [6], Xia [7], Stavrulaki [8], Zhao et al. [9, 10], Chen and Chang [11], Zhang et al. [12] and Zhao et al. [13].

In addition, there are many research results and achievements of green product modeling in China and any other country. Chialin [14] pricing model of green product based on product quality. Zugang et al. [15] a supply chain model with dynamic participants was built by taking consumer's environmental awareness into consideration.

However, to the best of our knowledge, none of these papers studied the price and carbon emission level decisions of green logistics distribution service and general logistics distribution service by maximizing profit of the whole supply chains and the two Chinese medicinal materials logistics distribution enterprises.

2 Model Symbols and Assumptions

Assume that there are two Chinese medicinal materials logistics distribution enterprises who provide two kinds of logistics distribution services that have a distinctive green degree and can be substituted. Both two kinds of logistics distribution services can be substituted, and are in liberal provide. Symbols and decision variables are used in this paper showed in Table 1.

Table 1. Model parameters and definition

Parameter	Definition
Δ	The total number of consumers in the logistics distribution service
t_1	The proportion of green consumers in the logistics distribution service
t_2	The proportion of general consumers in the logistics distribution service
Q_1	The quantity of green logistics distribution service
Q_2	The quantity of general logistics distribution service
C_1	The cost of green logistics distribution service
C_2	The cost of general logistics distribution service
λ	Consumers price sensitivity
δ	Consumer awareness
Decision variable	
P_1	The price of green logistics distribution service
P_2	The price of general logistics distribution service

Assumption 1. Total market demands are fixed, and total number of consumers in the market is. For convenient calculation we order $\Delta = 1$.

Assumption 2. The green logistics distribution service market of Chinese medicinal materials is divided into green logistics distribution service market and general logistics distribution service market. Consumer groups are separated into green consumers and consumers. Green consumers have higher environmental protection consciousness, and are not sensitive to price. General consumers have lower environmental protection consciousness, and are sensitive to price. Assume that, the proportion of green consumers in the logistics distribution service of Chinese medicinal materials is t_1, and the proportion of general consumers in the logistics distribution service of Chinese medicinal materials is t_2, and $t_1 + t_2 = 1$.

Assumption 3. Green logistics distribution enterprise of Chinese medicinal materials has an obligation to improve the green degree of logistics distribution services. We assume that the cost of green logistics distribution services of Chinese medicinal materials is higher than that of general logistics distribution services of Chinese medicinal materials, mean $C_1 > C_2$.

Assumption 4. Green consumer values the green logistics distribution service at $V_{\mathrm{G}}^{\mathrm{G}} = V$, assumed to be uniformly distributed between 0 and 1. General consumer values green logistics distribution service at $V_{\mathrm{G}}^{P} = \lambda V$. Green consumer values general logistics distribution service at $V_{\mathrm{P}}^{\mathrm{G}} = \delta V$, and general consumer values general logistics distribution service at $V_P^P = \lambda\delta V$. Assume valuation parameter V_j^i whose superscript represents customer type, and whose subscript represents logistics distribution service type. i = {Green consumers, general consumers}, j = {Green logistics distributions, general logistics distributions}. λ represents the general consumer's price sensitivity, and δ represents environmental protection consciousness of consumer, where $0 < \lambda < 1$ and $0 < \delta < 1$.

Different value of logistics distribution services are caused by many reasons. Marcus B et al. (2008) consider service level is the cause of different value [16]. In this study, valuation difference is the result of consumers' price sensitivity. As the government is proceeding with more and more environmental protection campaign, the environmental awareness of consumers has been improved. This paper assumed green consumers have higher environmental awareness, and general consumers have lower environmental awareness. Nevertheless, green consumers are not sensitive to price. That means green consumers are willing to pay higher to buy green logistics distribution services. General consumers are sensitive to price, and their willing to pay is lower. So green consumers' evaluation V of green logistics distribution service is higher, and general consumers' value λV for general logistics distribution service is lower. General consumers have the lowest evaluation $\lambda\delta V$ for the general logistics distribution service because they are sensitive to prices.

3 Demand Functions Analyses

Given the assumptions and the definitions in Table 1, a green consumer gets utility $\left(U_G^G = V - P_1\right)$ from green logistics distribution service and utility $\left(U_P^G = \delta V - P_2\right)$ from general logistics distribution service. A general consumer gets utility $\left(U_G^P = \lambda V - P_1\right)$ from green logistics distribution service and utility $\left(U_P^P = \lambda\delta V - P_2\right)$ from primary logistics distribution service. Green consumers purchase the green logistics distribution service if $U_G^G > 0$ and $U_G^G > U_P^G$. Otherwise, green consumers purchases the general logistics distribution service if $U_P^G > 0$ and $U_P^G > U_G^G$. Similarly, the general consumers purchase the green logistics distribution service if $U_G^P > 0$ and $U_G^P > U_P^P$. Otherwise, general consumers purchase the general logistics distribution service if $U_P^P > 0$ and $U_P^P > U_G^P$. Throughout our analysis we will consider cases where the Chinese medicinal material services logistics distribution enterprise prices green logistics distribution services higher than general, i.e., $P_1 > P_2$, which is the relevant case practice. So we got quantity of both green logistics distribution services and general logistics distribution services in Table 2.

Table 2. The quantity of green logistics distribution services and general logistics distribution services

P_1	Q_1	Q_2
$\left[C_1, \frac{P_2}{\delta}\right]$	$t_1(1 - P_1) + t_2\left(1 - \frac{P_1}{\lambda}\right)$	0
$\left[\frac{P_2}{\delta}, P_2 + \lambda V(1 - \delta)\right]$	$t_1\left(1 - \frac{P_1 - P_2}{1-\delta} + t_2\left(1 - \frac{P_1 - P_2}{\lambda - \lambda\delta}\right)\right)$	$t_1\left(\frac{P_1 - P_2}{1-\delta} - \frac{P}{\delta}\right) + t_2\left(\frac{P_1 - P_2}{\lambda - \lambda\delta} - \frac{P_2}{\lambda\delta}\right)$
$[P_2 + \lambda V(1 - \delta), 1]$	0	$t_1\left(1 - \frac{P_2}{\delta}\right) + t_2\left(1 - \frac{P_2}{\lambda\delta}\right)$

4 Model Building

This is a static game problem. The two Chinese medicinal materials logistics distribution enterprises set the service price based on consumer preferences. By analysis of the process of consumer decision-making we can know if the price of the general logistics distribution is not low enough. The green consumer will prefer green logistics distribution service to general logistics distribution service. Likewise, if the price of the green logistics distribution service is not low enough general consumer will prefer general logistics distribution service to green logistics distribution service.

The two Chinese medicinal materials logistics distribution enterprises will set the price to maximize their profits based on consumer preferences. The optimal pricing strategy of green logistics distribution service is making the green consumers buy green logistics distribution service, and also can capture some general consumers. The purpose of pricing strategy of general logistics distribution service is to enlarge both the green consumer market and general consumer market meanwhile at the lowest price. There are three kinds of pricing strategy of green logistics distribution service as follows.

Case 1. Green logistics distribution services are sold at a lower price $(C_1 < P_1 < \frac{P_2}{\delta})$. Only green logistics distribution service are sold under this pricing strategy. Only general logistics distribution enterprise of Chinese medicinal materials also sells its item at a low price can be the sold of general logistics distribution service are not zero. Then green logistics distribution enterprise of Chinese medicinal materials will cut down the price follow. There will be a vicious price competition. This malignancy competition will only make the consequence of harm them as well. So this pricing strategy is not scientific.

Case 2. Green logistics distribution services are sold at the market price $(\frac{P_2}{\delta} \leq P_1 < P_2 + \lambda V(g_1 + \delta g_2))$. Both green logistics distribution services and general logistics distribution services have sales under this pricing strategy, and the two manufacturers are profitable. So this pricing strategy is feasible.

Case 3. Green logistics distribution services are sold at a higher price $(P_1 > P_2 + \lambda V(g_1 + \delta g_2))$. There is no doubt that there is only general logistics distribution service has sales under this pricing strategy. And green logistics distribution enterprise of Chinese medicinal materials will not take this kind of pricing strategy.

We assume the two logistics distribution enterprises are rational, and the case 1 and 3 will be not allowed to happen. So, we are research the pricing strategy of case 2 in the following decision models.

To simplify expression, we use a series of constant replacement in following analysis:

$$A = 2C_1 - \delta C_1 - C_2,\ B = \delta C_1 + \delta C_2 - 2C_2,\ M = \lambda t_1 + t_2,\ T = C_1(2 - \delta) - C_2,$$
$$N = \delta C_1(2 - \delta) - C_2(4 - 3\delta).$$

The above assumptions of symbols are applicable to full-text.

4.1 Integrated Supply Chain

Under centralized decision-making, two logistics distribution enterprises reach an alliance. The objective of the alliance is to maximize the profit of the whole supply chain. Under the condition of $\frac{P_2}{\delta} < P_1 < P_2 + \lambda V(1-\delta)$, the profit function of the whole supply chain under centralized decision-making is:

$$\begin{aligned}\Pi^I &= \Pi_1 + \Pi_2 \\ &= t_1\left(1 - \frac{P_1 - P_2}{1-\delta}\right)(P_1 - C_1) + t_2\left(1 - \frac{P_1 - P_2}{\lambda(1-\delta)}\right)(P_1 - C_1) \\ &\quad + t_1\left(\frac{P_1 - P_2}{1-\delta} - \frac{P_2}{\delta}\right)(P_2 - C_2) + t_2\left(\frac{P_1 - P_2}{\lambda(1-\delta)} - \frac{P_2}{\lambda\delta}\right)(P_2 - C_2)\end{aligned} \tag{1}$$

Proposition 1. Under the condition of $C_2 < \delta C_1$, the optimal pricing strategies of the two logistics distribution enterprises are:

$$P_1^I = \frac{C_1}{2} + \frac{\lambda}{2M} \tag{2}$$

$$P_2^I = \frac{C_2}{2} + \frac{\lambda\delta}{2M} \tag{3}$$

Prove: Structure the Lagrange function of objective function (1):

$$\begin{aligned}L = {} & t_1\left(1 - \frac{P_1 - P_2}{1-\delta}\right)(P_1 - C_1) + t_2\left(1 - \frac{P_1 - P_2}{\lambda - \lambda\delta}\right)(P_1 - C_1) \\ & + t_1\left(\frac{P_1 - P_2}{1-\delta} - \frac{P_2}{\delta}\right)(P_2 - C_2) + t_2\left(\frac{P_1 - P_2}{\lambda - \lambda\delta} - \frac{P_2}{\lambda\delta}\right)(P_2 - C_2) \\ & + x_1(\delta P_1 - P_2) + x_2[P_2 + \lambda V(1-\delta) - P_1]\end{aligned}$$

According to Kuhn-Tucker:

$$\begin{cases} \frac{\partial L}{\partial P_1} = \frac{\partial L}{\partial P_2} = 0 \\ x_1(\delta P_1 - P_2) = x_2[P_2 - P_1 + \lambda V(1-\delta)] = 0, \ x_i \geq 0, i = 1, 2, \end{cases}$$

Then we get these two prices into objective function (1), and concluded the optimal profit of the whole supply chain under an integrated supply chain.

$$\Pi^I = \frac{M[\delta C_1^2 - 2\delta C_1 C_2 + C_2^2] - \lambda\delta C_1(1-\delta)}{4\lambda\delta(1-\delta)M} + \frac{\lambda - C_1 M}{4M} \tag{4}$$

4.2 Decentralized Supply Chain

In decentralized supply chain, two logistics distribution enterprises make pricing independently at the same time. The two logistics distribution enterprises' goals are maximizing their own profit. So under the condition of $\frac{P_2}{\delta} < P_1 < P_2 + \lambda V(1-\delta)$, the two logistics distribution enterprises' profits as follows:

$$\Pi_1 = t_1\left(1 - \frac{P_1 - P_2}{1-\delta}\right)(P_1 - C_1) + t_2\left(1 - \frac{P_1 - P_2}{\lambda(1-\delta)}\right)(P_1 - C_1) \tag{5}$$

$$\Pi_2 = t_1\left(\frac{P_1 - P_2}{1-\delta} - \frac{P_2}{\delta}\right)(P_2 - C_2) + t_2\left(\frac{P_1 - P_2}{\lambda(1-\delta)} - \frac{P_2}{\lambda\delta}\right)(P_2 - C_2) \tag{6}$$

Proposition 2. On the condition of $C_2 < \frac{\delta C_1}{2-\delta} + \frac{\lambda\delta(1-\delta)}{(2-\delta)M}$ the optimal pricing strategy of these two logistics distribution enterprises are:

$$P_1^D = \frac{2C_1 + C_2}{4-\delta} + \frac{2\lambda(1-\delta)}{M(4-\delta)} \tag{7}$$

$$P_2^D = \frac{\delta C_1 + 2C_2}{4-\delta} + \frac{\lambda\delta(1-\delta)}{M(4-\delta)} \tag{8}$$

Then we can get the optimal pricing of the two logistics distribution enterprises under decentralized supply chain. The optimal profits of the two logistics distribution enterprises are also getting by plugging the above into (5) and (6).

$$\Pi_1^D = \frac{[AM - 2\lambda(1-\delta)]^2}{\lambda(1-\delta)(4-\delta)^2 M} \tag{9}$$

$$\Pi_2^D = \frac{[BM + \lambda\delta(1-\delta)]^2}{\lambda\delta(1-\delta)(4-\delta)^2 M} \tag{10}$$

The proving of Proposition 2 is same as Proposition 2.

Theorem 1 In the integrated supply chain, by the equation in (2) and (3) we can know there is a positive correlation between the two logistics distribution enterprises' pricing and the consumer price sensitivity. Pricing of green logistics distribution enterprises is irrelevant to consumer environmental awareness. And the pricing of general logistics distribution enterprises is positively related to environmental awareness.

Prove:

$$\frac{\partial P_1^I}{\partial \lambda} = \frac{1}{2M} > 0,\ \frac{\partial P_2^I}{\partial \lambda} = \frac{\delta}{2M} > 0,\ P_1^I = \frac{C_1}{2} + \frac{\lambda}{2(\lambda t_1 + t_2)},\ \frac{\partial P_2^I}{\partial \delta} = \frac{\lambda}{2M} > 0.$$

Theorem 2 In the decentralized supply chain, by the equation of (7) and (8) we can know there is a positive correlation between the two logistics distribution enterprises pricing and the level of consumption of consumers λ, and there is a positive correlation between these two logistics distribution enterprises pricing and consumer environmental awareness δ when these conditions of $2MC_1 > 6\lambda - MC_2$ and $\delta MC_1 > -\delta^2 + 8\delta - 2MC_2 - 4$ are satisfied.

Prove:

$$\frac{\partial P_1^D}{\partial \lambda} = \frac{2(1-\delta)}{M(4-\delta)} > 0, \frac{\partial P_2^D}{\partial \lambda} = \frac{\delta(1-\delta)}{M(4-\delta)} > 0, \frac{\partial P_1^D}{\partial \delta} = \frac{M(2C_1+C_2)-6\lambda}{M(4-\delta)^2}, \frac{\partial P_2^D}{\partial \delta} = \frac{M(\delta C_1+2C_2)+(\delta^2-8\delta+4)}{M(4-\delta)^2}.$$

5 Conclusion

Consumer heterogeneity determines different consumer estimate of different logistics distribution services, and there be different behavior characteristics in the process of the purchase behavior. This paper studied the pricing of two monopoly Logistics distribution enterprises. These two logistics distribution enterprise provide two kinds of different green degree logistics distribution services, and can be substituted. We divided consumers into green consumer and general consumers. Main work and conclusions are as follows:

A. Setting up two profit functions of two logistics distribution enterprises by constructing the utility function of these two kinds of logistics distribution services. Then get customers purchase decision and give the optimal pricing strategies of the two logistics distribution enterprises.
B. The numerical simulation shows the influence of the proportion of green consumers to logistics distribution enterprise's profit. The result shows that these two logistics distribution enterprises' pricing strategies on the two supply chain strategy increase with consumer spending levels, specially increasing more rapidly in the integrated supply chain.
C. Through numerical simulation found that as consumers improve the price sensitivity of general consumers, two logistics distribution enterprises of price increases. With the decrease of the consumer environmental awareness, the pricing of these two logistics distribution enterprises is not gradually reduced on integrated supply chain, but the pricing is decreasing in decentralized supply chain.

From the above analysis we can know. Logistics distribution enterprise can increase green logistics distribution services propaganda work thus improving the environmental awareness of consumers. Or with new technology to promote green logistics distribution services cost, so can increase the proportion of green consumers. And realize the purpose of maximizing profits finally.

Acknowledgements. This work is supported by Anhui higher school natural fund key project No. KJ2015A415.

References

1. J.R. Birge, J. Drogosz, I. Duenyas, Setting single-period optimal capacity levels and prices for substitutable products. Int. J. Flex. Manuf. Syst. **10**(4), 407–430 (1998)
2. K. Rajaram, C.S. Tang, The impact of product substitution on retail merchandising. Eur. J. Oper. Res. **135**(3), 582–601 (2001)
3. W.J. Hopp, X. Xu, Product line selection and pricing with modularity in design. Manuf. Serv. Oper. Manag. **7**(3), 172–187 (2005)
4. H.A. Kuyumcu, I. Popescu, Deterministic price–inventory management for substitutable products. J. Revenue Pricing Manag. **4**(4), 354–366 (2006)
5. C.S. Tang, R. Yin, Joint ordering and pricing strategies for managing substitutable products. Prod. Oper. Manag. **16**(1), 138–153 (2007)
6. M. Karakul, L.M.A. Chan, Joint pricing and procurement of substitutable products with random demands—a technical note. Eur. J. Oper. Res. **201**(1), 324–328 (2010)
7. Y. Xia, Competitive strategies and market segmentation for suppliers with substitutable products. Eur. J. Oper. Res. **210**(2), 194–203 (2011)
8. E. Stavrulaki, Inventory decisions for substitutable products with stock-dependent demand. Int. J. Prod. Econ. **129**(1), 65–78 (2011)
9. J. Zhao, W. Tang, J. Wei, Pricing decision for substitutable products with retail competition in a fuzzy environment. Int. J. Prod. Econ. **135**(1), 144–153 (2012)
10. J. Zhao, W. Tang, R. Zhao, J. Wei, Pricing decisions for substitutable products with a common retailer in fuzzy environments. Eur. J. Oper. Res. **216**(2), 409–419 (2012)
11. J.-M. Chen, C.-I. Chang, Dynamic pricing for new and remanufactured products in a closed-loop supply chain. Int. J. Prod. Econ. **146**(1), 153–160 (2013)
12. J. Zhang, B. Shou, J. Chen, Postponed product differentiation with demand information update. Int. J. Prod. Econ. **141**(2), 529–540 (2013)
13. J. Zhao, J. Wei, Y. Li, Pricing decisions for substitutable products in a two-echelon supply chain with firms′ different channel powers. Int. J. Prod. Econ. **153**, 243–252 (2014)
14. C. Chialin, Design for the environment: a quality-based model for green product development. Manag. Sci. **47**, 250–263 (2001)
15. L. Zugang, D.A. Trisha, M.C. Jose, Consumer environmental awareness and competition in two-stage supply chains. Eur. J. Oper. Res. **218**, 602–613 (2012)
16. B. Marcus, C.K. Anderson, Revenue management for low-cost providers. Eur. J. Oper. Res. **188**, 258–272 (2008)

Research on Profit Allocation of Aquatic Products Cold Chain Logistics Alliance Based on Cooperative Game Theory

Yan Chen, Ya-wen Xie(✉), Lu Wang, and Nai-xu Song

Department of Management Engineering, Qingdao University of Technology, Qingdao, China
1323021797@qq.com

Abstract. With the people's increasing demands for aquatic products, the efficiency and service quality of aquatic products cold chain logistics are facing new challenges. The paper chooses cooperative game theory as the basic research method to study the profit allocation of aquatic products cold chain logistics alliance which is composed of four major participating enterprises conducting processing, warehousing, transportation and retailing activities. Based on Shapley value, the paper constructs a scientific profit allocation model for the above mentioned cold chain logistics alliance, providing an equitable and reasonable allocation plan for the coordination of interests among the main participating enterprises.

Keywords: Aquatic products · Cold chain logistics alliance · Profit allocation Cooperative alliance · Shapley value

1 Introduction

With the continuous development of economy and society, the consumption quantity of aquatic products is increasing. However, due to the nature of aquatic products (high moisture and active enzyme contents), it is easy for aquatic products to deteriorate in the process from processing to the ultimate consumer, which seriously affects the quality of fresh products and greatly reduces the utility value and economic value of fresh products. At present, the development of China's aquatic products cold chain logistics is still in the initial stage, and there is no integrated system for aquatic products cold chain logistics. Different participating enterprises in the cold-chain independently conduct their own logistical activities, which seriously affects the operation efficiency of the supply chain and damages the overall interests of participating enterprises as well. Cold chain logistics includes four different value-creation activities including processing, warehousing, transportation and retailing, and there are four major node enterprises accordingly. Each participating enterprise is a individual property right body. Sometimes, its own interests and targets may conflict with those of the entire supply chain, which affects the efficiency of the entire cold chain. Therefore, how to

G. Q. Huang et al. (eds.), *Proceeding of the 24th International Conference on Industrial Engineering and Engineering Management 2018*,
https://doi.org/10.1007/978-981-13-3402-3_55

effectively solve the problem of uncoordinated benefits of cold-chain logistics can help reduce the production and circulation costs of aquatic products. And the fair and reasonable profit allocation among participating enterprises of cold chain logistics is the key to ensure the effective operation of cold chain logistics. Meanwhile it plays an important part in improving the competitiveness of fresh aquatic products in our country.

The research of cold chain logistics in China is in its infancy, but it has also achieved some achievements. Zhu combined the characteristics of cold chain logistics to build a cold chain logistics distribution path model based on the vehicle routing problem under the traditional time window constraint [1]. Osvald and Stirn added the constraint conditions of the customer time service scope in the transportation process, and solved the problem of cold chain logistics distribution path by minimizing the total cost of distribution as the objective function [2]. Gong and Gao put forward the problems in the system construction, talent cultivation and technological development of China's water products cold chain logistics, and put forward corresponding countermeasures and Suggestions [3]. Based on the development model of cold chain logistics in North America, east Asia and Western Europe. Lu put forward the construction of the fourth party aquatic products cold chain logistics development mode which contains four main bodies-the government, Banks, transportation organization, quality inspection institutions and other departments [4]. Qin and Tian took the development of the "One Belt and One Road" and "Internet Plus" as the research background. They analyzed the problems existing in policy support, restriction of enterprise development, unsound network system, backwardness of logistics standards and related technologies in Guangxi cold chain logistics of aquatic products, and then put forward corresponding suggestions to promote aquatic products cold chain logistics development [5]. Liu and Zhu used FAHP empirical methods to identify the key factors affecting the development of cold chain logistics of aquatic products, and provided reference basis for focusing on limited resources to resolve major conflicts and accelerate the development of aquatic products cold chain logistics [6].

At present, there is a certain amount of literature on the research of profit allocation. Nishizaki and Sakawa described the linear production model with the mathematical programming problem with fuzzy parameters to realize the distribution of alliance profit and cost among members [7]. Zhou et al. pointed out that the rational profit allocation among the alliance enterprises is the basis for the coordinated development of alliance. And from the perspective of cooperative game, Shapley value method is used to calculate the reasonable return of each firm in the alliance enterprise [8]. Wu established a two-stage process model for the profit allocation of dynamic logistics alliance [9]. On the basis of Shapley value method, Nash equilibrium method and MCRS method, Yu set up the model of weighted group center of gravity by introducing heavy coefficient to distribute the profit of alliance [10]. Xu introduced risk factors as the method of income distribution in Shapley value method [11]. Du and Wang used game theory and AHP to study the profit allocation of logistics enterprise alliance [12]. Ren et al. studied the supply chain of agricultural products using Shapley value method, and the benefit of each participator on the basis of cooperation was much larger than that of non-cooperation [13]. Wang applied the Shapley value method to distributing the initial benefits of the alliance, and adjusted the profit allocation model by using TOPSIS method [14]. Wei analyzed the profit function to study the fresh agricultural products supply chain e-commerce on the profit distribution problems, and concluded that it is possible to balance the profits of the company itself by coordinating the behavior of each member in the supply chain [15]. Based on the alliance relationship between cold chain logistics node enterprises, this paper constructs a model of profit allocation of cold chain logistics alliance based on Shapley value, so as to distribute the benefits within the main body of the alliance.

2 Model Establishing

As far as game theory is concerned, the non-cooperative game between enterprises will eventually lead to the "prisoner's dilemma" of non-cooperation, which will not maximize the benefits. Similarly, the cold chain logistics of aquatic products generally includes four major participant enterprises of processing, warehousing, transportation and retailing, only through the establishment of a stable logistics enterprise alliance can the four enterprises achieve the multi-win situation. Logistics enterprise alliance, as a coalition parties to pursue a common economic interests, through the formation of a contract and an agreement to form a corporate cooperative collective, the interests of everyone is the original appeal and also the ultimate appeal. The desire to obtain such benefits will bring about dual effects. The rational profit allocation will promote the healthy development of the alliance. Therefore, in order to guarantee the stability of the alliance, the internal interests of the alliance must abide by certain principles:

(1) Equity principle. Although the differences in strength within aquatic products cold chain logistics federation may cause different status within the alliance, all participators in the alliance contribute to the operation of the supply chain. Without any one alliance enterprise, the operation of the whole cold chain will be affected. Therefore, the principle of fairness must be reflected in the distribution method and process.
(2) Benefit principle. The profit allocation of alliances must be distributed in the principle of fairness and to ensure that all participators in the best interests. Alliances should avoid the waste of resources caused by unnecessary internal fighting among internal members and affect the efficiency of the entire supply chain.
(3) The principle of mutual benefit. It asks the members of the alliance tao give more help to the partners within their capabilities so as to achieve the "win-win" goal.
(4) The principle of matching the share of interests and risk taking. This is a function that takes into account the influence of risk factors, and that interest acquisition should be positively related to the amount of risk.

Shapley value is based on the number of people's contribution to the alliance, so it is a fair and reasonable way of distribution. The income distribution model based on Shapley value is constructed as follows.

2.1 Basic Assumptions of the Model

Considering a problem of profit allocation on a cold chain logistics consisting of processing, warehousing, transportation and retailing, we use P, W, T and R for the enterprises of processing, warehousing, transportation and retailing of cold chain logistics respectively. In the four links of cold chain logistics, there are three types of cooperation: The participators in the game are not aligned with each other, some players form alliances, and all players are allied. In the first case, companies across the cold chain do not cooperate and are independent stakeholders; In the second case, it is the two-tiered master-slave countermeasures that the cooperation among the members of some bureaus constitutes between the alliance and the enterprises outside the alliance; The third scenario is the optimal decision-making problem for large coalitions formed by all players.

2.2 The Characteristic Function of Model

In the game of n players, use set $N = \{1, 2, ..., n\}$ to represent the participant set, S is any subset of N, S could be an empty set and a universal set, Such S we call a alliance. Use the feature function $V(S)$ to represent the maximum benefit that a number of participants can gain when they join together as an alliance. When S is an empty set, its maximum profit is zero. The feature function $V(S)$ of multiplayer cooperation model is super additive, That is, for any alliance S1, S2 belongs to N. When the intersection of S1 and S2 is empty, there is $V(S_1 \cap S_2) \geq V(S_1) + V(S_2)$, which is a necessary condition for forming alliances.

2.3 The Solution X of Multiplayer Cooperative Game Model

The solution of multi-player cooperative game model is the profit distribution of various subjects in multiplayer cooperation model. Using $X = \{X_1, X_2, ..., X_n\}$ to represent the profit and loss vectors of n participators. Where *i* represents the interest of the *i* participant. The solution of multi-player cooperative game model should satisfy two conditions: One is that the sum of the profits from the distribution of each subject from the alliance is the largest total revenue of all forms of alliances. The second is that the income derived from the distribution of each participator from the alliance should be greater than that of the individual operating income. That is formulas (1) and (2).

$$\sum_{i=1}^{n} x_i = v(N), i \in N \tag{1}$$

$$x_i \geq v(\{i\}), i \in N \tag{2}$$

2.4 The Solution of the Multiplayer Model

There are many methods for solving multiplayer cooperation models, such as Shapley value method, CGA method, MCRS method, etc. The Shapley value method allocates the benefits of each participant according to the amount of contribution that the participant makes to the cooperation, which has a certain fairness, and the calculation is simple and efficient. Therefore, the Shapley value method is used in this paper to solve the solution of the multi-participator cooperation model of aquatic product cold chain logistics. In the Shapley value method, the distribution of benefits to the participants in the cooperative enterprise is unique, called the Shapley value, denoted as $\varphi(v) = (\varphi_1(v), \varphi_2(v), \ldots, \varphi_n(v))$. Where $\varphi_i(v)$ represents the profit distribution of the i participant in the alliance, the formula is calculated as (3), (4), (5).

$$\varphi_i(v) = \sum_{i \in S_i}^{n} w(|S|)[v(S) - v(S \backslash \{i\})] \quad i = 1, 2, \ldots n \tag{3}$$

$$S \backslash \{i\} = S - \{i\} \tag{4}$$

$$w(|S|) = \frac{(n - |n|)!(|S| - 1)!}{n!} \tag{5}$$

where $|S|$ represents the number of members contained in subset S. $w(|S|)$ is a weighting factor, and it is actually the probability that the game subject i participates in the association S in a random manner, and the probability that i contributes to the alliance; $v(S \backslash \{i\})$ represents the total income when i does not participate in subset S. $v(S) - v(S \backslash \{i\})$ indicates the contribution of participant i to Alliance S.

3 Example Analysis

It is assumed that the annual profits of the four major enterprises of processing (P), warehousing (W), transportation (T), and retailing (R) in the first case operating independently are 6 million, 3 million and 4 million and 2 million yuan respectively. And the maximum benefit of the four logistics subjects in the second and third cooperation game, as shown in Table 1.

Table 1. Revenue of alliance cooperation (unit: rmb 10,000 yuan)

Method of cooperation	(P, W)	(P, T)	(P, R)	(W, T)	(W, R)	(T, R)	(P, W, T)	(P, W, R)	(P, T, R)	(W, T, R)	(P, W, T, R)
Alliance profit	1200	1400	1000	900	800	900	1600	1300	1600	1200	2000

The Shapley value method is used to calculate the income distribution of four logistics subjects in different cooperation modes, as shown in Tables 2, 3, 4 and 5.

Table 2. Processing enterprise (unit: 10,000 yuan)

S	$v(S)$	$v(S\backslash\{i\})$	$v(S) - v(S\backslash\{i\})$	$\|S\|$	$\omega(\|S\|)$	$\omega(\|S\|)[v(S) - v(S\backslash\{i\})]$
(P)	600	0	600	1	1/4	150
(P, W)	1200	300	900	2	1/12	75
(P, T)	1400	400	1000	2	1/12	1000/12
(P, R)	1000	200	800	2	1/12	800/12
(P, W, T)	1600	900	700	3	1/12	700/12
(P, W, R)	1300	800	500	3	1/12	500/12
(P, T, R)	1600	900	700	3	1/12	700/12
(P, W, T, R)	2000	1200	800	4	1/4	200

Table 3. Warehousing enterprise (unit: 10,000 yuan)

S	$v(S)$	$v(S\backslash\{i\})$	$v(S) - v(S\backslash\{i\})$	$\|S\|$	$\omega(\|S\|)$	$\omega(\|S\|)[v(S) - v(S\backslash\{i\})]$
(W)	300	0	300	1	1/4	75
(P, W)	1200	600	600	2	1/12	50
(W, T)	900	400	500	2	1/12	500/12
(W, R)	800	200	600	2	1/12	50
(P, W, T)	1600	1400	200	3	1/12	200/12
(P, W, R)	1300	1000	300	3	1/12	25
(W, T, R)	1200	900	300	3	1/12	25
(P, W, T, R)	2000	1600	400	4	1/4	100

Table 4. Transportation enterprise (unit: 10,000 yuan)

S	$v(S)$	$v(S\backslash\{i\})$	$v(S) - v(S\backslash\{i\})$	$\|S\|$	$\omega(\|S\|)$	$\omega(\|S\|)[v(S) - v(S\backslash\{i\})]$
(T)	400	0	400	1	1/4	100
(P, T)	1400	600	800	2	1/12	800/12
(W, T)	900	300	600	2	1/12	50
(T, R)	900	200	700	2	1/12	700/12
(P, W, T)	1600	1200	400	3	1/12	400/12
(P, T, R)	1600	1000	600	3	1/12	50
(W, T, R)	1200	800	400	3	1/12	400/12
(P, W, T, R)	2000	1300	700	4	1/4	175

Table 5. Retailing enterprise (unit: 10,000 yuan)

S	$v(S)$	$v(S\backslash\{i\})$	$v(S) - v(S\backslash\{i\})$	$\|S\|$	$\omega(\|S\|)$	$\omega(\|S\|)[v(S) - v(S\backslash\{i\})]$
(R)	200	0	200	1	1/4	50
(P, R)	1000	600	400	2	1/12	400/12
(W, R)	800	300	500	2	1/12	500/12
(T, R)	900	400	500	2	1/12	500/12
(P, W, R)	1300	1200	100	3	1/12	100/12
(P, T, R)	1600	1400	200	3	1/12	200/12
(W, T, R)	1200	900	300	3	1/12	25
(P, W, T, R)	2000	1600	400	4	1/4	100

According to formula (3), it can be calculated that the profits of processing (P), warehousing (W), transportation (T) and retailing (R) enterprises of aquatic product cold chain logistics after cooperation are $\varphi_1(v) = 733$, $\varphi_2(v) = 383$, $\varphi_3(v) = 567$, $\varphi_4(v) = 317$ respectively, and $\varphi_1(v) + \varphi_2(v) + \varphi_3(v) + \varphi_4(v) = 2000$.

Through comparison with related data, it can be found that $\varphi_1(v)$ $\varphi_2(v)$ $\varphi_3(v)$ $\varphi_4(v)$ are all greater than the benefits obtained by the individual business operators, the two-to-two cooperation, and the three-three cooperation.

Through the analysis of the above game simulation model, it can be seen that the alliance cooperation between each participator will obtain more benefits than any other single alliance cooperation. The participating companies will improve the efficiency of the entire cold chain logistics through the cooperation method of integrating resources and sharing information with the internet of things technology and cloud computing. In terms of current trends, Logistics alliance cooperation among the participators is the development direction of cold chain of aquatic products in the future. Alliance cooperation can improve the efficiency of the whole cold chain, it is also the important premise to ensure the quality of aquatic products. The rationality of alliance profit allocation has an important influence on the stability of the cooperative relationship among logistics participants, concerning the stable operation of the entire aquatic product cold chain. Coordinating the interests of the main participators of the cold chain logistics alliance with the aid of Shapley value method can ensure that each participator's benefit is not less than the benefit produced by individual operation and any other cooperation methods. And this is the key to guarantee the effective operation of aquatic products cold chain.

4 Conclusion

Based on the analysis of the principle of alliance members' interest distribution, the cooperation game is applied to the cold chain logistics of aquatic products. Using the Shapley value method to coordinate the interests of the main body of the cold chain logistics alliance, we can draw the following conclusions:

(1) No matter whether they are self-serving or altruistic, the fully-coordinated logistics alliance will get the maximum benefit from all the nodes of the cold-chain logistics. Therefore, by integrating resources such as Internet of things and cloud computing, node enterprises can reduce the cost of logistics and improve the efficiency of the whole cold chain logistics.

(2) There is a conflict of interest distribution in cooperation between node enterprises. The revenue distribution model based on Shapley value effectively realizes the potential profit of cooperation and resolves conflicts of interest in cooperation. It fully embodies the cooperative relationship of "revenue sharing and risk sharing", which ensures the long-term stable and efficient operation of the whole cold chain logistics.

Acknowledgements. Fund Project: Shandong Provincial Social Science Planning Research Project, Project Number: 16CGLJ07.

References

1. H. Zhu, Study on the distribution management of food cold chain logistics (in Chinese), Shanghai Jiao Tong University (2008)
2. A. Osvald, L.Z. Stirn, Vehicle routing algorithm for the distribution of fresh vegetables and similar perishable food. J. Food Eng. **85**(2), 285–295 (2008)
3. M. Gong, H. Gao, Research on the optimization of cold chain logistics of aquatic products in China (in Chinese). Market Mod. **28**, 71 (2013)
4. Y. Lu, Research on the development mode of cold chain logistics of aquatic product (in Chinese), Beijing Jiao Tong University (2014)
5. X.H. Qin, H.X. Tian, Problems and countermeasures of development of cold chain logistics in Guangxi aquatic products under new situation (in Chinese). China Market **49**, 47–48 (2016)
6. D.H. Liu, P.Y. Zhu, Analysis of affecting factors of aquatic products cold chain logistics in Fujian based on FAHP (in Chinese). J. Qingdao Agric. Univ. **1**, 35–39 (2016)
7. I. Nishizaki, M. Sakawa, Solutions based on fuzzy goals in fuzzy linear programming games. J. Jpn. Soc. Fuzzy Theory Syst. **115**(1), 105–119 (2000)
8. X.F. Zhou, C. Zhang, Y. Guo, Discussion on the railway logistics alliance and the problem of profit distribution (in Chinese). Railw. Transp. Econ. **6**, 74–76 (2009)
9. L. Wu, The method of profit allocation of dynamic logistic alliance under the mode of product share contract (in Chinese). Syst. Eng. **5**, 25–29 (2009)
10. H.W. Yu, Research on the logistics alliance distribution of interests based on weighted group barycenter algorithm (in Chinese), Hebei University of Engineering, p. 62 (2012)
11. Z.P. Xu, Evolutionary game analysis and revenue distribution of joint distribution in cold chain of urban logistics (in Chinese), Beijing Jiao Tong University, p. 114 (2012)
12. Z.P. Du, P. Wang, Study on benefit allocation of vehicle-transport enterprise alliances based on interest game (in Chinese). Logist. Technol. **13**, 122–125 (2015)
13. X.Y. Ren, Q.Q. Feng, S. Wang, X. Wen, Profit distribution of agricultural supply chain based on Shapley value. Adv. J. Food Sci. Technol. **7**(7), 479–483 (2015)
14. W.M. Wang, Research on profit allocation of cold chain logistics distribution alliance (in Chinese). Manag. Adm. **12**, 126–128 (2016)
15. L. Wei, Distribution of the benefit of fresh produce E-commerce supply chain (in Chinese). Logist. Eng. Manag. **2**, 47–50 (2018)

Evolutionary Game Analysis of Supply Chain Subjects' Information-Sharing from the Perspective of Building Industrialization

Yue Li(✉) and Yong-qing Li

Department of Management, Xi'an University of Science and Technology, Xi'an, China
357933242@qq.com

Abstract. The recent introduction of industry 4.0 and the promotion of the "One Belt One Road" initiative have had a positive impact on China's efforts towards architectural industrialization. This has been an important change for the traditional construction industry and an inevitable trend for the future development of the construction industry. Currently, the industrialization development of Chinese architecture is at an important stage, and it requires a scientific and efficient construction supply chain management to add value. This paper builds a model based on an evolutionary game theory to analyze the information-sharing process of construction supply chain companies that possess a solid background of building industrialization, and provides a theoretical basis for the development of supply chain management for main enterprises.

Keywords: Construction supply chain · Construction industrialization Evolutionary game · Information-sharing

1 Introduction

In recent years, relevant departments have successively issued a number of regulations and plans for the promotion of construction industrialization and housing industrialization, providing an opportunity for the development of the building industrialization. However, at present, there are many problems that will hinder the development in the early stages of the development of China's industrialization of buildings. And it is urgent to be solved [1]. Because of the characteristics of building industrialization itself, its information system has a certain degree of complexity with respect to the traditional construction industry, which makes its construction supply chain also have a certain degree of complexity. The purpose of this study is to address the issue of information sharing between supply chain entity companies in the process of the development of China's building industrialization, building models based on evolutionary game theory and perform simulation analysis. And to provide a preliminary theoretical basis for the main part of the supply chain to participate in the decision-making of enterprises in the process of the industrialization of construction.

G. Q. Huang et al. (eds.), *Proceeding of the 24th International Conference on Industrial Engineering and Engineering Management 2018*,
https://doi.org/10.1007/978-981-13-3402-3_56

2 Basic Principles of Model Building

2.1 System Background

Compared with the traditional building construction model, the construction production mode under the background of the industrialization of the building has undergone tremendous changes [2]. It uses the prefabrication components and the prefabrication construction as the main mode of production. This greatly enhances the role of building component manufacturers and suppliers in the supply chain [3]. The participants in the construction supply chain mainly include: Proprietor, Component producer, Designers, Builders, Component suppliers, Construction Materials and Equipment Suppliers, etc [4]. Due to the complexity and relevance of the supply chain, any decision made by each participant will have an impact on the decisions and development of other subjects.

2.2 System Specification

In the construction process under the background of building industrialization, the smooth construction supply chain is an important condition for ensuring the increase of value, and it is also the most important prerequisite for the smooth progress of assembly construction. Practice has shown that information interaction between building supply chain entities plays an important role in this process. The forms of information owned by participants in the construction supply chain can be divided into corporate proprietary information and common information. For each participating entity in the supply chain, proprietary information is the basic information that constitutes the core competitiveness of the enterprise. Through the sharing of proprietary information, the proprietary information of the other party can be continuously obtained and internalized to some extent. That can help them to improve the value of their own information. However, the sharing of proprietary information has two sides. While participating entities acquire and absorb information from other parties and increase their own value, they will also weaken their power in the entire supply chain because of their sharing, and may even cause the spillover of core corporate information resources. This will result in a certain sharing cost.

Some scholars have pointed out that the optimal governance structure of the building industrialized supply chain is a strategic alliance, which can reduce the risk costs arising from economic interference and information asymmetry [5]. This shows that information sharing between companies has an indispensable impact on the value-added of the construction supply chain. Information sharing in the building supply chain is a dynamic process that is constantly changing [6]. The difference between the strategic objectives and the responsibility orientation of the two information-sharing parties (A and B) will directly restrict the degree of information sharing [7]. At the same time, because of the information sharing party knows more about their own information holding situation than the other side, and because of participants' rationality and information asymmetry, participants can easily adopt the "free rider" approach to maximize their utility. Once the participants in the supply chain adopted the "not sharing" strategy, then the information-sharing process will fall into the "prisoner's dilemma". Basic conditions based on the dynamics of the supply chain system and the limited rationality of the participants in the system, this paper chooses the tool of evolutionary game theory to construct and analyze the model of information sharing process [8].

3 Evolutionary Game Analysis

3.1 Hypothesis

Based on the preliminary analysis and interpretation of the system, in order to more thoroughly analyze the strategies of the participating parties on the supply chain, the impact on the supply chain's value-added and architectural industrialization projects. The proposed hypotheses as follows:

H1: The strategies adopted by Enterprise A and Enterprise B in the information-sharing process are "sharing" (Y) and "not sharing" (N).
H2: Both Enterprise A and Enterprise B maintain bounded rationality and make decisions based on their actual conditions.
H3: The probability of "sharing" strategies for Enterprise A and Enterprise B based on their information holding situation, enterprise development status, different information importance and cost is x and y.
H4: Based on the dynamic nature of supply chain system, it is considered that information-sharing process of supply chain participants is a dynamic process.
H5: Assuming that there is no policy intervention or social impact in the information-sharing process, only the participants in the supply chain information-sharing system are the foundation of the research.

3.2 Evolutionary Game Model Establishment

In the process of information-sharing, taking part in "sharing" or "not sharing" by participating parties will bring a certain influence on the decision-making and development of enterprises. When information is sharing, the sharer will receive incentives and rewards for contributing information during the information-sharing process. On the other hand, due to the change of enterprise's income and cost, the relationship between the penalty cost of choosing "free rider" behavior without information sharing and the income generated by absorbing information from the other party, as well as some other external factors will directly affect the enterprise's decision-making [9].

In order to facilitate the analysis of different strategies and the evolution of the system in the supply chain's main business under different circumstances, setting the following parameter variables as shown in Table 1.

Table 1. Variable symbol setting table

Variable symbol	Definition
Π	General income without sharing process
π	Share revenue when both parties share
η	Information sharing earnings index
α	Information sharing incentive coefficient
β	Penalty coefficient when not sharing
γ	The amount of information-sharing
θ	Absorption coefficient of shared information
ε	Income absorption coefficient of shared information
C	The cost of information sharing
ω	Share cost sharing ratio of Enterprise A ($0 < \omega < 1$)
E	Expected revenue value of the main body of information sharing

Based on the above analysis and hypothesis, according to the correlation coefficient set in Table 1, the income matrix of information sharing game between supply chain principal enterprises is calculated as shown in Table 2.

Table 2. Supply chain enterprises' information-sharing revenue matrix

		B	
		Y	N
A	Y	$\Pi_1 + \pi_1 + \alpha_1\theta_1\gamma_2\varepsilon_1 - \omega C$, $\Pi_2 + \pi_2 + \alpha_2\theta_2\gamma_1\varepsilon_2 - (1-\omega)C$	$\Pi_1 - \omega C$, $\Pi_2 + (1-\beta_2)\gamma_1\theta_2\varepsilon_2$
	N	$\Pi_1 + (1-\beta_{1)}\gamma_2\theta_1\varepsilon_1$, $\Pi_2 - (1-\omega)C$	Π_1, Π_2

3.3 Evolution Path and Result Analysis

According to the revenue matrix of Table 2, calculate the expected return E_{AY} when company A adopts the "sharing" strategy, and the expected return E_{AN} when adopting the "not sharing" strategy, the average return E_A, and the replication dynamic equation are as follows:

$$\begin{aligned} E_{AY} &= y[\Pi_1 + \pi_1 + \alpha_1\theta_1\gamma_2\varepsilon_1 - \omega C] + (1-y)[\Pi_1 - \omega C] \\ &= \Pi_1 + y(\pi_1 + \alpha_1\theta_1\gamma_2\varepsilon_1) - \omega C \end{aligned} \tag{1}$$

$$\begin{aligned} E_{AN} &= y[\Pi_1 + (1-\beta_1)\gamma_2\theta_1\varepsilon_1] + (1-y)\Pi_1 \\ &= \Pi_1 + y(1-\beta_1)\gamma_2\theta_1\varepsilon_1 \end{aligned} \tag{2}$$

$$\begin{aligned} E_A &= xE_{AY} + (1-x)E_{AN} \\ &= \Pi_1 + xy\pi_1 - x\omega C + y[x\alpha_1 + (1-x)(1-\beta_1)]\theta_1\gamma_2\varepsilon_1 \end{aligned} \tag{3}$$

$$F_A(x) = \frac{dx}{dt} = x(E_{AY} - E_A) \\ = x(1-x)\{y[\pi_1 - (1-\alpha_1-\beta_1)\theta_1\gamma_2\varepsilon_1] - \omega C\} \tag{4}$$

Similarly, we can calculate the expected return E_{BY} of the company B when adopting the "sharing strategy", the expected profits E_{BN} when adopting the "not sharing" strategy, the average income E_B, and the replication dynamic equation are as follows:

$$E_{BY} = x[\Pi_2 + \pi_2 + \alpha_2\theta_2\gamma_1\varepsilon_2 - (1-\omega)C] + (1-x)[\Pi_2 - (1-\omega)C] \\ = \Pi_2 + x(\pi_2 + \alpha_2\theta_2\gamma_1\varepsilon_2) - (1-\omega)C \tag{5}$$

$$E_{BN} = x[\Pi_2 + (1-\beta_2)\gamma_1\theta_2\varepsilon_2] + (1-x)\Pi_2 \\ = \Pi_2 + x(1-\beta_2)\gamma_1\theta_2\varepsilon_2 \tag{6}$$

$$E_B = yE_{BY} + (1-y)E_{BN} \\ = \Pi_2 + xy\pi_2 - x(1-\omega)C + x[y\alpha_2 + (1-y)(1-\beta_2)]\theta_2\gamma_1\varepsilon_2 \tag{7}$$

$$F_B(y) = \frac{dy}{dt} = y(E_{BY} - E_B) \\ = y(1-y)\{x[\pi_2 - (1-\alpha_2-\beta_2)\theta_2\gamma_1\varepsilon_1] - (1-\omega)C\} \tag{8}$$

Analyze by copying the dynamic equation: $F_A(x)$, $F_B(y)$ [10].

$F_A(x) = 0$. The stable states of the replication dynamic equation of enterprise A are:

$$x* = 0,\ x* = 1,\ y^* = \omega C/[\pi_1 - (1-\alpha_1-\beta_1)\theta_1\gamma_2\varepsilon_1], (I = [0, 1])$$

$F_B(y) = 0$. The stable states of the replication dynamic equation of enterprise B can be obtained as:

$$y* = 0,\ y* = 1,\ x^* = (1-\omega)C/[\pi_2 - (1-\alpha_2-\beta_2)\theta_2\gamma_1\varepsilon_2], (I = [0, 1])$$

It can be seen that there are five evolutionary equilibrium points in this information sharing system:

$$O(0,0), A(0,1), B(1,1), C(1,0), D(x*, y*).$$

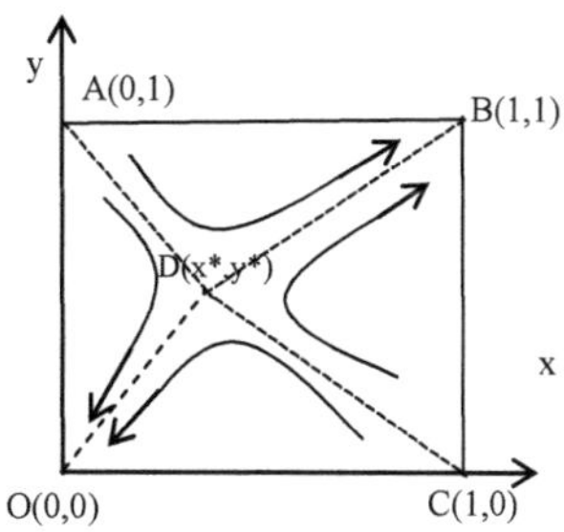

Fig. 1. Supply chain information-sharing system evolutionary game phase diagram

Based on this, the evolutionary phase diagram of this information-sharing system is shown in Fig. 1.

(1) $\begin{cases} \pi_1 - (1-\alpha_1-\beta_1)\theta_1\gamma_2\varepsilon_1 \succ 0, \\ \pi_2 - (1-\alpha_2-\beta_2)\theta_2\gamma_1\varepsilon_1 \succ 0 \end{cases}$. That means when $\begin{cases} \pi_1 + \alpha_1\theta_1\gamma_2\varepsilon_1 \succ (1-\beta_1)\theta_1\gamma_2\varepsilon_1, \\ \pi_2 + \alpha_2\theta_2\gamma_1\varepsilon_2 \succ (1-\beta_2)\theta_2\gamma_1\varepsilon_2 \end{cases}$ there are two ESS in this system: O (0, 0) and B (1, 1). It shows that when the participants of this information-sharing system are sharing information, the rewards and shared benefits are greater than the difference between the benefit and the penalty cost of the participants in the "hitch-hiker" behavior, the participants of information-sharing both adopt "sharing" or both "not sharing" decision results as long-term stability. In this case, there are two participating subject management strategies: a. Both parties share information through collaborative development. b. Nor parties sharing and they all do independent development based on their existing information.

(2) All game decision-making combinations of this information-sharing system will fall within the OABC area shown in Fig. 1. When the initial decision point falls in the area OADC, the system converges to (0, 0), that means the ESS is (0, 0). At this point, participants in the information-sharing process tend to "not sharing" strategy, and adopt strategies that develop with unique information. When the initial point of the decision falls in the area DABC, the system converges to (1, 1), that means the ESS is (1, 1). The information-sharing system participants tend to adopt "sharing" strategy and collaborative development for both parties.

(3) The area of regional AOCD indicates the probability that both sides of the information sharing system share the "no sharing" strategy. The area of the regional ABCD indicates that the information sharing system of the information sharing both parties choose the "sharing" strategy to test the probability of collaborative management of supply chain information [11]. The magnitude of the probability is directly related to the area of each area. The final evolution result and the result of the strategy selection are determined by the position of the D point. When the area AOCD is equal to the area of the region ABCD($S_{OADC}=S_{ABCD}$), both parties of the information-sharing system tend to have the same probability that the information is not shared and the probability that both parties share the information.

(4) After analysis: $S = S_{OADC} = (x^* + y^*)/2$. Calculate the partial derivative of S with respect to the relevant parameters: $\partial S/\partial\beta_i \prec 0, \partial S/\partial\varepsilon_i \prec 0, \partial S/\partial\alpha_i \prec 0$. It can be obtained that the increase of α, β and ε, S decreases. This shows that the area OADC area and the information of both parties do not share the "individual" development of supply chain participants. The probability of sharing with the *Information sharing incentive coefficient* (α), *Penalty coefficient of "not sharing"* (β), and *Income absorption coefficient of shared information* (ε) are positively correlated.

In summary, in the information-sharing system, only one party of the participating adopts the "sharing", and the other side adopts the "free rider" behavior is an extremely unstable state of the information-sharing system and cannot be developed for a long time. Eventually, it will evolve into two modes due to their own characteristics and external environment that is coordinated development or independent development.

4 Simulation Verification

Set initial value: $\Pi_1 = \Pi_2 = 50$, $\pi_1 = \pi_2 = 40$, $\eta_1 = \eta_2 = 5, \alpha_1 = 1, \alpha_2 = 2$, $\beta_1 = \beta_2 = 4, \gamma_1 = \gamma_2 = 5, \theta_1 = \theta_2 = 1, \varepsilon_1 = \varepsilon_2 = 3, \omega = 0.5, C = 50$. We obtained four evolution results as shown in Fig. 2.

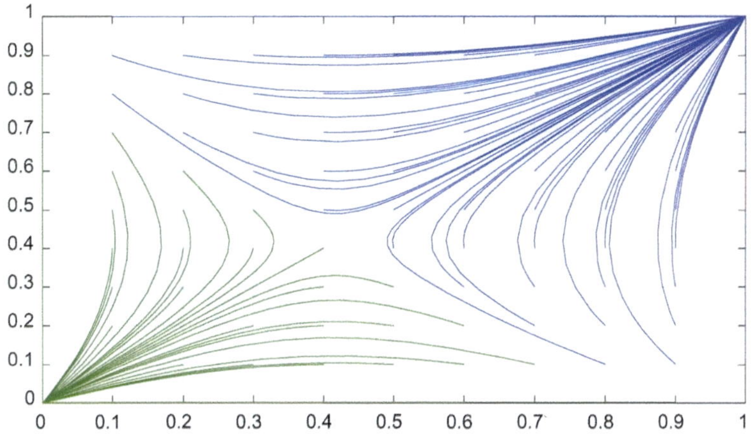

Fig. 2. Initial value evolution path simulation diagram

4.1 Impact of Information-Sharing Cost (C)

For this information-sharing system, on the premise that other parameters remain unchanged, set the value of C: C = 70, C = 80, C = 90. Using MATLAB to analyze the evolution path of this information-sharing system, we obtained four evolution results as shown in Fig. 3.

Through the analysis of the shared results in the three cases in Fig. 3, we can see that With the increase of sharing cost, the state and trajectory of the initial point of the independent development mode tend to increase, and the probability of "sharing" between both sides is decreasing. This shows that, with the increase of information sharing cost (C) in a certain range, the probability of "sharing" is reduced, and the probability of "sharing" is negatively correlated with the cost of information sharing (C).

4.2 Impact of Incentive Coefficient (α)

For this information-sharing system, on the premise that other parameters remain unchanged, set the value of information sharing incentive coefficient (α): $\alpha_1 = 2\,\alpha_2 = 3$, $\alpha_1 = 3\,\alpha_2 = 4$, $\alpha_1 = 4\,\alpha_2 = 5$. Using MATLAB to analyze the evolution path of this shared system, four evolution results are shown in Fig. 4.

The Fig. 4 shows that in the information-sharing system, within a certain range as the value of the information sharing incentive coefficient (α) increases, the evolutionary

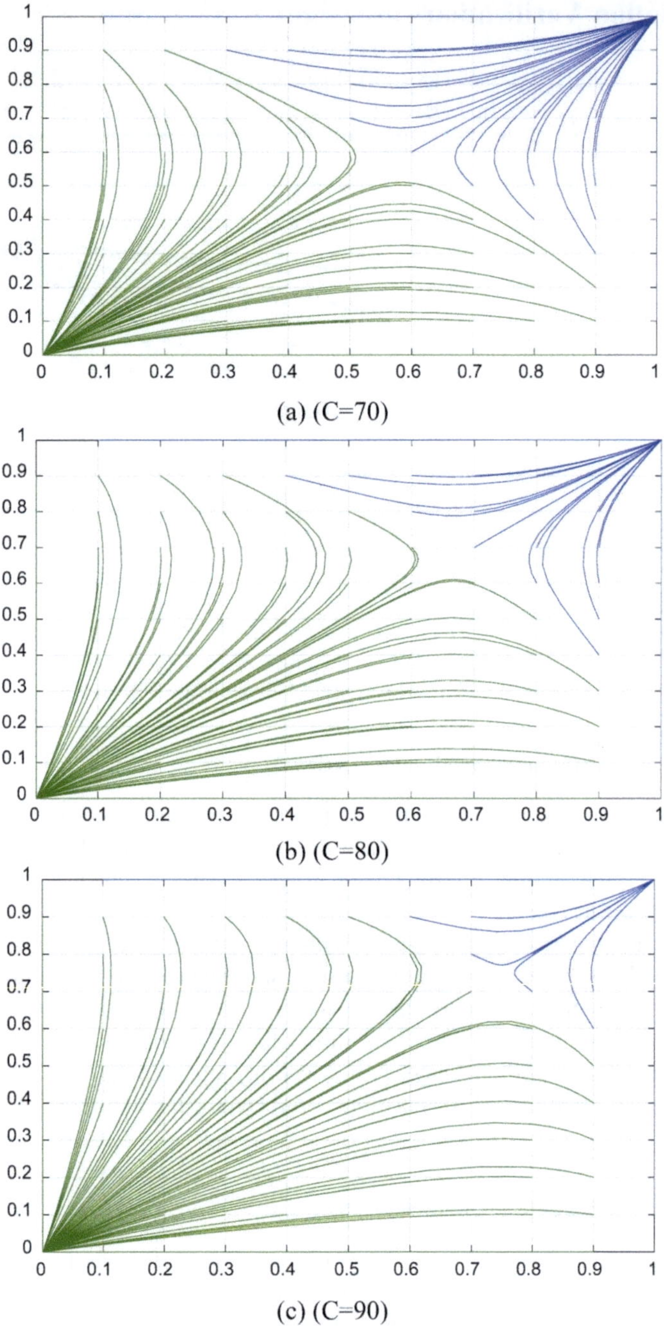

Fig. 3. Sharing cost (C) effect on sharing status

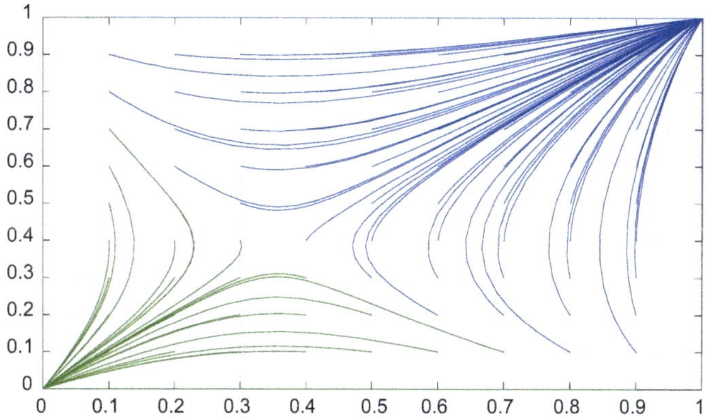

(a) (α_1=2 α_2=3)

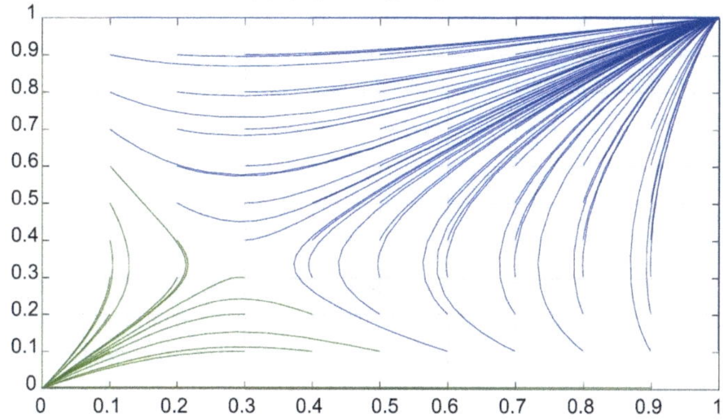

(b) (α_1=3 α_2=4)

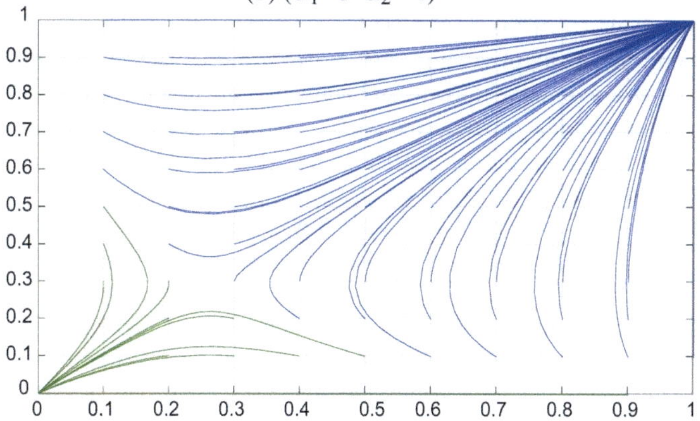

(c) (α_1=4 α_2=5)

Fig. 4. Incentive coefficient (α) impact on sharing status path curve of the system tends to steady state (1, 1) increases, and more initial point will evolve towards ESS (1, 1). This means that with the increase of the information sharing incentive coefficient (α), the probability of the information sharing subject adopting the "sharing" strategy increases. That means information sharing additional revenue is positively correlated with information sharing probability. To a certain extent, the additional benefit of information sharing will constrain the evolution of information sharing system and affect the final evolution result.

4.3 Impact of Other Factors

By assigning values to other parameters (π, η, β, γ, θ, ε) and using MATLAB to analyze the evolution path, we conclude that the above parameters affect the shared state as follows: π, η, β, γ, θ, ε all directly affect information sharing status and show a positive correlation. Within a certain range, with the increase of the above parameters' value, the probability of adopting "sharing" strategy will gradually increase, and the state of information-sharing will eventually evolve from an independent development state in which both parties "not sharing" strategy to a steady state of coordinated development where both parties "sharing".

5 Conclusion

The evolutionary game theory was applied in this paper, to the information-sharing system of supply chain enterprises from the perspective of architectural industrialization. This evolutionary game model was built based system theories and verified the results of the model analysis by using MATLAB simulation. The evolution results obtained in the model verification process through the parameter assignment simulation in the model were consistent with the results of the model analysis, which proves the feasibility and accuracy of the model. This study draws the following conclusions: a. The participants in the supply chain must actively participate in information sharing process to seek common development and actively adjust the development strategy. b. In the information sharing process, recipients should increase the reward and reward recipients of information sharing parties, when absorbing efficient and proprietary information resources. The information senders must also consider cooperation and development, and not only publish public low efficiency information. c. From the perspective of the overall system, it is also necessary to increase the punishment for "free rider" behavior through related policies to promote the collaborative development of construction supply chain enterprises. The main focus of this study is to provide a theoretical basis of supply chain management for the main enterprises in the construction supply chain business using the country's vigorous development in the construction industry as a backdrop. It also seeks make it easy for supply chain companies to analyze their macro and micro environment peculiar to their businesses and use that to adapt to the new era effectively. This process adds value to their services, creates better working conditions and provides an overall developmental opportunity for the companies.

Acknowledgements. First, I want to thank my tutor for his hard work in the study and writing guidance, and the inspiration for life planning and development. I would like to express my deep gratitude to my teachers for their care and guidance.
Second, I want to thank my fellow practitioners for their selfless help during my study, and many problems have been solved.
In addition, thank my family for supporting and encouraging me from time to time.
Finally, I would like to thank the experts for reviewing my paper in a busy schedule and I will continue my efforts.

References

1. K. Zheng, Q. Zhao, Problems in the development of building industrialization and analysis of countermeasures (in Chinese, Periodical style). Constr. Eng. Technol. (31) (2016)
2. N. Jia, Research on pricing of monolithic prefabricated house construction (in Chinese, Thesis or Dissertation style). Southeast University (2016)
3. K. Lei, R. Zhao, X. Wei, A review of the research on quality control of precast concrete components (in Chinese, Presented Conference Paper style), in *The 1st national engineering quality inspection, accreditation and reinforcement academic exchange conference* (2015)
4. Y. Shan, Z. Li, Evolutionary game of knowledge sharing in construction supply chain facing building industrialization (in Chinese, Periodical style). J. Eng. Manag. **2**, 1–5 (2015)
5. Y. Yue, J. Yin, Analysis of the economic motives of supply chain strategic alliance in the industrialization of new buildings (in Chinese, Periodical style). J. Hunan Univ. (Social Science) **28**(5), 45–49 (2014)
6. X. Yu, Maturity evaluation, model optimization and value-added analysis of information sharing in construction supply chain (in Chinese, Thesis or Dissertation style). Zhejiang University (2010)
7. K. Chen, Evolutionary game analysis of platform sharing mechanism of science and technology basic condition without external supervision (in Chinese, Periodical style). Sci. Decis. **2**, 45–53 (2011)
8. S. Ye, L. Yang, J. Zhang, Research on dynamic evolutionary game of supply chain knowledge sharing (in Chinese, Periodical style). J. Inf. (9), 126–130 (2012)
9. Y. Liu, P. Fan, Research on information sharing in internet of things environment based on game theory (in Chinese, Periodical style). J. Party Sch. Communist Party China **5**, 20–25 (2015)
10. X. Yi, Game analysis of the behavior of strategic alliance in real estate enterprises (in Chinese, Periodical style). J. East China Jiaotong Univ. **4**, 96–101 (2012)
11. X. Mi, Research on information sharing in decentralized supply chain network (in Chinese, Thesis or Dissertation style). Xiamen University (2006)

Research on Logistics Platform Resource Integration Based on UTAUT Model

Wei-hua Gan(✉), Chen Fu, Zhi-peng Chen, and Ren Zhong

School of Transportation and Logistics, East China Jiaotong University, Nanchang City, China
727595249@qq.com

Abstract. China's logistics platform has experienced a blowout development under the impetus of "Internet +", and attracted a lot investment from many organizations and enterprises. Consequently thousands of logistics platform companies appeared on the market. However, nowadays logistics platform market is facing a death suffering. This paper studies the resource integration of logistics platform based on the theory of acceptance and use of technology. Firstly, it describes the formation and elements of logistics platform. Then a conceptual framework of logistics platform resource integration is set up. Finally, it puts forward UTAUT model in order to accomplish the resource integration of the logistics platform.

Keywords: Logistics platform · Resource integration · Unified theory of acceptance and use of technology (UTAUT model)

1 Introduction

The concept of logistics platform came from France in 1960s. Nowadays there are more than 60 logistics platforms in developed countries, e.g. FIRST in US, FCPS in UK, TRADEGATE in Australia, PORTNET in Singapore, etc. [1, 2]. Especially some logistics platforms in these countries also include logistics parks or freight villages [3], these platforms play an important role in improving operation efficiency and economy development. Table 1 shows the definitions from foreign scholars.

Researches on logistics platform in China began with 1999. Prof. Zhitai Wang firstly proposed to establish logistics platform based on standardization and infrastructure, etc. in 2000 [10]. Before 2010 domestic scholars studied the logistics platform from the perspective of information system structure. But after 2010, their research interests focus on the combination with online and offline operation [11].

Although the concept of logistics platform can be understood and analyzed in different manners because of different factors like language translations, research areas, etc. Nevertheless, logistics platform must consider concepts like logistics operations, physical structures, process and its activities as well as information and communications technology (ICT) needed for design, operations and reporting and work as a base for logistics processes.

G. Q. Huang et al. (eds.), *Proceeding of the 24th International Conference on Industrial Engineering and Engineering Management 2018*,
https://doi.org/10.1007/978-981-13-3402-3_57

Table 1. Definitions of logistics platform

Authors	Definition
EAFV for EUROPLATFORMS (1997)	A defined area with which all activities relating to transport, logistics and the distribution of goods, both for national and international transit, are carried out by various operators. It is run by a single body, either public or private, and is equipped with all the public facilities to carry out the above mentioned operations
Abrahamsson Mats et al. (2003)	A homogenous part of the logistics system in the supply chain [4] that is centrally controlled and designed by focal organizations in a way that it is a resource-base for new marketing channel positions [5]
A. F. Dubke (2006)	Network coordination of all logistics services, providing greater competitiveness through lower costs of management and operation of the logistics chain [6]
Carolina Correa de Carvalho et al. (2010)	Where different agents of the supply chain can be integrated in the same physical place assisting logistics flows and acting as strategic interfaces between networks of global and regional dimensions with an aim to improve supply chain efficiency [7]
Juan Pablo Antúna, Rodrigo Alarcón (2014)	A specialized area with the infrastructure and services required for co-modal transportation and added value of the products making use of the infrastructure [8]
R. M. Silva et al. (2015)	The provision of a complex service for communicating and evaluating all relevant information that may affect the operation of supply channels [9]

The purpose of this article is to understand the key element of logistics platforms operation, especially discover the current relationship between users or providers and the logistics platforms through users' behavior. Therefore, this paper is composed as follows: The second section of this paper describes the status quo of logistics platforms in China, the third section explains the resource integration of logistics platforms, and the fourth section establishes the conceptual framework of logistics platform's resource integration. Finally the UTAUT model is given to explore the logistics platform's resource integration based on users' behavior.

2 Status Quo of Logistics Platform in China

During the recent years, China's logistics platform has experienced a blowout development under the impetus of "Internet +", and attracted a lot investment from many organizations and enterprises. In terms of APP matching with trucks and cargoes, nearly 200 APP can be available (see Table 2).

Table 2. Definitions of logistics platform

No.	Name	No.	Name
1	Luoji logistics	41	51Yundao
2	Haoduoche	42	Yushuzhijia
3	Lvmayi	43	Gonglu e-station
4	Fuyoukache	44	Gongluyizhan
5	Yunmanman	45	Duduhaoyun
6	Logistics QQ	46	Hongkong kuaigou
7	Yunce logistics	47	Huolala
8	Logistics xiaomi	48	CC Huodi
9	Shengshenghuitouche	49	No. 1 gui
10	Yundongxi	50	Sudiyi
11	Yipeihuo	51	Diyi
12	Haosiji	52	E Xiangui
13	Zhuanxinabao	53	Dada
14	Wo logistics	54	Shequ 001
15	Maeryi	55	Maowu
16	Logistics bangbang	56	Mr. Feng
17	First logistics	57	Kuaishou
18	Yiyunbao	58	Shansong
19	Sougulahuo	59	Xingfuyizhan
20	Logistics jingling	60	Sunwukong
21	Supaide	61	Anneng
22	58daojia	62	Manyitongda
23	Jingdongdaojia	63	Luxiantu
24	Lanxiniu	64	Tianjiao
25	Shendunkuaiyun	65	Chewing
26	Yuninao distribution	66	Beidou logistics
27	No.1 Huodi	67	Hechengzhida
28	OTMS	68	Yizhan
29	Zhuzhu	69	Logistics fangzhou
30	eTMS	70	Duoduohuoche
31	Logistics doudou	71	Huiyi TMS
32	Dianjue	72	Weizhineng logistics
33	Chemanman	73	Zhogjiyitongwang
34	Miyang	74	Huoyuanjia
35	56 Pinhuo	75	Yunwuliuzhushou
36	Babasupei	76	Tinatianhuichengche
37	Kuaidaowang	77	Kuaidaowang
38	Kuaizhahuoche	78	Kuaizhaohuoche
39	Tiantianyouhuo	79	Feiyiwang
40	Pinhuolang	80	Falama

Source collected by the author

It is obvious the fierce competition arouse among the logistics platforms, the function, target clients and marketing strategy of these logistics platforms are so homogeneous that the loyalty of users is very low, the users' habit is under cultivation. It is necessary to integrate so many logistics platforms.

3 Contents of Logistics Platform Resource Integration

3.1 Goal for Logistics Platform Resource Integration

The resource integration of logistics platforms refers to the users integration, since the accumulation of the users will bring about added trucks resource, cargoes resource, warehousing facilities, and information, etc., therefore the logistics platforms can realize the capacity integration and information integration.

The basic goal of logistics platforms integration is to improve the operation efficiency, make good use of the resources and enhance the overall competence. While the economic goal is to reduce the costs, maximize their profits and increase the stability among the platform members, see Formula (1).

$$\Delta_{LogisticsPlatform} = \sum \Delta_{members} + \Delta_{Co} \tag{1}$$

$\Delta_{LogisticsPlatform}$	The platform's overall economic effect after resource integration.
$\sum \Delta_{members}$	The amount of all members' economic effects.
Δ_{Co}	Value added after platform's resource integration.

3.2 Key Objects of Logistics Platform Resource Integration

Owing that the resource integration of logistics platforms refers to the users' integration, the users can be classified into four types, see Formula (2).

$$f(u) = f(u1) + f(u2) + f(u3) + f(u4) \tag{2}$$

(1) u1: Governance users, i.e., system administrators such as ICT managers.
(2) u2: Logistics service provider users, i.e., transportation companies, warehousing companies, freight forwarders, distribution centers.
(3) u3: Logistics service demander users, i.e., manufacturers, trade companies, etc.
(4) u4: Logistics service support users, i.e., government offices, finance and insurance sectors, etc.

It should be mentioned that u1 and u4 are the guarantees of logistics platforms operation, u2 and u3 are the real users from the supplier and demand side depending on the logistics platform.

4 Conceptual Framework of Logistics Platform's Resource Integration

In order to integrate the resources of logistics platforms successfully, the conceptual framework is established as follows, see Fig. 1.

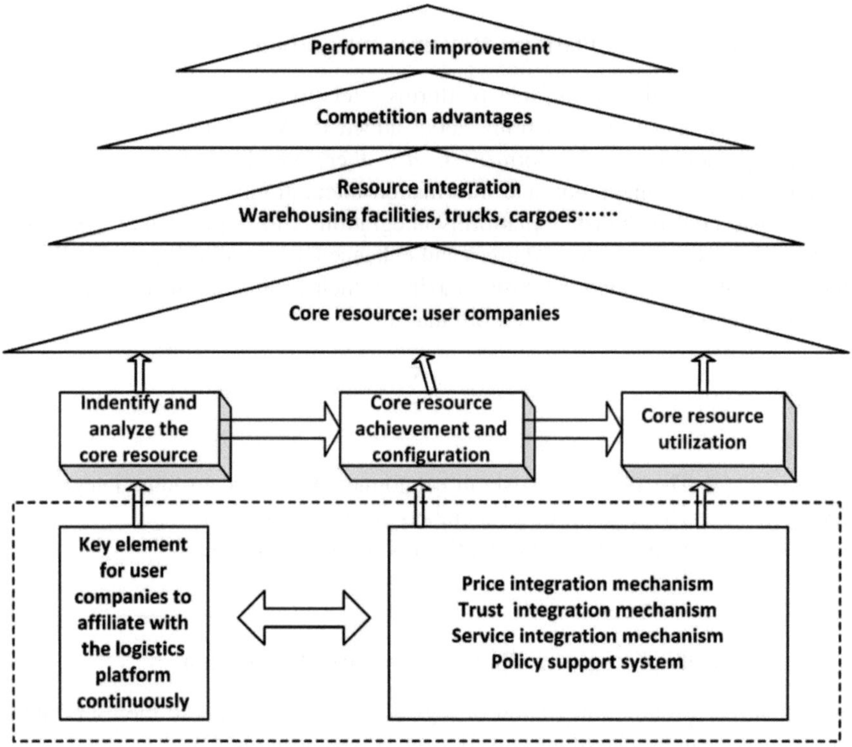

Fig. 1. Framework of logistics platform's resource integration

The first step is to identify and analyze the core resource of the integration objects, the second step is to obtain the core resource, and the third step is to utilize the core resources.

Moreover, in accordance with the platforms' integration, four factors should be considered, i.e. price integration, trust integration, service integration and policy system. In the end, the logistics platform can achieve the competition advantages and performance improvement.

5 UTAUT Model

Unified Theory of Acceptance and Use of Technology (UTAUT) combines eight theories and models such as Theory of Reasoned Action (TRA), Technology Acceptance Model (TAM), Motivational Model (MM), Theory of Planned Behavior (TPB), etc., which can explain the acceptance and usage behavior as high as 69%, more than the rate of 17–53% as eight theories and models. The variables in UTAUT model include social influence (SI), effort expectancy (EE), performance expectancy (PE) and facilitating conditions (FC). In addition, gender, age, experience and voluntariness, are the adjust variables of UTAUT model, see Fig. 2.

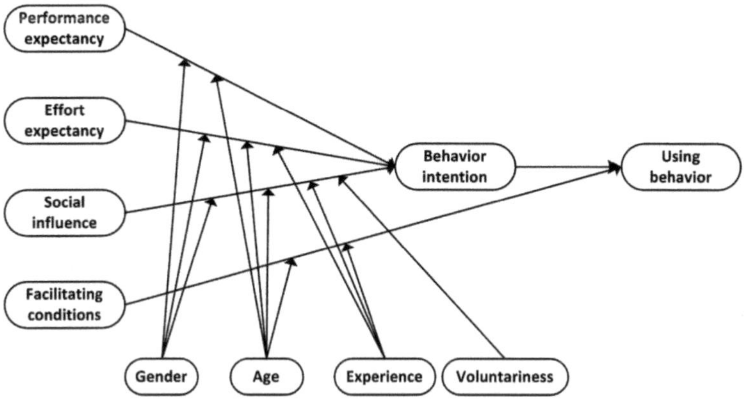

Fig. 2. UTAUT model

Logistics platform and the interaction with the users can influence its sustainable development. Based on the analysis of UTAUT model, essential factors can be deeply segmented and how the factors influence the acceptance and use behavior of users is worthy keeping studying since it can help logistics platform to turn potential users into real users.

6 Conclusion

Starting with its definition, logistics platforms, are understood as those spaces in which packaging, storage, transportation and distribution of merchandise can be carried out, both online and offline, taking into account that one or several logistics operators can be attended, who can be owners or tenants of the facilities and equipment that are there.

The essence of the logistics platforms is integration. Based on the study of the factors affecting the resource integration of logistics platform and according to the internal logic, the paper puts forward a conceptual model of logistics platform resource integration mechanism: price integration mechanism, trust integration mechanism and service integration mechanism, as well as a system: government support system, these four elements interact to form a joint force.

In order to verify the effectiveness of the logistics platform resource integration mechanism, the paper establishes the Unified Theory of Acceptance and Use of Technology (UTAUT) model in the perspective of users' behavior.

It is interesting to observe how the key factors, such as EE, SI, PE, FC decide the users' habit depending on the logistics platform. Further research will focus on this topic.

Acknowledgements. Thanks for National Social science Foundation of China (17BJY140): Research on the Cause and Effect between Logistics Platforms and User Companies in the Perspective of Symbiosis and Social science Foundation of Jiangxi province (16GL02): Research on the Vertical Integration Strategy of Logistics platforms Facing Supply Side Reform (16GL02).

References

1. J.C. Rochet, J. Tirole, Platform competition in two-sided markets. J. Eur. Econ. Assoc. **1**(4), 990–1029 (2003)
2. M. Armstrong, Competition in two-sided markets. Rand J. Econ. **37**(3), 668–691 (2006)
3. N.M.P. Bocken, S.W. Short, P. Rana, S. Evans, A literature and practice review to develop sustainable business model archetypes. J. Clean. Prod. **65**, 42–56 (2014)
4. M. Abrahamsson, N. Aldin, F. Stahre, Advantages of intermodal logistics platforms: insights from a Spanish platform. Int. J. Logist. Res. Appl. **6**(3), 85–106 (2003)
5. M. Abrahamsson, N. Aldin, F. Stahre, Logistics platforms for strategic flexibility. Int. J. Logist. Res. Appl. **6**(3), 85–106 (2003)
6. A.F. Dubke, F.R.N. Ferreira, N.D. Pizzolato, Logistics platform: characteristics and tendencies in Brasil. XXIV ENEGEP (2006)
7. Carolina Correa de Carvalho et al., Efficient logistics platform design: the case of campinas platform, in *XVI International Conference on Industrial Engineering and Operations Management*, pp. 1–11 (2010)
8. J.P. Antúna, R. Alarcón, Bases for feasibility analysis of logistics platforms at borders. Procedia Soc. Behav. Sci. 162, 6–14 (2014)
9. R.M. Silva et al., A framework of performance indicators used in the governance of logistics platforms: the multiple-case study. J. Transp. Lit. **9**(1), 5–9 (2015)
10. Z. Wang, Establish China's logistics platform. China Mater. Distrib. **01**, 33–34 (2000)
11. Z. Wang, Develop the specific logistics platform. China Distrib. Econ. **11**, 24–27 (2010)

Dynamic Pricing with Reference Quality Effects in Closed-Loop Supply Chain

Zhi-chao Zhang(✉), Lei Guo, Qian-qian Wu, Lin-mei Gong, Ting-ting Ni, and Pu-hao Guo

School of Management, Anhui Science and Technology University, Chuzhou, China
ekmzzc@163.com

Abstract. With the development of economic progress, dynamic pricing, as a powerful strategy to improve company profits, has been extensively adopted by companies worldwide. However, there are little researches which pay close attention to the effects of dynamic pricing on closed-loop supply chain (CLSC) performance. We consider a Stackelberg game between one manufacturer and one retailer where the manufacturer, as the leader, decides the product quality and the wholesale price, and the retailer, as the follower, sets the retail price. By analyzing, we obtain the following conclusions: It is the best for the manufacturer, the retailer and the whole supply chain to adopt dynamic pricing strategies compared with static pricing strategies both in integrated and decentralized case. In dynamic pricing strategies, the manufacturer should take a higher recycling rate where the recovery cost is low while should take a lower recycling ratio where the recovery cost is high.

Keywords: Static price case · Dynamic price case · Dynamic pricing strategy · Reference quality effects · CLSC

1 Introduction

With the rapid development of social economy and e-commerce, people's pursuit of products has gone from the price to the pursuit of quality, price and environmental protection. Therefore, enterprises need to comprehensively consider the price, quality and environmental protection of the products in the decision-making process so as to gain more benefits. Product price affects consumer demand, directly reflected in its quality improvement and product perfection, is one of the important product elements that customers value.

Product quality is one of the criteria to measure enterprise development and core competitiveness. Many literature are focused on the reference effect, including the reference price effect [1]; and reference quality effects [2]. The quality of reference plays a crucial role in influencing market demand. Zhang et al. [3] examine the relationship between price and reference quality and their impact on profit, based on price and reference quality. Kopalle and Winer [4] conclude the stability ratio between quality and price is influenced by reference quality effect and the relation of reference quality effect.

G. Q. Huang et al. (eds.), *Proceeding of the 24th International Conference on Industrial Engineering and Engineering Management 2018*,
https://doi.org/10.1007/978-981-13-3402-3_58

The product price represents the market position of the enterprise, reflecting its sensitivity to the market. The price of the enterprise is not static, but changes with the market environment, product life cycle and the development stage of the enterprise. The dynamic adjustment of price reflects the change of enterprise market and customer demand. The relative impact of quality and price was obtained in [5, 6] by relaxing the structural dependencies between quality and price. Their analysis shows that the steady-state price and overall quality are higher than those without these variables in the context of constant reference quality [7].

2 Problem Definition

Constructing a supply chain consisted of one manufacturer (M) and one retailer (R). We consider a Stackelberg game between the manufacturer, as the leader determines the wholesale price (w) and quality strategy (q), the retailer, as the follower decides the retail price (p).

Consider a finite planning horizon $[0, T]$, and let $R(t)$ denote the goodwill and $q(t)$ express the current quality of the products. According to [8], we use an exponential smoothing process of historical quality to model the dynamics of reference quality as follows:

$$\dot{R}(t) = \theta(q(t) - R(t)), R(0) = R_0 \tag{1}$$

where $\theta > 0$ captures the depreciation of the goodwill stock, and $R_0 > 0$ denotes the initial stock of goodwill.

We assume that the demand $D(t)$ is dependent on price, quality and the difference between product quality and reference quality, and adopt the following demand function

$$D(t) = \alpha - \beta p(t) + \gamma_1 q(t) + \gamma_2 (q(t) - R(t)) \tag{2}$$

where $\alpha > 0$ represents the basic market potential, $0 \leq \beta \leq 1$ denotes the price discount coefficient. In accordance with a plenty of previous literature (DeGiovanni and Roselli 2012; Feng et al. [9]; Prasad and Sethi [10]; Zhang et al. [11]), the quality cost function is assumed to be quadratic in the quality rate, i.e.

$$C(q) = \frac{1}{2} k_1 q^2(t) \tag{3}$$

According to a great deal of previous researches (Ferguson and Toktay [12]; Ovchinnikov [13]; Atasu et al. [14]; Guide and Van Wassenhove [15]), the product recycled cost function is assumed to be quadratic in the recycle rate, i.e.

$$C(\tau) = \frac{1}{2} k_2 \tau^2 \tag{4}$$

where $k_1 > 0$ is a measure value of manufacture cost, and $k_2 > 0$ is a measure value of recycle cost.

The objective functions of the manufacturer and retailer are expressed respectively as

$$J_M = \int_0^T \begin{pmatrix} (w - c_m + \Delta\tau)D(t) \\ -\frac{1}{2}k_1 q^2(t) - \frac{1}{2}k_2\tau^2 \end{pmatrix} dt \tag{5}$$

$$J_R = \int_0^T ((p - w)D(t))dt \tag{6}$$

Taking the dynamic relationships (1)–(6) together, the differential game as follows:

$$\begin{aligned} \max_{w(\cdot),q(\cdot)} J_M &= \int_0^T \begin{pmatrix} (p - c_m + \Delta\tau)D(t) \\ -\frac{1}{2}k_1 q^2(t) - \frac{1}{2}k_2\tau^2 \end{pmatrix} dt \\ \max_{p(\cdot)} J_R &= \int_0^T ((p - w)D(t))dt \\ s.t.\ \dot{R}(t) &= \theta(q(t) - R(t)), R(0) = R_0 \end{aligned} \tag{7}$$

3 Solutions

In the centralized scenario, the centralized channel's profit maximization problem can be given by

$$\begin{aligned} &\max_{w(\cdot),q(\cdot)} \int_0^T \begin{pmatrix} (p - c_m + \Delta\tau)D(t) \\ -\frac{1}{2}k_1 q^2(t) - \frac{1}{2}k_2\tau^2 \end{pmatrix} dt \\ &s.t.\ \dot{R}(t) = \theta(q(t) - R(t)), R(0) = R_0 \end{aligned} \tag{8}$$

Consider optimization problems through dynamic pricing (8). The optimal pricing and quality strategy of the product is proposed in the case of the centralized manufacturer adopting the dynamic pricing strategy.

We can get the Hamiltonian function of the centralized scenario by applying the maximum principle as below:

$$H^{cd} = (p - c_m + \Delta\tau)[\alpha - \beta p + \gamma_1 q + \gamma_2(q - R)] - \frac{k_1}{2}q^2 - \frac{k_2}{2}\tau^2 + \lambda^{cd}\theta(q - R) \tag{9}$$

where λ^{cd} is the adjoint variable associated with R, according to the first-order condition for optimality $\frac{\partial H^{cd}}{\partial q} = 0$, $\frac{\partial H^{cd}}{\partial p} = 0$ and adjoint equation $\dot{\lambda}^{cd} = -\frac{\partial H^{cd}}{\partial R}$, we can gain

$$p^{cd} = \frac{k_1(\alpha + \beta(c_m - \Delta\tau) - \gamma_2 R)}{2k_1\beta - (\gamma_1 + \gamma_2)^2} + \frac{(\Delta\tau - c_m)(\gamma_1 + \gamma_2)^2 + \theta\lambda_s^{cd}(\gamma_1 + \gamma_2)}{2k_1\beta - (\gamma_1 + \gamma_2)^2} \tag{10}$$

$$q^{cd} = \frac{2\beta\theta\lambda^{cd} - (\gamma_2^2 + \gamma_1\gamma_2)R}{2k_1\beta - (\gamma_1 + \gamma_2)^2} + \frac{(\gamma_1 + \gamma_2)(\alpha + (\Delta\tau - c_m)\beta)}{2k_1\beta - (\gamma_1 + \gamma_2)^2} \tag{11}$$

$$\dot{\lambda}^{cd} = \gamma_2(p - c_m + \Delta\tau) + \theta\lambda_S^{cd} \tag{12}$$

Substituting Eq. (10) into Eq. (12) and substituting Eq. (10) into Eq. (1), we have

$$\dot{\lambda}_S^{cd} = \frac{k_1\gamma_2(\alpha + \beta(\Delta\tau - c_m)) - k_1\gamma_2^2 R}{2k_1\beta - (\gamma_1 + \gamma_2)^2} + \frac{(2k_1\beta - \gamma_1^2 - \gamma_1\gamma_2)\theta\lambda_S^{cd}}{2k_1\beta - (\gamma_1 + \gamma_2)^2} \tag{13}$$

$$\dot{R}(t) = \frac{(\gamma_1 + \gamma_2)\theta(\alpha + (\Delta\tau - c_m)\beta)}{2k_1\beta - (\gamma_1 + \gamma_2)^2} + \frac{2\beta\lambda_S^{cd}\theta^2 + (\gamma_1^2 + \gamma_1\gamma_2 - 2k_1\beta)\theta R}{2k_1\beta - (\gamma_1 + \gamma_2)^2} \tag{14}$$

Rewriting them, we can get

$$\begin{bmatrix} \dot{\lambda}^{cd} \\ \dot{R} \end{bmatrix} = A \begin{bmatrix} \lambda^{cd} \\ R \end{bmatrix} + b \tag{15}$$

$$A = \begin{bmatrix} \frac{(2k_1\beta - \gamma_1^2 - \gamma_1\gamma_2)\theta}{2k_1\beta - (\gamma_1 + \gamma_2)^2} & \frac{-k_1\gamma_2^2}{2k\beta - (\gamma_1 + \gamma_2)^2} \\ \frac{2\beta\theta^2}{2k_1\beta - (\gamma_1 + \gamma_2)^2} & \frac{(\gamma_1^2 + \gamma_1\gamma_2 - 2k_1\beta)\theta}{2k_1\beta - (\gamma_1 + \gamma_2)^2} \end{bmatrix} \tag{16}$$

$$b = \begin{bmatrix} \dfrac{k_1\gamma_2(\alpha+\beta(\Delta\tau - c_m))}{2k\beta - (\gamma_1+\gamma_2)^2} \\ \dfrac{(\gamma_1+\gamma_2)\theta(\alpha+(\Delta\tau - c_m)\beta)}{2k\beta - (\gamma_1+\gamma_2)^2} \end{bmatrix} \tag{17}$$

The eigenvalues of A are given by

$$\begin{bmatrix} r_1 \\ r_2 \end{bmatrix} = \begin{bmatrix} \frac{\theta\sqrt{N_1^2 - 2k_1\gamma_2^2\beta}}{2k_1\beta - (\gamma_1+\gamma_2)^2} \\ \frac{-\theta\sqrt{N_1^2 - 2k_1\gamma_2^2\beta}}{2k_1\beta - (\gamma_1+\gamma_2)^2} \end{bmatrix} \tag{18}$$

where $N_1 = 2k_1\beta - \gamma_1^2 - \gamma_1\gamma_2$
and the matching matrix of eigenvectors of A is

$$P = \begin{bmatrix} \dfrac{N_1 + \sqrt{N_1^2 - 2k_1\gamma_2^2\beta}}{2\beta\theta} & \dfrac{N_1 - \sqrt{N_1^2 - 2k_1\gamma_2^2\beta}}{2\beta\theta} \\ 1 & 1 \end{bmatrix} \tag{19}$$

Accordingly we have

$$\begin{bmatrix} \lambda^{cd} \\ R \end{bmatrix} = P\begin{bmatrix} e^{r_1 t} & 0 \\ 0 & e^{r_2 t} \end{bmatrix}\begin{bmatrix} h_1 \\ h_2 \end{bmatrix} - A^{-1}b \tag{20}$$

According to the boundary conditions condition

$$R^{cd}(0) = R_0 \tag{21}$$

$$\lambda^{cd}(T) = 0 \tag{22}$$

the analytical solutions of h_1, h_2 can be gained as

$$R^{cd}(t) = h_1 e^{r_1 t} + h_2 e^{r_2 t} - B_1 \tag{23}$$

$$\lambda^{cd}(t) = h_1 \frac{N_1 + \sqrt{N_1^2 - 2k_1\gamma_2^2\beta}}{2\beta\theta} e^{r_1 t} + h_2 \frac{N_1 - \sqrt{N_1^2 - 2k_1\gamma_2^2\beta}}{2\beta\theta} e^{r_2 t} - A_1 \tag{24}$$

where

$$A_1 = \frac{k_1(\gamma_2^2(\gamma_1+\gamma_2) - N_1\gamma_2)(\alpha+\beta(\Delta\tau - c_m))}{\theta(2\beta k_1\gamma_2^2 - N_1^2)}$$
$$B_1 = \frac{(N_1(\gamma_1+\gamma_2) - 2\beta k_1\gamma_2)(\alpha+\beta(\Delta\tau - c_m))}{2\beta k_1\gamma_2^2 - N_1^2}$$
$$h_1 = \frac{(R_0+B_1)(N_1 - \sqrt{N_1^2 - 2k_1\gamma_2^2\beta})e^{r_2T} - 2\beta\theta A_1}{(N_1 - \sqrt{N_1^2 - 2k_1\gamma_2^2\beta})e^{r_2T} - (N_1 + \sqrt{N_1^2 - 2k_1\gamma_2^2\beta})e^{r_1T}}$$
$$h_2 = \frac{(R_0+B_1)(N_1 + \sqrt{N_1^2 - 2k_1\gamma_2^2\beta})e^{r_1T} - 2\beta\theta A_1}{(N_1 + \sqrt{N_1^2 - 2k_1\gamma_2^2\beta})e^{r_1T} - (N_1 - \sqrt{N_1^2 - 2k_1\gamma_2^2\beta})e^{r_2T}}$$

Substituting Eqs. (23) and (24) into (10) and (11), separately, the optimal pricing and quality strategies of the centralized firm can be respectively procured as

$$R^{cd}(t) = h_1e^{r_1t} + h_2e^{r_2t} - B_1 \tag{25}$$

$$\begin{aligned} p^{cd} &= \frac{k_1\alpha + (c_m - \Delta\tau)(\beta k_1 - (\gamma_1+\gamma_2)^2)}{2k_1\beta - (\gamma_1+\gamma_2)^2} \\ &+ \frac{(\gamma_1+\gamma_2)\theta}{2k_1\beta - (\gamma_1+\gamma_2)^2}(h_1\frac{N_1 + \sqrt{N_1^2 - 2k_1\gamma_2^2\beta}}{2\beta\theta}e^{r_1t} \\ &+ h_2\frac{N_1 - \sqrt{N_1^2 - 2k_1\gamma_2^2\beta}}{2\beta\theta}e^{r_2t} - A_1) \\ &- \frac{k_1\gamma_2}{2k_1\beta - (\gamma_1+\gamma_2)^2}(h_1e^{r_1t} + h_2e^{r_2t} - B_1) \end{aligned} \tag{26}$$

$$\begin{aligned} q^{cd} &= \frac{(\gamma_1+\gamma_2)(\alpha + (\Delta\tau - c_m)\beta)}{2k\beta - (\gamma_1+\gamma_2)^2} \\ &+ \frac{2\theta\beta}{2k\beta - (\gamma_1+\gamma_2)^2}(h_1\frac{N_1 + \sqrt{N_1^2 - 2k_1\gamma_2^2\beta}}{2\beta\theta}e^{r_1t} \\ &+ h_2\frac{N_1 - \sqrt{N_1^2 - 2k_1\gamma_2^2\beta}}{2\beta\theta}e^{r_2t} - A_1) \\ &- \frac{(\gamma_2^2 + \gamma_1\gamma_2)}{2k\beta - (\gamma_1+\gamma_2)^2}(h_1e^{r_1t} + h_2e^{r_2t} - B_1) \end{aligned} \tag{27}$$

4 Numerical Study

In this section, we need numerical examples to demonstrate the effectiveness of the results and sensitivity analysis of key parameters to test the impact of different results on solutions.

Let T = 10 and the following parameters setting: $\alpha = 50$, $\beta = 0.6$; Purchasing desire parameters: $R_0 = 10$; Profit parameters: $\gamma_1 = 0.7, \gamma_2 = 0.5, \theta = 0.2, R_0 = 10$, $k_1 = 15, k_2 = 12$. According to the above proposition and MATLAB, we get the manufacturer's profit, the retailer and the supply chain of two modes: we need a numerical example shows the validity of the results and some key parameters by assessing the sensitivity analysis of the factors of the solution.

In the case of concentration, the retailer can take J as an example for manufacturers: $J = 1437.4$.

Next, we assess the sensitivity analysis of some important parameters, such as β, θ, and r_1, r_2. To check the impacts on profit and supply chain efficiency in each case, which are shown in the next table, respectively. Additionally, to illustrate the comparison of profits and supply chain efficiency in different cases more clearly, we take the profits and supply chain efficiency with different proportion of the traditional retail price and the whole sale price as an example.

According to Table 1, then we can get some results.

Table 1. Impacts on profit

		J			J
β	0.2	5934.2	γ_1	0.2	1390.1
	0.4	2443.3		0.4	1403.9
	0.6	1437.4		0.6	1419.7
	0.8	963.48		0.8	1437.4
θ	0.2	1437.4	γ_2	0.2	1534.0
	0.4	1473.9		0.4	1479.5
	0.6	1505.8		0.6	1437.4
	0.8	1531.5		0.8	1385.2

(1) We found that the profits of manufacturers and retailers were declining as the value of β and r_2 increased. (2) We found that the profits of manufacturers and retailers showed an upward trend as the value of θ and r_1 increased.

Particularly, we discuss the influence the recover fraction coefficient have on the whole supply chain and its members.

The reason can be found in Fig. 1, when the cost is zero, as the recovery ratio of the manufacturer rises, the profit of the manufacturer will be higher and its maximum value will occur at 100% of the recovery ratio. In fact, the change in the cost of the recovery rate has little effect on the profit of the manufacturer. At the time of higher cost, the recovery ratio also rises, while the profit rises first and then decreases. The recovery ratio is 100% and the recovery cost is the maximum when the recovery costs are zero.

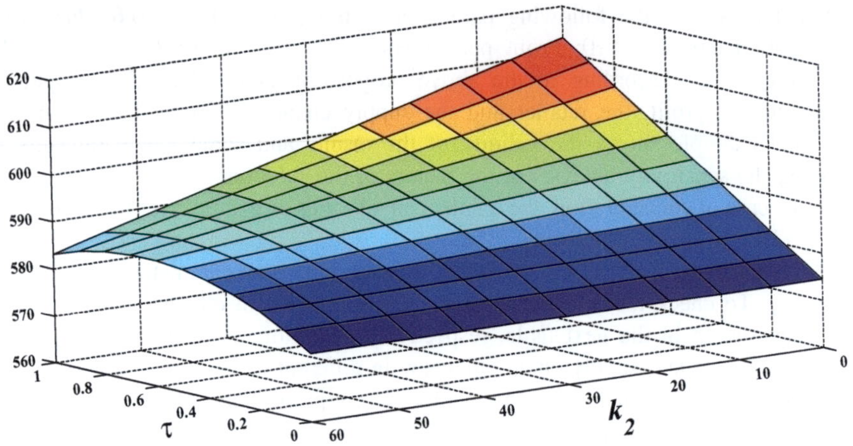

Fig. 1. Recovery rate, recovery cost and profit relationship

5 Conclusion

In this paper, we construct a closed-loop supply chain consisting of manufacturer and retailer, and compare the enterprise dynamic and static pricing strategy based on centralized model.

We designed a game of Stackelberg, where the manufacturer, as the leader, decided on product quality and wholesale price. Retailers, as followers, set retail prices. And the difference equation is used to discuss the optimal solution of enterprise pricing in dynamic and static situations. In addition, the sensitivity analysis of related parameters is discussed. In the dynamic pricing strategy, the recovery part depends on the recovery cost of the product and the recovery rate of the manufacturer. A lower recovery ratio and higher recovery costs should be accepted when economic recovery costs are low. However, when the recovery cost takes a fixed value, as the recovery rate increases, the profit of the enterprise increases first and then decreases.

References

1. F. El Ouardighi, K. Kogan, Dynamic conformance and design quality in a supply chain: an assessment of contracts' coordinating power. Ann. Oper. Res. **211**(1), 137–166 (2013)
2. D. Krass, T. Nedorezov, A. Ovchinnikov, Environmental taxes and the choice of green technology. Prod. Oper. Manag. **22**(5), 1035–1055 (2013)
3. Q. Zhang, J. Zhang, W. Tang, A dynamic advertising model with reference price effect. RAIRO Oper. Res. **49**(4) (2015)
4. Q. Zhang, J. Zhang, W. Tang, A dynamic advertising model with reference price effect. Oper. Res. **49**(4), 669–688 (2015)
5. P.K. Kopalle, R.S. Winer, A dynamic model of reference price and expected quality. Mark. Lett. **7**(1), 41–52 (1996)

6. P. De Giovanni, M. Roselli, Overcoming the drawbacks of a revenue-sharing contract through a support program. Ann. Oper. Res. **196**(1), 201–222 (2012)
7. F. Toyasaki, T. Boyacı, V. Verter, An analysis of monopolistic and competitive take-back schemes for WEEE recycling. Prod. Oper. Manag. **20**(6), 805–823 (2011)
8. G.P. Cachon, M.A. Lariviere, Supply chain coordination with revenue-sharing contracts: strengths and limitations. Manage. Sci. **51**(1), 30–44 (2005)
9. J. Guo, Z. Wen, Y. Zhou, K. Ji, The competitive strategies between the traditional and online retailers, in *2015 12th International Conference on Service Systems and Service Management (ICSSSM)*. IEEE, pp. 1–5
10. L. Feng, J. Zhang, W. Tang, A joint dynamic pricing and advertising model of perishable products. J. Oper. Res. Soc. **66**(8), 1341–1351 (2015)
11. A. Prasad, S.P. Sethi, Competitive advertising under uncertainty: a stochastic differential game approach. J. Optim. Theory Appl. **123**(1), 163–185 (2004)
12. M.E. Ferguson, L.B. Toktay, The effect of competition on recovery strategies. Prod. Oper. Manag. **15**(3), 351–368 (2006)
13. A. Ovchinnikov, Revenue and cost management for remanufactured products. Prod. Oper. Manag. **20**(6), 824–840 (2012)
14. A. Atasu, V.D.R. Guide, L.N. Wassenhove, Product reuse economics in closed-loop supply chain research. Prod. Oper. Manag. **17**(5), 483–496 (2008)
15. V.D.R. Guide Jr., L.N. Van Wassenhove, The evolution of closed-loop supply chain research. Oper. Res. **57**(1), 10–18 (2009)

Research on Coordination of Closed-Loop Supply Chain Revenue Sharing Contract Under Disruption

Qiao-na Song(✉)

College of Zhongshan, University of Electronic Science and Technology, Zhongshan 528402, Guangdong, China
sqn0912@163.com

Abstract. For a closed-loop supply chain consisting of one manufacturer and one retailer facing stochastic market demand, the coordination function with revenue sharing contract is analyzed firstly. Then, the coordination under disruptions with revenue sharing contract is studied. The result shows that the closed-loop supply chain coordination may be broken off by a disruption. Thus, an optimal strategy for the closed-loop supply chain to the disruptions is given, and an adjusted revenue sharing contract is proposed which has anti-disruption ability.

Keywords: Closed-loop supply chain coordination · Revenue sharing contract Disruption · Anti-disruption ability

1 Introduction

Emergencies refer to devastating events that occur beyond people's expectations, bring serious harm to social stability, economic development, life and property, and need to be dealt with immediately. Since the beginning of the 21st century, emergencies happen frequently. Which will make the overall stability of the supply chain broken even may collapse. Relevant enterprises will suffer the risk of business failures. The research on the coordination mechanism of supply chain under the influence of emergent events is a key to improve the emergency management capabilities of supply chain. In recent years, studies on the issues related have become the focus, attracted the attention of academia. Hui et al. [1], Cao and Lai [2] and other researchers find that when the market demands can't be ascertained, if the emergence of emergencies leads to the change of market scale, people can use the buyback contract and wholesale price contract to ensure the steady operation of supply chain. Hu and Wang [3] takes the three-stage supply chain as the research object, analyzes the consequences caused by emergencies, and makes some modification of the prices discount contract. Due to his contributions, the coordination of supply chain is realized, and new supply chain has a strong power to deal with emergencies. Teng et al. [4] studies the supply chain system that consists of multiple manufacturers and multiple retailers and points that when

G. Q. Huang et al. (eds.), *Proceeding of the 24th International Conference on Industrial Engineering and Engineering Management 2018*,
https://doi.org/10.1007/978-981-13-3402-3_59

emergencies lead to the changes of market scale, people can use quantity discounts contract to realize the coordination of supply chain. Tsay and other scholars [5] summarize the three functions of the supply chain contract—coordinating and improving overall revenue, sharing risks and establishing cooperation relationships. Cachon [6] makes a summary of the contracts related to the coordination of supply chain.

Savaskan and other researchers [7] explore the efficiency of closed-loop supply chain under three common recovery modes. Zuqing [8] studies the loss of supply chain efficiency when retailers and manufactures have different decision-making rights in the environment of the market demands have been ascertained. Toktay and others [9] study how to get the biggest profits for new products and remanufactured goods on the same market. Galbreth [10] thinks that under the situation of ascertaining demands and random demands, people can decide the recycling quantity of products according to the demands information. However, there are few articles about how do the closed-loop supply chain cope with emergencies by income sharing contract.

Based on the understandings to traditional closed-loop supply chain, this paper explores analyzes the coordination capacity of closed-loop supply chain under the influence of income sharing contract.

2 The Impacts of Emergencies on the Closed-Loop Supply Chain

For the closed-loop supply chain that consists of a single manufacturer and retailer, we need to consider the condition that the manufactures will ask the retailer to recycle these products that have a short service term, need to be ordered early, have less residual values but have a certain re-use space when they expired. At the end of the sale season, manufactures will recycle them, and re-manufacture. There is no difference in quality of re-manufactured products and new products. All products including re-manufactured will be put on the market in the nest sale season to meet the needs of customers. Based on the information symmetry, producers and sellers can know the costs of the two parties clearly and profit function, accurately grasp the demands of market, and pursue the highest benefits. If manufactures need to spend the capitals—c_m developing a new product, before the sale season, manufactures will give the wholesale prices of the product—w; the sale prices of retailers are p, the scale costs of a product are c_r, the purchase quantities are q. According to above data, manufactures will arrange production. If the demand amounts are more than the productions, the retailer's unit loss opportunity cost is h, the residual values of rest unit product are s. The retailer's recycling prices of wasted and old products are p_0, its unit costs are c_{r0}; the manufacture will purchase the wasted products from retailers, the purchasing prices are w_0, then the manufacture start to use them to create new products, the investments of unit product are c_{m0}; however, these expired products still have a certain value, the value is v_0, total amounts of wasted and old products used are D_0 that is a fixed value. Let's assume that the probability density function related to market demands of a product is $f(x)$, cumulative distribution function is $F(x)$, average values are μ, $F(x)$ belongs to differentiable strictly increasing function. In addition, we give the orders that

$F(0) = 0$, and $\bar{F}(x) = 1 - F(x)$, then according to the mathematical statistics, we can deduce the result—$S(N) = N - \int_0^N F(x)dx$. When the recycling prices of wasted and old products are p_0, the quantity of wasted and old products purchased by retailers is $H(p_0)$, and $H'(p_0) > 0$ and $H''(p_0) < 0$. This paper introduces income sharing contract, breaks traditional profits earning methods that manufactures simply earn the difference between the selling price and the cost. In new model, manufactures sell products to retailers with a lower price (The price is lower than traditional price), retailers try to sell and recycle; after a sales period, retailers get φ of sales income, manufactures get $1-\varphi$ of sales income to make up for the loss caused by lower prices. Based on the agreement, manufactures and retailers shall work together to realize the overall coordination of supply chain.

If manufactures combine the optimal order quantity to conduct corresponding production activity, before the sale season comes, the emergence of emergencies will make the demands of the product change. Let's assume that at the moment, average demand quantities are μ_G, its probability density function is $g(x)$, cumulative distribution function is $G(x)$. Here, $G(x)$ belongs to differentiable strictly increasing function. In addition, we give orders that $G(0) = 0$ and $\bar{G}(x) = 1 - G(x)$. When emergencies occur, if new order quantities are more than previous plan, newly increased production quantities are $q - q^*$, at the same time, the costs of each product increase, we assume the increasing amounts—k_1; on the contrary, the extra production quantities are $q^* - q$, and the corresponding treatment costs of each product are k_2.

Under this situation, the expected profits of the retailer are π_{rG}, the expected profits of the manufacture are π_{mG}, and the expected profits of the whole closed-loop supply chain are π_{cG}. The specific values of profits separately are:

$$\begin{aligned}
\pi_{rG} &= pS_G(q) + sI_G(q) - hL_G(q) - T - H(p_0)(p_0 + c_{r0}) - c_r q \\
&= [\varphi(p - s) + h]S_G(q) - (w + c_r - \varphi s)q \\
&\quad + [(1 - \varphi)v_0 + w_0 - p_0 - c_{r0}]H(p_0) - \mu_G h \\
\pi_{mG} &= T - c_m(q - D_0) - c_{m0}D_0 + v_0 H(p_0) - k_1(q - q^*)^+ \\
&\quad - k_2(q^* - q)^- \\
\pi_{cG} &= \pi_r + \pi_m \\
&= (p + h - s)S_G(q) - h\mu_G - (c_m - s)q + (c_m - c_{m0})D_0 \\
&\quad + H(p_0)(v_0 - p_0 - c_{r0}) - k_1(q - q^*)^+ - k_2(q^* - q)^-
\end{aligned}$$

Emergencies have impacts on the market scale. It means that if $x \geq 0$, then $\bar{G}(x) \geq \bar{F}(x)$ or $\bar{G}(x) \leq \bar{F}(x)$. Based on this point, researchers explore the impacts of the changes of market scale on production quantities and the optimal recycling prices of wasted and old products in the closed-loop supply chain.

When emergencies occur, the optimal production quantities of the closed-supply chain are $\bar{q}$ ($\bar{q} = \arg\max_{q>0} \pi_{cG}$). At this time, the optimal recycling prices of wasted and old products are $\bar{p}_0$ ($\bar{p}_0 = \arg\max_{p_0>0} \pi_{cG}$).

Lemma 1 Assuming that emergencies make the market scale expand significantly, it means that $\bar{G}(x) \geq \bar{F}(x)$, then as long as $\bar{q} \geq 0$, we can get $\bar{q} \geq q^*$ and $\bar{p}_0 = {p_0}^*$, if emergencies make the market scale decrease, it means that $\bar{G}(x) \leq \bar{F}(x)$, then as long as $\bar{q} \geq 0$ we can get $\bar{q} \leq q^*$ and $\bar{p}_0 = {p_0}^*$.

Theorem 1 Under the situation that emergencies have impacts on the market scale, the optimal production quantities of closed-loop supply chain are:

$$\bar{q} = \begin{cases} q^{\wedge} & \textit{When the market scales expands} \\ q^* & \textit{others} \\ q^{\odot} & \textit{When the market scale reduces} \end{cases}$$

$q^{\wedge}$ is the solution of the equation of $\bar{G}(x) = \frac{c_m - s + k_1}{p + h - s}$, $q^{\odot}$ is the solution of the equation of $\bar{G}(x) = \frac{c_m - s - k_2}{p + h - s}$.

Proof: If emergencies make original market scale expand, according to Lemma 1, we can get that $\bar{q} \geq q^*$. Under this situation, expected revenue function of closed-loop supply chain is:

Here, $\pi''_{cG} = -(p + h - s)g(q) < 0$, so we can know π_{cG} belongs to strict concave function, the first order optimal solution of q is $q^{\wedge}$, and $\bar{G}(q^{\wedge}) = \frac{c_m - s + k_1}{p + h - s}$.

If emergencies make the market scale expand slightly, then $q^{\wedge} \leq q^*$. Because π_{cG} belongs to strict concave function, for $[q^{\wedge}, +\infty]$, π_{cG} is strictly reducing function, and $q \geq q^*$. Under this situation, $\bar{q} = q^*$, its values belong to the optimal production quantities of closed-loop supply chain. If emergencies make the market expand significantly, then $q^{\wedge} > q^*$. Under this circumstance, the upper limit of the function π_{cG} is $q^{\wedge}$ that is in the interval of the collection $q > q^*$. Therefore, the optimal production quantities of the closed-loop supply chain are $q = q^*$.

If emergencies make the market scale reduce, then according to the Lemma 1, we can get that $\bar{q} \leq q^*$, under this circumstance, expected revenue function of closed-loop supply chain is:

$$\begin{aligned} \pi_{cG} = {} & (p + h - s)S_G(q) - h\mu_G - (c_m - s)q \\ & + (c_m - c_{m0})D_0 + H(p_0)(v_0 - p_0 - c_{r0}) - k_1[q - q^*] \end{aligned}$$

Under this situation, $\pi''_{cG} = -(p + h - s)g(q) < 0$, so π_{cG} belongs to strict concave function. The first order optimal solution of q is $q^{\odot}$, at the same time, $\bar{G}(q^{\odot}) = \frac{c_m - s - k_2}{p + h - s}$.

If emergencies don't make the market scale reduce greatly, then $q^{\odot} \geq q^*$. Because π_{cG} belongs to strict concave function, π_{cG} is also the strict concave function in $[0, q^{\odot}]$. Simultaneously, according to the limited precise—$q \leq q^*$, under this situation, $\bar{q} = q^*$ belongs to the optimal production quantities of closed-loop supply chain. If emergencies make the market scale reduce significantly, then $q^{\odot} < q^*$, the upper limit of the function π_{cG} is $q^{\odot}$ that is in the interval of the collection $q \leq q^*$, its optimal production quantities are $\bar{q} = q^{\odot}$.

Theorem 2 When emergencies occur, we still choose original income sharing contract (φ, p_0), then if the market scale has big changes, it can't ensure the coordinated operation of closed-loop supply chain.

Proof If emergencies make the whole market scale change a lot, under the limiting of original income sharing contract of (φ, p_0), expected revenues of the retailer are:

$$\begin{aligned}\pi_{rG} &= pS(q) + sI(q) - hL(q) - T - H(p_0)(p_0 + c_{r0}) - c_r q \\ &= \frac{(p+h-s)(w+c_r) - h(c_m - s)}{pc + sh - c_m s}[(p-s)S_G(q) + sq + v_0 H(p_0)] \\ &\quad + hS_G(q) - (w+c_r)q + (v_0 + w_0 - p_0 - c_{r0})H(p_0) - \mu_G h\end{aligned}$$

For the formula above, let's calculate the partial derivative about q, and make the first order derivative be zero, then the optimal production quantities $q_r^{\wedge}$ meets $\bar{G}(q_r^{\wedge}) = \frac{c_m - s}{p+h-s}$.

If emergencies make the market scale expand significantly, expected revenues of closed-loop supply chain are:

$$\begin{aligned}\pi_{cG} &= (p+h-s)S_G(q) - h\mu_G - (c_m - s)q + (c_m - c_{m0})D_0 \\ &\quad + H(p_0)(v_0 - p_0 - c_{r0}) - k_1[q - q^*]\end{aligned}$$

According to the Theorem 1, we can get that all conditions must meet the formula $\frac{c_m - s + k_1}{p+h-s} = \frac{c_m - s}{p+h-s}$ in order to realize the coordination of closed-loop supply chain. Because $k_1 > 0$, the income sharing contract used in the past can't ensure the coordinated cooperation of closed loop supply chain. If emergencies lead to serious shrink of market, expected revenues of closed-loop supply chain are:

$$\begin{aligned}\pi_{cG} &= (p+h-s)S_G(q) - h\mu_G - (c_m - s)q \\ &\quad + (c_m - c_{m0})D_0 + H(p_0)(v_0 - p_0 - c_{r0}) \\ &\quad - k_2[q^* - q]\end{aligned}$$

According to the Theorem 1, we can get that all conditions must meet the formula $\frac{c_m - s - k_2}{p+h-s} = \frac{c_m - s}{p+h-s}$ to realize the coordination of closed-loop supply chain. Because $k_2 > 0$, the past income sharing contract—$(\varphi,\ p_0)$ cannot remain the coordination of closed-loop supply chain.

3 Coordinated Closed-Loop Supply Chain Coping with the Emergencies

According to the Theorem 2, we can get that if emergencies make the market scale change, previous income sharing contract can't guarantee the coordinated operation of supply chain. Aiming at this point, this paper puts forward the suggestions.

Relevant literature represents that if a certain contract can ensure the coordinated operation of supply chain under any circumstances including the occurrence of emergencies, then the contract has anti-emergencies capacity. Based on this point, for closed-loop supply chain, if a certain contract can guarantee that closed-loop supply can work well under any situations including the occurrence of emergencies, then this contract has the capacity of anti-emergencies.

Theorem 3 After the income sharing contract—(φ, p_0) is improved, this contract can ensure the coordinated operation of closed-loop supply chain under the situation that emergencies occur.

$$\varphi_{anti} = \frac{\left[(p+h-s)\frac{w+c_r}{c_m-s} - h\right](c_m - s + k_1)}{pc + sh - c_m s}\min\left[1, [q-q^*]^+\right] + \frac{\left[(p+h-s)\frac{w+c_r}{c_m-s} - h\right](c_m - s - k_2)}{pc_m + sh - c_m s}\min\left[1, [q^*-q]^+\right]$$

Proof If emergencies occur, income sharing contract improved—(φ, p_0) is put into practice, then expected revenues of retailers:

$$\begin{aligned}\pi_{rG} &= pS_G(q) + sI_G(q) - hL_G(q) - T - H(p_0)(p_0 + c_{r0}) - c_r q \\ &= [\varphi(p-s)+h]S_G(q) - (w + c_r - \varphi s)q \\ &\quad + [(1-\varphi)v_0 + w_0 - p_0 - c_{r0}]H(p_0) - \mu_G h\end{aligned}$$

So, we can know that the order quantities of retailers must meet:

$$\bar{G}(q_r^{\wedge}) = \frac{w + c_r - \varphi_{anti} s}{\varphi_{anti}(p-s) + h}$$

When emergencies make the market scale expand significantly, for the coordinated operation of closed-loop supply chain, we can get that according to the Theorem 1: $\frac{c-s+k_1}{p+h-s} = \frac{w+c_r-\varphi_{anti}s}{\varphi_{anti}(p-s)+h}$ must be established, so

$$\varphi_{anti} = \frac{\left[(p+h-s)\frac{w+c_r}{c_m-s} - h\right](c_m - s + k_1)}{pc + sh - c_m s}$$

If emergencies make the market scale reduce greatly, for the coordinated operation of closed-loop supply chain, we can get that according to the Theorem 1: the formula $\frac{c-s-k_2}{p+h-s} = \frac{w+c_r-\varphi_{anti}s}{\varphi_{anti}(p-s)p+h}$ must be established, so

$$\varphi_{anti} = \frac{\left[(p+h-s)\frac{w+c_r}{c_m-s} - h\right](c_m - s - k_2)}{pc_m + sh - c_m s}$$

Combining the Lemma 1, we can know if

$$\varphi_{anti} = \frac{\left[(p+h-s)\frac{w+c_r}{c_m-s} - h\right](c_m - s + k_1)}{pc + sh - c_m s}\min\left[1, [q - q^*]^+\right] + \frac{\left[(p+h-s)\frac{w+c_r}{c_m-s} - h\right](c_m - s - k_2)}{pc_m + sh - c_m s}\min\left[1, [q^* - q]^+\right]$$

income sharing contract improved (φ, p_0) can ensure the coordinated operation of closed-loop supply chain even emergencies occur, so this income sharing contract (φ, p_0) has the capacity of anti-emergencies.

4 Conclusions

Thus, the closed-loop supply chain, under the incentive of the income sharing contract, will show the robustness characteristics: in the early period of emergencies, emergencies don't have huge impacts on the recycling activities of wasted and old products, at this time, we don't need to adjust the closed-loop supply chain; however, if emergencies make the market scale change a lot, we need to adjust and optimize previous income sharing contract to cope with emergencies. Based on that, we can find that the loss of revenues of manufactures caused by the emergencies needs other enterprises to bear according to the requirements in order to ensure the coordinated operation of closed-loop supply chain.

Acknowledgements. Natural Science Foundation of Guangdong Province (2015A030313681).

References

1. Y. Hui, C. Jian, Y. Gang, How do the supply chain cope with emergencies. Syst. Eng. Theory Pract. **7**, 9–16 (2005)
2. E. Cao, M. Lai, How do the supply chain cope with emergencies with the assistance of revenue sharing contract. J. Wuhan Univ. Sci. Technol. **30**(5), 557–560 (2007)
3. J. Hu, H. Wang, Study on price discount contracts in three-level supply chain dealing with emergencies. China Manag. Sci. **15**(3), 103–107 (2007)
4. C. Teng, Y. Hu, Y. Zhou, Supply chain network having random remands cope with emergencies. Syst. Eng. Theory Events **29**(3), 16–20 (2009)
5. A.A. Tsay, S. Nahmias, N. Agrawal, Modeling supply chain contracts: a review, in *Quantitative Models for Supply Chain Management*, Chap. 10, ed. by S. Tayur, M. Magazine, R. Ganeshan (Kluwer, Boston, MA, 1999)
6. G.P. Cachon, Supply chain coordination with contracts, in *Handbooks in Operations Research and Management Science: Supply Chain Management* (University of Pennsylvania, 2002)
7. R.C. Savaskan, S. Bhattacharya, L.N.V. Wassenhove, Closed-loop supply chain models with product remanufacturing. Manag. Sci. **50**(2), 239–252 (2004)

8. H. Zuqing, Efficiency analysis of linear re-manufacturing supply chain structure. J. Manag. Sci. **9**(4), 51–56 (2006)
9. L. Debo, B. Toktay, Market segmentation and technology selection for remanufacturable products. Manage. Sci. **51**(8), 1193–1205 (2005)
10. M.R. Galbreth, *Managing condition variability in remanufacturing* (Vanderbilt University., Nashville, Tennessee, 2006)

Quality and Reliability, Ergonomics and Human Factors and Data Mining

Quality Analysis of Rear Axle Assembly Based on AHP

Yun-rui Wang(✉), Juan Li, and Hong-Yu Ge

School of Mechanical Engineering, Xi'an University of Science and Technology, Xi'an, China
2002wyr@sina.com

Abstract. Product quality is an important factor for the survival and development of enterprise. Under the premise of ensuring the quality of design, the assembly process also plays a crucial role in the formation of product quality. Through field investigation and analysis, combining with the current situation of product quality of an axle factory, using the pareto chart, researchers found quality fluctuation was mainly caused by abnormal sound at rear axle and rough surface. According to the principle and step of AHP, a three-level evaluation system was established. Combining with the actual assembly process, it was calculated that the influence of assemble rear wheel hub and brake hub were the greatest. This method changed the status of the enterprises' experience as quality judgment method, and provided a reliable theoretical basis for the implementation of modern quality management.

Keywords: Rear axle assembly · Quality · AHP · Analysis

1 Overview

With the progress of technology and market competition becoming more and more fierce, the quality requirement from consumers are higher than before. To ensure the market product quality and quantity are perfect, we must pay attention to the production assembly process [1]. Rear axle, an important part of the automobile, supporting the bearing and driving of the car, its assembly process has its unique process flow. The components of the rear axle are: main reducer, differential, through shaft, rear axle housing, rear wheel hub, brake drum, half axle, etc. Shaft accessories have relatively higher qualified rate. According to the investigation of the production site of an axle factory, the raw material and accessory will be sent to the workshop warehouse directly, and the warehouse personnel will conduct spot-checking, and then return the unqualified products, and put the qualified products in the designated position, and send them to the assembly line when necessary. The axle assembly process is conventionally push mode of production, and the semi-finished product will be sent to the next procedure after the previous operation is completed, until the final assembly is completed, through the test and inspection. Qualified axle is packaged in accordance with standard operation, and the axle is rejected as waste disposal [2].

The assembly quality is one of the main factor which determine the quality of the rear axle, and understanding of quality characteristics and main influencing factors of

G. Q. Huang et al. (eds.), *Proceeding of the 24th International Conference on Industrial Engineering and Engineering Management 2018*,
https://doi.org/10.1007/978-981-13-3402-3_60

the assembly is basis of quality control and reducing the waste rate. At present, there are not too many quality prevention and monitoring methods in the enterprise, and the quality management has been used thoughts of the treatment after the unqualified product is inspected which do not meet the requirements of modern quality management, and which cannot improve the quality of assembly. Thus, using scientific methods of assembly quality analysis, find the key link of quality become the premise of modern quality management.

2 Analysis of Main Causes of Quality Fluctuation

The quality of the product depends on the quality chain formed in the product process, and the quality chain is directly relate to the procedure key chain. Therefore, it is necessary to find the key chain that forms the quality from quality characteristics, namely the key process [3, 4].

2.1 Analysis of Assembly Quality Characteristic

The Pareto chart is an effective way to find the main factor in many factors affecting the product. Quality problem can be expressed in the form of quality loss, and most of the loss is often caused by a few quality problems, which is caused by a few reasons. Therefore, as long as the key minority is identified, the resource can be centralized to solve the key problem and avoid the resulting loss. After field investigation, the quality characteristic of 200 unqualified products are shown in Table 1.

Get the arrangement shown in Fig. 1.

Class A factor: the rear axle has abnormal sound, and the surface is rough.
Class B factors: oil spill, brake instability.
Class C factors: other (crack, scratch).

It can be seen, that the abnormal sound and rough surface are the main quality characteristic of the rear axle assembly process.

2.2 Analysis of Main Factors of Assembling Quality Fluctuation

The quality factors can be summed up as "5M1E" which are man, machine, materials, process method, test means and environmental condition. In quality analysis to "5M1E" as the tool to expand, through the comprehensive observation and analysis of the factors affecting the quality characteristic, and then find out the causal relationship between the quality factor and quality characteristic. According to the actual production

Table 1. Unqualified product statistics

Quality characteristics	Oil spill	Rear axis has abnormal sound	Brake instability	Rough surface	Crack	Scratch
Unqualified number	25	108	21	41	3	2

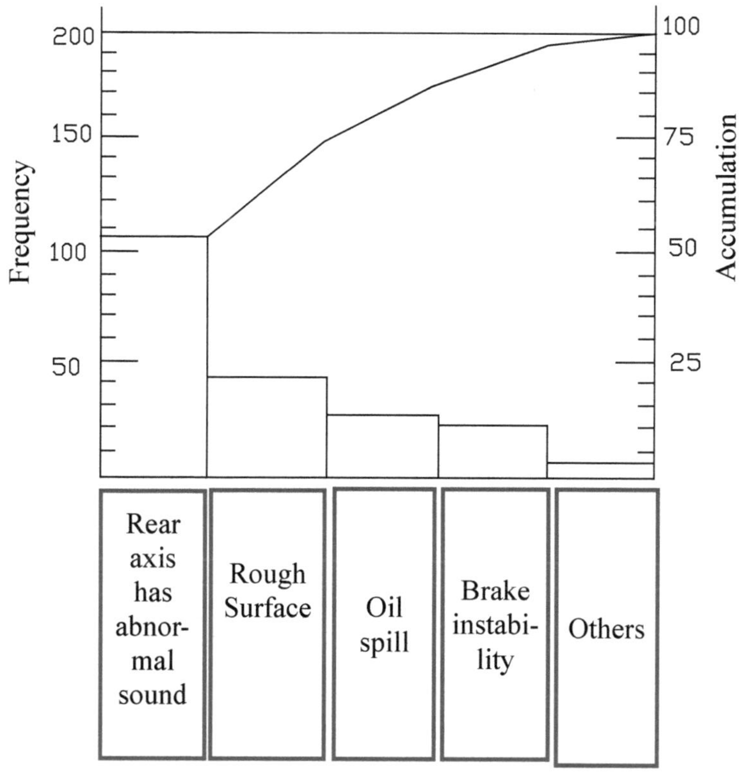

Fig. 1. Analysis of assembly quality characteristic

status of the enterprise, combining with the analysis result of quality characteristic analysis, the main reasons for the assembly quality fluctuation from the five factors of man, machine, material, method and environment are shown in Fig. 2 [5, 6].

This can be seen, the main factors of quality fluctuation include: fastening bolts loose, mechanical abrasion, stomata accessory is incorrect, lack sense of responsibility, high-intensity work, raw material is not cleaned, insufficient lubricating oil, material has burr and sand eye, the gear clearance is too large or too small, improper combination of brake friction pieces and so on.

3 The Analysis of the Main Process of the Assembly Quality

AHP is often used in the evaluation process of multi-objective and multi-criteria [7], and the analytic hierarchy process (AHP) is used to evaluate the system. The main steps are as follows:

(1) Analyzing the relationship among the elements in the evaluation system, and establish the hierarchical structure of the system.

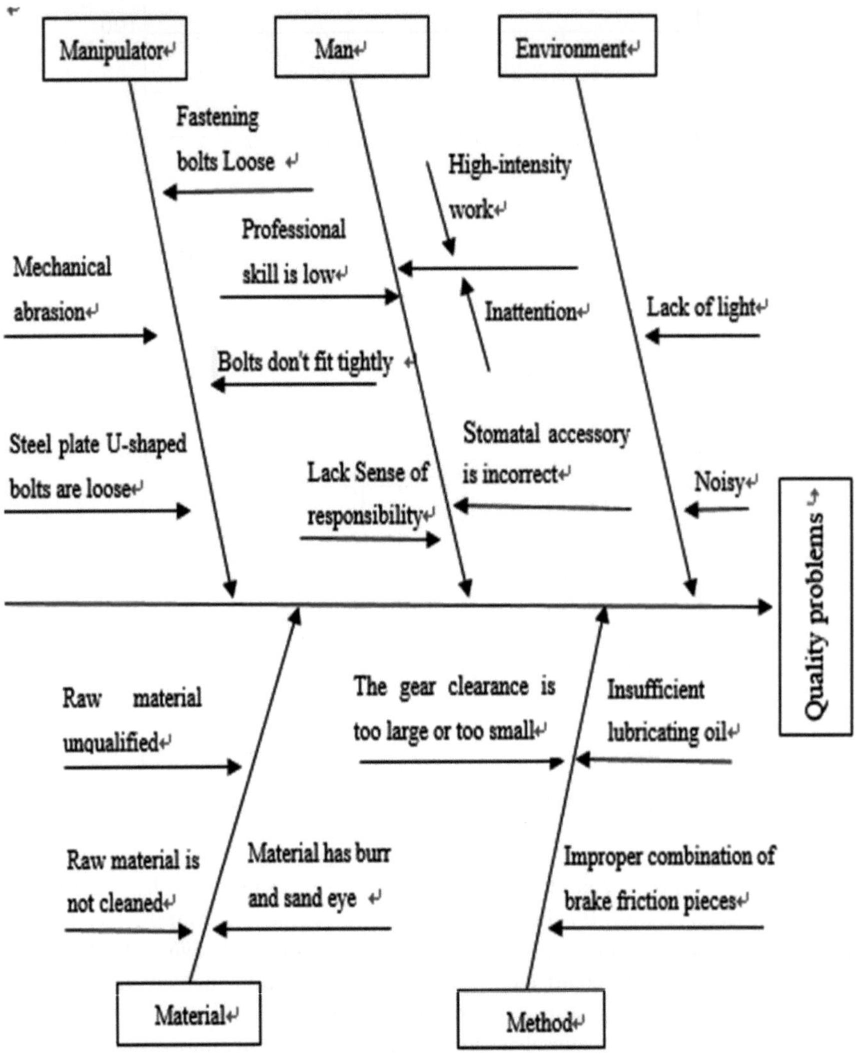

Fig. 2. Analysis of main factors of assembling quality

(2) Comparing the elements of the same level with the upper one, the relative importance is determined according to the evaluation scale, and two or two judgment matrices are constructed, and the consistency test is carried out.
(3) Calculate the relative importance of comparison element of the judgment matrix to criterion.
(4) Calculate the elements of each layer and sort various alternatives.

Based on the analysis of quality characteristic, the three-layer evaluation system is established. The first layer "A" takes the total quality as the goal. The second layer is four main quality characteristic indexes, which are D1, D2, D3, D4. The third layer has six main processes which are F1, F2, F3, F4, F5, F6 related to the assembly quality forming process, and the multi-level hierarchical structure is established by using the analytic hierarchy process, as shown in Fig. 3 [8–10].

Establish the judgment matrix. The element a_{ij} in the matrix represents the ratio of i to the relative importance of the j element, and has the following relationship.

$$a_{ij} = \frac{1}{a_{ji}}, \qquad a_{ii} = 1, \qquad i, j = 1, 2, \ldots, n$$

Obviously, the larger the ratio, the higher the degree of importance of i is. For convenience, the provision of 1, 3, 5, 7, 9 respectively, said i element and j element as same important, more important, important, very important, much more important. 2, 4, 6, 8, respectively, said the middle level. Through the investigation of the site responsible for the staff, the A layer matrix is described as below:

A = importance	D_1	D_2	D_3	D_4
D_1	1	5	6	7
D_2	1/5	1	3	5
D_3	1/6	1/3	1	3
D_4	1/7	1/5	1/3	1

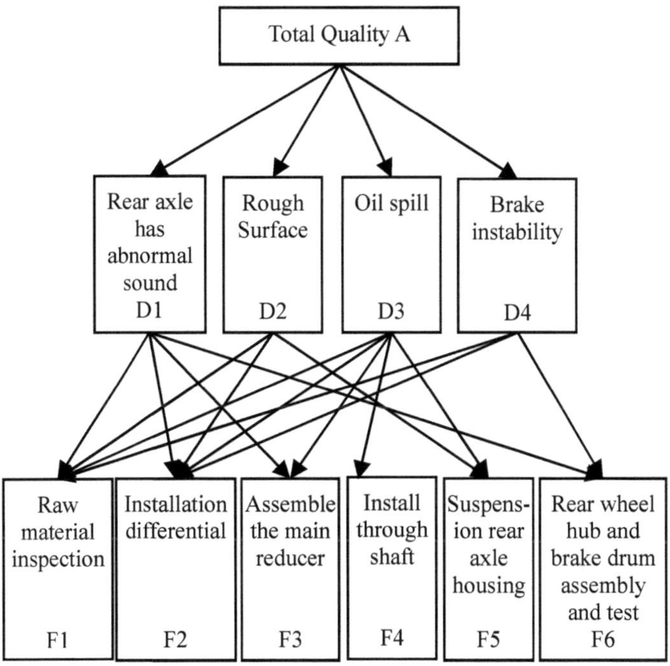

Fig. 3. Multilevel hierarchical structure diagram

By using the summation method, the matrix is normalized by column (even if the sum of column is 1):

$$b_{ij} = \frac{a_{ij}}{\sum a_{ij}} \tag{1}$$

Get the matrix:

$$B = \begin{bmatrix} 0.662 & 0.765 & 0.581 & 0.438 \\ 0.132 & 0.153 & 0.290 & 0.312 \\ 0.111 & 0.051 & 0.097 & 0.188 \\ 0.095 & 0.031 & 0.032 & 0.062 \end{bmatrix}$$

To sum by row:

$$v_i = \sum_{j}^{i} b_{ij} \tag{2}$$

Get the matrix:

$$V = \begin{bmatrix} 2.446 \\ 0.887 \\ 0.447 \\ 0.220 \end{bmatrix}$$

normalization of V:

$$\omega_i^0 = \frac{v_i}{\sum v_i} \qquad i = 1, 2, \ldots, n \tag{3}$$

The result $\omega_i^0(1, 2, \ldots, n)$ is an approximation of the eigenvector of D_i.

$$W = \begin{bmatrix} 0.611 \\ 0.223 \\ 0.111 \\ 0.055 \end{bmatrix}$$

Next, the consistency check is executed. In the actual evaluation, many evaluation problems of the evaluation object attributes are diverse, and structures are complex. Thus, the evaluator can only make a rough judgment of A, and sometimes even makes an inconsistent mistake. In order to test the consistency of the judgment matrix, according to the principle of AHP, the difference between $\lambda_{\max}$ and n can be used to check the consistency. At that time, $C.R. < 0.1$, the consistency of the judgment matrix was acceptable. Here, the consistency check is as follow:

$$R.I. = 0.89$$

$$AW = \begin{bmatrix} 1 & 5 & 6 & 7 \\ 1/5 & 1 & 3 & 5 \\ 1/6 & 1/3 & 1 & 3 \\ 1/7 & 1/5 & 1/3 & 1 \end{bmatrix} \begin{bmatrix} 0.677 \\ 0.223 \\ 0.111 \\ 0.055 \end{bmatrix} = \begin{bmatrix} 2.777 \\ 0.953 \\ 0.452 \\ 0.224 \end{bmatrix}$$

$$\lambda_{\max} = \frac{1}{4}\left[\frac{2.777}{0.611} + \frac{0.953}{0.223} + \frac{0.452}{0.111} + \frac{0.224}{0.055}\right] = 4.241$$

$$C.I. = \frac{4.241 - 4}{4 - 1} = 0.080$$

$$C.R. = \frac{0.080}{0.890} = 0.090 < 0.1$$

Consistency can be accepted.

Therefore, the important degree of D_1, D_2, D_3, D_4 to A is 0.611, 0.223, 0.111, 0.055.

Apply the same method to the D layer (layer 2),

D_1 the influencing processes are: F_1, F_2, F_3, F_6

$$D1 = \begin{bmatrix} 1 & 1 & 1/3 & 1/7 \\ 1 & 1 & 1/3 & 1/7 \\ 3 & 7 & 1 & 1/3 \\ 7 & 7 & 3 & 1 \end{bmatrix} \quad W_{D1} = \begin{bmatrix} 0.077 \\ 0.076 \\ 0.277 \\ 0.570 \end{bmatrix}$$

D_2: the influencing processes are: F_1, F_2, F_5

$$D2 = \begin{bmatrix} 1 & 1 & 1/3 \\ 1 & 1 & 1/3 \\ 3 & 3 & 1 \end{bmatrix} \quad W_{D2} = \begin{bmatrix} 0.2 \\ 0.2 \\ 0.6 \end{bmatrix}$$

D_3 the influencing processes are: F_1, F_2, F_3, F_4, F_5.

$$D3 = \begin{bmatrix} 1 & 1 & 1/5 & 1 & 1 \\ 1 & 1 & 1/5 & 1 & 1 \\ 5 & 5 & 1 & 5 & 5 \\ 1 & 1 & 1/5 & 1 & 1 \\ 1 & 1 & 1/5 & 1 & 1 \end{bmatrix} \quad W_{D5} = \begin{bmatrix} 0.111 \\ 0.111 \\ 0.556 \\ 0.111 \\ 0.111 \end{bmatrix}$$

D_4: the influencing processes are: F_1, F_2, F_6.

$$D4 = \begin{bmatrix} 1 & 1 & 1/3 \\ 1 & 1 & 1/3 \\ 3 & 3 & 1 \end{bmatrix} \quad W_{D4} = \begin{bmatrix} 0.2 \\ 0.2 \\ 0.6 \end{bmatrix}$$

After obtaining the relative importance of the components of the same layer, comprehensive importance of overall quality of elements can be calculated from upper and lower level of the correlation matrix. D level has four elements D_1, D_2, D_3, D_4 and their importance to the overall quality is in turn $\omega_1, \omega_2, \omega_3, \omega_4$; Its subordinate has six elements F_1, F_2, F_3, F_4, F_5, F_6, F_i, about the relative importance of D_i is v_{ij}. Thus the comprehensive importance of factor F_i in the F level is:

$$W_i = \sum_j \omega_j v_{ij} \tag{4}$$

The results of the calculation are shown in Table 2.

It can be seen that the rear wheel hub and the brake assembly and the inspection process have the greatest impact on product quality, assembly main reducer, lifting rear axle housing, raw material inspection, and installation differential impact are relatively large, and the installation of the shaft axis impact is small.

Table 2. Overall importance

D_i	D_1	D_2	D_3	D_4	W_i
ω_i	0.611	0.223	0.111	0.055	
F_1	0.077	0.2	0.111	0.2	0.115
F_2	0.076	0.2	0.111	0.2	0.114
F_3	0.277	0	0.556	0	0.231
F_4	0	0	0.111	0	0.012
F_5	0	0.6	0.111	0	0.146
F_6	0.570	0	0	0.6	0.381

4 Conclusion

From the analysis of the quality factor of a rear axle assembly process, it can be found that the assembly is the main process of product quality formation, and any process will influence quality. From 5 aspects of impact quality which are namely man, machine equipment, material, process method, test mean and environmental condition, to find the main factors of assembly quality fluctuation, and establish three-layer index system, qualitative knowledge quantification. By combining the actual assignment with scientific calculation, finally, it is found that the rear wheel hub and brake assembly and inspection process are the main process of assembly quality, and change the status of enterprise experience as quality judgment method, which could use for the implementation of modern quality management.

Acknowledgements. Project supported by the youth scientist fund of NSFC (No. 51705417).

References

1. G. Zhang, H. Zhao, Assembly quality characteristic prediction method under influence of multiple assembly features. Appl. Res. Comput. **32**(3), 709–712 (2015)
2. T. Yunfei, *Basic Craft of Axle* (Baoji Hande Axle Factory, March 2016)
3. F. Qian, Simulation of Quality Fluctuation Model of Key Process Chain Using QFD Method. Electron Quality (June 2016), pp. 53–56
4. Z. Jia, J. Ma, F. Wang, Characteristics forecasting method of assembled product based on multiple part geometric elements. J. Mech. Eng. **45**(7), 168–173 (2009)
5. C. Wen, Z. Li, J. Jianjun, An assembly quality improvement of turbocharger based on PDCA plus fishbone diagram analysis. Ind. Eng. J. **18**(1), 55–59 (2015)
6. B. Meng, F. Zhou, X. Wang, L. Sun, Research on depth analysis method for aircraft assembly quality data. Sci. Technol. Eng. **15**(34), 235–242 (2015)
7. J. Han, Decision analysis of maintenance strategy of trackless rubber vehicle based on AHP. Science and Technology Innovation Herald, pp. 81–82 (November 2015)
8. Y. Wang, *System Engineering* (Machinery Industry Press, Beijing, 2008)
9. Y. Wang, Analysis of influence factors of coal mine safety based on AHP. Energy Educ. Sci. Technol. Part A Energy Sci. Res. **32**(2), 1225–1230 (2014)
10. H. Li, Q. Jin, Q. Wang, R. Yue, Evaluation of performance of mechanical system of tourist vehicle based on grey/AHP theory, Mach. Build. Autom. **44**(4), 63–66 (August 2015)

Study on Productivity Improvement and Quality Control at a Smart Card Company Based on Six Sigma

Yue Wu[1(✉)], Chao Xu[1], Chen-xi Li[2], and Yi-jing Wan[3]

[1] School of Mechanical Engineering, Xi'an University of Science and Technology, Xi'an, China
wuyue1992@xust.edu.cn
[2] School of Architecture and Civil Engineering, Xi'an University of Science and Technology, Xi'an, China
[3] Department of Mechanical and Manufacturing Engineering, St. Cloud State University, St. Cloud, MN, USA

Abstract. The development of smart card is stepping into the era of high-speed as the improvement of producing techniques becomes more mature. The application of smart cards is widely used in multiple fields such as transportation, telecommunication, medical treatment, financials, etc. A study regarding six sigma was conducted at a smart card company which was undergoing downturn times due to the bottleneck process on assembly line and the market competition pressure. A very important methodology of Six Sigma, DMAIC, was used to improve the productivity without adding any unvalued equipment or additional labor, at the same time ensure the standard quality of smart cards to satisfy customers. This study finally presented the ideal improvement plan and put it into practical application, which brought the desirable achievements and considerable benefits to the company.

Keywords: Bottleneck process · DMAIC · DOE · Hypothesis testing
Smart card · Six Sigma management

1 Introduction

The study was conducted at a smart company based on Six Sigma. This smart card company owns fully automatic equipment for production, with perfect card detection methods, testing equipment, professional graphic design and printing equipment [1]. However, as the competitive trend of the smart card market becomes severe, increasingly small companies and workshops divide up the market share depending on the low price advantage [2]. In addition, the market for smart cards is in a status of demand exceeds supply. Therefore the company wanted to increase the yield of smart cards by improving the bottleneck process based on Six Sigma, meanwhile to fulfill the needs from the market under the prerequisite of believable and eligible quality and get more economic profits [3].

The main producing process of smart cards is showed in Fig. 1, from printing, milling, slotting card to embedding, grooving and personalization, it contains seven

G. Q. Huang et al. (eds.), *Proceeding of the 24th International Conference on Industrial Engineering and Engineering Management 2018*,
https://doi.org/10.1007/978-981-13-3402-3_61

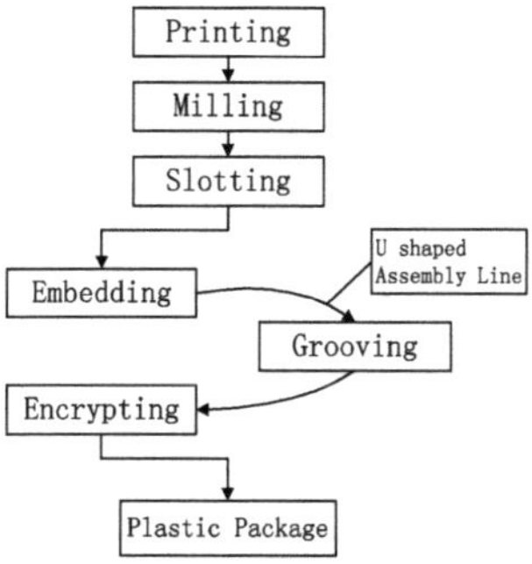

Fig. 1. Main producing flow chart of the smart card

main procedures. The U shaped assembly line which was observed particularly consists of embedding, grooving and encrypting. Their corresponding productivities could be observed directly. After analyzing the collected data, the grooving station was found as the bottleneck process in the whole assembly line, which caused the lower yield and undesired productivity for the production. Thus these three procedures were regarded as the start of problems as the critical keys.

2 Methodology

2.1 Six Sigma Management

Sigma is called the standard deviation in statistics, which is used as an indicator of the dispersion degree of the data [4]. In quality management it is used to describe the level of quality fluctuation. It is also an indicator for perfect improvement specifically associated with statistical modeling of manufacturing processes. The maturity of a process can be described by a sigma rate indicating its yield or the percentage of defect-free products it creates [5]. A Six Sigma process is one in which 99.99966% of all opportunities to produce some features of a part are statistically expected to be free of defects (3.4 defective features/million opportunities), although, consider of 1.5 sigma shift, this defect level corresponds to only a 4.5 sigma capability. Six Sigma management is the main thread and research basis of this study [6].

2.2 DMAIC

Six Sigma is precise technique and principle of implementation to mine the essence of the problem efficiently and to give suggestions for improvement. DMAIC process, as one of the most important methodologies in Six Sigma management, was the mainline and route in this study [7]. This process includes define, measure, analysis, improve and control these five phases [8]. The study followed the DMAIC process to solve the problems without influencing the customer satisfaction, to identify the significant factors and consolidate the ways to improve, and to control and maintain the expected results [9]. The data in the process was a basis for analyzing, some statistics tools and charts were

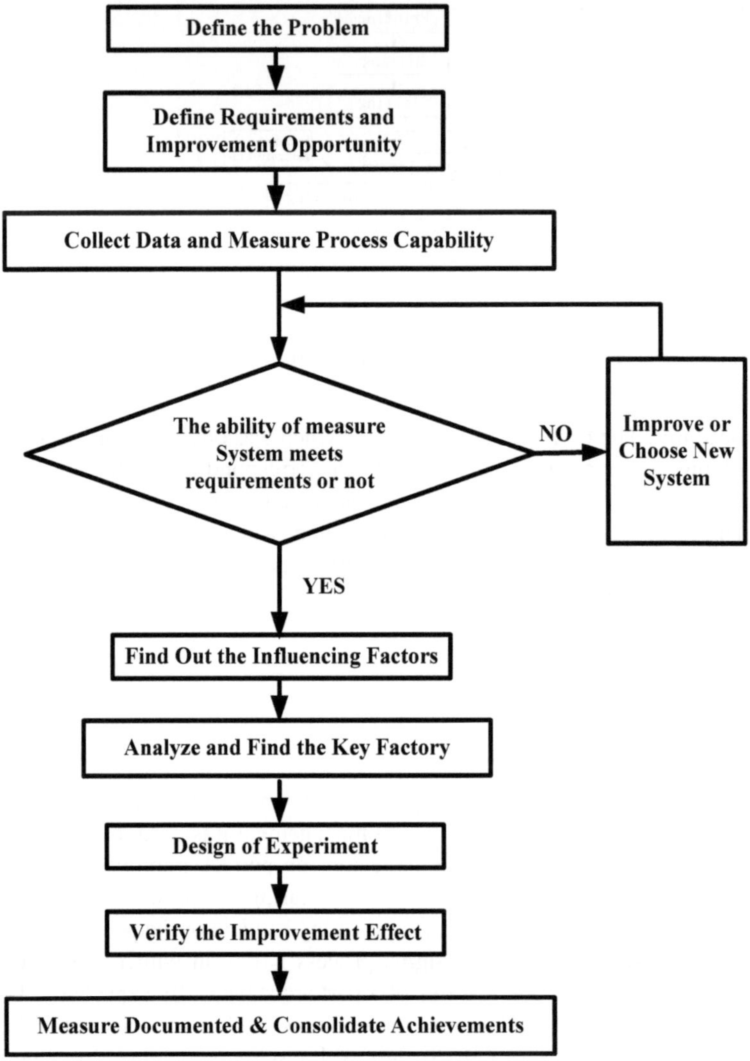

Fig. 2. DMAIC framework

applied to define, measure, analyze, improve, and control the process, such as process capability analysis, variance analysis, FMEA, DOE, Pareto chart, Fishbone map, etc. Figure 2 is the framework of this study.

(1) *Define phase*: First of all, the improvement opportunity and customers' requirement need to be defined. The equipment used for embedding, grooving, encrypting were in a U shaped assembly line layout. Their productivities were 3500 pieces per hour, 2800 pieces per hour, and 3200 pieces per hour. Grooving equipment contained two sets of machines, which had the lowest productivity of 1400 p/c among three working stations. The smart card company manufactured a set of detective device

of simulating artificial card-breaking activity. They received an average power value for the connecting force between the metal chip and the plastic part of the smart cards. The smart card company unified the technical specification requirements for connecting force to be controlled within a range of 28–40 N. If they could guarantee this condition for the force value, both the internal and external groups' requirements could be met.

(2) *Measure phase*: The goal of this phase was to gather the data that described the nature and extent of the problem. Two variables Y1 and Y2 were chose as the measured objects, Y1 stood for the speed of the grooving equipment. Two types of reference data were used to describe this speed. One is the producing cycle time for the single smart card, which could be used to make compare on the changes of speed during improvement. Another one is the monthly yield of the smart cards. Y2 stood for the defect-free rate of smart cards in the grooving process. The expectation was to improve Y1 while not bringing negative effects on Y2.

The defect-free rate belonged to attribute data, which presented whether the quality characteristic of the products meet the specification or not. In this company, every operator worked at process station had to conduct self-check and rated the quality level before the products going to the next process. In order to verify the operators' abilities of judging and classifying on the defective products, attribute agreement analysis is an ideal tool to figure out how likely the measurement system was to misclassify a part. There were 5 operators in this measuring experiment. They were picked up randomly from the assembly line, except the most inexperienced and the most sophisticated operators. These 5 operators made one time classification on the 50 smart cards including all kinds of typical defective sorts. Then the tools of Minitab software were used to analyze the data of judgment results.

Besides, T-test method for measuring devices was indispensable because measuring results of the connecting force must be controlled in a required range. It was use to ensure there were no differences between the measuring devices and the results of measurement were specific and reliable [10]. Experiment process was to ask one operator grooved one smart card sample from its two edges, then measured the two forces by using device A and device B, in this way could the operator maintain the same experiment conditions and material, only leaving the connecting force as the only variable. The sample size was 20.

(3) *Analysis phase*: In this phase, the potential causes that made effects on the most active factor should be figured out. FMEA (Failure modes and effects analysis) is a methodical approach for preventing defects by prioritizing the problems. It is also used to identify the contingency plans to eliminate or reduce the probability or severity of the problem. Calculating the RPN (risk priority number) value, which was the multiplying products of three indexes: severity, occurrence and detection levels; to determine the importance and priority level of all the causes and find the significant factors.

(4) *Improve phase*: This phase was to make some specific changes to counteract the causes after understanding what were the influencing factors in the previous phase. DOE (design of experiments) was the lean tool used in this project. Factorial design was adopted in this phase [11]. Here Minitab was a very popular software for conducting DOE to get important analysis graphs such as main effects plot, contour plot, and surface plot etc., which helped to get the response optimizer for the researching objects.

(5) *Control phase*: The last phase of control was very important to sustain the implemented improvements and achieve the desired results continuously [12]. The communication of the new methods, procedures, and responsibilities should be integrated into a training program for the process personnel.

3 Results and Discussion

3.1 Attribute Agreement Analysis

Five operators accepted the measuring experiment. By using Minitab, the analyzed results of their judgment data was showed in Fig. 3.

From the results, the appraise ability of operator "d" was the best; operator "c" was in the next place. The entire assessment agreement rate was 86% based on 95% confidence interval, which was greater than the usual acceptable rate of 80%. Also, the assessment agreement between appraisers was good. Therefore the judgment abilities and assessment agreement of operators were credible.

```
Each Appraiser vs Standard

Assessment Agreement

Appraiser  # Inspected  # Matched  Percent       95% CI

a504896.00  (86.29,  99.51)

b504896.00  (86.29,  99.51)

c504998.00  (89.35,  99.95)

d5050100.00  (94.18,  100.00)

e504896.00  (86.29,  99.51)

# Matched: Appraiser's assessment across trials agrees with the known standard.

Between Appraisers

Assessment Agreement

# Inspected  # Matched  Percent      95% CI

50 43  86.00  (73.26, 94.18)

# Matched: All appraisers' assessments agree with each other.

All Appraisers vs Standard

Assessment Agreement

# Inspected  # Matched  Percent      95% CI

50 43  86.00  (73.26, 94.18)

# Matched: All appraisers' assessments agree with the known standard.
```

Fig. 3. Attribute agreement analysis figure

3.2 T-Test for Measuring Devices

The first step of conducting two-sample T-test was to raise the null hypothesis and alternative hypothesis.

H_0: *There is no significant difference between device A and device B;* $\mu 1 - \mu 2 = 0$
H_1: *There are differences between device A and device B;* $\mu 1 - \mu 2 \neq 0$.

In T-test, the T-value measured the size of the difference relative to the variation in the sample data. The greater the magnitude of T-value is, the greater the evidence against the null hypothesis; the closer T-value is to 0, the more likely there is not a significant difference. The results showed in Fig. 4 indicated T-value was 0.02, and P-value was 0.985, which indicated the null hypothesis could not be rejected. There was no significant difference between device A and device B, which also proved that the measurement system was reliable.

So far all the preconditions got qualified for the design of experiment in the following phases.

3.3 Failure Modes and Effects Analysis

The severity for failure modes, the probability of occurrence and the likelihood of detection were marked by FMEA. The eligible and stable connecting force between the metal chip and the plastic part of the smart cards was the critical factor that determines the progress of all the producing activity, so how to maintain the connecting force became the problem to be solved next. The first five significant factors that would be influencing the connecting force are shown in Fig. 5. They are lower mold, stocking cutter, lower mold, molds match-up and molds-operating distance.

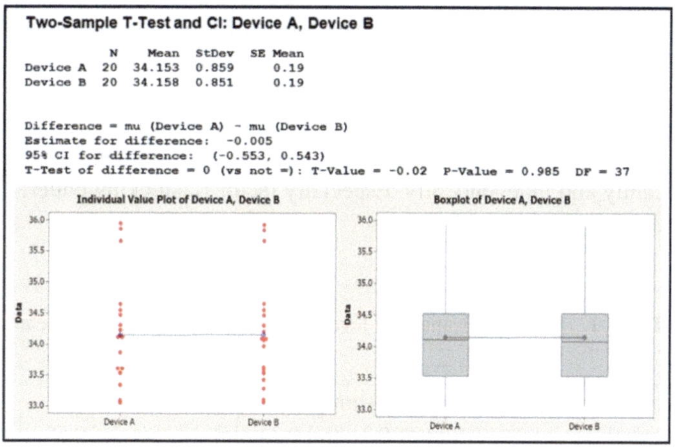

Fig. 4. T-test results figure

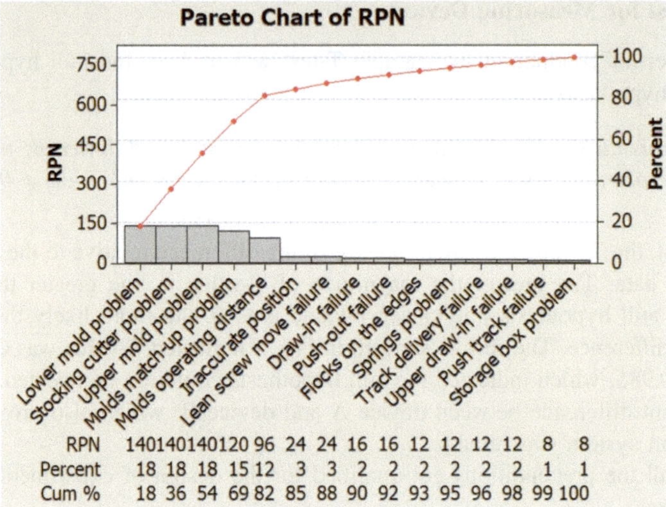

Fig. 5. Pareto chart of RPN value

3.4 Design of Experiment

To conduct the further experiment and to find the significant factor, these five high-RPN influencing factors were quantized by different levels [13]. The original sizes of stocking cutter were 0.31 and 0.34 mm. The adjusted sizes were 0.55 and 0.60 mm. The new size was not an exact value, it was the relatively high value used to determine whether the factor really has main effect. The new set of stocking cutters were provided by the supplier and the sizes were designed depend on their processing experience. Because the levels of molds match-up mainly depended on the operators' subjective working methods, which needed to be unitized and standardized, it would not be designed as a factor in the experiment. The levels of the remaining four factors were defined in Table 1.

This was a 4-factor and 2-level model of DOE. It contained 16 sets of experiments. The data of different connecting forces got analyzed by using Minitab. The results showed in Fig. 6 indicated that all the four factors were influencing the connecting force significantly and independently. Especially factor B, stocking cutter, was the most significant factor with no doubt.

Table 1. Four-factor and two-level model of DOE (mm: millimeter)

Factor	Level 1	Level 2
A: Lower mold	Old and worn part	New part
B: Stocking cutter	Size of 0.31 and 0.34 mm	Size of 0.55 and 0.60 mm
C: Upper mold	Old and worn part	New part
D: Mold-operating distance	Low position	High position

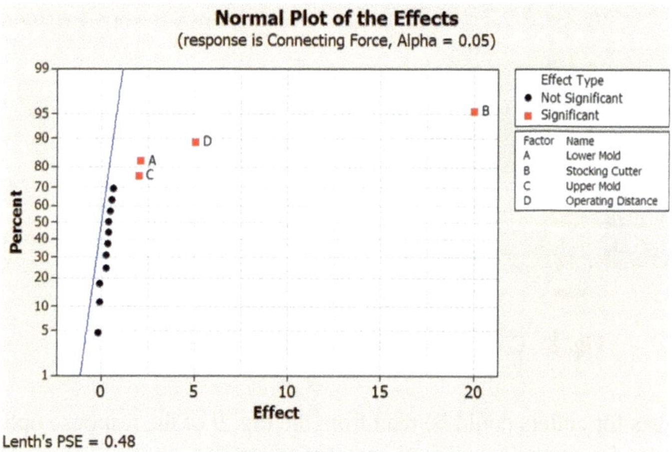

Fig. 6. Normal plot of the effects figure

Table 2. Two-factor and two-level model of DOE (mm: millimeter)

Factor	Level 1 (mm)	Level 2 (mm)
Cutter A	0.31	0.55
Cutter B	0.34	0.60

Another experiment for stocking cutter was designed in order to determine the best size of cutters. Table 2 was a 2-factor and 2-level model.

Figure 7 showed that the main effects plot and interaction plot for connecting force, both cutter A and cutter B had effects toward connecting force and they had interaction to each other. To find the optimal sizes of cutter A and cutter B became the critical task.

The optimal sizes of cutters for getting an average connecting force within required range could be distinguished by contour plot and surface plot in Fig. 8. While the

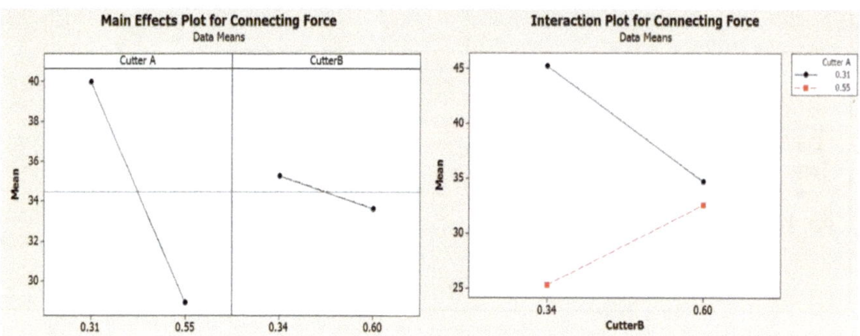

Fig. 7. Effects plots for connecting force

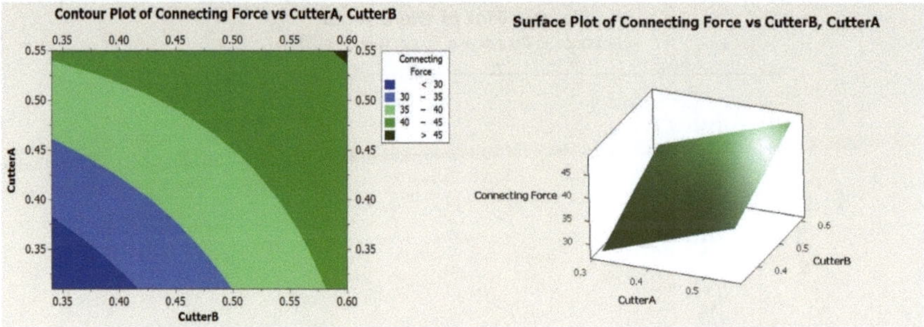

Fig. 8. Contour plot and surface plot of connecting force

specific values for cutters could be read from the Fig. 9 of the response optimizer. They were 0.39 mm for cutter A and 0.41 mm for cutter B.

3.5 Capability Analysis for New Connecting Force

Capability analysis was conducted towards the new connecting force [14]. Figure 10 showed that the new process was normally distributed. The histogram did not exceed the upper specification limit or the lower specification limit. The mean value of connecting force was 34.18 N, which was perfectly met standards. Additionally, some statistics that can be used to measure the capability are on the right side of the histogram.

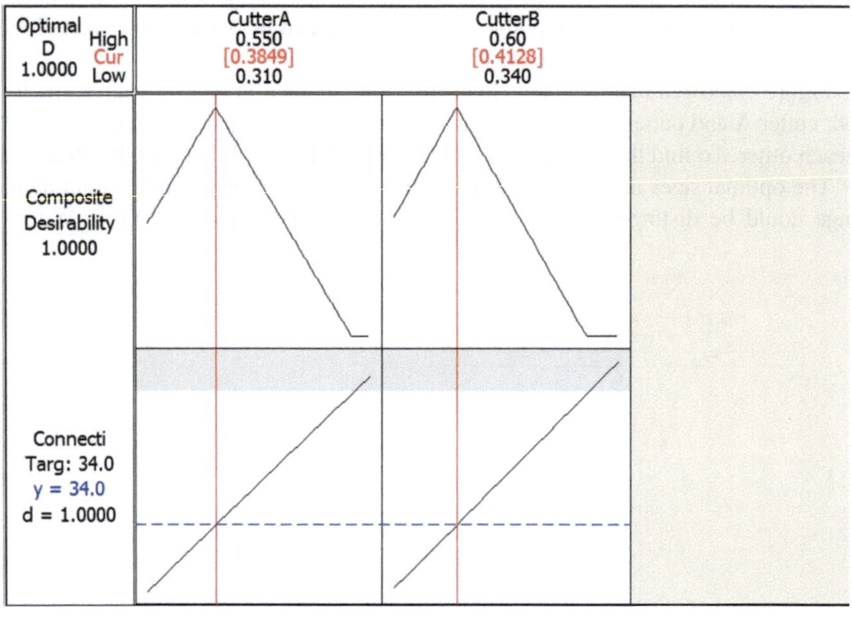

Fig. 9. Response optimizer figure

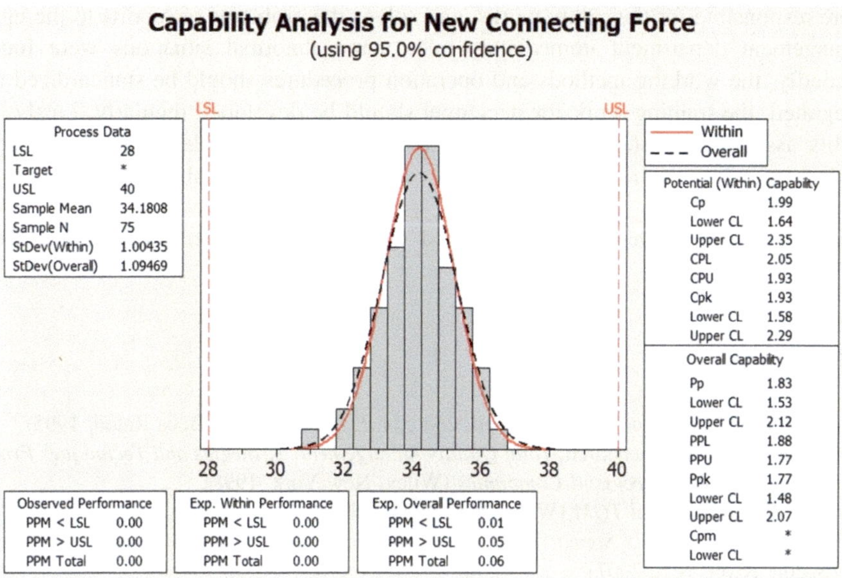

Fig. 10. Figure of capability analysis for new connecting force

Generally, Cp determines the spread of the process while Cpk determines the shift in the process. Both of indexes provide information about how the process is performing with respect to the specification limits. Here Cp was 1.99 and Cpk was 1.93, which stood that the current process capability was perfect and there was no need to adjust anymore. It also meant the new sets of cutters fit the standards perfectly. As for the capability analysis of new defective ratio, the present Sigma capability was 3.43, which increased a lot compared with the previous Sigma capability 3.36. PPM now was 298, which meant the defective parts per million decreased compared with the precious PPM 387. Moreover, after the improvement the cycle time of producing single card decreased from 2.49 to 2.19 s; the productivity of grooving had a 12% increase.

4 Conclusion

This study conducted the quality management systematically and applied DMAIC process in sequence. Obviously, the results were desirable and met the objectives and expectations of the company. The productivity of smart cards increased dramatically by improving the bottleneck, while the defect-free rate of smart cards still maintained well during improvement, which meant the quality of smart cards was perfectly in the control. What is the most important is that this study would eventually benefitting the company to become more competitive in the market. It was worthy of referring for the future control. There are some recommendations for the smart card company [15]. Firstly of all, the smart card company has to update and share information regarding improving methods with every department, meanwhile every department should be more responsible for their duty works, and solves the problems or reports to the upper management department immediately once some abnormal situations were found. Secondly, the working methods and operation procedures should be standardized and integrated, the training work for personnel should be developed regularly. Lastly, the ability assessment system and supervision system must be implemented in the smart card company to ensure all the processes could be in the control in the future.

Acknowledgements. This work is supported by the natural science foundation of China (51875451).

References

1. J.E. Ross, *Total Quality Management*, 2nd edn. (St. Lucie Press, Boca Raton, 1995)
2. S. George, A. Weimerskirch, *Total Quality Management: Strategies and Techniques Proven at Today's Most Successful Companies* (Wiley, New York, 1998)
3. K.J. Zink, *Successful TQM* (Wiley, New York, 1997)
4. D.C.S. Summers, *Six Sigma: Basic Tools and Techniques* (Pearson/Prentice Hall, Upper Saddle River, N.J., 2007)
5. H. Dreifus, *Smart Cards: A Guide to Building and Managing Smart Card Applications* (Wiley, New York, 1998)
6. J. Ferrari, S. Poh, R. Mackinnon, L. Yatawara, Smart Cards: A Case Study, Internet Article, 1998. Retrieved from http://www.redbooks.ibm.com/redbooks/pdfs/sg245239.pdf
7. S. Liu, China Smart Card Development Association Journal: Smart Cards Market Analysis, V1
8. Y. Han, *Application of Six Sigma to Improve Grass-Roots Ceramic Capacitor Yield Research* (Tianjin University, April 2008)
9. J. Liker, *The Toyota Way: 14 Management Principles form the World's Greatest Manufacturer* (McGraw-Hill Education, January 2004)
10. J.R. Evans, W.M. Lindsay, *The Management and Control of Quality*, 6 edn. (South-Western College Publish, April 2004)
11. S. Nahmias, *Production and Operations Analysis*, 6 edn. (McGraw-Hill/Irwin, March 2008)
12. P. Graupp, R.J. Wrona, *Implementing TWI: Creating and Managing a Skills-Based Culture* (Productivity Press, November 2010)
13. Z. He, M. Zhang, Y. Dong, Study on the method of cutting tool wear process capacity analysis. Chinese Agricultural Mechanization, V6
14. N. Hong, J. Hou, *MINITAB Tutorial* (Publishing House of Electronics Industry, Beijing)
15. China Smart Card Development Association Journal, V6 (2014)

Using Six Sigma to Optimize Fan Blade Structure

Cai-feng Chen, Bin Cheng(✉), Xi-yang Li, Jie Mo, and Yang Liu

Industrial Engineering, University of Shihezi, Xinjiang, China
1847513395@qq.com

Abstract. In this paper take the wind turbine blade as an example, to expound Six Sigma theories which apply to optimize fan blade structure. Based on the Six Sigma optimal design, the assembling dimension tolerance accumulative analysis is carried out, so that the blade design can improve the quality of blade design and the success rate of the assembly, then reduce the cost and improve the enterprise efficiency.

Keywords: Six Sigma · Fan blades · Structural optimization · ANSYS Workbench

1 Introduction

Design for Six Sigma (FSS) is a way of taking the Six Sigma level as the goal to design products [1]. It began in the 20th century, 90 years. Design for Six Sigma uses scientific methods to accurately understand and grasp the need of customers. In addition, it applies statistical methods to quantify the relationship between system performance and related to design parameters. The most important thing is to design the key customer's needs into the product, in order to achieve the product at Six Sigma quality levels at low cost [2]. Six Sigma theories not only improve product quality and reliability, but also reduces costs and shorten the development cycle, it has a high practical value [3].

The blades of wind turbines are the primary components that convert wind energy into mechanical energy and determine the efficiency of energy conversion for wind turbines. It is made up by airfoils of different twist angles. However, the aerodynamic characteristics of the airfoil affect the wind energy utilization directly. Designed and researched fan blade is a high-tech and complex work [4]. This article will be based on Six Sigma design methods to identify the products key parameters, and set the appropriate size and tolerance, so that the final product to meet the application requirements and achieve the Six Sigma quality level.

G. Q. Huang et al. (eds.), *Proceeding of the 24th International Conference on Industrial Engineering and Engineering Management 2018*,
https://doi.org/10.1007/978-981-13-3402-3_62

2 Wind Turbine Blade 3D Modeling

2.1 Calculate Parameters

The design parameters of the fan blades including follows: Airfoil, the number of blades, blade diameter, length, the tip speed ratio, chord length, twist angle, etc. In this paper, the fan blades are divided into 19 sections. By calculation, each section parameters are shown in Table 1.

Table 1. The parameters of each section

Section	Rotation radius (mm)	Twist angle (°)	Chord length (mm)
1	0	0	1600
2	1000	0	1600
3	2500	28.3	2220
4	3000	22.3	2340
5	5000	18.3	2450
6	7000	14.1	2700
7	10000	8.3	3040
8	13000	5.8	2730
9	15000	3.3	2640
10	17000	2.0	2270
11	20000	0.3	2000
12	23000	−0.6	1934
13	25000	−1.7	1850
14	27000	−1.93	1820
15	30000	−2.2	1780
16	33000	−2.9	1747
17	35000	−3.7	1685
18	37000	−4.3	1588
19	39240	−4.7	1468

2.2 Establish 3D Model

The two-dimensional airfoil of each section is closed by CAXA. The concrete steps are as follows:

Step1: Selecting "spline" in the toolbar.
Step2: Selecting "direct mapping".
Step3: Opening the spline data file.
Step4: Select "straight line" in the toolbar.

Since the coordinates are split into X, Y_{up}, and Y_{under}, the upper and lower two lines corresponding to each airfoil should be imported at the same time. Finally, each stitched section airfoil is stored as DXF (Figs. 1 and 2).

In SolidWorks, selecting the "Insert DXF" for the toolbar, and then import 19 sections into SolidWorks as showed in Fig. 4. Finally, the three-dimensional model of the fan blades is obtained by the lofting command, as showed in Fig. 5.

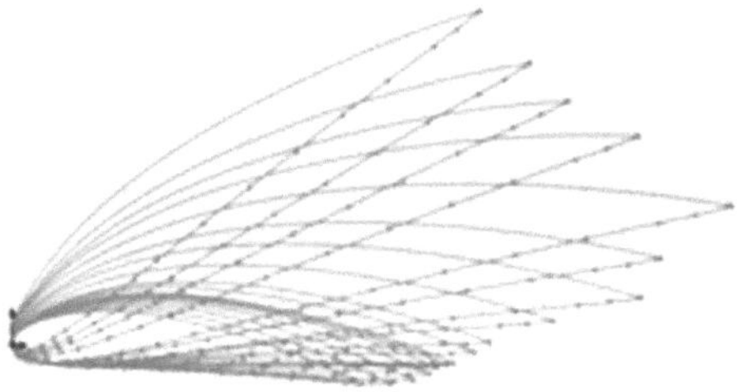

Fig. 1. The change of twist

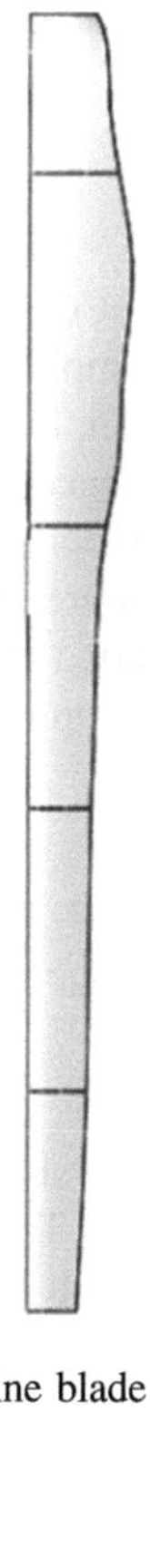

Fig. 2. Wind turbine blade 3D modeling

3 Optimization Process Based on Six Sigma

Six Sigma optimization based on six standard error theory [5], to assess the product reliability probability, and then determine whether the product meets the Six Sigma guidelines. The traditional structural optimization design is provided by the designer with several different design schemes, and then we compare numerous schemes and pick out the optimal scheme. This approach is usually based on the experience of the designer, due to the limited of time, the number of options available, and often is not necessarily the best solution. The number of options available is compact and often not necessarily the best solution. If you want to take the best solution, it is necessary to provide more design to compare, which requires a lot of resources. In addition, it's hard to be done by manpower alone. But computers can do this efficiently. So far, there is little software to do structural optimization, ANSYS Workbench software as a universal finite element analysis tool, not only has a strong front and rear processor, but also a very powerful optimization design function. It can do both structural size optimization and topology optimization, and its own algorithm can respond to the need of engineering. The purpose of structural optimization is tantamount to design a scheme to meet all design requirements, with minimal cost [5]. In order to ensure structural stiffness, and meet the strength of the requirements. We change some design variables, so that the weight of the structure of the most reasonable, which not only save structural supplies. In the transport installation also provides a convenient, lower transport cost.

3.1 Optimization Parameters

After the completion of the 3D model design and the tolerance calculation, only to complete the preliminary design is completed and the structural characteristics of the product must be taken into consideration. In this paper, ANSYS Workbench software is used to analyze the structural performance of the fan blades. 1.5 MW fan blade solid model generated in SolidWorks is imported into ANSYS Workbench for structural static analysis. The setting parameters are shown in Table 4. For finite element analysis, the blade is assumed to be simplified as a cantilever beam, then the full constraint is applied at the root of the blade, and applies a 7 MPa stress in the 45 direction on the upper surface of the blade (Table 2).

Table 2. Fan blade material properties [6]

Elastic modulus (GPa)	Poisson's ratio	Density (kg/s^3)
42.6	0.22	1950

In this paper, the finite element analysis is properly the result file is generated. The total deformation is illustrated in Fig. 3. The equivalent stress is shown in Fig. 4, and the safety factor is illustrated in Fig. 5. It can be seen from the figure that the deformation of the blade increases along the direction of the blade and reaches the maximum on the free end face of the blade. Through the equivalent stress map can be found that the maximum stress appears at one third of the root. When the load increases, the surface of the blade begins to tear here. There is stress concentration at the intersection of the inner surface of the blade, the intersection of the force plate and the free end face, the local stress reaches the maximum value, and the stress value of the other part is very small. So the blade structure design is safe and reliable, to meet the strength requirements of the circumstances, may be adapted to reduce the length and quality of the blade.

Fig. 3. Total deformation

Fig. 4. Equivalent stress

Fig. 5. Safety factor

Through the static structural analysis, the maximum deformation of the blade is 2297. The maximum equivalent stress is 270.6 MPa and the minimum safety factor is 0.924613. The structural statics analysis module and the Six Sigma optimization analysis module are established association as showed in Fig. 6. The Six Sigma optimization is mainly used to evaluate the reliability of the product, and its technology is based on six standard error theories, such as material attributes, geometrical dimensions, load and so on the effect of the probability distribution of uncertain input variables on product performance (stress, strain, etc.). A total of eight optimization parameters were set up at Six Sigma optimization, which were P1, P2, P3, P4, P5, P6, P7 and P8, where P1, P2, P3, P4 indicate the length of Sects. 1, 2, 3 and 4, P5 for blade quality, P6 for maximum deformation, P7 for maximum effect, P8 indicates minimum safety factor.

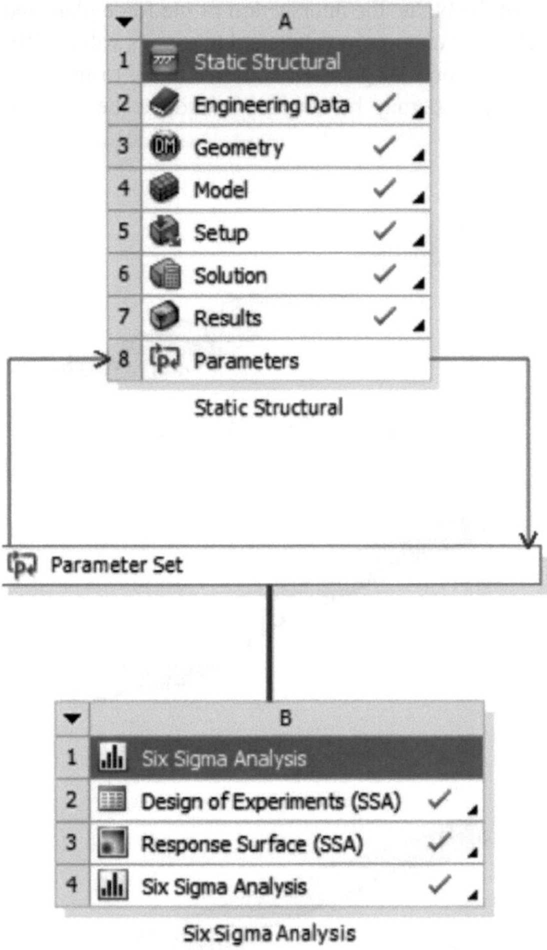

Fig. 6. Relationship between Six Sigma optimization and structural static analysis

3.2 Analysis Results

After six sigma optimization, given 25 design points, as showed in Table 3, the overall deformation of the design point of the relationship between the curve shown in Fig. 7.

Spider and local sensitivity reflect changes in the output parameters corresponding to one or two input parameters. Figure 8 detects the value and change of all output parameters, and local sensitivity. Figure 9 detects the component of each parameter in the vicinity of the response point. Depending on the change of these two parameters, change the parameters into the response surface, get the need to optimize the response surface showed in Fig. 10. The comparison of the parameters after the Six Sigma optimization is showed in Tables 4 and 5. Therefore, the use of Six Sigma principles of the fan blade structure optimization, both to shorten the design time, but also cost savings.

Table 3. Optimization results

Name	P1 (m)	P2 (m)	P3 (m)	P4 (m)	P5 (kg)	P6 (mm)	P7 (MPa)	P8
1	2500	500	2000	2000	116043.1	2297.5	271.6	0.920578
2	2497.5	500	2000	2000	116043.1	2297.5	274.5	0.910909
3	2502.5	500	2000	2000	116043.1	2297.5	272.8	0.916486
4	2500	497.5	2000	2000	116033.1	2297	272.5	0.917511
5	2500	502.5	2000	2000	116047.5	2298	270.9	0.92269
6	2500	500	1997.5	2000	116033.5	2297	271.2	0.921872
7	2500	500	2002.5	2000	116052.7	2298	272	0.919071
8	2500	500	2000	1997.5	116032.6	2297	271.2	0.921892
9	2500	500	2000	2002.5	116053.5	2298	271.9	0.919441
10	2498.3	498.3	1998.3	1998.3	116020.7	2296.4	272	0.919078
11	2501.7	498.3	1998.3	1998.3	116020.7	2296.4	271.4	0.921093
12	2498.3	501.7	1998.3	1998.3	116033.7	2297.1	272.4	0.917856
13	2501.7	501.7	1998.3	1998.3	116033.7	2297.1	272	0.918967
14	2498.3	498.3	2001.7	1998.3	116037.1	2297.2	272	0.918979
15	2501.7	498.3	2001.7	1998.3	116037.1	2297.2	273.1	0.915297
16	2498.3	501.7	2001.7	1998.3	116045.1	2297.9	269.9	0.926302
17	2501.7	501.7	2001.7	1998.3	116045.1	2297.9	271.4	0.92125
18	2498.3	498.3	1998.3	2001.7	116038.4	2297.1	273.2	0.914934
19	2501.7	498.3	1998.3	2001.7	116038.4	2297.1	272.0	0.918958
20	2498.3	501.7	1998.3	2001.7	116048.5	2297.8	271.8	0.919844
21	2501.7	501.7	1998.3	2001.7	116048.5	2297.8	272.3	0.91794
223	2498.3	498.3	2001.7	2001.7	116051	2297.8	272.1	0.918893
23	2501.7	498.3	2001.7	2001.7	116051	2297.8	272.2	0.918584
24	2498.3	501.7	2001	2001.7	116059.8	2298.6	272.6	0.917231
25	2501.7	501.7	2001.7	2001.7	116059.8	2298.6	235.9	1.059669

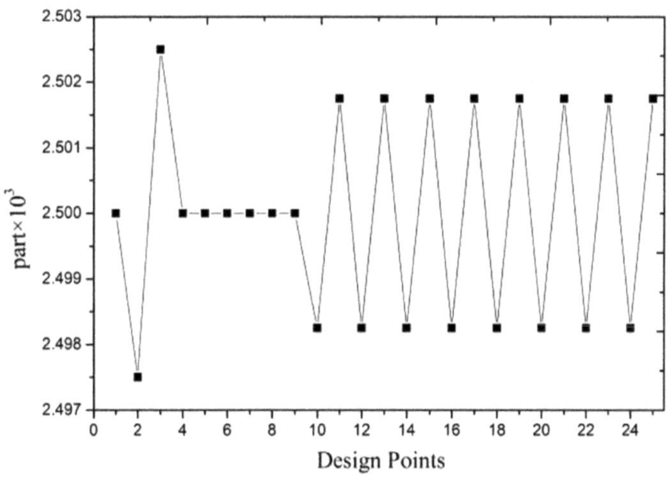

Fig. 7. Design points versus parameter

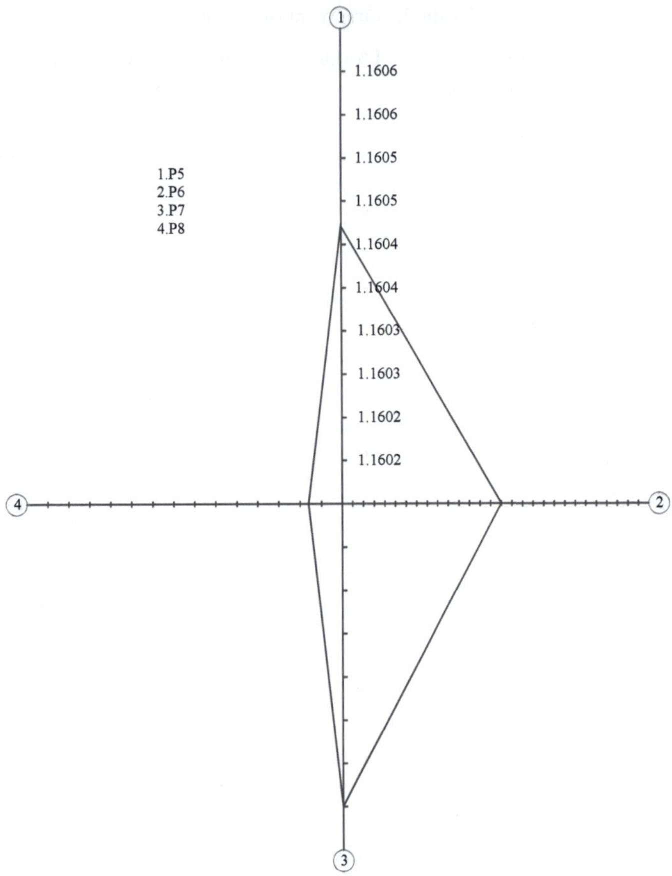

Fig. 8. Spider chart

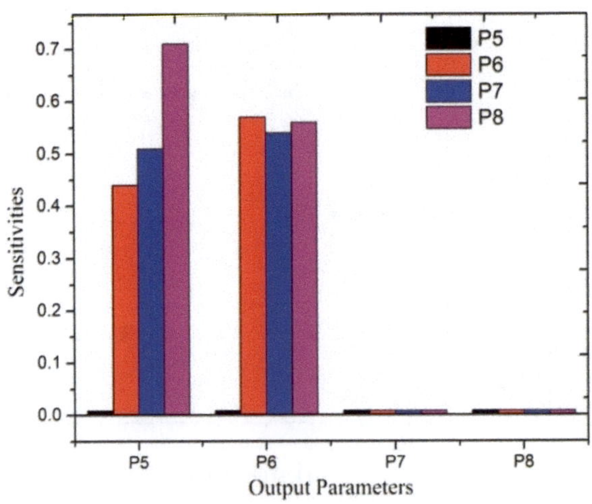

Fig. 9. Sensitivities

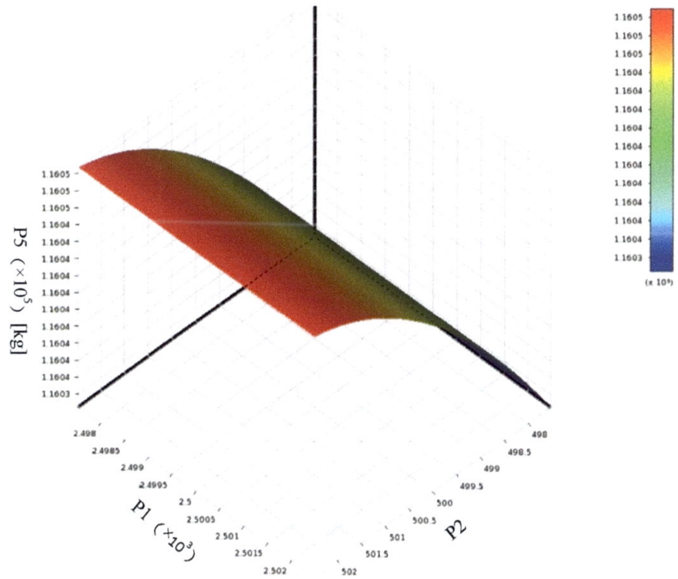

Fig. 10. Response chart for P5

Table 4. Optimization results (unit: mm)

Optimization parameters	P1	P2	P3	P4
Value	2500	500	2000	2000
Lower bound	2497.5	497.53	1997.5	1997.5
Upper bound	2502.5	502.47	2002.5	2002.5

Table 5. Optimization results

Objective	P5 (kg)	P6 (mm)	P7 (MPa)	P8
Calculated minimum	116020	2296.4	235.92	0.91091
Calculated maximum	116060	2298.6	274.45	1.0597

4 Conclusion

This paper describes how to use ANSYS workbench to analyze the structural characteristics of the fan blades and optimize them. The blade safety factor is taken into account when defining the load and calculating the analysis. From the equivalent strain diagram of the blade, it can be seen that the deformation of the blade reaches the maximum at the free end face of the blade, and the strain and stress values before and after the optimization are much smaller than the tensile strength and compressive strength of the material. Therefore, the blade structure design is safe reliable, optimized design is feasible. It is found that it is practicable to optimize the structure of fan blades by using Six Sigma optimization principles.

References

1. X. Cai, *Study on Six Sigma Optimization Design of Faucet Structure* (Xiamen University, 2008)
2. Optimization design of aeroengine products based on Six Sigma design. Gas Turbine Exp. Res. (04), 2 (2015)
3. H. Ji, Y. Qu, Y. Tang, Optimization of the locking structure of plastic parts based on Six Sigma design. Mach. Des. Manuf. (10), 104–106 (2009)
4. Y. Wu, *Information Security Risk Assessment* (Tsinghua University Press, Beijing, 2007), pp. 58–59
5. Z. Huang, C. Liu, *ANSYS Workbench14.0 Super Learning Manual* (People Post Press, Beijing, 2013)
6. X. Zhang, Finite element modeling and lightweight design of 1.5 MW fan blade. J. Liaoning Tech. Univ. (2015)

Analysis on the Strategy of Promoting the Quality of Shared Bicycle Service Based on Evolutionary Game

Yan-liang Zhang(✉) and Liu-yu Wen

School of Management Engineering, Zhengzhou University, Zhengzhou 450001, China
474510734@qq.com

Abstract. In order to enhance the service quality of shared bicycles under the background of popular bicycle market, the evolution path and evolutionary equilibrium of government and enterprise choice game strategy is analyzed by constructing the evolutionary game model of government and enterprise strategy from the perspective of management under the attention of sharing the bike in this paper. Then the evolution strategy of government and enterprise is analyzed by using the evolutionary game theory and the government and enterprise as the main body of the game. Moreover, the stable conditions that can promote the government and enterprise decision-making are discussed. The results show that strengthening the cooperation between government and enterprises, increasing the sharing of bicycle propaganda, strengthening government management awareness, promote the development of consumer standards, and standardize user behavior will help to enhance the quality of shared bicycles.

Keywords: Evolutionary game · Service quality · Shared cycling Strategy selection

1 Introduction

In the present era, the rapid development of social economy increases the environmental pressure, and makes the traffic situation gradually deteriorated. As the product of the "Internet +" era, shared bikes quickly get the favor of the public and quickly throughout the major cities and domestic universities because of its convenience. The emergence of shared bicycles not only meets the need of short trip, but also helps to solve the traffic congestion, environmental pollution and other urban problems, which is more fit the concepts such as green travel, healthy China and public entrepreneurship with innovation and so on.

However, with the popularity of shared bicycles, a series of social problems which have become increasingly prominent are triggered. The emergence of these problems not only exposes the individual quality, moral bottom line and the sense of the rules of the citizens, but also reveals the request of shared bicycle technology and management

G. Q. Huang et al. (eds.), *Proceeding of the 24th International Conference on Industrial Engineering and Engineering Management 2018*,
https://doi.org/10.1007/978-981-13-3402-3_63

[1]. The government and enterprises play a leading role in the management of shared bicycles and improve the rules of the system. At the same time, enterprises need to strengthen technical support. In addition, users should use vehicles more civilized and complying with the moral bottom line. Therefore, the shared bicycle can play a greater role in the public and bring more convenience to travel.

As a complex social system project, sharing the effectiveness of cycling promotion is influenced by government and enterprises [2]. In order to increase their own interests, the government and enterprises tend to deviate from the balance strategy because they only have limited rationality, which means the adjustment of the behavior of both parties will lead to changes in the equilibrium state of the system [3]. In addition, the sharing of bicycles rise time is short, the existing researches seldom concern about interaction the behavior of both government and business, and they also lack the accurate description of the system equilibrium state [4]. Based on this, this paper analyzes the influence factors of the game strategy selection from the perspective of the management of shared cycling, the use of evolutionary game theory, and the government and the enterprise as the main body of the game. Then this paper discusses the measures to improve the service quality of the shared bicycles. Finally, this paper proposes the corresponding suggestions and solutions.

2 Evolutionary Game Model Construction

2.1 Model Basic Assumptions

In the process of sharing bicycles share, it is assumed that the participants of the game are government and enterprise respectively, and assume that both sides are rational economic people, that is, to maximize their own interests. All of them, the government is an important policy maker, guides and advocates, and is responsible for supervision, the government through effective and effective incentives to stimulate other stakeholders to promote the full implementation of shared bicycles [5]. Its goal is to achieve policy objectives and to maximize social benefits through policy development to impact Enterprise decision-making. Business is a shared practice in the implementation of cycling.

2.1.1 Game Analysis of Government and Enterprise Double Matrix

Combined with the actual situation of the model to make the following assumptions:

(1) Both sides of the game are the government (P) and the enterprise (E); and both sides are bounded rational.
(2) The strategy choice is: the government chooses to regulate the cycling market to develop a shared bicycle market system or chooses not to standardize bicycle market; enterprises choose whether to implement the market system or not. The probability of government to choose the regulation is y, the probability of the government chose not to regulation is $1 - y$; the probability of enterprises choose

to implement is x, the probability of enterprises choose not to execute is $1 - x$; x, y are located within the range of $[0, 1]$.

(3) Suppose that the government chooses to subsidize the shared bicycle market to pay the subsidy is B, in which part of the subsidy provided by the government is paid to the enterprise, part of the payment to users, the subsidy paid to users is kB, the payment to the enterprise is $(1 - k)B$, the supervision cost is C_1, the government chooses not to monitor the shared bicycle market when the loss of social resources is D.

Assuming the cost of the firm's implementation is C_2, the proceeds are A_1, the social benefits is A_2; the cost of the enterprise's non-execution is C_2', the proceeds are A_1', and the government chooses to supervise the firm which does not obey the market system will be punished F. In this assumption $C_2' > C_2$.

(4) Supposing that government subsidies are available only when the government chooses to regulate the market and the firm chooses to implement the market system.

(5) Assuming that the government chooses to supervise, the net proceeds from the implementation of the market system will be less than the gains from the firm's implementation of the market system, which means $A_1' - F < A_1$.

2.2 Payment Matrix Construction

Based on the above assumptions, we can get the business and government game payment matrix as the Table 1 shows.

Table I. Government and enterprises on the sharing of the implementation of cycling matrix

		Government strategy	
		Supervise (y)	No supervise $(1 - y)$
Enterprise strategy	Carry out (x)	$A_1 + (1 - k)B - C_2$	$A_1 - C_2$
		$A_2 - C_1 - B$	$A_2 - D$
	No carry out $(1 - x)$	$A_1' - C_2' - F$	$A_1' - C_2'$
		$F - C_1 - kB$	$-D$

Analysis of the matrix in Table 1, available.

2.2.1 Stable Analysis of Enterprise Evolution Strategy

Enterprises take to "carry out" strategy for the effectiveness is:

$$U_{e1} = y[A_1 + (1 - k)B - C_2] + (1 - y)(A_1 - C_2)$$
$$= (1 - k)yB + A_1 - C_2$$

Enterprises take to "no carry out" strategy for the effectiveness is:

$$\begin{aligned} U_{e2} &= y(A_1' - C_2' - F) + (1 - y)(A_1' - C_2') \\ &= -yF + A_1' - C_2' \end{aligned}$$

The average expected return for an enterprise is:

$$\begin{aligned} U_e &= xU_{e1} + (1 - x)U_{e2} \\ &= x[(1 - k)yB + A_1 - C_2] + (1 - x)(-yF + A_1' - C_2') \end{aligned}$$

The dynamic equation of the firm's choice of "carry out" strategy is:

$$\begin{aligned} F(x) &= \frac{dx}{dt} = x(U_{e1} - U_e) \\ &= x(1 - x)\{y[(1 - k)B + F] + A_1 - C_2 - A_1' + C_2'\} \end{aligned} \quad (1)$$

If $.F(x) = 0.$, then we can get that $x_1' = 0$, $x_2' = 0$, and $y = \frac{C_2 + A_1' - A_1 - C_2'}{(1-k)B + F}$.

According to the stability theorem of differential equations, the evolution stabilization strategy can be expressed as satisfying $F(x') = 0$ and $F'(x) < 0$ in steady state [6].

$$F'(x) = (1 - 2x)\{y[(1 - k)B + F] + A_1 - C_2 - A_1' + C_2'\} \quad (2)$$

The evolutionary stabilization strategies of the enterprise group game are analyzed by the following Eqs. (1) and (2).

If $y = y'$, then $F(x) \equiv 0$, and all the equilibrium points are stable state.

If $y \neq y'$, then $x_1' = 0$, $x_2' = 1$ are the two stable state points, the classification of the different situation to discuss, because $0 < k < 1$, so $1 - k > 0$, and because $B > 0$, $F > 0$, so $(1 - k)B + F$ constant than F, So the direct discussion of the different situations is $C_2 + A_1' - A_1 - C_2'$.

(1) If $A_1' - C_2' < A_1 - C_2$, then $y' < 0$, $y > y'$ constant set up, $F'(x_1') > 0$, $F'(x_2') < 0$, so $x_2' = 1$ is a stable equilibrium point.
(2) If $A_1' - C_2' > A_1 - C_2$ then $0 < y' < 1$ because $A_1' - F < A_1$, $C_2' > C_2$, so $C_2 + A_1' - A_1 - C_2' < F$. At this point, there are two cases to discuss:

   ① When $y > y'$, $F'(x_1') > 0$, $F'(x_2') < 0$, so $x_2' = 1$ is a stable equilibrium point.
   ② When $y < y'$, $F'(x_1') < 0$, $F'(x_2') > 0$, so $x_1' = 0$ is a stable equilibrium point.

2.2.2 An Analysis of Government's Evolutionary Stabilization Strategy

The effectiveness of the government's "supervise" strategy is:

$$\begin{aligned} U_{p1} &= x(A_2 - C_1 - B) + (1 - x)(F - C_1 - kB) \\ &= x(A_2 - B + kB - F) + F - C_1 - kB \end{aligned}$$

The effectiveness of the government's "no supervise" strategy is:

$$U_{p2} = x(A_2 - D) + (1 - x)(-D) = xA_2 - D$$

The government's average expected earnings is:

$$\begin{aligned} U_p &= yU_{p1} + (1 - y)U_{p2} \\ &= y[x(A_2 - B + kB - F) + F - C_1 - kB] + (1 - y)(xA_2 - D) \end{aligned}$$

The dynamic equation of the government's choice of "supervise" strategy is:

$$\begin{aligned} F(y) &= \frac{dy}{dt} = y(U_{p1} - U_p) \\ &= y(1 - y)[x(kB - B - F) + F - C_1 - kB + D] \end{aligned} \tag{3}$$

$$F^{'}(y) = (1 - 2y)[x(kB - B - F) + F - C_1 - kB + D] \tag{4}$$

Let $F(y) = 0$, available $y_1' = 0$, $y_2' = 1$, $x' = \frac{C_1 + kB - F - D}{(k-1)B - F}$, using the formula (3), formula (4) on the evolution of the government group game stability strategy as follows:

If $x = x'$, then $F(y) \equiv 0$. It means that all the equilibrium points are stable state.

If $x \neq x'$, then $y_1' = 0$, $y_2' = 1$ are the two stable state points of y. Different cases of x' are discussed.

(1) If $C_1 + kB < D$, then $x' > 1$, $x < x'$ constant set, $F'(y_1') > 0$, $F'(y_2') < 0$, so $y_1' = 1$ is a stable equilibrium point.
(2) If $C_1 + kB > F + D$, then $x' < 0$, $x > x'$ constant set, $F'(y_1') < 0$, $F'(y_2') > 0$, so $y_1' = 0$ is a stable equilibrium point.
(3) If $D < C_1 + kB < F + D$, then $0 < x' < 1$, at this point, is divided into two cases to discuss:

① When $x > x'$, $F'(y_1') > 0$, $F'(y_2') < 0$, so $y_2' = 1$ is a stable equilibrium point.
② When $x < x'$, $F'(y_1') < 0$, $F'(y_2') > 0$, so $y_1' = 0$ is a stable equilibrium point.

2.3 Strategy Dynamic Evolution Trend Analysis

The dynamic relationship between government and enterprise strategy is shown with a coordinate plane, which shows the dynamic and stability of replication between the two sides as shown in Fig. 1 and the dynamic evolution of government and enterprise interaction as shown in Fig. 2. Divided into the following three cases:

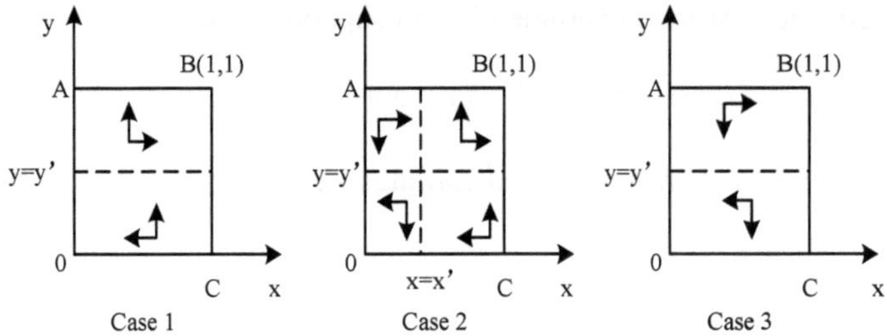

Fig. 1. Government and enterprise replication dynamics and stability

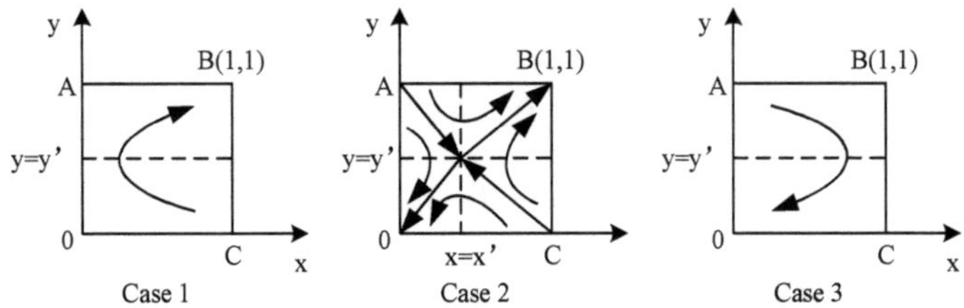

Fig. 2. Dynamic evolution of government and enterprise interactions

Case 1 When $x' > 1$, that is $C_1 + kB < D$, the government's cost of supervision and the subsidy given to the user can not compensate for the loss of government resources.

At this point, B (1, 1) is the stability of business and government game point, that is, the end result of the game for enterprises and the government, respectively, choose to implement and monitor as the best strategy.

Case 2 When $0 < x' < 1$ that is $D < C_1 + kB < F + D$, the loss of social resources at this time is not enough to make up for the cost of government supervision and sharing of bicycles and the amount of subsidies given to users, because the profits of enterprises need to pay a fine.

At this point 0 (0, 0) and B (1, 1) are the stability of the business and government game point, that is, the end result of the game is the government neither monitor the shared bicycle market, nor does the enterprise implement the government to develop the market system; Sharing the bicycle market, the implementation of the government to develop the market system. From the Case 2 in Fig. 2, when the initial state in the A0C area, the system will stabilize the zero point, that is, in this region, the government and enterprises choose not to monitor and choose not to implement as the best strategy; State in the ABC area, the system will stabilize the point B (1, 1), that is, in this region, the government and enterprises were selected to monitor and implement as the best strategy.

Case 3 When $x' < 0$, that is $C_1 + kB > F + D$, the sum of the fines paid by the enterprise and the loss of the social resources is not enough to cover the cost of the government regulation Shared bicycle market and the subsidy given to the user.

At this point, 0 (0, 0) is the stability of business and government game point, that is, the end result of the game for enterprises and the government is concerned, respectively, choose not to implement and not to monitor is the best strategy.

3 Analysis of Game Evolution and Suggestions on Quality Improvement

Based on the above analysis, you can get the following conclusions and recommendations:

(1) When $A_1' - C_2' < A_1 - C_2$, $x_2' = 1$ is the evolutionary stability strategy solution.

The strategy shows that if the firm chooses to implement a bicycle market system. The profit is higher than that the firm chooses not to execute the bicycle market system. The firm will choose the "implementation" strategy. However, in order to encourage the cooperation between the government and the enterprises, the following measures should be taken as a result of the fact that the shared bicycle market is not perfect in the emerging industries.

① To reduce the cost of maintaining the bicycle, that is, the government departments need to do a good job supervision, for those graffiti, deliberate destruction of the behavior must be stopped and make the appropriate punishment to monitor user behavior, in order to ensure the safe operation of bicycles.
② The government should actively cooperate with the enterprise, sharing the bicycle wants to develop by leaps and bounds, must solve the credit system and service standards. The shared bicycle enterprises should be encouraged to share data with government regulators, thereby reducing unwanted contradictions due to information asymmetry.
③ The first priority of government management is to enhance service awareness. Such as the original road planning to ignore the existence of the bike, leading to a lot of cycling chaos. Therefore, the government can manage the cycle of sharing the bike, re-prepared bike lane planning, and improve the infrastructure.

(2) When $C_1 + kB > F + D$, $y_1' = 0$ is a stable equilibrium point.

The strategy shows that if the sum of the fines paid by the firm and the loss of the social resources is not sufficient to cover the cost of the government to monitor the shared bicycle market and the sum of the subsidies given to the consumer. The government will choose "no supervision" strategy. In order to enable the government to benefit from the shared cycling market and play a role. The following measures should be taken:

① The government should increase the publicity of cycling, increase the awareness of the masses, promote the residents of green travel and strengthen the use of shared bikes guidance, thereby reducing the loss of social resources.
② Improve the cost of punishment for enterprise's opportunistic behavior. Enterprises should actively cooperate with the government to standardize the parking point, clear the rights and responsibilities of cycling users, build and improve the

user credit system, and the violation should bring in the government traffic illegal platform. Second, enterprises should improve services, strengthen the quality of cycling control, improve operational service levels and enhance the user experience.

③ The government and enterprises should develop the consumer's own standard for the use of bicycle standards. Encounter man-made damage, is the most common problem encountered in sharing bicycles. Therefore, the government and enterprises should strengthen the restrictions on consumer violations, allowing users can use shared bicycles be more standardized to radically reduce the monitoring cost.

4 Conclusion

This paper firstly uses the evolutionary game method to analyze the shared cycling market background, and then constructs the evolutionary game model of government and enterprise strategy selection, finally analyzes its evolution path and evolutionary equilibrium. The results of the evolution show that strengthening the cooperation between government and enterprises will increase the sharing of bicycle propaganda, enhance the sense of government management, develop consumer standards and standardize user behavior, which are conducive to sharing the quality of bicycle service. In addition, the information asymmetry in the process of government and enterprise cooperation, conflict of interest and moral hazard and other issues are difficult to avoid. In order to reduce the occurrence of these contradictions, both government and enterprises need to strengthen cooperation which can be realized through the specific approach including two steps. First, the government should play a dominant market, and give full play to its regulatory functions. Second, enterprises need to improve the quality of cycling complying with the government to develop the market system. Through the promotion of government and business cooperation, the quality of shared bicycles can be continuously improved to be better for the public.

Acknowledgements. This research is supported by Humanity and Social Science Youth foundation of Ministry of Education of China under Grant #11YJC630291 to Yanliang Zhang.

References

1. Z. Zhang, W. Wu, The status, problems and their development countermeasures and suggestions of sharing bicycles. Mod. Bus., 162–163 (2017). (Chinese)
2. X. Wang, R. Du, S. Ai, Z. Zhanu, The evolution analysis of the community hospitals and patients' behavior selection under the background of telemedicine. Ind. Eng. Manag. **20**(2), 130–137 (2015). (Chinese)
3. X. Pu, Y. Wu, L. Lu, L. Jiang, The analysis of evolution game model and simulation between the productive behavior of the firms and the supervision of the officials. Chin. J. Manag. Sci. (21), 390–396 (2013). (Chinese)

4. W. Zhang, G. Zhou, J. Cao, Analysis on evolutionary game of government regulation mode and enterprise pollution emissions. China Popul. Resour. Environ. **24**(11), 108–113 (2014). (Chinese)
5. M. Cheung, J. Zhuang, Regulation games between government and competing companies: oil spills and other disaster. Decis. Anal. **9**(2), 156–164 (2012)
6. Q.H. Zhu, Y.J. Dou, Evolutionary game model between governments and core enterprises in greening supply chains. Syst. Eng. Theory Pract. **27**(12), 85–89 (2007)

The Study and Construction of Quality Credit Contract Model Between the Enterprise and the Consumer Based on Game Theory

Li Zhou(✉), Bisong Liu, Yan Zhao, and Xianghua Li

China National Institute of Standardization, Beijing, China
zhouli@cnis.gov.cn

Abstract. Quality credit can essentially reflect a contractual relationship of product quality commitments between the enterprise and the consumer. By studying and constructing the evolutionary game model between the enterprise and the consumer, the evolutionary stable strategy in game between the enterprise and the consumer had been analyzed. And through numerical experiments under different situations, the impact of different behaviors on evolutionary results had been analyzed. Accordingly, recommended measures of reference had been proposed to enhance quality integrity management and improve credit level of enterprises.

Keywords: Evolutionary game model · Quality credit · Strategy analysis

1 Introduction

Quality credit is the willingness and ability of enterprise to fulfill its quality commitments. Quality credit can essentially reflect a contractual relationship of product quality between the enterprise and the consumer [1]. Strengthening the construction of quality credit system is a sound strategy for rectifying and standardizing the order of the market economy, and promoting the quality improvement, efficiency increase and upgrading of China's economy and society. It has been proved with facts that, depending on a good image of high quality and high credit, the products have played an irreplaceable role in improving the people's consumer confidence and stimulating the economic growth [2]. Due to the imperfection of the market economy order, some enterprises lack the quality credit consciousness, so that the behaviors of breaching quality promises have occurred from time to time in all fields, e.g. production and sale of counterfeit goods, poor construction, selling poisonous foods, breach of service contract and so on.

As the main body of quality credit, in the face of huge profits brought from cheap costs, the enterprise may become reckless in desperation to sacrifice its own quality credit [3]. Any behaviors of breaching quality promises may not only seriously interfere with the objective law of market economy operation, harm the macroeconomic environment in many ways, but also may seriously affect the safety of the people's lives and properties, directly affect the interests of consumers, and restrict the sound development of socialist market economy. Therefore, in order to strengthen the

G. Q. Huang et al. (eds.), *Proceeding of the 24th International Conference on Industrial Engineering and Engineering Management 2018*,
https://doi.org/10.1007/978-981-13-3402-3_64

construction of quality credit system, it is necessary to objectively analyze a relationship between the enterprise quality credit and the consumer, so as to take any targeted measures for guidance, in order to achieve the effective progress.

2 Basic Assumptions and Parameter Settings of the Game Model Between the Enterprise and the Consumer

Through the analysis of a quality credit relationship between the enterprise and the consumer, this paper studies on the establishment of the game strategy model between the enterprise and the consumer, and analyzes the strategy selection and equilibrium problems on the behavior and interest interdependence between the two main bodies of enterprise and consumer from the evolutionary game model of both parties, so as to improve the policy system and enhance the economic efficiency. To make it easier to study on the game model below, this paper presents the basic assumptions of model, and sets up the core parameters of model.

(1) Game players: There are the three-party players in the game model, i.e. government regulator, enterprise and consumer. In the evolutionary game model, both parties of the game hope to maximize their own gains by minimizing the cost based on the prediction of each other's behaviors and the selection of their own behaviors.
(2) The main functions of government regulator are: To strengthen the supervision of market players; maintain a sound and good market development environment; build a quality credit system; develop the rules and regulations; and supervise and manage the quality credit of enterprises [4]. The cost of government regulation is expressed by C_g. When the government regulates the enterprise with high quality credit, the income obtained by government is expressed by E_g. When the government fails to regulate the enterprise with low production quality, the government's economic loss caused by low quality credit of enterprise is expressed by L_g.
(3) The strategic space for enterprise is expressed by (high quality credit, low quality credit). When the enterprise sells its products at the price V_f, the extraneous income E_f can be obtained by maintaining a high quality credit level. The cost for maintaining the high quality credit level is expressed by C_h; the cost for maintaining the high quality credit level is expressed by C_l, and $C_h > C_l$. When the enterprise is at a low quality credit level, if the government fails to exercise control and supervision, any additional cost is required to be paid. If the government exercises control and supervision, the enterprise will be penalized by the government due to the enterprise's behaviors of breaching quality promises P_g (if any).
(4) The strategic space for consumer is expressed by $\gamma = (\gamma_1, \gamma_2) =$ (complaints, non-complaints). The income obtained by the consumer due to the enterprise's high quality credit is expressed by E_c. The loss caused by the enterprise's low quality credit is expressed by L_c. The cost for complaining to the government regulator about the enterprise's low quality credit problems is expressed by C_c. When the

government exercises control and supervision, if the consumer lodges a complaint, the enterprise will pay the cost R_c to the consumer as required, and $R_c > C_c$. When the government fails to exercise control and supervision, the consumer will be dissatisfied with the government's behavior, so the loss of reputation suffered by the government is expressed by S_g.

3 Onstruction and Analysis of the Evolutionary Game Model Between the Enterprise and the Consumer

Under actual operating conditions of the market, the enterprise and consumer have a certain proportion of groups to choose the same strategy, respectively [5]. The evolutionary stable strategy in the game between the enterprise and the consumer is to study the evolution of behavior between the groups in enterprise and consumer. Solving the evolutionary stable strategy in the game is to replicate the dynamic state of stability.

3.1 Construction of the Game Model Between the Enterprise and the Consumer

The proportion of high-quality credit level maintained by the enterprise is expressed by $\rho(0 \leq \rho \leq 1)$. The proportion of low-quality credit level maintained by the enterprise is expressed by $1 - \rho$. The proportion of complaints from customers is expressed by $\sigma(0 \leq \sigma \leq 1)$. The proportion of non-complaints from customers is expressed by $1 - \sigma$. Therefore, the income matrix of evolutionary game strategy between the enterprise and the consumer is constructed as follows (Table 1).

Table 1. Income matrix of the evolutionary game strategy between the enterprise and the consumer

Income matrix		Enterprise	
		High production quality credit	Low production quality credit $1 - \rho$
Consumer	Complaints	$(E_c - C_c, V_f + E_f - C_h)$	$(R_c - L_s - C_c, V_f - C_l - (R_c + P_g))$
	Non-complaints $1 - \sigma$	$(E_c, V_f + E_f - C_h)$	$(-L_s, V_f - C_l - P_g)$

3.2 Replicated Dynamic Equation of Game Between the Enterprise and the Consumer

The expected returns of strategies for "high production quality credit" and "low production quality credit" as selected by the enterprise are expressed by U_{β_1} and U_{β_2}, respectively. The average expected return is expressed by $\overline{U_\beta}$. The expression of U_{β_1} and U_{β_2} and is as follows:

$$U_{\beta_1} = \sigma\left(V_f + E_f - C_h\right) + (1-\sigma)\left(V_f + E_f - C_h\right) = \left(V_f + E_f - C_h\right)$$
$$U_{\beta_2} = \sigma\left(V_f - C_l - \left(R_c + R_g\right)\right) + (1-\sigma)\left(V_f - C_l - P_g\right) = -\sigma R_c + V_f - C_l - P_g$$
$$\overline{U} = \rho U_{\beta_1} + (1-\rho)U_{\beta_2} = \rho\left(V_f + E_f - C_h\right) + (1-\rho)\left(-\sigma R_c + V_f - C_l - P_g\right)$$

The growth rate of strategic quantity of enterprise's "high production quality credit" is equal to $U_{\beta_1} - \overline{U}_\beta$. t represents the time. Therefore, the replicated dynamic equation of enterprise is as follows:

$$\begin{aligned} F(\rho) &= \frac{d\rho}{dt} = \rho\left(U_{\beta_1} - \bar{U}_\beta\right) \\ &= \rho(1-\rho)\left(\sigma R_c + E_f - C_h + C_l + P_g\right) \end{aligned}$$

Similarly, the expected returns of strategies for "complaints" and "non-complaints" selected by the consumer are expressed by and, respectively. The average expected return is expressed by $\overline{U}_\gamma$. The expression of U_{γ_1} and U_{γ_2} and $\overline{U}_\gamma$ is as follows:

$$\begin{aligned} U_{\gamma_1} &= \rho(E_c - C_c) + (1-\rho)(R_c - L_c - C_c) \\ U_{\gamma_2} &= \rho E_c + (1-\rho)(-L_c) = \rho(E_c + L_c) - L_c \\ \overline{U}_\gamma &= \sigma U_{\gamma_1} + (1-\sigma)U_{\gamma_2} \\ &= \sigma[\rho(E_c - C_c) + (1-\rho)(R_c - L_c - C_c)] + (1-\sigma)[\rho(E_c + L_c) - L_c] \end{aligned}$$

The growth rate of strategic quantity of "complaints" from the consumer is equal to $U_{\gamma_1} - \overline{U}_\gamma$ t represents the time. Therefore, the replicated dynamic equation of consumer is as follows:

$$\begin{aligned} F(\sigma) &= \frac{d\sigma}{dt} = \sigma\left(U_{\gamma_1} - \overline{U}_\gamma\right) \\ &= \sigma(1-\sigma)[(1-\rho)R_c - 2L_c - C_c] \end{aligned}$$

According to the above analysis, a two-dimensional power system (L) can be, namely:

$$\begin{cases} \frac{d\rho}{dt} = \rho(1-\rho)\left(\sigma R_c + E_f - C_h + C_l + P_g\right) \\ \frac{d\gamma}{dt} = \sigma(1-\sigma)[(1-\rho)R_c - 2L_c - C_c] \end{cases}$$

3.3 Analysis of Evolutionary Stable Strategy in Game Between the Enterprise and the Consumer

- Analysis of equilibrium point and stability of evolutionary stable strategy

First of all, it can be solved by the replicated dynamic equation that there are the five local equilibrium points existing in the game between the enterprise and the consumer, i.e. $A_1(0,0), A_2(0,1), A_3(1,0), A_4(1,1)$, $A_5\left(\frac{C_h-C_l-E_f-P_g}{R_c}, \frac{R_c-2L_c-C_c}{R_c}\right)$.

Combined with the Jacobian matrix, the a_{11}, a_{12}, a_{21} and a_{22} values can be obtained at the five local equilibrium points (as shown in Table 2):

Table 2. A_{11}, A_{12}, A_{21} and A_{22} values obtained at the local equilibrium points

Local equilibrium points	a_{11}	a_{12}	a_{21}	a_{22}
$A_1(0,0)$	$E_f - C_h + C_l + P_g$	0	0	$R_c - 2L_c - C_c$
$A_2(0,1)$	$R_c + E_f - C_h + C_l + P_g$	0	0	$-(R_c - 2L_c - C_c)$
$A_3(1,0)$	$-(E_f - C_h + C_l + P_g)$	0	0	$-2L_c - C_c$
$A_4(1,1)$	$-(R_c + E_f - C_h + C_l + P_g)$	0	0	$2L_c + C_c$
$A_5\left(\frac{C_h-C_l-E_f-P_g}{R_c}, \frac{R_c-2L_c-C_c}{R_c}\right)$	0	M	N	0

where

$$M = -\frac{(C_h-C_i-E_f-P_g)[R_c-(C_h-C_i-E_f-P_g)]}{R_c},$$
$$N = \frac{-(R_c-2L_c-C_c)(2L_c+C_c)}{R_c}$$

If $R_c + E_f - C_h + C_l + P_g < 0$ and $R_c - 2L_c - C_c < 0$, the equilibrium point of evolutionary stable strategy between the enterprise and the consumer is (0, 0)

If $R_c + E_f - C_h + C_l + P_g < 0$ and $R_c - 2L_c - C_c > 0$, the equilibrium point of evolutionary stable strategy between the enterprise and the consumer is (0, 1).

If $E_f - C_h + C_l + P_g > 0$, the equilibrium point of evolutionary stable strategy between the enterprise and the consumer is (1, 0).

- Model parameter analysis and system phase diagram

If $E_f - C_h + C_l + P_g > 0$ and $R_c - 2L_c - C_c > 0$, the equilibrium point of evolutionary stable strategy between the enterprise and the consumer is (1, 0), namely, all the enterprises in the market can maintain the high quality credit level; by this time, the consumer is not required to lodge a complaint, and the market can be operated under optimum conditions [6]. In this case, the stability analysis of equilibrium point of evolutionary stable strategy between the enterprise and the consumer is shown in Table 3.

Table 3. Stability analysis results of equilibrium point of evolutionary stable strategy

Equilibrium point	trJ	$\det J$	Results
$A_1(0,0)$	+	+	Unstable point
$A_2(0,1)$	+	−	Unstable point
$A_3(1,0)$	−	+	ESS
$A_4(1,1)$	+	+	Unstable point
$A_5\left(\frac{C_h-C_l-E_f-P_g}{R_c}, \frac{R_c-2L_c-C_c}{R_c}\right)$	0	+	Saddle point

The dynamic process of interaction between the enterprise and the consumer is shown in Fig. 1.

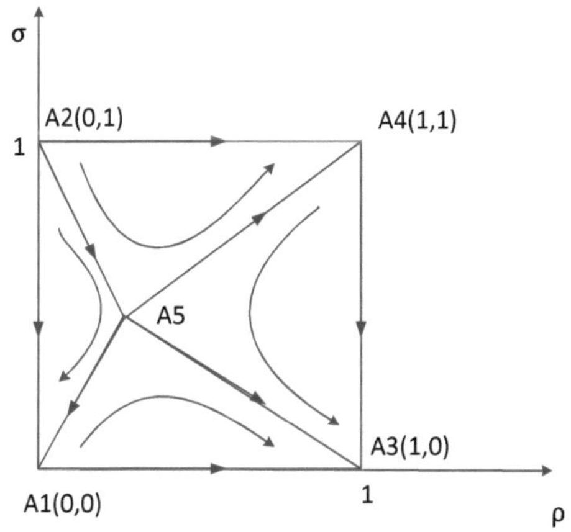

Fig. 1. Dynamic process of interaction between the enterprise and the consumer

From the expression of saddle point, the relevant parameters will be changed, causing the displacement of saddle point, and then regulating and controlling the evolutionary direction of strategy between the enterprise and the consumer.

4 Numerical Experiment and Result Analysis of Evolutionary Model Between the Enterprise and the Consumer

According to the constraint conditions and replicated dynamic equations in the evolutionary stable strategy in game between the enterprise and the consumer, the numerical experiment can be conducted for the analysis of evolutionary stable strategy in game between the enterprise and the consumer by using the Matlab tool, in order to analyze the impact of parameter changes on evolutionary results, e.g. the proportion of initial groups in certain strategies selected between the enterprise and the consumer (ρ, σ), the compensation paid by the enterprise to the consumer (R_c), the cost for maintaining the high quality credit level by the enterprise (C_h), the cost for maintaining the low quality credit level by the enterprise (C_l), the income obtained by the consumer resulted from the high quality credit of enterprise (E_f), etc.

4.1 Impact of the Cost for Maintaining the High Quality Credit Level by the Enterprise (C_h) on Evolutionary Results

The cost for maintaining the high quality credit level by the enterprise (C_h) is taken as 1.0 or 3.5, respectively. It is necessary to contrastively analyze the evolution of cost for maintaining the high quality credit level by the enterprise (C_h) in different cases of initial values. The numerical experimental results are shown in Fig. 2. With an increase in the cost for maintaining the high quality credit level by the enterprise (C_h), the longer it takes for the curve to converge to the stable state, namely, the time required for both parties to reach a steady state is increased, and the increase in the cost for maintaining the high quality credit level by the enterprise has shown an adverse effect [7].

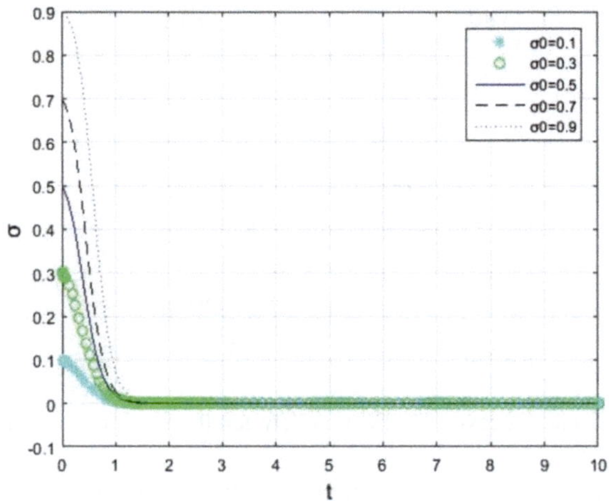

(a) $C_h = 1.0$

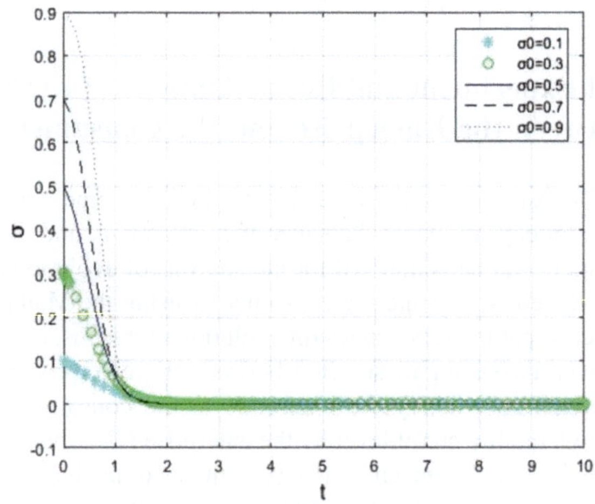

(b) $C_h = 3.5$

Fig. 2. Impact of the cost for maintaining the high quality credit level by the enterprise (C_h) on evolutionary results

4.2 Impact of the Cost for Maintaining the Low Quality Credit Level by the Enterprise on Evolutionary Results

The cost for maintaining the low quality credit level by the enterprise (C_l) is taken as 0.5 or 3.0, respectively. It is necessary to contrastively analyze the evolution of cost for maintaining the low quality credit level by the enterprise (C_l) in different cases of initial values. The numerical experimental results are shown in Fig. 3. With an increase in the cost for maintaining the low quality credit level by the enterprise (C_l), the shorter it takes for the curve to converge to the stable state, contrary to the state presented by the curve with the increase in cost for maintaining the high quality credit level by the enterprise (C_l), namely, the time required for both parties to reach a steady state is shortened, and the increase in the cost for maintaining the low quality credit level by the enterprise has shown a positive effect [8].

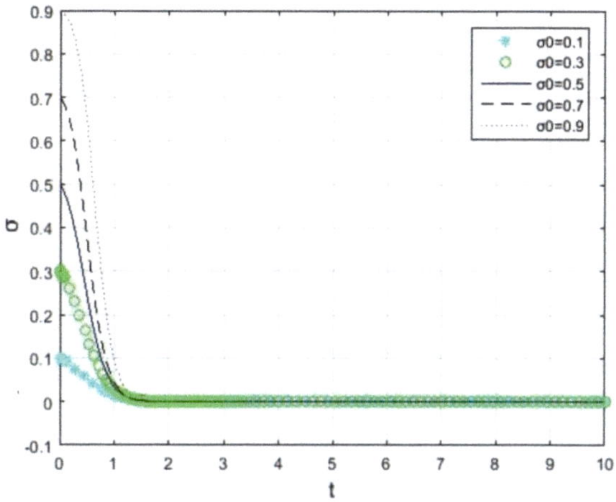

(a) $C_l = 0.5$

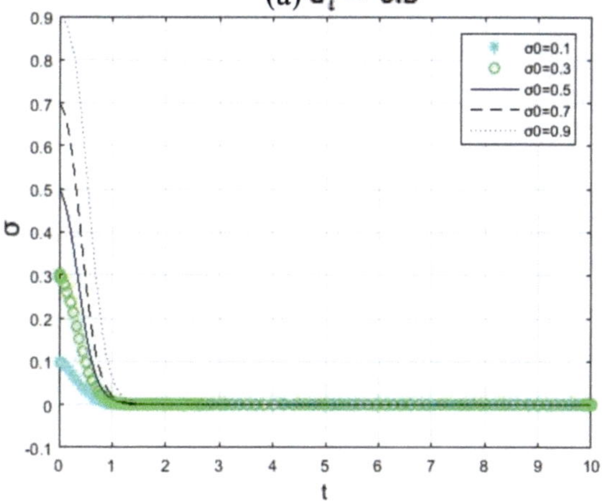

(b) $C_l = 3.0$

Fig. 3. Impact of the cost for maintaining the low quality credit level by the enterprise (C_l) on evolutionary results

4.3 Impact of the Income Obtained by the Consumer Resulted from the High Quality Credit of Enterprise on Evolutionary Results

The income obtained by the consumer resulted from the high quality credit of enterprise (E_f) is taken as 0.3 or 10.3, respectively. It is necessary to contrastively analyze the evolution of income obtained by the consumer resulted from the high quality credit of enterprise (E_f) in different cases of initial values. The numerical experimental results are shown in Fig. 4. An increase in the income obtained by the consumer resulted from the high quality credit of enterprise will promote both parties to reach a steady state in game, and showing a good positive effect [9].

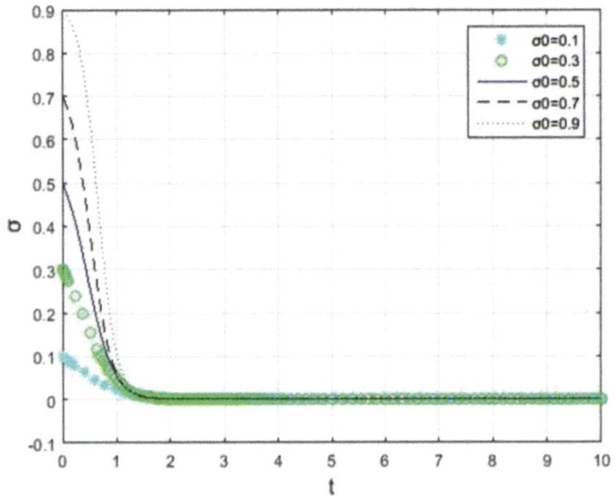

(a) $E_f = 0.3$

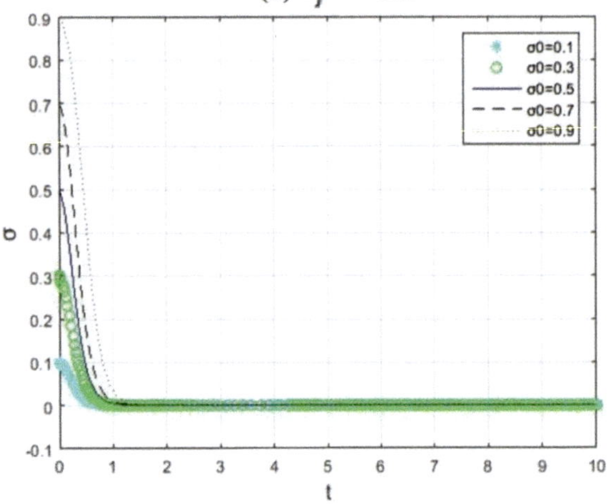

(b) $E_f = 10.3$

Fig. 4. Impact of the income obtained by the consumer resulted from the high quality credit of enterprise (E_f) on evolutionary results

4.4 Impact of the Government's Punishment for the Enterprise's Behaviors of Breaching Quality Promises (P_g) on Evolutionary Results

The government's punishment for the enterprise's behaviors of breaching quality promises (P_g) is taken as 0.1 or 10.0, respectively. It is necessary to contrastively analyze the evolution of government's punishment for the enterprise's behaviors of breaching quality promises (P_g) in different cases of initial values. The numerical experimental results are shown in Fig. 5. With an increase in the government's punishment for the enterprise's behaviors of breaching quality promises (P_g), the shorter it takes for the curve to converge to the stable state, namely, the time required for both

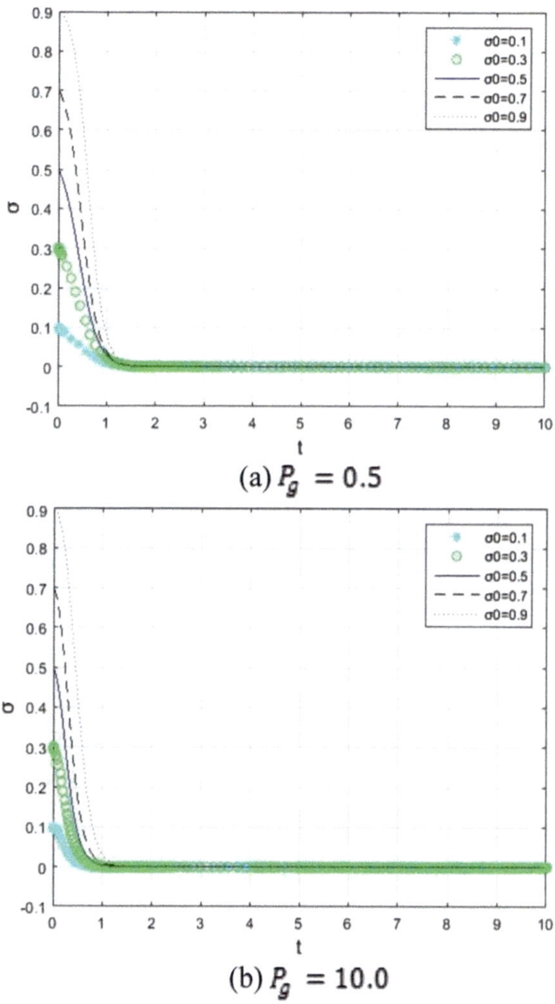

Fig. 5. Impact of the government's punishment for the enterprise's behaviors of breaching quality promises (P_g) on evolutionary results

parties to reach a steady state is shortened. Therefore, increasing the intensity of government's punishment for the enterprise's behaviors of breaching quality promises will help to promote the game between the enterprise and the consumer to achieve the ideal state [10].

4.5 Impact of the Loss to the Consumer Caused by the Enterprise's Low Quality Credit on Evolutionary Results

The loss to the consumer caused by the enterprise's low quality credit (L_c) is taken as 0.3 or 2.8, respectively. It is necessary to contrastively analyze the evolution of loss to the consumer caused by the enterprise's low quality credit (L_c) in different cases of initial values. The numerical experimental results are shown in Fig. 6. In the case of a

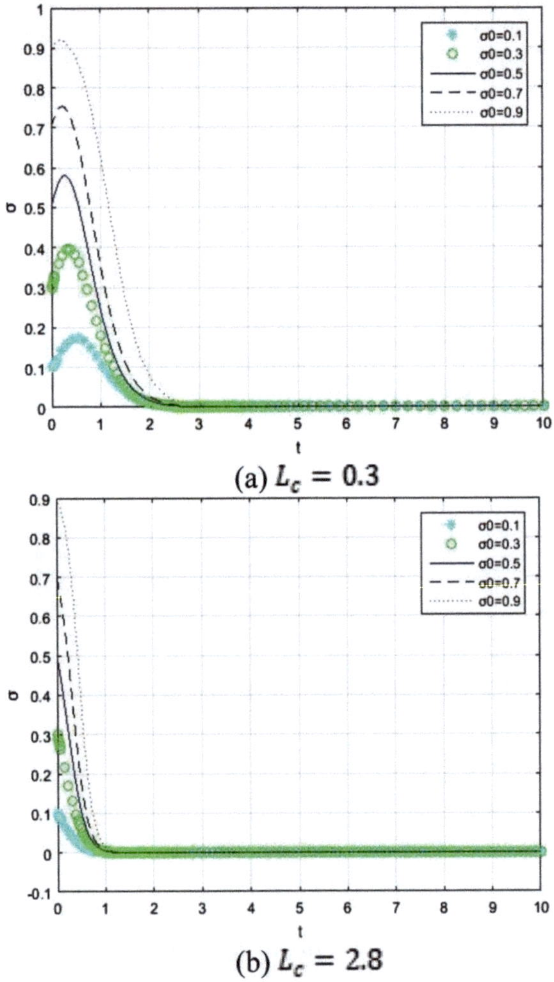

Fig. 6. Impact of the loss to the consumer caused by the enterprise's low quality credit (L_c) on evolutionary results

relatively low proportion of enterprise's high quality credit, the consumer will not quickly cancel the complaint against the enterprise because of the loss to the consumer caused by the enterprise's low quality credit. On the contrary, the consumer will more worry about the overall quality credit problems, showing a lack confidence and trust in the enterprise's improvement of its own quality credit problems [11]. And the higher the initial proportion of choosing the complaints among the consumers, the longer it takes for the curve to reach a stable state.

5 Summary

From the point of view of evolutionary analysis in different cases of taking values by building the evolutionary game model between the enterprise and the consumer, the enterprise isn't positive about the cost expenditure for maintaining the high-quality credit. However, an increase in the income obtained by the consumer resulted from the high quality credit of enterprise will promote both parties to reach a steady state in game, and showing a good positive effect. Meanwhile, increasing the intensity of government's punishment for the enterprise's behaviors of breaching quality promises will help to promote the game between the enterprise and the consumer to achieve the ideal state [12]. From this, in order to strengthen the enterprise quality credit, it is necessary to fundamentally make the enterprise aware of the long-term significance and sustainable market-oriented effect of construction of social credit system to/on the development of enterprise; increase the government's supervision and guidance functions/effects; jointly create a social atmosphere of honesty and trustworthiness together with all circles of the society and customers; form a long-term social governance and joint defense mechanism; make the enterprise aware of a large "development dividend" brought about by maintaining the high quality credit; thus form a virtuous circle; and fundamentally improve the achievements in the construction of quality credit system.

References

1. Z. Li, R. Ye, K. Xian, Y. Wang, *Theory and Practice of Quality Credit System Construction*, 1st edn. (China Zhijian Publishing House, 2010)
2. G. Li, X. Wei, *Definition, Measurements and Economic Consequences of Corporate Social Responsibility: A Survey on Theories of Corporate Social Responsibility, Accounting Research*, vol. 08 (2014), pp. 33–40
3. W. Li, H. Xiao, Logics for Corporate Social Responsibility, China Industrial Economics, vol. 10 (2011), pp. 87–97
4. R. Narasimhan, S. Ghosh, D.A. Mendez, Dynamic model of product quality and pricing decisions on sales response. Decis. Sci. **24**(5), 893–908 (2007)
5. Z. Li, Technology research on quality credit evaluation of the enterprise with real-name system, in *Standard Science*, vol. 05 (2013), pp. 77–81
6. S.A. Starbird, Designing foodsafety regulations: the effect of inspection policy and penalties for non-compliance on food processor behavior. J. Agric. Resour. Econ. **25**(3), 615–635 (2006)

7. K.J. Arrow, *Benefits-Cost Analysis in Environmental Health and Safety Regulation: A Statement of Principles* (Washington D.C.: The AEI Press, 1996)
8. E. Millstone, P.V. Zwanenberg, The evolution of food safety policy–making institutions in the UK, EU and Codex Alimentarius. Soc. Policy Adm. **36**(6), 593–609 (2003)
9. Coates, Cost-benefit analysis of financial regulation: case studies and implications. Yale Law J., 124–125 (2015)
10. J.T. Bennett, Introduction to cost-benefit analysis. Mon. Labor Rev. **96**(8), 94–95 (2016)
11. GB/T 31863-2015, Evaluation index of enterprise quality credit
12. L. Yang, J. Wu, A research on the credit rating to SMEs in China based on the perspective of statistics and clustering. in *Journal of Sichuan University* (Social Science Edition), vol. 06 (2014) pp. 89–97

Patent Quality Evaluation Model Based on Hesitant Fuzzy Soft Set

Li Zhang, Qian Li(✉), and Jian-li Yu

School of Management Engineering, Zhengzhou University of Aeronautics, Zhengzhou 450046, China
894483752@qq.com

Abstract. The existing evaluation methods of patent quality take expert scoring as the principal thing. Each expert is required to score all the indexes and to assign the corresponding weight according to the relative importance of the index. But index system generally involves multiple indicators in different fields. Experts can not make accurate evaluation of indicators beyond their research fields, and fuzzy indicators also make experts hesitate and wander between several evaluation values. Hesitant fuzzy soft sets can more accurately describe the fuzzy nature of things, have no restrictions on optional objects, and take full account of the degree of hesitation of expert decision making. Accordingly, this paper proposes a evaluation model of patent quality based on hesitant fuzzy soft sets and gives specific evaluation steps.

Keywords: Evaluation index system · Hesitant fuzzy soft set
Patent · Patent quality

1 Introduction

As the core of intellectual property, patent contributes particularly to economic growth and technological progress. In terms of quantity, China has become the patent great country, but the phenomenon of garbage patent, problem patent and patent monster has come out in an endless stream, which has aroused great concern to the quality of patent in all circles of society [1]. According to a large number of patent information shows that the distribution of patent value is highly non-uniform. Most patents just make little improvement on the basis of the original technology, only a few patents have a great innovation, and hold tremendous economic value. This makes people realize that the production of economic benefits is not due to the number of patents but the quality of patents, so the demand for quality evaluation of patents has increased dramatically [2].

Funded projects: 1. A quality monitoring economic model based on feedforward adjustment and its performance evaluation (172102210078); 2. Economic model of self correlation process quality monitoring and its performance evaluation (17A630069); 3. Multiple objective optimization design based on the economic statistics of PLD process quality control (71672209); 4. The smart phone surface of defect monitoring and real time quality diagnosis of manufacturing process (U1604262).

G. Q. Huang et al. (eds.), *Proceeding of the 24th International Conference on Industrial Engineering and Engineering Management 2018*,
https://doi.org/10.1007/978-981-13-3402-3_65

On the study of the evaluation index system, some research institutions and scholars abroad have taken the lead in establishing some index system for evaluating the quality of patent. For example, The United States CHI and the National Science Foundation (NSF) jointly developed the first patent evaluation index system all over the world [3]. Pantros IP has made a comprehensive assessment of the quality of single piece patent from three aspects of law, commerce and technology [4]. In comparison, the research on the quality evaluation of patent in China started late. Liu et al. [5] appraised the quality of patent from the patent technology. Zhang and Hu [6] took into account the influence of technology, competition, market, law and other factors on the quality of patents. Feng et al. [7] evaluated the quality of single piece patents by using the analytic hierarchy process of multiobjective decision based on qualitative analysis and quantitative analysis. In order to reduce the error caused by the qualitative analysis and fuzzy problems, Ren et al. [8] combined Delphi method with fuzzy comprehensive evaluation to evaluate the patent quality. State Intellectual Property Office and the China Technology Exchange calculated patent value by using expert scoring method [9].

Under literature research, finding that the evaluation index system generally covers a number of indicators in different aspects, some of which can be accurately quantified according to the relevant patent information. The high complexity of the patent makes most of the indicators vague, so these indicators are difficult to describe with accurate results. Therefore, the fuzzy evaluation method based on expert scoring is mainly used in the evaluation of patent quality at present. But it can not take into account the hesitancy of the expert scoring and the selectivity of the expert for the evaluation index. In 2013, Babitha et al. [10] proposed the hesitant fuzzy soft set theory on the basis of the related study of the fuzzy set, and gave the definition and basic operation properties of the hesitant fuzzy soft set. It is widely used to solve the hesitant fuzzy decision problem with multiple attributes, and receives good performance. Inspired by the above application research, this paper attempts to apply the theory and method of hesitant fuzzy soft set to the process of patent quality evaluation, and proposes a patent quality evaluation model based on hesitant fuzzy soft set. The membership value of elements in hesitant fuzzy soft set is more than one, which can more accurately describe the fuzziness of indicators, and allow experts to choose indicators according to their research fields, and give more than one hesitant score for each index. The model makes the evaluation process more practical and the evaluation results are more credible.

2 Evaluation Index System

Technology is the core and essence of the patent, and it can reflect the intrinsic value and quality of a patent. Some scholars believe that high quality patents must achieve essential innovation on the basis of current technology. Other scholars think that the quality of the patent means the progressiveness and importance of the technology in the patent invention itself. So the patent quality must be explored to the technical level of the patent.

The economic value of a patent refers to the sum of the economic value produced by a patent technology in the process of market use. It is the core content of the state and the enterprise pay attention to. The original purpose of applying for a patent is to obtain the authorization of the patent, but the ultimate goal is to put the patent technology into practice, to transform into productivity, and to generate economic benefits, and contribute to the society. If a patent has high technology level, but its implementation is difficult and its economic efficiency is low, and it is difficult to achieve industrialization, such a patent can not be called a high-quality patent. Therefore, the consideration of the economic value of the patent is an indispensable part of the evaluation of the patent quality [11].

The protection of the patent right is mainly embodied in the actual legal effect of the patent right. The patent right is the monopoly enforcement power within a certain period of time granted by the state to the owner of the patent according to law. It can ensure that the patented technology is protected reasonably and effectively in the process of transfer and diffusion, and provide strong legal protection for the innovation subjects or patent holders. But the authorized patent does not always have legal effect, it is faced with the risk of ineffectiveness, infringement and so on at any time. When the period of protection is reached, the technology can be used without compensation, and the value of the transaction is lost [12].

By analyzing and studying the relevant literature of evaluation index system at home and abroad and based on the requirements of the above index selection, this paper sets up an evaluation index system including 3 dimensions of technical level, economic value and protection of rights, and a total of 19 two level indicators, specific as Table 1.

Table 1. Evaluation index system

Evaluation object	First level index	Two level index
Patent value	A Technological level	A1 Innovation degree (m)
		A2 Technical life cycle (q)
		A3 Application scope of Technology (m)
		A4 Technical correlation coefficient (q)
		A5 Scientific correlation coefficient (q)
		A6 Substitutability (m)
		A7 Technical defensive forces (m)
	B Economic value	B1 Technical maturity (m)
		B2 Market demand (m)
		B3 The remaining validity period of a patent (q)
		B4 Conform to the degree of policy orientation (m)
		B5 Supply basis of raw materials (m)
	C Right protection	C1 Independence of patent (m)
		C2 The number of Patent family (q)
		C3 The number of independent claim (q)
		C4 The situation of instructions support claims document (m)
		C5 Patent infringement litigation (q)
		C6 Patent invalidity litigation (q)
		C7 Status of license implementation (q)

“q” means: clear conceptual indicators, “m” means: fuzzy conceptual indicators

3 Patent Quality Evaluation Model Based on Hesitation Fuzzy Soft Set

3.1 Basic Concepts of Hesitant Fuzzy Soft Sets

Definition 1 [13] Given a fixed set X, then the hesitant fuzzy set H on the X is defined as:

$$H = \{\frac{h_H(x)}{x} | x \in X\} \tag{1}$$

The set $h_H(x)$ is made up of several different values on the interval [0, 1], which indicates that the element x in X belongs to a number of possible membership degrees of the set H.

$h_H(x)$ is called a hesitant fuzzy element, and the number of values in different hesitant fuzzy elements may be different and disorderly. In the hesitant fuzzy element $h_H(x)$, the number of values is expressed with $l(h_H(x))$, and arrange these values in ascending order. Counting from the back to the j, this value is recorded as $h_H^{\sigma(j)}(x)$.

Definition 2 [13] Set $h(x), h_1(x), h_2(x)$ as three hesitant fuzzy sets, then:
$h_1(x) \cup h_2(x) = \{h \in (h_1(x) \cup h_2(x)) | h \geq \max(h_1^-, h_2^-)\}$; 其中,
$h_1^- = \min h_1(x), h_2^- = \min h_2(x)$.

Definition 3 [13] Let U be an initial universe set, E be a set of parameters. Let $HF(U)$ represents be all hesitant fuzzy sets over U. Set $A \subseteq E, \tilde{F} : A \to HF(U)$ be a mapping, and then a pair $(\tilde{F}, A)$ is called a hesitant fuzzy soft set on the U.

In other words, a hesitant fuzzy soft set of U is an attribute cluster of hesitant fuzzy sets on the domain U.

Definition 4 [14] For two fuzzy soft sets $(\tilde{F}, A)$ and $(\tilde{G}, B)$ over a common universe (U, E), and $A, B \subseteq E$. Define $\tilde{H} : A \cup B \longrightarrow HF(U)$ is a mapping, as long as $e \in A \cup B \neq \varphi$, then

$$\tilde{H}(e) = \begin{cases} \tilde{F}(e), & if \quad e \in A - B, \\ \tilde{G}(e), & if \quad e \in B - A, \\ \tilde{H}(e), & if \quad e \in A \cap B. \end{cases}$$

If $\tilde{H}(e) = \tilde{F}(\mathrm{e}) \cup \tilde{G}(e)$, $(\tilde{H}, A \cup B)$ is the generalized combination of $(\tilde{F}, A)$ and $(\tilde{G}, B)$, written $(\tilde{F}, A) \tilde{\cup} (\tilde{G}, B)$.

The generalized union and generalized intersection of hesitant fuzzy soft set can be extended to n hesitant fuzzy soft sets.

The generalized union of n hesitant fuzzy soft sets is recorded as:

$$(\tilde{F}, A) = \tilde{\cup}_{i=1}^{n} (\tilde{F}_i, A_i),\ A = \overset{n}{\underset{i=1}{U}} A_i,\ \tilde{F}(e) = \underset{1 \leq i \leq n, e \in A_i}{U} \tilde{F}_i(e);$$

Definition 5 [14] The characteristic function of setting $x, y \geq 0$, $x \geq y$ is defined as:

$$f(x,y) = \begin{cases} 1, & if \;\; x > y, \\ 0.5, & if \;\; x = y, \\ 0, & others. \end{cases}$$

Definition 6 [14] Let h_1 and h_2 are two hesitant fuzzy elements, $l(h_1)$ and $l(h_2)$ respectively represent the number of values in h_1 and h_2. Let $l_{\max} = \max\{l(h_1), l(h_2)\}$, and the probability that h_1 is greater than h_2 is:

$$p(h_1 \succeq h_2) = \frac{1}{l_{\max}} \sum_{j=1}^{l_{\max}} f(h_1^{\sigma(j)}, h_2^{\sigma(j)}) \tag{2}$$

Definition 7 [14] Let $A = \{e_1, e_2, \ldots e_m\}$, $(\tilde{F}, A)$ is a hesitant fuzzy soft set over $U = \{u_1, u_2, \ldots, u_n\}$. Generalized comparison table of $(\tilde{F}, A)$ has the equal number of rows and the number of columns. The elements in the table are expressed as C_{ij}, and the calculation formula is as follows:

$$\begin{aligned} & C_{ij} = \sum_{k=1}^{m} p(h_{\tilde{F}(e_k)}(u_i) \succeq h_{\tilde{F}(e_k)}(u_j)), \quad i, j = 1, 2, \ldots, n. \\ & 0 \leq C_{ij} \leq m, \; C_{ii} = 0.5m \end{aligned} \tag{3}$$

m is the number of attributes in the $(\tilde{F}, A)$.

Definition 8 [14] Let $\lambda > 0$, h is a hesitant fuzzy element, and the results of the multiplication of h and λ are recorded as:

$$\lambda \cdot h = \cup_{\gamma \in h} \{\lambda \gamma\} \tag{4}$$

When two hesitant fuzzy elements $H_1 = \{\frac{h_{H_1}(x)}{x} | x \in X\}$ and $H_2 = \{\frac{h_{H_2}(x)}{x} | x \in X\}$ are compared, the number of their values must be the same. But in most cases, $l(h_{H_1}(x)) \neq l(h_{H_2}(x))$. In order to make the length of H_1 and H_2 equal, it is necessary to extend the hesitant fuzzy element with short length. The extension rule [15, 16] is adding the minimum of the short hesitant fuzzy element.

3.2 The Concrete Steps of the Evaluation Method

Setting $U = \{u_1, u_2, \ldots, u_n\}$ is a collection of patents that need to be evaluated, $E = \{e_1, e_2, \ldots, e_m\}$ is a collection of the evaluation index of patent quality, $w = (w_1, w_2, \ldots, w_m)$ is the weight vector of index, and $D = \{d_1, d_2, \ldots, d_s\}$ is a collection of experts who evaluate the quality of the patent. s experts use multiple attribute group decision making methods to evaluate the quality of n patents with m attributes. According to the conceptual features of the two level index, this paper divides it into clear concept index and fuzzy concept index. In Table 1, the clear concept index is labeled with "q", and the fuzzy concept index is labeled with "m".

1. Score of clear concept index

Clear concept indexes scoring method is divided into two steps. First of all, a defined number of H_{ij}^{k} is used to evaluate the clear concept. For example, technical relevance is used to evaluate its technological creativity, it can be represented by the number of related patents (H_{ij}^{k}) that are referenced in the patent to be evaluated; Then, according to the actual situation, the quantity H_{ij}^{k} is fuzzed, and get the membership value $h_{ij}^{k}(i = 1, 2, \ldots, n;\ k = 1, 2, \ldots, s)$. j is the sequence number of clear indicators in $E = \{e_1, e_2, \ldots, e_m\}$.

2. The score of fuzzy concept index

Set $A_k \subseteq E$, and the number of evaluation indexes in A_k is expressed with $l(A_k)$. Each evaluation expert $d_k \in D(k = 1, 2, \ldots, s)$ could select a number of fuzzy concepts $A_k = \left\{e_1^k, e_2^k, \ldots, e_{l(A_k)}^k\right\}$ $(k = 1, 2, \ldots, s)$ in the index system according to his professional expertise, and then evaluate the quality of a single piece of patent. Every expert $d_k(k = 1, 2, \ldots, s)$ evaluates each fuzzy index in the attribute set A_k of the patent $u_i(i = 1, 2, \ldots, n)$ and gives more than one score. Then the hesitant fuzzy decision matrix $R = \left(h_{ij}^{k}\right)_{m \times n}$ $(k = 1, 2, \ldots, s)$ is obtained, which can be expressed by the hesitant fuzzy soft set $\left(\tilde{F}_k, A_k\right) = \left\{\tilde{F}_k\left(e_j^k\right) \middle| e_j^k \in A_k,\ j = 1, 2, \ldots, l(A_K)\right\}$.

$$\tilde{F}_k(e_j^k) = \left\{ \frac{h^k_{\tilde{F}_k(e_j^k)}(u_j)}{u_j} \middle| h^k_{\tilde{F}_k(e_j^k)}(u_i) = h_{ij}^k,\ u_i \in U \right\}.$$

3. All the single hesitant fuzzy soft sets $\left(\tilde{F}'_K, A_K\right)(k = 1, 2, \ldots, s)$ are generalized and then combined with the score of clear concepts to get a comprehensive hesitant fuzzy soft set $\left(\tilde{F}, A\right) = \tilde{U}_{k=1}^{s}\left(F'_k, A_K\right)$, and $A = U_{k=1}^{s} A_k$.
4. According to the formula (4), the weighted hesitant fuzzy soft set $\left(\tilde{F}, A\right)$ of $\left(\tilde{F}, A\right)$ is calculated.
5. According to the formula (3), the C_{ij} is calculated and then the generalized comparison table of the hesitant fuzzy soft set $\left(\tilde{F}, A\right)$ is obtained.
6. Calculating the row sum $P_i = \sum_{j=1}^{n} C_{ij}$ and column sum $Q_i = \sum_{i=1}^{n} C_{ij}$ of u_i in the generalized comparison table, and then calculating the value $S_i = P_i - Q_i$ of each patent u_i to be evaluated. The patent u_t with the highest score S_t is the best quality.

4 Conclusion

Based on the analysis and research of patent evaluation index system at home and abroad, this paper establishes the scientific, effective and operable quality evaluation index system of single piece patent, there are 19 two level indicators in 3 fields of technology, economy and law. Taking into account the hesitancy and selectivity of the expert evaluation, this paper puts forward to apply hesitant fuzzy soft set to the process of evaluating the quality of patent and establishes the decision model of patent quality evaluation based on hesitant fuzzy soft set. The model solves the problems of fuzziness of evaluation indexes, hesitation and selectivity of expert evaluation, and thus improves the quality of patent evaluation. The accurate evaluation results can provide scientific reference for companies or government departments to select high quality and potential patents among numerous authorized patents and supply the reliable guarantee for enterprise innovation and economic benefit, so that its theoretical value can be fully demonstrated in practice.

References

1. X.L. Wan, X.Z. Zhu, Status and trends of patent quality indicator research in international perspective. J. Intell. **28**(7), 49–54 (2009). (Chinese)
2. Y.T. Zhang, W.C. Du, M.S. Jia et al., Research on the evaluation of enterprise patents quality based on the adaptive analytic hierarchy process. Libr. Inf. Serv. **7**, 110–115 (2016). (Chinese)
3. F. Narin, Patent as indicators for the evaluation of industrial research output. Scientometrics **34**(3), 489–496 (1995)
4. Q.H. Li, Y. Liu, S.Z. Wu, et al., The overview and hierarchical analysis of the evaluation index of patent value. Stud. Sci. Sci. **25**(2), 281–286 (2007). (Chinese)
5. C. Liu, J.P. Jing, J. Yu, Analysis on the definition and composition factor of patent quality in intellectual property rights. Inf. Sci. **11**, 1710–1713 (2009). (Chinese)
6. X. Zhang, Y.J. Hu, A study on the patent evaluation method without market bench marking: theoretical basis, empirical research and future challenges. Soft Sci. **24**(9), 142–144 (2010). (Chinese)
7. J. Feng, J.Z. Zhou, Y. Du, Research on quality evaluation index system of single patent. Sci. Technol. Manag. Res. **32**(23), 166–170 (2012). (Chinese)
8. P.M. Ren, Y.H. Chen, B. Jiang et al., Research on evaluation index system of patent pledge financing for small and medium sized enterprises. J. Shandong Agric. Univ. (Social Science Edition) **4**, 55–60 (2012). (Chinese)
9. China Technology Exchange Organization Writing, China. Operation manual of the patent value analysis index system. Intellectual Property Press, 2012. (Chinese)
10. K.V. Babitha, S.J. John, Hesitant fuzzy soft sets. J. New Results Sci. **3**, 98–107 (2013)
11. H. Mao, Economic significance and practical use of patent indicators. Intellect. Prop. **07**, 72–79 (2015). (Chinese)
12. T. Jiang, The cornerstone of strict intellectual property protection: good patent authorization and the quality of the right. Intellect. Prop. **12**, 65–70 (2016). (Chinese)
13. V. Torra, Hesitant fuzzy sets. Int. J. Intell. Syst. **25**(6), 529–539 (2010)

14. X.Q. Zhou, *Soft set and hesitant fuzzy set with their application in decision making*. Hunan University (2014). (Chinese)
15. Z. Xu, M. Xia, Distance and similarity measures for hesitant fuzzy sets. Inf. Sci. **181**(11), 2128–2138 (2011)
16. Z. Xu, M. Xia, On distance and correlation measures of hesitant fuzzy information. Int. J. Intell. Syst. **26**(5), 410–425 (2011)

The Research of Affected Factors of Musculoskeletal Pain of Video Display Terminal Operator Based on ISM

Lin-hui Sun[1,2(✉)], Ying Lv[1], Zhi-kui Zhang[1], and Wei Zhang[1]

[1] Department of Industrial Engineering, University of Science and Technology, Xi'an, China
45642572@qq.com

[2] The State Key Laboratory for Manufacturing Systems, Xi'an, China

Abstract. The present study was conducted to systematically analyze the associations among work-related factors that cause musculoskeletal disorders of video display terminal operator. Around the "man-machine-materials-method-environment" five elements, by the method of literature searching, questionnaire survey and interview, extract the related impacts of 6 dimension, ultimately 16 factors was selected as the potential factors from the 6 dimension, which were taken qualitative analysis, and then a five-level structure stratum model was established by conducting Interpretative Structural Modeling (ISM) theory. S2, S6, S7 are direct surface factors; indirect middle factors mainly involve S5, S8, S9, S10, S3, S4, S11, S13, S14, S15, S16; deep fundamental factors include S1, S12. The established hierarchical model provide some theoretical reference for occupational health and safety manager to improve work condition that in accordance with ergonomics, so that protect the video display terminal operator health.

Keywords: Musculoskeletal disorders · Interpretative structural modeling Video display terminal operator · Work posture · Occupational health

1 Introduction

In 2016, The State Council put forward the strategic target—"Build and share, national health"—in the outline of "Healthy China 2030", it indicated that improve health status of population has risen to central strategic level. With the popularization of information-based office, VDT (video display terminals) operators attribute to long hours at the computer in a relatively enclosed space, on the one hand, they bear high work intensity and iteratively use mouse and keyboard with a static work. On the other hand, they don't have a balance break work during a certain time. Just as the Chinese proverb goes "put a fire under a pile of faggots-a hidden danger." Above all of these phenomenon are easily lead to WMSDs (musculoskeletal disorders) and influence employees' occupational health. The related literature show that [1–4] the overwhelming majority of employees

G. Q. Huang et al. (eds.), *Proceeding of the 24th International Conference on Industrial Engineering and Engineering Management 2018*,
https://doi.org/10.1007/978-981-13-3402-3_66

with typical VDT mode have been at high risk in suffering from WMSDs (include following body regions: neck; shoulder; elbow; low back; etc.). As for WMSDs, researchers pay most of their attention to the pain of manual worker (such as aircraft mechanic, assembly worker, col miner, etc.). In the existed literatures [5–7] and mainly focus on the basic evaluation of injury degree by questionnaire and experimental method, there are few of scholars emphasis on mental worker who work in video display terminal. Unlike others occupational diseases, such as pneumoconiosis, vocational fluorosis, etc. The WMSDs will not immediately lethal harm operators, but exert a subtle influence on operators by the accumulation of awkward posture over many years. So it's too easy to lose sight of. There is no systematic improvement strategy yet, it's necessary to do further research in this area and establish the hierarchical diagram among factors and find out deep-seated factor so that health and safety managers carry out effectively work of preventing WMSDs.

2 Methodology

Interpretive Structure Modeling (ISM for short), was developed by Professor John Warfield in 1973. It has been mainly used to disassemble complex system into sub-systems. The following steps are required to build the ISM of occupational factors that result in WMSDs on VDT operators:

(1) Ensure the critical influencing factors in system.
(2) Define the binary relationship among the critical influencing factors, and establish adjacency matrix; as specified, when S_i have an impact on S_j, then matrix element a_{ij} is 1, when S_i have no impact on S_j, then matrix element a_{ij} is 0 [8]. That is to say if $A = (a_{ij})_{n \times n}$, the definition is:

$$a_{ij} = \begin{cases} 0, & S_i \text{ have no impact on } S_j; \\ 1, & S_i \text{ have no impact on } S_j \end{cases}$$

(3) Calculating the reachable matrix based on adjacency matrix by means of Boolean algebra method, and then the reachable matrix M can be obtained.

$$i.e.\ M = (A + I)k + l = (A + I) \quad (1)$$

(4) According to the resulting reachable matrix M, establishing the reachable set $R(S_i)$, advanced set $A(S_i)$, and common set $C(S_i)$ about various factors. Divide the level of influencing factors that resulting WMSDs on condition that $R(S_i) \cap Q(S_i) = R(S_i)$.
(5) Drawing multilevel hierarchical structure, form a hierarchical model.

3 Results

3.1 Ensure the Critical Influencing Factors

In order to make sure the effectiveness of its research, the present paper used a large-scale e-commerce company employees as the sample. The person who ever suffered from (1) surgery; (2) MUSDs pains were excluded. Employees who work directly related to computer use more than 7 h per day were invited to participate in the study. The research group went to Jiang Su to take videos about work posture in workplace from April to June in 2017, then take the screenshots from video at regular intervals, every screenshots were taken one-to-one pain evaluation by the Rapid Upper Limb Assessment (as shown in Fig. 1). In the subjective study, the total of 54 Maastricht upper extremity questionnaires [9] were distributed among employees, and 50 completed the questionnaire properly (response rate, 92.6%). We invited 4 expert in human factor engineering to judge the relevance of the risk factors, Removing and supplement the specific risk factors on the basis of representativeness and significance, finally, the research group concluded that there are 6 dimension and 16 representative factors (see Table 1).

Fig. 1. Screenshots from video

RULA SCORE:

Upper arm: 40°; shoulder is raised	+3
Lower arm: 70°	+1
Wrist position: 10° wrist is bend from the middle	+3
Wrist twist: wrist is twisted in mid-range	+1

Look-up posture score in table A:	4
Add muscle use score:	+1
Add force:	+0
Neck Position: 20°	+3
Legs: legs and feet not be supported and balanced	+1
Look-up posture score in table B:	5
Add muscle use score: action 4/min or more	+1
Ass force:	+0
Find column in table C:	7

Table 1. The factors of work-related MUSD

Dimension	Factors	Dimension	Factors
Demographic characteristic	Working-age S_1	Work intensity	Computer screen time S_9
	Body mass index S2		Work proficiency S_{10}
Work facilities	The position of chairs and Table S3	Psychological factors	Work attitude S_{11}
	Workspace volume S_4		Health consciousness S_{12}
	Using-frequency of mouse and keyboard S_5	Organizational environment	Work atmosphere S_{13}
	Lighting and screen S_6		Leadership care S_{14}
Work posture	The bending degree of head S_7		Working institution S_{15}
	Support of elbow and legs S_8		Coffee break S_{16}

Final score: high risk, investigate and change immediately.

The concrete analysis of potential factors that cause WMSDs is as follows:

Demographic characteristic: foreign scholars Kaliniene et al. [10] survey found that gender, body mass index, and working-age are three significant contributors to WMSDs by conducting a questionnaire investigation for 513 public service office workers. Namely working-age recorded as S1, body mass index is recorded as S2.

Work facilities: It's well known that computer mouse and keyboard are the main office appliance for VDT operator, the position of chairs and table decide the work posture. Yu et al. [11] put forward the mode of workplace facility layout problem on the basis of ergonomics. Namely the position of chairs and table recorded as S3, workspace volume recorded as S4, using-frequency of mouse and keyboard recorded as S5, lighting and screen recorded as S6.

Work posture: Liu et al. [12] take a human factors analysis on the musculoskeletal disorder of us drivers by the method of OWAS, the analysts reveals that awkward posture of neck and back in driving will increase the risk. Zhang and Kong [13] constructed local fatigue function and find that poor posture for a long time will increase pain risk and less productive in their jobs. Namely the bending degree of head recorded as S7; support of elbow and legs recorded as S8.

Work intensity: Tornqvist et al. [14] analysed statistically numerous factors and illustrate that onerous and heavy work and the work proficiency will lead to work stress, which is significant to neck pain. In additional, using computer with longer time, the potential WMSDs symptom will aggravate. Namely computer screen time recorded as S9; work proficiency recorded as S10.

Psychological factors: One side, people is a social being with independency and consciousness, job turnout will have an effect to cause unconsciously irregular operation, the other side, ego to protect consciousness is an important factors to avoid WMSDs. Namely work attitude recorded as S11; health consciousness recorded as S12.

Organizational environment: Wu et al. [15] analyzed the relationship between personal factors and occupational factors and neck WMSDs, the result reveals that a positive correlation exists between coffee break and neck pain. Namely work atmosphere S13; leadership care S14; working institution S15; coffee break S16.

3.2 Adjacency Matrix Building

Through the above mentioned analysis can abstain the mutual relation of {S1, S2, S3, S4, …, S16}, then establish the adjacent matrix A (see Table 2).

Table 2. Adjacency matrix

$$A = \begin{bmatrix}
0 & 1 & 0 & 0 & 0 & 0 & 0 & 0 & 0 & 1 & 0 & 1 & 0 & 0 & 0 & 0 \\
0 & 0 & 0 & 0 & 0 & 0 & 0 & 0 & 0 & 0 & 0 & 0 & 0 & 0 & 0 & 0 \\
0 & 0 & 0 & 1 & 0 & 0 & 1 & 1 & 0 & 0 & 0 & 0 & 0 & 0 & 0 & 0 \\
0 & 0 & 1 & 0 & 0 & 0 & 1 & 1 & 0 & 0 & 0 & 0 & 1 & 0 & 0 & 1 \\
0 & 0 & 0 & 0 & 0 & 0 & 1 & 0 & 0 & 1 & 0 & 0 & 0 & 0 & 0 & 0 \\
0 & 0 & 0 & 0 & 0 & 0 & 1 & 0 & 0 & 0 & 0 & 0 & 0 & 0 & 0 & 0 \\
0 & 0 & 0 & 0 & 0 & 0 & 0 & 0 & 0 & 0 & 0 & 0 & 0 & 0 & 0 & 0 \\
0 & 0 & 0 & 0 & 0 & 0 & 1 & 0 & 0 & 0 & 0 & 0 & 0 & 0 & 0 & 0 \\
0 & 0 & 0 & 0 & 1 & 0 & 1 & 0 & 0 & 1 & 0 & 0 & 0 & 0 & 0 & 0 \\
0 & 0 & 0 & 0 & 1 & 0 & 0 & 1 & 0 & 0 & 0 & 0 & 0 & 0 & 0 & 0 \\
0 & 0 & 0 & 0 & 1 & 0 & 1 & 0 & 1 & 1 & 0 & 0 & 1 & 1 & 0 & 1 \\
0 & 1 & 0 & 0 & 0 & 1 & 1 & 1 & 1 & 0 & 1 & 0 & 0 & 0 & 0 & 1 \\
0 & 0 & 0 & 0 & 0 & 0 & 0 & 0 & 0 & 1 & 1 & 0 & 0 & 0 & 0 & 0 \\
0 & 0 & 1 & 1 & 0 & 1 & 0 & 1 & 1 & 1 & 1 & 0 & 1 & 0 & 1 & 1 \\
0 & 0 & 0 & 0 & 1 & 0 & 0 & 0 & 1 & 0 & 1 & 0 & 1 & 0 & 0 & 1 \\
0 & 1 & 0 & 0 & 0 & 0 & 0 & 0 & 1 & 0 & 1 & 0 & 0 & 0 & 0 & 0
\end{bmatrix}$$

3.3 Reachable Matrix Calculating

Adjacent matrix plus Identity matrix equals A + I, then calculating the reachable matrix based on Boolean algebra Method, the resulting M as Table 3 shows.

Table 3. Reachable matrix

$$
M = \begin{bmatrix}
1 & 1 & 1 & 1 & 1 & 1 & 1 & 1 & 1 & 1 & 1 & 1 & 1 & 1 & 1 & 1 \\
0 & 1 & 0 & 0 & 0 & 0 & 0 & 0 & 0 & 0 & 0 & 0 & 0 & 0 & 0 & 0 \\
0 & 1 & 1 & 1 & 1 & 1 & 1 & 1 & 1 & 1 & 1 & 0 & 1 & 1 & 1 & 1 \\
0 & 1 & 1 & 1 & 1 & 1 & 1 & 1 & 1 & 1 & 1 & 0 & 1 & 1 & 1 & 1 \\
0 & 0 & 0 & 0 & 1 & 0 & 1 & 0 & 1 & 1 & 0 & 0 & 0 & 0 & 0 & 0 \\
0 & 0 & 0 & 0 & 0 & 1 & 1 & 0 & 0 & 0 & 0 & 0 & 0 & 0 & 0 & 0 \\
0 & 0 & 0 & 0 & 0 & 0 & 1 & 0 & 0 & 0 & 0 & 0 & 0 & 0 & 0 & 0 \\
0 & 0 & 0 & 0 & 0 & 0 & 1 & 1 & 0 & 0 & 0 & 0 & 0 & 0 & 0 & 0 \\
0 & 0 & 0 & 0 & 1 & 0 & 1 & 0 & 1 & 1 & 0 & 0 & 0 & 0 & 0 & 0 \\
0 & 0 & 0 & 0 & 1 & 0 & 1 & 0 & 1 & 1 & 0 & 0 & 0 & 0 & 0 & 0 \\
0 & 1 & 1 & 1 & 1 & 1 & 1 & 1 & 1 & 1 & 1 & 0 & 1 & 1 & 1 & 1 \\
0 & 1 & 1 & 1 & 1 & 1 & 1 & 1 & 1 & 1 & 1 & 1 & 1 & 1 & 1 & 1 \\
0 & 1 & 1 & 1 & 1 & 1 & 1 & 1 & 1 & 1 & 1 & 0 & 1 & 1 & 1 & 1 \\
0 & 1 & 1 & 1 & 1 & 1 & 1 & 1 & 1 & 1 & 1 & 0 & 1 & 1 & 1 & 1 \\
0 & 1 & 1 & 1 & 1 & 1 & 1 & 1 & 1 & 1 & 1 & 0 & 1 & 1 & 1 & 1 \\
0 & 1 & 1 & 1 & 1 & 1 & 1 & 1 & 1 & 1 & 1 & 0 & 1 & 1 & 1 & 1
\end{bmatrix}
$$

Table 4. Hierarchical relation

i	R(Si)	A(Si)	C(Si)
1	1,2,3,4,5,6,7,8,9,10,11,12,13,14,15,16	1	1
2	2	1,2,3,4,11,12,13,14,15,16	2
3	2,3,4,5,6,7,8,9,10,11,13,14,15,16	1,3,4,11,12, 13,14,15,16	3,4,11,13,14,15,16
4	2,3,4,5,6,7,8,9,10,11,13,14,15,16	1,3,4,11,12, 13,14,15,16	3,4,11,13,14,15,16
5	5,7,9,10	1,3,4,5,9,10,11,12,13,14,15,16	5,9,10
6	6,7	1,3,4,6,11,12,13,14,15,16	6,7
7	7	1,3,4,5,6,7,8,9,10,11,12,13,14,15,16	7
8	7,8	1,3,4,8,11,12,13,14,15,16	8
9	5,7,9,10	1,3,4,5,9,10,11,12,13,14,15,16	5,9
10	5,7,9,10	1,3,4,5,9,10,11,12,13,14,15,16	5,9
11	2,3,4,5,6,7,8,9,10,11,13,14,15,16	1,3,4,11,12,13,14,15,16	3,4,11,12,13,14,15,16
12	2,3,4,5,6,7,8,9,10,11,12,13,14,15,16	1,12	12
13	2,3,4,5,6,7,8,9,10,11,13,14,15,16	1,3,4,11,12,13,14,15,16	3,4,11,13,14,15,16
14	2,3,4,5,6,7,8,9,10,11,13,14,15,16	1,3,4,11,12,13,14,15,16	3,4,11,13,14,15,16
15	2,3,4,5,6,7,8,9,10,11,13,14,15,16	1,3,4,11,12,13,14,15,16	3,4,11,13,14,15,16
16	2,3,4,5,6,7,8,9,10,11,13,14,15,16	1,3,4,11,12,13,14,15,16	3,4,11,13,14,15,16

3.4 Hierarchical Relation Judging

According the resulting reachability matrix M, establish the reachability set namely R (Si), advanced set namely A(Si) and common set namely C(Si). Judging hierarchical relation of the occupational factors that result in WMSDs on VDT operators on the condition that R(Si) ∩ Q(Si) = R(Si) (as Table 4 shows).

3.5 Constrict the ISM

See Fig. 2.

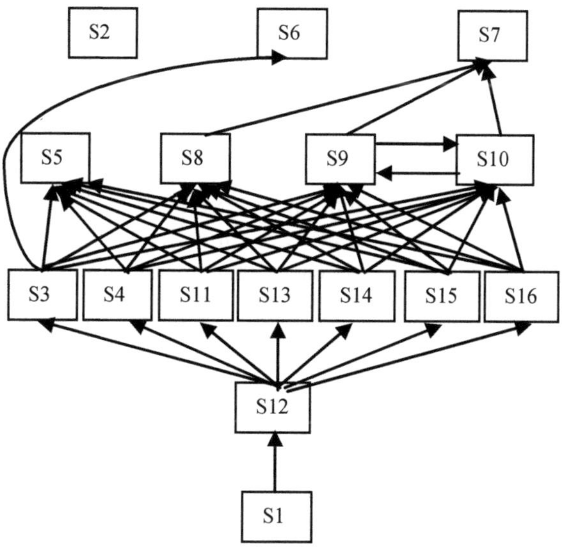

Fig. 2. ISM of factors of WMSDs

4 Discussion

The as seen in Table 4, there is a five-level structure stratum model which reflect the logical relationship among theses occupational factors. Concrete analysis is as follows:

4.1 Surface Layer Factors

The surface layer factors can be divided into two levels, include 7 elements: body mass index S2; lighting and screen S6; bending degree of head S7; using-frequency of mouse and keyboard S5; support of elbow and legs or not S8; computer screen time S9; work proficiency S10, they directly influence the occupational health of VDT operators.

4.2 Middle Layer Factors

Middle layer factors involve the position of chairs and table S3; workspace volume S4; work attitude S11; work atmosphere S13; leadership care S14; working institution S15; coffee break S16.

From this we may certainly infer that middle layer factors focus on the soft environment and work facilities about VDT operators. Among them, the manager play a significant role.

4.3 Deep Layer Factors

Deep layer factors concentrate on working-age S1 and health consciousness S12. This conclusion is in accordance with the previous studies. The WMSDs is the accumulation of awkward posture over many years, with more work life, more symptom that carpal tunnel and tendinitis syndrome will anabatic. The health consciousness is the most fundamental factor, manager may perform improvement from this aspect.

5 Conclusion

Regarding to the occupational health of WMSDs aspect, making off a five-level structure stratum model. From the ISM, we can easily find that if employees the on selves lack of occupational health awareness, then no matter how perfect external work facilities and management system, cannot play a good role in the prevention of musculoskeletal pain, managers should start with the employees thinking to strengthen the awareness of health and safety, at the same time, cultivate a self-protection awareness. In view of the middle factors, managers should layout reasonably the work facilities (e.g. tables, chairs) and work space with reference to the relevant ergonomic knowledge before going to work, improve the office environment and hardware facilities, such as add wrist rest, foot pedal and other auxiliary facilities to relieve upper limb musculoskeletal pain, while improving work efficiency and promote the better development of individuals and companies. At the same time, in order to avoid static sitting posture more than 2 h, managers should arrange the appropriate break time and reasonable work tasks, create a happy and harmonious working environment for the employees. Start from the essence of preamble, the results of the study provide a new way to improve the improvement of the occupational health and safety in the future.

Acknowledgements. We are grateful to an anonymous referee for his/her helpful comments on earlier version of this paper. This work was supported by the National Natural Science Foundation of China [grant number 71673220].

References

1. J. He, L. Zhou, J. Li, Z. Zhou, W. Du, Ergonomic evaluation of motions associated with computer operation in office staff. Chin. J. Ind. Med. **18**(2), 80–83 (2005) (in Chinese)
2. P. Ranasinghe, Y.S. Perera, D.A. Lamabadusuriya, S. Kulatunga, N. Jayawardana, S. Rajapakse et al., Work related complaints of neck, shoulder and arm among computer office workers: across-sectional evaluation of prevalence and risk factors in a developing country. Environ. Health (2011)
3. G. Kaliniene, R. Ustinaviciene, L. Skemiene, V. Januskevicius, Associations between the neck musculoskeletal complaints and work related factors among public service computer workers in Kaunas. Int. J. Occup. Med. Environ. Health **26**(5), 67–81 (2013)
4. Y. Huang, Research progress on ergonomic preventive measures for Musculoskeletal injuries of VDT operators. China Occup. Med. **38**(2), 167–170 (2011) (in Chinese)

5. L. Zhang, X. Yuan, Application of rapid upper limb assessment in aircraft maintenance. China Saf. Sci. J. **14**(7), 34–37 (2004) (in Chinese)
6. H. Huang, F. Kong, R. Xue, Research on the work fatigue conditions of multi-functional employees under lean production. Value Eng. 224–228 (2016) (in Chinese)
7. L. Wang, D. Shi, S. Chai, H. Wang, G. Lu, Improvement on drilling posture in tunneling. Coal Eng. **47**(4), 75–78 (2015) (in Chinese)
8. G. Ye, S. Duan, H. Wang, Investigation on factors influencing contractor's decision on safety investment. China Saf. Sci. J. 24(11), 23–30 (2014) (in Chinese)
9. N. Jia, G. Li, R. Qin et al., Reliability and validity of maastricht upper extremityquestionnaire in Chinese version. Ind. Health Occup. Dis. **39**(3), 129–135 (2013) (inChinese)
10. G. Kaliniene, R. Ustinaviciene et al., Associations between musculoskeletal pain and work-related factors among public service sector computer workers in Kaunas County, Lithuania. BMC Musculoskelet. Disord. **17**, 420–432 (2016)
11. R. Yu, Y. Wang, Y. Zhu, Optimizing research on facility layout problem in workplace based on human factors. Syst. Eng.-Theory Pract. **3**, 40–46 (2004) (in Chinese)
12. Q. Liu, L. Sun, J. Liu, Human factor analysis on the musculoskeletal disorder of bus drivers. Ind. Eng. Manag. **17**(2), 118–123 (2012) (in Chinese)
13. J. Zhang, Q. Kong, Study of working posture on assembly line based on fatigue analysis. Modern Manuf. Eng. **10**, 58–62 (2009) (in Chinese)
14. E.W. Tornqvist, M. Hagberg et al., The influence of working conditions and individual factors on the incidence of neck and upper limb symptoms among professional computer users. Int. Arch. Occup. Environ. Health **82**, 689–702 (2009)
15. S. Wu, L. He, J. Li et al., A study of the impact of working with computer on musculoskeletal disorders in the neck among office workers. Ind. Health Occup. Dis. **42**(4), 294–298 (2016) (in Chinese)

Factors Affecting Decision on Mobile Phone Types for Consumer Purchase Based on Discrete Choice Model

Yu-meng Yan[1,2(✉)], Xin-yu Sun[1,2], and Zi-xian Liu[1,2]

[1] College of Management & Economics, Tianjin University, Tianjin, China
helen941224@163.com
[2] College of Management, Xi'an Jiaotong University, Xi'an, China

Abstract. This paper starts from the purchase of mobile phone, on the basis of effectively collecting and summarizing the large user data of a operator in Xi'an, Shaanxi Province, summarizes the factors that affect the selection behavior of users when their mobile phone is purchased, include the type attributes of mobile phone and individual level attribute of users. Then with the support of Expected Utility Theory and Discrete Choice Model, a Multinomial Logit model is established to explore how these attributes affect the users' preference of mobile phone type, from the perspective of the mobile phone manufacturers. According to the results of the study, consumers can be roughly classified, and a precision marketing scheme of mobile phone based on operator's large data can be ultimately provided to the manufacturers, to help them deepen their cognition of the regularity of the consumers' choice behavior, and identify as well as divide potential customers effectively, provide customized services, precise marketing of new phones, enhance market competitiveness.

Keywords: Discrete choice model · Precision marketing · Smart phones Operators big data

1 Introduction

Logit Model, also known as Logistic Regression Model, is a kind of Discrete Choice Model which has been growing since 1950. Discrete Choice Model is one of the main research tools to study and analyze the individual's choice behavior. In the market research, it is matched with the Utility Theory and the Conjoint Analysis in the research topic and model, and through the discussion and judgment of the consumers' choice tendencies to different products and services, to measure, verify, and estimate the market demand.

At present, the research on Discrete Choice Model mainly focuses on the choice of traffic travel mode and the analysis of bank customer default risk [1, 2]. In the field of marketing, a series of studies have shown that the Logit Model can be used to explain the consumer's decision-making process from the micro level and to explain many of

G. Q. Huang et al. (eds.), *Proceeding of the 24th International Conference on Industrial Engineering and Engineering Management 2018*,
https://doi.org/10.1007/978-981-13-3402-3_67

the similar problems in marketing research and provide more valuable quantitative information, which contributes to better aware of the preference of consumer choice as well as behavior and provide a scientific basis for effective prediction of the market [3–9].

However, in China, the research on the selection behavior of the mobile phone types in the process of consumers' purchase is still in the embryonic stage. A study is to obtain the data through the telephone questionnaire, then based on this, constructs a Discrete Choice Model to explore the influence of demographic factors on consumers' mobile phone brand decision-making, and put forward some guidance recommendations about selecting target market [10]. At the same time, after years of operation, operators have accumulated a wide range of customer groups, and thus have a large amount of user behavior data, but the value of these data has not been fully utilized. Another study on the basis of counting on a large number of real user data which comes from an operator, then use the Dominance-based Rough Set Approach to find mobile user's intention of updating terminal [11].

This paper is composed based on Utility Theory and Discrete Choice Model, firstly constructs a multinomial logit model by counting on large user data of an operator in Xi'an, Shaanxi Province, and summarizing the factors that affect the selection behavior of users when they purchased their mobile phone, include the type attributes of mobile phone and individual level attribute of users. Then with the help of Stata software to obtain the regression results of the model parameters, explain the meaning of the parameters, and according to its symbol, size and significance to discuss and determine which elements really affect users' behavior and their effect, thus revealing the laws of consumers' behavior in the purchase process, it can be as a basis for dividing consumer groups simply. Finally, based on the research results of this model, this paper puts forward the precise marketing scheme for mobile phone based on the big data of operator, and expounds the value that it can bring to the mobile phone manufacturers.

2 Methodology

2.1 Utility Theory

Utility Theory is used to measure the consumer in the purchase or consumption of a group of goods or enjoy the service can get the satisfaction and pleasure. In the study of customer choice behavior, utility maximization behavior is the most basic hypothesis, but also the theoretical basis of all discrete choice model.

McFadden [12] established the random utility theory, and stochastic utility-based discrete choice model considered that the utility value U can be perceived by the consumer is the total utility value, and the utility value obtained by the researcher is fixed.

For the consumer n, he can accurately measure and know the utility value U_{nj} for the program j, but the researchers can not accurately measure the consumer n on the j program perceived utility, V_{nj}. This leads to the existence of errors, ε_{nj}, in the measure of true utility U_{nj}. Thus, the total utility of consumer perception can be divided into two parts:

$$U_{nj} = V_{nj} + \varepsilon_{nj} \quad (1)$$

In the formula,

V_{nj} the fixed utility, that is, the observable portion in the total utility. It mainly includes the attribute variables of the decision maker, such as age, sex, position and so on, and the related attributes of the product in the research process.

ε_{nj} the random utility, the unobservable part of the total utility. Mainly including the effects of difficult parts and errors that are difficult to observe in the utility. The stochastic utility lays the theoretical basis for the calculation of probability afterwards.

From the perspective of the utility of consumers who purchases a mobile phone, to build the general form of mobile phone type selection behavior model. According to the utility theory, assuming that the customer n chooses the scheme i in the selection scheme set J and the utility is U_{ni}, then the probability that the consumer n chooses the scheme i from the set J is:

$$P_{ni} = P\left(U_{ni} > U_{nj}, \forall j \neq i\right) \quad (2)$$

P_{im} obeys the cumulative distribution, the integral form is:

$$P_{ni} = \int I\left(\varepsilon_{nj} - \varepsilon_{ni} < V_{ni} - V_{nj}, \forall j \neq i\right) = f(\varepsilon_n) d\varepsilon_n \quad (3)$$

2.2 MNL Model

The discrete choice model has a variety of categories, depending on the distribution of the random utility ε_{nj}. If the random utility ε i is obey the Gumbel distribution, then the discrete choice model is the Logit model; if ε is obey the normal distribution, then the Probit model. In addition, Nested Logit, Mixed Logit and so on, depending on the application. This paper is used in the Multinomial Logit Model, which is one of the logit model that option $J \geq 3$, it also is the most widely used model.

The probability density function for ε_{nj} following Gumbel distribution is shown below:

$$f\left(\varepsilon_{nj}\right) = e^{-\varepsilon_{nj}} e^{e^{-\varepsilon_{nj}}} \quad (4)$$

The distribution function is:

$$F\left(\varepsilon_{nj}\right) = e^{e^{-\varepsilon_{nj}}} \quad (5)$$

After derivation, the logit model can be expressed as:

$$P_{nj} = \frac{e^{V_j}}{\sum_{i=1}^{I} e^{V_i}}, j = 1, 2, \ldots, I \quad (6)$$

In the formula, P_{nj} represents the probability that decision maker n chooses scheme j.

Based on the utility function is a linear assumption, the utility function of the selection can be expressed as:

$$P_{nj} = \frac{1}{\sum_{i=1}^{I} e^{\beta^T (x_{ni} - x_{nj})}} \quad (7)$$

In the formula, x—attribute vector; β—attribute parameter vector.

3 Data and Model Establishment

3.1 Basic Statistical Description of the Data

First of all, this paper collects and summarizes the online public data, sort out a data table which formed by property parameters from total 115 types of mobile phones, including time to market, CPU brand, screen size, pixels, ROM, RAM, and so on. It also includes the prices from May to October 2016. The proportion of the brand distribution is shown in Fig. 1.

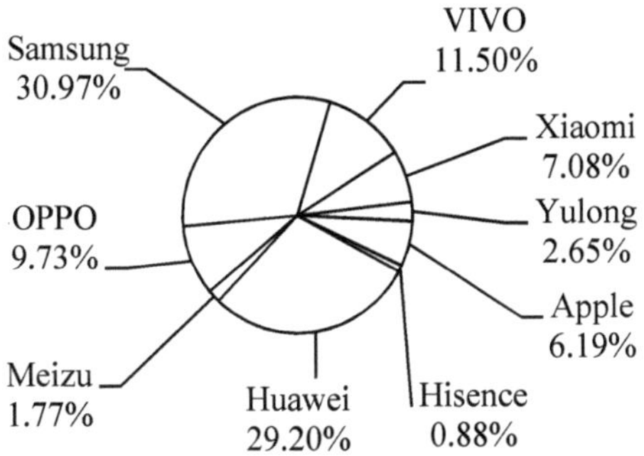

Fig. 1. The proportion of the brand distribution

User data is from the user database of an operator in Shaanxi Province, we randomly selected 30,000 users, because each type of mobile only provides the price from May to October 2016, we tracked them back for their monthly mobile terminal data for this period, including the gender, age, the first landing time of user, the terminal brand, mobile type, the first landing time of the mobile terminal and monthly traffic and ARPU, etc., and then screened out the user data of the mobile terminal included in the 115 types described above, deleted the data for missing user information at the same

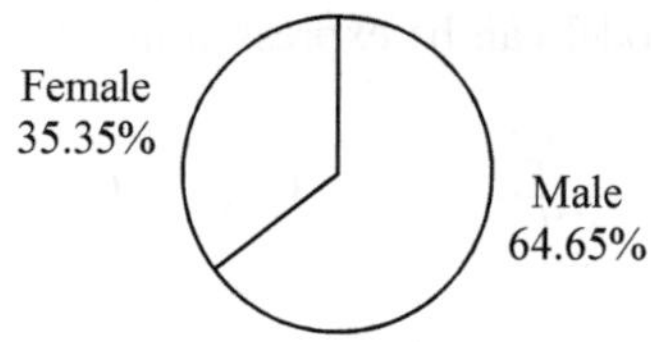

(a) The proportion of the gender distribution

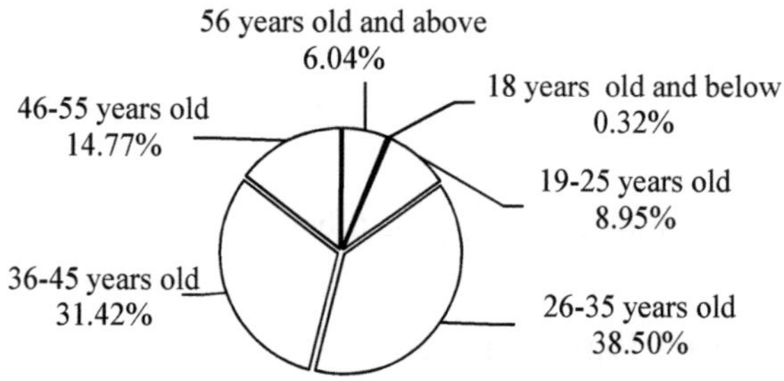

(b) The proportion of the age distribution

Fig. 2. The proportion of the gender and age distribution of the sample users

time. The final valid samples included mobile terminal data for 17,582 users during the period from May to October 2016, the proportion of the gender and age distribution of the sample users is shown in Fig. 2. But because it cannot confirm that the price of the mobile terminal when it first land is the user's initial purchase price, so the price is not added to the analysis model. The descriptive statistics of variables is show in Table 1.

3.2 Variable Selection

From the purpose of this study, the dependent variable represents different type of mobile phone, so it is a variable with multiple choices. In the long-term life and consumption practice, the individual characteristics of consumers interact with mobile phone parameters, will inevitably act on the decision-making process. At the same time, the independent objective performance indicators of the mobile phone itself will also have an impact on the purchase options. It is worth noting that in the probability model, explanatory variables are generally divided into two categories, one is the demographic variables such as age, race and ethnic background, or gender and socioeconomic variables such as education, income and occupation, it does not change due to individual choices, its values only change between people and people; Another type of explanatory variable will be specific to the choice, based on the different options, even if the same person, the value of the variable is also different. Thus, in this paper, the choice of independent variables includes both the consumer characteristic variables (age, sex, last month data traffic, last month calls, brand loyalty), which belong to the individual level, and also include the phone's performance parameters

Table 1. Descriptive statistics of variables

Variable	Mean	Std. Dev.	Min	Max
Consumer characteristics				
Age	37.552	10.397	9	115
Male	0.681	0.466	0	1
Loyalty	0.782	0.413	0	1
Lagged flow	1.009	2.542	0	395.37
Lagged fee	61.126	67.108	−38.826	1536.44
Cell phone attributes				
No. net type	5.148	1.709	2	9
Screen size	5.046	0.637	3.3	6.1
Pixel	1.073	0.393	0.3	1.6
RAM	2.14	1.179	0.5	4
ROM	24.923	23.952	2	128
Model age	23.746	15.477	1	74
Attribute dummies				
Large size	0.524	0.499	0	1
High pixel	0.585	0.493	0	1
Big RAM	0.628	0.483	0	1
Big ROM	0.334	0.472	0	1
New	0.423	0.494	0	1
CPU dummies				
CPU—Apple	0.069	0.254	0	1
CPU—Hisilicon	0.159	0.366	0	1
CPU—Mediatek	0.047	0.211	0	1
CPU—Qualcomm	0.621	0.485	0	1
CPU—Samsung	0.104	0.305	0	1
Brand dummies				
Brand—Apple	0.062	0.254	0	1
Brand—HS	0.009	0.093	0	1
Brand—HW	0.292	0.453	0	1
Brand—MZ	0.018	0.13	0	1
Brand—OB	0.097	0.293	0	1
Brand—SCH	0.31	0.46	0	1
Brand—VIV	0.115	0.333	0	1
Brand—XMP	0.071	0.245	0	1
Brand—YL	0.027	0.159	0	1

variables (phone age, network type, screen size, CPU manufacturer dummy variable, brand dummy variable, RAM, ROM, and pixels), which belong to the explanatory variable determined by the selection.

3.3 Establishment of Selection Model for Mobile Phone Types

Assuming that consumers follow the principle of maximizing utility in the selection process, whatever the option they select is the result of maximizing utility. According to McFadden (1973), we set the random utility equation of the consumer's choice behavior during the purchase process. Assuming a total of J options, if the consumer i chooses option j, the utility function is:

$$U_{ij} = \beta_{ji}x_i + \alpha_k z_{jk} + \varepsilon_{ij} \tag{8}$$

In the formula,

U_{ij} the utility when consumer i chooses option j;
x_i the consumer characteristic variables;
z_{jk} the phone's performance parameters variables;
ε_{ij} the stochastic error.

If customer i chooses the scheme j in the selection scheme set J, which means that in the total of J options, option j bring the greatest utility to him, U_{ij} is maximum. Then the probability of the statistical model selection option j is:

$$Prob(Y_i = j) = Prob\left(U_{ij} > U_{im}, \forall m \neq j, m = 1, 2, 3, \ldots, J\right) \tag{9}$$

Assume that ε_{ij} follows the i.i.d, Gumbel distribution, and that the selections between different mobile types is independent of each other. Under the above assumptions, $Y_i = j$ means customer i chooses the mobile type j, so can infer the probability of the customer i chooses the mobile type j:

$$Prob(Y_i = j) = \frac{e^{U_{ij}}}{\sum_{j=1}^{J} e^{U_{ij}}} = \frac{e^{\sum_{i=1}^{n} \beta_{ji}x_i + \sum_{k=1}^{K} \alpha_k z_{jk}}}{\sum_{j=1}^{J} e^{\sum_{i=1}^{n} \beta_{ji}x_i + \sum_{k=1}^{K} \alpha_k z_{jk}}} \tag{10}$$

The formula is the probability formula of Multiple Logit Models, and the solution of model parameters α and β is obtained by maximum likelihood estimation method. In this paper, the Stata software's clogit command is used to estimate the model.

4 Results and Discussion

4.1 The Results of Model Test

The results of the model test show that the likelihood ratio, $Prob > chi2$ is 0, indicating that the model is significant. But the value of Pseudo R2 is only 0.0798, indicating that the model's goodness of fit is not very high. However, the focus of this paper model is applied to explain rather than predict, in this model, the number of mobile phone models is very large, however, the focus of this paper model is applied to explain rather than predict, in this model, the number of mobile phone models is very large, to calculate and predict the probability of each type selected is not much practical significance. And the focus of this paper is to effectively estimate the user's preference for mobile phone's brand and attribute through the model, so as to help the mobile phone manufacturers to divide the potential customer base. Therefore, the final regression results for this model are more focused on the analysis of the symbols, size and significance of α, β to discuss and determine which elements really affect users' behavior and their effect, thus revealing the laws of consumers' behavior in the purchase process, therefore, compared to the significance, the requirements of the Goodness of Fit is not very high.

4.2 The Analysis of Regression Results

The regression results of model parameter is show in Table 2. By giving the practical significance to the regression results, the results show that under the same conditions, the user preferred to choose mobile phone type with various network type, big screen size and high RAM. But users do not like to use the type with higher Pixel and ROM, one possible reason is that, relative to the type with basic configuration, higher Pixel and ROM lead to higher prices. In addition, a very interesting phenomenon is the age of the type, that is, the time of the type to enter the market, will increase the user's choice probability. About the CPU brand, we found that the type configured with HISILICON and QUALCOMM brand CPU will reduce the probability of consumers' choice, the specific reasons need to analyze the user experience and evaluation. About the mobile phone brand, we found that the four most popular consumer brands followed by Apple, VIVO, Huawei and OPPO (because we only samples from Shaanxi, the results are not general).

Then by analyzing the differences in preference between the consumer groups for the type parameters, the results show:

(1) the users including the young people, female, and who with large data traffic and low monthly telephone charge, like to use large-screen mobile phone, which may correspond to low-income young fashion female consumer groups.
(2) the users including the older, male and high monthly telephone charge prefer high-pixel mobile phone.
(3) Similar to the screen selection, young, female and low monthly telephone charge, users prefer to use large RAM phones. This may be related to the users' awareness of running speed and data traffic have a certain relationship, such as high traffic users do not have high requirements on the running speed.
(4) the users including the young people, male, and who with low data traffic and high monthly telephone charge, like to use large ROM mobile phone, may correspond to the business male consumer groups with higher income.

Finally, we analyze the user's preference for new mobile phone type which is less than 18 months old. First of all, young people, female and high data traffic users prefer to buy a new type phone, which is consistent with the choice of large-screen mobile phone groups, because the new mobile phone is often configured large screen. In addition, we found that the brand loyalty of this consumer group is low. This group have low probability to keep using the same brand of the mobile phone when they

Table 2. The regression result of model paramater

Variables	Coef.	Std. Err.	z	P > z	Odds ratio	Std. Err.
No. net type	0.1637	0.003	54.39	0	1.1779	0.0035
Screen size	0.3372	0.011	30.65	0	1.401	0.0154
Pixel	−0.1161	0.0246	−4.72	0	0.8904	0.0219
RAM	0.088	0.0105	8.37	0	1.092	0.0115
ROM	−0.0193	0.0005	−40.84	0	0.9809	0.0005
Model age	0.0044	0.0005	9.43	0	1.0045	0.0005
CPU dummies (base: Samsung)						
CPU—Apple	2.1499	0.0336	63.95	0	8.584	0.2886
CPU—Hisilicon	−0.328	0.0214	−15.29	0	0.7204	0.0154
CPU—Mediatek	0.3211	0.0252	12.74	0	1.3786	0.0347
CPU—Qualcomm	−0.0558	0.0154	−3.62	0	0.9457	0.0146
Brand dummies (base: YL)						
Brand—HS	−0.1355	0.0666	−2.04	0.042	0.8733	0.0581
Brand—HW	0.7892	0.0299	26.37	0	2.2016	0.0659
Brand—MZ	0.2237	0.0451	4.96	0	1.2506	0.0564
Brand—OB	0.7874	0.0318	24.77	0	2.1976	0.0699
Brand—SCH	0.5514	0.0299	18.47	0	1.7357	0.0518
Brand—VIV	1.6248	0.0657	24.75	0	5.0773	0.3334
Brand—XMP	0.1783	0.0328	5.43	0	1.1952	0.0392
Large size = 1 if screen size > 5						
Large size × age	−0.0051	0.0004	−12.43	0	0.9949	0.0004
Large size × male	−0.2019	0.016	−12.61	0	0.8172	0.0131
Large size × lagged flow	0.0694	0.0054	12.79	0	1.0718	0.0058
Large size × lagged fee	−0.0011	0.0001	−8.09	0	0.9989	0.0001
High pixel = 1 if pixel > 1200						
High pixel × age	0.0118	0.0008	14.1	0	1.0119	0.0008
High pixel × male	0.1088	0.0349	3.12	0.002	1.1149	0.0389
High pixel × lagged flow	−0.0054	0.0046	−1.19	0.235	0.9946	0.0046
High pixel × lagged fee	0.0021	0.0003	7.2	0	1.0021	0.0003
Big RAM = 1 if RAM ≥ 2 GB						
Big RAM × age	−0.0082	0.0008	−10.38	0	0.9919	0.0008
Big RAM × male	−0.0512	0.0328	−1.56	0.119	0.9501	0.0312
Big RAM × lagged flow	−0.0077	0.0051	−1.53	0.127	0.9923	0.005
Big RAM × lagged fee	−0.0012	0.0003	−4.23	0	0.9988	0.0003

(*continued*)

Table 2. (*continued*)

Variables	Coef.	Std. Err.	z	P > z	Odds ratio	Std. Err.
Big ROM = 1 if ROM > 16 GB						
Big ROM × age	−0.0046	0.0006	−7.41	0	0.9954	0.0006
Big ROM × male	0.1632	0.0234	6.97	0	1.1773	0.0276
Big ROM × lagged flow	−0.0777	0.0066	−11.74	0	0.9253	0.0061
Big ROM × lagged fee	0.0025	0.0002	14.57	0	1.0025	0.0002
New = 1 if model age ≤ 18						
New × age	−0.0026	0.0005	−4.98	0	0.9974	0.0005
New × male	−0.0174	0.0205	−0.85	0.396	0.9827	0.0201
New × loyalty	−0.8056	0.0151	−53.22	0	0.4468	0.0068
New × lagged flow	0.1326	0.0064	20.64	0	1.1418	0.0073
New × lagged fee	0.0001	0.0002	0.47	0.641	1.0001	0.0002
LR chi2(43)	79268.45					
Prob > chi2	0					
Log likelihood	−456891.69					
Pseudo R2	0.0798					

purchase a new one. This also to an extent explained that the current mobile phone market is highly competitive and high substitutability.

Mobile phone itself is a combination of brand, price and other attributes, so if the merchant launched a new type of mobile phone, according to the mobile phone type parameters and combined with our analysis, they can find the consumer groups who most likely to buy this phone, then based on the analyze results to deliver accurate advertising for precise marketing.

5 The Precise Marketing Scheme

5.1 The Precise Marketing Scheme for a Mobile Phone Type

Based on the operator's large data and the characteristics of the smart phone industry, combined with the research methods adopted in this paper, we give the precise marketing scheme for a mobile phone type based on the big business data.

(1) To target the potential user
Based on the operator data and user purchase forecasting model, to estimate the user's preferences for mobile phone brands, prices and attributes, then classify users according to their preferences, which delineate the potential target users the consumer groups who most likely to buy this mobile phone type.

(2) To target the potential use
According to the user's basic information, consumption information, terminal behavior, replacement cycle, the behavioral habits of using, we can analyze users'

characteristics, type preferences, consumption quota, activity range, and so on, then draw multi-dimensional portraits for potential target customers, refine the customer's user characteristics, layout of the market core accurately, multi-point control operating costs.

(3) Implementation of targeted precision marketing
Use large data mining technology to obtain effective information from the user behavior data and terminal manufacturers channel system and resource allocation, then multi-dimensional define the customer portraits, capture the customer's consumption opportunity, according to LBS information to determine distribution characteristics of potential target customers' daily location, match up policies/products and consumer attributes/preferences. Based on the library of marketing scene, with SMS, telephone outbound system, channel maintenance system and LBS precision advertising, implement the marketing of publicity for different target groups efficiently.

(4) Trace the marketing effect
Multi-channel tracking and monitoring the status of target customer before and after accepting precise marketing, according to consumer feedback timely adjust and optimize the process of precise marketing, to achieve continuous improvement in marketing effectiveness.

5.2 The Value of the Precise Marketing Scheme

In the era of large data, the competition among the mobile phone manufacturers will be more and more from the front-end transfer to the back-end, who first make full use of large data to understand the user, who is more timely to understand the needs and expectations of the target market effectively. Therefore, mobile phone manufacturers to choose the precise marketing scheme which based on the operator's large data as one of its marketing model, will effectively help their enterprises to reduce marketing costs and improve marketing efficiency, while improving the user experience to explore wider development space for its own products.

(1) Solve the ROI problem
Traditional marketing model mainly uses advertising, push bundles for promotion, etc., because of its human cost investment, promotional materials, investment and promotional costs, resulting in a higher total cost. And through the operator's large data for precision marketing is largely reduced consumer satisfied-costs and channel circulation time, while also reducing the flow of links, saving storage costs, so that a significant reduction in marketing costs, improve customer experience but also for its own products to provide a broader space for development.

(2) Improve the user experience
The precise marketing scheme based on the operator's large data, use multi-dimensional portraits to achieve multi-dimensional user insight analysis, which support the brand mobile phone personalized marketing. Make customizable mobile advertising be given to the accurate consumer group in the accurate time and accurate location.

(3) Support business decisions with users' big data
Operators of large data directly reflect the using habits of mobile phone users, is the most real and accurate consumer data, avoid the distortion of data which is due to spread and filter through a number of links in the traditional channel, help business decision-making transform from the traditional business-experience driven into the data quantitative drive.

6 Conclusion

This paper refers to some domestic and international research about large data, precision marketing [13, 14] and mobile phone industry, tourism industry, transportation, then summarizes the commonly used methods of studying behavior preference, including utility theory and discrete selection model. Constructs a multinomial logit model by counting on large user data of an operator in Xi'an, Shaanxi Province, sums up the preferences of consumers' selection behavior when they purchased their mobile phone, for example, under the same conditions, the user preferred to choose mobile phone type with various network type, big screen size and high RAM, but users do not like to use the type with higher Pixel and ROM. Then roughly classifies he consumer groups based on the consumer preferences, for example, the users including the young people, female, and who with large data traffic and low monthly telephone charge, like to use large-screen mobile phone, which may correspond to low-income young fashion female consumer groups. Mobile phone itself is a combination of brand, price and other attributes, so if the merchant launched a new type of mobile phone, according to the mobile phone type parameters and combined with our analysis, they can find the consumer groups who most likely to buy this phone, then based on the analyze results to deliver accurate advertising for precise marketing. At the same time, this paper provides the precise marketing scheme for mobile phone based on the big data of operator, which is of great importance to identify as well as divide potential customers effectively, provide customized services, precise marketing of new phones and enhance market competitiveness.

References

1. S. Wu, Q. Shi, H. Lu, Review of error structure extension of discrete choice models in travel mode choice. Highway Eng. **32**(6), 92–97 (2007) (Chinese)
2. J.K. Dow, J.W. Endersby, Multinomial probit and multinomial logit: a comparison of choice models for voting research. Elector. Stud. **23**(1), 107–122 (2004) (English)
3. H. Zhou, X. Liao, Study on logit combined forecasting of the market demand. Syst. Eng.-Theory Pract. **23**(7), 63–69 (2003) (Chinese)
4. K. Chen, X. Wang, M. Zhu, The application of logit model to marketing econometric research on consumer's brand choice of domestic appliances. Commer. Res. **24**, 195–199 (2006) (Chinese)

5. S. Yang, F. Xu, J. Chen, Identification of dynamic loyalty of online consumer brand choice—based on the consumer panel data of a famous E-tailer in China. Ind. Eng. Manag. **14**(3), 112–117 (2009) (Chinese)
6. Z. Hu, F. Huang, X. Lu, P. Zhao, Multinomial logit model in consumer choice: an application in the car purchasing in China. Chin. J. Manag. Sci. **15**(z1), 35–40 (2007) (Chinese)
7. Q. Li, A study on the choice behavior of rural residents to the retail terminal based on logit model—taking Fujian Province as an example. J. Sichuau Univ. Sci. Eng.: Soc. Sci. Ed. **6**, 62–65 (2011) (Chinese)
8. J. Guo, Y. Li, Analysis of customer's choice preference based on discrete choice model. Technoecon. Manag. Res. **8**, 13–16 (2012) (Chinese)
9. H. Zhang, J. Zhang, J. Cao, C. Liu, C. Shi, Y. Yang, A study on random coefficient logit model about tourists' destination choice based on their travel motivation. Tour. Tribune **6**, 43–47 (2008) (Chinese)
10. G. Zhai, Y. Cao, Research on the consumer choice of mobile phone brand in China—empirical analysis based on discrete choice model. Modern Bus. Trade Ind. **20**(1), 55–56 (2008) (Chinese)
11. L. Liu, G. Wang, W. Deng, Prediction of mobile users for updating terminal method based on dominance-based rough set approach. J. Chin. Comput. Syst. **36**(8), 1789–1794 (2015) (Chinese)
12. D. Mcfadden, Conditional logit analysis of qualitative choice behavior. Front. Econom. 105–142 (1972) (English)
13. H. Wang, Q. Sun, K. Hu, Precision marketing model based on technology of signaling big data. J. Beijing Univ. Posts Telecommun. (Soc. Sci. Ed.) **18**(4), 70–76, 107 (2016) (Chinese)
14. H. Chen, Mobile precision marketing era. E-Bus. J. **10**, 32–33 (2013) (Chinese)

The Spatial Econometric Analysis of Regional Economic Development in Jiangxi Province

Chang-bin Xu[1], Ze-yun Wei[2], Ke-xin Cheng[3], and You-yuan Wang[2](✉)

[1] College of Aviation and Tourism, Jiangxi Teachers College, Yingtan, China
[2] School of Economics and Management, Nanchang Hangkong University, Nanchang, China
yywnc@sina.com
[3] Science and Technology Information Institution of Jiangxi Province, Nanchang, Jiangxi, China

Abstract. By using the R language, this paper analyzes the spatial relationship between the value's added value of various cities in Jiangxi province and the imbalance of regional economic development in Jiangxi province by using the Moran's I index. The results show that there is an obvious imbalance between the economic development of Nanchang and the surrounding cities. The paper puts forward some Suggestions on how to implement the regional economy, adjust the economic development center of the region, improve the structural tax reduction, and use the location advantage to develop the economy.

Keywords: Regional economy · Spatial measurement · R language

1 Introduction

With the development of large data and geostatistics, spatial data acquisition has become convenient. The development of economic geography also provides scholars with a new path to explore economic issues, how to develop valuable information from complex spatial and economic data, which is an urgent need to address issues related to economic geography scholars.

R is an open source software system, scholars around the world can publish their own statistical methods in the software, making the R language has a small system, full-featured advantages. It provides users with a lot of statistical and graphical display functions, such as variance analysis, clustering analysis, spatial statistical analysis, machine learning and so on. This paper will use the spatial statistics function of R language to analyze the value of GDP growth and spatial information synthesis in Jiangxi Province from 2014 to 2016.

Regional economic differences have been the concern of academia, government and people. regional economic imbalances are common all over the world [1–4]. Raising the development level of regional economic is the fundamental way to balance the spatial gap of regional economic. However, the development of urban economy must be based on the scientific understanding of the spatial pattern of regional economy. By exploring the spatial differences in the economic characteristics of the region,

G. Q. Huang et al. (eds.), *Proceeding of the 24th International Conference on Industrial Engineering and Engineering Management 2018*,
https://doi.org/10.1007/978-981-13-3402-3_68

find out the reasons for the formation of regional economic spatial patterns. Based on the existing resources, then formulate strategies that are suitable for the coordinated development of the regional economy.

However, at present, most scholars' researches on the spatial structure of China's regional economy are concentrated in the larger areas of the province and the region. There is relatively little research on the economic differences of municipal cities in each province [5–8].

Jiangxi Province in the wave of reform and opening up, economic strength has also been rapid development, but the regional economic imbalances have been there, long-term serious economic imbalance will lead to social stratification of all levels of the problem.

This paper analyzes the economic development differences between different cities in Jiangxi Province and their spatial relations, and uses the results of the calculation to analyze the imbalance of economic development in different cities in Jiangxi Province [9, 10]. The relevant conclusions can provide reference for the Jiangxi provincial government to formulate regional economic policy.

2 Spatial Statistical Analysis

2.1 Local Moran's I Index

The local Moran's I-index is the degree of similarity in terms of the representation of spatial units that are adjacent or contiguous in space. The local autocorrelation formula is:

$$I_i = \frac{(X_i - X)}{S^2} \sum_j w_{ij}(X_j - \bar{X}) \tag{1}$$

In the formula (1), w_{ij} is the neighborhood relation of the spatial regions i and j, x_i and x_j are the observation values of the spatial regions i and j.

When i and j are adjacent spatial regions, $w_{ij} = 1$; otherwise, $w_{ij} = 0$.

When I_i is greater than zero, the region i is positively correlated with the surrounding area properties.

When I_i is less than zero, the region i is negatively correlated with the surrounding area properties.

When I_i is zero, the region i and the surrounding area attribute are randomly distributed.

In the local Moran's I index, the existence of the spatial autocorrelation relationship between the test space region i and the surrounding space region is achieved by normalizing the statistic Z, and Z is calculated as:

$$Z(I_i) = \frac{I_i - E(I_i)}{\sqrt{VAR(I_i)}} \tag{2}$$

$E(I_i)$ and VAR (I_i) are their theoretical and theoretical variance. Where the statistic Z is positive and through the significance test, then the spatial region i has the same direction of spatial autocorrelation as the surrounding spatial region attribute.

When the statistic Z is negative and through the significance test, the spatial autocorrelation of the spatial region i in the opposite direction to the surrounding spatial region attribute. If the result of the statistic Z value is zero, then the spatial region i and the surrounding spatial region attribute are randomly distributed.

3 Empirical Analysis

The data of this article are derived from the official website of Jiangxi Provincial Bureau of Statistics, 2014–2016 the province's GDP growth is calculated by the year's GDP minus the previous year's GDP derived. The specific values are shown in Table 1.

Table 1. The added value of the GDP of each city from 2014 to 2016

Nanchang	331.96	332.05	354.98
Jingdezhen	58.21	33.85	68.09
Pingxiang	66.95	47.44	85.89
Jiujiang	178.96	122.72	193.45
Xinyu	55.27	46.53	81.37
Yingtan	53.98	32.28	56.09
Ganzhou	170.59	130.28	220.47
Ji'an	119.11	86.41	132.85
Yichun	135.99	98.03	149.36
Fuzhou	96.77	68.37	105.77
Shangrao	149.24	100.57	160.24

Note The data from the Jiangxi Provincial Bureau of Statistics [11–13]

According to the data of increasing the GDP of Jiangxi Province in 2014–2016 and the spatial information of administrative map of Jiangxi Province, we can measure the local Moran's index and related statistics of the added value of GDP in the province through the function of maptools, ggplots and plry in *R* language and the functions of (1) and (2). The results are shown in Table 2.

It can be seen from Table 2 that the *I*, *Z* and *P* values of Nanchang in the period from 2013 to 2014 are −3.250, −1.864 and 0.062 respectively.

When the significance level is 0.1, it indicates that the economic development of Nanchang and the surrounding area is high and low imbalance, that is, the economic development level of Nanchang is higher and the level of economic development in the surrounding cities is low. The *I*, *Z*, *P* values of Jiujiang City were 1.860, 1.833 and 0.067.

Table 2. Local Moran Ii index and statistics Z (Ii)

	2013–2014			Type	2014–2015			Type	2015–2016			Type
	I_i	$Z(I_i)$	P		I_i	$Z(I_i)$	P		I_i	$Z(I_i)$	P	
Ganzhou	0.006	0.116	0.908		−0.003	0.122	0.903		−0.012	0.086	0.932	
Ji'an	−0.009	0.227	0.820		0.025	0.275	0.783		0.004	0.234	0.815	
Xinyu	1.057	0.957	0.338		0.626	0.722	0.470		0.859	0.812	0.417	
Fuzhou	0.016	0.209	0.835		−0.160	0.084	0.933		0.098	0.266	0.790	
Yingtan	−0.208	0.143	0.887		−0.493	−0.042	0.967		0.024	0.289	0.773	
Yichun	0.113	0.389	0.697		0.129	0.401	0.689		0.087	0.372	0.710	
Nanchang	−3.250	−1.864	0.062	High, low	−3.819	−2.343	0.019	High, low	−3.850	−2.244	0.025	High, low
Shangrao	−0.520	−0.289	0.773		−0.178	0.004	0.997		−0.440	−0.216	0.829	
Jingdezhen	0.219	0.351	0.726		0.314	0.508	0.611		0.339	0.448	0.654	
Jiujiang	1.860	1.833	0.067	High, high	1.171	1.414	0.157		1.617	1.581	0.114	
Pingxiang	0.327	0.458	0.647		0.220	0.421	0.674		0.238	0.380	0.704	

Note When the P value is less than 0.1 in the statistic, the local Moran Ii index and the statistic Z (Ii) are significant, and the spatial distribution type can be judged

When the significance level is 0.1, it indicates that the economic development level of Jiujiang City and the surrounding area is high.

In 2014–2015, the I, Z and P values of Nanchang were −3.819, −2.343 and 0.019 respectively. When the significance level is 0.05, it indicates that the economic development of Nanchang and surrounding areas continues to appear high and low imbalance.

In the 2015–2016 Nanchang City I, Z, P values were −3.850, −2.244, 0.025. When the significance level is 0.05, it still shows that the economic development of Nanchang and the surrounding area is high and low imbalance.

In the past three years, the economic development of Nanchang and the development of the surrounding cities are imbalanced, and the absolute value of the local Moran's I index has been increasing, indicating that Nanchang City and the surrounding city economic imbalance is increasing.

4 Suggestions

Aiming at the above studies on the development of local economy in Jiangxi Province and the analysis of spatial effects, this paper puts forward some suggestions on how to reduce the local economic differences in Jiangxi Province.

(1) The implementation of the development of regional economic differentiation
Through the analysis, it can be seen that there are significant differences in the economic development of Jiangxi Province, and there are development differences not only between the economic regions but also the development gap in the region.
Therefore, structural adjustment should be carried out according to the economic development advantages. But the current economic structure of the industry is unreasonable, Xinyu, Nanchang and other regions should focus on the development of leasing and business economy, the financial industry, give full play to regional advantages, making the economic distribution is reasonable, and thus improve the overall level of the economy. Pingxiang in culture, sports and entertainment have significant advantages in development, it should focus on the development of Pingxiang cultural industries for improving cultural soft power, thereby improving the overall level of the economy. Ji'an should vigorously develop the tourism economy, cultivate Jing Gangshan, Luling culture, green eco-tourism competitiveness. Improve tourism propaganda, improve public facilities, strengthen safety management, and then attract tourists, promote economic development.

(2) Adjust the focus of economic development
First, vigorously develop productive services. In the service industry, the productive service industry is human capital, knowledge and technology-intensive service industry, highly dependent on talent and knowledge, so the relevant departments can provide the corresponding supporting measures to introduce talent, provide a smooth employment platform, Job seekers to communicate, reduce corporate recruitment costs.

Second, the government can increase the support of the economy, the appropriate introduction of the field of leading technology, while the technical training of personnel to improve the application of advanced technology in the region, and thus improve the ability to innovate.

(3) Improve the protection of structural tax cuts

First of all, increase financial investment to improve the people's livelihood security system. Improve the social security system, improve the level of economic development. The total amount of social retailing is also a significant influence on the increase of the economy. The government could reduce the taxation of retail items and improve the vitality of the retail items and increase the economic value.

Finally, improve the level of industrial development, urbanization. Industrial development level and the level of urbanization are the important factors to promote the level of economic development, the Jiangxi provincial government should follow the premise of the laws of the market economy, optimize the upgrading of industrial structure, improve the rate of urbanization, these initiatives will effectively promote the local economic level.

(4) Utilize the Regional Advantages to Develop the Economy

According to the above analysis, Nanchang's Moran index is negative for three consecutive years, indicating that Nanchang and the surrounding areas of economic development is uneven and are high and low distribution.

Therefore, Nanchang as a highland economic development should establish economic development center in the conditional areas, while leading the capital city to the surrounding areas of the diffusion effect, and then drive the surrounding cities into the high economic level stage.

5 Concluding Remarks

The economic strength of Jiangxi Province was rapidly developed during "12th Five-Year" period, but the regional economic imbalances have been there, long-term economic imbalance will lead to social equality of all classes.

In this paper, by studying the economic development difference between different cities in Jiangxi Province and its spatial relationship, the paper analyzes the imbalance of regional economy in Jiangxi Province. And put forward the proposals that implementation of differentiated regional economic development, adjust the regional economic development center, improve the protection of structural tax cuts and the use of location advantages of economic development.

Acknowledgements. This work was financially supported by Jiangxi Academy of Social Science "13th Five-Year" (2016) planning project (No. 16YJ16).

References

1. L. Qiu, C. Fang, Beijing economic spatial agglomeration comprehensive measurement. Geogr. Res. **01**, 99–110 (2013)
2. A. Fan, H. Pan, Study on the difference of economic region and its influencing factors—an empirical analysis based on Yunnan data. J. Yunnan Natl. Univ. (Nat. Sci. Ed.) **02**, 109–114 (2013)
3. F. Yang, J. Ye, China's economic development and spatial distribution. Trop. Geogr. **02**, 178–186 (2013)
4. L. Minghua, G. Hui, Beijing's economic growth and its spatial distribution characteristics. Urban Dev. Res. **04**, 65–71 (2012)
5. N. Lin, Analysis of china's regional economic optimization and remodeling development trend. Macroecon. Manag. **06**, 39–43 (2017)
6. G. Yingzhao, Y. Chengrong, C. Jia, The effect of government governance mechanism on regional economic convergence in China. J. Shandong Univ. (Philos. Soc. Sci.) **04**, 8–18 (2017)
7. W. Jun, A. Wei, The trend decomposition of spatial correlation of regional economy and analysis of its influencing factors: policy suggestions of "one belt and one road" strategy. Econ. Surv. **01**, 1–10 (2018)
8. W. Song, The convergence of regional economic growth and knowledge spillover effects in China, Ph.D. dissertation, Quantitative Economics, Jilin University, Jilin Province (2016)
9. J. Zhu, Industrial agglomeration effect of the industry difference analysis—based on the Guangdong economy empirical research. Central Univ. Finance Econ **06**, 74–79 (2012)
10. L. Zhao, Study on economic spatial difference of Fujian province. J. Shenyang Univ. Technol. (Soc. Sci. Ed.) **02**, 141–148 (2012)
11. Jiangxi Provincial Bureau of Statistics, *Jiangxi Statistical Yearbook* (China Statistics Press, Jiangxi, 2015)
12. Jiangxi Provincial Bureau of Statistics, *Jiangxi Statistical Yearbook* (China Statistics Press, Jiangxi, 2016)
13. Jiangxi Provincial Bureau of Statistics, *Jiangxi Statistical Yearbook* (China Statistics Press, Jiangxi, 2017)

Segmentation of Human Motion Sequence Based on Motion Primitives

Yu-ting Zhao[1], Yao Wang[1], Shuang Wu[1], Xiao-jun Li[2], Hua Qin[3], Taijie Liu[4], Linghua Ran[4], and Jianwei Niu[1](✉)

[1] School of Mechanical Engineering, University of Science and Technology Beijing, Beijing, China
niujw@ustb.edu.cn

[2] Xi'an Research Institute of Hi-Tech, Xi An, China

[3] School of Mechanical-electronic and Automobile Engineering, Beijing University of Civil Engineering and Architecture, Beijing, China

[4] China National Institute of Standardization, Beijing, China

Abstract. Motion capture data could describe human motion precisely. It's anticipated some uncapturable human motions may be generated through easily capturable motions. A method was introduced to segment motion captured data, which could cut a long motion sequence into some unique motion primitives. This method involved 14 joints of body hierarchy. Singular Value Decomposition (SVM) was used to determine how motion data dimension changed, which could identify the segmentation frame. The validity of method was verified with an 8,401 frames motion sequence from Carnegie Mellon University Motion Capture Database, and it turned out to be valid in accordance with human intuitive judgment.

Keywords: Motion capture · Motion segment · Motion primitive
Singular value decomposition (SVM)

1 Introduction

In motion capture technology, research on human motion needs to be based on motion data. However, if we want to explore the laws of human motion in a broad sense, we can't just stop at the idea of "make action 1—capture data 1—make action 2—capture data 2". On the one hand, in the real world, the human movements are always changing. Even under the same scenario, the differences in the human body and the differences in the movement amplitudes may cause differences in the motion data. On the other hand, the constraints between the various joints in the human body under normal conditions cannot make some certain actions, and therefore the total number of human movement states which can be represented by motion captured data is quite limited. In order to improve the flexibility, efficiency, and to reduce the cost of motion data applications, the motion data can be processed to obtain the motion data segments, and subsequent human motion studies can be conducted on this basis motion primitives.

G. Q. Huang et al. (eds.), *Proceeding of the 24th International Conference on Industrial Engineering and Engineering Management 2018*,
https://doi.org/10.1007/978-981-13-3402-3_69

Data segmentation is a basic means for processing motion sequence data. Its purpose is to segment a long motion sequence into several segments of small motion segments according to different requirements. To address this issue, many scholars have conducted relevant research at this stage. Pomplun and Mataric proposed a joint space motion segmentation and decomposition algorithm based on experiments on human arm [1]. Arikan et al. first artificially labeled the sample motion data, and then used Support Vector Machine (SVM) to repeatedly mark and segment the motion data, resulting in a smooth and natural motion [2]. Forbes and Fiume proposed using the weighted Principal Component Analysis (PCA) method to represent the human pose, and then efficiently search and segment the motion sequence by calculating the similarity distance between the poses [3]. Wang et al. adopted Hidden Markov Models (HMM) to segment and cluster motion data in video, inspired by algorithms in natural language processing [4]. Kanav et al. argued the human skeletal structure can be represented by a hierarchical, named Dynamic Hierarchical Layered Structure. In this method, the body of each layer (Hierarchy) is divided into a plurality of sections (layer), based on which the movement split-level behavior (Hierarchical Activity Segmentations) algorithm and a simple Bayesian classifier on the motion sequences of the human body motion segmentation could be implemented [5]. Wang introduced Kernel Dynamic Texture (KDT) into the analysis of human motion data. KDT modeling based on Martin distance was used as a measure of the similarity between KDTs to achieve segmentation of motion capture data [6].

This article proposes a method for segmenting motion sequences according to the type of motion. The motion data are high-dimensional, and the inherent dimensions of the motion data can be obtained through dimensionality reduction methods. Due to the different types of motion of the human body, the corresponding intrinsic dimensions will also be different. Thus, the internal dimensions may vary according to a data segment length in the motion sequence. We can analyze the changes during a period of a motion sequence to segment the motion sequence into fragments of different types of motion.

2 Methodology

2.1 Fundamentals

The movement state of human beings has certain regularity. Any movement is composed of some simple primitive movements that can be superimposed on each other; these movement primitives are generally the movement units that describe different movement characteristics, such as running, jump, walk, etc. The concept of motion primitives may refer to the concept of motion template proposed by Meinard Müller [7]. They considered the motion templates can reflect the nature of a certain type of movement with similar logic, and the data may be described by data matrix which has clear, specific, and independent semantic meaning. It can be seen that the motion primitive is the basis of human movement. It not only can accurately describe the characteristics and changes of human movement, but also can be used for data reuse in the subsequent creation of virtual human movement.

For a period of motion data $x_i (i = 1, 2, \ldots, n)$, all the motion data is distributed in the space formed by the dimension. Motion frames with similar motion types can form a relatively concentrated cluster; therefore, if the motion data in the segment contains multiple types of motion, then the motion data will be distributed in the entire space to form a corresponding number of clusters. The center of the cluster can be represented by the mean of the motion sequence:

$$\overline{x} = \frac{1}{n} \sum_{i=1}^{m} x_i \tag{1}$$

Each movement cluster can be represented by its mean of motion, say $\bar{x}$. Think of $\bar{x}$ as a plane and cast each one in $x_i (i = 1, 2, \ldots, m)$ onto this plane to get $x_i^{'}$, represented by:

$$x_i^{'} = \overline{x} + \alpha_{i1} v_1 + \alpha_{i2} v_2 + \cdots + \alpha_{ir} v_r \tag{2}$$

where $v_1, v_2, \ldots, v_r$ are orthogonal vectors, and also the main components of the optimal subspace for $\bar{x}$, $\alpha_{i1}, \alpha_{i2}, \ldots, \alpha_{ir}$ can be used to determine the coefficients of $x_i^{'}$ of the subspace, the coefficients of r means the intrinsic dimension in the subspace of $x_i^{'}$, and the intrinsic dimensions of x_i in the motion cluster.

Based on the above analysis, the types of motion represented by different kinematic primitives are also different. The intrinsic dimensions of motion frames within a single type of kinematic primitive are approximately constant, and it can therefore be concluded that the number of intrinsic dimensions of the sequence of motions containing a single kinematic primitive is smaller than that of containing multiple types of motion primitives.

2.2 Characteristic of Data

When segmenting the motion data, in order to reduce unnecessary calculations, it is not necessary to consider the change of the root node in the spine. At the same time, the movements of joints in the AMC file that do not affect the overall effect of human motion, such as fingers and toes, are ignored. In this algorithm, 14 individual segments other than some motion ends are selected as feature data extraction objects, and the 14 individual segments have a greater impact on human motion. The distribution of body segments is shown in Fig. 1.

2.3 Singular Value Decomposition

For the motion data matrix $A_{n \times q}$, we calculate the Singular-Value Decomposition (SVD) matrix decomposition to get three matrix, i.e., US and V, as shown in Eq. 3

$$A = USV^T \tag{3}$$

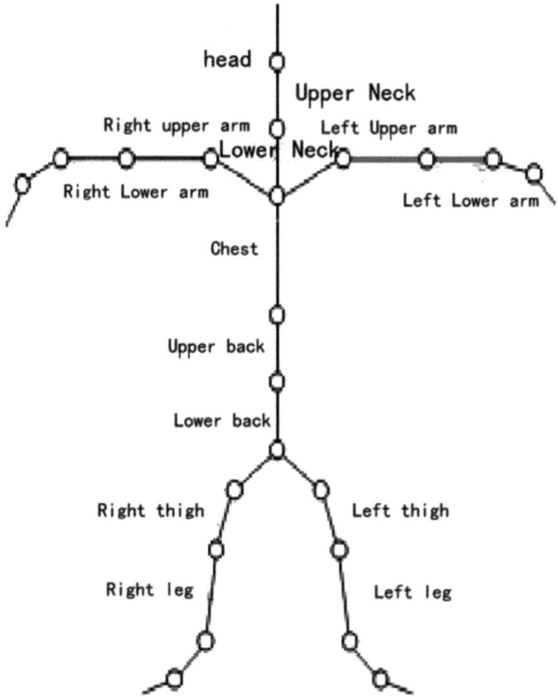

Fig. 1. Distribution of 14 individual segments of digital human body

In the matrix $S_{n \times q}$, the fore part with r dimensions is a diagonal matrix composed of the feature values of the motion data, indicating that the motion data can be represented by the r dimensions. In other words, this r dimensions can represent the optimal space of data matrix $A_{n \times q}$, marked as $v_1, v_2, \ldots, v_r$. Based on this optimal space, two concepts are proposed [8]:

(1) Projection error: When all values in the matrix are projected onto the optimal space, some of the data is lost (i.e., in the S matrix, feature values other than the fore r dimensions) and this will produce a projection error. The projection error is calculated as follows:

$$e = \sum_{i=1}^{n} \left\| x_i - x_i^{'} \right\|^2 = \sum_{j=r+1}^{q} \sigma_j^2 \tag{4}$$

(2) Projection Contribution Rate: It refers to the eigenvalues that remain in the projection after the values in the matrix are projected into the optimal space. These values ensure the trueness of the description of the original matrix. The projection contribution rate is calculated as follows:

$$E_r = \frac{\sum_{j=1}^{r} \sigma_j^2}{\sum_{j=1}^{q} \sigma_j^2} \tag{5}$$

By setting a threshold of 0.95 for the contribution rate of projection, when $E_r > 0.95$ it outputs the dimension of optimal space.

3 Splitting Frame Detection

For a motion sequence, get a dimension r from the initial timing window. With the expansion of the timing window, the data in the new timing window is projected to the r, a series of projection error values can be obtained. By recording the change of the projection error value, whether the new motion primitive is covered in the detection timing window can be obtained.

The projection error sequence is marked as $E = e_T, e_{T+1}, e_{T+2}, \ldots$, record the projection error change sequence as $D = d_i, d_{i+1}, \ldots$, among which $d_i = e_i - e_{i-1}$. When a certain motion frame meets the condition $d_i > d\_mean + k_\sigma \times d\_sd$. ($d\_mean$ *is t*he mean of the change value of the projection error sequence; d_sd is the standard deviation of the variation of the projection error sequence, k_σ is a threshold specified in the experiment), the motion frame will be marked as a motion segmentation frame, namely splitting frame.

4 Results

The experiment was conducted on a PC with i3–3.40 GHz and 4 GB of memory. The programming environment was Windows/Matlab R2010b. In order to verify the effectiveness of this segmentation method, we used SVD segmentation method for a relatively long motion sequence.

The experimental data comes from the CMU motion capture database [9]. We used the motion sequence #86_03 as our sample. In this data, 8401 frames of motion are included. The types of motion include six types of motion, including walking, running, jumping forward, kicking, jumping on one foot, and expanding the upper limb.

We set the parameters as follows: K value is 240 (initialization timing window is 240 frames), tau value is 0.9 (the threshold tau is 0.9, and when the current time window has a projection error of $E_r > 0.9$, the dimension of the current timing window will be outputted). The SVD segmentation method divides the motion sequence by determining the number of frames of the segmentation segment. After segmentation of the #86_03 sequence, all segmentation points can be obtained, as shown in Table 1.

It should be noted that the SVD based method treats forward jumps and walks as the same kind of movement primitive, which is inconsistent with subjective cognition. Meanwhile, the action of kicking the right foot is divided into two fragments. This is in accordance with the fact that the sequence of motion does show a clear change of posture in kicking the right foot. SVD based method is not a perfect segmentation method, it still needs further improvement through the optimization of parameters.

Table 1. Results of SVD segmentation on human motion data

File No	Sports primitive	Split frame	Sports interval	Number of motion frames
1	Walk	866	[1, 866]	866
2	Run	1907	[867, 1907]	1041
3	Jump + walking	3558	[1908, 3558]	1651
4	Kick right foot	4044	[3559, 4044]	486
5	Kick right foot	4690	[4045, 4690]	646
6	Left foot hop	5622	[4691, 5622]	932
7	Right foot hop	6323	[5412, 6323]	701
8	Upper limb extension	7104	[6324, 7104]	781
9	Walk	8401	[7105, 8401]	1297

However, for simpler action sequences, SVD segmentation method can achieve the segmentation of the motion primitives in the sequence quite well.

5 Conclusion

In reality, the basic movement units (motion primitive) of human motion are multifarious, and it is difficult to obtain all kinds of motion data by using motion capture. It is an effective means to generate some new movements with existing motion sequences. Therefore, it is necessary to study the segmentation of motion data. This paper introduces a motion segmentation method based on the motion primitives by using SVD based dimension reduction. The change of the intrinsic dimension of a long motion sequence has the following characteristics: the number of dimensions of the optimal subspace of a motion sequence containing a single motion primitive is smaller than that of a plurality of motion primitive data. Based on this, to segment a motion sequence into a number of motion primitives, it is first necessary to determine the variation of dimensions in this motion sequence, and then detect the variation of the projection error when the dimensions change. If the projection error changes beyond the determination threshold, the corresponding motion frame is treated as a segmentation point.

However, this method has limitations: It has better recognition and segmentation effects for the human body's macro overall movement, such as walking, running, jumping, etc. But it is not applicable for microscopic motions of the human body, such as the activities of the big arms or the lower legs. The ability to identify and segment transitional data needs to be strengthened. Through the examples, we can see that after the segmentation, we would get the transition motion before or after the motion splitting frame. This shows that the segmentation algorithm misunderstands the transitional motion as a motion primitive, and thus we need to further study the parameters to avoid this problem.

Acknowledgements. This research is supported by National Key R&D Program of China (2017YFF0206602) and Special funds for the basic R&D undertakings by welfare research institutions (522016Y-4680). The authors also appreciate the support from General Administration of Quality Supervision, Inspection and Quarantine of the People's Republic of China (201510042), the State Scholarship Fund from China Scholarship Council (201208110144), the National Natural Science Foundation of China (51005016), and Fundamental Research Funds for the Central Universities, China (FRF-TP-14-026A2).

References

1. M. Pomplun, M. Mataric, Evaluation metrics and results of human arm Movement imitation, in *Proceedings of the First IEEE-RAS International Conference on Humanoid Robots* (2000)
2. O. Arikan, D. Forsyth, Interactive motion generation from examples, in *Proceedings of the 29th Annual Conference on Computer Graphics and Interactive Techniques*, vol. 21, no. 3 (2002), pp. 483–490
3. K. Forbes, E. Fiume, An efficient search algorithm for motion data using weighted PCA, in *Proceedings of the 2005 ACM SIGGRAPH/Eurographics Symposium on Computer animation* (2005), pp. 67–76
4. T.S. Wang, N.N. Zheng, Y.X. Xu, X.Y. Shen, Unsupervised cluster analysis of human motion. J. Softw. **14**(2), 209–213 (2003)
5. K. Kanav, Gesture segmentation in complex motion sequences, in *Proceedings IEEE International Conference on Image Processing* (1996), pp. 94–99
6. M.J. Wang, Virtual person motion synthesis technology and its engineering application research. *Thesis of National University of Defense Technology* (2010)
7. M. Müller, T. Röder, Motion templates for automatic classification and retrieval of motion capture data, in *Eurographics/ACM SIGGRAPH Symposium on Computer Animation* (2006), pp. 137–146
8. J. Barbič, A. Safonova, Jia-Yu Pan, C. Faloutsos, Segmenting motion capture data into distinct behaviors, in *Proceedings of Graphics Interface* (2004), pp 185–194
9. https://www.mocap.cs.cmu.edu/. Accessed 31 Mar 2018

Human Motion Simulation and Analysis in Microgravity Environment Based on ADAMS

Chao-zhong Guo[1], Shi An[2], Ji-hong Yan[1(✉)], and Shi-cheng Qing[1]

[1] Department of Industrial Engineering, Harbin Institute of Technology, Harbin, China
jyan@hit.edu.cn

[2] School of Management, Harbin Institute of Technology, Harbin, China

Abstract. Microgravity has significant influence on the astronaut's extravehicular tasks in space environment. This paper proposes an ADAMS based human motion simulation method to analyze the force and torque changes due to microgravity. According to basic theory of human anatomy, a human rigid model is built using SolidWorks 3D modeling software considering the human bones, joints and muscles structures, as well as the functions and pattern of motion of the body. The model refers to the 90 percentile of Chinese human body data, and keep all the main joints, in which the degree of freedom is considered to make sure the model design is reasonable and practical. Then the rigid model is imported into the ADAMS software by setting constraint, load and the motion angle to perform the dynamic analysis of the human body during one walking cycle. The force and torque of joints of shoulder, hip, knee and ankle are compared in normal gravity and microgravity environment. This paper provides necessary dynamic data for astronaut's outer space training to optimize the operation ability when accomplishing space missions.

Keywords: ADAMS · Human body model · Microgravity · Motion simulation

1 Introduction

With the construction of the international space station and the "heavenly palace" space station in China, manned space activities develop much faster. Astronauts play more important role in the utilization and development of space environment. However, due to the microgravity, the operation of the astronauts outside the cabin is very different from that on the ground. Since there is more space interference in outer space without support, astronauts often suffer from overexertion, disorientation and muscle tension during exercise or operation, which leads to decline of working ability. Therefore, the motion states of the astronauts under microgravity environment should be analyzed and the relevant motion and mechanics data can be derived in order to provide technical support for the precise operation of astronauts in space.

Human dynamics is the science of the study of the relationship between the motion of the human body and the force acting on the body, which is a branch of sports

G. Q. Huang et al. (eds.), *Proceeding of the 24th International Conference on Industrial Engineering and Engineering Management 2018*,
https://doi.org/10.1007/978-981-13-3402-3_70

biomechanics, and an interdisciplinary subject between sports science, bionics, mechanics and dynamics [1]. It focuses on the gravity, support reaction force, friction, fluid resistance, internal force, etc. Hara et al. carried out research on the ergonomics evaluation strength, activity range, touch perception, flexibility and fatigue of manual operation on the ground by taking account of the space flight situation [2]. Wang et al. from the Chinese Astronaut Research and Training center developed a biomechanical model of a human-extra vehicular cloth system, and joint torque in elbow and shoulder under unsuited and suited conditions was measured and compared [3]. Szczesna [4] proposed a quaternion blobby model for constraining the joint range of motions based on real captured data. The boundary of the feasible region is modeled using a geometric approach. Draganich et al. [4] developed an analytical model to simulate a trip and fall during gait. The human body was modeled as a 12 degree-of-freedom linkage system. The motion simulation was animated using 3D studio. Song et al. [5] discussed the force and stress distribution within the anteromedial and posterolateral bundles of the anterior cruciate ligament (ACL) in response to an anterior tibial load with the knee at a full extension a 3D finite element model of a human (ACL). Liu et al. [6] used a human biochemical mechanics software to building up a human body model. And a bushing constraint was applied to knee joint. The ADAMS was used to obtain the curves of the acting force and movement on human knee joint. Similar work has been performed in [7–9]. Previous literature provides a good foundation for extravehicular operations of astronauts. However, in most of the literature, the dynamics analysis was performed in normal gravity environment.

In this paper, human dynamics in microgravity environment is discussed. A rigid model considering the Chinese human characteristic is built by using the SolidWorks software, in which all the main joints are maintained. The ADAMS software is employed to perform the simulation of the rigid body model, and the force and torque of the joints during walking process are analyzed. Comparisons are made between the simulation results of normal gravity and microgravity.

2 Structure Analysis of Human Body

The human skeleton is made up of 206 bones, the functions of which are to provide support, give our bodies shape, provide protection to other systems and organs of the body, to produce movement and so on as shown in Fig. 1. The main bones of the human skeleton are the skull, shoulder gridle, arm, hand, chest, spine, leg, ankle and foot. These bones can be classified into 5 types which are long bones, short bones, flat bones, irregular bones and sesamoid bones.

A system of rigid bodies interconnected by joints is called kinematic chain. Joints are movable connections between skeleton elements (rigid bodies) and allow relative motion between them with imposed constraints [4]. There are two main types of joints: cartilaginous (partially moveable) and synovial (free moveable) joint. Cartilaginous joints occur where the connection between the articulating bones is made up of cartilage for example between vertebrae in the spine. Synovial joints are the most common classification of joint with the human body which can be classified as hinge joint (e.g. Elbow, Knee), pivot joint (e.g. Neck), ball and socket joint (e.g. Shoulder, Hip), saddle joint (e.g.

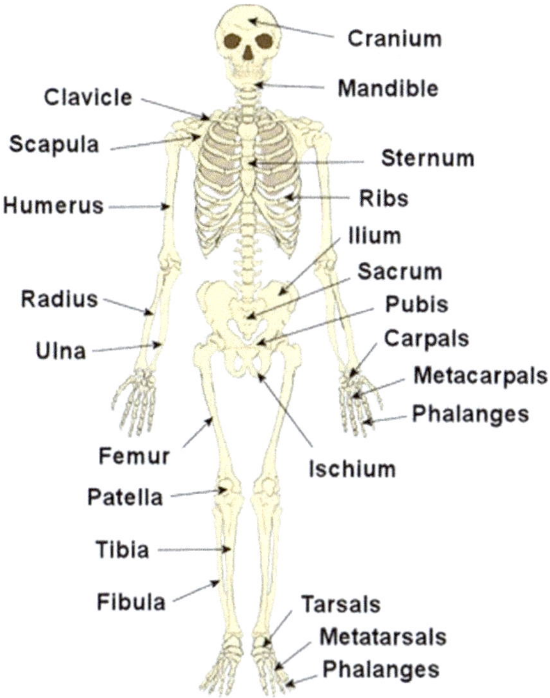

Fig. 1. Diagram of the human skeleton

CMC joint of the thumb), condyloid joint (e.g. Wrist), gliding joint (e.g. Intercarpal joints of hand) [10]. Based on these bones and joints, the whole body can be built according to the 90 percentile of the national standard GB10000-88 of Chinese adult human body data. The main scales of Chinese human body can be listed in Table 1.

Table 1. Main scales of chinese human body

Item	Height	Weight (mm) (kg)	Upper arm length	Forearm length (mm)	Thigh length (mm)	Shin length (mm)
Scale	1755	70	333	253	495	397
Item	Shoulder height (mm)	Foot length (mm)	Max. hand breadth (mm)	Crotch height (mm)	Tibial height (mm)	Thigh clearance height (mm)
Scale	1435	260	108	840	472	146
Item	Lower leg-foot length (mm)	Chest breadth (mm)	Chest depth (mm)	Shoulder breadth (mm)	Waistline (mm)	Hand length
Scale	439	307	237	397	859	193

The above analysis provides necessary information for the human body modeling.

3 Rigid Body Modelling Based on Solidworks

Currently, the most effective design method is 3D modeling. Geometrically accurate 3D models enable one to evaluate and check operation of the future mechanism, make strength and kinematics calculations [11]. The SolidWorks software is one of the most widely used 3D modelling software produced by Dassault System Corporation. In this paper, the SolidWorks software is utilized to build the 3D model for the human body.

In order to connect different parts of the body, two main types of joints are used, which are the ball/socket joint, and hinge joint. Here, it is noted that the rigid body is a simplified model of the real body. Therefore, not all the joints are used. The rigid body is made up of head, two upper arms, two forearms, two hands, torso, two thighs, two shines, and two feet. The ball/socket joints are utilized in the connection of head and torso, upper arm and torso, thigh and torso. The hinge joints are used for upper arm and forearm, forearm and hand, thigh and shin, and shin and foot.

According to the national standard of Chinese human body, 3D models of different parts of the multi rigid body are established and the final rigid model is shown in Fig. 2.

Fig. 2. SolidWorks model

In the upper part of the human body, there are 3 rotation degrees of freedom (DOFs) of the head, 3 DOF of the shoulder, 1 DOF of elbow, wrist and knuckle, separately. In the low limb, the hip joint has 3 DOFs, and the knee and ankle have 1 DOF separately. Hence, the whole rigid model can perform most of the human actions such as walking, squatting, lying down, grabbing, nodding, shaking head, picking up, etc., which satisfies the basic requirement for human body simulation.

4 Dynamic Simulation of Human Body Using Adams

The human body model built in the SolidWorks environment cannot be directly loaded into the ADAMS software. A format transform is accomplished by exporting the body model to a '.Parasolid(x_t)' file. The rigid human body in ADAMS is shown in Fig. 3.

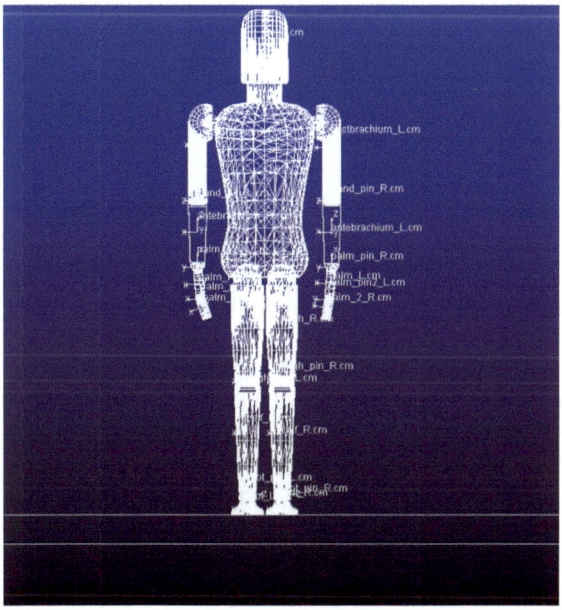

Fig. 3. ADAMS model

In order to perform the dynamics simulation, parameter definitions of all the parts in the rigid body model are necessary. The main parameters and the corresponding values are listed in Table 2.

Table 2. Main parameters of body parts

Item	Torso	Head	Thigh	Shin	Foot
Mass (kgs)	30.835	6.034	9.933	2.569	1.036
I^x/G (kg cm^2)	10177.14	3.704	1527.58	369.45	48.8
I^y/G (kg cm^2)	8720.24	3.242	1527.58	369.45	42.9
I^z/G (kg cm^2)	28806.64	1.917	225.52	34.03	14.88
Item	Upper arm	Forearm	Hand	Finger	
Mass (kgs)	1.701	0.875	0.310	0.138	
I^x/G (kg cm^2)	122.922	27.8	4.228	1.0	
I^y/G (kg cm^2)	122.922	27.8	4.228	1.0	
I^z/G (kg cm^2)	15.800	2.973	2.209	0.6	

In the human body, some parts are constraints to each other due to the connections by the joints. Based on the analysis of human body skeleton in the above section, the constraints are built in the ADAMS model. In order to carry out dynamic analysis, the definitions of load and driver force are necessary which are another type of constraints.

For the upper limb, the shoulder connecting with the body is a main constraint, which is simplified as a ball joint and the other constraints are all treated as the hinge joint. The constraint design of the upper limb is shown in Table 3.

Table 3. Constraint design of upper limbs

Item	Shoulder joint			Ulnar joint	Wrist joint	Ankle
Coordinate axis	X	Y	Z	Z	Y	Y
Maximum angle	−100	−90	−40	−140	−79	0
Minimum angle	+50	+90	+100	0	+70	+90

The lower limb model is critical for simulation accuracy of human walking motion. The hip joint is one of the most important parts which connects legs with the body and has 3 DOFs of rotation. The other joints are knee joint and ankle join, and each of them has one rotation DOF. The constraint design of lower limbs is listed in Table 4. The human model with constraints are built in ADAMS as shown in Fig. 4. It can be seen that several main constraints are included at the connection joints. In addition, the motion control strategy is set in order to simulate the walking motion and working action in real environment.

Table 4. Constraints design of lower limbs

Item	Hip joint			Ulnar joint	Wrist joint
Coordinate axis	X	Y	Z	Z	Y
Maximum angle (degree)	−120	−40	−30	−135	−30
Minimum angle (degree)	+50	+50	+60	0	+30

In ADAMS, there are many function makers which can create mathematical functions, data elements, constant and variables, as well as self-defined subfunctions. In order to accomplish the motion control, action breakout is performed and the current coordinate position is input into the function to control the motion path. During this process, the velocity, acceleration, angular velocity, etc. should be set for the motion control. Another important factor is the environment where the worker stays such as with and without gravity. In ADAMS, motion control function can be shown as Fig. 5. After setting the suitable parameter values, simulation can be performed for certain actions. For example, the pull action is shown in Fig. 6. A pull force of 5 Newtons is added in the right hand from 8th to 12th second, and the stress, torque, displacement, relative velocity, relative angular velocity, relative accelerate velocity, etc. are derived under normal gravity and zero gravity. The force of the shoulder and wrist joint are shown in Figs. 7 and 8. In these figures, the red solid curve indicates the profile with normal gravity, while the blue dash curve indicates that with zero gravity.

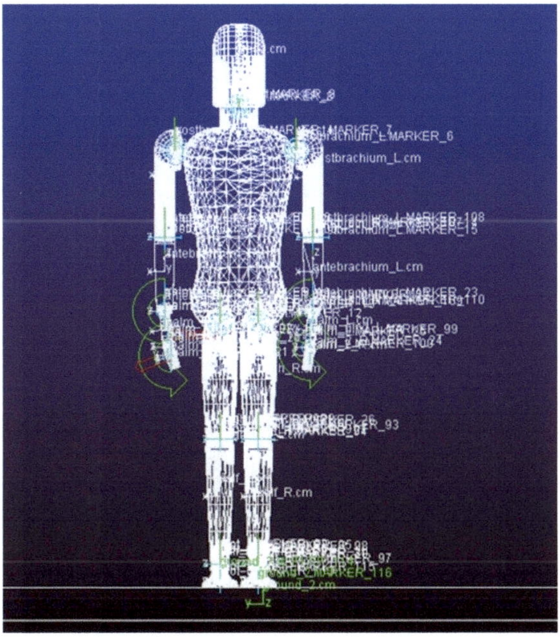

Fig. 4. Human model with constraints in ADAMS

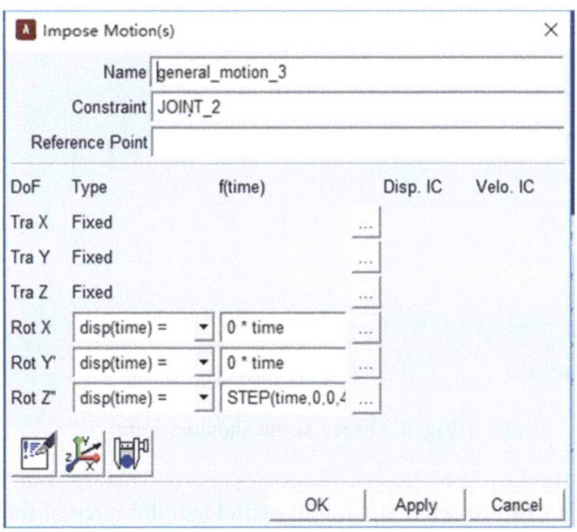

Fig. 5. Motion control function in ADAMS

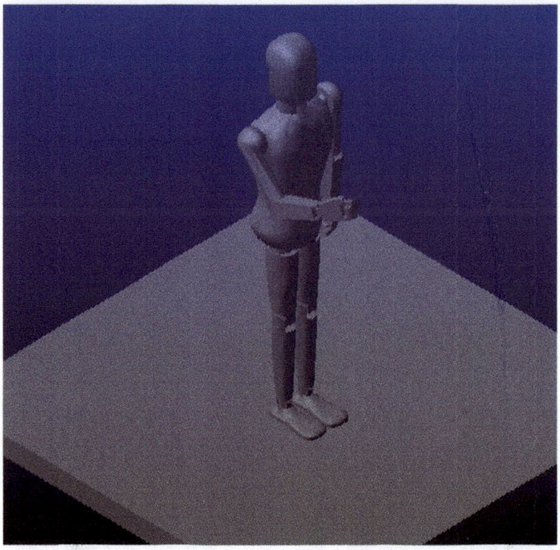

Fig. 6. Full action model

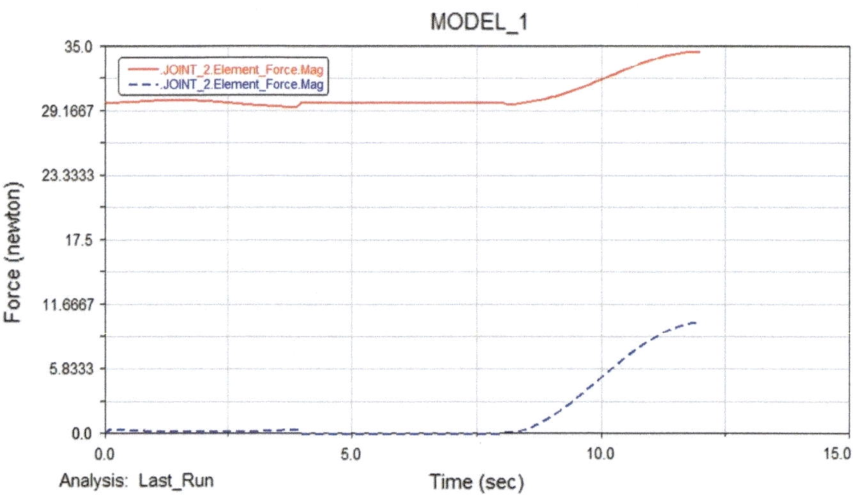

Fig. 7. Force at the shoulder joint

From Fig. 7, it can be seen that without extra load, the force of the shoulder joint in zero gravity environment is 29.9 Newtons less than that in normal gravity environment which is exactly the weight of the upper arm. However, in Fig. 8, there is no great difference of the wrist joint torque. Therefore, it can be concluded that gravity affects the force in a more obviously way than torque.

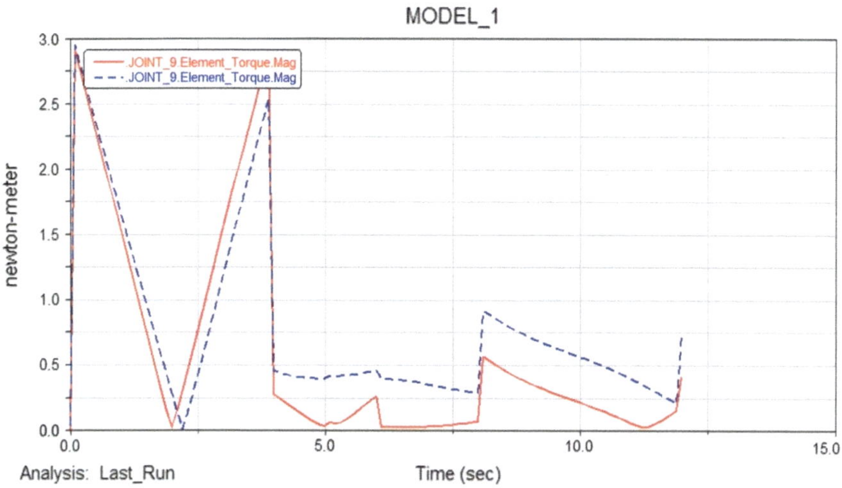

Fig. 8. Torque at the wrist joint

Fig. 9. Walking model

The next simulation is about walking motion as shown in Fig. 9. The force and torque analysis of knee, ankle and hip joint are accomplished. The simulation time is set as 2 s. The results are shown in Fig. 10. It can be seen that in Fig. 10a, the hip force also shows an obvious reduction while Fig. 10b, c do not show so obvious variation.

Finally, another more complicated simulation about a continuous working action is finished. The whole screw fixing process is: (1) pick up a tool using right hand, (2) walk to the working place, (3) squat and hold the work piece with left hand, (4) fix

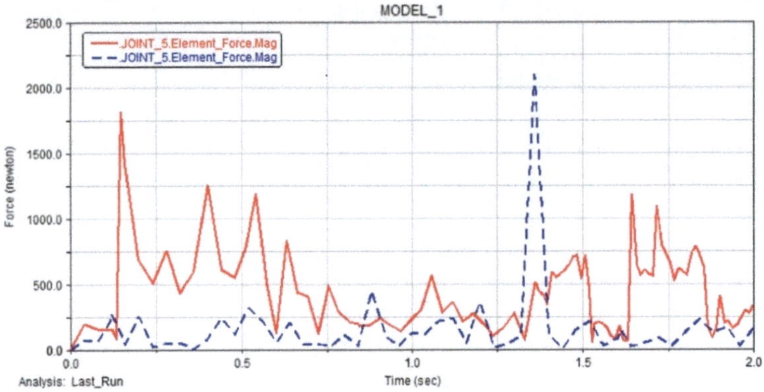

(a) Force of hip joint

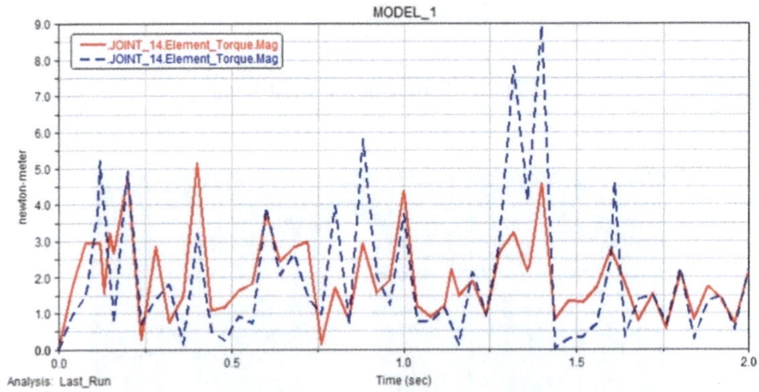

(b) Torque of knee joint

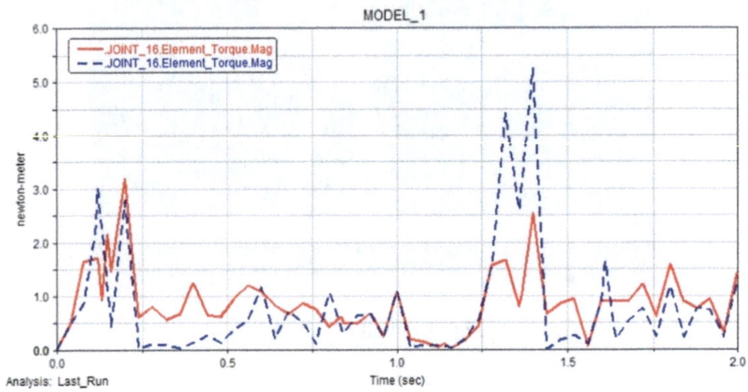

(c) Torque of ankle joint

Fig. 10. Simluation results of walking motion

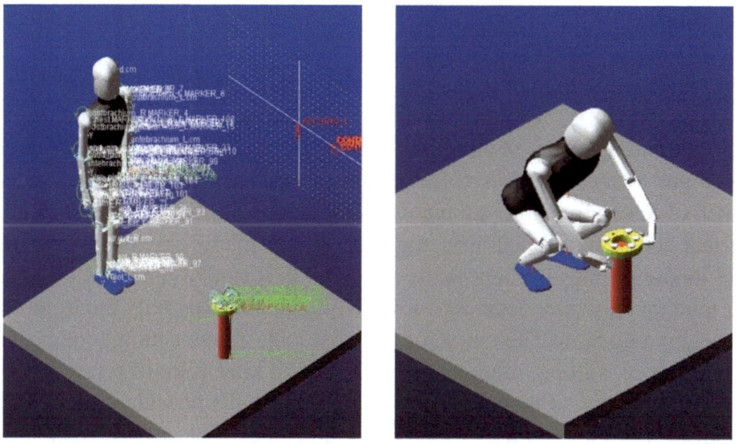

(a) Constraints **(b)** Screw fixing

Fig. 11. Screw fixing model

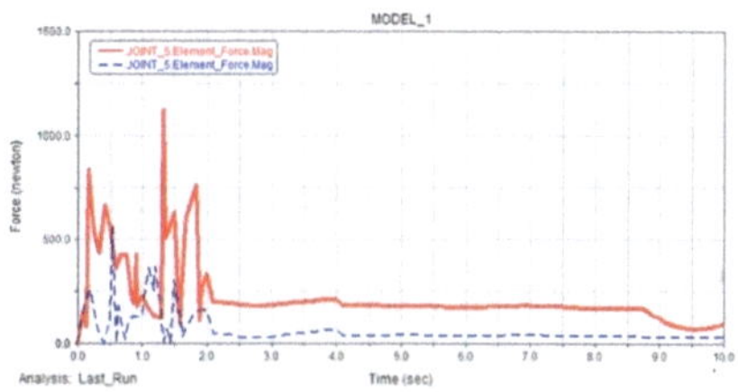

(a) Force of hip joint

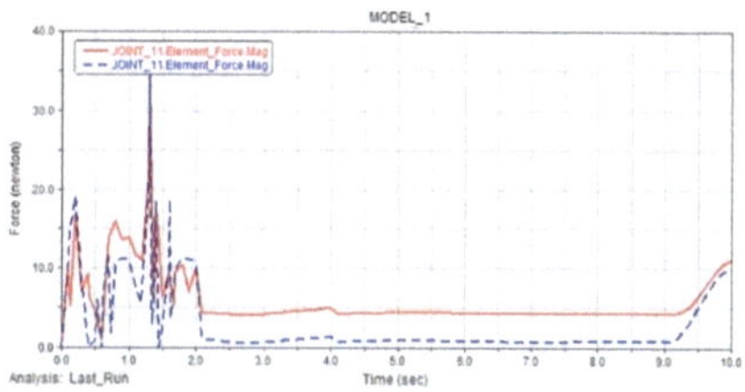

(c) Force of left wrist joint

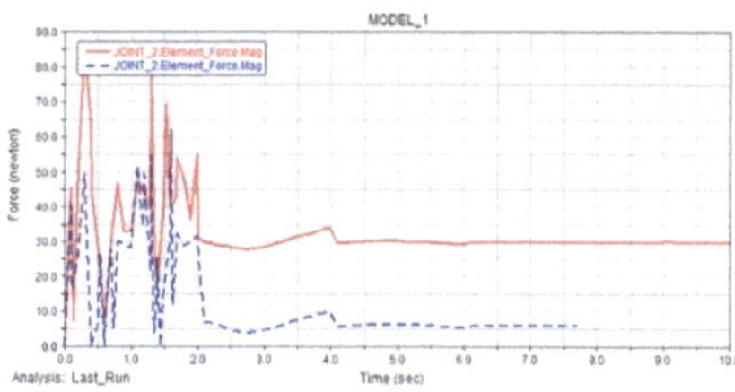

(b) Force of right shoulder joint

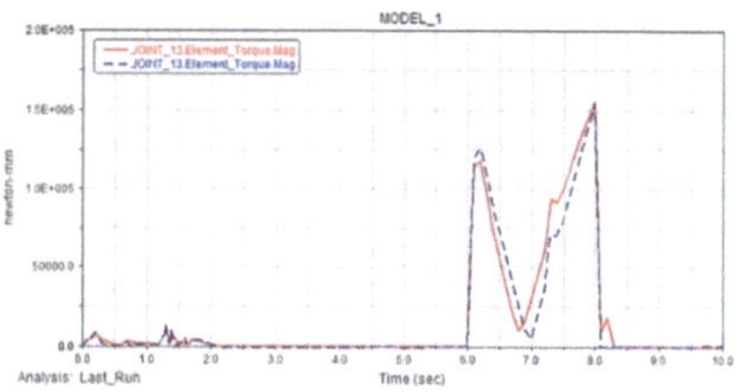

(d) Torque of left knee joint

Fig. 12. Simulation results of screw fixing

the screw using the tool. The constraint model and screw fixing action model are shown in Fig. 11. The simulation results of different joints in microgravity and normal gravity are shown in Fig. 12. It can be seen that the force of hip joint is very large, and decreases more obviously compared with that of the upper limb joints. The reason is that the low limb supports the body weight which is reduced a lot in the microgravity environment. The results provide a good guidance for the astronaut extravehicular task design.

5 Conclusion

This paper studies the human dynamics by using the ADMAS software. The Chinese human body model is built in SolidWorks environment considering the national standard data and appropriate simplification of the main joints. The 3D human body model is input to the ADAMS software and necessary constraints are added to the proper parts. Several action simulations are accomplished including pull action, walking and screw fixing action in the normal and micro gravity environment. The force and torque profiles of main joints are derived. It can be seen that the force of the shoulder joint and hip joint decreases obviously in the microgravity environment due to the self-weight reduction. The torque of the wrist joint of pull action, toques of the knee and ankle joints of walking model, and the torque of knee joint of screw fixing are not affected by the gravity so much like the force. The results provide critical information for the astronaut training on the ground to simulate the space environment and the task design so that the proper actions can be planned to avoid muscle injury and task failure.

Acknowledgements. This paper is supported by the National Natural Foundation of China (Grants No. 51505099 and No. 71390522).

References

1. M. Vukobratovic, V. Potkonjak, S.G. Tzafestas, Human and humanoid dynamics: from the past to the future. J. Int. Robot. Syst. **41**(1), 65–84 (2004)
2. J.M. Ohara, M. Briganti, J. Cleland, W. Dan, Extravehicular activities limitations study. Volume 2: Establishment of physiological and performance criteria for EVA gloves, in *NTRS* (1988)
3. X.D. Wang, Z. Wang, H. Li, C.H. Wang, Research on method of extravehicular spacesuit joint torque measurement and modeling. Space Med. Med. Eng. **28**(3), 195–202 (2015)
4. A. Szcz̨esna, *Verification of the Blobby Quaternion Model of Human Joint Limits*, vol. 39, pp. 130–138 (2018)
5. X. Zhou, L. Draganich, F. Amirouche, A dynamic model for simulating a trip and fall during gait. Med. Eng. Phys. **24**, 121–127 (2002)
6. J. Liu, J. Zheng, Q. Wu, ADAMS-based human knee joint kinematics and dynamics research. Comput. Appl. Softw. **20**(6), 202–204 (2012)
7. X. Li, G. Wang, G. Jin, Human modeling and motion simulation based on ADAMS. Mech. Eng. Autom. **4**, 11–13 (2009)
8. C. Ma, Y. Liu, X. Zhu, Research of human modeling and motion simulation based on ADAMS. Adv. Mater. Res. **1016**, 292–297 (2014)
9. L. Stirling, D.J. Newman, K. Willcox, Self-rotations in simulated microgravity: performance effects of strategy training. Aviat. Space Environ. Med. **80**, 5–14 (2009)
10. http://www.teachpe.com/anatomy/joints.php
11. A.N. Loginovsky, L.I. KhmarovaV, 3D Model of geometrically accurate Helical-Gear Set. Procedia Eng. **150**, 734–741 (2016)

Cloud Computing Knowledge Domains Mining Based on Patent Networks

Cheng Cheng[1], Peng-hui Lyu[1,2(✉)], and Chao Fu[2]

[1] School of Management, Hefei University of Technology, Hefei 230009, People's Republic of China
lyuph@uwm.edu
[2] Department of Management and Marketing, Hong Kong Polytechnic University, Hung Hom 999077, Hong Kong, People's Republic of China

Abstract. In this paper several technical areas was selected to analyze the research fronts and hotspots of cloud computing. From the perspective of patent networks of cloud computing in global patent database, we summed up the development of global cloud computing. In the view of patents analysis, this paper puts forward the development directions of cloud computing as "one hard and one soft", in two-way development which is the combination of information technology or new generation of information technology industry, and then provided the implementation of the cloud computing industry develop plan for China.

Keywords: Cloud computing · Informatics analysis · New generation of information technology · Patent network

1 Introduction

The concept of "cloud computing" was first proposed by Google, it's most basic concept is splitting the huge computing program into numerous smaller subroutine automatically by network, and then by searching and analysis of the huge system which is made up by a number of servers, after the analysis, the results will be passed back to the user [1]. Through this technology, network service providers can handle tens of millions of or even hundreds of information within a few seconds to achieve the same powerful performance of supercomputer [2].

Cloud computing is the personal computer, the Internet, the third representative of the information technology revolution [3], is a kind of through the Internet to provide customers with hardware and software services such as the new information technology service model, but also the next generation of technical architecture trends [4]. Cloud Computing is Distributed Computing [5], Parallel Computing [6], Utility Computing, Network Storage Technologies, Virtualization, Load Balancing, High Available and other traditional computer and network technology to develop the product of integration. Compared with the traditional technology, cloud computing has a wide range of network access [7], flexibility [8], super-scale [9], scalability [10], resource sharing

G. Q. Huang et al. (eds.), *Proceeding of the 24th International Conference on Industrial Engineering and Engineering Management 2018*,
https://doi.org/10.1007/978-981-13-3402-3_71

[11], on-demand and other technical features. These characteristics change the patterns that people use and give fundamental changes to people's lifestyles and methods of production.

Understand the development trend of cloud computing, for the country to develop industrial policy, innovation and development services for enterprises, providing the necessary scientific basis. The formation of knowledge network and its evolution law can effectively provide scientific methods and corresponding tools for knowledge development and innovation trend forecasting. At present, there is not much research on the knowledge network of cloud computing, and the development trend of cloud computing is mainly based on the literature content analysis [12]. With the information and data as the center, through the information measurement and data mining technology analysis cloud computing technology trends become hot. The first is the use of patent database in the inventor, citation and other rules to study [13], the most commonly used is the citation analysis and clustering analysis [14]; the other is the text mining, mainly on the description of the study [15]. Based on the patent network of knowledge network, this paper analyzes the development trend of cloud computing, expands the details of cloud computing technology and its related relationship.

2 Methodology

The data were retrieved on July 31st, 2017 from Derwent Innovation Index (DII) database offered by the producer by Clarivate Analytics. In order to improve the relevance of the search topic, the "cloud comput*" was used as the search term to retrieve the data of the global cloud computing patent data from 1963 to 2016, and 10372 data were obtained.

This paper chooses Derwent Innovation Index's database as a retrieval platform, and patent analyzes and studies data mining with Thomson Data Analyzer and Origin as the main analytical tools. From the perspective of patent results measurement analysis, the global patent results of the annual distribution, distribution and other aspects of the analysis and research were shown in this paper.

3 Data Analysis

3.1 Cloud Computing Patent Trend Study

From Fig. 1, we can learn that from 2000 to 2007, cloud computing patent results are fewer, indicating that in this time cloud computing has not yet as developed, technical supports were not enough. From 2008 to 2016, the number of cloud computing patent applications is in steadily creasing, reaching its peak in 2014. The priority year and the number of applications year yields are very close; there are two major gaps under the main reasons. Firstly, the patent application does not require priority after the request in that year. Secondly, the appearance of patents applied does not include priority

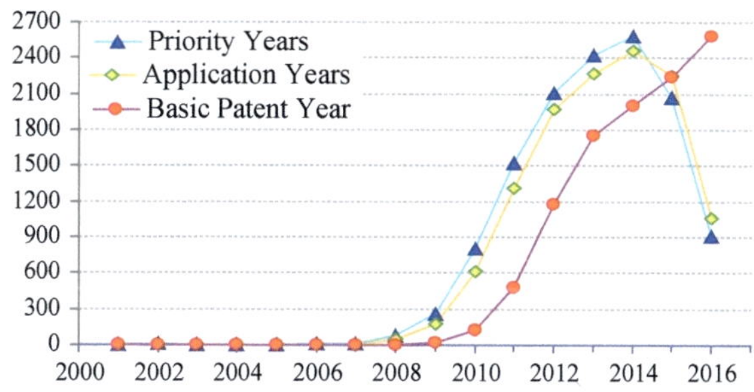

Fig. 1. Global production trends of cloud computing patent

requirements. So on the whole, the number of cloud computing patents has been growing always, indicating that cloud computing gathered more and more attentions from all countries, this emerging area has great value to develop.

As shown from Fig. 2, the number of cloud computing patents and the corresponding number of applicants grow rapidly since 2000, until now cloud computing research is still on its patents baby stage.

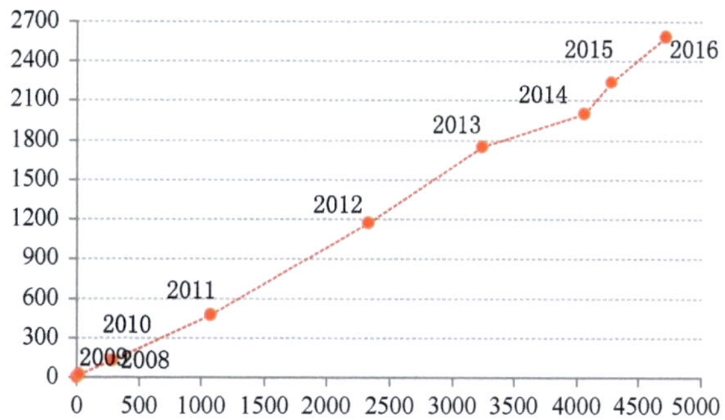

Fig. 2. Technology life cycle of cloud computing patents

To show patent technology life cycle clear, the detailed patents and inventor were provided in Fig. 3, the number of cloud computing applicants is greater than the cloud computing patent, which shown in the diagram that the cloud computing research still kept in its early development, as in fast increasing stage these inventors may apply for more cloud computing patents, but not a patent with multiple applicants as now shown in Fig. 3.

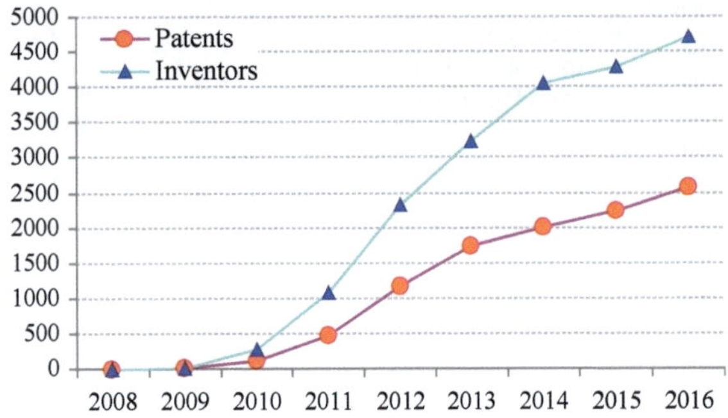

Fig. 3. Cloud computing patents and inventors

3.2 Cloud Computing Patent Assignee Network

Table 1 shows the code, the number of patents, their productive countries as well as the meanings of cloud computing patent code. In order to ensure the uniqueness of the organization and the accuracy of the data records, the standard patentee code were chosen. As seen in Table 1, the top ten companies included eight US companies and two Chinese companies, they contributed more than about 70 patents in the cloud computing, the contribution of Chinese companies is not that strong, which need more research efforts in future.

Table 1. Productive patent assignees in cloud computing

No.	Patent assignee	Records	County
1	IBMC-C	1891	USA
2	MICT-C	351	USA
3	INEI-C	264	China
4	HEWP-C	135	USA
5	ECEM-C	129	USA
6	GOOG-C	112	USA
7	REDH-C	84	USA
8	ZTEC-C	81	China
9	ORAC-C	74	USA
10	AMAZ-C	69	USA

Figure 4 show the collaboration network of patent assignees, which shows a network of productive partners, including patent individuals, standard companies and non-standard companies. Obviously, from the perspective of personal patents, we can find that the three individuals located in the center of the cloud computing cooperation network. From the company's view, IBM and Microsoft enter into the central of the cooperative network already.

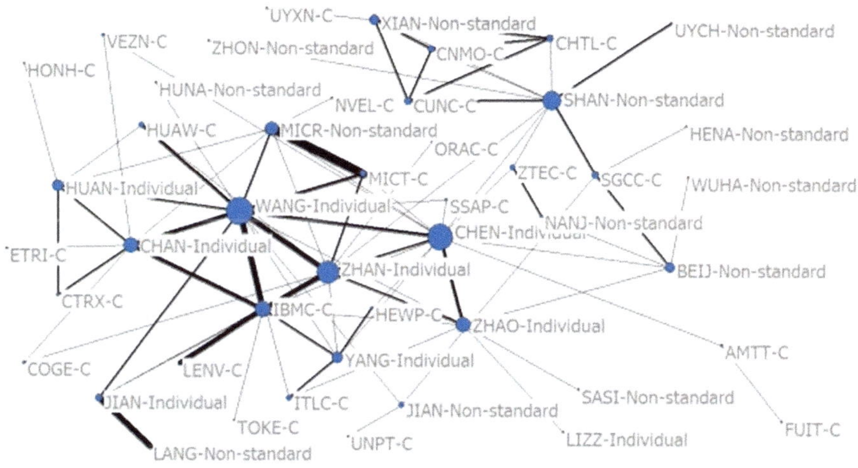

Fig. 4. The collaboration network of patent assignees

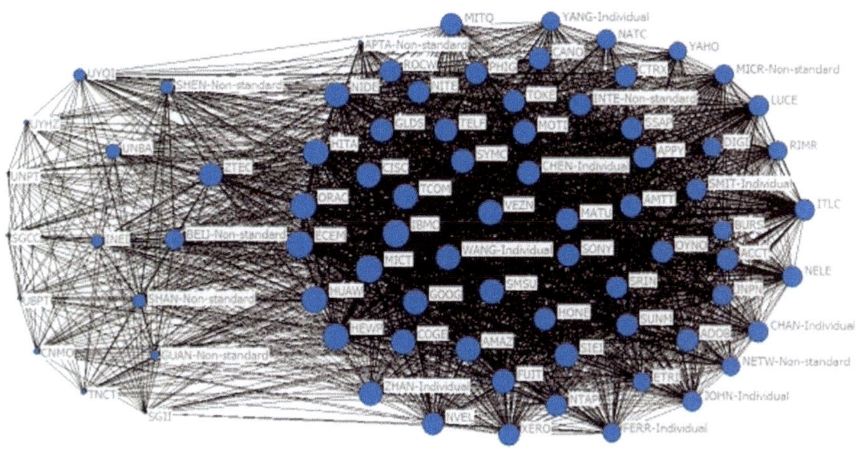

Fig. 5. The co-citation network of patents assignees cited

From Fig. 5, it can be seen that in the field of cloud computing two large productive patents assignees came into the same two groups. Located in the right side, led by the international companies or camp institutions of the United States, is the group such as IBM, Microsoft, Google, Amazon. While in the other side patents from more mature international companies, such as Samsung, Huawei and Sony were cited more. From the left side of the graph shows us the emerging camps of Chinese companies or institutions, including ZTE, Bei-hang University, Anhui Electric Power and so on, they will be the strengthened contributions in the next stage in cloud computing research.

3.3 Cloud Computing Patent Classifications

The cloud computing field of the top research topics with the records more than 100 patents was shown in Table 2. In which digital data processing and digital information transmission possessed most shares in all the classifications.

Table 2. The meanings of the international classification of cloud computing

No.	International classifications	Records	Meanings
1	G06F	5672	Digital data processing
2	H04L	4786	Digital information transmission
3	G06Q	1483	Data processing systems or methods that are specific; other categories do not include processing systems or methods that are specifically applicable
4	H04W	432	Wireless communication network
5	H04N	399	Image communication
6	G05B	260	A general control or adjustment system; a functional unit of such a system; a monitoring or testing device for such a system or unit
7	G06K	244	Data identification; data representation; record carrier; record carrier processing
8	H04M	164	Telephone communication
9	G06T	161	General image data processing or generation
10	G06N	144	Computer system based on specific computational model
11	G08G	116	Traffic control system
12	A61B	112	Diagnosis; surgery

The cloud computing international classification code occurrences in a same patent, which can be used to show the research hot issues, just as Fig. 6 do. From the figure we can see that the current research hotspots of cloud computing are digital data processing and digital information transmission as patent results shows, the number of records recorded by Derwent Innovation Index database is 5,672 and 4,786, far beyond the other research directions. On the whole, the current research directions of cloud computing are still focused on information transmission, data identification, processing and wireless communication network, which is mainly due to the rapid development of big data technology.

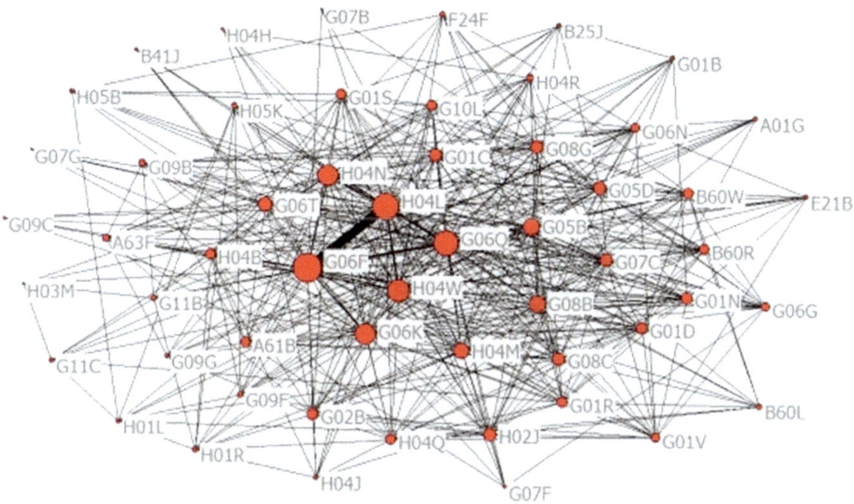

Fig. 6. The cooccurrence network of IC in cloud computing

3.4 Cloud Computing Inventor Network Study

The inventors and their co-occurrence relationships were shown in Table 3, in which the main numbers were given. And then the collaboration network of cloud computing patents inventors can be formed and more vivid information can be obtained as in the Fig. 7. From the inventor co-occurrence matrix of the cloud computing patents, on can learn the more the value, the relationship more closely, and the little, the less. From the Table 3 we can conclude, LI J J and WANG Y become the center of the cooperative network and the research cooperation between Zhang Y and Wang Y close, the correlation coefficient of them is 0.095. In addition, Zhang J M and Liu Y are close, with the correlation coefficient of 0.086.

From the cloud computing patent inventor network we can also clearly get that most of the potential inventor cooperation still can be kept in developing. Several isolated points on the edge of the figure also show that the inventors and the other producers have basically no cooperation, which can be developed as the future relationships.

Table 3. The co-occurrence relationships between inventors of cloud computing research

	Zhang Y	Li Y	Chen Y	Li Z	Wang Y	Li J J	Zhang X Y	Zhang J M	Wang W J	Liu Y	Li X Y
Zhang Y	1	0.006	0.027	0	0.095	0.047	0.01	0.019	0.036	0.012	−0.003
Li Y	0.006	1	0.021	0.015	0.024	0.003	0.066	0.012	0.005	0.022	0.022
Chen Y	0.027	0.021	1	−0.006	0.009	0.01	0.019	−0.003	−0.011	0.013	0.022
Li Z	0	0.015	−0.006	1	0.033	0.011	0.011	−0.003	0.022	0.005	0.014
Wang Y	0.095	0.024	0.009	0.033	1	0.021	−0.011	0.006	−0.002	0.024	0.025
Li J J	0.047	0.003	0.01	0.011	0.021	1	0.014	0.007	0.016	0.016	0.072
Zhang X Y	0.01	0.066	0.019	0.011	−0.011	0.014	1	−0.001	−0.01	0.008	0.008
Zhang J M	0.019	0.012	−0.003	-0.003	0.006	0.007	−0.001	1	−0.01	0.086	0.009
Wang W J	0.036	0.005	−0.011	0.022	−0.002	0.016	−0.01	−0.01	1	0	0.02
Liu Y	0.012	0.022	0.013	0.005	0.024	0.016	0.008	0.086	0	1	0.05
Li X Y	−0.003	0.022	0.022	0.014	0.025	0.072	0.008	0.009	0.02	0.05	1

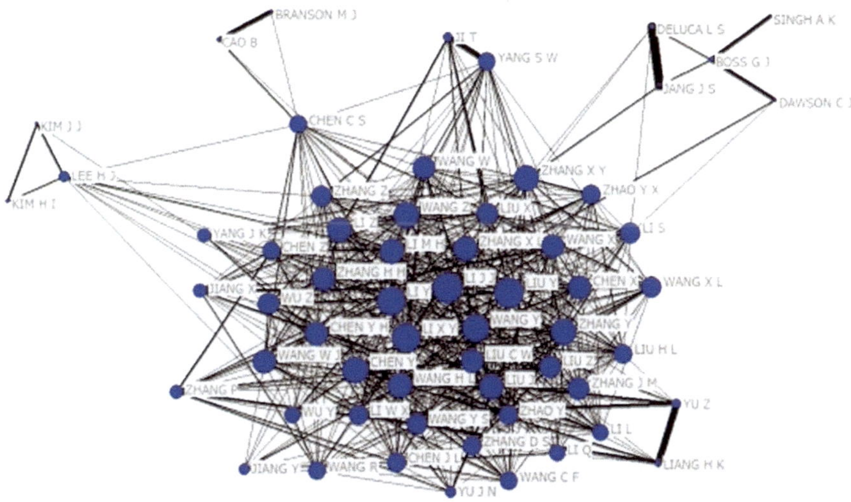

Fig. 7. The inventors' collaboration network of cloud computing research

4 Conclusion

Through the cloud computing patent data analysis, the conclusions can be summed up as the following points: Firstly, cloud computing patents developed rapidly, as in its fast rising period, the main countries in the world still need to increase their investment and construction efforts.

Secondly, cloud computing development is still quiet uneven, the United States of America is still very obvious have advantage in the field of cloud computing, followed by the contribution of P.R. China, the production results are not very prominent, so there is room for improvement.

Thirdly, the cooperation between famous companies in the research field of cloud computing is very important. The large international companies' patents from the United States of America are still need collaborating with others. Their continuous investments are very urgent and their outputs are amazing. Chinese enterprises need to strengthen international cooperation continually, and constantly improve their own international corporations.

Lastly, from the classification of cloud computing patents, information transmission and data processing are important directions, for those enterprises from the field of cloud computing to grow bigger; it is not only new technologies but also a part of their future strategy.

Acknowledgements. Thanks for the support from the National Science Foundation of China (Program No. 71571060).

References

1. M. Armbrust, A. Fox, R. Griffith et al., A view of cloud computing. Commun. ACM **53**(4), 50–58 (2010)
2. C. Yang, Q. Huang, Z. Li, Big Data and cloud computing: innovation opportunities and challenges. Int. J. Digit. Earth 1–41 (2016)
3. S. Marston, Z. Li, S. Bandyopadhyay, A. Ghalsasi, Cloud computing—the business perspective. Decis. Support Syst. **51**(1), 176–189 (2011)
4. C. Hutchinson, J. Ward, K. Castilon, Navigating the next-generation application architecture. IT Prof. **11**(2), 18–22 (2009)
5. M.D. Dikaiakos, D. Katsaros, P. Mehra, G. Pallis, A. Vakali, Cloud computing: distributed internet computing for IT and scientific research. IEEE Internet Comput. **13**(5), 10–13 (2009)
6. T. Gunarathne, B. Zhang, T.L. Wu, Scalable parallel computing on clouds using Twister Azure [J]. Futur. Gener. Comput. Syst. **29**(4), 1035–1048 (2013)
7. R. Buyya, C.S. Yeo, S. Venugopal, Market-Oriented cloud computing: vision, hype, and reality, in *IEEE International Conference on High Performance Computing and Communications* (2008), pp. 5–13
8. V. Chang, Y.H. Kuo, M. Ramachandran, Cloud computing adoption framework: a security framework. Futur. Gener. Comput. Syst. **57**, 24–41 (2016)
9. A. Bateman, M. Wood, Cloud computing. Bioinformatics **25**(12), 1475 (2009)
10. C. Hewitt, ORGs for scalable, robust, privacy-friendly client cloud computing. IEEE Int. Comput. **12**(5), 96–99 (2008)
11. M. Armbrust, A. Fox et al., in *Above the Clouds: A Berkeley View of Cloud Computing*, vol. 53(4) (EECS Department University of California Berkeley, 2009), pp. 50–58
12. M. Bayramusta, V.A. Nasir, A fad or future of IT: a comprehensive literature review on the cloud computing research. Int. J. Inf. Manag. **36**(4), 635–644 (2016)
13. T.U. Daim, G. Rueda, H. Martin, Forecasting emerging technologies: use of bibliometrics and patent analysis. Technol. Forecast. Soc. Change **73**(8), 981–1012 (2006)
14. S.A. Morris, Z. Wu, G. Yen, A SOM mapping technique for visualizing documents in a database, in *Proceedings of the 2001 International Joint Conference on Neural Networks IJCNN'01*, vol. 3 (2001), pp. 1914–1919
15. S. Lee, B. Yoon, Y. Park, An approach to discovering new technology opportunities: keyword-based patent map approach. Technovation **29**(6–7), 481–497 (2009)

Big Data-Driven Simulation Analysis for Inventory Management in a Dynamic Retail Environment

Haixia Sang[1(✉)], Shingo Takahashi[1], and Rie Gaku[2]

[1] Faculty of Science and Engineering, Department of Industrial and Management, Waseda University, Tokyo, Japan
sanghaixia@aoni.waseda.jp; Shingo@waseda.jp

[2] Graduate School of Business Administration, St. Andrew's University, Osaka, Japan
r-gaku@andrew.ac.jp

Abstract. Inventory management is one of the most important factors in logistics operations. However, real-world inventory systems are complexly intertwined with related elements, and determining the optimal parameters and identifying the determining factors that influence inventory changes are complex problems. In this paper, using real POS data, we propose a simulation-based algorithm to optimize automated refreshment systems in a retail environment. The inventory system is modeled and simulated, which then returns the performance functions. The expectations of these functions are then estimated by an algorithm and the optimal combination result is obtained. Based on the sensitivity analysis, the determining factor that influences inventory changes is identified. The results show that the proposed simulation-based algorithm is powerful and effective.

Keywords: Inventory management · Simulation · Optimization

1 Introduction

In today's competitive environment the automated refreshment system is becoming a widely accepted inventory method, especially within the retail industry. Currently automated refreshment systems have been adopted by the majority of retail establishments in Japan [1]. As this study found, the reason for the popular use of such method is that many managers realize that an effective automated refreshment system can maintain inventory level while it does not require high levels of running costs. However, automated refreshment systems bring with them many challenges. For example, inventory control includes a high level of demand uncertainty, contradictory objectives, constraints, and many decision variables, [2] and such decision variables may be complexly intertwined with each other. Generally many settings, such as reorder point and order quantity, are usually based on the manager's intuition and

G. Q. Huang et al. (eds.), *Proceeding of the 24th International Conference on Industrial Engineering and Engineering Management 2018*,
https://doi.org/10.1007/978-981-13-3402-3_72

experience, hence, overstock or shortages tend to happen. To achieve increased satisfaction in service levels, the determining factor that influences inventory changes must be identified and the optimal parameters found.

Big data used to be a technical problem, but is now a business opportunity and enterprises are exploring big data to discover previously unknown fact [2]. Now almost all retail chains collect big data from point of sales system (POS), however, how to use this accumulated data to improve business performance is still an important problem. Simulation has been used as a decision-making tool to solve complicated real-world problems. However, simulation models do not provide the capability to define the optimal set of decision variables related to a predefined objective function. Instead this is made possible by optimization models that allow decision makers to find the best possible alternatives. Existing literature proves that Optimization via Simulation (OvS) is relatively easy to develop regardless of the complexity of the problem and provides a much more realistic solution methodology without assumption [3–5]. Many papers have demonstrated that simulations are also a powerful and pragmatic tool for analyzing and optimizing inventory problems. Chou and You proposed a novel simulation-based optimization framework to optimize a multi-echelon inventory system under uncertainties [6]. Zheng et al. simulated a client's finished product inventory at its manufacturing sites and warehouses [7]. Sang and Takakuwa developed a simulation-based approach to obtain the optimal order quantities of short-expiration-date items, which are delivered and scrapped three times a day [8]. The purpose of this paper differs from previous research in that the study aims to develop a simulation-based algorithm within a retail environment. Further, the approach is taken from a practical point of view and is not only used to optimize the inventory problem, but also to identify the determining factors that influence inventory changes.

2 Problem Statement

In this work we focus on the retail inventory system. Recently, the majority of retail chains in Japan have introduced an automated refreshment system to enhance order performance. Demand is forecast, then, based on the reorder point and order policy, the order quantity is determined.

Different from manufacturing inventory and distribution inventory problems, the retailer's customer is the end consumer who may be influenced by stock levels in a shop. Therefore, a concept called "Face" was introduced to set the reorder pint and the order quantity. "Face" is a row of a particular item on a shelf as shown in Fig. 1. The item's overall display amount will be determined by the number of goods per row and the number of "Face". For example, if the reorder point is set as 70% of the item's overall display amount, then as Fig. 1 shows, item A and item B's reorder points would be set as 16 and 21, respectively. In the automated refreshment system the order quantity can be calculated by using the following function:

$$O = F + V - I$$

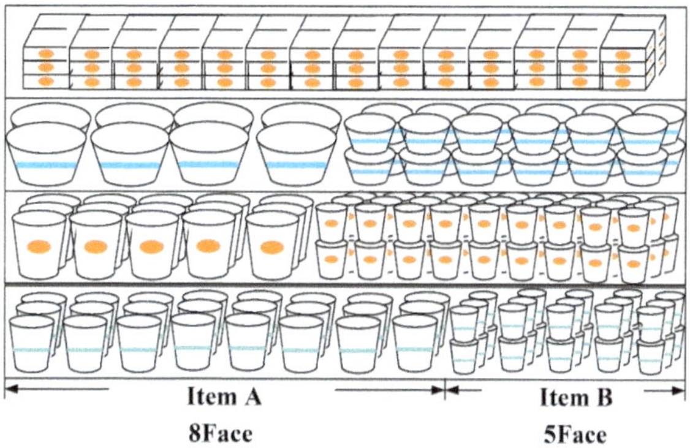

Fig. 1. The concept of 'face'

where
F: forecasted order quantity (pcs.)
V: reorder point (pcs.)
I: The present inventory level (pcs.)

From the function, it can be found that the reorder point and order quantity are influenced by the demand prediction method, the item's overall display amount. In addition, when the goods are shipped from the distribution center or manufacturer, the lot-sizes may also impact the retailer's inventory level. In the automated refreshment system, the relationship among the inventory change, the reorder point, and lot-sizes can be represented as shown in Fig. 2.

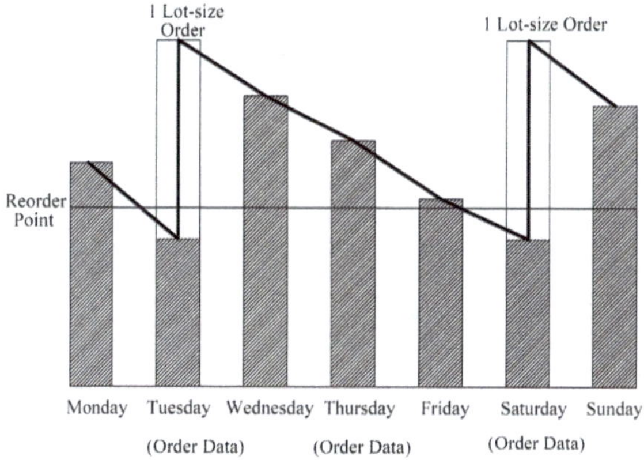

Fig. 2. The relationship among the reorder point, order quantity and lot-size

3 Case Study

3.1 An Automated Refreshment System

This paper takes as a case study an automated refreshment system used by Japan's third largest convenience store franchise chain. To satisfy diverse demands from different customers, 3000 different items are sold in the store. The items, such as cup noodles, beverages and general merchandise have a large item quantity and lower demand variability and can be ordered using the automated refreshment system. Although the economic benefit of these items is not crucial to the entire system they are essential necessity items at convenience stores, and therefore how to make inventory orders of these items more efficient is important to store managers.

3.2 Simulation Model Construction

In this paper, item A is selected as the research object. Based on the one week actual data, the demand for 1 year was forecast using the moving average method. The item A is ordered on Tuesday, Thursday and Saturday. If the item is ordered before 10:00 a.m., then it would ship by 8 p.m. on the same day.

Based on analyses of real POS data from January 1, 2012 to December 31, 2012 an original simulation model was constructed to analyze the current automated refreshment system. This model is termed the AS-IS model. This study used the Excel VBA simulation platform to develop the AS-IS model, the simulation logic flow and the simulation result as shown in Figs. 3 and 4.

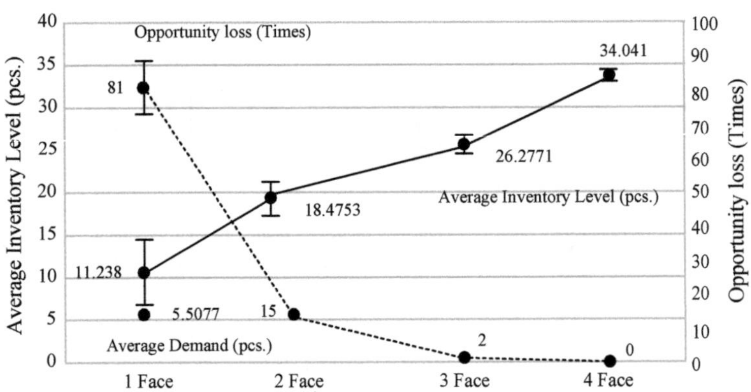

Fig. 3. The results of the AS-IS model

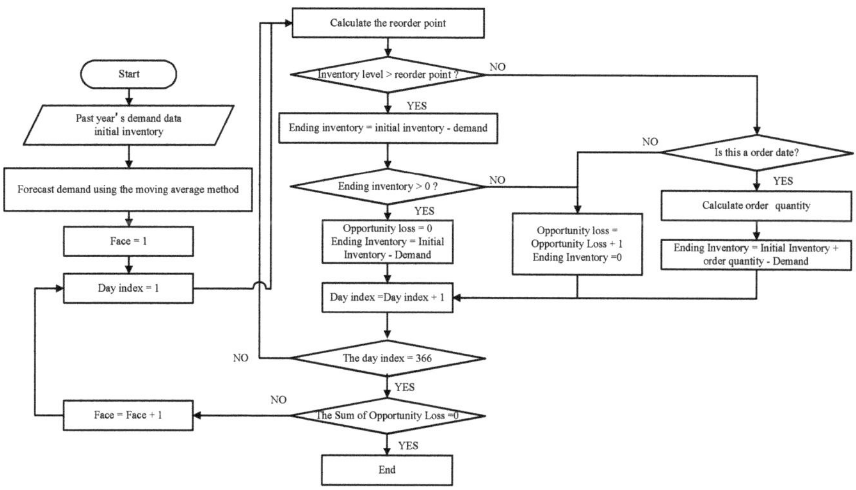

Fig. 4. The AS-IS model logic

By using the AS-IS model, the average inventory level and opportunity loss times for each "Face" can be output. As the results show, the opportunity loss is inconsistent with the average inventory level. Although one "Face" can achieve a low average inventory level, the opportunity loss time is 81 times a year, which shows a poor service level. No opportunity loss appears when the "Face" number is increased up until 4, however, the average inventory level is 34.041, which is too high compared with the average demand.

3.3 Optimal Algorithm Model Construction

The objective of inventory optimization is to identify the optimal parameters for a given inventory control policy to minimize inventory levels while maintaining acceptable service levels [9]. In this study, we set the acceptable service level for opportunity loss as 0, and consider forecast method, lot-sizes, and the number of goods per "Face" and the ratio of the item's overall display amount as most relevant factor for our investigations.

Different forecast methods will produce different prediction results for different order quantities. Hence, in this paper, the moving-average method, exponential smoothing method, regression analysis and multiple linear regression analysis were introduced in the system to evaluate the impact of forecasting results on the inventory level.

In the automated refreshment system, although the order quantity is calculated in the system, due to the given lot-sizes, the amount of goods that arrive is always more than the quantity that the system wants.

In this paper, we set the item's overall display amount ratio at 0.1–0.9 to simulate the inventory level changes and opportunity losses that can occur. When the ratio of the item's overall display amount is set as a low value, which means the reorder point is low due to the late timing of an order, there is a high probability that opportunity loss may happen. In contrast, if the amount is set as a high value this means the order is placed frequently, which makes it a high inventory level.

As we mentioned in Sect. 2, the main driver in the automated refreshment system is the item's overall display amount, which is determined by the number of goods per "Face" and the number of "Faces". Because the optimal number of "Faces" is the system output, thus the number of goods per "Face" must be evaluated carefully. The input forecasting method and parameters that we used in this paper are shown in Table 1. The problem surrounds how to determine the optimal parameter combination that server an opportunity loss of 0 for a lower inventory level. Though a simulation-based optimization method is applicable to a complicated inventory system, this procedure only provides black-box functions and no analytical expressions characterizing the input-output relationship. To solve the problem, an algorithm was constructed that allows the input parameter combinations and the corresponding results to be output to an Excel sheet, which can make the problem visible. The algorithm's logic is shown in Fig. 5. After running the optimal procedure, the optimal parameter combination was identified. By comparing the AS-IS model and the optimal algorithm model's result, it was found that the service level was maintained and at the same time, the inventory problem was improved.

Table 1. Comparison of the AS-IS model and optimal model

	Element factors				Results	
	Forecast method	Lot-sizes (pcs.)	The numbers of goods for per column (pcs.)	The ratio of item's overall display amount (pcs.)	Average inventory level (pcs.)	Opportunity loss (times)
AS-IS model	Moving-average method	12	9	0.5	34.041	0
Optimal model	Regression analysis	12	15	0.7	20.40274	0

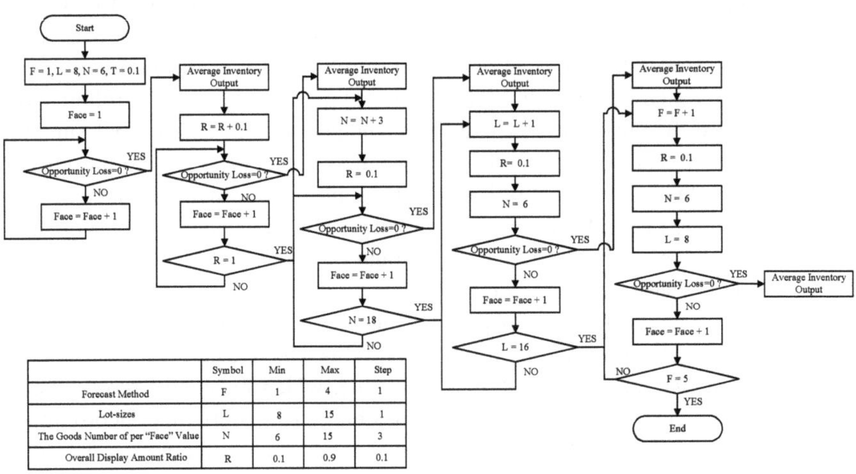

Fig. 5. The optimal procedure algorithm

3.4 Determination Factor Identification

An optimal solution was achieved following optimization analysis. Based on the optimal solution result, the determining factor was identified using sensitivity analysis. In this study, we used four factors to simulate function change to achieve the optimal results. By fixing three factors and only changing the value of one factor, the inventory level and also opportunity loss change could be simulated as shown in Fig. 6. Compared with the forecast results and lot-sizes, the inventory level and opportunity loss were heavily affected by the numbers of goods per "Face" and the ratio of the item's overall display amount, which are critical factors that should be treated with the utmost care.

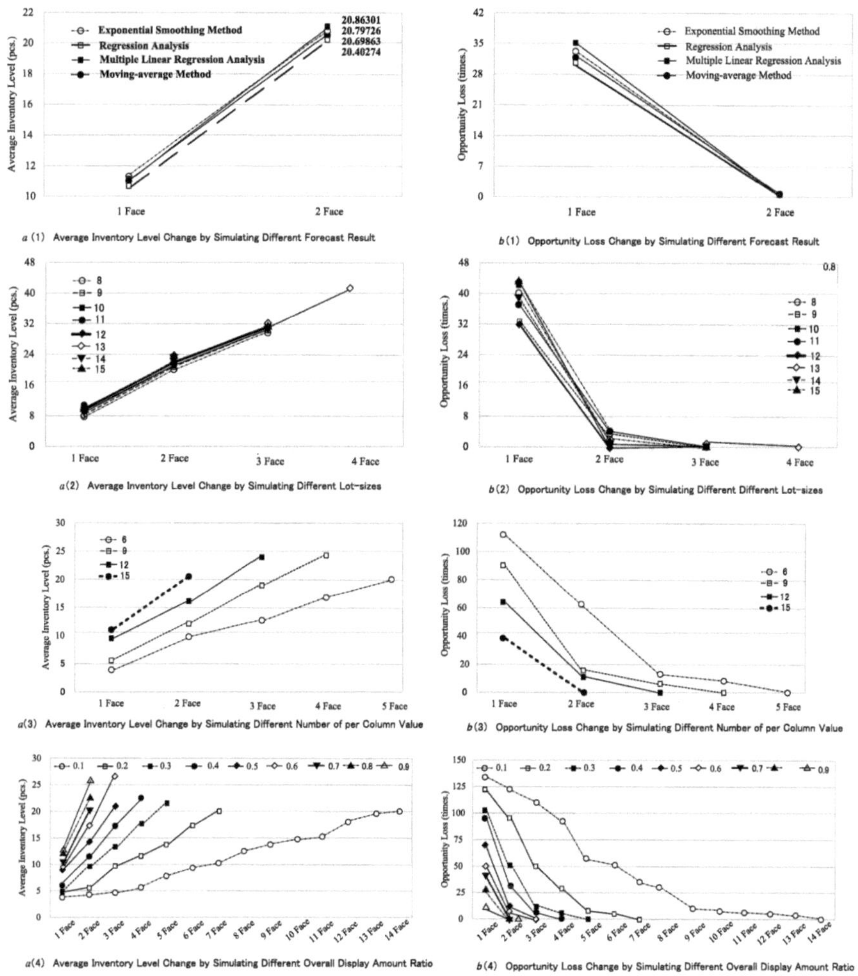

Fig. 6. Sensitivity analysis result

4 Conclusions

A simulation model was constructed and used to identify the inventory problem for an automated refreshment system. An optimal algorithm was described that reduced the inventory level and at the same time maintained the service level. After the optimal procedure, the exterminating factors that heavily influenced the inventory were identified. The study was applied to a real-world case and the results show that the inventory problem can be solved easily and the proposed algorithm is both practical and powerful in its ability to assist logistics mangers in their inventory planning efforts. In future research, experiments will be designed to employ the algorithm to find ways to make the simulation procedure more efficient.

References

1. ARIMA models applied to distribution operations. IT Solut. Front. **10**, 6–9 (2011)
2. A. Arisha, W.A. Hamad, Simulation optimization methods in supply chain applications. Ir. J. Manag. 90–124 (2010)
3. A.M. Law, M.G. McComas, Simulation optimization: simulation-based optimization, in *Proceedings of the 2002 Winter Simulation Conference*, ed. by E. Yucesan, C.H. Chen, J.L. Snowdon, J.M. Charnes (Institute of Electrical and Electronics Engineers, Inc, Piscataway, NJ, 2002), pp. 41–44
4. J.R. Swisher, D.H. Paul, H.J. Sheldon, W.S. Lee, A survey of simulation optimization techniques and procedures, in *Proceedings of the 2000 Winter Simulation Conference*, ed. by J.A. Joines, R.R. Barton, K. Kang, P.A. Fishwick (Institute of Electrical and Electronics Engineers, Inc, Piscataway, NJ, 2000), pp. 119–128
5. X. Wan, J.F. Pekny, G.V. Reklaitis, Simulation-based optimization with surrogate models—application to supply chain management. Comput. Chem. Eng. **29**(6), 1317–1328 (2005)
6. Y. Chu, F. You, Simulation-based optimization for multi-echelon inventory systems under uncertainty, in *Proceedings of the 2014 Winter Simulation Conference*, ed. by A. Tolk, S.Y. Diallo, I.O. Ryzhov, L. Yilmaz, S. Buckley, J.A. Miller (Institute of Electrical and Electronics Engineers, Inc, Piscataway, NJ, 2014), pp. 385–394
7. X. Zheng, M. He, L. Tang, C. Ren, B. Shao, A multiple-purpose simulation-based inventory optimization system: applied to a large detergent company in China, in *Proceedings of the 2015 Winter Simulation Conference*, ed. by L. Ylimaz, W.K.V. Chan, I. Moon, T.M.K. Roeder, C. Macal, M.D. Rossetti (Institute of Electrical and Electronics Engineers, Inc, Piscataway, NJ, 2015), pp. 1218–1229
8. H. Sang, S. Takakuwa, A simulation-based approach for obtaining optimal order quantities of short-expiration date items at a retail store, in *Proceedings of the 2012 Winter Simulation Conference*, ed. by C. Laroque, J. Himmelspach. R, Pasupathy, O. Rose, A.M. Uhrmacher (Institute of Electrical and Electronics Engineers, Inc, Piscataway, NJ, 2012), pp. 1466–1477
9. D.J. Yue, F.Q. You, Planning and scheduling of flexible process networks under uncertainty with stochastic inventory: MINLP models and algorithm. AIChE J. **59**, 1511–1532 (2013)

Research on Application of PCA and K-Means Clustering in Enterprise Human Resources

Hong-hua Sun(✉) and Qing-yang Li

Department of Industrial Engineering, Inner Mongolia University of Technology, Hohhot, China
1990864705@qq.com

Abstract. For enterprises, the effective allocation of human resources is of utmost importance. In order to solve the problems existing in the recruitment and assignment of employees, the application of principal component analysis is used to reduce the dimension of evaluation indicators, and K-means clustering analysis of data after dimension reduction. By comparing the size of weighted comprehensive evaluation values of the cluster centers, the four classes are sorted among classes. By comparing the size of principal components scores, achieve within the class sorting, so as to achieve the total ranking results. Case studies show that enterprises can use this method to classify recruiters according to the actual situation, and the evaluation results are conducive to arrange reasonable jobs for new hires.

Keywords: Comprehensive assessment · PCA · K-means

1 Introduction

With the continuous development of modern society, the competition among enterprises has become increasingly fierce. The focus of enterprise competition has also shifted from the competition of capital and material resources to the competition of human resources. Therefore, human resources management has become an important part of corporate management. The rise and fall of companies and their success or failure largely depend on their capabilities and qualities. How to select high-quality and excellent talents and how to allocate the corresponding talents to the right positions has become an increasingly acute problem that plagues corporate decision makers and human resource managers. In view of the above problems, the paper proposes the method of principal component analysis and K-means combination to classify and sort enterprise personnel, and provides an effective reference for managers of enterprises to reasonably allocate personnel. Principal component analysis is a statistical method widely used in unsupervised dimension reduction; K-means clustering algorithm and self-organizing neural network are two commonly used data clustering methods in unsupervised learning tasks, but this clustering algorithm directly when it comes to multidimensional data sets, the efficiency of clustering algorithms is extremely low. For this reason, the paper proposes that after the principal component analysis reduces the dimensions and then clusters. This method reduces the computational complexity and also improves the recognition accuracy.

G. Q. Huang et al. (eds.), *Proceeding of the 24th International Conference on Industrial Engineering and Engineering Management 2018*,
https://doi.org/10.1007/978-981-13-3402-3_73

Through the PCA, the research carried out comprehensive evaluation and principal component scores of the research objects, combined with other observation information of the research objects, and finally established a geological environment safety evaluation model [1]. Dynamic PCA also applies to different engineering fields [2–4]. An adaptive PCA algorithm for HGS is proposed for process monitoring and anomaly detection of hydroelectric generating units [5]. This study evaluates accounting information' effect on listed companies' capital cost at Stock Exchange based on PCA-BP method [6]. This study is specifically focused on the PCA-based face recognition techniques [7–9]. A PCA (principal component analysis)-SVDD (support vector data description)-based method was presented for chiller fault detection [10]. In addition, the dimensionality reduction of the PCA algorithm is also widely used [11, 12]. This study adds weight metrics to the standard K-means algorithm to form a new clustering algorithm that can better solve load balancing [13]. This study improves the K-means algorithm by improving the stability and accuracy of the K-means algorithm [14–16]. The principle and flow of K-means algorithm are expounded, and the problems existing in the application of K-means algorithm are analyzed [17]. This study proposes a K-means method for large-scale data sets, which can better solve the big data clustering problem [18]. Focuses on the analysis based on the clustering and the classification method of fatigue strain signals [19]. This study proposes a new clustering method, which mainly improves the performance of traditional clustering by combining K-means and particle swarm [20]. The following documents have improved the K-means method from different angles [14, 15, 21–23]. This study proposes an improved K-means algorithm, mainly for robotic control of the action space and discrete continuous state [24]. This study uses principal component analysis to guide K-means search to solve the K-means clustering centroid search problem. The research shows that the method is more efficient [25]. The study used a combination of PCA and K-means to study big data [26]. The study proposes a method for estimating ore grade by combining principal component analysis with K-means [27]. In this paper, the author analyse the design of wireless sensor intelligent traffic control system based on K-means algorithm. Make the traffic flow more reasonable and balanced, reducing travel time of vehicle delay and number of stops, so that he can make full use of the efficiency of the road traffic system [28].

2 Basic Principles and Steps of Principal Component Analysis

Principal component analysis is a very useful mathematical method. It is an optimal transformation in the sense of least mean square. The purpose is to remove the correlation between the input random vectors and highlight the implicit properties in the original data.

Principal component analysis uses a method of transforming variables to merge or eliminate redundant and correlated variables into a small number of unrelated composite indicator variables. From the perspective of algebra, it is to convert the covariance matrix of the original variable into a diagonal matrix. From a geometric point of view, principal component analysis is equivalent to an extraction method based on minimum mean square error. The basic steps of the principal component analysis method are as follows:

(1) Calculate the sample mean $\overline{Y_j}$ and sample standard deviation S_j for each indicator, standardize the data $X_{ij} = \frac{Y_{ij}-\overline{Y_j}}{S_j}$.
where $\overline{Y_j} = \frac{1}{m}\sum_{i=1}^{n} Y_{ij}$, $S_j = \sqrt{\frac{1}{m-1}\sum_{i=1}^{n}\left(Y_{ij} - \overline{Y_j}\right)^2}$.
Normalization of the obtained data to a standardized matrix.

$$X = \begin{bmatrix} x'_1 \\ x'_2 \\ \vdots \\ x'_m \end{bmatrix} = \begin{bmatrix} x_{11} & x_{12} & \cdots & x_{1n} \\ x_{21} & x_{22} & \cdots & x_{2n} \\ \vdots & \vdots & \ddots & \vdots \\ x_{m1} & x_{m2} & \cdots & x_{mn} \end{bmatrix}$$

The role of standardized processing is to eliminate the dimension between the original data indicators, so that the indicators can be compared, so that the standardized sample meets $E(X) = 0$.

(2) CalculateX'sindexcorrelationmatrix$R = \left(r_{ij}\right)_{n\times n}$,$r_{ij} = \frac{1}{m-1}\sum_{k=1}^{m}\frac{(x_{ki}-\overline{x_i})}{S_i}\frac{\left(x_{kj}-\overline{x_j}\right)}{S_j}$.
(3) Find the eigenvalue λ and eigenvector matrix B of R. If the index correlation matrix can be transformed by the orthogonal matrix Q, then

$$Q^T RQ = \begin{bmatrix} \lambda_1 & 0 & \cdots & 0 \\ 0 & \lambda_2 & \cdots & 0 \\ \vdots & \vdots & \ddots & \vdots \\ 0 & 0 & \cdots & \lambda_n \end{bmatrix}$$

When n feature values $\lambda_1 \geq \lambda_2 \geq \cdots \geq \lambda_n$ are obtained, each column $a_j = \left[a_1 \quad a_2 \quad \cdots \quad a_{nj}\right]^T$ of Q is a characterization vector of λ_j. And its composition of the orthogonal matrix is

$$B = \begin{bmatrix} B_{11} & B_{12} & \cdots & B_{1n} \\ B_{21} & B_{22} & \cdots & B \\ \vdots & \vdots & \ddots & \vdots \\ B_{n1} & B_{n2} & \cdots & B_{nn} \end{bmatrix}$$

(4) Determine the main component:
According to $Z_k = a_k X$, the n principal components of the k-th sample are:

$$\begin{bmatrix} Z_{k1} \\ Z_{k2} \\ \vdots \\ Z_{kn} \end{bmatrix}_1 = \begin{bmatrix} a_{11} & a_{12} & \cdots & a_{1n} \\ a_{21} & a_{22} & \cdots & a_{2n} \\ \vdots & \vdots & \ddots & \vdots \\ a_{n1} & a_{n2} & \cdots & a_{nn} \end{bmatrix} \begin{bmatrix} x_{k1} \\ x_{k2} \\ \vdots \\ x_{kn} \end{bmatrix}.$$

For all m samples there are

$$\begin{bmatrix} Z_{11} & Z_{12} & \cdots & Z_{m1} \\ Z_{12} & Z_{22} & \cdots & Z_{m2} \\ \vdots & \vdots & \ddots & \vdots \\ Z_{1n} & Z_{2n} & \cdots & Z_{mn} \end{bmatrix} = \begin{bmatrix} a_{11} & a_{12} & \cdots & a_{1n} \\ a_{21} & a_{22} & \cdots & a_{2n} \\ \vdots & \vdots & \ddots & \vdots \\ a_{n1} & a_{n2} & \cdots & a_{nn} \end{bmatrix} \begin{bmatrix} x_{11} & x_{12} & \cdots & x_{m1} \\ x_{12} & x_{22} & \cdots & x_{m2} \\ \vdots & \vdots & \ddots & \vdots \\ x_{1n} & x_{2n} & \cdots & x_{mn} \end{bmatrix}.$$

Several representative indicators are used as the main components, and reasonable choices are needed. Introduce the principal component contribution rate here: If λ_i is the i-th characteristic root of the covariance matrix σ, then $\lambda_k / \sum_{i=1}^{n} \lambda_i$ is the contribution rate of k principal components. $\sum_{i=1}^{r} \lambda_i / \sum_{i=1}^{n} \lambda_i$ is the cumulative contribution rate of the preceding R principal component. Finally, select the number of principal components according to the criterion of $\sum_{i=1}^{r} \lambda_i / \sum_{i=1}^{n} \lambda_i \geq 80\%$.

(5) After the original data matrix is multiplied by the eigenvector matrix B containing the principal components, the dimensionality reduction matrix C is obtained.

3 K-Means Clustering Principle and Steps

When applying an algorithm for cluster analysis, the number of classifications k needs to be given in advance. However, in many cases, it is not known in advance how much this k value should be given. This is also a defect of the K-means algorithm but it can also be used as needed. Sample space distribution, specifying the number of classifications. K-means clustering algorithm has fast calculation speed and good clustering performance. Therefore, K-means clustering algorithm has been widely used in engineering practice.

The core idea of K-means clustering is that for a data set, k number of data is randomly selected as the initial center. Then calculate the distance from each data to the clustering center according to the principle of similarity measurement and form k clusters. Then find the latest cluster centers for the k clusters and compare the new cluster centers with the cluster centers in the previous round. If the new cluster center does not change, the result can be output directly. If the clustering center has changed, the data needs to be re-clustered according to the new clustering center. Through repeated iteration until the center of the cluster does not change, the result can be output.

(1) Select k samples from the m samples as the initial clustering center.
(2) Calculate the distance from each sample to the cluster center using the similarity measure principle, the nearest sample of the distance clustering center is classified as a class.
(3) Calculate the error sum of squares: $E = \sum_{j=1}^{K} \sum_{x \in C_j} \left\| x - m_j \right\|^2$.
(4) Recalculate the cluster center, calculate the average value of all the data in this cluster, and get a new cluster center. Then calculate the error sum of squares: E_1.
(5) Compare E with E_1. If $E = E_1$, proceed to step 6. If $E \neq E_1$, proceed to step 2.
(6) Output the result and get k clusters.

4 PCA and K-Means Combined Application

This paper proposes a method to classify and rank the PCA algorithm and the K-means method. The original data is first dimension-reduced, the main components are extracted, and then the K-means algorithm is used for classification. The flow chart is shown in Fig. 1.

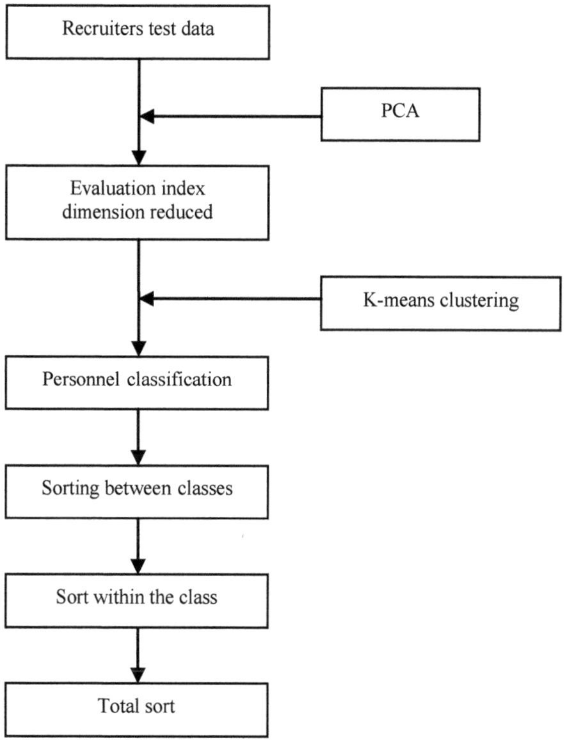

Fig. 1. PCA and K-means combination flow chart

5 Case Study

The following is the data of 28 employees recruited by a company. The company is going to assign these persons to four different posts. The score of each person on 11 evaluation indicators is shown in Table 1. Among them, A1–A11 represents 11 indicators for evaluation.

Table 1. Type sizes for camera-ready papers

Sample	A1	A2	A3	A4	A5	A6	A7	A8	A9	A10	A11
1	85	75	85	75	60	75	78	60	75	60	3.6
2	95	75	75	95	60	85	87	75	85	60	3.6
3	85	85	95	95	60	95	89	60	95	60	3.5
4	85	60	75	75	60	60	65	45	60	45	3.4
5	60	60	75	85	45	75	87	60	75	60	3.5
6	85	60	85	85	60	60	81	75	85	60	3.6
7	85	60	75	75	75	85	85	75	75	60	3.4
8	95	85	95	95	85	85	77	75	85	60	3.4
9	85	60	75	75	75	85	85	75	75	60	3.5
10	85	45	75	75	45	85	83	60	75	45	3.4
11	85	60	85	85	75	75	90	75	95	60	3.5
12	75	75	95	85	60	85	92	75	75	45	3.4
13	75	60	75	85	60	85	85	75	75	60	3.6
14	75	75	75	75	60	75	67	60	75	60	3.5
15	85	45	75	95	60	95	93	75	95	45	3.4
16	60	85	85	95	85	85	87	60	95	60	3.5
17	60	45	85	95	60	60	67	60	85	60	3.5
18	60	45	75	75	45	45	83	60	75	45	3.4
19	95	60	85	95	60	85	90	85	85	60	3.5
20	75	75	75	75	60	60	88	60	75	60	3.6
21	85	75	75	75	60	75	87	60	85	60	3.6
22	85	60	75	75	60	75	68	60	85	60	3.4
23	95	60	85	85	75	85	80	60	60	60	3.4
24	85	60	85	85	75	75	85	60	95	45	3.4
25	95	60	95	95	75	75	85	60	95	60	3.6
26	75	60	85	75	60	60	71	60	85	60	3.5
27	75	45	75	75	60	60	92	60	75	60	3.4
28	75	60	75	75	45	75	76	45	85	45	3.6

5.1 Results After Principal Component Analysis

According to the principal component analysis method, using the princomp function of the statistical software SPSS tool, after the principal component analysis of the data in Table 1 is performed with the SPSS software, the principal component analysis results are shown in Table 2, from which it can be seen that the total interpretation of the first six principal components The variance is 83.900% (cumulative contribution rate), which shows that the first six principal components provide enough information of the original data. Therefore, the first six components are extracted to build a classification index system, i.e., the first six principal components are selected.

$$C = \begin{bmatrix} 226.314 & -4.409 & -4.076 & 7.129 & 15.512 & 3.570 \\ 247.211 & -17.815 & 6.269 & 15.094 & 15.669 & 9.985 \\ 256.494 & -10.208 & 7.415 & -4.960 & 27.757 & 0.201 \\ 197.220 & -6.922 & -12.115 & 0.659 & 13.609 & 13.548 \\ 212.886 & -13.618 & 21.828 & 4.235 & 7.531 & -6.377 \\ 229.443 & -12.784 & 9.146 & 3.721 & -4.502 & 15.122 \\ 234.124 & -19.236 & -2.230 & 20.501 & 3.356 & 3.360 \\ 263.412 & -6.910 & -13.296 & 1.069 & 10.751 & 5.561 \\ 234.133 & -19.182 & -2.185 & 20.524 & 3.379 & 3.389 \\ 210.454 & -30.627 & 10.022 & 11.421 & 18.714 & 16.841 \\ 245.343 & -18.547 & 9.503 & 4.047 & -1.851 & 9.910 \\ 239.798 & -22.217 & 7.162 & 2.200 & 18.568 & -7.302 \\ 230.185 & -19.999 & 9.840 & 15.324 & 4.874 & 0.343 \\ 217.516 & -0.248 & -3.392 & 6.138 & 12.577 & 0.323 \\ 240.816 & -41.250 & 21.115 & 6.711 & 11.541 & 16.781 \\ 251.302 & -2.878 & 8.684 & -12.365 & 9.361 & -19.008 \\ 214.215 & -8.291 & 15.457 & -16.516 & -13.680 & 11.287 \\ 189.672 & -17.560 & 23.769 & -4.693 & -5.203 & 5.852 \\ 250.455 & -26.981 & 9.772 & 14.910 & 4.938 & 16.043 \\ 217.336 & -3.636 & 7.250 & 8.155 & 7.435 & -4.922 \\ 228.086 & -8.253 & 5.524 & 10.339 & 16.584 & 2.646 \\ 219.147 & -8.089 & -0.247 & 5.771 & 8.244 & 16.094 \\ 233.463 & -15.033 & -16.048 & 14.329 & 9.554 & 9.397 \\ 235.850 & -20.470 & 5.525 & -8.030 & 10.366 & 12.580 \\ 249.249 & -14.266 & 1.894 & -7.825 & 4.533 & 19.990 \\ 215.407 & -3.315 & 4.925 & -4.573 & -0.904 & 10.942 \end{bmatrix}$$

Table 2. Total variance of interpretation

Ingredient	Initial eigenvalues	Contribution rate (%)	Cumulative contribution rate (%)
1	3.568	32.436	32.436
2	1.608	14.623	47.058
3	1.286	11.688	58.747
4	1.212	11.022	69.769
5	0.853	7.759	77.528
6	0.701	6.372	83.900
7	0.525	4.773	88.673
8	0.505	4.591	93.264
9	0.314	2.851	96.115
10	0.253	2.300	98.415
11	0.174	1.585	100.000

According to the original data and feature vector matrix B multiplied to obtain a dimensionality reduction matrix C.

5.2 Cluster Analysis Results

K-means clustering analysis was performed using SPSS software, and the number of clusters k was assumed to be 4 as required. The specific data is shown in Table 3.

The calculated cluster centers are shown in Table 4.

Table 3. The number of cases included in the cluster

Cluster	Cases	Number of cases
1	6, 17, 18, 25, 26, 27	6
2	2, 10, 11, 13, 15, 19	6
3	3, 5, 12, 16, 24, 28	6
4	1, 4, 7, 8, 9, 14, 20, 21, 22, 23	10

Table 4. Cluster center

Cluster center	Number of clusters k			
	1	2	3	4
Z_{i1}	−0.58407	0.47151	0.26153	−0.08938
Z_{i2}	0.23306	−1.24095	0.05585	0.57122
Z_{i3}	0.60109	0.55498	0.48294	−0.98341
Z_{i4}	−0.79436	0.73431	−0.85354	0.54815
Z_{i5}	−1.31640	0.09385	0.86798	0.21274
Z_{i6}	0.58381	0.57489	−0.92295	−0.14145

Z_{ip}: denotes the p-th component value of the i-th cluster center

5.3 Sorting Between Classes

Calculate the weights of the six principal components. Each principal component value is divided by the sum of the six principal components as the weight. The result is $\omega_1 = 0.387$, $\omega_2 = 0.174$, $\omega_3 = 0.139$, $\omega_4 = 0.131$, $\omega_5 = 0.092$, $\omega_6 = 0.076$.

The formula for calculating the weighted comprehensive evaluation values for each cluster center is as follows $LJ_j = \omega_1 z_{i1}(m) + \omega_2 z_{i2}(m) + \cdots + \omega_p z_{ip}(m)$. Where $i = 1, 2, \ldots, K$; ω_k represents the weight in the principal component. $z_{ik}(m)$ represents the k-th component of the $i(1, 2, \ldots, k)$-th cluster center after m iterations (the final cluster center). After taking the data into the calculation, get $LJ_1 = -0.283$, $LJ_2 = 0.192$, $LJ_3 = 0.076$, $LJ_4 = 0.009$. By comparing four worth sizes, get $LJ_2 > LJ_3 > LJ_4 > LJ_1$. So the result of sorting between classes is $2 \succ 3 \succ 4 \succ 1$. Sorting between classes divides people with similar characteristics into a class. The same class of people is suitable for the same kind of work and provides a reference for the distribution of company personnel.

5.4 Sort Within the Class

When sorting inside the class, only the first principal component score of the sample needs to be sorted (matrix *C*). So, the sorting result of class 3 is $3 \succ 16 \succ 12 \succ 24 \succ 5 \succ 28$. The sorting result of class 1 is $25 \succ 6 \succ 26 \succ 17 \succ 27 \succ 18$. The sorting result of class 2 is $19 \succ 2 \succ 11 \succ 15 \succ 13 \succ 10$. The sorting result of class 4 is $8 \succ 9 \succ 7 \succ 23 \succ 21 \succ 1 \succ 22 \succ 20 \succ 14 \succ 4$. The sort within the class implements the sorting of individuals in the same class. Provide reference for the distribution of personnel in the enterprise.

6 Conclusion

This paper focuses on the problems existing in the recruitment and assignment of personnel in the company, and uses the principal component analysis and K-means clustering method to sort and divide the personnel. First, the steps and principles of principal component analysis and K-means clustering are introduced, and then a flow chart combining the two methods is drawn. Finally, combine a business instance, based on the staff's 11 evaluation index scores, through principal component dimension reduction and K-means cluster analysis. after the classification of staff, the sort between classes, and the sort within the class, the overall ranking of 28 new staff was obtained. Providing decision-making basis for enterprises to use personnel rationally, it also plays a more effective role in the personnel management of the company. It is conducive to discovering and promoting truly outstanding talents and promoting the better development of the enterprise.

Acknowledgements. This study was supported by the Inner Mongolia Natural Science Found (Grant No. 2014BS0707).

References

1. F.S. Wang, Q.B. Rong, H.J. Zhang, PCA-based safety evaluation model on geological environment. J. Comput. Methods Sci. Eng. **15**(3), 559–568 (2015)
2. J.G. Cui, X. Yan, X.P. Pu, Y.W. Qi, L.Y. Jiang, J.Q. Shi, Aero-engine fault diagnosis based on dynamic PCA and improved SVM. Meas. Diagn. **35**(1), 94–99 (2015)
3. C.D. Tong, T. Lan, X.H. Shi, Fault detection by decentralized dynamic PCA algorithm on mutual information. Huagong Xuebao/CIESC J. **67**(10), 4317–4323 (2016)
4. Y.J. Wang, F.M. Sun, M.X. Jia, Online monitoring method for multiple operating batch processes based on local collection standardization and multi-model dynamic PCA. Can. J. Chem. Eng. **94**(10), 1965–1976 (2016)
5. F.W. Chu, Anomaly detection of hydropower generating set using operation condition and adaptive PCA. Int. J. Power Energy Syst. **37**(1), 34–44 (2017)

6. X.H. Peng, PCA-BP based analysis of the evaluation of computerized accounting information. Int. J. Hybrid Inf. Technol. **9**(8), 277–288 (2016)
7. X.G. Xiang, J. Yang, Q.P. Chen, Color face recognition by PCA-like approach. Neurocomputing **152**, 231–235 (2015)
8. C.G. Turhan, H.S. Bilge, Class-wise two-dimensional PCA method for face recognition. IET Comput. Vis. **11**(4), 286–300 (2017)
9. J.J. Hu, G.Z. Tan, F.G. Luan, A.S.M. Libda, 2DPCA versus PCA for face recognition. J. Cent. South Univ. **22**(5), 1809–1816 (2015)
10. G.N. Li, Y.P. Hu, H.X. Chen, H.R. Li, PCA-SVDD-based chiller fault detection method. J. Huazhong Univ. Sci. Technol. (Natural Science Edition) **43**(8), 119–122 (2015)
11. M.A. Belarbi, S. Mahmoudi, G. Belalem, PCA as dimensionality reduction for large-scale image retrieval systems. Int. J. Ambient Comput. Intell. **8**(4), 45–58 (2017)
12. X. Zhong, D. Enke, Forecasting daily stock market return using dimensionality reduction. Expert Syst. Appl. **67**, 126–139 (2017)
13. Y.L. Dai, Y. Wang, G.J. Wang, Package balancing k-means algorithm for physical distribution. Int. J. Comput. Sci. Eng. **14**(4), 349–358 (2017)
14. G. Zhang, C.C. Zhang, H.Y. Zhang, Improved K-means algorithm based on density Canopy. Knowl. Based Syst. **145**, 1–14 (2018)
15. W. Li, J.Y. Zhao, T.S. Yan, Improved K-means clustering algorithm optimizing initial clustering centers based on average difference degree. Kongzhi yu Juece/Control Decis. **32** (4), 759–762 (2017)
16. G. Zhao, The application of user behavior analysis by improved K-means algorithm based on Hadoop. Revista de la Facultad de Ingenieria **31**(7), 111–120 (2016)
17. Q.X. Zhang, J.X. Li, Application of K-means clustering algorithm in information security problem. Revista de la Facultad de Ingenieria **32**(11), 462–467 (2017)
18. T.S. Xu, H.D. Chang, G.Y. Liu, C.W. Tan, Hierarchical K-means method for clustering large-scale advanced metering infrastructure data. IEEE Trans. Power Deliv. **32**(2), 609–616 (2017)
19. M.F.M. Yunoh, S. Abdullah, M.H.M. Saad, Z.M. Nopiah, M.Z. Nuawi, K-means clustering analysis and artificial neural network classification of fatigue strain signals. J. Brazilian Soc. Mech. Sci. Eng. **39**(3), 757–764 (2017)
20. G.K. Patel, V.K. Dabhi, H.B. Prajapati, Clustering using a combination of particle swarm optimization and K-means. J. Intell. Syst. **26**(3), 457–469 (2017)
21. Z.X. Fan, Y. Sun, H. Luo, Clustering of college students based on improved K-means algorithm. J. Comput. (Taiwan) **28**(6), 195–203 (2017)
22. S.Y. Xiao, B. Wang, J. Zhang, Q.F. Zhang, J.Z. Zou, Automatic sleep stage classification based on an improved k-means clustering algorithm. Shengwu Yixue Gongchengxue Zazhi/J. Biomed. Eng. **33**(5), 847–854 (2016)
23. Y.F. Jin, G. Li, Application of improved k-means algorithm in evaluation of network resource allocation. Boletin Tecnico/Tech. Bull. **55**(5), 284–292 (2017)
24. X.B. Guo, Y. Zhai, K-means clustering based reinforcement learning algorithm for automatic control in robots. Int. J. Simul. Syst. Sci. Technol. **17**(24) (2016)
25. Q. Xu, C. Ding, J.P. Liu, B. Luo, PCA-guided search for K-means. Pattern Recogn. Lett. **54**, 50–55 (2015)
26. P.A. Farhad, B. Stephen, Preconditioned data sparsification for big data with applications to PCA and K-mean. IEEE Trans. Inf. Theory **63**(5), 2954–2974 (2017)

27. N.M. Mohammadi, H. Ardeshir, M. Abbas, Application of K-means and PCA approaches to estimation of gold grade in Khooni district (central Iran). Acta Geochim. **37**(1), 102–112 (2018)
28. D.X. Liu, Design of wireless sensor intelligent traffic control system based on K-means algorithm. Revista de la Facultad de Ingenieria **32**(12), 354–362 (2017)

A Multi-objective Optimization Approach for Influence Maximization in Social Networks

Jian-bin Guo[1], Fu-zan Chen[2(✉)], and Min-qiang Li[1,2]

[1] College of Management and Economics, Tianjin University, Tianjin, China
[2] State Key Laboratory of Hydraulic Engineering Simulation and Safety, Tianjin University, Tianjin, China
fzchen@tju.edu.cn

Abstract. Influence maximization (IM) is to select a set of seed nodes in a social network that maximizes the influence spread. The scalability of IM is a key factor in large scale online social networks. Most of existing approaches, such as greedy approaches and heuristic approaches, are not scalable or don't provide consistently good performance on influence spreads. In this paper, we propose a multi-objective optimization method for IM problem. The IM problem is formulated to a multi-objective problem (MOP) model including two optimization objectives, i.e., spread of influence and cost. Furthermore, we develop a multi-objective differential evolution algorithm to solve the MOP model of the IM problem. Finally, we evaluate the proposed method on a real-world dataset. The experimental results show that the proposed method has a good performance in terms of effectiveness.

Keywords: Influence maximization · Multi-objective differential evolution algorithm · Multi-objective optimization model · Social network

1 Introduction

With the rapid development of social networks, more and more people exchange information on social networks. Social networks provides a broader platform for information propagation. The commercial value embedded in social networks gradually emerges (such as 'virtue marketing').

Influence maximization (IM) problem is a task of finding a set of seed nodes (i.e., seed set) that make these nodes have the broadest influence spread based on a specific propagation model. It has attracted great attention of scholars and industrial practice.

With the increasing expansion of social networks, searching a seed set on a large-scale social network is a NP-hard problem. Most of existing approaches fail to get the optimal solution for a large-scale social network in a reasonable time.

Furthermore, most of previous researches only pursue the maximal of influence spread. Few of them consider cost of spread i.e., payment for delivering targeted information.

To solve the problem of IM, we propose a multi-objective optimization method. The IM problem is formulated to a multi-objective optimization problem (MOP) model

G. Q. Huang et al. (eds.), *Proceeding of the 24th International Conference on Industrial Engineering and Engineering Management 2018*,
https://doi.org/10.1007/978-981-13-3402-3_74

including two optimization objectives, i.e., influence spread and cost. Furthermore, we develop a multi-objective differential evolution algorithm to solve the MOP model of IM. Finally, the proposed approach is validated on a real-world dataset.

The rest of the paper is organized as follows. Section 2 provides a review of related work. In Sect. 3, we present the proposed multi-objective optimization model for the IM problem. Section 4 develops a multi-objective differential evolution algorithm to solve the MOP model of IM. Section 5 reports experimental analysis on a real-world dataset to validate the proposed method. We draw a conclusion and discuss the future work in Sect. 6.

2 Related Work

Existing approaches for the IM can be divided into two types, i.e., greedy approaches and heuristic approaches. However, greedy approaches are not scalable and heuristic approaches do not provide consistently good performance on influence spreads.

Domingos and Richardson firstly considered IM problem as an algorithmic problem, using *Markov random field* modeling to simulate the influence propagation process, and proposed a heuristic method to solve this problem [1].

Kempe et al. considered the influence maximization problem as a discrete optimization problem [2]. His work focused on two propagation models: *IC (independent cascade model)* and *LT (linear threshold model).* Based on the above two propagation models, a greedy algorithm is proposed to solve the IM problem.

In the above work, *Monte Carlo simulation* is used to estimate the propagation range, but the use of *Monte Carlo simulation* increases the running time of the algorithm and makes it difficult to apply to large-scale social networks. Therefore, most of the scholars put a lot of effort into improving the efficiency of the algorithm. Leskovec et al. used the submodularity of propagation to propose that *CELF* algorithm can greatly improve the algorithm running speed [3]. Chen et al. proposed algorithm *CGA* its main idea is to find the optimal set of seed nodes from the newly constructed sub-social network diagram [4]. Goyal et al. proposed a *CELF++* algorithm for further improvement of *CELF* [6].

Some scholars proposed the use of evolutionary algorithms to solve this NP-hard problem. Gong et al. proposes using particle swarm optimization algorithm (PSO) to solve IM problem [11]. Bucur et al. proposes the use of genetic algorithms (GA) to solve IM problem [12]. Jiang Q et al. proposed to use simulated annealing algorithm (SA) to solve IM problem [13]. The use of evolutionary algorithms greatly shortens the time for solving IM problems.

Node influence measure is based on some characteristics of the network to construct the corresponding formula to calculate the node's global influence. Kempe et al. propose to use *Degree Centrality* and *Closeness Centrality* to measure the node's global influence according to the node's influence definition in social networks [1]. Cha et al. proposed using the '*Retweet*', '*Comment*' and '*Mention*' in social networks to measure the influence of nodes in twitter [7]. Romero et al. use the *Hirsch Index* to estimate the global influence of nodes in social networks [8]. Gayo-Avello et al.

proposed a physics-based variable mass system's influence metric [9]. Kitsak et al. proposed a measure of the influence of *k-shell* decomposition on the nodes of influence in dynamic propagation [10].

3 Multi-object Optimization Model for the IM

In practice, enterprises releasing information in social networks usually pay money or give coupons to individuals according to their influence in the social network in order to encourage them to retweet the targeted information. As mentioned, existing literatures of IM pursuit of the maximal spread of influence, but few of them consider spread cost, i.e., payment or coupons. Besides of the spread of influence, we consider the spread cost are considered. The problem of IM is formulated as a MOP and a multi-objective optimization model is proposed in this section.

3.1 Spread of Influence

The *independent cascade (IC) model* is used in this paper to simulate the propagation process of the targeted information [2]. Suppose a given social network is represented by a directed graph $G = (V, E, P)$. The nodes $v \in V$ in the directed graph represent the users in the social network, the edges $(u, v) \in E$ represent the relationships between users, and the weights P_{uv} on the edges (u, v) represent the probability of influence between users.

There are a number of calculation methods defined for the evaluation of the probability of activation between nodes. Although the traditional heuristic calculation method is simple to calculate, the final probability is difficult to fit the true propagation probability.

Zhang et al. proposes a method for calculating the probability of influence between nodes [14], which considers that the more frequent the interaction between two nodes, the more likely the two nodes will influence each other, and the greater the probability of activation between nodes. They define that the activation probability of node u to activate node v is P_{uv}. ω_{uv} denotes the weight on the edge from u to v. $p \in [0, 1]$ is a designated propagation probability. The probability of activation between nodes is defined as:

$$p_{uv} = 1 - (1 - p)^{\omega_{uv}} \tag{1}$$

The probability of influence between nodes calculated by the above method are more fitting the actual propagation probabilities than the heuristic activation probabilities of the nodes in the universal *IC* model, so that the actual propagation of information can be more accurately simulated.

The *independent cascade model (IC)* simulates the random propagation of information. We represent the social network as a directed graph $G = (V, E, P)$. The nodes v contained in V represent users in the social network. The edges e contained in E represent the relationships between nodes and are defined for each edge in E. P_{uv} represents the influence probability of the edge (u, v).

The state of the node in this model includes two types: active, inactive, and some nodes (called seed nodes) are pre-activated at the initial stage $t = 0$ to form a node set S. In any step $t > 1$, if node u has been activated at step $t - 1$, then node u has only one chance to attempt to activate its inactivated neighbor node v with probability P_{uv}, and the node cannot be activated once it is activated. The process terminates when no new node is activated.

3.2 Cost of Influence

In practice, discounts or coupons issued by companies in order to encourage initial users to deliver targeted information are the major components of the cost of information spread.

And the greater the influence of the initial user, the greater the cost of activating the user. Therefore, in this paper we assume that the user's activation cost is positively related to the user's influence.

So we define the activation cost of the seed node based on the global influence of the node itself. The cost function is defined as follows:

$$\cos t(S) = \sum_{u \in S} \sum_{v \in V} p_{uv} \tag{2}$$

S represents the seed set, and P_{uv} represents the probability that node u activates node v.

We define the sum of the activation probabilities between u and all its *outgoing degree node* as the global influence of node u on the social network.

3.3 Multi-objective Optimization Model for the IM

In this paper, we consider the two objectives of maximizing the influence spread and minimizing the cost of influence, and propose a multi-objective optimization based influence maximization model. Two objective function formulas are defined as follows:

(1) Maximization of Influence spread

We use the *IC* model to simulate the final number of nodes activated by the seed set as the final influence spread. Previous scheme often used *Monte Carlo simulations* to simulate the final range of propagation, but multiple *Monte Carlo* simulations were very time- consuming, so we used a more simple *LIE* function to calculate the number of nodes what were finally activated [11].

$$S^* = \arg \max_{|S|=k, S \in V} LIE(S) \tag{3}$$

S represents the selected seed set, and k represents the number of seed nodes.

(2) Minimization of cost

Minimizing the cost of influence is equivalent to minimizing the activation cost of the initial set of nodes. Therefore, the objective function is to minimize the activation cost of the initial seed set. So cost minimization objective function:

$$S^* = \arg \max_{|S|=k, S \in V} \cos t(S) \tag{4}$$

3.4 Mathematical Model of Multi-objective Optimization

Because the MOP model is suitable for solving the minimization problem. And our objective function contains a maximum spread of influence. So we need to convert the objective function. D represents the number of nodes in the social network. So we turn the optimization goal of the maximum spread of influence into the minimum number of inactive nodes.

$$\begin{cases} f_1(S) = D - LIE(S) \\ f_2(S) = \cos t(S) \end{cases} \tag{5}$$

The multi-objective optimization mathematical model constructed in this paper is as follows:

$$\begin{aligned} &\min(f(S) = \min(f_1(S), f_2(S)) \\ &s.t.\ |S| = k; \\ &S \in V \end{aligned} \tag{6}$$

4 Solving Algorithm for the MOP Model

In this section, we develop a multi-objective differential evolutionary algorithm (*MODEA*) to solve the MOP model for IM.

For the above multi-objective optimization model, we adopt the multi-objective evolutionary algorithm to optimize the solution. In view of the complexity of the problem of maximizing influence, we have improved the optimization algorithm so that it can converge more quickly to the Pareto frontier.

Differential evolution algorithm is an evolutionary algorithm based on evolutionary theory of genetic algorithm. The essence is a multi-objective optimization algorithm (*MOEAs*) for solving the global optimal solution in multi-dimensional space. Has the advantages of easy to use, simple structure, fast convergence, and robustness. We extend the DE algorithm to a multi-objective form (*MODEA*) and improve the selection strategy of the DE algorithm so that it can quickly converge to the Pareto frontier under the premise of ensuring the diversity of the optimal solution distribution.

The principles in *DE* were simplicity, efficiency, and the use of floating-point encoding instead of binary numbers. As with traditional evolution, the *DE* algorithm owns an initial population and is promoted by selection, mutation and crossover during the iteration.

4.1 Population Initialization

In the solution space, N individuals are randomly generated, each of which is represented as n-dimensional vectors such as:

$$X_i(0) = (X_{i1}(0), X_{i2}(0), \ldots, X_{iN}(0)), i = 1, 2, 3, \ldots, N; \tag{7}$$

The i th individual's j th dimension value is as follows:

$$\begin{aligned} &X_{ij}(0) = L_{j-min} + rand(0,1) * \left(L_{j-max} - L_{j-min}\right) \\ &i = 1, 2, 3, \ldots, N; \\ &j = 1, 2, 3, \ldots, k; \end{aligned} \tag{8}$$

$L_{j\text{-}min}$ represents the minimum value on the j-th dimension, and $L_{j\text{-}max}$ represents the maximum value in the j-th dimension.

4.2 Mutation

In the *g*th iteration, 3 individuals were randomly selected from the population $X_{p1}(g)$, $X_{p2}(g)$, $X_{p3}(g)$ and p1 $\neq$ p2 $\neq$ p3. The resulting mutation vector is:

$$H_i(g) = X_{p1}(g) + F * \left(X_{p2}(g) - X_{p3}(g)\right) \tag{9}$$

where $X_{p2}(g) - X_{p3}(g)$ is the difference vector, F is the differential weight, for the differential weight F, generally choose between [0, 2], usually take 0.5.

4.3 Crossover

$$V_{i,j} = \begin{cases} h_{i,j}(g), rand(0,1) \leq cr \\ X_{i,j}(g), else \end{cases} \tag{10}$$

where $cr \in [0, 1]$ is the crossover probability.

4.4 Selection

To expand *DE* algorithm into multi-objective optimization algorithm, we need to improve the selection operation in *DE* algorithm.

Selection in *MODE*A is based on the following rules:

(1) When there is a Pareto dominance relationship between two solution vectors, we choose a better solution vector based on Pareto dominance to enter the next generation population.
(2) When there is no Pareto dominance relationship between two solution vectors, we choose all the two solution vectors into the next generation population.

After an iteration, the size of the population may increase. If the population size grows to a pre-set threshold, we use a selection operation similar to the one in *NSGA-II* to resize the population to the original size. The solution vector is sorted according to the indexes of non-domination and crowding degree, and delete the solution vectors with poor performance to reduce the population size to the initial size.

The increment ε for each iteration size of the population is between *N* and *2N* (*N* is assumed to be the initial population size), since both vectors can enter the next generation without any dominance between the vectors. We assume that the population size threshold is *2N*. When the population size increases to *2N* after many times of iterations, a selection operation with non-dominated sorting is called to reduce the population size to the original population size.

From the above process, we know that the improved algorithm does not have to call selection operations with non-dominated sorting every time, and the number of invocations is much smaller than that of the *NSGA-II* algorithm. However, the time complexity of a select operation with non-dominated sorting is $O(g * N^2)$, which is a major part of the computational complexity of the algorithm. Reducing the number of calls can significantly shorten the algorithm's running time. In the improved algorithm, the increase of the population size within the preset threshold makes the algorithm not easy to be trapped in the local optimum, which is beneficial to increase the diversity of the optimal solution. The pseudocode of the above algorithm is as follows:

Algorithm 1. MODEA

```
Input: social graph G=(V,E,P), M and n ∈N⁺,cr=0.2,
population size threshold: T .
Output: a series of seed set S={s1,s2,...,sM}
For i =1,2,...,M:
    For j=1,2,...,n:
        Xi,j(0) ←Lj_min+rand(0,1)*(Lj-max – Lj-min)
    end
    Xi(0) ←{Xi,1(0), Xi,2(0),...,Xi,n(0)}
end
X←{X1(0), X2(0),..., XM(0)}
Repeat:
    For i =1,2,...,M:
        {p1,p2,p3}←rand(1,M) // Randomly select three
        different integers from 1 to M.
        Hi(g) ←Xp1(g)+F*[Xp2(g) – Xp3(g)]
    end
    For i=1,2,...,M:
        For j=1,2,...,n:
            If rand(0,1)<cr:
                Vi,j(g) ← Hi,j(g)
            else:
                Vi,j(g) ← Xi,j(g)
        end
    end
    For i=1,2,...,M:
        if f [Vi(g)] <f [Xi(g)]:
            Xi(g+1) ←Vi(g)
        else:
            Xi(g+1) ←Vi(g)
            Xi+1(g+1) ←Xi+1(g)
    end
    If pop[ X(g+1)]>T:
        Non-dominant sorting[X(g+1)]
    else:
        pass
Until convergence
```

5 Experimental Analysis

In this section, we construct several experiments on a public real-world dataset used in [5] to evaluate the performance of the proposed method.

Dataset is a collection of data from academics and scholarly works obtained from the *ArnetMiner* Academic Search System. The collection contains 2,162 academic authors and 2,555 scholarly articles. And based on academic articles to establish a reference of 19,875 times between scholars.

We argue that the more times Scholar *A* refers to Scholar *B* in a certain field, the more influence Scholar *B* has on Scholar *A* in a certain field, which means Scholar *B* can activate Scholar *A* with a greater probability in this field.

We use real-coded instead of binary code. Each population individual represents a set of candidate seed nodes. The coding of the individual population consists of the coding of the nodes it contains.

In experiments, the weights of the edges between nodes in the network represent the number of interaction between the nodes. The default parameter p has a value of 0.5.

For our proposed new multi-objective differential algorithm, we compared the performance of the algorithm with the most commonly used *NSGA-II* algorithm for solving multi-objective optimization problems. In the experiment, the population size $N = 30$, the number of iterations $g = 300$, the size of the seed node set $k = 20$, and the crossover probability $cr = 0.2$. The final experiment result are Fig. 1. From the experiment result, we can see that the new proposed multi-objective differential algorithm can obtain more excellent frontiers than the NSGA-II algorithm.

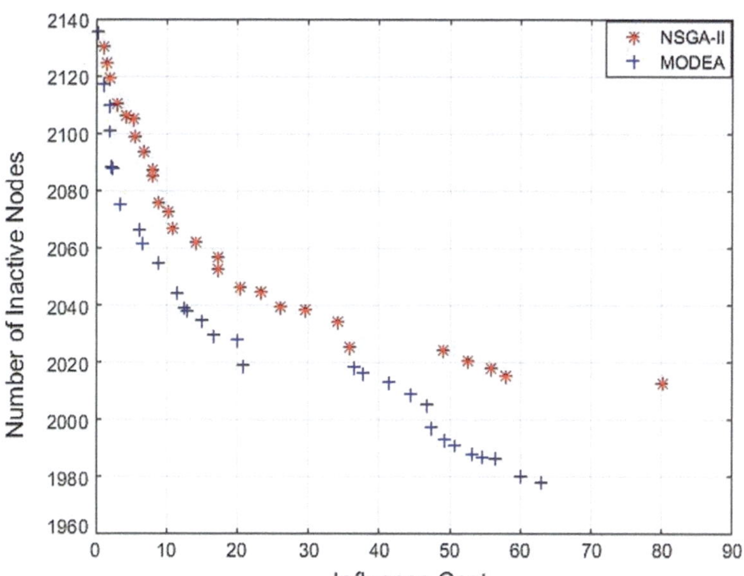

Fig. 1. Comparison of NSGA-II and MODEA

From the above figure, we can obtain a smooth frontier surface by using the multi-objective optimization model to solve the problem of maximizing the influence. Increased diversity of feasible solutions, closer to actual business analysis, providing more direct and comprehensive information for business decisions.

6 Conclusion

The influence maximization problem based on multi-objective optimization proposed in this paper can quickly and accurately (using improved multi-objective evolutionary algorithm) find the optimal set of initial seed nodes for particular targeted information so that the final targeted information has the largest spread of influence. This not only saves costs but also enables more intuitively support for actual decisions and reduces the difficulty of applying research to practical business.

However, there are some limitations to the solution proposed in this paper. For example, the cost function constructed in this paper is only a linear function of the global influence of the node. In reality, the effect of fitting the actual cost may not be very good. The introduction of a more accurate pricing model is an exploration of our future work.

Acknowledgements. The work was supported by the Key Program of National Natural Science Foundation of China (No. 71631003) and the General Program of National Natural Science Foundation of China (No. 71771169).

References

1. P. Domingos, M. Richardson, Mining the network value of customers, in *KDD* (2001), pp. 57–66
2. D. Kempe, J. Kleinberg, Maximizing the spread of influence through a social network, in *Proceedings of the Ninth ACM SIGKDD International Conference on Knowledge Discovery and Data Mining*, KDD '03 (ACM, 2003), pp. 137–146
3. J. Leskovec, A. Krause, C. Guestrin, C. Faloutsos, J.M. Van Briesen, N.S. Glance, Cost-effective outbreak detection in networks, in *KDD* (2007), pp. 420–429
4. W. Chen, Y. Wang, S. Yang, Efficient influence maximization in social networks, in *KDD* (2009), pp. 199–208
5. J. Tang, J. Sun, C. Wang et al., Social influence analysis in large-scale networks, in *ACM SIGKDD International Conference on Knowledge Discovery and Data Mining* (ACM, 2009), pp. 807–816
6. A. Goyal, W. Lu, L.V. Lakshmanan, Celf++: optimizing the greedy algorithm for influence maximization in social networks, in *Proceedings of the 20th International Conference Companion on World Wide Web* (ACM, 2011), pp. 47–48
7. M. Cha, H. Haddadi, F. Benevenuto, P.K. Gummadi, Measuring user influence in twitter: the million follower fallacy, in *Proceedings of the Fourth International Conference on Weblogs and Social Media*, ed. by W.W. Cohen, S. Gosling, ICWSM 2010, Washington, DC, USA, May 23–26, 2010 (The AAAI Press, 2010)

8. D.M. Romero, W. Galuba, S. Asur et al., Influence and passivity in social media, in *Joint European Conference on Machine Learning and Knowledge Discovery in Databases* (Springer, Berlin, Heidelberg, 2011), pp. 113–114
9. D. Gayo-Avello, D.J. Brenes, D. Fernández-Fernández, M.E. Fernández-Menéndez, R. García-Suárez, De retibus socialibus et legibus momenti. EPL (Europhysics Letters) **94**(3), 38001 (2011)
10. M. Kitsak, L.K. Gallos, S. Havlin, F. Liljeros, L. Muchnik, H.E. Stanley, H.A. Makse, Identification of influential spreaders in complex networks. Nat. Phys. **6**, 888–893 (2010)
11. M. Gong, J. Yan, B. Shen et al., Influence maximization in social networks based on discrete particle swarm optimization. Inf. Sci. **367**(C), 600–614 (2016)
12. D. Bucur, G. Iacca, Influence maximization in social networks with genetic algorithms, in *European Conference on the Applications of Evolutionary Computation* (Springer, Cham, 2016), pp. 379–392
13. Q. Jiang, G. Song, G. Cong et al., Simulated annealing based influence maximization in social networks, in *AAAI Conference on Artificial Intelligence* (AAAI Press, 2011), pp. 127–132
14. X. Zhang, J. Zhu, Q. Wang et al., Identifying influential nodes in complex networks with community structure. Knowl. Based Syst. **42**(2), 74–84 (2013)

A Collaborative Filtering Method for Cloud Service Recommendation via Exploring Usage History

Fang-fang Wang[1], Fu-zan Chen[1(✉)], and Min-qiang Li[1,2]

[1] College of Management and Economics, Tianjin University, Tianjin, China
fzchen@tju.edu.cn

[2] State Key Laboratory of Hydraulic Engineering Simulation and Safety, Tianjin University, Tianjin, China

Abstract. With the rapid development of cloud computing technique, cloud service recommendation has attracted significant research interest nowadays. The main task of cloud service recommendation is to provide a list of new cloud services for users to satisfy their requirements and the key of accurate recommendation is the complete and accurate identification of users' preference. In this study, we propose a cloud service recommendation approach based on collaborative filtering via exploring user usage history. This approach first computes user similarity using an improved cosine similarity method, which is adjusted by cloud service popularity. Then several similar users are selected as user neighbors. At last, it predicts the possibility that active users invoke candidate services according to user neighbors. Experimental results show that the proposed approach performs better than existing approaches in terms of recommendation accuracy.

Keywords: Cloud computing · Collaborative filtering · Recommendation Usage history

1 Introduction

As a new computing paradigm, cloud computing technique has been rapidly developed in recent years [1]. Due to the advantages of cloud service such as low cost, high reliability and scalability, cloud service has been widely used in the construction of distributed applications in heterogeneous environments and a great number of cloud services have been published by cloud service providers. With the rapid increase of the number of available cloud services on the Internet, however, it has become a difficult and time-consuming task for users to find the cloud services that can satisfy users' functional and non-functional requirements. To solve this problem, cloud service recommendation has attracted significant research interest [2].

Most prior web service recommendation approaches focus on the non-functional properties, QoS (Quality of Service, such as response time, throughput, availability, reliability, etc.). The basic idea of these approaches is to recommend cloud services with superior QoS values to users. And the main task is to predict the missing QoS values using collaborative filtering technique [3–8]. However, the recommendation

G. Q. Huang et al. (eds.), *Proceeding of the 24th International Conference on Industrial Engineering and Engineering Management 2018*,
https://doi.org/10.1007/978-981-13-3402-3_75

result may not be accurate since cloud services with superior QoS values may not satisfy users' functional requirements.

To recommend cloud services that can meet users' functional requirements, some works propose content-based approaches [9–12], which mining users' personalized preference by analyzing the description information of cloud services. However, the recommendation accuracy rate is low because of the deviation between the function of cloud services and description information and the complexity of analyzing instructed textual information [13]. Collaborative filtering technique is an effective technique to identify users' potential interests by directly exploring user usage history [14].

To completely and precisely mine user preference and generate accurate recommendation results for users, we propose a cloud service recommendation approach based on collaborative filtering via exploring user usage history in this paper. It is composed of three phases: user similarity computation, user neighbor selection, and possibility prediction. A key enabler to getting accurate recommendation in this approach is the identification of user neighbors, which is determined by the method of user similarity computation and the strategy of user neighbor selection. We introduce a weighting factor about cloud service popularity into cosine similarity to improve user similarity computation and set a parameter to filter user neighbors. The evaluation in this study shows that our proposed approach outperforms other state-of-the-art cloud service recommendation approaches in terms of F-Measure. The performance improvement of our proposed approach can be mainly attributed to the method of user similarity computation and the strategy of user neighbor selection.

The remainder of the paper is organized as follows: Sect. 2 summarizes related work in previous research. Section 3 elaborates our proposed approach in detail. Experiments and evaluations are presented in Sect. 4. Section 5 concludes this study and discusses the future work.

2 Related Work

Cloud service recommendation has been extensively studied in recent years. Most prior works recommend cloud services with superior QoS values to users and they concentrate on the accurate prediction of missing QoS values using collaborative filtering technique. Memory-based approaches [3], model-based approaches [7] and hybrid approaches [8] have been proposed. In practice, however, cloud services are usually used to implement specific functions of distributed applications. Therefore, the recommended cloud services should not only perform excellently in QoS but more importantly they should satisfy users' functional requirements.

In order to recommend cloud services that satisfy users' functional requirements, some works extract functional preference of a specific user by analyzing the description information (such as WSDL, OWL-S) of cloud services that he used and use content-based method to generate results [9–12]. However, description information always requires tedious, error-prone manual work and contains redundant information that is entirely unrelated to the function of cloud services so that users' functional preference extracted based on the description information of services have large deviation [13]. Besides, the analysis of untrusted textual information is extremely complicated, which increases the difficulty of accurately extract users' preference. Content-based approaches are based on users' own usage history, so they can only recommend cloud services that are similar with the cloud services that they have used but have no ability to find the potential interests of users. As a result, the accuracy rate of recommendation results is low.

Collaborative filtering technique is effective for the identifying of users' potential interests. Reference [14] proposed a collaborative filtering based approach for cloud service recommendation by exploring user usage history. In this approach, the cloud services invoked by similar users will be recommended to the active user. It is composed of two phases: user similarity computation and user preference prediction. However, since this approach treat each cloud services without distinction in user similarity computation and only use the most similar user to predict the preference of active user, although it performs better than content-based method, due to a lack of personalization treatment of cloud services and appropriate neighbor selection, it could be improved.

3 Recommendation Approach

We first define some notations that will be used throughout this paper. Let $U = \{u_1, u_2, \ldots, u_m\}$ denote the set of m users and $S = \{s_1, s_2, \ldots, s_m\}$ denote the set of n cloud services. We also define a user-service matrix $R = \{r_{i,j}\}_{m \times n}$, where $r_{i,j} = 1$ if user u_i invoked s_j, otherwise $r_{i,j} = 0$.

The problem of cloud service recommendation in this paper can be formally described as follows: given a user, called active user, with his usage records and a list of cloud service that he has not invoked, called candidate services, the system will rank the candidate services in descending order of the possibility that the active user invoke them and then a predefined number of top-ranked candidate services will be recommended to the active user. We use collaborative filtering technique to calculate the possibility that active users invoke candidate services. Our proposed approach is composed of three phases: user similarity computation, user neighbor selection, and possibility prediction. In the following sections, we will introduce the details of this approach.

3.1 User Similarity Computation

Cosine similarity has been employed for similarity computation in a large number of recommender systems [13], since it could be implemented easily and achieves good performance. Similarity between two users u_i and u_j based on cosine similarity computation method can be calculated by the following equation:

$$sim(u_i, u_j) = \frac{\sum_{k=1}^{n} r_{i,k} \cdot r_{j,k}}{\sqrt{\sum_{k=1}^{n} r_{i,k}^2} \cdot \sqrt{\sum_{k=1}^{n} r_{j,k}^2}}. \tag{1}$$

In above equation, each cloud service has the same weighting in similarity computation, but it may be not reasonable [15, 16]. In the field of sociology, there is a famous theory named "Matthew Effect", which means strong the stronger, weak the weaker. In recommendation system, long tail problem is reflection of Matthew Effect and Matthew Effect contributes to the long tail problem in some extension. Popular cloud services are easy to be found and invoked by the users that have no specific requirements and then become more popular, which implies that some usage records may be generated because of the popularity of cloud service but not users' personalized preference and unpopular cloud services relatively reflect more personalized preference of users than popular services. To improve the reliability of user similarity, we should fully mine users' personalized preference and reduce the influence of data that is rarely related to users' personalized preference. Under such a consideration, we introduce a weighting factor about cloud service popularity as following:

$$w(s_i) = avgpop/pop(s_i), \tag{2}$$

Therein $pop(s_i)$ is the total number of usage records of cloud services s_i and $avgpop$ is the average number of usage records of all cloud services in system.

In our proposed cloud service recommendation approach, user similarity is calculated using the following equation:

$$sim(u_i, u_j) = \frac{\sum_{k=1}^{n} (w(s_k) \cdot r_{i,k}) \cdot (w(s_k) \cdot r_{j,k})}{\sqrt{\sum_{k=1}^{n} (w(s_k) \cdot r_{i,k})^2} \cdot \sqrt{\sum_{k=1}^{n} (w(s_k) \cdot r_{j,k})^2}}. \tag{3}$$

From above definition, the similarity of two users, $sim(u_i, u_j)$, is in the interval of [0, 1], where a larger similarity indicates that two users u_i and u_j are more similar. If two users commonly invoked an unpopular cloud service, they are more similar than two users who commonly invoke a popular cloud service.

3.2 User Neighbor Selection

Before predicting the possibility that active users invoke candidate services, user neighbor, which includes a set of similar users, need to be identified. The quality and quantity of user neighbors have significant influence on recommendation results since two few neighbors may lead to lose some valuable neighbors that help active users to find his potential interests and too many neighbors may introduce noise to possibility prediction.

For user u_i, a set of similar users $S(u_i)$ can be found by the following equations:

$$S(u_i) = \{u_a | u_a \in T(u_i), i \neq a, sim(u_i, u_a) > 0\}. \tag{4}$$

We rank the similarity between user u_i and each other user in the system in descending order and $T(u_i)$ are the k users whose similarities with user u_i are larger than other users. Specially, if the number of users whose similarity with user u_i is greater than 0 is less than k, we choose the users whose similarity with user u_i is greater than 0 as neighbors of user u_i. By this method, valuable users that are beneficial to find user potential interests will be selected and invaluable users that may introduce noise will be filter out.

3.3 Possibility Prediction

After identifying neighbors for the active user u_i, we apply the neighbor set $S(u_i)$ of user u_i to predict the possibility that user u_i will invoke each candidate service. The possibility $P(u_i, s_j)$ that user u_i will invoke the candidate service s_j are computed by the following equation:

$$P(u_i, s_j) = \frac{\sum_{u_a \in (u_i)} sim(u_i, u_a)}{|S(u_i)|}. \quad (5)$$

After predicting the possibility that the active user u_i invoke each candidate service, recommendation can be generated for user u_i. The candidate services of user u_i is ranked in descending order of the predicted possibility and top-*N* cloud services that are more likely to be invoked by the active user u_i will be recommended to him.

4 Experiment and Evaluation

4.1 Dataset and Experimental Setup

Programmable Web is a well-known cloud service community where many cloud service providers public cloud services. It provides the register information of services, including name, web description, provider, URL, tags and so on. To evaluate our proposed approach, we collected data from the website, in the range from 2005 to 2016, as experimental dataset. In the dataset, 3917 usage records of 600 users who have invoked services are collected and 500 services are involved by these users. The average usage records of each user are approximately 6. Furthermore, we collected description documents and usage records of each cloud services. Details of the data we collected are illustrated in Table 1.

Table 1. Dataset from programmable web

Statistics	Values
Total # of users	600
Total # of cloud services	500
Total # of usage records	3917
Total # of cloud services descriptions	500

We conducted a K-folder cross-validation (K = 10) on the overall dataset. 600 users in the dataset were divided into 10 groups. The experiments were performed 10 times. In each experiment, the usage records of 9 groups composed training set and the other one group composed validation set. For each user in the test set, 20% of usage records are omitted in order to evaluate the effect of recommendation.

4.2 Evaluation Metrics

We evaluated the recommendation performance using F-Measure which is widely used in recommendation system. F-Measure is the weighted harmonic mean of precision and recall.

Precision and Recall are calculated as following equations:

$$Precision = \frac{\sum_u |R(u) \cap I(u)|}{\sum_u |R(u)|}, \tag{6}$$

$$Recall = \frac{\sum_u |R(u) \cap I(u)|}{\sum_u |I(u)|}. \tag{7}$$

F-Measure is calculated by Precision and Recall as following equation:

$$F - Measure = \frac{2 \cdot Precision \cdot Recall}{Precision + Recall}. \tag{8}$$

In above three equations, $R(u)$ is the subset of recommended cloud services for user u and $I(u)$ is the subset of invoked cloud services of user u. We evaluated F-Measure on the test set for each user and then averaged the metric over the set of test users.

4.3 Performance Comparison

To validate the effectiveness of the approach (named as CFR) in this paper, we compare it to several existing approaches, including:

(1) Popularity: this approach recommends popular cloud services to users. Popularity of each service is measured by the times it has been invoked by users.
(2) Content-based: this approach extracts preference of users and feature of cloud services by analyzing description information of cloud services using TF-IDF technique. It computes the similarity of the preference of active users and feature of candidate services and services that get high similarity are recommended [9].
(3) CF-based: this approach computes user Jaccard similarity and predicts the possibility that users invoked services by the most similar user [14].
(4) CFR: it computes user similarity by improved cosine similarity and predicts the possibility that active users will invoke candidate services by user neighbors.

We conducted K-folder cross-validation experiments on our dataset for above four approaches. The number of cloud services that system would recommend to each user N is set to 1, 5, 10 and 15. For CFR, the number of user neighbor k is set to 4. The results are shown in Table 2.

Table 2. Experimental results (Avg ± Stdev)

Approach	N = 1	N = 5	N = 10	N = 15
Popularity	0.0033** ±0.0105	0.0027** ±0.0028	0.0035** ±0.0019	0.0041** ±0.0013
Content-based	0.0033** ±0.0105	0.0021** ±0.0028	0.0029** ±0.0022	0.0035** ±0.0021
CF-based	0.0673** ±0.0289	0.0545** ±0.0110	0.0418** ±0.0065	0.0345** ±0.0049
CFR	**0.1817** ±0.0232	**0.0863** ±0.0844	**0.0541** ±0.0068	**0.0405** ±0.0044

**Indicates extremely significant difference ($P < 0.01$)

Bold figures in Table 2 indicate the best F-Measure among four approached with one specific *N* values. It indicates that the F-Measure of CFR outperforms all the other approaches consistently over each *N* values. We also performed the paired t-test on the experimental results and it is shown that the differences between the CFR and each other approach are always extremely significant. Benefiting from the improved similarity computation and user neighbor selection, the F-Measure of the CFR is 26 times higher than content-based approach and 60% higher than based CF-based approach, respectively.

4.4 Impact of Cloud Service Popularity

Cloud service popularity is considered as a weighting factor in user similarity computation. This adjustment can increase the similarity between two users if they invoked unpopular cloud services commonly and decrease the similarity between two users if they invoked popular cloud services commonly. To study the impact of cloud service popularity, we implement two versions of CFR. One version (CFR-pop) employs cloud service popularity to adjust user similarity computation and the other version (CFR-base) does not. In the experiments, the number of cloud services that the system would recommend to each user *N* is set to 5 and the number of user neighbor *k* is set to 4. Figure 1 shows the experimental results.

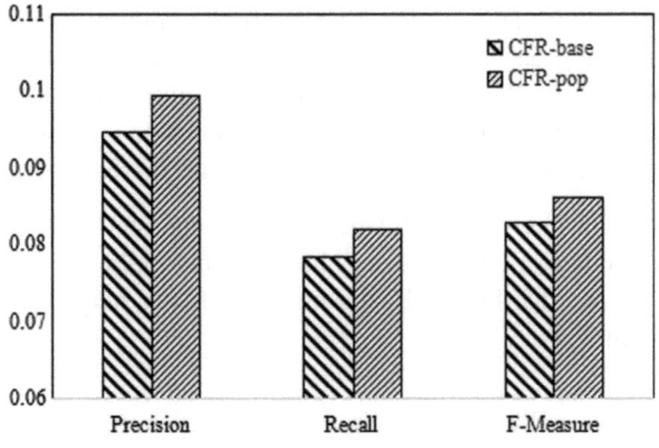

Fig. 1. Impact of cloud service popularity

Figure 1 indicates that the approach CFR-pop adjusted by cloud service popularity achieves better results in term of F-Measure than the approach CFR-base. Such an improvement verifies that taking cloud service popularity as weighting factor in user similarity is effective. We can also infer that unpopular cloud services imply more users' personalized preference. If two users invoke unpopular cloud services commonly, they are more similar than two users who invoke a popular cloud service commonly.

4.5 Impact of Neighbor Size *k*

In our proposed CFR approach, the parameter k determines the size of the user neighbor set. We suppose that a too small k (two few neighbors) may lead the approach lose some valuable neighbors that help the active user find potential interest, but a too large k (too many neighbors) may introduce noise to possibility prediction. In this experiment, we study the impact of neighbor size k. The number of cloud services that the system would recommend to each user N is set to 1, 5, 10, 15 and the parameter k is set in the range of 1–10. The experimental results are shown in the following Fig. 2.

Figure 2 demonstrates that under the predefined N, as k increase, the F-Measure of CFR first increases then reach the maximum point at $k = 4$ and continues to decrease to a limited extent. This trend indicates that suitable neighbor size benefits to achieve better recommendation accuracy.

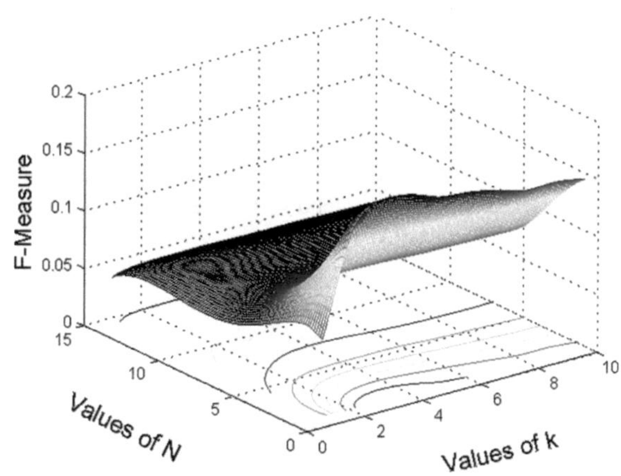

Fig. 2. Impact of neighbor size *k*

5 Conclusion and Future Work

In this paper, we have proposed a cloud service recommendation approach based on collaborative filtering via exploring user usage history, which is composed of three phases: user similarity computation, user neighbor selection, and possibility prediction. It first computes user similarity using the method of cosine similarity, which is improved by a weighting factor about cloud service popularity, then it identifies neighbors for active user and at last it generates recommendation results for a specific active user by predicting the possibility that the user will invoke candidate services based on neighbors. In contrast to the state-of-the-art approaches [9, 14], which failed to completely and precisely identify users' potential interests, our proposed approach can select appropriate user neighbors to identify users' preference and potential interests. Experimental results show that it performs better than existing approaches in terms of F-Measure. We trust that the effective cloud service recommendation approach will help cloud service users find services they demand and build distributed applications more efficiently.

However, the proposed approach has some limitations. Collaborative filtering based recommendation suffers from the cold start problem, which means new published cloud services that have no usage records have no opportunity to be recommended. Future work will address this problem and improve recommendation performance by hybrid recommendation techniques, such as content-based recommendation.

Acknowledgements. The work was supported by the Key Program of National Natural Science Foundation of China (No. 71631003) and the General Program of National Natural Science Foundation of China (No. 71771169).

References

1. T. Erl, R.S. Puttini, Z. Mahmood et al., Cloud computing: concepts, technology & architecture, in *Cloud Computing* (2013)
2. X.W. Zhang, H.E. Ke-Qing, J. Wang et al., A survey of personalized web service recommendation. Comput. Eng. Sci. (2013)
3. Z. Zheng, H. Ma, M.R. Lyu et al., QoS-aware cloud service recommendation by collaborative filtering. IEEE Trans. Serv. Comput. **4**(2), 140–152 (2011)
4. M. Chen, Y. Ma, A hybrid approach to cloud service recommendation based on QoS-aware rating and ranking. Comput. Sci. (2015)
5. J. Liu, M. Tang, Z. Zheng et al., Location-aware and personalized collaborative filtering for cloud service recommendation. IEEE Trans. Serv. Comput. **9**(5), 1–1 (2016)
6. X. Wang, J. Zhu, Z. Zheng et al., A spatial-temporal QoS prediction approach for time-aware cloud service recommendation. ACM Trans. Cloud **10**(1), 1–25 (2016)
7. Y. Xu, J. Yin, S. Deng et al., Context-aware QoS prediction for web service recommendation and selection. Expert. Syst. Appl. **53**(C), 75–86 (2016)
8. Y.M. Afify, I.F. Moawad, N.L. Badr et al., Enhanced similarity measure for personalized cloud services recommendation. Concurr. Comput. Pract. Exp. **29** (2017)
9. G. Kang, J. Liu, M. Tang et al., AWSR: active cloud service recommendation based on usage history, in *IEEE International Conference on Cloud Services* (IEEE, 2012), pp. 186–193
10. X. Zhang, K. He, C. Wang et al., Interest-driven cloud service recommendation based on MFI-7, in *IEEE International Conference on Services Computing* (IEEE, 2013), p. 438
11. L. Liu, F. Lecue, N. Mehandjiev, Semantic content-based recommendation of software services using context, in *ACM Transactions on the Cloud* (2013)
12. B. Cao, J. Liu, M. Tang et al., Cloud API recommendation based on user usage history and reputation evaluation. Chin. J. Polym. Sci. **48**(2), 77–98 (2015)

13. X. Zhang, K. He, J. Wang et al., Cloud service recommendation based on watchlist via temporal and tag preference fusion, in *IEEE International Conference on Cloud Services* (2014), pp. 281–288
14. G. Kang, M. Tang, J. Liu et al., Diversifying cloud service recommendation results via exploring service usage history. IEEE Trans. Serv. Comput., 1–1 (2015)
15. L.Y. Hao, J. Wang, Collaborative filtering TopN-recommendation algorithm based on item popularity. Comput. Eng. Des. **34**(10), 3497–3501 (2013)
16. W. Xia, L. He, L. Ren et al., A new collaborative filtering approach utilizing item's popularity, in *IEEE International Conference on Industrial Engineering and Engineering Management* (IEEE, 2008), pp. 1480–1484

MIX
Papier aus verantwortungsvollen Quellen
Paper from responsible sources
FSC® C105338

If you have any concerns about our products,
you can contact us on
ProductSafety@springernature.com

In case Publisher is established outside the EU,
the EU authorized representative is:
Springer Nature Customer Service Center GmbH
Europaplatz 3, 69115 Heidelberg, Germany

Printed by Libri Plureos GmbH
in Hamburg, Germany